T0189488

Advances in Intelligent Systems and Computing

Volume 891

Series editor

Janusz Kacprzyk, Systems Research Institute, Polish Academy of Sciences,
Warsaw, Poland
e-mail: kacprzyk@ibspan.waw.pl

The series "Advances in Intelligent Systems and Computing" contains publications on theory, applications, and design methods of Intelligent Systems and Intelligent Computing. Virtually all disciplines such as engineering, natural sciences, computer and information science, ICT, economics, business, e-commerce, environment, healthcare, life science are covered. The list of topics spans all the areas of modern intelligent systems and computing such as: computational intelligence, soft computing including neural networks, fuzzy systems, evolutionary computing and the fusion of these paradigms, social intelligence, ambient intelligence, computational neuroscience, artificial life, virtual worlds and society, cognitive science and systems, Perception and Vision, DNA and immune based systems, self-organizing and adaptive systems, e-Learning and teaching, human-centered and human-centric computing, recommender systems, intelligent control, robotics and mechatronics including human-machine teaming, knowledge-based paradigms, learning paradigms, machine ethics, intelligent data analysis, knowledge management, intelligent agents, intelligent decision making and support, intelligent network security, trust management, interactive entertainment, Web intelligence and multimedia.

The publications within "Advances in Intelligent Systems and Computing" are primarily proceedings of important conferences, symposia and congresses. They cover significant recent developments in the field, both of a foundational and applicable character. An important characteristic feature of the series is the short publication time and world-wide distribution. This permits a rapid and broad dissemination of research results.

More information about this series at http://www.springer.com/series/11156

Pavel Krömer · Hong Zhang
Yongquan Liang · Jeng-Shyang Pan
Editors

Proceedings of the Fifth Euro-China Conference on Intelligent Data Analysis and Applications

Editors
Pavel Krömer
Department of Computer Science
VSB-Technical University of Ostrava
Ostrava, Czech Republic

Hong Zhang
School of Automation
Xi'an University of Posts
and Telecommunications
Xi'an, China

Yongquan Liang
College of Computer Science
and Engineering
Shandong University of Science
and Technology
Qingdao, China

Jeng-Shyang Pan
College of Information Science
and Engineering
Fujian University of Technology
Fuzhou, Fujian, China

ISSN 2194-5357 ISSN 2194-5365 (electronic)
Advances in Intelligent Systems and Computing
ISBN 978-3-030-03765-9 ISBN 978-3-030-03766-6 (eBook)
https://doi.org/10.1007/978-3-030-03766-6

Library of Congress Control Number: 2018960434

This Springer imprint is published by the registered company Springer Nature Switzerland AG
The registered company address is: Gewerbestrasse 11, 6330 Cham, Switzerland

Preface

This volume composes the Proceedings of the Fifth Euro-China Conference on Intelligent Data Analysis and Applications (ECC2018), which is hosted by Xi'an University of Posts and Telecommunications, and is held in Xi'an, China, on 12–14 October 2018. ECC2018 is technically co-sponsored by Springer, Xi'an University of Posts and Telecommunications, VSB-Technical University of Ostrava in Czech, Fujian University of Technology, Fujian Provincial Key Laboratory of Digital Equipment, Fujian Provincial Key Laboratory of Big Data Mining and Applications and Shandong University of Science and Technology in China.

ECC puts a special emphasis on promoting research and scientific collaboration between Europe and China, strengthening research partnerships and providing an opportunity for joint efforts leading to higher quality fundamental and applied research. The aim of ECC2018 is to bring together researchers, engineers and policymakers to discuss the intelligent computational and data analysis techniques, to exchange related research ideas and to make friends.

Ninety-four excellent papers were accepted for the final proceeding. We would like to thank the authors for their tremendous contributions. We would also express our sincere appreciation to the reviewers, program committee members and the local committee members for making this conference successful. Finally, we would like to express special thanks for the financial support from Fujian University of Technology, China, and VSB-Technical University of Ostrava in Czech in making ECC2018 possible, and we also appreciate the great help from Xi'an University of Posts and Telecommunications for locally organizing the conference.

September 2018

Pavel Krömer
Hong Zhang
Yongquan Liang
Jeng-Shyang Pan

Organizing Committee

Honorary Chairs

Jiulun Fan	Xi'an University of Posts and Telecommunications, China
Vaclav Snasel	VSB-Technical University of Ostrava, Czech Republic

Advisory Committee Chair

Chongzhao Han	Xi'an Jiaotong University, China

Conference Chairs

Wenqing Wang	Xi'an University of Posts and Telecommunications, China
Jeng-Shyang Pan	Fujian University of Technology, China
Vaclav Snasel	VSB-Technical University of Ostrava, Czech Republic

Program Chairs

Jiamin Gong	Xi'an University of Posts and Telecommunications, China
Pavel Krömer	VSB-Technical University of Ostrava, Czech Republic
Jeng-Shyang Pan	Fujian University of Technology, China

Publication Chairs

Wenxue Chen	Xi'an University of Posts and Telecommunications, China
Pavel Krömer	VSB-Technical University of Ostrava, Czech Republic

Local Organization Chairs

Wenqing Wang	Xi'an University of Posts and Telecommunications, China
Xiao Qiang Xi	Xi'an University of Posts and Telecommunications, China
Hong Zhang	Xi'an University of Posts and Telecommunications, China

Finance Chairs

Naili Tong	Xi'an University of Posts and Telecommunications, China
Jeng-Shyang Pan	Fujian University of Technology, China

Program Committees

Abdel hamid Bouchachia	University of Klagenfurt, Austria
Aihong Ren	Baoji University of Arts and Sciences, China
Bo Wang	Xi'an University of Posts and Telecommunications, China
Brijesh Verma	Central Queensland University, Australia
Chang-Shing Lee	National University of Tainan, Taiwan
Chao-Chun Chen	Southern Taiwan University, Taiwan
Chia-Feng Juang	National Chung Hsing University, Taiwan
Chien-Ming Chen	Harbin Institute of Technology (Shenzhen), China
Chin-Chen Chang	Feng Chia University, Taiwan
Chuan-Kang Ting	National Chung Cheng University, Taiwan
Chuan-Yu Chang	National Yunlin University of Science and Technology, Taiwan
Chu-Hsing Lin	Tunghai University, Taiwan
Fatemeh Afghah	Northern Arizona University, USA
Feng Feng	Xi'an University of Posts and Telecommunications, China
Feng-Cheng Chang	Tamkang University, Taiwan
Han Peng	Northern Arizona University, USA

Han Zhuang	Peking University, China
Haoyang Tang	Xi'an University of Posts and Telecommunications, China
Hsiang-Cheh Huang	National University of Kaohsiung, Taiwan
Jan Martinovic	VSB-Technical University of Ostrava, Czech Republic
Jana Heckenbergerova	University of Pardubice, Czech Republic
Jianwei Wang	Xi'an University of Posts and Telecommunications, China
Jimmy Min-Tai Wu	Shandong University of Science and Technology, China
Junjie Fu	Southeast University, China
Jyh-Horng Chou	National Kaohsiung First University of Science and Technology, Taiwan
Leon Wang	National University of Kaohsiung, Taiwan
Liang-Cheng Shiu	National Pingtung University, Taiwan
Liangpeng Zhang	University of Birmingham, UK
Lin Xu	Fujian Normal University, China
Lingping Kong	VSB-Technical University of Ostrava, Czech Republic
Miao Ma	Shaanxi Normal University, China
Michael Blumenstein	Griffith University, Australia
Michal Kratky	VSB-Technical University of Ostrava, Czech Republic
Michal Musilek	University of Hradec Kralove, Czech Republic
Milos Kudelka	VSB-Technical University of Ostrava, Czech Republic
Na Huang	Hangzhou Dianzi University, China
Peihu Duan	Peking University, China
Qingping Zhang	Xi'an University of Posts and Telecommunications, China
Qunsuo Qu	Xi'an University of Posts and Telecommunications, China
Roman Neruda	Institute of Computer Science, Czech Republic
Sebastian Basterrech	Czech Technical University, Czech Republic
Shaojie Tang	Xi'an University of Posts and Telecommunications, China
Tieshuang Hou	Xi'an University of Posts and Telecommunications, China
Ting-Ting Wu	National Yunlin University of Science and Technology, Taiwan
Tzung-Pei Hong	National University of Kaohsiung, Taiwan
Vaclav Snasel	VSB-Technical University of Ostrava, Czech Republic
Wei-Chiang Hong	Oriental Institute of Technology, Taiwan

Weiyi Zhang	Xi'an University of Posts and Telecommunications, China
Wen-Yang Lin	National University of Kaohsiung, Taiwan
Wenyu Zhang	Xi'an University of Posts and Telecommunications, China
Xiang Ren	Beihang University, China
Xiaoping Zhang	North China University of Technology, China
Xiaoqiang Xi	Xi'an University of Posts and Telecommunications, China
Xiumei Cai	Xi'an University of Posts and Telecommunications, China
Yu Zhao	Northwestern Polytechnical University, China
Yueh-Hong Chen	Far East University, Taiwan
Yuh-Yih Lu	Minghsin University of Science and Technology, Taiwan
Yunsheng Li	Xi'an University of Posts and Telecommunications, China
Yuqing Hao	Beihang University, China
Zhigang Pan	Xi'an University of Posts and Telecommunications, China

Contents

Energizing Topics and Applications in Computer Sciences

Robot and Intelligent Control

Pattern Recognition Technologies and Applications

Complex Network Control

Truncation Error Correction for Dynamic Matrix Control Based on RBF Neural Network

Youming Wang[1(✉)], Jugang Li[1], and Feng Ji[2]

[1] Xi'an University of Posts and Telecommunications, Xi'an 710121, China
wangyouming@xupt.edu.cn

[2] Shaanxi Environmental Protection Research Institute, Co., Xi'an 710100, China

Abstract. In this paper, a correction method for truncation error of dynamic matrix control is studied. A truncation error correction method is designed by using the law of the dynamic change of control system. Firstly, the predictive initial value is calculated using different compensation parameters and the difference between the last two components of the predicted initial value vector is obtained. Secondly, the compensation parameter is fitted based on RBF neural network. Finally, the compensation parameter is used to correct error during feedback correction. Numerical experiments show that the proposed method can improve the overshoot and hysteresis characteristics of the system.

Keywords: Dynamic matrix control · Truncation error · RBF neural network

1 Introduction

Dynamic Matrix Control (DMC) is a new type of computer control algorithm that emerged in the 1970s and has been successfully applied in many fields such as advanced manufacturing, energy, environment, aerospace, medical, etc. [1–3]. It has the characteristics of simple algorithm, less computation and strong robustness. However, DMC also has some shortcomings, one of which is the sampling period Ts and the modeling time domain N. In order to solve this problem, it is necessary to compensate for the truncation error of the prediction model. Many researchers have proposed a method for correcting the truncation error using compensation parameters, which can reduce the error caused by the truncation of the model time domain [5–7]. Although the limitations of traditional algorithms have changed, the truncation error is not well compensated because β is a fixed value. The above method does not quantitatively study the truncation error and the exact compensation parameter expression cannot be obtained.

RBF neural network is a forward network with simple structure and fast convergence speed. It has global approximation performance and strong self-learning and self-adaptive ability [8, 9]. It is based on the above advantages of neural network.

In this paper, a neural network based on truncation error correction method is designed for typical control system and the expression of the correction parameters is fitted by the neural network. The RBF neural network model is established by taking the difference in the predictive initial value as the input and the compensation parameter as the output. The exact expression of the compensation parameter can be represented

P. Krömer et al. (Eds.): ECC 2018, AISC 891, pp. 3–9, 2019.
https://doi.org/10.1007/978-3-030-03766-6_1

by the fitted result. This method can improve the overshoot and hysteresis characteristics of the system.

2 Dynamic Matrix Control Algorithm

Dynamic matrix control is an advanced control algorithm based on prediction model, rolling optimization and feedback correction. At each sampling k, the predictive model and the initial value $y_{N0}(k)$ are used to predict the output of controlled object at P moments in the future. $y_{N0}(k)$ determines the future M control increments $\Delta u_M(k) = [\Delta u(k), \Delta u(k+1), \ldots, \Delta u(k+M-1)]$ so that the predictive output $y_{N1}(k+i/k)(i = 1, 2, \ldots, P)$ is as close as possible to the given expected value $w(k+i)(i = 1, 2, \ldots P)$. The $\Delta u_M(k)$ inflicts the first control increment $\Delta u(k)$ at $t = kT$. However, the second control increment $\Delta u(k+1)$ is always not implemented at $t = (k+1)T$. The prediction error between the actual output of the system $y(k + 1)$ and the first component $y_1(k+1|k)$ in $y_{N1}(k)$ can be expressed as

$$e(k+1) = y(k+1) - y_1(k+1|k) \tag{1}$$

The output predictive value is corrected by weighting $e(k + 1)$ in the form of

$$y_{cor}(k+1) = y_{N1}(k) + he(k+1) \tag{2}$$

where h is the error correction vector, $y_{cor}(k + 1)$ is the predictive output after correcting.

As the time changes, $y_{cor}(k + 1)$ is usually shifted to be the initial predictive value of time $k + 1$.

$$y_0(k+1+i|k+1) = y_{cor}(k+1+i|k+1), i \in [1, N-1] \tag{3}$$

For a stable control system, $y_0(k+1+N|k+1)$ can be approximated by $y_{cor}(k+N|k+1)$. The last component of the initial predictive value $y_0(k+1+N|k+1)$ should be obtained by $y_1(k+1+N|k)$. In traditional algorithm, the predicted value $y_1(k+N|k+1)$ derived from the predictive model is $y_1(k+1+N|k+1)$ only if there is no truncation error in the predictive model. However, when there is a truncation error, the error will replace the predictive output with this approximation.

3 Truncation Error Correction Based on Neural Network

3.1 Analysis of Truncation Error

The truncation error is corrected by the coefficient β and the predictive value is approximately replaced by the predictive value at times $k + N$ and $k + N - 1$ [5–8]. The initial value of using the compensation coefficient is written as follows

$$y_0(k+1+N/k+1) \approx y_{cor}(k+N/k+1) + \beta\left[y_{cor}(k+N/k+1) - y_{cor}(k+N-1/k+1)\right] \tag{4}$$

The correction coefficient β is a fixed parameter for the error compensation. However, the single correction coefficient cannot satisfy the accuracy of the compensation. The correction coefficient μ is designed to replace the fixed correction coefficient β so that it can reduce the fluctuation of the control system. The correction coefficient is output and the prediction difference is input. RBF neural network is constructed for fitting and the expression of dynamic compensation parameter is obtained.

3.2 Radial Basis Function Neural Network (RBF) Modeling

Radial basis functions in neural network are generally based on Gaussian functions. The difference in the predictive value is input and the compensation parameter is the output. The network input vector is

$$x(k) = \left[x_1, x_2, \ldots, x_n\right]^T = \left[\mu_1, \mu_2, \ldots, \mu_N\right]^T \tag{5}$$

The hidden layer output is

$$h_j(x) = \exp\left(\frac{\left\|x - c_j\right\|^2}{2S_j^2}\right) = \exp\left(\frac{\left(\mu_1 - c_{j1}\right)^2 + \left(\mu_2 - c_{j2}\right)^2 + \ldots + \left(\mu_N - c_{jN}\right)^2}{2S_j^2}\right) \tag{6}$$

where $(j = 1, 2, \ldots, N)$

The network output is

$$y_1(k+1) = \Delta y_{cor}(k) = \sum_{i=1}^{N} w_j h_j(x(k)) \tag{7}$$

The objective function is defined as

$$E_m = \frac{1}{2}\left[y(k+1) - y_1(k+1)\right]^2 \tag{8}$$

The weight adjustment formula using gradient descent method is

$$w_j(k+1) = w_j(k) - \alpha_j \frac{\partial E_m}{\partial w_j(k)} \tag{9}$$

The RBF neural network is established by compensation parameter and prediction difference. It's trained by MATLAB toolbox. The fitting result is shown in Fig. 1.

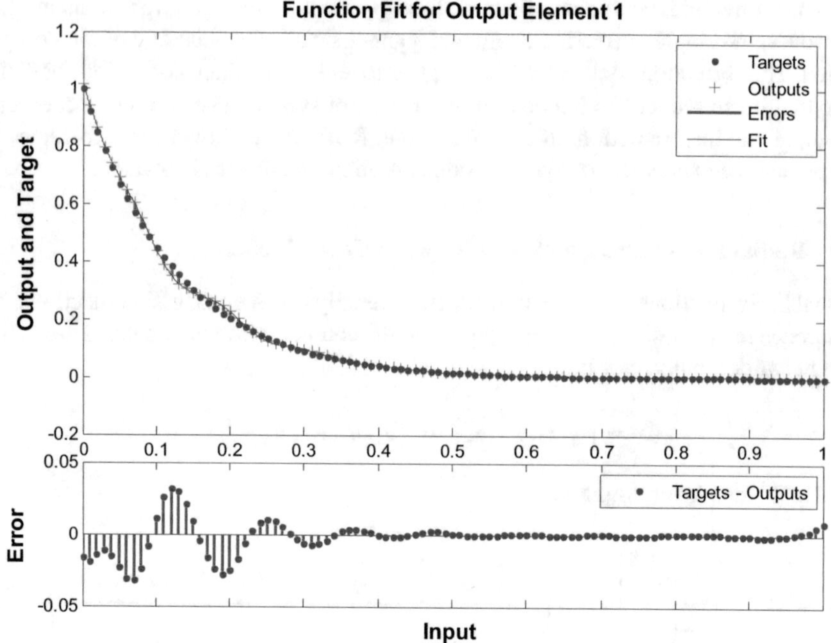

Fig. 1. The fitting results.

4 The Simulation Analysis

In order to verify the efficiency of dynamic matrix control based on RBF neural network, comparison of traditional DMC algorithm with improved DMC algorithm is implemented in MATLAB under the typical system. The controlled object selects the common object model in production practice. Its transfer function is

$$F(z) = \frac{1}{1 + 0.5z^{-1}}$$

The parameters of DMC algorithm can be selected as follow

$$T = 1; N = 80; M = 30; P = 5;$$

$$Q = \begin{bmatrix} 1 \\ \vdots \\ 1 \end{bmatrix}_{P\times1} R = \begin{bmatrix} 1 \\ \vdots \\ 1 \end{bmatrix}_{M\times1} h = \begin{bmatrix} 1 \\ \vdots \\ 1 \end{bmatrix}_{N\times1};$$

It can be seen from Fig. 2 that the improved DMC algorithm reduces the overshoot by 50%, while the difference decreases between the set value and the predicted value. At the same time, the system hysteresis characteristics, overshoot and system error are improved. The performance of the control system is improved to some extent by the correction parameters of subsection compensation. Within a certain range of values, the relative overshoot of the control system is decreasing and the adjustment time is also greatly reduced. DMC algorithm based on neural network to correct truncation error has achieved good control effect in adjusting time and relative overshoot.

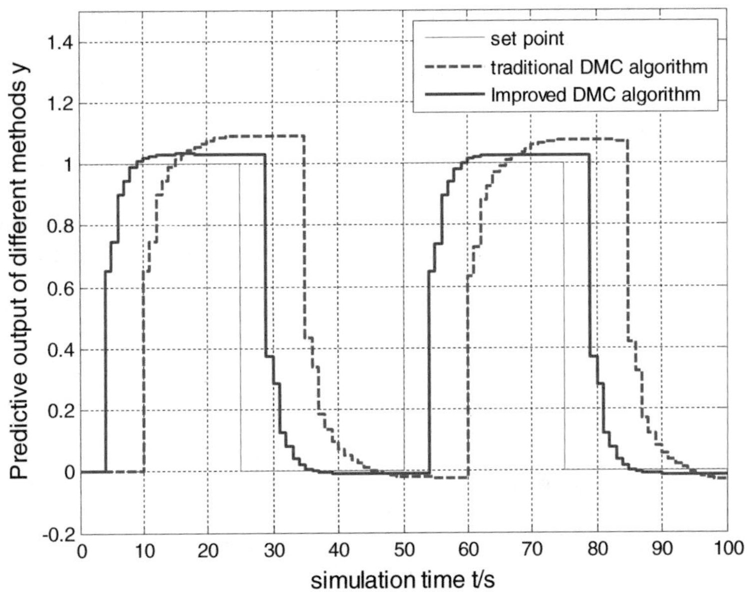

Fig. 2. The simulation results

5 Conclusion

In this paper, a correction method for truncation error of dynamic matrix control is proposed. A method of truncation error correction based on RBF neural network is designed by using the rule of dynamic change of calibration parameters. Simulation results of typical control objects show that, compared with the traditional DMC algorithm, this method can improve the system's overshoot and hysteresis characteristics.

Acknowledgements. This work was supported by the Key Research and Development Program of Shaanxi Province of China (2017GY-168) and the New Star Team of Xi'an University of Posts and Telecommunications. It was also supported by the graduate student innovation fund of Xi'an University of Posts and Telecommunications (CXL2016-20) and the Department of Education Shaanxi Province, China, under Grant 2013JK1023.

References

1. Hudson, R.A.: Model predictive heuristic control: applications to industrial processes. Automatica **14**(5), 413–428 (1978). https://doi.org/10.1016/0005-1098(78)90001-8
2. Qin, S.J., Badgwell, T.A.: A survey of industrial model predictive control technology. Control Eng. Pract. **11**(7), 733–764 (2003). https://doi.org/10.1016/S0967-0661(02)00186-7
3. Clarke, D.W., Mohtadi, C., Tuffs, P.S.: Generalized predictive control—part I. The basic algorithm. Automatica **23**(2), 149–160 (1987). https://doi.org/10.1016/0005-1098(87)90087-2
4. Zhong, Q.C., Xie, J.Y.: Dynamic matrix control with model truncation error compensation. J. Shanghai Jiaotong Univ. **33**(5), 623–625 (1999). https://doi.org/10.16183/j.cnki.jsjtu.1999.05.032
5. Liang, Y., Ping, L.I.: A new method for compensating model truncation error in dynamic matrix control. Comput. Simul. **22**(1), 84–99 (2005). https://doi.org/10.3969/j.issn.1006-9348.2005.01.025
6. Zhang, X.Y., Huang, J.T., Guo, X.F., Cao, Z.: An adaptive method for compensating model truncation error in DMC. Control Eng. China **24**(11), 2308–2313 (2017). https://doi.org/10.14107/j.cnki.kzgc.C2.0403
7. Gao, H., Pan, H., Zang, Q., Liu, G.: Dynamic matrix control for ACU position loop. Lect. Notes Electr. Eng. **323**, 173–184 (2015). https://doi.org/10.1007/978-3-662-44687-4_16
8. Wang, C.Y., Zhang, F., Han, W.D.: A study on the application of RBF neural network in slope stability of bayan obo east mine. Adv. Mater. Res. **1010–1012**(1), 1507–1510 (2014). https://doi.org/10.4028/www.scientific.net/AMR.1010-1012.1507
9. Liu, Z.B., Yang, X.W.: Assessment of the underground water contaminated by the leachate of waste dump of open pit coal mine based on RBF neural network. Adv. Mater. Res. **599**, 272–277 (2012). https://doi.org/10.4028/www.scientific.net/AMR.599.272
10. Moon, U.C., Lee, Y., Lee, K.Y.: Practical dynamic matrix control for thermal power plant coordinated control. Control Eng. Pract. **71**, 154–163 (2018). https://doi.org/10.1016/j.conengprac.2017.10.014
11. Ramdani, A., Grouni, S.: Dynamic matrix control and generalized predictive control, comparison study with IMC-PID. Int. J. Hydrog. Energy **42**(28), 17561–17570 (2017). https://doi.org/10.1016/j.ijhydene.2017.04.015
12. Klopot, T., Skupin, P., Metzger, M., Grelewicz, P.: Tuning strategy for dynamic matrix control with reduced horizons. ISA Trans. **76**, 145–154 (2018). https://doi.org/10.1016/j.isatra.2018.03.003
13. Koo, M.S., Choi, H.L.: Dynamic gain control with a matrix inequality approach to uncertain systems with triangular and non-triangular nonlinearities. Int. J. Syst. Sci. **47**(6), 1453–1464 (2014). https://doi.org/10.1080/00207721.2014.934749
14. Bagheri, P., Sedigh, A.K.: Robust tuning of dynamic matrix controllers for first order plus dead time models. Appl. Math. Model. **39**(22), 7017–7031 (2015). https://doi.org/10.1016/j.apm.2015.02.035
15. Fu, Y., Chang, L., Henson, M.A., Liu, X.G.: Dynamic matrix control of a bubble-column reactor for microbial synthesis gas fermentation. Chem. Eng. Technol. **40**(4), 727–736 (2017). https://doi.org/10.1002/ceat.201600520

16. Li, Y., Wang, G., Chen, H.: Simultaneously regular inversion of unsteady heating boundary conditions based on dynamic matrix control. Int. J. Therm. Sci. **88**, 148–157 (2015). https://doi.org/10.1016/j.ijthermalsci.2014.09.013
17. Lima, D.M., Normeyrico, J.E., Santos, T.L.: Temperature control in a solar collector field using filtered dynamic matrix control. ISA Trans. **62**, 39–49 (2015). https://doi.org/10.1016/j.isatra.2015.09.016

Research on Temperature Compensation Technology of Micro-Electro-Mechanical Systems Gyroscope in Strap-Down Inertial Measurement Unit

Ying Liu[1(✉)], Cong Liu[1], Jintao Xu[2], and Xiaodong Zhao[2]

[1] Xi'an University of Posts and Telecommunications, Xi'an 710121, China
Ly676@163.com
[2] Xi'an Institute of Optics and Precision Mechanics of CAS, Xi'an 710119, China

Abstract. Due to the characteristics of MEMS gyroscope and the influence of the peripheral driving circuit, the MEMS gyroscope is easily affected by temperature and the accuracy is deteriorated. The compensation delay is caused by the complexity of the model in practical engineering applications. A second-order polynomial compensation model for temperature-divided regions is proposed by analyzing the mechanism of gyroscope zero-bias temperature drift. The Model first divides the temperature region of the gyroscope work, and then uses the least squares method to solve the parameters according to multiple linear regression analysis. Finally, the model was verified by experiments. The results show that the model can effectively reduce the drift temperature drift caused by temperature changes, which can reduce the temperature drift after compensation by 73.3%.

Keywords: MEMS gyroscope · Zero-bias temperature drift
Multiple linear regression analysis

1 Introduction

With the rapid development of inertial technology, the Inertial Measurement Combination Unit of Micro-Electro-Mechanical Systems (MEMS-IMU) has become an important development direction of national defense technology with the advantages of fast start-up, small size, low power consumption and easy maintenance in actual use. How to improve MEMS-IMU accuracy has become the focus of current research. The errors of MEMS-based Strap-down Inertial Navigation Systems (INS) can be divided into two categories: one is the overall machine error caused by integrated MEMS-IMU; the other is the inherent error of the MEMS inertial sensor used. The error of MEMS inertial devices is mainly caused by MEMS gyroscopes. Due to the influence of ambient temperature, the material properties of the MEMS gyroscope and the electrical properties of its peripheral circuits will change with temperature, and bring thermal noise affecting the accuracy of the gyroscope, thus affected the gyroscope's zero offset and scale factor, so the accuracy of the gyroscope is lowered and the performance is

P. Krömer et al. (Eds.): ECC 2018, AISC 891, pp. 10–16, 2019.
https://doi.org/10.1007/978-3-030-03766-6_2

deteriorated. In order to meet the performance requirements of the multi-strap-type gyroscope in the application environment, it is necessary to adopt an effective and feasible method to improve the accuracy of the gyroscope and compensate the drift of the gyroscope.

In recent years, the research on random drift of MEMS gyroscope has received extensive attention and achieved certain results. Aiming at the modeling of the relationship between zero bias and temperature, many methods are proposed, such as grey model and the RBF neural network, Wavelet theory, Fuzzy logic, polynomial fitting of Multiple regression analysis and so on [1–5]. However, in engineering applications, these models will increase the computational load of the processor and cause delay compensation. Based on this, the polynomial fitting of multivariate regression analysis is improved, The working temperature region was divided first and then fitted according to polynomials. The complexity of the traditional polynomial modeling is reduced, the computational load of the processor is reduced, and the compensation accuracy is improved on the basis of the traditional polynomial fitting.

2 Error Analysis of MEMS Gyroscope

The Errors of MEMS-IMU can be divided into two major categories: one is the overall machine error caused by integrated MEMS-IMU. The other is own error of the MEMS Inertial sensor used. The machine error mainly includes installation error, Rod arm effect error and Structural deformation error. Rod arm effect errors and Structural deformation errors are generally negligible, and installation error can be eliminated by calibrating the MEMS-IMU. Therefore, improving the accuracy of MEMS inertial sensors has become the focus. At present, MEMS inertial devices can't reach the inertia level, which severely limits its development prospects. There are two main reasons for this phenomenon: one is the inevitable processing error, and the other is complex and variable environmental factors, such as vibration, temperature changes, and so on.

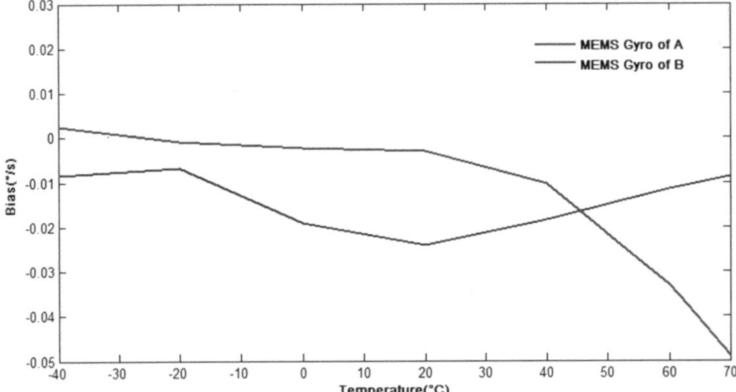

Fig. 1. Relationship between two types of gyroscope zero bias and temperature change characteristics

Temperature is the main factor that causes the accuracy of the gyroscope to deteriorate. MEMS gyroscopes are very sensitive to temperature and belong to temperature sensitive devices. When the temperature changes, the change of the gyroscope zero bias is very obvious, which seriously affects the gyroscope index. The error caused by temperature on the gyroscope output is not due to the temperature directly affecting the output but the error caused by affecting the properties of itself and the peripheral circuits. Moreover, the zero bias and temperature variation characteristics of the MEMS gyroscope are nonlinear, as shown in Fig. 1.

The influence of temperature on the MEMS gyroscope is mainly reflected in two aspects: on the one hand, when the temperature changes, the gyroscope's own silicon material will undergo thermal expansion and contraction, which causes the gyroscope to deform in the structural size, and the detected Coriolis Force changes, and finally the output changes; the driving and detecting circuits inside the gyroscope are also silicon materials, and the temperature changes also cause electrical parameters to change, causing errors. On the other hand, the electrical characteristics of the drive circuit around MEMS gyroscope change with temperature. The peripheral driving circuit is similar to the material of gyroscope. It will affect the temperature of gyroscope and affect the output of gyroscope, causing the drift of the gyroscope.

3 MEMS Gyroscope Temperature Drift Modeling Compensation

According to the self-characteristic and working characteristics of MEMS gyroscope, the established MEMS gyroscope error model is shown in Eq. (1)

$$\tilde{\omega}_g = S_g \omega_{ib}^b + \Delta data + N_g \tag{1}$$

Where: $\tilde{\omega}_g = \left[\tilde{\omega}_{gx}, \tilde{\omega}_{gy}, \tilde{\omega}_{gz}\right]^T$ represents the output of gyroscope; $S_g = \left[S_{gx}, S_{gy}, S_{gz}\right]^T$ is the scale factor of gyroscope; $S_g = \left[S_{gx}, S_{gy}, S_{gz}\right]^T$ is the rate of input angular of the gyroscope; $\Delta data$ is the drift caused by the temperature; N_g is the random noise.

Gyroscope temperature modeling is to find out the relationship between temperature-related factors. In combination with the actual needs of the project, the linear or polynomial model has a simple structure, and its operation speed can meet the application requirements. Therefore, according to the relationship between the MEMS gyroscope bias and the characteristics of temperature variation, the temperature range is divided into four temperature zones: [−40.0, −10.0); [−10.0, 20.0); [20.0,50.0); [50.0,70.0]. According to the polynomial modeling, the gyroscope and temperature variation characteristics of each temperature region are separately modeled and compensated.

$$\Delta data = \begin{cases} a_{01}T^n + \ldots + a_{(n-1)1}T + a_{n1} & A \\ a_{02}T^n + \ldots + a_{(n-1)2}T + a_{n2} & B \\ a_{03}T^n + \ldots + a_{(n-1)3}T + a_{n3} & C \\ a_{04}T^n + \ldots + a_{(n-1)4}T + a_{n4} & D \end{cases} \quad (2)$$

Where: A, B, C, and D are drifts in four temperature regions of [− 40.0, −10.0); [−10.0, 20.0); [20.0, 50.0); [50.0, 70.0] Model; a_{ii} are polynomial parameters, where $i = 1, 2, 3 \cdots$.

The parameters of the model were analyzed by multiple linear regression analysis and solved by the least squares method, which were realized by Matlab tool. In order to obtain a suitable and streamlined compensation model, it is not necessary to excessively pursue the minimum fitting error, because the more variables introduced, the higher the complexity of the model, not only the improvement of the accuracy after compensation is not obvious, but also increases the amount of processor operations. Therefore, combined with Eq. (2), and select the gyroscope sensor output stable data as the fitting data, and use the principle of optimal curve approximation to estimate the parameters of different order models to obtain various order model indexes, as shown is in Table 1. Where: $\overline{\Delta e}$ indicates the average of the error between the fitted curve in the four temperature regions and the actual value; σ_e represents the standard deviation of the error between the fitted curve in the four temperature regions and the actual value.

Table 1. n Choose 1–4 polynomial fitting model effect

N	$\overline{\Delta e}$	σ_e	After compensation
1	0.0658	0.0523	12.2040 °/h
2	0.0303	0.0234	3.6084 °/h
3	0.0173	0.0236	3.4982 °/h
4	0.0157	0.0223	3.4651 °/h

From Table 1, it can be seen that as the order n increases, the average and standard deviation of the error show the same trend, indicating that fitting phenomenon has not occurred. The larger the value of n, the better the compensation effect. When n = 2, the compensation accuracy and model complexity match the best. On the basis of this, increasing the order of the variables does not improve the accuracy.

According to the above modeling scheme, the temperature-induced drift of the gyroscope in the MEMS-IMU is.

$$data = \begin{cases} A_{01}T^2 + A_{11}T + A_{21} & A \\ A_{02}T^2 + A_{12}T + A_{22} & B \\ A_{03}T^2 + A_{13}T + A_{23} & C \\ A_{04}T^2 + A_{14}T + A_{24} & D \end{cases} \quad (3)$$

Where: $A_{0i} = diag(a_{0\mathrm{xi}}, a_{0\mathrm{yi}}, a_{0\mathrm{zi}})$, $A_{1i} = diag(a_{1\mathrm{xi}}, a_{1\mathrm{yi}}, a_{1\mathrm{zi}})$, $A_{2i} = diag(a_{2\mathrm{xi}}, a_{2\mathrm{yi}}, a_{2\mathrm{zi}})$ $i = 1, 2, 3\ldots$.

4 Experimental Results and Analysis

The compensation experiment is carried out on the high and low single-axis turntable, and the controllable range of the thermostat is −60 ~ 85 °C, which fully meets the experimental requirements. At −40 °C, −30 °C, −20 °C, −10 °C, 0 °C, 20 °C, 30 °C 40 °C, 50 °C, 60 °C, 70 °C ambient temperature, keeping each temperature point for two hours is to ensure internal temperature of the IMU reaches the ambient temperature. After each temperature point is insulated, first, collect the temperature and gyroscope output measurement information and preprocess the collected information; then polynomial fitting is performed according to Eq. (3); finally, the fitted parameters are programmed to the processor.

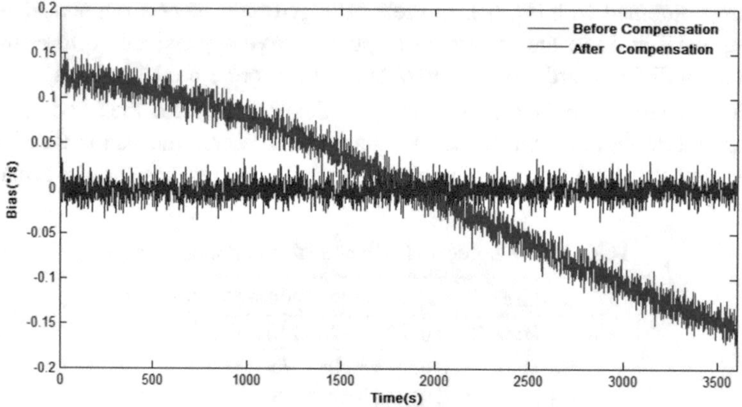

Fig. 2. The data before and after compensation of X-axis gyroscope

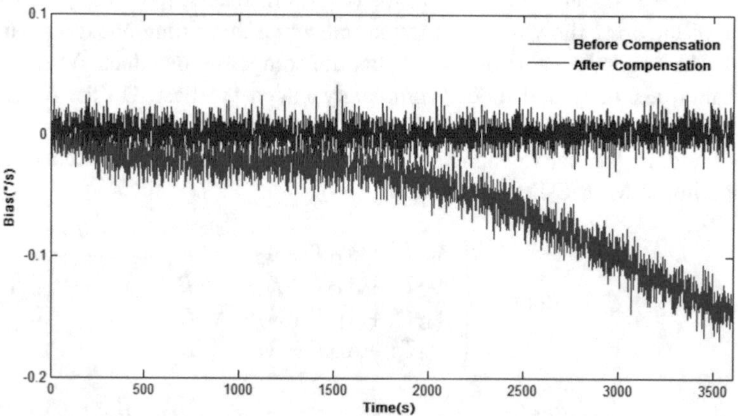

Fig. 3. The data before and after compensation of Y-axis gyroscope

The temperature test is performed on the compensated MEMS-IMU. The test procedure is as follows: Put the IMU in the high and low temperature box; Cool the temperature box to −40 °C, and keep it for two hours; collect the measurement information of the gyroscope output and take the temperature of 2 °C/min raises the temperature to 70 °C; finally, the measurement information of the gyroscope output is averaged by 1 s as Figs. 2, 3 and 4:

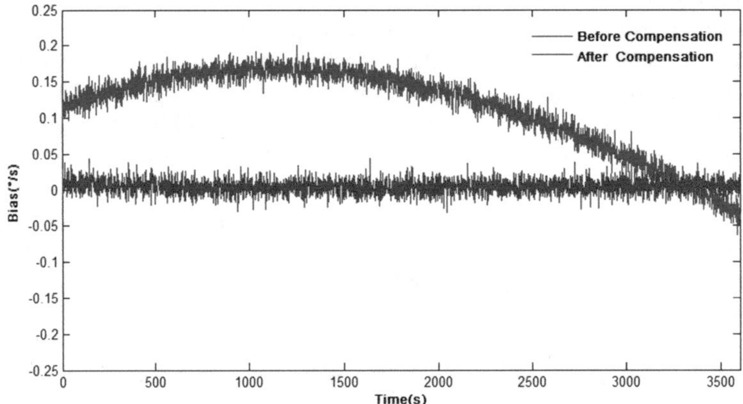

Fig. 4. The data before and after compensation of Z-axis gyroscope

According to Figs. 2, 3 and 4, it can be concluded that the gyroscope output trend before compensation is obvious, indicating that the gyroscope is easily affected by temperature changes, and the zero bias has a nonlinear relationship with the temperature change characteristics; the compensated gyroscope output trend is relatively flat. It is indicated that the compensation method can effectively suppress the drift caused by the temperature change. Further, the IMU is subjected to constant temperature static testing of high and low temperatures and the zero-bias temperature drift before and after temperature compensation is as shown in Table 2.

Table 2. Zero-temperature drift of gyro before and after temperature compensation

	Before compensation	After compensation
X	1085.76 °/h	18.72 °/h
Y	852.12 °/h	34.92 °/h
Z	438.48 °/h	56.52 °/h

It can be seen from Table 2 that the temperature drift after compensation is reduced by an order of magnitude compared with the temperature drift before compensation. The zero drift is reduced by 73.3%. It is indicated that the compensation method can effectively suppress the drift caused by temperature.

5 Conclusion

In order to meet the performance requirements of the multi-strap-type gyro in the application environment, it is necessary to adopt an effective and feasible method to improve the accuracy of the gyro and compensate the drift of the gyroscope. Due to the characteristics of MEMS gyroscope and the influence of the peripheral driving circuit, the MEMS gyroscope is easily affected by temperature and the accuracy is deteriorated. In order to avoid the traditional compensation method, the compensation delay is caused by the complexity of the model in practical engineering applications. In this paper, a second-order polynomial compensation model for temperature-divided regions is proposed by analyzing the mechanism of gyroscope zero-bias drift. The method firstly divides the ambient temperature of the MEMS gyro working; then separately models the zero offset of each region separately; finally solves the parameters by the least squares method and burns the model and parameters into the processor. The model can effectively reduce the drift temperature drift caused by temperature change, and can reduce the temperature drift after compensation by 73.3%, indicating that the compensation method can effectively suppress the drift caused by temperature.

Acknowledgements. This work was supported by the Shaanxi Natural Science Foun-dation (2016JQ5051) and the National Science Foundation for Young Scientists of China (51405387).

References

1. Li, S., Wang, X., Weng, H., et al.: Temperature compensation of MEMS gyroscope based on grey model and RBF neural network. J. Chin. Inert. Technol. **18**(6), 742–746 (2010). https://doi.org/10.13695/j.cnki.12-1222/o3.2010.06.002
2. Cheng, L., Wang, S., Ye, P.: Research on bias temperature compensation for micromachined vibratory gyroscope. J. Chin. Sens. Actuators **21**(3), 483–485 (2008)
3. Qin, W., Fan, W., Chang, H., et al.: Zero drift temperature compensation technology of MEMS gyroscope based on fuzzy logic. J. Proj. Guides **31**(6), 19–22 (2011). https://doi.org/10.15892/j.cnki.djzdxb.2011.06.019
4. Qin, W., Zeng, Z., Liu, G., et al.: Modeling method of gyroscope's random drift based on wavelet analysis and LSSVM. J. Chin. Inert. Technol. **16**(6), 721–724 (2008). https://doi.org/10.13695/j.cnki.12-1222/o3.2008.06.013
5. Chen, W., Chen, Z., Ma, L., et al.: Temperature characteristic analysis and modeling of MEMS micromachined gyroscope. J. Chin. Sens. Actuators **27**(2), 194–197 (2014). https://doi.org/10.3969/j.issn.1004-1699.2014.02.009
6. Xu, P., Wang, F., Dong, B., et al.: A low-cost adaptive MEMS gyroscope temperature compensation method. Micronanoelectron. Technol. **53**(8), 535–540, 562 (2016). https://doi.org/10.13250/j.cnki.wndz.2016.08.007
7. Zhao, X., Su, Z., Ma, X., et al.: Study on MEMS gyroscope zero offset compensation in large temperature difference application environment. J. Chin. Sens. Actuators **25**(8), 1079–1083 (2012). https://doi.org/10.3969/j.issn.1004-1699.2012.08.012
8. Sun, T., Liu, J.: A New method for modeling and compensating temperature error of MEMS gyroscope. Piezoelectrics Acoustooptics **39**(1), 136–139 (2017)

A Network Security Situation Awareness Model Based on Risk Assessment

Yixian Liu(✉) and Dejun Mu

School of Automation, Northwestern Polytechnical University,
Xi'an 710072, China
liu-yi-xian@xupt.edu.cn

Abstract. Network Security Situation Awareness (NSSA) can provide holistic status to administrator, and most related works rely on real time packet inspection technique to detect the security attacks which are happening and may already have caused some damage. In this paper, we propose the Risk Assessment NSSA model which collects the vulnerabilities information and uses corresponding risk level to qualitatively represent the security situation. This model is easy to apply and conveniently helps the administrator to monitor the whole network and be alerted to possible threat in future.

Keywords: Network security · Situation Awareness · Risk assessment Vulnerability

1 Introduction

Network technology plays a very important role in modern life, meanwhile network security is also endangered by multiple threats. To cope with these security issues, some methods can be implemented to enhance the security level of the network, such as firewall, intrusion detection and biometric authentication, etc. Most of these methods target the specific security problem and unable provide holistic security status of the network. Therefore Network Security Situation Awareness (NSSA) arises as more appropriate way to tackle different problems in the network, and it has already been researched in many different scenarios and aspects [1–3]. The traditional methodology of NSSA is to gather all kinds of information like log files on the server or packets through the router to detect potential attack in real time. Because of the inherently vulnerabilities of the assets, the network is still facing security risk without been attacked. In this paper, we propose a network security awareness model, Risk Assessment NSSA (RA-NSSA), which collects the vulnerabilities of the network and assesses the corresponding risk which indicates the security situation of the whole network.

In the rest of this paper, we introduce the related works and the motivation of our work in Sect. 2, Sect. 3 describes the modeling process of RA-NSSA in detail, Sect. 4 uses an example to demonstrate how the proposed model works, and Sect. 5 concludes the work and states the future research.

P. Krömer et al. (Eds.): ECC 2018, AISC 891, pp. 17–24, 2019.
https://doi.org/10.1007/978-3-030-03766-6_3

2 Related Works and Motivation

The concept of situation awareness (SA) was first mentioned by Endsley [4], and the main purpose is to help design the aircraft system. After years evolvement, SA is also great helpful in various fields for decision making [5–9] as well as in network security. Since Bass [10] applied NSSA in building more effective IDS, there have been a lot of works achieved significant progress in this topic. For example, Zhao and Liu [11] proposed a method which uses particle swarm optimization algorithm in big data environment. Zhang et al. [12] use DS evidence theory to fuse data submitted from heterogeneous network sensors like firewall, NIDS and HIDS etc., to infer the security situation. Another novel work was proposed in [13], which uses semantic ontology to define the essential object in the network, and follow the user-defined reasoning rule to automatically generate the current situation value.

All the works we mentioned above and some other NSSA works, mainly rely on the real time network traffic inspection technique, and cannot present the security situation of the network to IT administrator before the attack happened. The goal of this paper is to propose a model which checks the vulnerabilities in the network, calculates risk level and present it in qualitative manner as the security situation. This model is also capable of assisting the practitioner to mitigate the risk before actual attack happened.

3 Modeling of RA-NSSA

The general SA model which proposed by Endsley has three levels [14], first level is perception of the elements in the environment, second level is comprehension of the current situation and the last one is projection of future status. Our RA-NSSA model complies with the same data manipulation process, and has more specific task in each level according to the network security scenario. Figure 1 illustrates the structure of the RA-NSSA model.

3.1 Function of Perception Level

Understand the behavior pattern of hackers definitely helps to understand the functionality of the RA-NSSA model. Usually, attacks lunched by the adversary have similar methodology. Gather information and scanning are the prior steps before the exploitation. Gather information means before the hacker lunches an attack, he needs to know who is victim. With help of command line, website and search engine, the victim's information can be easily retrieved, including the target network's namespace server, web server, IP address ranges, etc. Among these tools and skills, Google hacking is really a powerful one and concerned the security researchers [15–17]. In the scanning stage, tools like Nmap, Nessus can be used to find the running assets and the corresponding vulnerabilities. After knowing the information of the vulnerabilities in the network, hacker can exploit the network purposely. According to the specific vulnerability, Attacker can send malicious data remotely to cause the target crashed or get the full access privilege of the target. The exploit tools can be acquired from the Internet and some are even been integrated into the operating system and security

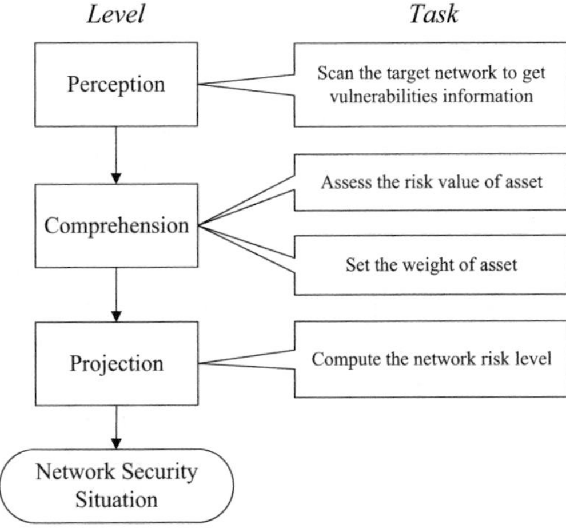

Fig. 1. Structure of RA-NSSA

testing framework like Kali Linux, Backtrack, Metasploit and so on, which are pre-installed with thousands embedded codes and tools. So in the perception level, our work is to find the vulnerabilities by using the same methodology to simulate the action before the real exploitation, which the attacker will do.

3.2 Function of Comprehension Level

Only have the vulnerability information cannot deduce the security situation of the network, even though most vulnerability severity can be queried from open source vulnerability database like National Vulnerability Database (NVD). Because the relationship of vulnerability, asset and risk are rather complicated due to particular network environment. Generally, the relationship can be illustrated in Fig. 2.

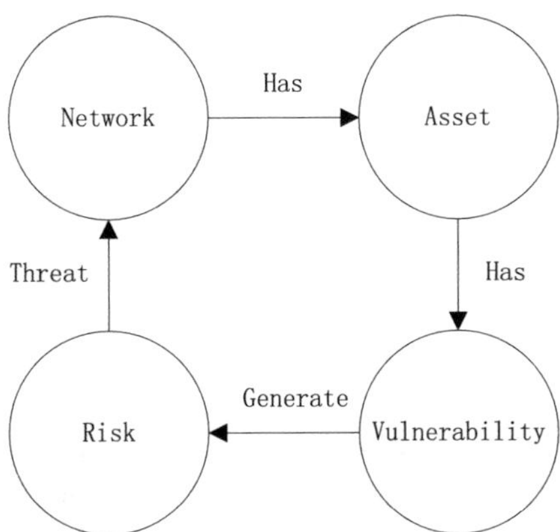

Fig. 2. Relationship of network, asset, vulnerability and risk

A network usually has multiple assets and every asset has multiple vulnerabilities. Same vulnerability on different assets may cause the impact to the network unequally. The reason is every asset has different importance in the network. For example, A workstation with no hard drive generates less risk than a server installed with database which contains confidential information of the company when they are both been compromised. So there are two issues need to be addressed in the second level i.e., comprehension level. The first one is every asset has multiple vulnerabilities has to be assigned a certain risk value, the second is every asset has to be endowed with appropriate weight to address the importance of the asset. For the first issue, if we have a set $\boldsymbol{V} = (v_1, v_2, \cdots, v_n)$, v_i represents the severity of ith vulnerabilities on a particular asset, n is the number of vulnerabilities on this asset, a function is needed to use the $\boldsymbol{V}$ as a parameter to calculate the risk value r of the asset. The function could be various, depending on the standpoint of what is the integrated impact of the vulnerabilities. If the administrator thinks the weakest loophole poses the most critical situation, the risk value r of the asset can be represented by the component which has maximum value in the $\boldsymbol{V}$, the function can be denoted as:

$$r = \max_{1 \le i \le n}(v_i), v_i \in \boldsymbol{V} \tag{1}$$

On the other hand, if the hacker chooses the objective randomly to exploit, the r can be represented by average value of all components in the $\boldsymbol{V}$, the function can be denoted as:

$$r = \frac{\sum_{i=1}^{n} v_i}{n}, v_i \in \boldsymbol{V} \tag{2}$$

These two functions are just the example or recommendation, other functions also can be applied in this process. For the second issue, the weight of each asset mainly needs the experts or specialists to determine according to their empirical analysis. Actually, there are some works in this topic which discuss how to systematically analyze the asset value for security management [18, 19]. These solution can be used in this process when there is lack of dependable expert opinion.

3.3 Function of Projection Level

At last, we have two sets, $\boldsymbol{R} = (r_1, r_2, \cdots, r_n)$ which stands for risk value of the assets in the network and $\boldsymbol{W} = (w_1, w_2, \cdots, w_n)$ stands for weight of the assets respectively, n is the number of assets. The risk value r_{ne} of the whole network can be denoted as:

$$r_{ne} = \sum\nolimits_i r_i \cdot w_i \tag{3}$$

So the risk value r_{ne} reflects the network security situation, and for intuitively to be observed, r_{ne} can be transformed to qualitative form by using linguistic expression like low, medium and high etc. One solution is divide the r_{ne} value range to corresponding continual sections, each section represents a specific situation. For example, as Table 1 demonstrated, range of r_{ne} is $[0, 10]$ and it is divided to four same length sections, the linguistic words low, medium, high and very high are the corresponding security situations. According which section the r_{ne} falls in, the security situation of the target network can be determined. More reasonable method for this issue is to apply the fuzzy logic if the original vulnerability emergency level is linguistic word, because it suits the human convention much better, and the result is easier for people to understand. Applying membership function is essential step of this process, and trapezoidal and triangular membership functions are commonly used which can be found in some security related analysis works [8, 20, 21].

Table 1. Risk value range and corresponding linguistic security situation

Risk value range	Security situation
[0,2.5)	Low
[2.5,5)	Medium
[5,7.5)	High
(7.5,10]	Very high

4 Implementation Example

In this section we will use an example to demonstrate how RA-NSSA model works. In traditional risk assessment, assets could be various, not only the physical device, the software, the operation manual and the personnel, etc., all could be the assets which influence the security situation of the organization. In our model, we only treat a node in the network as a distinct asset which could be routers, workstations and application server, etc. In the example, assuming the target network has 6 nodes i.e., 6 assets. In the perception level, the task is to collect the vulnerabilities information. To simulate the

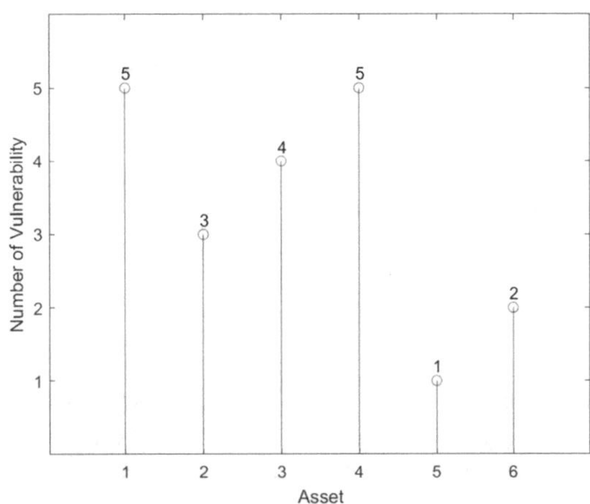

Fig. 3. Number of vulnerability on each asset

scanning result we randomly generate vulnerabilities information for each asset. The count of the vulnerability on each asset between 0 and 5, the emergency level of the vulnerabilities are randomly generated. Figure 3 shows the number of vulnerabilities on each asset and Fig. 4 shows the average severity of vulnerabilities on each asset.

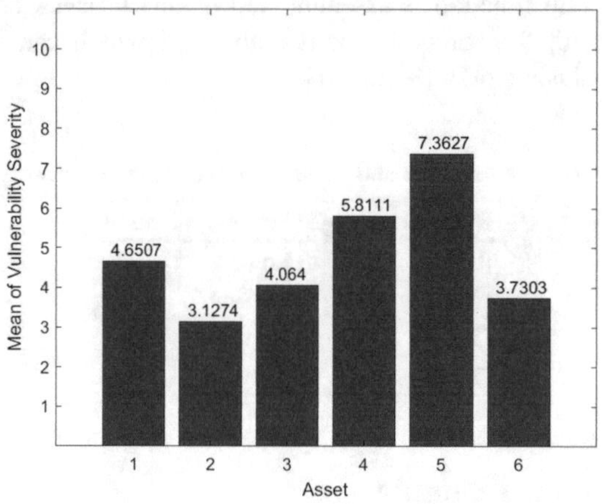

Fig. 4. Average severity of vulnerability on each asset

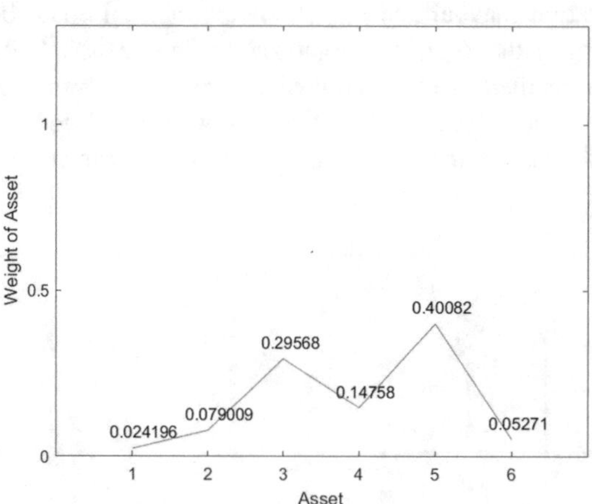

Fig. 5. Weight of each asset

In the comprehension level, we chose to use average value of vulnerabilities severity on each asset to represent the overall risk value of the particular asset. For each asset, we randomly assign the weight to simulate the process of expert empirical involvement. The asset weights are illustrated as Fig. 5.

Finally, we can compute the risk value of the whole network and convert the real number to qualitative manner according to the mapping relationship listed in Table 1. The security situation of the network is illustrated as Fig. 6.

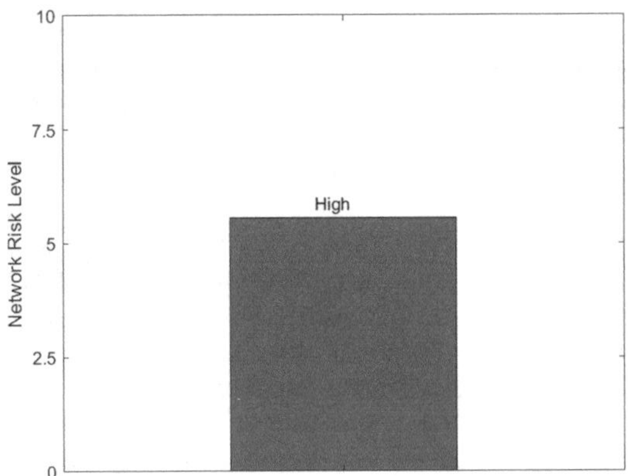

Fig. 6. Security situation of target network

5 Conclusion

NSSA is capable of providing holistic security view of the target Network, which other security techniques are rarely to achieve. Most NSSA approaches depend on real time network traffic inspection and detect the attack when it happens. In this paper we propose our RA-NSSA model which gathers the vulnerability information in the network and deduces the risk level of the target network in qualitative form. This model can help the administrator to master the network security status in the near future and take precaution against potential threat. For further research, how to measure the aggregate impact of the multiple vulnerabilities is our main direction, which will definitely enhance the performance of our model.

Acknowledgement. This research is supported by the China Natural Science Foundation (61672433).

References

1. Tao, H., Zhou, J., Liu, S.: A survey of network security situation awareness in power monitoring system. In: IEEE Conference on Energy Internet and Energy System Integration, pp. 1–3. IEEE, Beijing (2017). https://doi.org/10.1109/ei2.2017.8245487
2. Singh, M., Bhandari, P.: Building a framework for network security situation awareness. In: International Conference on Computing for Sustainable Global Development, pp. 2578–2583. IEEE, Delhi (2016)
3. Evangelopoulou, M., Johnson, C.W.: Attack visualisation for cyber-security situation awareness. In: IET International Conference on System Safety and Cyber Security, pp. 1–6. IEEE, London (2015). https://doi.org/10.1109/dese.2016.20

4. Endsley, M.R.: Design and evaluation for situation awareness enhancement. In: Human Factors Society-32nd Annual Meeting, CA, Santa Monica, pp. 97–101 (1988)
5. Vlahakis, G., Apostolou, D., Kopanaki, E.: Enabling situation awareness with supply chain event management. Expert Syst. Appl. **93**, 86–103 (2018). https://doi.org/10.1016/j.eswa.2017.10.013
6. Kalloniatis, A., Ali, I., Neville, T., La, P., Macleod, I., Zuparic, M., Kohn, E.: The situation awareness weighted network (SAWN) model and method: theory and application. Appl. Ergon. **61**, 178–196 (2017). https://doi.org/10.1016/j.apergo.2017.02.002
7. Wolf, F., Kuber, R.: Developing a head-mounted tactile prototype to support situational awareness. Int. J. Hum. Comput. Stud. **109**, 54–67 (2017). https://doi.org/10.1016/j.ijhcs.2017.08.002
8. Naderpour, M., Lu, J., Zhang, G.: A situation risk awareness approach for process systems safety. Saf. Sci. **64**, 173–189 (2014). https://doi.org/10.1016/j.ssci.2013.12.005
9. Yong, K.C.: Assessment of operator's situation awareness for smart operation of mobile cranes. Autom. Constr. **85**, 65–75 (2017). https://doi.org/10.1016/j.autcon.2017.10.007
10. Bass, T., Gruber, D.: A glimpse into the future of ID. In: The Magazine of USENIX & SAGE, vol. 24, pp. 40–45 (1999)
11. Zhao, D., Liu, J.: Study on network security situation awareness based on particle swarm optimization algorithm. Comput. Ind. Eng. (2018). https://doi.org/10.1016/j.cie.2018.01.006
12. Zhang, Y., Huang, S., Guo, S., Zhu, J.: Multi-sensor data fusion for cyber security situation awareness. Procedia Environ. Sci. **10**, 1029–1034 (2011). https://doi.org/10.1016/j.proenv.2011.09.165
13. Xu, G., Cao, Y., Ren, Y., Li, X., Feng, Z.: Network security situation awareness based on semantic ontology and user-defined rules for internet of things. IEEE Access **5**, 21046–21056 (2017). https://doi.org/10.1109/access.2017.2734681
14. Endsley, M.R.: Toward a theory of situation awareness in dynamic systems. Hum. Factors **37**, 32–64 (1995)
15. Munir, R., Mufti, M.R., Awan, I., Hu, Y.F., Disso, J.P.: Detection, mitigation and quantitative security risk assessment of invisible attacks at enterprise network. In: 2015 International Conference on Future Internet of Things and Cloud, FiCloud 2015 and 2015 International Conference on Open and Big Data, pp. 256–263. IEEE, Rome (2015). https://doi.org/10.1109/ficloud.2015.24
16. Mansfield-Devine, S.: Google hacking 101. Netw. Secur. **2009**(3), 4–6 (2009). https://doi.org/10.1016/s1353-4858(09)70025-x
17. Abdelhalim, A., Traore, I.: The impact of Google hacking on identity and application fraud. In: IEEE Pacific Rim Conference on Communications, Computers and Signal Processing, pp. 240–244. IEEE, Victoria (2007). https://doi.org/10.1109/pacrim.2007.4313220
18. Beaudoin, L., Eng, P.: Asset valuation technique for network management and security. In: IEEE International Conference on Data Mining Workshops 2006, ICDM Workshops, pp. 718–721. IEEE, Hong Kong (2006)
19. Loloei, I., Shahriari, H.R., Sadeghi, A.: A model for asset valuation in security risk analysis regarding assets' dependencies. In: 20th Iranian Conference on Electrical Engineering (ICEE2012), pp. 763–768. IEEE, Tehran (2012). https://doi.org/10.1109/iraniancee.2012.6292456
20. Deng, Y., Sadiq, R., Jiang, W., Tesfamariam, S.: Risk analysis in a linguistic environment: a fuzzy evidential reasoning-based approach. Expert Syst. Appl. **38**(12), 15438–15446 (2011). https://doi.org/10.1016/j.eswa.2011.06.018
21. Sendi, A.S., Jabbarifar, M., Shajari, M., Dagenais, M.: FEMRA: fuzzy expert model for risk assessment. In: Fifth International Conference on Internet Monitoring & Protection, pp. 48–53. IEEE, Barcelona (2010). https://doi.org/10.1109/icimp.2010.15

An Algorithm of Crowdsourcing Answer Integration Based on Specialty Categories of Workers

Yanping Chen[1,2], Han Wang[1], Hong Xia[1,2(✉)], Cong Gao[1,2], and Zhongmin Wang[1,2]

[1] School of Computer Science and Technology, Xi'an University of Posts and Telecommunications, Xi'an, China
{chenyp, xiahong, cgao, zmwang}@xupt.edu.cn, wanghan_0115@163.com

[2] Shannxi Key Laboratory of Network Data Analysis and Intelligent Processing, Xi'an University of Posts and Telecommunications, Xi'an, China

Abstract. The effective integration of crowdsourcing answers has become research hot spots in crowdsourcing quality control. Taking into account the influence of the specialty categories of workers on the accuracy of crowdsourced answers, a crowdsourced answer integration algorithm based on the specialty categories of workers is proposed(SCAI). Firstly, SCAI use the crowdsourced answer set to determine the difficulty of the task. Secondly calculate the accuracy of each crowdsourced answer, then obtain the professional classification of the workers and update the professional accuracy. Experiments were conducted on real data sets and compared with classical majority voting method(MV) and expectation maximization evaluation algorithm(EM). The results show that the proposed algorithm can effectively improve the accuracy of crowd-sourced answer.

Keywords: Crowdsourcing · Quality control · Answers integration
Specialty categories of workers

1 Introduction

Crowdsourcing is a distributed computing model that is solved by outsourcing tasks to a large number of workers. It has become a method to solve complex problems by using crowd's intelligence [1, 2]. At present, crowdsourcing has achieved marked successes in correlation evaluation [3], CrowdDB system [4] and data mining [5]. Workers may be unreliable when performing crowdsourcing tasks, and the quality of their tasks depends heavily on their skills, expertise, and behavior. It is acknowledged that crowdsourcing systems have been widely subjected to malicious activities such as fake reviews posted to online markets [6]. With the wide application of crowdsourcing, how to control the quality of results of crowdsourcing tasks is an important challenge in crowdsourcing applications [7].

Most of the existing algorithms employ the same accuracy that the same worker provides to all types of tasks to model the quality of the worker. In fact, the accuracy

P. Krömer et al. (Eds.): ECC 2018, AISC 891, pp. 25–35, 2019.
https://doi.org/10.1007/978-3-030-03766-6_4

for different types of task workers varies greatly. We proposes an algorithm of crowdsourcing answer integration based on specialty categories of workers, which introduce specialized accuracy to distinguish the quality of workers for different types of tasks. First, we use Bayesian probability formula to calculate the accuracy of the result. Then, we find the specialty categories of workers according to their historical records, and update the accuracy in different kinds task, so as to achieve efficient and accurate evaluation of crowdsourcing task results.

2 Related Work

In crowdsourcing quality control, a common technique is redundancy. Hence, an aggregating method majority vote [8, 9] to infer the final result of the task. But it does not consider the difference in accuracy between workers. Therefore, gold standard data is used for quality control to estimate the quality of worker, such as the MTurk system [10]. However, this method does not refer to the performance of workers in previous tasks. Liu et al. [11] obtained the accuracy of workers' answers in this way. Over time, iterative algorithms are increasingly used in crowdsourcing quality control. The most admired is the expectation maximization evaluation algorithm, in which a mixing matrix is used to reflect the quality of workers. Through the continuous iteration to evaluate the actual results of the task and the quality of the workers, until the convergence of the task results [12, 13].The iterative algorithm is difficult to set, resulting in the algorithm cannot converge.

[14] proposed a probability model based on factor graph to reflect the accuracy of workers. Zhang [15] proposed a phased strategy for evaluating crowdsourcing results through iterative operations of performance evaluation and worker replacement. [16] designed a new type of worker model to dynamically reflect the quality of the worker's answer. There is a problem that the accuracy of the different types of tasks performed by workers varies. Therefore, this paper proposes crowdsourcing quality control based on specialty categories of workers.

3 Crowdsourcing Answer Integration Based on Specialty Categories of Workers

This paper focuses on the problem of crowdsourcing answers integration based on the specialty categories of workers. The goal is to obtain the accuracy difference in different types of tasks based on the task answers given by the worker, and then to obtain the final answer of the task based on this difference. The formal definition is as follows: for a crowdsourcing task set $T = \{t_1, t_2, \ldots, t_k\}$ and the final result v_i of the crowdsourcing task t_j is obtained according to the result set $V(t_j) = \{v_1, v_2, \ldots, v_n\}$ of the task t_j given by the worker set $W = \{w_1, w_2, \ldots, w_m\}$, where t_j represents the *j-th* task, and n represents the number of different result values in the result set given by m workers.

The algorithm is divided into four parts: evaluating the difficulty of the task, calculating the score of result accuracy, detecting specialty categories of workers and updating the specialized accuracy. First to complete the evaluation of the difficulty of the task, if the result provided by each worker is concentrated in a small part of the answer that the task is less difficulty, the impact of the task on the accuracy of the worker is small; then calculate the accuracy of result values, the specialty categories of workers, and the specialized accuracy. The higher the accuracy of the worker, the more likely the result is to be the final result, and the more correct result given by the worker, the higher accuracy of the worker. Suppose the following conditions:

Suppose 1. A task has and only one final true answer and each task belongs to only one category.
Suppose 2. Workers are independent of each other to provide a result value for the task, and each worker can only provide one answer for the same task.
Suppose 3. The categories of tasks that different workers can accomplish is the same.

Table 1 show the important notations used in this paper and their meanings.

Table 1.

Notations	Meaning
$T = \{t_j \vert j \in [1,k]\}$	Task set
$W = \{w_i \vert i \in [1,m]\}$	Worker set
$V(t) = \{v_i \vert i \in [1,n]\}$	Different result value sets for tasks
$diff(t_j)$	Task difficulty
$A(v_i)$	The accuracy of the *i-th* answer
$T(w_i, c)$	The accuracy of worker w_i completes c-category tasks

3.1 Task Difficulty

The difficulty $diff(t_j)$ of a task t_j is the difficulty of providing the correct result for the task. For a task, if all the workers who participated in the task give a result focused on a small part of the answers, it means that the task is less difficult.

$$diff(t_j) = \frac{|V(t_j)|}{\max\{|V(t_l)| \,|\, l \in [1,k]\}} \tag{1}$$

However, the above method only considers the number of different result values of the task and does not consider the distribution of the result values. Therefore, the balance degree is used to describe the consistency of the result value distribution:

$$\mathrm{B}^{-1} = \frac{n}{\sum\limits_{v_i=1}^{n} (AR(v_i) - \overline{AR})^2} \tag{2}$$

where $AR(v_i)$ is the approval rating for the *i-th* result in $V(t_j)$, and $\overline{AR}$ is the average value of $AR(v_i)$:

$$AR(v_i) = \frac{M(v_i)}{\sum_{v_i=1}^{n} M(v_i)} \tag{3}$$

where $M(v_i)$ is the total number of supporters of the *i-th* result value in $V(t)$.

If the result set $|V(t_j)|$ of a task t_j is smaller, the difference in the number of proponents between each result is greater, the task is less difficult. Therefore, we use the sigmoid function method to evaluate the difficulty of the task.

$$diff(t_j) = \frac{1}{1 + e^{-\frac{\mu n}{\mathrm{B}}}} \tag{4}$$

3.2 Answer Accuracy

According to the assumptions 2, the Bayesian method can be used to calculate the probability $P(v_i)$ that each result value is the true result of the task. According to the Bayesian formula, we first obtain the probability $p(V(t_j)|v_i)$ of $V(t_j)$ under the condition that v_i is the real result. $p(V(t_j)|v_i)$ indicates that all worker $W(t_j, v_i)$ providing the result value of v_i for task t_j provides the correct answer, and that the result value provided for task t_j provides the probability of error result for all worker $W(t_j, \neg v_i)$ except v_i.

$$p(V(t_j)|v_i) = \prod_{w \in W(t_j, v_i)} T(w, c) * \prod_{w \in W(t_j, \neg v_i)} \frac{1 - T(w, c)}{|V(t_j)|} \tag{5}$$

where c is the category of task t_j.

In addition, calculate the probability $p(V(t_j))$ of $V(t_j)$, suppose that the prior probability $p_r(v_i)$ that v_i is true is the same in each result set $V(t_j)$, and set as γ:

$$p(V(t_j)) = \sum_{v_i \in V(t_j)} (\gamma * \prod_{w \in W(t_j, v_i)} T(w, c) * \prod_{w \in W(t_j, \neg v_i)} \frac{1 - T(w, c)}{|V(t_j)|}) \tag{6}$$

Therefore, the accuracy $P(v_i)$ of the result values is:

$$P(v_i) = p(v_i|V(t_j)) = \frac{p(V(t_j)|v_i) * p_r(v_i)}{p(V(t_j))}$$
$$= \frac{\prod_{w \in W(t_j, v_i)} T(w, c) * \prod_{w \in W(t_j, \neg v_i)} \frac{1 - T(w,c)}{|V(t_j)|} * \gamma}{\sum_{v_i \in V(t_j)} (\gamma * \prod_{w \in W(t_j, v_i)} T(w, c) * \prod_{w \in W(t_j, \neg v_i)} \frac{1 - T(w,c)}{|V(t_j)|})} \tag{7}$$

For all result values v_i in $V(t_j)$, $p(V(t_j))$ is the same, so it can be neglected.

$$P(v_i) = \prod_{w \in W(t_j, v_i)} T(w, c) * \prod_{w \in W(t_j, \neg v_i)} \frac{1 - T(w, c)}{|V(t_j)|} \tag{8}$$

In addition, in order to prevent the overflow, take the logarithm of $P(v_i)$ to obtain the accuracy score of the result value v_i:

$$A'(v_i) = \sum_{w \in W(t_j, v_i)} \ln(T(w, c)) + \sum_{w \in W(t_j, \neg v_i)} \ln(\frac{1 - T(w, c)}{|V(t_j)|}) \tag{9}$$

Further standardization, we obtain the accuracy score of v_i:

$$A(v_i) = \frac{A'(v_i)}{\sum_{v \in V(t_j)} A'(v)} \tag{10}$$

3.3 Detecting Specialty Categories of Workers

With the accumulation of workers' knowledge, their accuracy of completing tasks is also changing. If the quality of the two categories of tasks is the same for the worker, it is not necessary to distinguish the difference in accuracy between these two categories. In this paper, we obtain the specialty categories of workers based on a fixed category detection method.

First, construct a hierarchical tree of the worker according to the category of historical tasks completed by the worker. Starting from the root node, we traverse all child nodes of the tree in turn using the breadth traversal method.

Then, we use the degree of dispersion of the accuracy of the result values of this category of tasks given by the worker to detect whether it needs to distinguish the credibility of its subclasses. We use the standard deviation to measure the degree of dispersion of the accuracy of the result values. if the degree of dispersion is less than the threshold ε, we delete all the subtrees of the current node. Otherwise, we push the root node of the node subtree onto the stack until the stack is empty. Thus, we can obtain specialty categories C of workers.

3.4 Specialized Accuracy

Worker specialized accuracy is the possibility of workers providing real results for a certain type of task. Therefore, the accuracy $T(w_i, c)$ is the average of all the provided task attribute values.

$$T'(w_i, c) = \frac{\sum_{t_j \cap v} A(v)}{M(w_i, c)} \tag{11}$$

where, $t_j \in t(w_i, c)$, $v = Value(w_i, t_j)$, $t(w_i, c)$ denotes the set of c-class tasks completed by worker w_i, $Value(w_i, t_j)$ denotes the result value provided by worker w_i for task t_j, and $M(w_i, c)$ denotes the total number of c-class tasks completed by worker w_i.

Because the worker's ability to perform on different difficulty tasks is different, the ability of the worker to handle the task with lower difficulty is relatively higher. Therefore, we have added the task difficulty in $R(w_i, c)$ to consideration.

$$T''(w_i, c) = \frac{\sum_{t_j \cap v} diff(t_j) * A(v)}{M(w_i, c)} \tag{12}$$

In addition, the total amount of tasks accomplished by workers also affects the credibility of workers. Therefore, improving the above formula yields the following worker accuracy calculation formula:

$$T(w_i, c) = \frac{M(w_i, c)}{M(-, c)} * \frac{\sum_{t_j \cap v} diff(t_j) * A(v)}{M(w_i, c)} \tag{13}$$

where $M(-, c)$ indicates the number of tasks for all c category.

3.5 Algorithm Steps

Algorithm 1: An algorithm of crowdsourcing answer integration based on specialty categories of workers.

Input: answer set, initial worker specialized accuracy $T_0(w, c)$, number of iterations r

Output: task result

1. Determine whether the number of different results n in the result set is 1. If yes, then it ends; if not, go to step 2.
2. Calculate the difficulty of the task according to formula (4), and get the difficulty value of the task $diff(t_j)$.
3. Calculate the accuracy score for each result $A(v_i)$ based on Equation (10).
4. Get specialty categories C of workers.
5. Calculate the specialized accuracy $T_i(w, c)$ of each type in C according to equation (13).
6. Determine whether the number of iterations reaches r. If yes, step 7; if not, step 3.
7. Return to the result value v_i with the highest accuracy score obtained in step 6.

4 Experimental Analysis

During the experiment, the true answer for each task in the dataset is known. By running the algorithm on the dataset to get the answer for each task and comparing it with the real answer, the accuracy of the algorithm is obtained. At the same time, the impact of the parameters on performance is judged by changing the value of the parameters. Finally, compare the accuracy and efficiency of the SCAI algorithm with EM algorithm and MV algorithm.

4.1 Dataset

We experimented on a real-world data set also used in [17]. We removed duplicate records and records provided by only one worker. In the data set there are 877 bookstores, 1263 books, and 24364 listings, each listing contains a list of authors on a book provided by a bookstore. We regard bookstores as workers, books as tasks and listings as answer of tasks provided by workers. And randomly selected 600 books, and each book randomly selected 9 listings. A total of 163 workers participated in. In addition, we obtain the classification information and author of these books from Amazon.com.

4.2 Parameter Sensitivity Analysis

4.2.1 μ

The algorithm of this paper uses the sigmoid function method to assess the difficulty of the task. When the total number of different results n in the result set of task t_j is greater or the task's balance B^{-1} is higher, the task is more difficult. The difficulty curve of the task changes with the change of the coefficient, that is, it affects the difficulty of the task, and then affects the accuracy of the algorithm. Figure 1 shows the effect of different values of μ on the accuracy of the results.

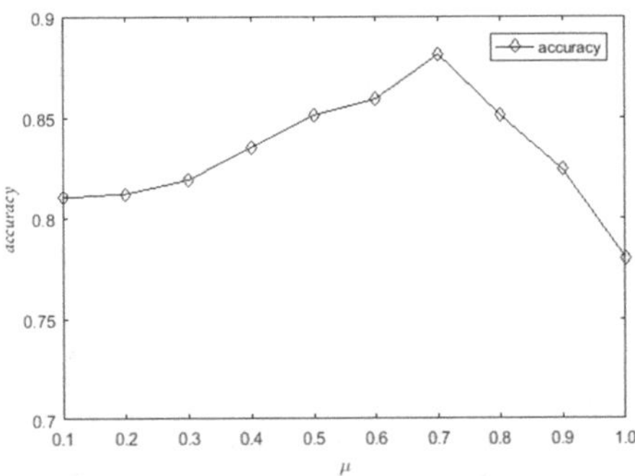

Fig. 1. The effect of the value of μ on the accuracy.

As can be seen from Fig. 1, when $\mu < 0.7$, with the increase of μ, the accuracy of the task results increased; $\mu > 0.7$, the accuracy of the task gradually decreased. This is because when $\mu < 0.7$, the difficulty distribution range of the task gradually increases, and the degree of distinction between tasks of different difficulty levels is more obvious, so the accuracy rate gradually increases; and when $\mu > 0.7$,the difficulty of all tasks is greatly increased, so the accuracy of the task drops at a faster rate.

4.2.2 ε

In the workers' specialty categories detection process of the algorithm, the threshold ε is used to judge whether worker needs to further distinguish its subclass nodes at a task classification node. Different ε values affect the specialty categories of workers, which in turn affects the specialized accuracy of workers. Therefore, the accuracy of the algorithm changes with ε changes. Figure 2 shows the effect of different values of ε on the accuracy of the results.

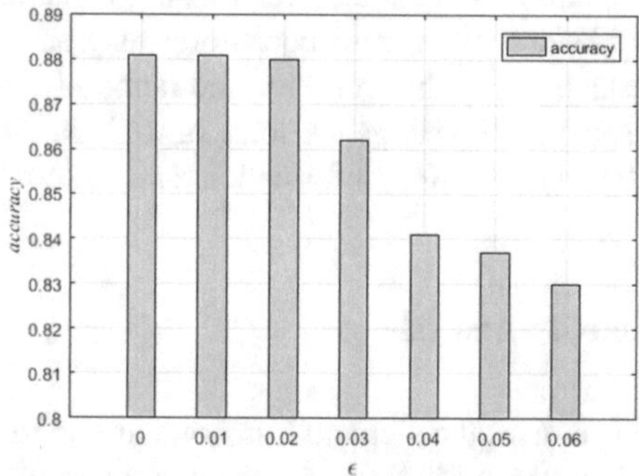

Fig. 2. The effect of the value of ε on the accuracy

In the specialty categories detection of workers, when the standard deviation of the accuracy of the result value is greater than ε, the detection is continued. Therefore, the larger ε, the lower the detection efficiency of specialty categories is and the lower the accuracy of the algorithm is. The smaller ε is, the higher the amount of calculation of the algorithm is. According to Fig. 2, the accuracy of the algorithm decreases as the value of ε increases. When $\varepsilon < 0.02$, the accuracy of the algorithm is basically stable; when $\varepsilon > 0.02$, the accuracy of the algorithm begins to decrease.

4.3 Algorithm Accuracy Comparison

According to the above parameter sensitivity experiment results, we can see that the accuracy of the SCAI algorithm reaches a peak when $\mu = 0.7$, $\varepsilon = 0.02$. Therefore, in following part of the experiment we set $\mu = 0.7$, $\varepsilon = 0.02$.

Figure 3 shows the algorithm accuracy of each algorithm in the case of inconsistent total number of task answers. It can be seen that the SCAI algorithm has the best results. With the increase of the number of answers, the accuracy of MV algorithm fluctuates little, while the accuracy of the other two algorithms is gradually improved. Among them, the accuracy of the EM algorithm is lower than that of the SCAI algorithm. This is because the MV algorithm directly uses the answer given by the worker to infer the task result, ignoring the influence of the worker's quality on the task result. The confusion matrix is used in the EM algorithm to estimate the quality of the overall answer for each worker. Therefore, when the number of worker answers is few, the confusion matrix is less accurate, which leads to the low quality of the result. But when all the workers' answers are received, the accuracy of the EM algorithm and the SCAI algorithm was very close.

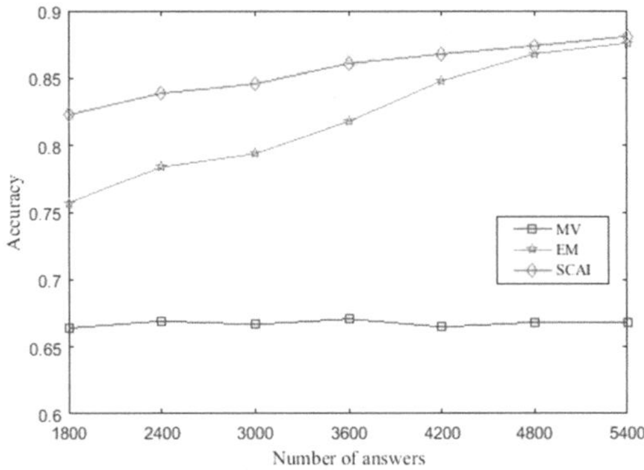

Fig. 3. The effect of the number of task answers on the accuracy

4.4 Algorithm Operation Efficiency Evaluation

Figure 4 shows the effect of the number of answer on the algorithm running time. Observation found that the more the number of task answers received, the more time the algorithm runs. Among them, the MV algorithm has the least running time, because the MV algorithm is simple in calculation, high in execution efficiency, but low in accuracy. Since the EM algorithm and the SCAI algorithm all adopt the iterative method, so the time complexity is high. When the number of task answers is small, the EM algorithm is superior to the SCAI algorithm, but as the number of answers increases, the running time of the EM algorithm increases obviously. This is because the EM algorithm first uses the accuracy of the existing worker to answer the question to estimate prior probability of the crowdsourcing task results, and then calculates the worker's whole answer accuracy. Therefore, the total amount of the task answer has a greater impact on its calculation. While the running time of the SCAI algorithm is constantly increasing, it shows obvious advantages when there are a large number of answers, and a large accuracy of results when the number of answers is small.

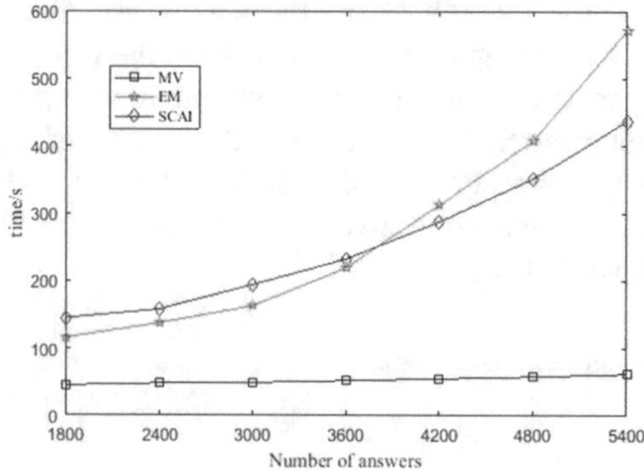

Fig. 4. The running time of the number of task answers

5 Conclusions

In this paper, we presented answer integration models that aim to predict the true labels from a set of labels gathered from workers for crowdsourced task. By detecting the worker's specialty categories, we can update the quality change of the workers to accomplish the task effectively and accurately evaluate the results of the crowdsourcing task. The experiment verifies that the task results are evaluated by introducing the specialty categories of the workers with high accuracy. In the follow-up work, we will study the relationship between the worker's response time and other implicit parameters and accuracy, so that the result of crowdsourcing is of higher quality.

Acknowledgements. This research is supported by the National Natural Science Foundation of China (61373116) and Science and the Technology Project in Shaanxi Province of China (Program No. 2016KTZDGY04-01) and the International Science and Technology Cooperation Program of the Science and Technology Department of Shaanxi Province of China (Grant No. 2018KW-049), and the Special Scientific Research Program of the Education Department of Shaanxi Province of China (Grant No. 17JK0711).

References

1. Allahbakhsh, M., Benatallah, B., Ignjatovic, A., Motahari-Nezhad, H.R., Bertino, E., Dustdar, S.: Quality control in crowdsourcing systems: issues and directions. IEEE Internet Comput. **17**(2), 76–81 (2013). https://doi.org/10.1109/MIC.2013.20
2. Feng, J.H., Guo-Liang, L.I., Feng, J.H.: A survey on crowdsourcing. Chin. J. Comput. **38**(9), 1713–1726 (2015). https://doi.org/10.11897/SP.J.1016.2015.01713
3. Alonso, O., Mizzaro, S.: Can we get rid of TREC assessors? using mechanical turk for relevance assessment. In: SIGIR Workshop on the Future of IR Evaluation, pp. 19–23 (2009). https://doi.org/10.1016/j.ipm.2012.01.004
4. Franklin, M.J., Kossmann, D., Kraska, T., Ramesh, S., Xin, R.: CrowdDB: answering queries with crowdsourcing. In: ACM SIGMOD International Conference on Management of Data, pp. 61–72. ACM (2011).https://doi.org/10.1145/1989323.1989331

5. Lease, M., Carvalho, V.R., Yilmaz, E.: Crowdsourcing for search and data mining. ACM SIGIR Forum **45**(1), 18–24 (2011). https://doi.org/10.1145/1988852.1988856
6. Alabduljabbar, R., Al-Dossari, H.: A task ontology-based model for quality control in crowdsourcing systems. In: International Conference on Research in Adaptive and Convergent Systems, pp. 22–28. ACM (2016). https://doi.org/10.1145/2987386.2987413
7. Li, G., Fan, J., Fan, J., Wang, J., Cheng, R.: Crowdsourced data management: overview and challenges. In: ACM International Conference on Management of Data, pp. 1711–1716. ACM (2017). https://doi.org/10.1145/3035918.3054776
8. Muhammadi, J., Rabiee, H.R., Hosseini, A.: A unified statistical framework for crowd labeling. Knowl. Inf. Syst. **45**(2), 271–294 (2015). https://doi.org/10.1007/s10115-014-0790-7
9. Yue, D.J., Ge, Y.U., Shen, D.R., Xiao-Cong, Y.U.: Crowdsourcing quality evaluation strategies based on voting consistency. J. Northeast. Univ. **35**(8), 1097–1101 (2014). https://doi.org/10.3969/j.issn.1005-3026.2014.08.008
10. Ipeirotis, P.G., Provost, F., Wang, J.: Quality management on Amazon Mechanical Turk. In: ACM SIGKDD Workshop on Human Computation, pp. 64–67. ACM (2010). https://doi.org/10.1145/1837885.1837906
11. Liu, X., Lu, M., Ooi, B.C., et al.: CDAS: a crowdsourcing data analytics system. In: Proceedings of the VLDB Endowment (2012). https://doi.org/10.14778/2336664.2336676
12. Ding, Y., Wang, P.: Quality control algorithm research of crowdsourcing based on social platform. Softw. Guide **16**(12), 90–93 (2017). https://doi.org/10.11907/rjdk.171970
13. Zheng, Z., Jiang, G., Zhang, D., et al.: Crowdsourcing quality evaluation algorithm based on sliding task window. Small Microcomput. Syst. **38**(09), 2125–2129 (2017). https://doi.org/10.3969/j.issn.1000-1220.2017.09.038. **5**(10), 1040–1051
14. Demartini, G., Difallah, D.E., Cudré Mauroux, P.: ZenCrowd: leveraging probabilistic reasoning and crowdsourcing techniques for large-scale entity linking. In: International Conference on World Wide Web, pp. 469–478. ACM (2012). https://doi.org/10.1145/2187836.2187900
15. Zhang, Z.Q.: Research on crowdsourcing quality control strategies and evaluation algorithm. Chin. J. Comput. **36**(8), 1636–1649 (2013). https://doi.org/10.3724/SP.J.1016.2013.01636
16. Feng, J., Li, G., Wang, H., Feng, J.: Incremental Quality Inference in Crowdsourcing. In: International Conference on Database Systems for Advanced Applications, vol. 8422, pp. 453–467. Springer (2014). https://doi.org/10.1007/978-3-319-05813-9_30
17. Yin, X., Han, J., Yu, P.S.: Truth discovery with multiple conflicting information providers on the web. In: ACM SIGKDD International Conference on Knowledge Discovery and Data Mining, vol. 20, pp. 1048–1052. ACM (2007). https://doi.org/10.1109/tkdE.2007.190745

A Dynamic Load Balancing Strategy Based on HAProxy and TCP Long Connection Multiplexing Technology

Wei Li(✉), Jinwei Liang, Xiang Ma, Bo Qin, and Bang Liu

Xi'an University of Posts and Telecommunications, Xi'an 710121, China
wli@xupt.edu.cn

Abstract. A load balancing strategy based on HAProxy and TCP long connection multiplexing technology is proposed to solve the waste of network resources caused by TCP requests in the data storage module of electrical energy management system. First, based on the load capacity and processing capacity of the cluster server, a server with the best performance is selected to establish a long connection. Then the proposed TCP long connection multiplexing technology is used to reuse different TCP requests to this long TCP connection. Experiments show that this strategy not only enhances the processing capacity of the server, but also dynamically selects the best server. Compared with the default load balancing strategy in HAProxy, it has higher feasibility and stability.

Keywords: Multiplexing of TCP · Load balancing · HAProxy

1 Introduction

With the development of the Internet of things technology and the rapid increase of sensor nodes, the traditional server mode cannot support the increasing concurrent requests on the back end of the data storage module of the electrical energy management system. How to reduce the load of servers and ensure High Availability of the electrical energy management system become an urgent problem to be solved [1]. Load balancing technology based on HAProxy builds a cluster system by connecting multiple servers together. From the inside of the cluster, the client requests are distributed to the servers by HAProxy through a certain load balancing policy. It solves the performance bottleneck and low efficiency of a single server particularly well [2].

Due to the device of the electric energy management system will send real-time datas to the data storage module, the device and the module frequently establish TCP connections. Every data transfer takes three connection handshakes and four disconnection handshakes. It is bound to consume resources and time [3]. In this paper, Combine with the electric energy management system, design a data storage module based on database cluster, use HAProxy as the TCP proxy server of the cluster. The optimized load balancing algorithm is used to dynamically establish long TCP connections with the back-end server cluster. This paper proposes a TCP long connection multiplexing technology, which adds session IDs and encapsulates TCP connection

P. Krömer et al. (Eds.): ECC 2018, AISC 891, pp. 36–43, 2019.
https://doi.org/10.1007/978-3-030-03766-6_5

requests from different clients on the HAProxy server, to Realize the reuse of TCP long connection to the back end and reduce the number of TCP long connection.

2 Data Storage Module

The overall architecture of the data storage module of the power management system is shown in Fig. 1. Database cluster is adopted for distributed storage of data, and HAProxy is used to distribute and forward different requests on the device access side. First, HAProxy uniformly receives the database connection request from the device access side. Then, based on the improved dynamic load balancing strategy, a long TCP connection is established with a server on the back-end, and the proposed TCP connection multiplexing technology is used to reuse requests from different devices to a long connection. Then, multiple sessions can share the same data channel.

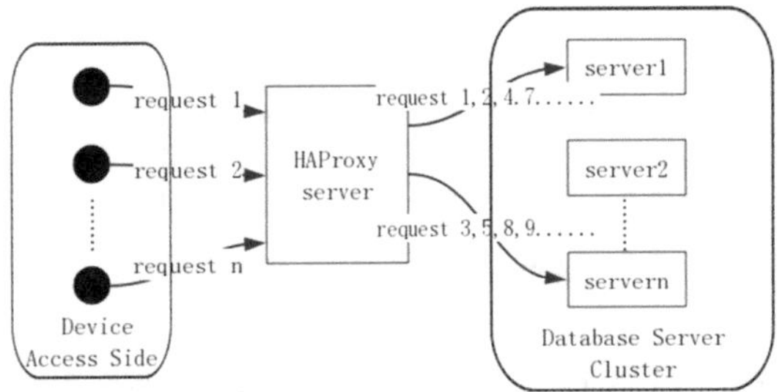

Fig. 1. The overall architecture of the data storage module

3 TCP Long Connection Multiplexing Technology

3.1 TCP Protocol

TCP is a reliable data transmission protocol, providing services for many applications [4]. With the diversification of network business, the use of TCP has many limitations. In the electrical energy management system, the device access module generates frequent data transmission services with the data storage module. There will be a lot of problems when a large number of TCP connections impact the server at the same time:

1. Every time a TCP connection is established, three handshakes are required, and resources are redistributed for each TCP connection, resulting in a waste of server resources [5].
2. The server has a limited number of connections, and it is easy to fail if a large number of connections are requested simultaneously.
3. In some cases, the network only allows one ip address to use a TCP connection, which imposes restrictions on the access module to transmit different data.

3.2 TCP Long Connection Multiplexing Technology

TCP Long connection multiplexing technology is a shared transmission of different TCP connections through the same TCP long connection to improve transmission efficiency and save network resources [6]. According to the existing theoretical support, the HAProxy is transformed into a load balancing server composed of the event-driven module, the TCP multiplexing module and the connection management module. The different TCP connections from the device access side are managed by the Haproxy event-driven module; the TCP Multiplexing module is responsible for managing and distinguishing the different sessions on the same TCP long connection; and the connection Management module is responsible for selecting a reasonable load balancing strategy and establishing and maintaining the TCP long connection with the upstream server. The concrete model diagram is shown in Fig. 2:

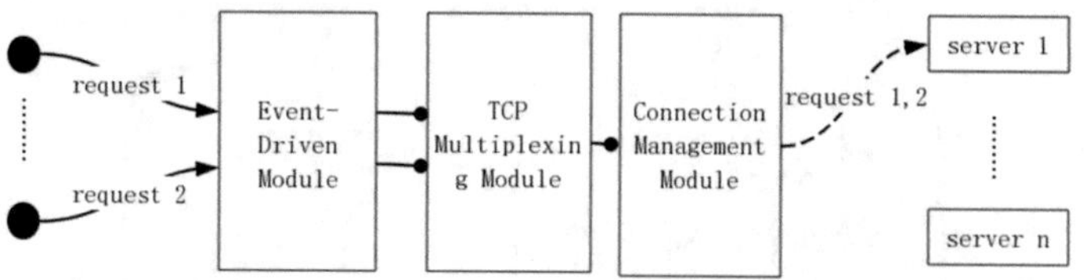

Fig. 2. The model diagram of TCP connection multiplexing technology

The key technology of TCP connection multiplexing is how to correctly distinguish multiple sessions when the back-end connection shares the same TCP long connection. For a traditional TCP protocol, a session produces a connection. The multiplexed TCP long connection transmits data from multiple clients at a time, so different data units need to be distinguished. Data units for TCP protocol are transmitted as messages, and the TCP segment are divided into the header and the data. The TCP header format is shown in Fig. 3.

<table>
<tr><td colspan="4">source port</td><td colspan="5">destination port</td></tr>
<tr><td colspan="9">serial number</td></tr>
<tr><td colspan="9">confirmation number</td></tr>
<tr><td>data offset</td><td>reserved bits</td><td>1</td><td>2</td><td>3</td><td>4</td><td>5</td><td>6</td><td>window</td></tr>
<tr><td colspan="4">checksum</td><td colspan="5">urgent pointer</td></tr>
<tr><td colspan="8">Options(variable-length)</td><td>filling</td></tr>
</table>

Fig. 3. TCP segment header format

The TCP header has a 6 bit reserved field that was not used at the beginning of the design and was all set to 0 States for scientific researchers to develop and use. By modifying this 6 bit reserved field, it is set to the session ID, and the TCP multiplexing module increases the session ID for each incoming TCP connection to differentiate different sessions that are transmitted on the same TCP long connection. Theoretically a TCP long connection can reuse up to 2^6 different sessions, but the number of reuses is much smaller.

When HAProxy receives a database connection request from the device access side, it quantifies the upstream servers and selects one of the servers that meets its load-balancing policy, then creates or uses a TCP long connection to the server, and creates a unique session ID for the current request. At this point, the TCP long connection between the Haproxy backend and the upstream server cluster transmits data with different session IDs, and the upstream server distinguishes the data sent by different devices according to the session IDs.

4 Dynamic Load Balancing Strategy

When Haproxy backend has a long TCP connection with the upstream server cluster, first, the servers in the cluster will be quantified and analyzed. Based on the analysis results, the HAProxy server selects the server with the best indexs to establish or reuse the TCP long connection of the server. Server performance quantification can well reflect the load status and processing capacity of the current cluster, providing feasibility analysis for selecting the appropriate upstream server.

4.1 Load Information Quantification

Server load informations includes CPU Utilization L(Ci), memory Utilization L(Mi), network bandwidth Utilization L(Ni), process Quantity Utilization L(Pi), disk I/O Utilization L(Di) [7]. Because the load informations has different influence factors on the server load status, the influence factor of different load information on the comprehensive load is indicated by introducing the parameter $\alpha = \{0.2, 0.25, 0.15, 0.25, 0.15\}$, which emphasizes the importance of CPU, Network bandwidth Utilization and disk to data transmission services in the electrical energy management system. After conversion, the indexs are synthesized to the integrated load information L(Si) of the current cluster server, and the transformation function is shown in the Eq. (1).

$$L(S_i) = [0.2, 0.15, 0.25, 0.15, 0.25] \begin{bmatrix} L(C_i) \\ L(M_i) \\ L(N_i) \\ L(P_i) \\ L(D_i) \end{bmatrix}, \ i = 0, 1, \cdots, n-1 \tag{1}$$

4.2 Processing Capacity Quantification

The main indicators for calculating the processing power of the server are CPU type C(Ci), number of CPUs mi, request response time C(ti), memory capacity C(Mi), network throughput C(Ni), maximum number of processes C(Pi), Disk I/O rate C(Di). Similarly, by introducing the parameters β = {0.1, 0.15, 0.15, 0.25, 0.15, 0.2}, the influence factors of different server parameters on the processing performance are indicated to emphasize the network throughput and the disk to in data transmission. The comprehensive processing capacity of the server C(Si) is as shown in Eq. (2).

$$C(S_i) = [0.1, 0.15, 0.15, 0.25, 0.15, 0.2] \begin{bmatrix} m_i C(C_i) \\ C(T_i) \\ C(M_i) \\ C(N_i) \\ C(P_i) \\ C(D_i) \end{bmatrix}, \; i = 0, 1, \cdots, n-1 \quad (2)$$

4.3 Comprehensive Quantitative Index of Server

According to the comprehensive load capacity and comprehensive processing performance of the server, the parameter γ = {0.6, 0.4} is introduced to convert the comprehensive load amount L(Si) and the comprehensive processing capacity C(Si) to obtain a preliminary comprehensive quantitative index of the server $Q'(S_i)$.

$$Q'(S_i) = [\gamma_1, \gamma_2] \begin{bmatrix} L(S_i) \\ C(S_i) \end{bmatrix}, \; i = 0, 1, \cdots, n-1 \quad (3)$$

$P(N_i)$ and $R(N_i)$ are introduced to represent the number of preset and actual multiplexes on the TCP long connection of the current server. The residual reuse rate is calculated as Re:

$$\mathrm{Re} = 1 - \frac{R(Ni)}{P(Ni)}, \; i = 0, 1, \cdots, n-1 \quad (4)$$

Then, convert to final comprehensive quantitative index of the server $Q(S_i)$. The conversion formula is as shown in Eq. (5):

$$Q(Si) = \mathrm{Re} * Q'(Si), \; i = 0, 1, \cdots, n-1 \quad (5)$$

5 Experimental Analysis

In order to verify the effectiveness of the dynamic load balancing strategy based on HAProxy and TCP long connection multiplexing technology, we apply it to the electrical energy management system of a company. We use weighted round-robin algorithm and dynamic load balancing strategy to experiment. By comparing the average weighted delay and the throughput of the system to validate this strategy.

The average response delay of different load balancing strategies is shown in Fig. 4. HAProxy server uses the weighted round-robin algorithm for its default load balancing policy. As can be seen from the figure, when the number of connections is small, the default weighted round-robin algorithm of HAProxy does not need to comprehensively quantify cluster server performance, and the speed is faster than the load balancing strategy using TCP long connection multiplexing. When the number of connections is greater than 150–200, the load balancing strategy proposed in this paper can significantly reduce the average response delay, which proves the advantage of TCP long connection multiplexing technology in high concurrent data environment.

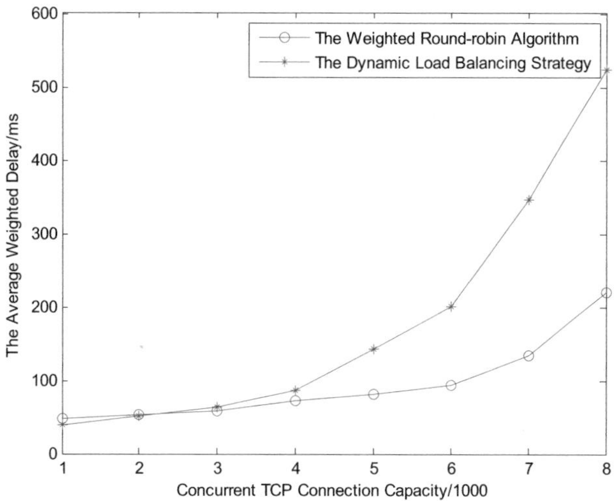

Fig. 4. Average response time delay comparison

The throughput comparison between the two load balancing strategies is shown in Fig. 5. It can be seen from the figure that the load balancing strategy using TCP long connection multiplexing technology can significantly improve the system throughput when the amount of concurrency is large. And within a certain range, the greater the number of concurrency, the more significant the effect. The default weighted round-robin algorithm of HAProxy server is gradually saturated after the concurrent quantity is greater than 5000–6000. Even with the increase of concurrency, server doesn't processing timely, resulting in the consumption of network resources, failure to successfully establish connections, and the situation of throughput decline.

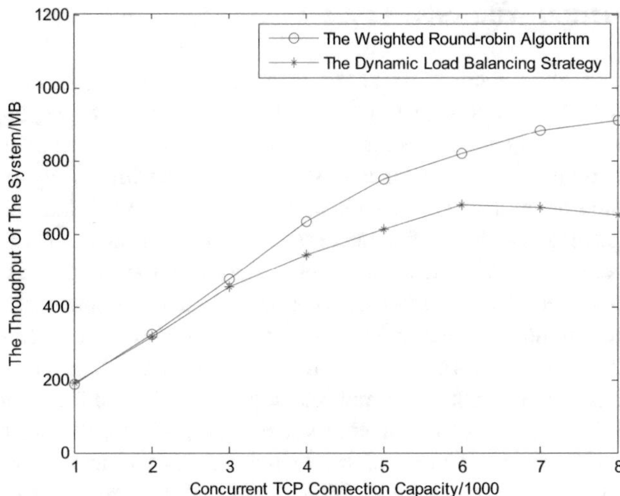

Fig. 5. Throughput comparison

6 Conclusion

This paper proposes a dynamic load balancing strategy based on HAProxy and TCP long connection multiplexing technology. HAProxy is used as the load balancer and TCP reverse proxy server of the power management platform. By modifying the reserved field of the TCP packet, the sessions is accurately distinguished. And through the comprehensive evaluation of the load information and processing capability of the back-end server, the comprehensive quantitative indicators of each server are further obtained. Experimental analysis proves that the dynamic load balancing strategy can significantly reduce the average response delay, greatly improve the throughput, and better cope with high concurrency requirements.

Acknowledgements. This work was supported by Shaanxi Province Technical Innovation Guide Special Project (2018SJRG-G-03) and Shaanxi education department industrialization project (16JF024).

References

1. Zhou, L., Wang, F.: Study on data load balancing method of railway passenger ticket system. Railw. Transp. Econ. **40**(5), 46–50 (2018). https://doi.org/10.16668/j.cnki.issn.1003-1421.2018.05.09
2. Liu, K.: HAProxy is used to realize the Web load balance of course selection system. Comput. Knowl. Technol. **7**(1), 35–36 (2011). https://doi.org/10.3969/j.issn.1009-3044.2011.01.015
3. Wang, W., Hao, X., Duan, G., et al.: The multiplexing mechanism of concurrent TCP in satellite network. J. Cent. South Univ. (Sci. Technol.) **48**(3), 712–720 (2017). https://doi.org/10.11817/j.issn.1672-7207.2017.03.020

4. Liu, M.: Factors influencing TCP transmission rate and optimization methods. China Comput. Commun. (8), 87–88 (2016). https://doi.org/10.3969/j.issn.1003-9767.2016.08.041
5. Dou, L., Lu, X., Duan, H.: Design and implementation of multiplexed network model based on TCP. Appl. Res. Comput. (6), 245–247 (2006). https://doi.org/10.3969/j.issn.1001-3695.2006.06.081
6. Zhou, S.: Research and implementation of TCP long connection multiplexing based on HAProxy. South China University of Technology (2011)
7. Liu, J., Xu, L., Zhang, W.: Load balancing algorithm based on dynamic feedback. Comput. Eng. Sci. **25**(5), 65–68 (2003). https://doi.org/10.3969/j.issn.1007-130x.2003

An Improved LBG Algorithm for User Clustering in Ultra-Dense Network

Yanxia Liang[1(✉)], Yao Liu[2], Changyin Sun[1], Xin Liu[3], Jing Jiang[1], Yan Gao[1], and Yongbin Xie[1]

[1] Shaanxi Key Laboratory of Information Communication Network and Security, Xi'an University of Posts and Telecommunications, Xi'an 710121, China
liangyanxia201@xupt.edu.cn
[2] Department of Computer Science and Engineering, University of South Florida, Tampa, FL 33647, USA
[3] School of Information Engineering, Xi'an Eurasia University, Xi'an 710065, China

Abstract. A novel LBG-based user clustering algorithm is proposed to reduce interference efficiently in Ultra-Dense Network (UDN). There are two stages, weight design and user clustering. Because a user could interfere and be interfered by other users at the same time, a balanced cooperative transmission strategy is utilized in weight design. The improved LBG algorithm is used for user clustering, which overcomes the shortcoming of local optimum of conventional LBG. Moreover, this algorithm is superior to conventional LBG in computational complexity. Simulation results show that the sum rate of cell-edge users increases a lot compared to the reference algorithm, and the average system throughput gets higher obviously.

Keywords: Ultra-Dense Network · Clustering · Throughput · Cell-edge user Interference

1 Introduction

For the variety of users' communication behaviors, applications of mobile communication become more multifarious. It's hard to cope with the explosive growth of data by the traditional mobile communication network which is based on macro cells. Future mobile cellular networks suffer from heavy data pressure. Ultra-Dense Network (UND) [1], as a solution, uses large-scale antenna and high-frequency communications. UND is considered to be the most innovative means to overcome the challenge [2], which composes of many low-power, small base stations. However, the intensive deployment of cells in UDN brings interference, which would reduce the network capacity and user experience, and result in low spectral utilization and cell-edge throughput.

According to the above, we need advanced interference suppression technology [3]. Coordinated multi-point (CoMP) or Network MIMO is the emerging technology which has been proposed to reduce interference and hence improve high data rate coverage

P. Krömer et al. (Eds.): ECC 2018, AISC 891, pp. 44–52, 2019.
https://doi.org/10.1007/978-3-030-03766-6_6

and cell edge throughout for future wireless networks [4]. Transmission occurs when the decision is made among networks, base stations (BSs) and users cooperatively. However, coordination between all cells in the network is a very complex task, due to precise synchronization requirement within coordinated cells, additional pilot overhead, additional signal processing, complex beamforming design and scheduling among all BSs [5]. To reduce this overhead, smaller size cooperation clusters are required where coordination only takes place within the cluster. Users are clustered at first, then cooperative transmission is carried out within or between clusters. Optimal CoMP clustering is one of the key challenges for CoMP implementation for future wireless networks. Selecting the right group of BSs or users for cooperation is key to maximize potential CoMP gains.

Some researchers have focused on coordination clustering in recent years. There is static clustering [7], Semi-dynamic clustering and Dynamic clustering [8]. Static clustering is mostly based on topological structure, which is less complex due to less signaling overhead. But this method is not responsive to changes in the network nodes or user locations, hence the performance gains are limited. And it is based on an assumption of hexagonal grid [9], which is not suitable for actual networks. Static clustering cannot meet the needs of system capacity in future networks due to inadequate spectral efficiency gain. This is the common shortcoming for most static clustering solutions. Multi-layer static clustering is designed in semi-dynamic clustering to avoid interference among clusters. But as an improved version, semi-dynamic clustering assumes ideal hexagonal grid in most solutions too, which makes it have the same disadvantage as static clustering. Dynamic clustering is a hybrid item based on network and users, which can improve the data rate and cell-edge throughput effectively.

It is presented in [10] that an adaptive semi-dynamic iterative clustering based on greedy algorithm. It includes scheduling initialization, collaborative user judgment and adaptive semi-dynamic iterative clustering scheduling. Whether the users can be cooperative depends on the Signal to Noise Ratio (SNR) in scheduling initialization. However, the calculation of SNR is large due to the complicated formula, which leads to great complexity in realization. The clustering algorithm proposed in [11] mainly aims to decrease the handover times. Handoff rate and failure times are important parameters which reflect the system performance in this Ref. It is deficient in obtaining the handoff failure times in this paper, so the result is untrustworthy. User density and channel noise are joint considered in one clustering algorithm in [12]. User signals are processed in the cluster individually. Although the effect of noise reduces, the computation increases. Moreover, the clustering results are different when the user's distribution density changes.

Considering the insufficiency of these algorithms and the complex environment in UDN, this paper focuses on simple, low-computation, relatively independent user clustering algorithm. There are two stages in this algorithm, interference weights generation and user clustering. In the first stage, weight balance strategy is taken into account due to the users' interference to and from other users. In this strategy, users' interference coefficients are adjusted, and the weights are designed for required signal to obtain gains. In the clustering phase, an improved algorithm based on vector quantization LBG [13] is proposed. Users are clustered by mining the sum of interference in each cluster. Users in the same cluster share the spectrum, which improves the spectrum utilization and cell-edge throughput. The algorithm has less computational complexity than the algorithms in the above-mentioned literatures. Moreover, the optimal user clusters are formed iteratively, and there's no need to know the handoff strategies and channel noise in this algorithm.

The clustering algorithm is introduced in Sect. 2, including interference weight design and improved LBG algorithm. This algorithm is simulinked in Sect. 3. And a summary is given at last.

2 User Clustering Algorithms

2.1 Interference Weight Design

In the mixed ultra-dense cooperative transmission network scenario, users in a cell interferes and are interfered by other users at the same time. Cell-edge users are interfered seriously. In order to improve the user experience at edge and reduce the impact caused by interference, multi-dimensional cooperation strategy is taken into account. Overlap between cells is considered in weight design. The overall performance of the system is dependent on multidimensional joint optimization of radio resources, power spectrum and space. Hence, the weight design takes channel amplitude and channel direction into account, which effectively reflect the mixed cooperative transmission gain. The vector angles of channels between users can be acquired from the selected channels.

We take user i and user j as candidates in one cluster, and assume the composite channel vector as: H_{i,C_i} (composite channel vector of user i in virtual cell C_i), H_{j,C_j} (composite channel vector of user j in virtual cell C_j). And the weight between user i in virtual cell C_i and user j in virtual cell C_j is noted as $W\text{ab}$ as follows:

$$\begin{aligned} W\text{ab}(i,j) = & \left[1+\alpha\frac{|H_{i,C_i}H_{i,C_j}^H|}{||H_{i,C_i}||_F||H_{i,C_j}||_F}\right]\cdot\left[1+\beta\frac{|H_{i,C_i}H_{i,C_i}^H|}{||H_{i,C_i}||_F||H_{j,C_i}||_F}\right]\cdot||H_{i,C_i}||_F \\ & +\left[1+\alpha\frac{|H_{j,C_i}H_{j,C_j}^H|}{||H_{j,C_i}||_F||H_{j,C_j}||_F}\right]\cdot\left[1+\beta\frac{|H_{i,Cj}H_{j,C_j}^H|}{||H_{i,C_j}||_F||H_{j,C_j}||_F}\right]\cdot||H_{j,C_j}||_F \end{aligned} \tag{1}$$

Where coefficients α and β denotes the proportions of power and space in multi-dimensional cooperation respectively. This formula is suitable for the overlapped cells scenario. Users in overlapped cells adopt joint transmission mode to obtain cooperative transmission gain. When the cells are non-overlap or part-overlap, this paper adopts the spatial coordinated transmission for users to eliminate the inter-user interference.

Table 1. Weights design in different scenarios

Scenarios	Weights
Channel vector cosine when user i and user j are in the same cell	$\cos(H_{i,C_j}, H_{j,C_i}) = \frac{\lvert H_{i,C_i} H^H_{j,C_i} \rvert}{\lVert H_{i,C_i} \rVert_F \lVert H_{j,C_i} \rVert_F}$
Channel vector cosine when user i and user j are in different cells	$\cos(H_{i,C_i}, H_{i,C_j}) = \frac{\lvert H_{i,C_i} H^H_{i,C_j} \rvert}{\lVert H_{i,C_i} \rVert_F \lVert H_{i,C_j} \rVert_F}$

The nearer the users' channels to orthogonal, the less interference between users and the higher energy and frequency efficiency gains we get. In this case, weights are designed as follows in Table 1:

Assumptions include

(1) base station a and b are in cell i in this model, and
(2) $H_i = \left[h_{1,a}h_{1,b}; h_{2,a}h_{2,b}\right]$ is the composite channel in cell i, where the subscript parameters 1 and 2 in H_i refer to the users from base station a and b respectively.

The relationship between these factors are shown in Fig. 1. The precoding matrix of cell i is:

$$W_1 = \frac{1}{\sqrt{C_i}} \cdot \frac{1}{(h_{1,a}h_{2,b} - h_{1,b}h_{2,a})} \cdot \begin{bmatrix} h_{2,b} & -h_{1,b} \\ -h_{2,a} & h_{1,a} \end{bmatrix} \tag{2}$$

The power division factor of each user is applied to precoding matrix, so parameter C_i can be written as:

$$\begin{aligned} C_i &= \max_j(\overline{W}_i \overline{W}_i^H)[j,j] \\ &= \max\left\{ \frac{|h_{1,b}|^2 + |h_{2,b}|^2}{h_{1,a}h_{2,b} - h_{1,b}h_{2,a}}, \frac{|h_{1,a}|^2 + |h_{2,a}|^2}{h_{1,a}h_{2,b} - h_{1,b}h_{2,a}} \right\} \end{aligned} \tag{3}$$

Without loss of generality, the SINR of user 1 after clustering is calculated as follows:

$$\begin{aligned} \Delta\text{SINR} &= \text{SINR}_i - \text{SINR}_a \\ &= \frac{P_{tx}}{\sigma_n^2} \cdot \frac{1}{C_i} - \frac{|h_{1,a}|^2 P_{tx}}{|h_{1,b}|^2 P_{tx} + \sigma_n^2} \\ &= \frac{P_{tx}}{\sigma_n^2} \cdot \frac{|h_1 \cdot w_1|^2 P_{tx}}{|h_{1,b}|^2 + |h_{2,b}|^2} - \frac{|h_{1,a}|^2 P_{tx}}{|h_{1,b}|^2 P_{tx} + \sigma_n^2} \\ &= \frac{P_{tx}}{\sigma_n^2} \cdot \frac{||h_1||^2 ||h_2||^2}{|h_{1,b}|^2 + |h_{2,b}|^2} \sin^2\langle h_1, h_2 \rangle - \frac{|h_{1,a}|^2 P_{tx}}{|h_{1,b}|^2 P_{tx} + \sigma_n^2} \end{aligned} \tag{4}$$

Where SINR_i represents the SINR of user 1 after clustering, and SINR_a denotes the SINR of user 1 under the service of base station a. Because UDN is an interference-limited system, the noise can be ignored. So the interference signal satisfies

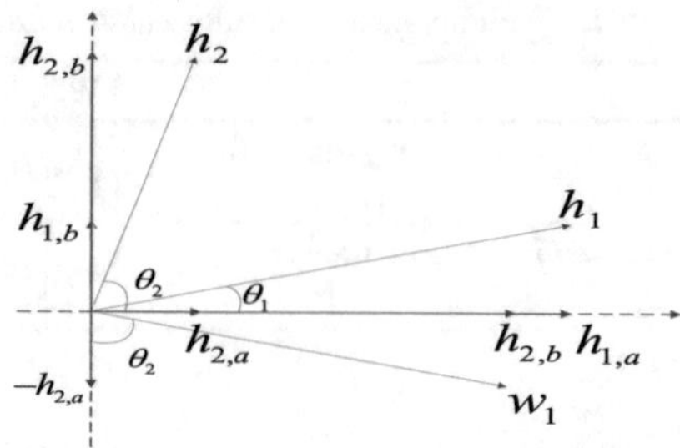

Fig. 1. Channel vector

$|h_{1,k}|^2 P_{tx} \gg s_n^2$. When user i is at the edge of an overlapped cell, the channel intensity satisfies $|h_{1,a}| \approx |h_{1,b}|$. Thus Eq. (4) can be simplified as:

$$\Delta\text{SINR} = \frac{P_{\text{tx}}}{\sigma_n^2} \cdot \frac{||h_1||^2 ||h_2||^2}{|h_{1,b}|^2 + |h_{2,b}|^2} \sin^2\langle h_1, h_2 \rangle - 1 \tag{5}$$

We can see from Eq. (5) that the improvement of the SINR of user 1 is closely related to the intensity of the cooperative channel h_2 and the orthogonality of Channel h_1 and h_2. Therefore, it is proved that the weight design of this paper is reasonable.

2.2 Improved LBG Algorithm in Clustering

The improved LBG algorithm is proposed in this paper on the basis of the interference weights designed above. *Wab* is a matrix. We use its elements to denote the interference between users. For example, the element of row i in column j, which is *Wab*(i, j), represents the interference between user i and user j. In the improved LBG algorithm, the initial cores are selected seriously to avert local optimization. Clusters are formed without any training sequence in this algorithm, the computational complexity of which is lower than traditional LBG algorithm.

In order to avoid local optimization in improved LBG algorithm, the initial cores are chosen carefully. There are two alternatives in this paper, which are called LBG_Ave and LBG_MDCI. In the former algorithm, users are selected as initial cores whose *Wab* is closest to the mean of all of the *Wab*. In LBG_MDCI, users with maximum *Wab* are chosen as the initial cores. Steps of the algorithm are shown in Table 2.

We take V_1 as a new set of users and repeat the procedure to split V_1 into two new sets. Keeping this way, we can get a power of 2 clusters.

Table 2. LBG_ave algorithm and LBG_MDCI algorithm

Algorithm 1 proposed improved LBG algorithm

INPUT: user's weight Wab, the set of users U, the number of users N_user

OUTPUT: Clusters of users V_i, $i = 1, 2$

```
Initialize:
    Calculate the mean of U's weight as Wab_center (LBG_Ave),
    Find the maximum value of U's weight as Wab_max (LBG_MDCI);
for ll = 1 : N_user
    for kk = 1 : N_user
      (for LBG_Ave)
      if |Wab(ll,kk) - Wab_center| is the smallest
            m = ll;   n = kk;
            V1 = {m}; V2 = {n};
      end
      (for LBG_MDCI)
      if Wab(ll,kk) is the maximum
            m = ll;   n = kk;
            V1 = {m}; V2 = {n};
      end
    end
end
for ll = 1 : N_user
    if ll is not m or n
      if Wab(m,ll) < Wab(n,ll)
          V1 = V1 ∪ {ll};
      else
          V2 = V2 ∪ {ll};
      end
    end
end
```

3 Simulation and Analysis

Weight design and improved LBG algorithm for user clustering are simulated by MATLAB in this paper. This algorithm is compared to K-Mean algorithm.

In the simulation, users are clustered by improved LBG algorithm based on virtual cells. In each cluster, users are scheduled among cells [14, 15] which are coordinated to adapt to beam forming [16] and power allocation [17]. Under the criterion of proportional fairness and rate maximization, greedy scheduling algorithm is used to select scheduled users. Reciprocity strategy is adopted for cooperative beamforming. For the BS of the overlapping virtual cell, power segmentation algorithm is administered based on the intensity of the instantaneous channel of the scheduled user. And water-filling algorithm is used for power control.

In the clustering process, the initial cores ensure the minimization of the sum of weights intra cluster and maximization of weights inter cluster. The simulation parameters are shown in Table 3.

Table 3. Simulation parameters

Parameters	Values
Number of users	36
Number of cells	6
Number of channels	2
Noise power/(dBm)	$N = -173.9 + 10 * \log_{10}(10.\wedge 7) + 9$
Propagating power of a Pico base station/(dBm)	20
Number of transmitting antennas	36
Number of receiving antennas	6

For fairness and authenticity of the simulation results, it is identical that parameter setting, power partition and power control algorithms in reference and improved LBG algorithms, except clustering algorithms. Figure 2 illustrates the cumulative distribution function (CDF) to system throughput. Here, the improvement of our proposal is notable against the reference algorithm. Targeting the 10th percentile of the CDF as QoS measure, up to 55% improvement of LBG_Ave is observable over K_Mean algorithm. Additionally, LBG_MDCI shows 80% enhancement over the reference algorithm. Targeting the 90th percentile, the improvement is also promising (40% over K-Mean of LBG_Ave, 46% of LBG_MDCI).

Table 4 shows significant improvements of our proposal. Compared to K-Mean algorithm, bigger sum rate of cell-edge users is obtained by LBG_Ave and LBG_MDCI, respectively. Users with the least mutual interference are put into the same cluster. The cell-edge effect is eliminated. The sum of interference in a cluster is as less as possible.

In addition, radio resource allocation has been considered in user clustering and weight design. Hence, the scheduled channels between users tend to be orthogonal, which is suitable for hybrid cooperative transmission. System performance, i.e. system throughput, is improved consequently.

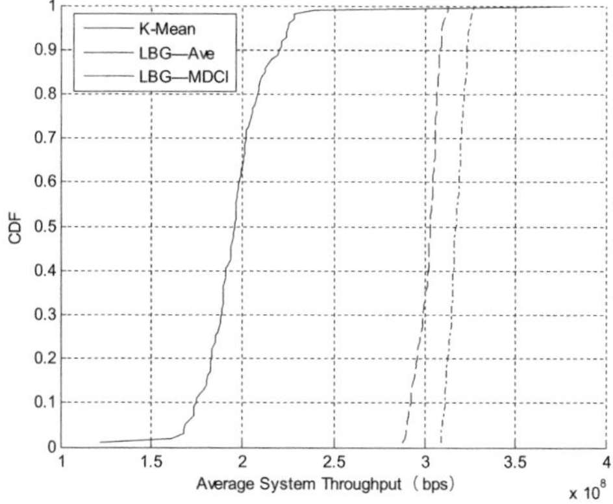

Fig. 2. Comparison of LBG_Ave and LBG_MDCI with K_Mean algorithm

Table 4. Comparation in sum rate of cell-edge users and system throughput

	K-Mean	LBG_Ave	LBG_MDCI
Sum rate of cell-edge users (bps)	0.53×10^8	0.78×10^8	0.94×10^8
System throughput (bps)	1.62×10^8	2.33×10^8	2.56×10^8

4 Conclusion

In this paper, an improved LBG algorithms is proposed, which is used to cluster the users under UDN in order to get good anti-interference effect. In view of the short-comings that the traditional LBG algorithm may fall into the local optimal, the initial cores are designed based on average (LBG_Ave) and maximum interference (LBG_MDCI). The simulation results show that the algorithm proposed in this paper improves the system performance and reduces interference. Moreover, LBG_MDCI is superior to LBG_Ave. The overall system throughput is improved, and the sum of throughput of cell- edge users increases distinctly.

Acknowledgement. This work was supported by National Science and Technology Major Project of the Ministry of Science and Technology of China (ZX201703001012-005), National Natural Science Foundation of China (61501371), Shaanxi STA International Cooperation and Exchanges Project (2017KW-011) and the Department of Education Shaanxi Province, China, under Grant 2013JK1023.

References

1. Yunas, S., Valkama, M., Niemela, J.: Spectral and energy efficiency of Ultra-Dense Networks under different eployment strategies. IEEE Commun. Mag. **53**(1), 90–100 (2015)
2. Wang, C., Hu, B., Chen, S., et al.: Joint dynamic access points grouping and resource allocation for coordinated transmission in user-centric UDN. Trans. Emerg. Telecommun. Technol. **29**(3), e3265 (2017)
3. Kunitaka, M., Tomoaki, O.: Orthogonal beamforming using Gram-Schmidt orthogonalization for downlink CoMP system. ITE Tech. Rep. **36**(10), 17–20 (2012)
4. Bu, H.W., Xu, Y.H., Yuan, Z., Hu, Y.J., Yi, H.Y.: An efficient method for managing CoMP cooperating set based on central controller in LTE-A systems. Appl. Mech. Mater. **719–720**, 721–726 (2015)
5. Bassoy, S., Farooq, H., Imran, M.A., Imran, A.: Coordinated multi-point clustering schemes: a survey. IEEE Commun. Surv. Tutor. **19**(2), 743–764 (2017)
6. Grebla, G., Birand, B., van de Ven, P., Zussman, G.: Joint transmission in cellular networks with CoMP-stability and scheduling algorithms. Perform. Eval. **91**(C), 38–55 (2015)
7. Du, T., Qu, S., Liu, F., Wang, Q.: An energy efficiency semi-static routing algorithm for WSNs based on HAC clustering method. Inf. Fusion **21**(1), 18–29 (2015)
8. Xu, D., Ren, P., Du, Q., Sun, L.: Joint dynamic clustering and user scheduling for downlink cloud radio access network with limited feedback. China Commun. **12**(12), 147–159 (2015)
9. Ali, S.S., Saxena, N.: A novel static clustering approach for CoMP. In: IEEE 7th International Conference on Computing and Convergence Technology (ICCCT), Seoul, South Korea, pp. 757–762. IEEE Press (2012)
10. Wan, Q.: Research on multi-cell clustering cooperative technology in CoMP scene. Beijing University of Posts and Telecommunications, Beijing (2015)
11. Meng, N., Zhang, H.T., Lu, H.T.: Virtual cell-based mobility enhancement and performance evaluation in Ultra-Dense Networks. In: IEEE Wireless Communications and Networking Conference, Doha, Qatar, pp. 1–6. IEEE Press (2016)
12. Kurras, M., Fahse, S., Thiele, L.: Density based user clustering for wireless massive connectivity enabling Internet of Things. In: Globecom Workshops (GCWkshps), San Diego, CA, USA, pp. 1–6. IEEE Press (2015)
13. Patané, G., Russo, M.: The enhanced LBG algorithm. Neural Netw. Off. J. Int. Neural Netw. Soc. **14**(9), 1219 (2001)
14. Wang, J., Tang, S., Sun, C.: Resource allocation based on user clustering in ultra-dense small cell networks. J. Xi'an Univ. Posts Telecommun. **21**(1), 16–20 (2016)
15. Gong, J., Zhou, S., Niu, Z., et al.: Joint scheduling and dynamic clustering in downlink cellular networks. In: Global Telecommunications Conference (Globecom), Houston, Texas, USA, pp. 1–5. IEEE Press (2011)
16. Ho, Z.K.M., Gesbert, D.: Balancing egoism and altruism on interference channel: the MIMO case. In: International Conference on Communications (ICC), Cape Town, South Africa, pp. 1–5. IEEE Press (2010)
17. Jindal, N., Rhee, W., Vishwanath, S., et al.: Sum power iterative water-filling for multi-antenna Gaussian broadcast channels. IEEE Trans. Inf. Theory **51**(4), 1570–1580 (2015)

Quadrotors Finite-Time Formation by Nonsingular Terminal Sliding Mode Control with a High-Gain Observer

Jin Ke, Kangshu Chen, Jingyao Wang(✉), and Jianping Zeng

Xiamen University, Fujian 361101, China
wangjingyao1@xmu.edu.cn

Abstract. This paper investigates the distributed finite-time formation problem of quadrotors using the information of relative position only. A high-gain observer is constructed to estimate the relative velocity through the relative position. Based on the estimated relative velocity, nonsingular terminal sliding mode (NTSM) protocols are designed for followers. The control protocols for the position subsystem of quadrotors are developed by the combination of the isokinetic trending law and the idempotent trending law, which guarantees the realization of finite-time formation accurately and quickly in the presence of the bounded external disturbances and internal uncertainties. Moreover, an idempotent term is introduced to the attitude subsystem, which eliminates the chattering caused by the isokinetic trending law. Finally, a numerical example is given to illustrate the effectiveness of the proposed method.

Keywords: Quadrotors · Finite-time formation · High-gain observer
Nonsingular terminal sliding mode control

1 Introduction

In recent years, formation control has received significant attention due to its wide application background, such as robots [1], satellites [2] and sensor networks [3], etc. Among those various control objects, each agent updates its states based on the state information of its local neighbors to keep the desired formation [4]. Up to now, there are roughly three popular approaches in the literatures for multi-agent coordination, namely the leader-following, the behavioral, and the virtual structures [5].

For practical multi-agent systems, both the asymptotical stability and the convergence rate are required. Implementing finite-time formation control for multi-agent systems is of great significance. An efficient approach for finite-time formation is sliding mode control (SMC) [6]. In [7], a continuous control law is proposed to achieve fast finite-time consensus tracking with the existence of uncertainties and bounded disturbances. Moreover, the problem of finite-time consensus tracking for multi-agent systems subject to input saturation is considered in [8, 9]. In [10], the integral-type nonsingular terminal sliding mode (NTSM) control and the extended-state observer are combined to achieve finite-time consensus. In [11], the distributed finite-time formation tracking protocols are given via the fast terminal sliding mode control (FTSMC)

P. Krömer et al. (Eds.): ECC 2018, AISC 891, pp. 53–64, 2019.
https://doi.org/10.1007/978-3-030-03766-6_7

scheme. It is worth noting that the above works are limited by the condition that both the relative position and the relative velocity should be known in advance. Unfortunately, the information of relative velocity for agents is unavailable in many practical situations because of the poor detection signal.

Regarding the above challenge, we consider the finite-time consensus tracking for quadrotors using the relative position only. A high-gain observer is constructed to estimate the relative velocity. Based on the prior relative position and estimated relative velocity, the dynamic NTSM protocols are designed for followers to achieve global consensus, which avoids the singularity problem at the same time. Moreover, considering the bounded external disturbances and internal uncertainties, the control protocols are developed by the combination of the isokinetic trending law and the idempotent trending law, which achieves finite-time formation with decent performance. For the attitude subsystem, an idempotent term is also introduced to eliminate the chattering caused by the isokinetic trending law.

The paper is organized as follows: the system description and some necessary preliminaries are given in Sect. 2. Section 3 states the finite-time formation control by NTSM scheme for quadrotors. The simulation results of quadrotors formation are demonstrated in Sect. 4. Finally, we draw the conclusion of this paper in Sect. 5.

2 Problem Formulation and Preliminaries

2.1 Graph Theory

Utilize a general directed graph $G = \{\mathrm{V}, E, D\}$ to describe the information exchange among quadrotors during the formation process. The adjacency matrix is defined as $A = [a_{ij}] \in R^{n \times n}$ where $a_{ij} = 1$ if the *jth* quadrotor can obtain the information from the *ith* quadrotor, otherwise $a_{ij} = 0$. The Laplacian matrix is denoted by $L = [l_{ij}] \in \mathrm{R}^{n \times n}$ where $l_{ij} = \sum_{j=1}^{N} a_{ij}$ if $i = j$, otherwise $l_{ij} = -a_{ij}$ [12]. Define $B = diag\{b_1, b_2 \cdots b_n\}$, where $b_i = 1$ if the *ith* quadrotor can access the information from the leader quadrotor.

Lemma 1 **[13].** *$L + B$ is of full rank if there exists a diagraph such that each member node except the leader node has a directed path from the leader.*

2.2 Problem Formulation

Denote $\sum_P$ as the north-east-down reference frame, $\sum_b$ the body frame.$p \in R^3$ and $v \in R^3$ are respectively the position and linear velocity of quadrotor expressed in $\sum_P$. $m \in R$ is quadrotor's mass, $T_d \in R^3$ is the control torque, g is the gravitational force, $w \in R^3$ is the angular velocity of $\sum_b$ with respect to $\sum_P$, and $e_3 = [0 \quad 0 \quad 1]^T$. $Q = [\eta, \vec{q}^T]$ is the unit-quaternion, which satisfies $\eta^2 + \vec{q}^T \vec{q} = 1$. $J \in R^{3 \times 3}$ is the inertial matrix, and $\tau \in R^3$ is the magnitude of the control torque. Then, the model of the *ith* quadrotor can be described as follows

$$\begin{cases} \dot{p} = v, \quad \dot{v} = ge_3 - \frac{T_d}{m} C_{pb} e_3, \quad \dot{\eta} = -\frac{1}{2} \vec{q}^T w \\ \dot{\vec{q}} = \frac{1}{2} [\eta I_3 + \vec{q} \times] w, \quad J\dot{w} = \tau - w \times Jw \end{cases}. \tag{1}$$

***Lemma 2* [14].** *$C_{pb} \in R^{3*3}$ is the rotational matrix representing the transformation from the body frame to the inertial frame, that is*

$$C_{pb} = (\eta^2 - \vec{q}^T \vec{q}) I_3 + 2\vec{q}\vec{q}^T + 2\eta \vec{q} \times, \tag{2}$$

where $\vec{q}\times$ is the skew-symmetric matrix.

Define $\xi_{id} = \xi_i - \xi_d \in R^3$ as the desired distance between the leader and the *ith* quadrotor. p^d is the trajectory of leader. $e_{pi} = p_i - p^d - \xi_{id}$ and $e_{vi} = v_i - \dot{p}^d$ are the relative position and the relative velocity, respectively. Moreover, let $Q_{id} = [\eta_{id}, \vec{q}_{id}]$ be the expected rotation quaternion, $Q_{ie} = [\eta_{ie}, \vec{q}_{ie}]$ be the rotation quaternion error of the *ith* quadrotor from the desired frame Σ_{b^*} to the body frame Σ_b, and $w_{id} \in R^3$ be the expected angular velocity in Σ_{b^*}. Then, the differential equation of tracking error for the *ith* quadrotor can be derived as

$$\begin{cases} \dot{e}_{pi} = e_{vi} \\ \dot{e}_{vi} = ge_3 - \ddot{p}^d - \frac{T_{id}}{m} C_{ipb^*} e_3 + \frac{T_{id}}{m} (C_{ipb^*} - C_{ipb}) e_3 \\ \dot{\eta}_{ie} = -\frac{1}{2} \vec{q}_{ie}^T w_{ie} \\ \dot{\vec{q}}_{ie} = \frac{1}{2} [\eta_{ie} I_3 + \vec{q}_{ie} \times] w_{ie} \\ \dot{w}_{ie} = J^{-1}(\tau_i - (w_{ie} + C_{ibb^*} w_{id}) \times J(w_{ie} + C_{ibb^*} w_{id})) + w_{ie} \times C_{ibb^*} w_{id} - C_{ibb^*} \dot{w}_{id} \end{cases}. \tag{3}$$

Let the virtual control input be $U = [\, u_1 \quad u_2 \quad u_3 \,] = T_d C_{pb^*} e_3$. Then the control torque and the desired angular velocity are given by

$$\begin{cases} T_d = m\|U\|, \qquad \vec{q}_d = \frac{m}{2T_d \eta_d} [-u_2 \quad u_1 \quad 0]^T \\ \eta_d = \sqrt{\frac{1}{2} + \frac{mu_3}{2T_d}}, \quad w_d = 2[\, \vec{q}_d \eta_d I_3 + \vec{q}_d \times \,] \cdot \dot{Q}_d \end{cases}. \tag{4}$$

For system (3), the algorithm of formation consensus can be expressed as

$$U_i = -\sum\nolimits_{j=1}^{N} a_{ij} \left[(p_i - p_j - \xi_{ij}) + \lambda_i (v_i - v_j) \right], \tag{5}$$

where $\lambda_i > 0$. Note that the formation can not be reached asymptotically among the N quadrotors unless $p_i - p_j \to \xi_{ij}$ and $v_i - v_j \to 0$ for all $p_i(t_0)$ and $v_i(t_0)$.

In order to facilitate subsequent research, the following assumption is introduced.

Assumption 1. *Let d_i be the sum of the external disturbances and the internal uncertainties. d_i is assumed to be bounded and slowly time varying, which satisfies* $|d_i| \leq D_i$ and $\dot{d}_i \simeq 0$.

Moreover, the absolute tracking error of the position subsystem can be described as

$$\begin{cases} \dot{e}_{pi} = e_{vi} \\ \dot{e}_{vi} = f_i - \dfrac{u_i}{m} - a_0 + \mathrm{d}_i \end{cases}, \tag{6}$$

where $e_{pi} = p_i - p^d - \xi_{id}$, $e_{vi} = v_i - \dot{p}^d$, $f_i = ge_3$, $a_0 = \ddot{p}^d$ and $u_i = T_{id}C_{ipb^*}e_3$.

The coupling tracking error is defined as

$$\begin{cases} \varepsilon_{pi} = \sum_{j=1}^{N} a_{ij}(p_i - p_j) + b_i(p_i - p^d - \xi_{id}) \\ \varepsilon_{vi} = \sum_{j=1}^{N} a_{ij}(v_i - v_j) + b_i(v_i - \dot{p}^d) \end{cases}, \tag{7}$$

which can be rewritten as

$$\begin{cases} \varepsilon_p = (L+B) \otimes I_3(e_p) \\ \varepsilon_v = (L+B) \otimes I_3(e_v) \end{cases}. \tag{8}$$

3 Main Results

In this section, for the *ith* quadrotor in (1), we consider the finite-time consensus tracking of N quadrotors.

3.1 Relative Velocity Measurement

Consider the situation that only the relative position is available. Without the information of e_v, the control input of (5) cannot be used directly. Thus, we design the high-gain observer to estimate the relative velocity e_v via the relative position e_p.

For the closed-loop system in (6), the distributed observer is

$$\begin{cases} \dot{\hat{e}}_{pi} = \hat{e}_{vi} + h_1(e_{pi} - \hat{e}_{pi}) \\ \dot{\hat{e}}_{vi} = f - a_0 - \dfrac{u_i}{m} + \hat{d}_i + h_2(e_{pi} - \hat{e}_{pi}) \\ \dot{\hat{d}}_i = h_3(e_{pi} - \hat{e}_{pi}) \end{cases}. \tag{9}$$

For simplicity, we take the x-direction state of the *ith* quadrotor as an example. The observation error of relative position and relative velocity are $\tilde{e}_{pix} = e_{pix} - \hat{e}_{pix}$, $\tilde{e}_{vix} = e_{vix} - \hat{e}_{vix}$, and the observation error of d_{ix} is $\tilde{d}_{ix} = d_{ix} - \hat{d}_{ix}$. Then, the state equation of observation error is derived as

$$\begin{bmatrix} \dot{\tilde{e}}_{pix} \\ \dot{\tilde{e}}_{vix} \\ \dot{\tilde{d}}_{ix} \end{bmatrix} = \begin{bmatrix} -h_1 & 1 & 0 \\ -h_2 & 0 & 1 \\ -h_3 & 0 & 0 \end{bmatrix} \begin{bmatrix} \tilde{e}_{pix} \\ \tilde{e}_{vix} \\ \tilde{d}_{ix} \end{bmatrix} + \begin{bmatrix} 0 \\ 0 \\ 1 \end{bmatrix} \dot{d}_{ix}. \tag{10}$$

Then the characteristic equation of system (10) is formulated as

$$\lambda^3 + h_1\lambda^2 + h_2\lambda + h_3 = 0. \tag{11}$$

Note that (11) is Hurwitz if $h_1 > 0,\ h_2 > 0,\ h_3 > 0$. Moreover, $\hat{e}_p,\ \hat{e}_v$ and $\hat{d}$ will converge to $e_p,\ e_v$ and d if $h_1,\ h_2,\ h_3$ are large enough.

3.2 NTSM Control of Position Subsystem

Consider the following second-order dynamic system

$$\begin{cases} \dot{x}_1 = x_2 \\ \dot{x}_2 = a(x) - b(x)u + d\,, \\ y = x_1 \end{cases} \tag{12}$$

where $x \in R^2$ is the state; u and y are the control input and control output, respectively; $a(x)$ and $b(x)$ are the functions of x.

A sliding mode surface can be designed as below

$$s = x_1 + \frac{1}{\beta} x_2^{r/q}. \tag{13}$$

Define $\hat{s}$ as the estimate of s, $\tilde{s}$ as the estimated error. Then the control strategy can be formulated as

$$u = b^{-1}(x)[a(x) + \beta \frac{q}{r} \hat{x}_2^{2-r/q} + (D + k|\hat{s}|^{\alpha})\mathrm{sgn}(\hat{s})], \tag{14}$$

where $1 < r/q < 2$, both r and q are prime numbers; $k > 0,\ \beta > 0$ and $0 < \alpha < 1$.

Theorem 1. Consider the system in (12). Choosing the sliding mode surface of (13) and the control strategy of (14), system (12) is asymptotically stable in finite time.

Proof: For the sliding surface of (13), its time derivation is given by

$$\begin{aligned} \dot{\hat{s}} &= \dot{\hat{x}}_1 + \frac{1}{\beta}\frac{r}{q}(\hat{x}_2)^{r/q-1}\dot{\hat{x}}_2 \\ &= \hat{x}_2 + \frac{1}{\beta}\frac{r}{q}\hat{x}_2^{r/q-1}[-\beta\frac{q}{p}\hat{x}_2^{2-r/q} - (D + k|\hat{s}|^{\alpha})\mathrm{sgn}(\hat{s}) + d]\,. \\ &= \frac{1}{\beta}\frac{r}{q}\hat{x}_2^{r/q-1}[-(D + k|\hat{s}|^{\alpha})\mathrm{sgn}(\hat{s}) + d] \end{aligned} \tag{15}$$

Choose the Lyapunov function $\hat{V}_1 = \frac{1}{2}\hat{s}^2$. The time derivation of $\hat{V}_1$ is

$$\begin{aligned} \dot{\hat{V}}_1 &= -\frac{1}{\beta}\frac{r}{q}\hat{x}_2^{r/q-1}[D\text{sgn}(\hat{s}) + k|\hat{s}|^{\alpha}\text{sgn}(\hat{s}) - d]\hat{s} \\ &\leq -\frac{1}{\beta}\frac{r}{q}\hat{x}_2^{r/q-1}k|\hat{s}|^{\alpha+1} \\ &= K\hat{V}_1^{(\alpha+1)/2} \end{aligned}, \tag{16}$$

where $K = -\frac{1}{\beta}\frac{r}{q}\hat{x}_2^{r/q-1}k$. If $\hat{x}_2 \neq 0$, $\dot{\hat{V}}_1 \leq 0$ and the equality holds unless $\hat{s} = 0$. Then the system is globally stable. Namely, $\hat{s} = 0$ can be reached. If $\hat{x}_2 = 0$, (12) can be denoted as $\dot{\hat{x}}_2 = -(D + k_2|\hat{s}|^{\alpha})\text{sgn}(\hat{s}) + d$. Note that $\hat{x}_2 = 0$ is not the equilibrium point if $\hat{s} \neq 0$. Meanwhile, we have

$$V_1 = \frac{1}{2}(\hat{s} + \tilde{s})^2 = \hat{V}_1 + f(\tilde{s}), \tag{17}$$

where $f(\tilde{s}) = \frac{1}{2}\tilde{s}^2 + \hat{s}\tilde{s}$ is a function of $\tilde{s}$. Both $\hat{V}_1$ and $f(\tilde{s})$ converge to zero. Furthermore, the convergence speed of $f(\tilde{s})$ is faster than $\hat{V}(t)$ because of the high-gain parameters of the state observer. The setting time can be estimated via (16) by

$$t_{1s} \leq \frac{2\hat{V}_1(t_0)^{\frac{1-\alpha}{2}}}{K(\alpha - 1)}.$$

Thus $V_1 = 0$ if $t \geq t_{1s}$. This indicates $s = 0$ can be reached in finite time. If the state variables reach the sliding mode surface, $x_1 + \frac{1}{\beta}\dot{x}_1^{r/q} = 0$. Then the system is stable with parameters which satisfy (14). Similarly, we get the convergence time as

$$t_{1r} = \frac{1}{\beta}\frac{r}{r - q}(-\beta x_1)^{1-\frac{q}{r}}.$$

For the position subsystem of quadrotors, the sliding mode surface can be denoted as $\hat{s}_i = \hat{\varepsilon}_{pi} + \frac{1}{\beta}\hat{\varepsilon}_{vi}^{r/q}$, and the control strategy of the *ith* quadrotor is

$$U_i = m(ge_3 - a_0 + (D + k_2|\hat{s}_i|^{\alpha})sgn(\hat{s}_i) + \beta\frac{q}{r}\hat{\varepsilon}_{vi}^{2-r/q}), \tag{18}$$

where $\frac{1}{\beta}\frac{p}{q}\hat{\varepsilon}_{vi}^{r/q-1}[-(D + k|\hat{s}|^{\alpha})\text{sgn}(\hat{s}) + d]$ is the trending law, which combines the isokinetic and the idempotent trending law. The idempotent trending law is faster than the isokinetic trending law when the state errors of the system are far away from the sliding surface. Therefore, quadrotors formation can be achieved via NTSM control with less time than using the isokinetic trending law only [15].

3.3 NTSM Control of Attitude Subsystem

Based on the attitude subsystem in (3), the NTSM surface can be designed as

$$s_i = w_{ie} + \kappa_{i1}\vec{q}_{ie} + \kappa_{i2}e^{-\lambda t}(\vec{q}_{ie}^T\vec{q}_{ie})^{-\alpha}\vec{q}_{ie}, \tag{19}$$

where $\kappa_{i1},\ \kappa_{i2} > 0$, $0<\alpha<1$ and $\lambda > 0$.

Choose the control torque of the *ith* quadrotor as

$$\begin{aligned}\tau_i =&(w_{ie} + C_{ibb^*}w_{id}) \times J(w_{ie} + C_{ibb^*}w_{id}) + JC_{ibb^*}\dot{w}_{id} - Jw_{ie} \times C_{ibb^*}w_{id} \\ &- \kappa_{i1}J\dot{\vec{q}}_{ie} - \kappa_{i2}JP - \rho\,\mathrm{sgn}(s_i)|s_i|^{\alpha}\end{aligned}, \tag{20}$$

where $\rho > 0$ and

$$P = (-\lambda)e^{-\lambda t}(\vec{q}_{ie}^T\vec{q}_{ie})^{-\alpha}\vec{q}_{je} + e^{-\lambda t}(\vec{q}_{ie}^T\vec{q}_{ie})^{-\alpha}\dot{\vec{q}}_{ie} + e^{-\lambda t}(-2\alpha)(\vec{q}_{ie}^T\vec{q}_{ie})^{-\alpha-1}(\vec{q}_{ie}^T\dot{\vec{q}}_{ie})\vec{q}_{ie}.$$

Theorem 2. For the system in (3), choosing the NTSM surface as (19) and the control torque as (20), if $\frac{1}{2}\alpha\kappa_{i1} - \lambda > 0$, the attitude subsystem can indeed be stabilized in finite time. And the equilibrium point of the system is $w_{ie} = 0,\ Q_{ie} = [1 \quad 0 \quad 0 \quad 0]^T$.

Proof: Construct the following Lyapunov function

$$V_{2i} = \frac{1}{2}s_i^T J s_i,\ i = 1, 2 \cdots N. \tag{21}$$

Taking the derivative of (21), we can get the following inequality

$$\begin{aligned}\dot{V}_{2i} =& s_i^T(J\dot{w}_{ie} + J\kappa_{i1}\dot{\vec{q}}_{ie} + J\kappa_{i2}P) \\ =& s_i^T(\tau_i - (w_{ie} + C_{ibb^*}w_{id}) \times J(w_{ie} + C_{ibb^*}w_{id}) - JC_{ibb^*}\dot{w}_{id} \\ &+ Jw_{ie} \times C_{ibb^*}w_{id} + \kappa_{i1}J\dot{\vec{q}}_{ie} + \kappa_{i2}JP) \\ =& -\rho|s_i|^{1+\alpha} \\ \le& -\sqrt{2}\rho(\lambda_{\min}^{-1}(J))^{(1+\alpha)/2}V_{2i}^{(1+\alpha)/2}\end{aligned}. \tag{22}$$

Similar to Theorem 1, the state variables of the attitude subsystem can reach the NTSM surface in finite time under the control law of (20). And the setting time is

$$t_{2s} \le \frac{\sqrt{2}V_{2i}(t_0)^{\frac{1-\alpha}{2}}}{(1-\alpha)\rho\lambda_{\min}^{-\frac{1+\alpha}{2}}(J)}. \tag{23}$$

When the subsystem operates in the proposed surface, (19) becomes

$$w_{ie} + \kappa_{i1}\vec{q}_{ie} + \kappa_{i2}e^{-\lambda t}(\vec{q}_{ie}^T\vec{q}_{ie})^{-\alpha}\vec{q}_{ie} = 0.$$

Define the Lyapunov function as

$$\begin{aligned} V_{3i} &= \vec{q}_{ie}^T \vec{q}_{ie} + (1 - \eta_{ie})^2 \\ &\leq \vec{q}_{ie}^T \vec{q}_{ie} + (1 - \eta_{ie})(1 + \eta_{ie}). \\ &= 2\vec{q}_{ie}^T \vec{q}_{ie} \end{aligned} \tag{24}$$

Differentiating V_{3i} with respect to the time, we can obtain

$$\begin{aligned} \dot{V}_{3i} &= \vec{q}_{ie}^T w_{ie} \\ &= -\kappa_{i1} \vec{q}_{ie}^T \vec{q}_{ie} - \kappa_{i2} e^{-\lambda t} (\vec{q}_{ie}^T \vec{q}_{ie})^{1-\alpha} \\ &\leq -\frac{1}{2}\kappa_{i1} V_{3i} - (\frac{1}{2})^{1-\alpha} \kappa_{i2} e^{-\lambda t} V_{3i}^{1-\alpha} \end{aligned}. \tag{25}$$

From (25), we know that $\dot{V}_{3i} \leq 0$. According to [16], the convergence time can be estimated by the following inequality

$$t_{2r} \leq \frac{\ln[1 + (1/2\alpha\kappa_{i1} - \lambda)V_{3i}^{\alpha}(t_0)/((1/2)^{1-\alpha}\alpha\kappa_{i2})]}{1/2\alpha\kappa_{i1} - \lambda}. \tag{26}$$

3.4 Stability Analysis

Theorem 3. Based on Theorems 1 and 2, it is possible to ensure that the state variables arrive at the sliding surface in finite time with the control law of (18).

Proof: According to Lemma 1, if $L + B$ is of full rank, the Lyapunov function can be constructed as

$$V = S^T \cdot (L + B)^{-1} \otimes I_3 \cdot S, \tag{27}$$

where $s_i = \varepsilon_{pi} + \frac{1}{\beta}\varepsilon_{vi}^{r/q}$. According to Theorem 2, if the control strategy satisfies (23), the attitude subsystem can achieve the desired orientation in finite time. Namely,

$$\forall \varepsilon_1 > 0,\ \exists t_{\varepsilon_1},\ \left\|(C_{ib^*p} - C_{ibp})e_3\right\| < \varepsilon_1,$$

then

$$\forall \varepsilon_2 > \varepsilon_1,\ \exists t_{\varepsilon_2},\ \dot{V} \leq 0,\ \|s_i\| < \varepsilon_2.$$

Thus s_i can converge to a compact set with the radius of ε_2 in finite time. This indicates that the tracking error of relative position converges to zero.

4 Simulation Results

Apply the proposed method to the finite-time formation of three quadrotors. The parameters are listed below:

$$m = 1\ kg,\ \ J = diag\{0.089 \quad 0.089 \quad 0.89\}\ kg \cdot m^2,\ \ g = 9.8\ N/kg.$$

The connection relationship among three quadrotors are shown in Fig. 1 by the form of the general directed graph

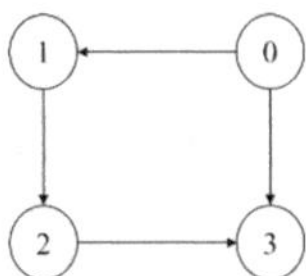

Fig. 1. The connection among three quadrotors.

The reference trajectories of leader along the directions of $x,\ y,\ z$ are

$$P^d = [-5\sin(t/2\pi),\ 2.5sin(t/2\pi),\ -0.5t]^T\ m.$$

The expected distances among the team are

$$\xi_{1d} = [0,2,0]^T\ m,\ \ \xi_{2d} = [-1,0,0]^T\ m,\ \ \xi_{3d} = [1,0,0]^T\ m,$$

and the initial positions are set as

$$e_{p1} = [0.5, 0.5, -0.3]^T\ m,\ e_{p2} = [-0.4, -0.6, -0.8]^T\ m,\ e_{p3} = [0.5, -1.5, -0.2]^T\ m.$$

Finally, we introduce the external disturbances and internal uncertainties as

$$d(t) = 0.1 * [\sin(0.01t),\ 0.3\sin(0.02t),\ 0.5\sin(0.01t)]^T\ N.$$

The simulation results are shown below

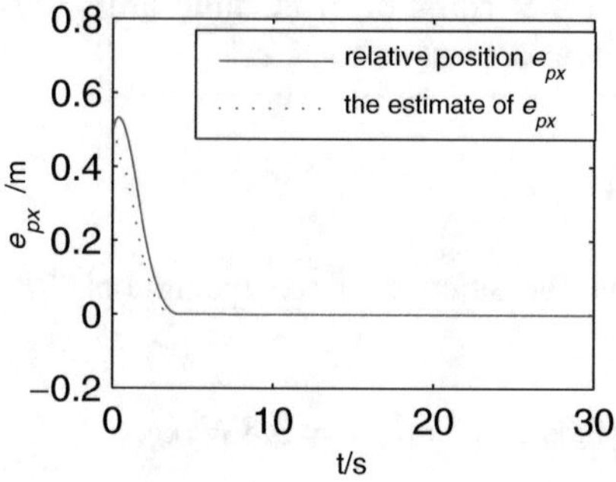

Fig. 2. The estimation of relative position

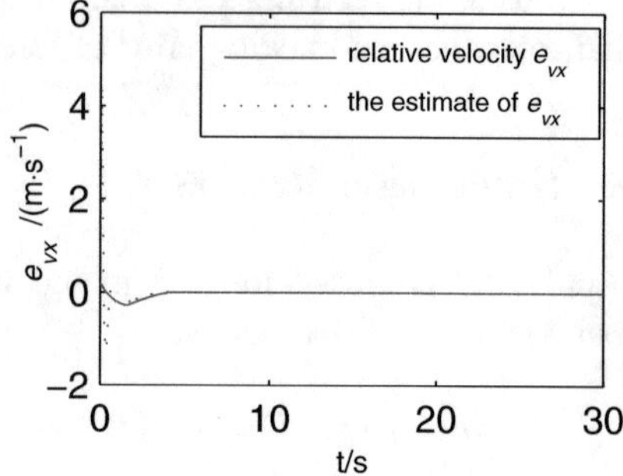

Fig. 3. The estimation of relative velocity

We use the high-gain observer to estimate the relative velocity via the relative position of quadrotors. As it can be seen from Figs. 2 and 3, the relative velocity of follower 1 can be estimated accurately with the proposed high-gain observer. Due to the space limitation, we do not list the estimates of the rest two followers, which achieve the same level of performance as follower 1.

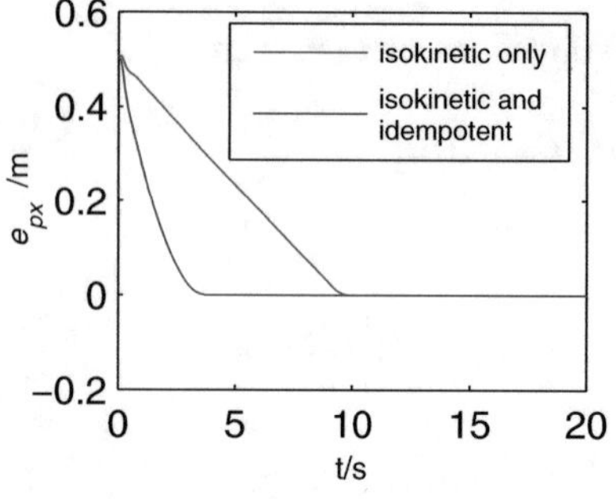

Fig. 4. The comparison of relative position

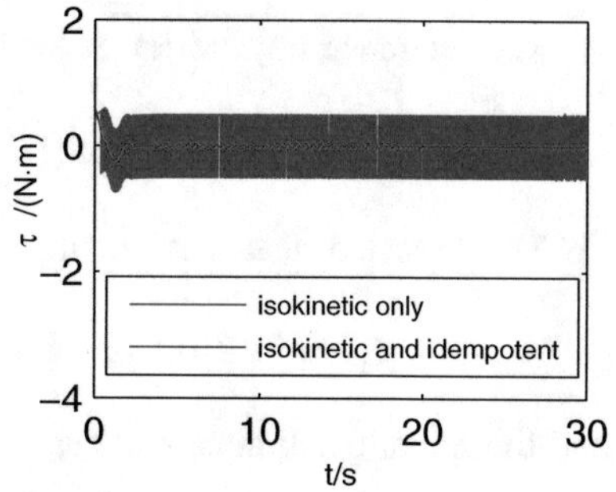

Fig. 5. The comparison of control torques

Figures 4 and 5 show the comparison of two different trending laws applied during the formation process, which indicates that the convergence rate is significantly improved via the combination of the isokinetic trending law and the idempotent trending law. Moreover, the chattering of control torques is greatly alleviated because of the introduced idempotent term in the attitude subsystem.

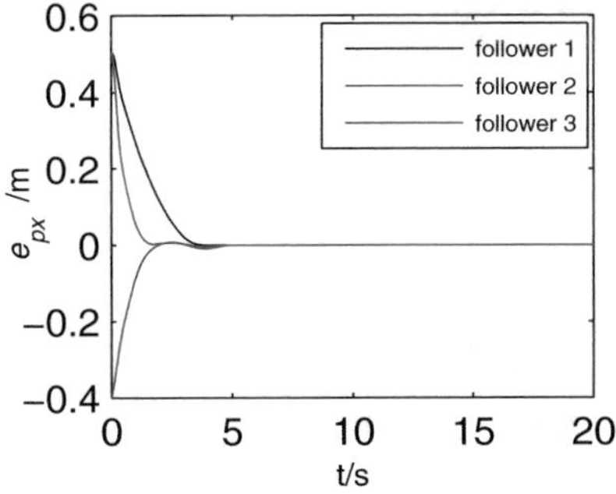

Fig. 6. The tracking error of relative position with three followers

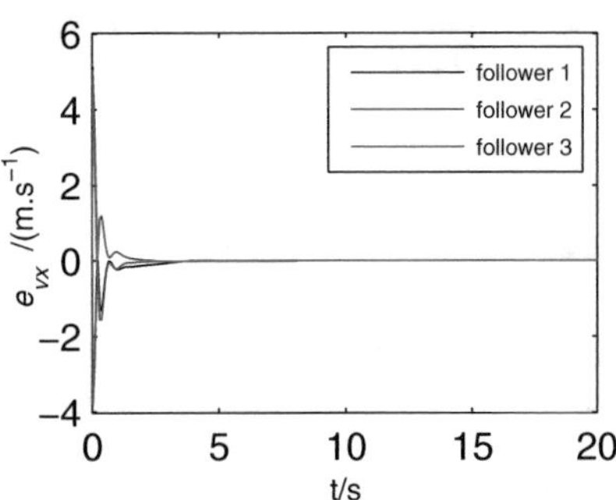

Fig. 7. The tracking error of relative velocity with three followers

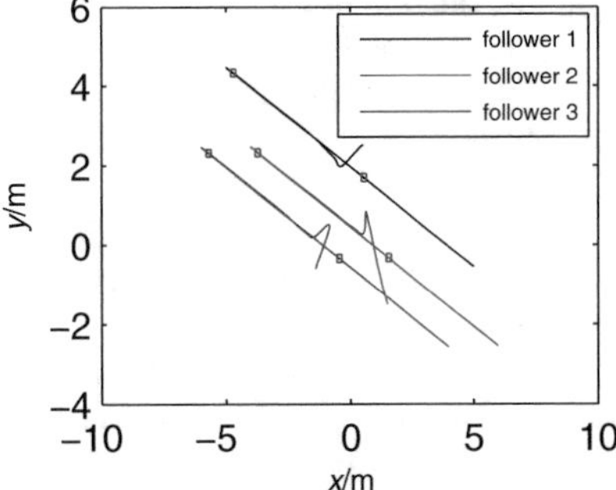

Fig. 8. Finite-time formation of quadrotors projected in two dimensions.

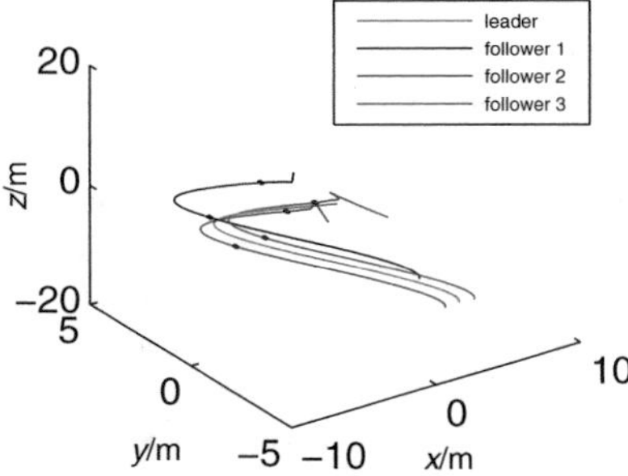

Fig. 9. Finite-time formation of quadrotors in three dimensions.

The tracking errors of relative position and relative velocity with three followers are shown in Figs. 6 and 7. We can see that both the tracking errors of relative position and relative velocity converge to zero after 5 s. Especially, Figs. 8 and 9 show that the finite-time consensus tracking of three quadrotors is achieved.

5 Conclusion

The problem of consensus tracking for quadrotors using the information of relative position only has been addressed. A high-gain observer has been constructed to estimate the relative velocity, which is used for the design of nonsingular terminal sliding mode protocols. Moreover, the isokinetic trending law and the idempotent trending law has been combined to achieve faster finite-time formation. Meanwhile, an idempotent term has been introduced in the attitude subsystem to eliminate the chattering caused by the isokinetic trending law. The simulation results have validated the effectiveness of the proposed method.

Acknowledgments. The authors would thank the National Natural Science Foundation of China (Grant No. U1713223 and 61673325) and the Chancellor Fund of Xiamen University (Grant No. 20720180090) for supporting this research.

References

1. Lei, C., Sun, W., Yeow, J.T.W.: A distributed output regulation problem for multi-agent linear systems with application to leader-follower robot's formation control. In: Proceedings of the 35th Chinese Control Conference, pp. 614–619. IEEE (2016)
2. Sabol, C., Burns, R., Mclaughlin, C.A.: Satellite formation flying design and evolution. J. Spacecraft Rockets **38**(2), 270–278 (2012)
3. Anderson, B.D.O., Yu, C.: Range-only sensing for formation shape control and easy sensor network localization. In: Proceedings of the 2011 Chinese Control and Decision Conference, pp. 3310–3315. IEEE (2011)
4. Hua, C., Chen, J., Li, Y.: Leader-follower finite-time formation control of multiple quadrotors with prescribed performance. Int. J. Syst. Sci. **48**(16), 1–10 (2017)
5. Beard, R.W., Lawton, J., Hadaegh, F.Y.: A coordination architecture for spacecraft formation control. IEEE Trans. Control Syst. Technol. **9**(6), 777–790 (2001)
6. Ding, S., Li, S.: A survey for finite-time control problems. Control Decis. **26**(2), 161–169 (2011)
7. Sun, C., Hu, G., Xie, L.: Fast finite-time consensus tracking for first-order multi-agent systems with unmodelled dynamics and disturbances. In: Proceedings of the 11th IEEE International Conference on Control & Automation, pp. 249–254. IEEE (2014)
8. Fu, J., Wang, Q., Wang, J.: Global saturated finite-time consensus tracking for uncertain second-order multi-agent systems with directed communication graphs. In: Proceedings of the 35th Chinese Control Conference, pp. 7684–7689. IEEE (2016)
9. Fu, J., Wang, Q., Wang, J.: Robust finite-time consensus tracking for second-order multi-agent systems with input saturation under general directed communication graphs. Int. J. Control, 1–25 (2017)
10. Wang, X., Li, S.: Consensus disturbance rejection control for second-order multi-agent systems via nonsingular terminal sliding-mode control. In: Proceedings of the 14th International Workshop on Variable Structure Systems, pp. 114–119. IEEE (2016)
11. Han, T., Guan, Z.H., Liao, R.Q., et al.: Distributed finite-time formation tracking control of multi-agent systems via FTSMC approach. IET Control Theor. Appl. **11**(15), 2585–2590 (2017)
12. Bondy, J.A., Murty, U.S.R.: Graph theory with applications. North Holland (1976)
13. Hu, J., Hong, Y.: Leader-following coordination of multi-agent systems with coupling time delays. Physica A Stat. Mech. Appl. **374**(2), 853–863 (2007)
14. Chen, W.C., Xiao, Y.L., Zhao, L.H., et al.: Kernel matrix of quaternion and its application in spacraft attitude control. Acta Aeronautica Et Astronautica Sinic **21**(5), 389–392 (2000)
15. Liu, J.K.: Sliding Mode Variable Structure Control and Matlab Simulation. 3rd edn. Tsinghua University press (2015)
16. Tran, M.D., Kang, H.J.: Nonsingular terminal sliding mode control of uncertain second-order nonlinear systems. Math. Probl. Eng. **2**, 1–8 (2015)

Flight Control of Tilt Rotor UAV During Transition Mode Based on Finite-Time Control Theory

Hang Yang, Huangxing Lin, Jingyao Wang, and Jianping Zeng(✉)

Xiamen University, Fujian 361101, China
jpzeng@xmu.edu.cn

Abstract. This paper focuses on the finite time convergence problem of system states in the course of the transition flight control for a small tilt rotor unmanned aerial vehicle (UAV). A controller design method using nonsingular terminal sliding mode surface and extended state observers (ESOs) is proposed. Due to the velocity and structure of tilt rotor UAV vary significantly with the variation of tilt angle, the transition mode is divided into two parts. To adapt to complex aerodynamic characteristics and maneuvering characteristics, and the vibrational control structure in different part of the transition mode, a nonsingular terminal sliding mode control method is applied to make the states converge to the reference trajectories in finite time. Moreover, ESOs are provided to enhance the robustness of the system for uncertainties. Finally, a numerical example is given to verify the effectiveness and robustness of the proposed approach. Regardless of disturbances, the aircraft can achieve the mode transition safely and smoothly.

Keywords: Tilt rotor UAV · Transition mode · Extended state observer Finite-time control

1 Introduction

With the capability of landing and hovering, vertical taking-off and good maneuverability, the tilt rotor unmanned aerial vehicles (UAVs) have become one of the most popular aircraft [1–4]. They have a wide range of application prospects in civil and military field, such as rescue, highway supervision, intelligence gathering and battlefield surveillance, etc. Without doubt, the rotor is an important feature of the tilt rotor UAV. It not only enables an aircraft to hover and ascend or descend in helicopter mode, but also makes an aircraft to high-speed fly and long-range cruise in airplane mode. The special configuration not only brings about excellent performance, but also produces many new technical problems. Especially during the transition mode between helicopter and airplane configurations, the air speed and aerodynamic characteristics change obviously with the variation of tilt angle, which leads to a complex system involving control redundancies, strong nonlinearities and strong couplings. Therefore, it has become a great attractive and challenging undertaking to design high-performance flight control systems in transition mode.

P. Krömer et al. (Eds.): ECC 2018, AISC 891, pp. 65–77, 2019.
https://doi.org/10.1007/978-3-030-03766-6_8

In the past, plenty of scholars have paid much attention to study the above questions and have proposed different kinds of control approaches, such as feedback linearization control, backstepping control, sliding mode control, etc. Moreover, the gain scheduling method for implementing the transition flight of tilt rotor UAV have been reported in [5, 6]. However, the gain scheduling has limited application since a large workload is needed and it is lacking in theoretical basis. As an effective control approach, PID is popular among many research groups in designing controllers for the flight control systems. The literatures [7–9] have combined the advantages of PID controllers which are simple and effective for the attitude control in transition mode. In order to realize the trajectory tracking in transition mode, the literatures [10, 11] have got solvability conditions via adopting the dynamic inversion method. However, the high-precision model is required. It is worth noting that there are no published results available in the literature for the tilt rotor UAV to track the reference trajectories in finite time.

In this paper, it takes the longitudinal model of a small tilt rotor UAV as the research object and focuses on its transition flight control. The control objective is to make the states converge to the reference trajectories in finite time by designing a non-singular terminal sliding mode controller. Firstly, the effect of uncertainties to the system is estimated and compensated by using ESOs. Then, in order to reduce the couplings between different channels and deal with the control redundancy problem, a control allocation scheme is presented which can enlarge the application scope of the tilt rotor UAV greatly. Finally, simulations are given to show the effectiveness of the proposed method.

2 Model of a Tilt Rotor UAV

The tilt rotor UAVs are equipped with rotatable grid plates on the wings and the rotors are mounted on grid plates (Fig. 1). The thrust line of each rotor can be changed as the grid plates tilt between the vertical and horizontal states, thus changing the flight mode [12].

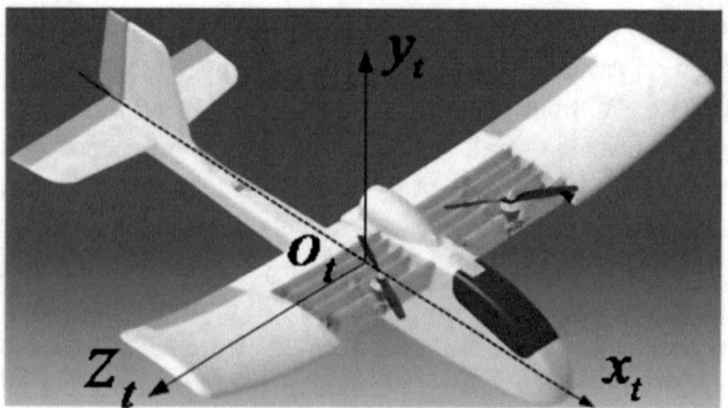

Fig. 1. Body axes of tilt rotor UAV

In transition mode, the tilt rotor UAV is a typical over-actuated system. Its control surface includes not only the conventional aerodynamic control surface, such as the elevator δ_z, but also the tension vector generated by the rotors. The size and direction of the tensions provided by both lateral sides can be changed independently, thus generating the other longitudinal control inputs, which are defined as follows

$$\begin{cases} \delta_{pe} = \dfrac{\delta_{pL} + \delta_{pR}}{2} \\ \delta_{te} = \dfrac{\tau_L + \tau_R}{2} \end{cases} \tag{1}$$

where δ_{pe} and δ_{te} are the mid-value of throttle and tilt angle, δ_{pL} and δ_{pR} are the left and right throttles, τ_L and τ_R are the left and right tilt angles, respectively. As in the longitudinal model, the left tilt angle is equal to the right tilt angle, which implies that $\tau := \tau_L = \tau_R$.

The longitudinal nonlinear model of the tilt rotor UAV can be expressed as

$$\begin{cases} m\dot{V} = F_{xt}\cos\alpha - F_{yt}\sin\alpha \\ mV\dot{\alpha} = mV\omega_z - F_{xt}\sin\alpha - F_{yt}\cos\alpha \\ \dot{\vartheta} = \omega_z \\ \dot{\omega}_z = \frac{M_z}{I_z} \\ \dot{H} = V\cos\alpha\sin\vartheta - V\sin\alpha\cos\vartheta \end{cases} \tag{2}$$

where V, m, α, ϑ, ω_z, H, I_z, F_{xt}, F_{yt} and M_z are the velocity, the mass of the aircraft, the angle of attack, the angle of pitch, the pitch angle rate, the altitude, the moment of inertia about the pitch axis, the component of thrust along body x_t axis, the component of thrust along body y_t axis, and the pitching moment, respectively.

Denote the states by $x = [V, \alpha, \vartheta, \omega_z, H]^T$, and the control inputs by $u = [\delta_z, \delta_{pe}, \delta_{te}]^{\mathrm{T}}$. The control objective is to ensure the states track the reference trajectories $x^* = [V^*, \alpha^*, \vartheta^*, \omega_z^*, H^*]^{\mathrm{T}}$ in finite time by designing admissible controller.

Let

$$\begin{aligned} \tilde{x} &= x - x^* \\ &= \begin{bmatrix} \tilde{V} & \tilde{\alpha} & \tilde{\vartheta} & \tilde{\omega}_z & \tilde{H} \end{bmatrix}^T \end{aligned} \tag{3}$$

$$\begin{aligned} \tilde{u} &= u - u^* \\ &= \begin{bmatrix} \tilde{\delta}_z & \tilde{\delta}_{pe} & \tilde{\delta}_{te} \end{bmatrix}^T \end{aligned} \tag{4}$$

where $\tilde{x}$ and $\tilde{u}$ are the tracking error and the increment of the control inputs, respectively.

Since α and ϑ remain small throughout the flight, we use the following approximations

$$\begin{aligned} \sin\alpha \approx \alpha,\ \cos\alpha \approx 1 \\ \sin\vartheta \approx \vartheta,\ \cos\vartheta \approx 1 \end{aligned} \tag{5}$$

Based on all the above conditions, the trajectory-tracking problem of the system (1) can be converted into the stabilization problem of the following error system

$$\begin{cases} \dot{\tilde{V}} = \frac{\tilde{F}_{xt}}{m} - \frac{F_{yt}^*}{m}\tilde{\alpha} - (\alpha^* + \tilde{\alpha})\frac{\tilde{F}_{yt}}{m} \\ \dot{\tilde{\alpha}} = \tilde{\omega}_z - \frac{F_{xt}^*}{mV}\tilde{\alpha} - \frac{\tilde{F}_{xt}}{mV}(\alpha^* + \tilde{\alpha}) - \frac{\tilde{F}_{yt}}{mV} \\ \dot{\tilde{\vartheta}} = \tilde{\omega}_z \\ \dot{\tilde{\omega}}_z = \frac{\tilde{M}_z}{I_z} \\ \dot{\tilde{H}} = V(\tilde{\vartheta} - \tilde{\alpha}) \end{cases} \tag{6}$$

Before we formally introduce the control law, the following lemmas are needed.

Lemma 1 [13]. *Consider a nonlinear system in the form of*

$$\dot{x} = f(x, t),\ f(0, t) = 0,\ x \in R^n \tag{7}$$

if there exists a continuous differentiable function $V(x)$, which is defined in the neighborhood of the origin $\hat{U} \subset R^n$, satisfying the following conditions

(1) *$V(x)$ is positive definite;*
(2) *there exists real number $\sigma > 0$, $0 < \lambda < 1$, such that*

$$\dot{V}(x) + \sigma V^{\lambda}(x) \le 0 \tag{8}$$

the equilibrium point of system (7) is finite-time stable. Moreover, if $\hat{U} = R^n$ and $V(x)$ is radially unbounded, i.e., $V(x) \to \infty$ as $\|x\| \to +\infty$, the equilibrium point of system (7) is globally finite-time stable.

Lemma 2 [14]. *Consider the nonlinear system (7), the equilibrium point $x = 0$ is globally finite-time stable for any given initial condition $x(0) = x_0$, if there exists a Lyapunov function $V(x)$ satisfying the following given inequalities*

$$\begin{aligned} \dot{V}(x) \le -\lambda_1 V(x) - \lambda_2 V^{\sigma}(x) \\ \lambda_1 > 0,\ \lambda_2 > 0,\ 0 < \sigma < 1, \end{aligned} \tag{9}$$

the settling time can be given by

$$T \le \frac{1}{\lambda_1(1-\sigma)} \ln \frac{\lambda_1 V^{1-\sigma}(x_0) + \lambda_2}{\lambda_2} \tag{10}$$

where $V(x_0)$ is the initial value of $V(x)$.

3 Transition Flight Control

During transition mode, the elevator δ_z and the mid-value of tilt angle δ_{te} control the pitching moment simultaneously, which leads to the control redundancy problem and complicates the design of flight controller. To solve the above disadvantages, a two-stage design scheme is adopted in this paper. First, substituting the virtual elevator δ_{zV} for the actual elevator δ_z and the actual mid-value of tilt angle δ_{te}, meanwhile, substituting the virtual mid-value of throttle δ_{pV} for the actual mid-value of throttle δ_{pe}. Next, the virtual control inputs $u_V = \left[\delta_{zV}, \delta_{pV}\right]^{\mathrm{T}}$ are allocated to the actual control inputs $u = \left[\delta_z, \delta_{pe}, \delta_{te}\right]^{\mathrm{T}}$ via some proper control allocation strategy.

Define the increment of virtual control inputs as follows

$$\begin{aligned} \tilde{u}_V &= \begin{bmatrix} \tilde{\delta}_{zV} & \tilde{\delta}_{pV} \end{bmatrix}^T \\ &= \begin{bmatrix} \delta_{zV} - \delta_{zV}^* & \delta_{pV} - \delta_{pV}^* \end{bmatrix}^T \end{aligned} \tag{11}$$

Then our aim is to prove that the resulting closed-loop system (6) under the state feedback control law $\tilde{u}_V = \begin{bmatrix} \tilde{\delta}_{zV} & \tilde{\delta}_{pV} \end{bmatrix}^T$ is asymptotically stable at the equilibrium point.

3.1 Control Strategies

Since the aerodynamic characteristics of the aircraft vary significantly with the variation of tilt angle, the control mechanism of the aircraft should also vary correspondingly. When the tilt angle is small and the air speed is high, the control mechanism of the aircraft is more similar to that of a turboprop airplane. When the tilt angle is larger and the air speed is lower, the control mechanism of the aircraft is more similar to that of a helicopter.

Therefore, the point $\tau = 40^\circ$ is tentatively chosen as the dividing point and thus the transition mode is divided into two parts [15]. When the tilt angle satisfies $\tau \in [0^\circ, 40^\circ)$, the virtual mid-value of throttle δ_{pV} controls the air speed, and the pitching movement which is controlled by the virtual elevator δ_{zV} controls the altitude. Correspondingly, when the tilt angle satisfies $\tau \in [40^\circ, 78^\circ]$, the virtual mid-value of throttle controls the altitude, and the pitching movement which is controlled by the virtual elevator controls the air speed. When the tilt angle is on the demarcation point, i.e. $\tau = 40^\circ$, the structure of the controller switches to adapt to the complex aerodynamic characteristics and operating characteristics of the transition mode.

In each part of the transition mode, selecting an operating point to design a sliding mode controller to ensure that the states converge to desired sliding surface in finite time and move along the sliding surface and finally converge to the equilibrium point. Moreover, ESOs are provided to enhance the robustness of the system for uncertainties, and to restrain the chattering phenomenon.

The control mechanism of the aircraft during transition flight are shown in the table below (Table 1).

Table 1. Control strategy for transition mode

State variable	Control input $\tau < 40°$	Control input $\tau \geq 40°$
Velocity	Virtual throttle	Pitch movement
Altitude	Pitch movement	Virtual throttle
Pitch movement	Virtual elevator	Virtual elevator

3.2 Non-singular Terminal Sliding Mode Control Law Based on ESO

A non-singular terminal sliding mode controller and an ESO are designed for each channel of the longitudinal model. The pitch angle channel is taken as an example.

By Eq. (6), the equation of the pitch angle channel is given by

$$\begin{cases} \dot{\tilde{\vartheta}} = \tilde{\omega}_z \\ \dot{\tilde{\omega}}_z = \dfrac{\tilde{M}_z}{I_z} \end{cases} \tag{12}$$

Let $x_1 = \tilde{\vartheta}$, $x_2 = \tilde{\omega}_z$. The Eq. (12) can be rewritten as

$$\begin{cases} \dot{x}_1 = x_2 \\ \dot{x}_2 = \dfrac{\tilde{M}_z}{I_z} \end{cases} \tag{13}$$

The performance of systems will be deteriorated in the presence of disturbances. The disturbances include the variation of the aerodynamic coefficients, modelling errors, unmodelled dynamics and external disturbances such as wind gust. Taking into account these factors, the Eq. (13) can be rewritten as

$$\begin{cases} \dot{x}_1 = x_2 \\ \dot{x}_2 = a(t) + b\tilde{\delta}_{zV} \end{cases} \tag{14}$$

$$a(t) = \frac{\tilde{M}_z}{I_z} - b\tilde{\delta}_{zV} \tag{15}$$

where $\tilde{\delta}_{zV}$ is the increment of the virtual elevator. b is the coefficient of the virtual elevator, which can be obtained by linearizing the system (2) at concerned operating points. $a(t)$ represents the total disturbance of the system, which can be compensated by using the ESO.

Define the extended state by $x_3 = a(t)$, one has

$$\begin{cases} \dot{x}_1 = x_2 \\ \dot{x}_2 = x_3 + b\tilde{\delta}_{zV} \\ \dot{x}_3 = h \end{cases} \tag{16}$$

where h is the derivative of $a(t)$.

A third-order ESO is applied to estimate the disturbance $a(t)$, which can be described as follows

$$\begin{cases} e = z_1 - x_1 \\ \dot{z}_1 = z_2 - \beta_{01} e \\ \dot{z}_2 = z_3 - \beta_{02} fal(e, 0.5, 0.01) + b\tilde{\delta}_{zV} \\ \dot{z}_3 = -\beta_{03} fal(e, 0.25, 0.01) \end{cases} \tag{17}$$

$$fal(e, m, n) = \begin{cases} \dfrac{e}{n^{1-m}}, & |e| \le n \\ |e|^m \mathrm{sgn}(e), & |e| > n \end{cases} \tag{18}$$

where e denotes the estimation error, z_1, z_2, z_3 are the observer states, and β_{01}, β_{02}, β_{03} are the observer gains to be designed. One can readily verify that, given appropriate values of the observer gains and the function $fal(\cdot)$, the observer state z_i tends to x_i and the error of observation $|z_i - x_i|$ converges to l_i, where l_i is a small positive number and $i = 1, 2, 3$.

Then an ESO-based control law is designed as

$$\tilde{\delta}_{zV} = \frac{u_c - z_3}{b} \tag{19}$$

where u_c represents the non-singular terminal sliding mode control law.

The non-singular terminal sliding surface is selected as

$$s = x_1 + \eta \cdot sig^r(x_2) \tag{20}$$

where $sig^r(x_2) = |x_2|^r \mathrm{sgn}(x_2)$, $\eta > 0$, $1 < r < 2$.

The control law $\tilde{\delta}_{zV}$ can be designed as

$$\tilde{\delta}_{zV} = -(z_3 + \eta^{-1} r^{-1} sig^{2-r}(x_2) + k_{01} s + k_{02} sig^{\rho}(s) + \varepsilon \mathrm{sgn}(s))/b \tag{21}$$

where $k_{01} > 0$, $k_{02} > 0$, $0 < \rho < 1$, $\varepsilon > l_3$.

Theorem 1. Consider the system (14) under the ESO-based non-singular terminal sliding mode control law (21), the states will converge to the desired sliding surface in finite time. On the sliding mode surface $s = 0$, it can be known from the characteristics of the non-singular terminal sliding mode that the states will converge to the equilibrium point in finite time.

Proof. Let $V(x)$ be a candidate Lyapunov function defined as

$$V(s) = s^2 \tag{22}$$

The time derivative of $V(s)$ is given by

$$\begin{aligned} \dot{V}(s) &= 2s\dot{s} \\ &= 2s(x_2 + \eta r|x_2|^{r-1}\dot{x}_2) \\ &= 2s(x_2 + \eta r|x_2|^{r-1}(a(t) + b\tilde{\delta}_{zV})) \end{aligned} \tag{23}$$

Substituting the control law (21) into Eq. (23), we obtain

$$\dot{V}(s) = 2\eta r|x_2|^{r-1}(d(t)s - k_{01}s^2 - k_{02}|s|^{\rho+1} - \varepsilon|s|) \tag{24}$$

$$|d(t)| = |a(t) - z_3| \le l_3 \tag{25}$$

Note that

$$d(t)s - \varepsilon|s| \le (l_3 - \varepsilon)|s| \le 0 \tag{26}$$

By Eq. (24), we obtain

$$\begin{aligned} \dot{V}(s) &\le 2\eta r|x_2|^{r-1}(-k_{01}s^2 - k_{02}|s|^{\rho+1} + (l_3 - \varepsilon)|s|) \\ &\le -\mu V(s) - \lambda V^{\frac{\rho+1}{2}}(s) \end{aligned} \tag{27}$$

where $\mu = 2k_{01}\eta r|x_2|^{r-1}$, $\lambda = 2k_{02}\eta r|x_2|^{r-1}$. When $x_2 \neq 0$, according to Lemma 2, the states will be driven onto the sliding surface $s = 0$ in finite time. When $x_2 = 0$ and $s \neq 0$, we can know that $x_1 \neq 0$ according to Eq. (20) and the system does not stay at $\dot{V} = 0$. Since $\dot{V} \le 0$, the states will be driven onto the sliding surface $s = 0$ in finite time. In summary, no matter what the initial states of the system are, they will be driven onto the sliding surface $s = 0$ in finite time ultimately.

After the states converge to the desired sliding surface $s = 0$, by Eq. (20), we obtain

$$\dot{x}_1 = -\eta^{-\frac{1}{r}}|x_1|^{\frac{1}{r}}\operatorname{sgn}(x_1) \tag{28}$$

Define the Lyapunov function candidate as

$$\bar{V}(t) = \frac{1}{2}x_1^2 \tag{29}$$

The time derivative of $\bar{V}(t)$ is given by

$$\begin{aligned}\dot{\bar{V}}(t) &= x_1\dot{x}_1 = -\eta^{-\frac{1}{r}}|x_1|^{\frac{1}{r}+1} \\ &= -2^{\frac{1+r}{2r}}\eta^{-\frac{1}{r}}\bar{V}^{\frac{1+r}{2r}}(t) \leq 0\end{aligned} \tag{30}$$

According to Lemma 1, the states will move along the sliding surface and finally converge to the equilibrium point in finite time. Solving the differential Eq. (28), the convergence time is

$$t_c = \frac{r\eta^{\frac{1}{r}}}{r-1}|x_1(t_r)|^{1-\frac{1}{r}} \tag{31}$$

This completes the proof.

The design methods for velocity and altitude channels are similar to that of the pitch angle channel.

3.3 Control Allocation

In transition mode, the movements of attitude and position are both directly affected by engine thrust and aerodynamic forces, which cause the control redundancy problem and renders the flight control design more difficult. In order to reduce the couplings between different channels and to tackle the control redundancy problem, the concepts of the virtual elevator and the virtual throttle are introduced. Considering the effectiveness of each control surface at different tilt angles, the mapping relationships between the virtual control inputs and the actual control inputs can be written as

$$\tilde{u} = K_\delta(\tau)\tilde{u}_V \tag{32}$$

where $\tilde{u} = [\,\tilde{\delta}_z \quad \tilde{\delta}_{pe} \quad \tilde{\delta}_{te}\,]^T$ and $\tilde{u}_V = [\,\tilde{\delta}_{zV} \quad \tilde{\delta}_{pV}\,]^T$ represent the increments of the actual control inputs and the virtual control inputs, and $K_\delta(\tau)$ is the control allocation matrix, respectively.

The control allocation matrix $K_\delta(\tau)$ is given by

$$K_\delta(\tau) = \begin{bmatrix} K_{\delta_z} & 0 \\ 0 & 1 \\ K_{\delta_{te}} & 0 \end{bmatrix} \tag{33}$$

$$K_{\delta_z} = \frac{B_{\omega_z}^{\delta_{zV}}}{B_{\omega_z}^{\delta_z}}K_z, K_{\delta_{te}} = \frac{B_{\omega_z}^{\delta_{zV}}}{B_{\omega_z}^{\delta_{te}}}K_{Tz}$$

where $B_{\omega_z}^{\delta_{zv}}$ represents the effectiveness of the virtual elevator, $B_{\omega_z}^{\delta_z}$ and $B_{\omega_z}^{\delta_{te}}$ are the effectiveness of $\tilde{\delta}_z$ and $\tilde{\delta}_{te}$ under different tilt angles, which can be obtained by the effectiveness matrix after linearizing the system (2). K_z and K_{Tz} are the allocation coefficients of $\tilde{\delta}_z$ and $\tilde{\delta}_{te}$, and can be expressed as

$$K_{Tz} = \begin{cases} 0, & 0^\circ \leq \tau \leq 15^\circ \\ (\tau - 15)/45, & 15^\circ < \tau \leq 60^\circ \\ 1, & 60^\circ < \tau \leq 78^\circ \end{cases} \tag{34}$$

$$K_z = 1 - K_{Tz} \tag{35}$$

4 Simulation and Analysis

$\tau = 40^\circ$ is tentatively chosen as the dividing point and thus the transition mode is divided into two parts. When the tilt angle is on the demarcation point, i.e. $\tau = 40^\circ$, the structure of the controller switches to adapt to the operating characteristics of the transition mode. Furthermore, selecting $\tau = 20^\circ$ and $\tau = 60^\circ$ as the operating points, and then the ESOs and sliding mode controllers are designed at these points.

The reference trajectories of the control inputs and the states can be written as

$$\begin{aligned} V^* = & -0.000000944493684\tau^4 + 0.000114201311023\tau^3 \\ & -0.006078837535368\tau^2 - 0.034872054994257\tau \\ & + 23.380059324866302 \end{aligned}$$

$$H^* = 20$$

$$\delta_z^* = \begin{cases} 1.142 \times 10^{-7}\tau^4 - 8.928 \times 10^{-6}\tau^3 + \\ 0.0001713\tau^2 + 0.01087\tau + 0.3895, & \tau \leq 60^\circ \\ 0.008692\tau^3 - 1.726\tau^2 \\ + 114\tau - 2503, & 60^\circ < \tau \leq 78^\circ \end{cases}$$

$$\begin{aligned} \delta_{pe}^* = & -0.000000566256427\tau^3 + 0.000141395965809\tau^2 \\ & -0.000206338598641\tau + 0.143911631016043 \end{aligned}$$

The initial states are selected as $V_0 = 3\,\text{m/s}$, $\alpha_0 = \vartheta_0 = 1.15^\circ$, $\omega_{z0} = 0\,\text{rad/s}$, and $H_0 = 20\,\text{m}$. Firstly, the aircraft is commanded to tilt from the helicopter mode (78°) to the airplane mode (0°) at the tilting angular velocity $1^\circ/\text{s}$. Then, the aircraft keeps flying in the airplane mode for 10 s. Finally, the aircraft converts from the airplane mode to the helicopter mode. The whole simulation time is 166 s.

Moreover, to test the robustness of the proposed approach, $\pm 30\%$ parameter perturbation of lift coefficient C_y is performed in the system. The simulation results are given in Figs. 2, 3, 4, 5, 6, 7 and 8.

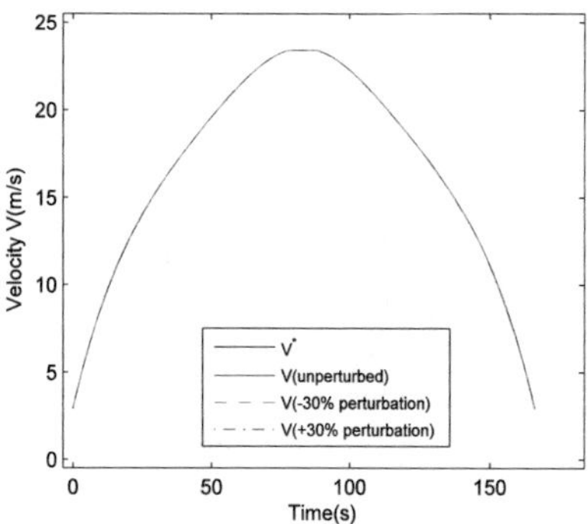

Fig. 2. Trajectories of the velocity

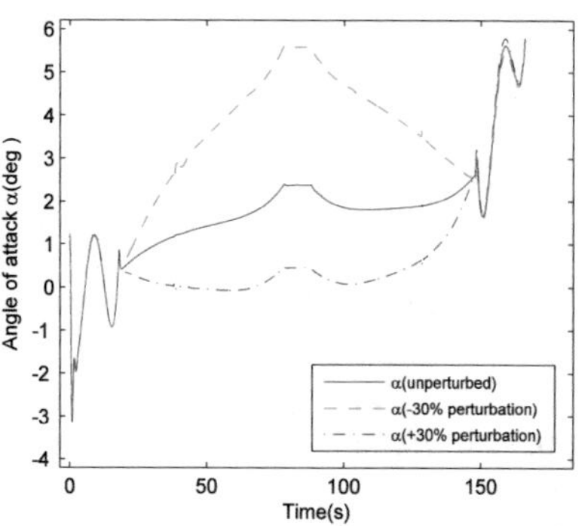

Fig. 3. Trajectories of the attack angle

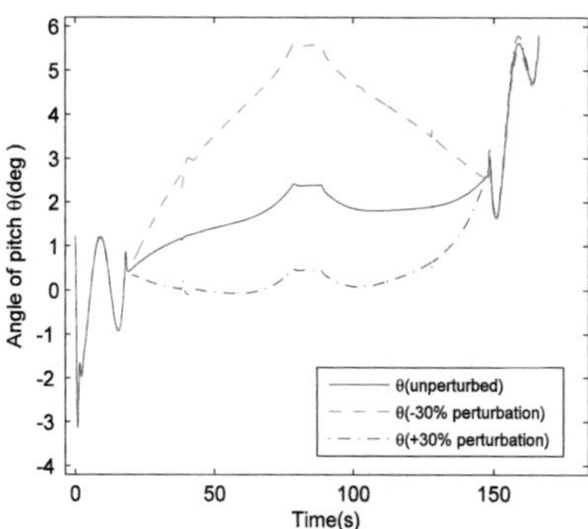

Fig. 4. Trajectories of the pitch angle

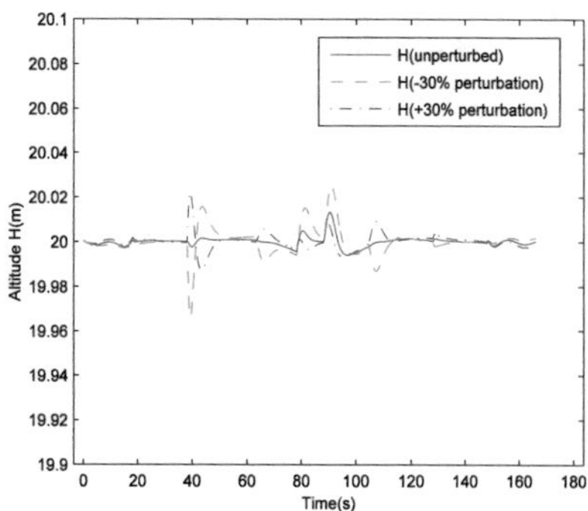

Fig. 5. Trajectories of the altitude

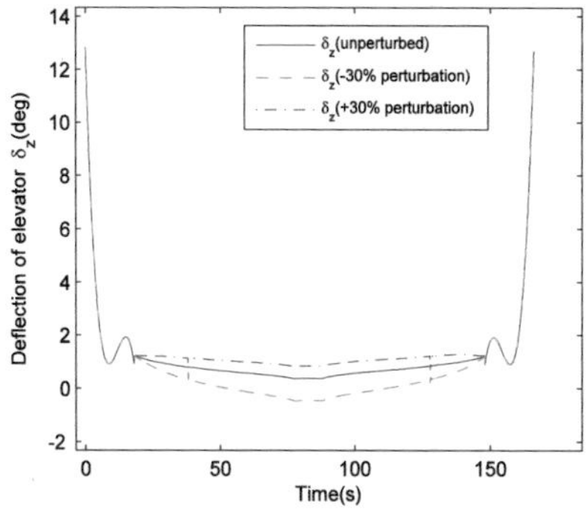

Fig. 6. Deflection of the elevator

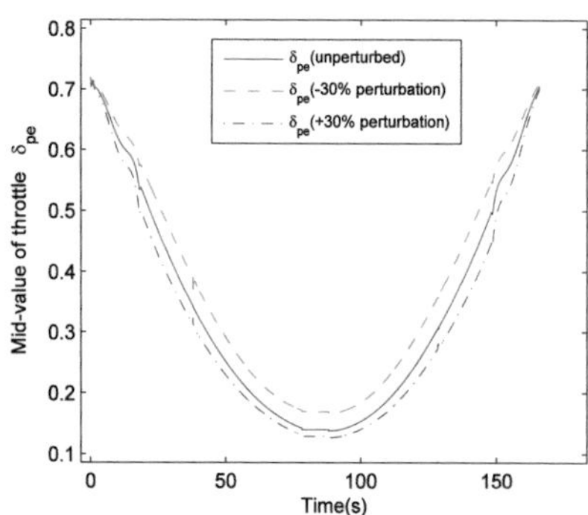

Fig. 7. Mid-value of the throttle

From the simulation results in Figs. 2, 3, 4, 5, 6, 7 and 8, we can make the conclusion that

(1) The system can track the reference trajectories rapidly, and its steady-state performance is good. Meanwhile, the states and the control surface change reasonably, and the transition flight can be completed according to the predetermined trajectories. All above show that the design of the control system is excellent.
(2) Under the influence of parameter perturbation, the pitch angle and angle of attack are correspondingly changed to generate sufficient aerodynamic force to ensure the smooth flight of the aircraft, and the changes are within a reasonable range. The entire transition mode is smooth and stable, which shows that the proposed control law has good robustness to the uncertainties.

In a word, the experimental results demonstrate that our method can realize the control objectives effectively. Furthermore, the UAV can still complete the transition flight according to the predetermined trajectories in the presence of parameter perturbation. Meanwhile, the system has rapid dynamic response and good robustness to the uncertainties, which shows that the system can satisfy various performance indicators.

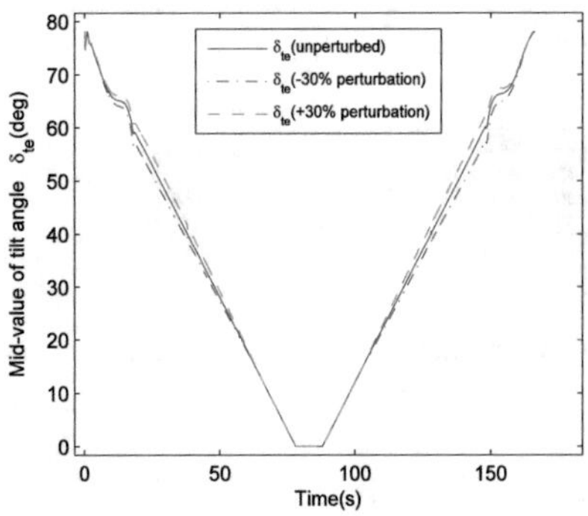

Fig. 8. Mid-value of the tilt angle

5 Conclusion

This paper focuses on the transition flight control for a small tilt rotor UAV. Considering that the maneuvering characteristics of the UAV will change under different tilt angles, a series of suitable control strategies for transition mode are proposed. Then, a non-singular terminal sliding mode controller is designed for each channel of the system to ensure that the states converge to the reference trajectories in finite time. And the robustness of the system is improved by using ESOs. Moreover, a simple and effective control allocation scheme is presented for the control redundancy problem. The simulation results show that the proposed control method has good performance and robustness. With or without parameter perturbation, the aircraft can achieve the mode transition safely and smoothly.

Acknowledgments. The authors would thank the National Natural Science Foundation of China (Grant No. 61673325 and U1713223) and the Chancellor Fund of Xiamen University (Grant No. 20720180090) for supporting this research.

References

1. Hirschberg, M.J.: An overview of the history of vertical and/or short take-off and landing (V/STOL) Aircraft. In: Proceedings (2006). www.vstol.org
2. Yeo, H., Johnson, W.: Performance and design investigation of heavy lift tilt-rotor with aerodynamic interference effects. J. Aircraft **46**(4), 1231–1239 (2009)
3. Ahn, O., Kim, J.M., Lim, C.H.: Smart UAV research program status update: achievement of tilt-rotor technology development and vision ahead. In: ICAS 2010, 27th Congress of International Council of the Aeronautical Sciences (2010)
4. Fu, R., Sun, H.F., Zeng, J.P.: Exponential stabilisation of nonlinear parameter-varying systems with applications to conversion flight control of a tilt rotor aircraft. Int. J. Control, 1–11 (2018)
5. Sato, M., Muraoka, K.: Flight controller design and demonstration of quad-tilt-wing unmanned aerial vehicle. J. Guid. Control Dyn. **38**(6), 1071–1082 (2014)
6. Cai, X.H., Fu, R., Zeng, J.P.: Robust H∞ gain-scheduling control for mode conversion of tilt rotor aircrafts. J. Xiamen Univ. (Nat. Sci.) **55**(3), 382–389 (2016)
7. Song, Y.G., Wang, H.J.: Design of flight control system for a small unmanned tilt rotor aircraft. Chin. J. Aeronaut. **22**(3), 250–256 (2009)
8. Chen, Y., Gong, H.J., Wang, B.: Research on longitudinal attitude control technology of tilt rotor during transition. Flight Dyn. **29**(1), 30–33 (2011)
9. Chen, Q., Jiang, T., Shi, F.M.: Longitudinal attitude control for a tilt tri-rotor UAV in transition mode. Flight Dyn. **34**(6), 49–53 (2016)
10. Rysdyk, R.T., Calise, A.J.: Adaptive model inversion flight control for tilt-rotor aircraft. J. Guid. Control Dyn. **22**(3), 402–407 (1999)
11. Rysdyk, R.T., Calise, A.J.: Adaptive nonlinear control for tiltrotor aircraft. In: Proceedings of the IEEE International Conference on Control Applications, pp. 980–984 (1998)
12. Lu, L.H., Fu, R., Wang, Y., et al.: Mode conversion of electric tilt rotor aircraft based on corrected generalized corridor. Acta Aeronautica et Astronautica Sinica **39**(7), 121900 (2018)
13. Zhou, H.B., Song, H.M., Liu, H.K.: Nonsingular terminal sliding mode guidance law with impact angle constraint. J. Chin. Inertial Technol. **22**(5), 606–611,618 (2014)
14. Yu, S., Yu, X., Shirinzadeh, B., Man, Z.: Continuous finite-time control for robotic manipulators with terminal sliding mode. Automatica **41**(11), 1957–1964 (2005)
15. Lin, H.X., Fu, R., Zeng, J.P.: Extended state observer based sliding mode control for a tilt rotor UAV. In: Proceedings of the 36th Chinese Control Conference, pp. 3771–3775. IEEE (2017)

Energizing Topics and Applications in Computer Sciences

A Minimum Spanning Tree Algorithm Based on Fuzzy Weighted Distance

Lu Sun[1,2], Yong-quan Liang[1,2,3(✉)], and Jian-cong Fan[1,2,3,4]

[1] Princeton University, Princeton, NJ 08544, USA
[2] College of Computer Science and Engineering, Shandong University of Science and Technology, Qingdao, China
lyq@sdust.edu.cn
[3] Provincial Key Laboratory for Information Technology of Wisdom Mining of Shandong Province, Shandong University of Science and Technology, Qingdao, China
[4] Provincial Experimental Teaching Demonstration Center of Computer, Shandong University of Science and Technology, Qingdao, China

Abstract. The traditional minimum spanning tree clustering algorithm uses a simple Euclidean distance metric method to calculate the distance between two entities. For the processing of noise data, the similarity can't be well described. In this regard, first of all, we integrate fuzzy set theory to improve, and propose a method with indeterminacy fuzzy distance measurement. In the distance metric method, fuzzy set theory is introduced to measure the differences between two entities. Moreover, the attributes are fuzzy weighted on this basis, which overcomes the shortcomings of the simple Euclidean distance measurement method. So, it not only has a good tolerance for data noise to solve the misclassification of noise information in the actual data, but also takes into account the difference of distinguishability contribution degree of attributes in classification. Thus, the accuracy of clustering is improved, and it also has significance in practical project application. Then, the proposed distance metric is applied into the traditional MST clustering algorithm. Compared with the traditional MST clustering algorithm and other classical clustering algorithms, the results show that the MST algorithm based on the new distance metric is more effective.

Keywords: Clustering algorithm · Minimum spanning tree algorithm Fuzzy set · Membership degree · Distance metric

1 Introduction

1.1 Minimum Spanning Tree (MST)

Minimum spanning tree (MST) [1] is one of the classical algorithms of graph theory. Minimum spanning tree clustering algorithm is also a typical clustering method [2]. Because its principle is simple and easy to implement, it has attracted more and more attention of researchers. [3] Traditional MST clustering algorithm is insensitive to noise data and can't use these uncertain information, which leads to the inaccuracy of MST segmentation and affects the accuracy of clustering. Therefore, the research of fuzzy

P. Krömer et al. (Eds.): ECC 2018, AISC 891, pp. 81–89, 2019.
https://doi.org/10.1007/978-3-030-03766-6_9

information processing has become an important topic. It has a great significance in the deeper research of MST clustering. Many methods of [4] graph theory are used to solve clustering problems. Clustering based on MST is a good way to cluster irregular data sets. [5] Zahn and others first obtained the minimum spanning tree of the data set through the adjacency matrix of the graph. [6] Zhong et al. proposed a clustering algorithm based on the minimum spanning tree of two rounds, which divides the clustering problems into separate clustering and contact clustering, and can automatically identify the two types. [7] March et al. proposed a dual tree structure based on KD tree and covering tree to construct MST. [8] Victor Olman and Fenglou Mao and others deal with the minimum spanning tree through the idea of dividing - parallel computing - merge. The algorithm improves computing efficiency at the expense of computer resources. [9] Wang and others put forward a fast MST clustering algorithm based on the divide and conquer thought. The above methods have improved the clustering method based on MST in varying degrees.

1.2 Indeterminacy

In 1999, the concept of indeterminacy [10] was proposed by Smarandache, an American scholar. The theory of indeterminacy is a generalization of the fuzzy set theory [11], which is closer to human thinking and can describe incomplete, uncertain and inconsistent information, while the intuitionistic fuzzy set can only deal with incomplete information. The minimum spanning tree clustering algorithm is easily affected by noise interference and data uncertainty. These factors increase the complexity of MST segmentation. However for the general MST clustering algorithm, it is difficult to solve these problems. And in fact, noise can be regarded as an uncertainty. Therefore, this paper introduces the indeterminacy theory into the clustering algorithm based on MST, which can effectively solve the uncertain information [12] in the MST segmentation, so as to improve the accuracy of the segmentation results. And it has a certain practical significance in the application of the actual project [13]. Compared with other clustering algorithms based on MST and other classical clustering algorithms, the results show that the algorithm in this paper is more effective.

2 Related Work

2.1 The Basic Concepts of the Indeterminacy Tet

The indeterminacy set is the collection of the true extent (T), the incertitude extent (I) and the false extent (F) of the element in the nonstandard unit interval. It is a summary of the fuzzy set, the intuitionistic fuzzy set and the paradoxes set [14].

Definition 1 Indeterminacy Set

[15] Sets X as an object set, x as any element of it, and an indeterminacy set A on X can be defined as

$$A = \{ <x, T_A(x), I_A(x), F_A(x) > |x \in X\},$$

Among them, $T_A(x)$, $I_A(x)$ and $F_A(x)$ respectively denote the function of the true extent, the incertitude extent and the false extent. $x \in X$, $T_A(x)$, $I_A(x)$, $F_A(x)$are the standard or nonstandard subset of $]0^-, 1^+[$, that is,

$$T_A(x) : X \to]0^-, 1^+[,$$
$$I_A(x) : X \to]0^-, 1^+[,$$
$$F_A(x) : X \to]0^-, 1^+[,$$

(in which non-standard finite number $1^+ = 1 + \mathcal{E}$. "1" is its standard part. "$\mathcal{E} > 0$" is an infinitesimal, and is its non-standard part. And $0^- \leq sup\ T_A(x) + sup\ I_A(x) + sup\ F_A(x) \leq 3^+$).

If $T_A(x)$, $I_A(x)$ and $F_A(x)$ are all real numbers in the closed interval [0, 1], then A is reduced to single valued indeterminacy set [16].

Definition 2 [17] Indeterminacy Distance
Sets $x_1 = \{T_1, I_1, F_1\}$ and $x_2 = \{T_2, I_2, F_2\}$ are two single valued indeterminacy number, the normalized Euclidean distance is:

$$\mathrm{D}x_1, x_2 = \sqrt{\left\{(T_1 - T_2)^2 + (I_1 - I_2)^2 + (F_1 - F_2)^2\right\}/3} \tag{1}$$

Definition 3 [18] Fuzziness Measure.
Sets $x_1 = \{T_1, I_1, F_1\}$ is a single valued indeterminacy number, the fuzzy measure of it is defined as:

$$\mu_1 = 1 - \sqrt{\left\{(1 - T_1)^2 + I_1^2 + F_1^2\right\}/3} \tag{2}$$

Definition 4 [19] Entropy of Indeterminacy
Sets $X = (x_1, x_2, \ldots, x_n)$, single valued indeterminacy set $A = \{<x, T_A(x), I_A(x), F_A(x) > | x \in X\}$, the entropy of indeterminacy of A is:

$$\mathrm{EA} = 1 - \frac{1}{n}\sum_{x \in X}(T_A(x_i) + F_A(x_i)) \cdot |I_A(x_i) - I_{A^*}(x_i)| \tag{3}$$

2.2 Traditional Clustering Algorithm Based on MST

The minimum spanning tree (MST) is an important data structure in graph theory. It has a wide range of applications in many fields, and its structure on the set has many advantages. In the construction method of MST, Kruskal algorithm and Prim algorithm are the most representative algorithms. Among them, the Prim algorithm is applicable to with the situation of that the MST with more sides [19]. The Prim algorithm is used in this paper.

In minimum spanning tree clustering, the first step is to build minimum spanning tree. Take the objects in the data set as the nodes of the graph and the distance measure between two objects as the edges of the graph. The minimum spanning tree of a graph is obtained on the basis of the above graph theory (a connected graph with the minimal weight accumulation) [20]. Delete those edges whose weights are greater than a given threshold, thus forming a forest. Each tree in the forest is a cluster. The algorithm steps are as follows (Table 1):

Table 1. Traditional clustering algorithm based on MST

Traditional clustering algorithm based on MST
Sets data set $X = (x_1, x_2, \ldots, x_n)$
(1) Construct the acyclic weighted graph $G\ (V, E, W)$; among them: $V = X = (x_1, x_2, \ldots, x_n)$; $W\ (x_i, x_j) = \mathrm{d}\ (x_i, x_j)$; $i, j = 1, \ldots, n$; $i \neq j$
(2) Find the minimum spanning tree MST of the complete graph G. $MST = \{(V,E)\mid E = \{e_1, e_2, \ldots, e_{n-1}\}\}$
(3) Remove the edges whose weights are bigger than the given threshold γ, get k sub trees and form forest F of X. $F = \{(V, E')\mid E' = T\text{-}\{e' \mid m(e') > \gamma\}\}$; The trees in the forest F is $T_i = \{(V_i, E_i)\mid i = 1, 2, \ldots, m\}$, $F = \bigcup_{i=1}^{m} \mathrm{T_i}$
(4) The result of clustering is the parts that MST still join together after these edges are deleted. Each tree T_i is a cluster, $C_i = T_i$.

3 MST Clustering Algorithm Based on Fuzzy Weighted Distance (FWMSTClust)

3.1 Fuzzy Weighted Distance Metric Based on Indeterminacy Set

Because the most commonly used experimental data sets have the characteristics of integrity and certainty, the distance measured by traditional distance metrics can often get better results. In real life, the environment is complex and uncertain, which makes the data set usually contains noise, and it is difficult to give accurate data. In recent years, these data information is often represented as fuzzy information such as fuzzy number and intuitionistic fuzzy number [21]. Therefore, based on the indeterminacy set theory, this paper proposes a new distance metric. This distance measure can better handle the noise data and weigh it according to the importance of each attribute. This algorithm improves the accuracy of distance metric and has better performance in the measurement of symbol attribute distance metric.

Firstly, according to the method of determining attribute weights based on intuitionistic fuzzy entropy, a method of determining attribute weights based on indeterminacy entropy is proposed. Let $F = (a_{ij})_{m*n}$ be a single valued indeterminacy set matrix, where a_{ij}= {T_{ij}, I_{ij}, F_{ij}} represent the *j* attribute of the *i* instance, and for any attribute,

(1) The indeterminacy entropy $E(a_{ij})$ for every instance can be calculated according to formula (3), The formula of determining indeterminacy entropy is as follows:

$$e_j = \frac{1}{m}\sum_{i=1}^{m} E\left(a_{ij}\right) \tag{4}$$

(2) Entropy indicates the uncertainty of attribute value. The greater the entropy and the greater the uncertainty [22]. The formula of determining attribute weight is as follows:

$$w_j = \frac{1 - e_j}{n - \sum_{j=1}^{n} e_j}, \quad j = 1, 2, \ldots, n \tag{5}$$

Then, based on the indeterminacy entropy weight and the indeterminacy distance, a distance measure method based on attribute fuzzy weighting is proposed.

Definition 5 Weighted fuzzy distance
Let $S = (U, A)$ be a classification information system, $A = \{a_1, a_2, \ldots, a_m\}$, for any *X*, $Y \in U$, the distance of *X* and *Y* is defined as:

$$d_\lambda(\mathrm{X,Y}) = \left\{\frac{1}{3}\sum_{j=1}^{m} w_j\left[\left|T_{xj} - T_{yj}\right|^\lambda + \left|I_{xj} - I_{yj}\right|^\lambda + \left|F_{xj} - F_{yj}\right|^\lambda\right]\right\}^{\frac{1}{\lambda}} \tag{6}$$

Among them, $\lambda > 0; X = (x_1, x_2, \ldots, x_m), x_j = (T_j, I_j, F_j), j = 1, 2, \ldots, m; Y = (Y_1, Y_2, \ldots,$ $Ym), y_j = (T_j, I_j, F_j), j = 1, 2, \ldots, m$.

3.2 The Improved MST Clustering Algorithm in This Paper

Let $S = (U, A)$ be a classification information system. Next, we apply the new distance metric to the traditional MST clustering algorithm. The steps of MST clustering algorithm based on the new distance measure is as follows:

The analysis shows that the time complexity of the MST clustering algorithm based on the new distance measure is O ($mn + n_2 + n_2 + n$). The algorithm mainly consists of four parts. The time complexity of solving the attribute weights is $m*n$, and the time complexity of solving the adjacency matrix is O (n_2). The time complexity of constructing the minimum spanning tree is O (n_2). The time complexity of the minimum spanning tree is just the time complexity of solving adjacency matrix, which time is O (n). Among them, n is the number of data points in the point set, and m is the number of attributes.

FWMSTClust algorithm can get arbitrary shape, arbitrary density class, and requires less input parameters. For high dimensional datasets, the algorithm can be solved by calculating the similarity matrix. And the algorithm has good extensibility (for example, if the constraint conditions are considered in solving the distance matrix, the algorithm can be extended to deal with the constrained clustering problem), there is no relation to the input order, the outlier can be found, and so on.

4 Experimental Analysis

The following three measures are used to analyze the clustering quality of the new algorithm that respectively are ACC (accuracy), AMI (adjusted mutual information) and ARI (adjusted Rand index). The upper bounds of the values of the three indexes are all 1, and the larger the values, the better the clustering results.

In order to test the effectiveness of the new algorithm, ten sets of artificial data sets are selected from UCI, and a new algorithm is used on each data set to verify the new algorithm. The algorithm is compared with K-means, AP and traditional minimum spanning tree algorithm respectively. Among them, the k-means algorithm calls the library function in Matlab, and other algorithms use the source code or program provided by the author. The ten sets of data sets are described as shown in Table 3.

Table 2. The MST clustering algorithm based on the new distance measure

The MST clustering algorithm based on the new distance measure is as follows:
Step 1. According to *formula (5)*, the weight of each attribute is calculated
Step 2. According to *formula (6)*, the distance matrix *G* is calculated by weighted fuzzy distance
Step 3. The minimum spanning tree *T* is calculated by prim algorithm, and the weight vector in the minimum spanning tree is *VT*
Step 4. Find the edge *e* with the maximum value in *VT*; remove e from *T*
Step 5. Delete the weight of the edge *e* from the *VT*
Step 6. According to the division of *T*, the *U* is divided into *K* clusters

Table 3. Description of data sets

Datasets	Samples/attributes	Clusters
Dermatology	366/34	6
Iris	150/4	3
Libras movement	360/91	15
Pima-Indians-diab	768/8	2
Seeds	210/7	3
Segmentation	2310/19	7
Waveform	5000/21	3
Waveform(noise)	5000/40	3
WDBC	569/30	2
Wine	178/12	3

Table 4. Comparison with Algorithms on the ACC Data

Datasets	AP	K-means	MST	FWMSTClust
Dermatology	0.814	0.691	0.787	0.950
Iris	0.907	0.825	0.892	0.972
Libras movement	0.450	0.443	0.561	0.628
Pima-Indians-diabetes	0.624	0.668	0.658	0.694
Seeds	0.895	0.890	0.890	0.910
Segmentation	0.670	0.602	0.698	0.832
Waveform	-	0.501	0.562	0.574
Waveform(noise)	-	0.512	0.523	0.642
WDBC	0.924	0.928	0.928	0.950
Wine	0.854	0.932	0.882	0.947

Table 5. Comparison with algorithms on the AMI data

Datasets	AP	K-means	MST	FWMSTClust
Dermatology	0.771	0.786	0.756	0.884
Iris	0.756	0.692	0.687	0.900
Libras movement	0.497	0.519	0.390	0.628
Pima-Indians-diabetes	0.045	0.050	0.050	0.023
Seeds	0.685	0.671	0.671	0.782
Segmentation	0.605	0.578	0.578	0.720
Waveform	-	0.269	0.269	0.350
Waveform(noise)	-	0.184	0.184	0.230
WDBC	0.602	0.411	0.411	0.685
Wine	0.686	0.687	0.687	0.826

Table 6. Comparison with algorithms on the ARI data

Datasets	AP	K-means	MST	FWMSTClust
Dermatology	0.717	0.654	0.564	0.910
Iris	0.757	0.660	0.760	0.932
Libras movement	0.277	0.304	0.314	0.420
Pima-Indians-diabetes	0.089	0.102	0.113	0.087
Seeds	0.715	0.705	0.705	0.824
Segmentation	0.502	0.483	0.483	0.595
Waveform	-	0.254	0.235	0.320
Waveform(noise)	-	0.252	0.254	0.213
WDBC	0.718	0.730	0.718	0.825
Wine	0.616	0.830	0.658	0.890

Through the analysis of Tables 2, 3 and 4, the FWMSTClust algorithm achieves better clustering results than other algorithms, on the ten sets of artificial datasets. The experimental results also show that the new distance measurement improved in this paper is effective, and the MST algorithm based on the new distance measure can obtain higher clustering accuracy in the uncertain data (Tables 5 and 6).

5 Conclusion

In this paper, a new distance metric is proposed based on the indeterminacy set theory. This algorithm can do better in data noise dealing and has better performance in symbol attribute processing. This distance metric method is applied to the traditional MST clustering algorithm. And compared with other classical clustering algorithms based on MST and other classical clustering algorithms, the results show that the proposed algorithm in this paper is better than other algorithms and can get higher accuracy in fewer iterations.

Acknowledgement. We would like to thank the anonymous reviewers for their valuable comments and suggestions. This work is supported by The State Key Research Development Program of China under Grant 2016YFC0801403, Shandong Provincial Natural Science Foundation of China under Grant ZR2018MF009 and ZR2015FM013, the Special Funds of Taishan Scholars Construction Project, and Leading Talent Project of Shandong University of Science and Technology.

References

1. Xin, C.: Research on clustering analysis method based on minimum spanning tree. A master's degree thesis. Chongqing University, Chongqing (2013)
2. Tang, F.: Data Structure Tutorial, pp. 158–276. Beihang University Press, Beijing (2005)
3. Barna, S., Pabitra, M.: Dynamic algorithm for graph clustering using minimum cut tree. In: Proceeding of Sixth IEEE International Conference on Data Mining (2006)
4. Wang, X., Wang, X., Wilkes, D.M.: A divide-and-conquer approach for minimum spanning tree-based clustering. IEEE Trans. Knowl. Data Eng. **21**(7), 945–958 (2009)
5. Caetano Jr., T., Traina, A.J.M., Faloutsos, C.: Feature selection using fractal dimension-ten years later. J. Inf. Data Manag. **1**(1), 17–20 (2010)
6. Zhong, C., Miao, D., Wang, R.: A graph-theoretical clustering method based on two rounds of minimum spanning trees. Pattern Recognit. **43**(3), 752–766 (2010)
7. Moore, A.W.: An Introductory Tutorial on KD-Trees. University of Cambridge, UK (1991)
8. Olman, V., Mao, F., Wu, H., Xu, Y.: Parallel clustering algorithm for large data sets with applications in bioinformatics. IEEE Computer Society Press **6**(2), 344–352 (2009)
9. Wang, H., Zhang, H., Ray, N.: Adaptive shape prior in graph cut image segmentation. Pattern Recogn. **46**(5), 1409–1414 (2013)
10. Smarandache, F.: A unifying field in logics. Neutrosophy: Neutrosphic probability, set and logic. American Research Press, Rehoboth, D E (1999)
11. Zadeh, L.A.: In there a need for fuzzy logic? Inf. Sci. **178**, 2751–2779 (2008)
12. Zhang, D.: Research on Motor Fault Diagnosis Method Based on Rough Set Theory. Bohai University, Jinzhou (2015)

13. Ye, J.: Single-valued neutrosophic similarity measures based on cotangent function and their application and their application in the fault diagnosis of steam turbine. Soft. Comput. **21**(3), 1–9 (2017)
14. Wang, H., Smarandache, F., Zhang, Y.Q., Sunderraman, R.: Interval neutrosophic sets and logic: theory and applications in computing. Hexis, Phoenix, A Z (2005)
15. Wang, H.B., Smarandache, F., Zhang, Y.Q., et al.: Single valued neutrosophic sets. Multispace Multructure **4**(10), 410–413 (2010)
16. Peng, J.J., Wang, J.Q., Zhang, H.Y., et al.: An outranking approach for multi-criteria decision-making problems with simplified neutrosophic sets. Appl. Soft Comput. **25**(25), 336–346 (2014)
17. Biswas, P., Pramanik, S., Giri, B.C.: TOPSIS method for multi- attribute group decision-making under single-valued netrosophic environment. Neural Comput. Appl. **27**(3), 727–737 (2016)
18. Majumdar, P., Damanta, S.K.: On similarity and entropy of neutrosophic sets. J. Intell. Fuzzy Syst. Appl. Eng. Technol. **26**(3), 1245–1252 (2014)
19. Yan, W., Wu, W.: Data structure. Tsinghua University Press, Beijing (1992)
20. Graham, R.L., Hell, P.: On the history of the minimum spanning tree problem. Ann. Hist. Comput. **7**(1), 43–57 (1985)
21. Atanassov, K.T.: Intuitionstic fuzzy sets. Fuzzy Syst. **20**(1), 87–96 (1986)
22. Shi, Y., Shi, W., Jin, F.: Entropy and its application in the study of uncertainty in spatial data. Comput. Eng. **31** (24), 36–37 (2005)

Cuckoo Search Algorithm Based on Stochastic Gradient Descent

Yuan Tian[1,2], Yong-quan Liang[1,2,3(✉)], and Yan-jun Peng[1,2,3,4]

[1] Princeton University, Princeton, NJ 08544, USA
[2] College of Computer Science and Engineering, Shandong University of Science and Technology, Qingdao, China
lyq@sdust.edu.cn
[3] Provincial Key Laboratory for Information Technology of Wisdom Mining of Shandong Province, Shandong University of Science and Technology, Qingdao, China
[4] Experimental Teaching Center of National Virtual Simulation for Security Mining, Shandong University of Science and Technology, Qingdao, China

Abstract. Cuckoo Search (CS) is a global search algorithm for solving multi-objective optimization problems. Cuckoo Search algorithm is easy to implement and has a few number of control parameters, excellent search path and strong optimization capability. It has been successfully applied to practical problems, such as engineering optimization. To improve the refining ability and convergence rate of CS algorithm, solve the problem of slow convergence rate and unstable search accuracy in later stage, this paper proposes a Cuckoo Search Algorithm based on Stochastic Gradient Descent (SGDCS). This algorithm uses Stochastic Gradient Descent to enhance the search of the local optimum, convergence process and algorithm adaptability, which improves the calculation accuracy and convergence rate of cuckoo search algorithm. The simulation experiments show that the proposed algorithm is simple and efficient, efficiently improves the performances on calculation accuracy and convergence rate on the basis of maintaining the advantages of the standard CS algorithm.

Keywords: Cuckoo Search Algorithm · Lévy flight · Function optimization Stochastic Gradient Descent

1 Introduction

Since the late twentieth Century, various modern metaheuristic algorithms have been proposed by studying the self-organizing behaviour of social animals like ants and birds, such as ant colony optimization (AC) [1], particles swarm optimization (PSO) [2] and so on. They solve function optimization problems by simulating natural phenomenon and animal behavior. These nature-inspired metaheuristic algorithms have become one of the research hotspots in intelligent computation field and been used in a wide range of practical problems.

With the rapid development of computational intelligence technology, all kinds of novel bionic optimization algorithms have been put forward by researchers. Cuckoo

P. Krömer et al. (Eds.): ECC 2018, AISC 891, pp. 90–99, 2019.
https://doi.org/10.1007/978-3-030-03766-6_10

Search Algorithm (CS) [3] is a global search method proposed by Xin-She Yang and others at University of Cambridge in 2009. It was inspired by the behavior of cuckoos for nesting and laying eggs. This new heuristic algorithm is used to solve the problem of function objective optimization. Cuckoo Search is easy to implement and has a few number of control parameters, excellent search path and strong optimization capability. It has been successfully applied to practical problems, such as engineering optimization [4, 5]. Cuckoo Search has been proved theoretically that it can converge to global optimization through the Markov chain model [6]. It has outperformed both GA and PSO in terms of convergence rate and robustness [3] because of less number of control parameters and a fine balance of randomization and intensification.

As a new bionic algorithm, CS needs further improvement in its convergence rate and accuracy of convergence results. Aiming at function optimization problems, most of the improved algorithms are based on the change of detection probability, self-adaptive step length and hybrid CS algorithm.

In the improved CS algorithms based on the change of detection probability, according to the reference [5], if the probability P is invariable, it will make the better or worse solutions be replaced at the same probability. Therefore, a dynamic method is proposed to adjust the detection probability P on the basis of the global solution in order to improve convergence rate and refining ability. In reference [7], an improved algorithm based on self-adaptive machine is proposed to control the scaling factor and detection probability. This algorithm improves the refining ability and convergence rate of CS algorithm for solving function optimization problems.

In the improved CS algorithms based on self-adaptive step length, according to reference [8], it is considered that the Lévy flight mode is lack of self-adaptability. A self-adaptive steps adjustment cuckoo search algorithm is proposed to accelerate search speed and improve its accuracy. Reference [9] uses a modified version based on population ranking to guide the step length of random walks, thus improving the adaptability of step length. The improved algorithm is superior to CS algorithm in terms of convergence rate and convergence accuracy.

In the improved hybrid CS algorithms, reference [10] optimizes local search and adds PSO into random walk and Lévy flight. PSO algorithm is used to optimize the population, and then CS algorithm is used to continue optimizing in optimal individuals, so that the performance of the algorithm is improved. Reference [11] combines with the framework of cooperative co-evolutionary, divides the solution vectors of population into several sub-vectors and constructs the corresponding sub-swarms. The solution vectors of each sub-population are updated by CS algorithm. This modified algorithm efficiently improves the performances of solving continuous function optimization problems.

CS algorithm has some shortcomings. It has the problem of premature convergence and its convergence rate is slow at later stage, the local search ability of this algorithm is weak. In the proposed improved methods, the algorithm complexity and the running memory are increased whether dynamic change of detection probability or combination of other frameworks. They cover up the advantages of the standard CS algorithm, such as less number of control parameters and simple operation. Aiming at the shortcomings of CS algorithm, this paper proposes an improved cuckoo search algorithm based on stochastic gradient descent (SGD Cuckoo Search, SGDSC).

This proposed algorithm uses gradient descent to optimize local search and improves the adaptability of CS algorithm. Through three sets of standard test functions (11 functions), the simulation experiments show that this improved algorithm is simple and efficient, and the convergence rate and optimization accuracy of the algorithm are improved on the basis of maintaining the advantages of the standard CS algorithm.

2 Cuckoo Search

In nature, certain species of cuckoos use a special breeding strategy of brood parasitism. Cuckoos lay eggs in the nests of host birds and host birds replace cuckoos to hatch and raise the next generation of cuckoos. Once the first cuckoo chick is hatched, it will blindly propel the eggs out of the nest, which increases the cuckoo chick's survival.

Cuckoo search algorithm adopts Lévy flight mode in the process of optimization. Lévy flight is a kind of random walk, the direction of each step is completely random and isotropic. Short walk with small step alternates with occasional long walk with large step, and the step size obeys the heavy-tail distribution. Shlesinger implanted this flight mode into swarm intelligence search algorithm [12]. Long walk is used to explore, expand the search scope and jump out of the local optimal situation. The greater the step size is, the easier it is to search for the global optimal. But it will reduce the search accuracy, sometimes there will be concussion. Short walk is used to converge to the global optimal solution in a small range. The smaller the step size is, the higher the accuracy of the solution is. But it will reduce the speed of search. Lévy flight is controlled by two factors per step: one is the direction of random walk, which generally selects a number that obeys the uniform distribution; the other is the step size, which obeys the Lévy distribution. Reynolds et al. have proved that when the target location presents random features and the target is in an irregular sparse arrangement, Lévy flight is the most effective and ideal search strategy for M mutual independent searchers [13]. Lévy flight can make an effective search maximally in uncertain areas, the foraging patterns of many animals show the typical feature of Lévy flight, and so does human behavior.

Cuckoos search for a suitable nest to lay eggs randomly. By simulating the cuckoo's parasitic brood behavior and birds' Lévy flight behavior, the following three idealized rules are used [3, 12]:

(1) Each cuckoo lays one egg at a time, and dumps its egg in randomly chosen nest;
(2) The best nests with high quality of eggs (solutions) will carry over to the next generations;
(3) The number of available host nests is fixed, and a host bird can discover an alien egg with a probability $p_a \in [0, 1]$. In this case, the host bird can either throw the egg away or abandon the nest, and build a completely new nest.

On the basis of these three idealized rules, when generating new solutions $x_i^{(t+1)}$ for, say a cuckoo i, a Lévy flight is performed

$$\mathrm{x}_i^{(t+1)} = \mathrm{x}_i^{(t)} + \alpha \oplus Levy(\beta) \tag{1}$$

where $\mathrm{x}_i^{(t)}$ represents the ith solution of the tth generation; $\oplus$ means entry-wise multiplications; $\alpha > 0$ is the step size which should be related to the scales of the problem of interest.

$$\alpha = \alpha_0 \left(x_i^{(t)} - x_{best} \right) \tag{2}$$

where α_0 is a constant ($\alpha_0 = 0.01$); x_{best} represents the current optimal solution.

$Levy(\beta)$ is a Lévy random search path, obeying Lévy probability distribution:

$$Levy \sim u = t^{-1-\beta}, (0 < \beta \leq 2) \tag{3}$$

Obviously, the generation of step size s samples is not trivial using Lévy flight. A simple scheme discussed in detail by Yang [14, 15] can be summarized as

$$s = \alpha_0 \left(x_j^{(t)} - x_i^{(t)} \right) \oplus Levy(\beta) \sim 0.01 \frac{u}{|v|^{1/\beta}} \left(x_j^{(t)} - x_i^{(t)} \right) \tag{4}$$

where u and v are drawn from standard normal distributions. That is

$$u \sim N\left(0, \sigma_u^2\right), v \sim N\left(0, \sigma_v^2\right) \tag{5}$$

with

$$\sigma_u = \left\{ \frac{\Gamma(1+\beta)\sin(\pi\beta/2)}{\Gamma[(1+\beta)/2]\beta 2^{(\beta-1)/2}} \right\}^{1/\beta}, \sigma_v = 1 \tag{6}$$

Here Γ is the standard Gamma function.

Combined with formula (1)–(6), in the Lévy flight random walk component, the formula for generating a new solution $x_i^{(t)}$ can be summed up as

$$x_i^{(t+1)} = x_i^{(t)} + \alpha_0 \frac{\phi \times u}{|v|^{1/\beta}} \left(x_i^{(t)} - x_{best} \right) \tag{7}$$

After the position is updated, compare the random number $r \in [0, 1]$ with the detection probability $p_a \in [0, 1]$. If $r > p_a$, change $\mathrm{x}_i^{(t+1)}$ randomly and abandon a small part of bad nests. Generate the same number of nests as abandoned nests according to random walk and mix with the nests that are not abandoned to get a new set of nests, which can avoid trapping in a local optimum. If $r \leq p_a$, no need to change. Finally, hold the set of nests positions with the best test value recorded as $x_i^{(t+1)}$:

$$x_i^{(t+1)} = x_i^{(t)} + r\left(x_j^{(t)} - x_k^{(t)}\right) \tag{8}$$

where random number $r \in [0, 1]$ is a scaling factor; $x_j^{(t)}$ and $x_k^{(t)}$ are random solutions.

In summary, the effect of random walk in Cuckoo Search is obvious. Some long walk with large step effectively avoids the algorithm falling into local optimum and short walk with small step searches for the optimal solution effectively. Consequently CS algorithm is very effective and efficient in searching the optimal solution. In addition, the number of control parameters needed to be adjusted is few in the algorithm. This makes CS algorithm more suitable for function optimization problems. Every set of nests in the algorithm iteration process can be regarded as a set of problem solution. Therefore, CS algorithm can also be extended to metapopulation algorithms.

3 Cuckoo Search Algorithm Based on Stochastic Gradient Descent

CS algorithm is an unconstrained global optimization method. Every generation refers to the current optimal nests, which makes CS algorithm highly efficient for optimization. In CS algorithm, Lévy flight generates random steps. The larger the step is, the easier it is to search for the global optimal, but it will reduce the search precision. The smaller the step is, the higher the solution precision is, but it will reduce the search speed. Therefore, the step length generated by Lévy flight is random, but it lacks self-adaptability. There are problems that the convergence rate is slow and the search accuracy is unstable in later period. To solve these problems, Stochastic Gradient Descent (SGD) is drawn into the process of random walk and combines with CS algorithm to obtain the improved algorithm called Cuckoo Search Algorithm Based on Stochastic Gradient Descent (SGDCS).

Gradient descent algorithm is the most commonly used optimization algorithm in machine learning and widely used in various optimization problems. SGD algorithm only uses one random-selected training sample for updating parameters every time. It is easy to operate and has low computation complexity. The calculation speed of SGD is very fast and there is no need to solve the error function in the operation process. The overall direction of the gradient points to the global optimal solution in the operation process. The randomicity in convergence process is the lowest. SGD algorithm has a good self-adaptive process. Cuckoo Search Algorithm Based on Stochastic Gradient Descent (SGDCS) combines with the advantages of cuckoo algorithm to break away from the local optimum and the advantages of SGD algorithm to find the optimal solution, which improves the refining ability, convergence rate and the self-adaptability of the algorithm.

Utilize the gradient obtained from SGD as the direction vector $\vec{\theta}$ for searching new solutions:

$$\vec{\theta}_{\mathrm{i}} = \vec{\theta}_{\mathrm{i-1}} - \alpha\left(h_{\theta}\left(x_0^j, x_1^j, \cdots, x_n^j\right) - y_i\right)x_i^j \tag{9}$$

Start from the current solution along the direction $\vec{\theta}$, combine with Lévy flight to get new solutions. Improve the efficiency of finding the optimum solution and avoid falling into the local optimum due to the randomness of Lévy flight. Get a new solution

$$x_i^{(t+1)} = x_i^{(t)} \cdot \vec{\theta}_i + \alpha \oplus Levy(\beta) \tag{10}$$

According to the analysis, the steps of the SGDCS algorithm are as follows:

Step 1. (Initialization) Generate n nests and detection probability p_a randomly. The search space dimension is nd, the upper and lower bound of the solution are Ub and Lb, the precision is set to *iteration*, the maximum number of iterations is set to *iteration*, initial nest position is $\mathrm{x}_i^{(0)}$, $\mathrm{i} \in \{1, 2, \cdots, n\}$ and the current optimal solution is $f\min$.

Step 2. (Iteration cycle) Keep the position $x_i^{(t)}$ of the best nests in the last generation, obtain the gradient by SGD as the direction vector $\vec{\theta}$, combine with Lévy flight to update the rest of nests along the direction $\vec{\theta}$ and get a set of solutions $x_i^{(t+1)}$. Test the position of the set of nests, compare this set of solutions with the solutions of previous generation, replace the bad solutions in previous generation with the new better solutions. Thus, a better set of nests positions is obtained record as $x_i^{(t+1)}$, $\mathrm{i} \in \{1, 2, \cdots, n\}$.

Step 3. Use the random number $r \in [0, 1]$ with uniform distribution as the possibility of discovering alien eggs by host birds and compare with p_a. If $r > p_a$, preserve the position of nests which have small detection probability, at the same time randomly change the nests positions with large detection probability. Test this new set of nests positions and compare the test value with nests positions before replacement. The set of nests positions with better test value is used as a new set of better nests positions.

Step 4. Preserve the best set of nests $x_i^{(t+1)}$, calculate the optimal solution $f\min$. Determine whether $f\min$ satisfies the requirement of accuracy. If so, output results; if not, return to step 2.

4 Experiment Results and Analysis

The operating environment of the simulation experiments is Intel(R) Core(TM) i5-4200U CPU, the main frequency is 1.60 GHz, the computer memory is 4 GB, the operating system is Windows 10 64 bit. The experiment simulation software adopts MatlabR2013b.

In order to observe the convergence rate and refining ability of the improved algorithm for solving function optimization problems, and prove the validity of the proposed algorithm in this paper, simulation experiments select 3 types of test functions, include 5 sets of unimodal benchmark functions, 3 sets of multimodal benchmark functions and 3 sets of fixed-dimensions multimodal benchmark functions [16, 17], which are shown partially in Table 1.

Table 1. A part of standard test functions.

Test function	Dimension	Search space	Optimal value
$f_1(x) = \sum_{i=1}^{n} x^2$	30	[−100, 100]	0
$f_2(x) = \sum_{i=1}^{n} \lvert x_i \rvert + \prod_{i=1}^{n} \lvert x_i \rvert$	30	[−10, 10]	0
$f_3(x) = \sum_{i=1}^{n} \left(\sum_{j=1}^{i} x_j \right)^2$	30	[−100, 100]	0
$f_4(x) = \max\{\lvert x_i \rvert, 1 \le i \le n\}$	30	[−100, 100]	0

In the experiments, CS algorithm and the improved SGDCS algorithm are used to run 20 times on each test function, do 200 times iterations every runtime. The average value of the results represents the accuracy of search results. The parameters in the experiments are set to the default: the size of nests n is 20, the detection probability p_a is 0.25. The results of experiments are shown in Tables 2 and 3. Compared with the original CS algorithm, the improved SGDCS algorithm has obvious improvement in convergence rate and refining ability.

Table 2. Average number of iterations to convergence.

Function	CS	SGDCS	Increase ratio	Function	CS	SGDCS	Increase ratio
f1	20.95	4.5	78.52%	f7	193.6	2	99.90%
f2	45.3	2	95.58%	f8	77.95	2	97.43%
f3	19.5	3.75	80.77%	f9	63.65	56.8	10.76%
f4	61.5	2	96.75%	f10	117.35	86.55	26.25%
f5	142.9	46.7	67.32%	f11	49.3	30.4	38.34%
f6	61.9	1	98.38%				

CS algorithm has the advantage of less number of control parameters. The improved SGDCS algorithm is the same as the original CS algorithm, only needs to initialize n and p_a. Choose test function f1, f5 and f9 for parameter sensitivity test. The results are shown in Tables 4, 5, 6 and 7. Through the parameter sensitivity test of SGDCS algorithm by single-parameter analysis, SGDCS algorithm is insensitive to initialized nest number n and detection probability p_a. The algorithm has high stability.

Table 3. Average convergence result.

Function	CS	Error of CS	SGDCS	Error of SGDCS	Accuracy improvement
f1	1.046E−19	1.046E−19	0	0	1.046E−19
f2	3.51E−10	3.51E−10	0	0	3.51E−10
f3	2.82E−19	2.82E−19	0	0	2.82E−19
f4	2.08348E−10	2.08348E−10	0	0	2.08348E−10
f5	0.0004638	0.0004638	0.0001019	0.0001019	0.0003619
f6	1.0823E−05	1.0823E−05	0	0	1.0823E−05
f7	12.86807	12.86807	0	0	12.80237
f8	1.56E−08	1.56E−08	0	0	1.56E−08
f9	0.39789	0	0.39789	0	0
f10	3	0	3	0	0
f11	−1.036	0	−1.036	0	0

Table 4. The average convergence times of test functions f1, f5 and f9 on n.

Function	n = 10	n = 15	n = 20	n = 25 (default)	n = 30	n = 35	n = 40	Standard deviation
f1	4.6	4.1	4.2	4.5	4	4.2	4.4	0.2193
f5	44.7	44.9	47.7	46.7	45.1	47.1	45.4	1.1998
f9	57.7	56.6	57.3	56.8	56.7	56.4	56.2	0.5210

Table 5. The average convergence results of test functions f1, f5 and f9 on n.

Function	n = 10	n = 15	n = 20	n = 25 (default)	n = 30	n = 35	n = 40	Standard deviation
f1	0	0	0	0	0	0	0	0
f5	0.00046	0.00013	0.00014	0.00010191	0.0001	9.5E−05	6.9E−05	1.3E−04
f9	0.39789	0.39789	0.39789	0.39789	0.39789	0.39789	0.39789	0

Table 6. The average convergence times of test functions f1, f5 and f9 on p_a.

Function	pa = 0.10	pa = 0.15	pa = 0.20	pa = 0.25 (default)	pa = 0.30	pa = 0.35	pa = 0.40	Standard deviation
f1	4.5	4.5	4.4	4.5	4.5	4.7	4.6	0.0949
f5	47.8	46.8	47.2	46.7	46.7	45.2	45	1.0238
f9	56.1	56.6	56.2	56.8	57.4	56.7	56.2	0.4572

Table 7. The average convergence results of test functions f1, f5 and f9 on p_a.

Function	pa = 0.10	pa = 0.15	pa = 0.20	pa = 0.25 (default)	pa = 0.30	pa = 0.35	pa = 0.40	Standard deviation
f1	0	0	0	0	0	0	0	0
f5	8.9E−05	7.9E−05	7.2E−05	0.000101905	8.1E−05	6.8E−05	6.8E−05	1.2e−05
f9	0.39789	0.39789	0.39789	0.39789	0.39789	0.39789	0.39789	0

5 Conclusions

CS Algorithm is a novel bionic optimization algorithm, which is easy to implement and has a few number of control parameters, excellent search path and strong optimization capability. It has been successfully applied to practical problems. Aiming at the problem of premature convergence, slow convergence rate at later stage and weak local search ability in CS Algorithm, this paper proposes an improved cuckoo search algorithm based on stochastic gradient descent (SGD Cuckoo Search, SGDSC). SGDCS algorithm uses gradient descent to optimize local search and improves the adaptability of CS algorithm. Through three sets of standard test functions (11 functions), the simulation experiments show that the improved algorithm has faster convergence rate and higher precision, especially for the multimodal benchmark functions, the improvement is very significant. Because of the good performance of the CS algorithm and the related improved algorithm on the optimization problem, it has great development prospects in the field of computational intelligence. It has great application space in the practical engineering problems and needs further study.

Acknowledgement. We would like to thank the anonymous reviewers for their valuable comments and suggestions. This work is supported by The State Key Research Development Program of China under Grant 2016YFC0801403, Shandong Provincial Natural Science Foundation of China under Grant ZR2018MF009 and ZR2015FM013, the Special Funds of Taishan Scholars Construction Project, and Leading Talent Project of Shandong University of Science and Technology.

References

1. Dorigo, M., Maniezzo, V., Colorni, A.: The ant system: optimization by a colony of cooperating agents. IEEE Trans. Syst. Man Cybern. Part B **26**(1), 29–41 (1996). https://doi.org/10.1109/3477.484436
2. Kennedy, J.,Eberhart, R.: Particle swarm optimization. In: Proceedings of the IEEE International Conference on Neural Networks, Perth, Australia: [s.n.], 1942–1948 (1995)
3. Yang, X.S., Deb, S.: Cuckoo search via Lévy flights. In: Abraham, A., Carvalho, A., Herrera, F., et al. (eds.) Proceedings of the World Congress on Nature and Biologically Inspired Computing (NaBIC 2009), pp. 210–214. IEEE Publications, Piscataway (2009). https://doi.org/10.1109/nabic.2009.5393690
4. Chen, L.: Modified cuckoo search algorithm for solving engineering structural optimization problem. Appl. Res. Comput. **31**(3), 679–683 (2014)
5. Wang, L., Yang, S., Zhao, W.: Structural damage identification of bridge erecting machine based on improved Cuckoo search algorithm. J. Beijing Jiaotong Univ. **37**(4), 168–173 (2013)
6. Wang, F.: Markov model and convergence analysis based on cuckoo search algorithm. Comput. Eng. **38**(11), 180–182 (2012)
7. Hu, X.: Improvement cuckoo search algorithm for function optimization problems. Comput. Eng. Des. **34**(10), 3639–3642 (2013)
8. Zheng, H.: Self-adaptive step cuckoo search algorithm. Comput. Eng. Appl. **49**(10), 68–71 (2013)
9. Andrew, W.: Modified cuckoo search. Chaos, Solitons Fractals **44**(9), 710–728 (2011)

10. Ghodrati, A.: A hybrid CS/PSO algorithm for global optimization. In: Intelligent Information and Database Systems, pp. 89–98 (2012)
11. Hu, X., Yin, Y.: Cooperative co-evolutionary cuckoo search algorithm for continuous function optimization problems. PR & AI **26**(11), 1041–1049 (2013)
12. Shlesinger, M.F.: Mathematical physics: search research. Nature **443**(7109), 281–282 (2006)
13. Reynolds, A.M., Smith, A.D., Menzel, R., et al.: Displaced honey bees perform optimal scale-free search flights. Ecology **88**(8), 1955–1961 (2007)
14. Yang, X.S., Deb, S.: Engineering optimisation by cuckoo search. Int. J. Math. Model. Numer. Optim. **1**(4), 330–343 (2010)
15. Yang, X.S.: Engineering Optimization: An Introduction with Metaheuristic Applications. Wiley, Hoboken (2010)
16. Chattopadhyay, R.: A study of test functions for optimization algorithms. J. Optim. Theor. Appl. **8**, 231–236 (1971)
17. Schoen, F.: A wide class of test functions for global optimization. J. Global Optim. **3**, 133–137 (1993)

Chaotic Time Series Prediction Method Based on BP Neural Network and Extended Kalman Filter

Xiu-Zhen Zhang(✉) and Li-Sang Liu

School of Information Science and Engineering, Fujian University of Technology, Fuzhou 350118, China
289126754@qq.com, liu_ls@163.com

Abstract. For neural networks, there are local minimum problems and slow convergence speeds. In order to improve the prediction accuracy of the BP neural network prediction model for chaotic time series, the EKF algorithm with BP neural network is used in the field of chaotic time series prediction. Namely, the use of the weight of its output of BP neural network is suitable for the state equation and observation equation of the Kalman filter, which gives the evolution of the Kalman filter algorithm suitable for nonlinear systems. Extended Kalman filter (EKF) algorithmtypical and Mackey-Glass chaotic time series were simulated. The simulation results show that the method of chaotic time series with nonlinear fitting better and higher prediction accuracy.

Keywords: BP neural network · Extended Kalman Filtering (EKF) Chaotic time series prediction

1 Introduction

There are chaotic phenomena everywhere in nature and human society, it is an irregular form of movement which is unique to nonlinear dynamic systems and shows the complexity of things [1]. In the analysis of the chaos, it is a very important field of study to predict the system based on the nonlinear time series extracted from the chaos system [2].

The prediction of the chaos time sequence is based on the development process and trend that the sequence reflects, and so on and so forth, to predict the future state of the system. There is a certain regularity in the chaotic time series, which is shown in the correlation in the time-delay state space of the series. Therefore, the prediction of nonlinear system must have certain intelligent information processing capabilities [3]. This capability is available for neural networks and is mainly used for the prediction of nonlinear system modeling, in which BP(back propagation) neural network, the error reverse propagation algorithm, is a multilayer feed-forward network of weight training for non-linear differential function [4]. The activation function of the neural network is an area composed of a nonlinear superplane, which is a relatively soft and smooth interface, so its classification is accurate, reasonable and fault tolerant [5]. It is another feature that that activation function is continuous and micro, which can be use strictly

P. Krömer et al. (Eds.): ECC 2018, AISC 891, pp. 100–108, 2019.
https://doi.org/10.1007/978-3-030-03766-6_11

by the gradient method, and the resolution formula of its weight correction is very clear. With its unique information processing characteristics, the chaotic time series can be learned and predicted the future state [6].

Mackey-Glass chaotic time series is one of the benchmark issues in time series prediction problems, and it has nonlinear characteristics [7]. In real life, there's a lot of typical Mackey-glass problems, and for these questions, the non-linear filter of sustainability is applied in the chaos time series prediction, and it has a better realistic significance [8, 9].

2 BP Neural Network Structure

The structure of BP neural network is shown in Fig. 1, the input layer has m nodes, the output layer has l nodes, and the hidden layer of the network has only one layer with n nodes. w_{ij} represents the connection weight between the input layer and the hidden layer neurons, as shown in Eq. (1).

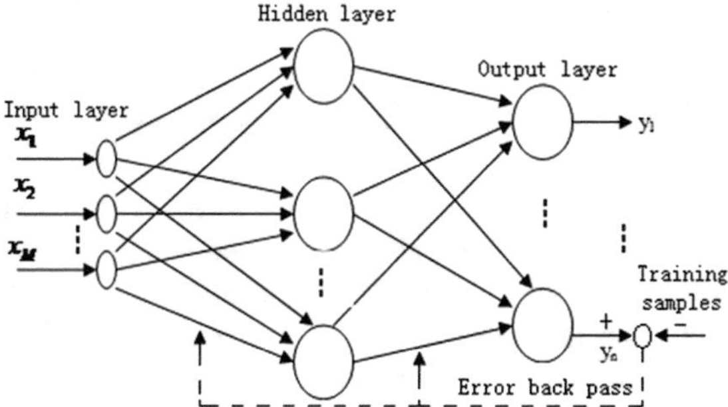

Fig. 1. BP neural network model structure

$$O_i = f(\sum_{j=1}^{m} w_{ij}x_j), i = 1, 2, \ldots, n \tag{1}$$

w_{jk} represents the connection weight between the hidden layer and the output layer neuron, the input of the hidden layer and output layer node is the weighted sum of the output of the previous layer node, and the output amplitude of each node is determined by its activation function, as shown in Eq. (2).

$$O_k = f(\sum_{j=1}^{n} w_{jk}O_j), k = 1, 2, \ldots, l \tag{2}$$

If the output layer does not get the expected output value d_k, the error signal will be returned along the original connection path, and the weighted coefficients of each layer neuron will be modified to get as close as possible to the desired output value of d_k.

Set the quadratic error function corresponding to the input and output modes of each pair as formula (3).

$$E_p = \frac{1}{2}\sum_{k=1}^{l}(d_k - O_k)^2 \tag{3}$$

And the cost function of the average error of the system is shown in Eq. (4).

$$E = \sum_{p=1}^{P} E_p \tag{4}$$

In (3) and (4), p and l are respectively the sample pattern logarithm and the number of network output nodes. Based on Eq. (3), a stepwise optimization method, namely the steepest descent method, can obtain the network connection weight adjustment Eq. (5).

$$w_{ij}(t+1) = w_{ij}(t) + \eta\delta_j O_i + \alpha[w_{ij}(t) - w_{ij}(t-1)] \tag{5}$$

In the Eq. (5), the smaller the learning rate parameter η is, the smaller the weight variation is, and the smoother the trajectory is in the weight space, but the learning speed is slowed down. Therefore, $\alpha(0 < \alpha < 1)$ is added as the smoothing factor, which is beneficial to the learning algorithm. Using the above algorithm and process sequence, adjust the weight coefficient of each layer of neurons, repeatedly input all training sample sequences, repeat calculation, until the output error is within the desired range, and the network training is finished.

3 Extended Kalman Filtering Method

Kalman filter is a least mean square error estimator based on a first-order Taylor series expansion of a nonlinear function. The new measurement value $y(k + 1)$ is added once every time, the new state filtering value $\hat{x}(k + 1|k + 1)$ and the new filtering error variance parray $p(k + 1|k + 1) = E[\tilde{x}(k + 1|k + 1)\tilde{x}(k + 1|k + 1)^T]$ can be calculated only by using the previously calculated preform filter value $\hat{x}(k|k)$ and the filter error variance matrix $p(k|k) = E\left[\tilde{x}\left(k|k\right)\tilde{x}\left(k|k\right)^T\right]$. Therefore, whatever the number of measurements is and however it increases, it is not necessary to calculate a large inverse matrix and store a large amount of historical survey data so as to meet the real time requirement. In the world, the system is basically nonlinear, so to expand the kalman filter algorithm, first you have to linearize the system model, and then use the linear optimal filter algorithm, which is the kalman filter.

3.1 Nonlinear Model Linearization

The state equation and the observation equation of the system are as per Eq. (6).

In the formula (6),$\{w(k)\}$ and $\{v(k)\}$ are zero-mean Gaussian white noise sequences that are not related to each other. The initial state $x(0)$ is a Gaussian random variable, $E[v(k)v^T(k)] = R_2(k)$, $E[w(k)w^T(k)] = R_1(k)$, the mean of $x(0)$ is $\bar{x}(0)$, and the covariance is $p(0)$.

$$\begin{cases} x(k+1) = f[x(k), k] + g[x(k), k]w(k) \\ y(k) = h[x(k), k] + v(k) \end{cases} \tag{6}$$

smooth enough, they can be Taylor-expanded along the $\hat{x}(k|k)$ and $\hat{x}(k|k-1)$ as shown in (7).

$$\begin{cases} f[x(k), k] = f[\hat{x}(k|k), k] + \varphi(k)[x(k) - \hat{x}(k|k)] + \cdots \\ g[x(k), k] = g[\hat{x}(k|k), k] + \cdots = G(k) \\ h[x(k), k] = f[\hat{x}(k|k-1), k] + H(k)[x(k) - \hat{x}(k|k-1)] + \cdots \end{cases} \tag{7}$$

In the formula (7),

$$\varphi(k) = \frac{\partial f(x,k)}{\partial x}|_{x=\hat{x}(k|k)}, H(k) = \frac{\partial h(x,k)}{\partial x}|_{x=\hat{x}(k|k-1)}, G(k) = g[\hat{x}(k|k), k]$$

Row i, column j of matrix $\varphi(k)$ is the partial derivative of the i element of the vector function $f(x, k)$ with respect to the j element of the state vector x, then substitute $x = \hat{x}(k|k)$ into the required formula and so on, you can get the matri $H(k)$.

In the Taylor series expansion (7), high subitem $x(k) - \hat{x}(k|k)$ is ignored, and only linear item is retained, the system Eq. (6) can be rewritten into the state equation and observation equation of (8).

$$\begin{cases} x(k+1) = \varphi(k)x(k) + G(k)w(k) + u(k) \\ y(k) = H(k)x(k) + v(k) + b(k) \end{cases} \tag{8}$$

In the Eq. (8), $u(k)$ and $b(k)$ are calculated as Eq. (9).

$$\begin{cases} u(k) = f[\hat{x}(k|k), k] - \varphi(k)\hat{x}(k|k) \\ b(k) = h[\hat{x}(k|k-1), k] - H(k)\hat{x}(k|k-1) \end{cases} \tag{9}$$

Through the above mathematical model transformation, the nonlinear model is simplified into linear form, and the nonlinear optimal estimation problem can be transformed into a state estimation by the kalman filtering algorithm.

3.2 Mathematical Calculation of the Extended Kalman Filter Algorithm

Through the analysis of Eqs. (8) and (9), combined with the above-mentioned linear optimal filtering algorithm, the extended kalman filtering mathematical expression is obtained from the general kalman filtering algorithm.

$$K(k) = P(k|k-1)H^T(k)[H(k)P(k|k-1)H^T(k) + R_2(k)]^{-1} \tag{10}$$

$$\hat{x}(k|k) = \hat{x}(k|k-1) + K(k)\{y(k) - h[\hat{x}(k|k-1), k]\} \tag{11}$$

$$\begin{aligned} P(k|k) = P(k|k-1) - P(k|k-1)H^T(k)[H(k)P(k|k-1)H^T(k) + R_2(k)]^{-1} \\ *H(k)P(k|k-1) \end{aligned} \tag{12}$$

$$\hat{x}(k+1|k) = f[\hat{x}(k|k), k] \tag{13}$$

$$P(k+1|k) = \varphi(k)P(k|k)\varphi^T(k) + G(k)R_1(k)G^T(k) \tag{14}$$

The initial value of recursion is given in Eq. (15).

$$\begin{cases} \hat{x}(0|-1) = \bar{x}(0) \\ P(0|-1) = P(0) \end{cases} \tag{15}$$

It can be seen from the above that *K*(*k*) is needed to calculate the state $\hat{x}(k|k)$ of the system which is to be estimated, that is to say, *H*(*k*) will be used. If the calculation of *H*(*k*) is known that the predicted state $\hat{x}(k|k-1)$ n the previous step is linearized as the standard value, it cannot be linearized with $\hat{x}(k|k)$ as the standard value like $\varphi(k)$.

4 Mathematical Models and Experimental Results of Chaotic Time Series Prediction

4.1 Mathematical Model of Chaotic Time Series Prediction

Mackey-Glass time series is a typical chaotic time series. The mathematical model of Mackey-glass differential equation is (16). It can be known that this differential equation has nonlinear characteristics. In real life, many phenomena and problems are Mackey-Glass problems.

$$\dot{x} = ax(t-\tau)/[1 + x^{10}(t-\tau)] - bx(t) \tag{16}$$

Let $\alpha = 0.2$, $b = 0.1$, when $\tau > 17$ begins to generate chaos, the greater τ is, the higher the chaos is. $\tau = 29$, x is the observed variable, generates 10,000 data, removes the first 5,000 transient points and takes the last thousand data points as the training set, then tests the predictive performance of the model, the data is shown in Fig. 2.

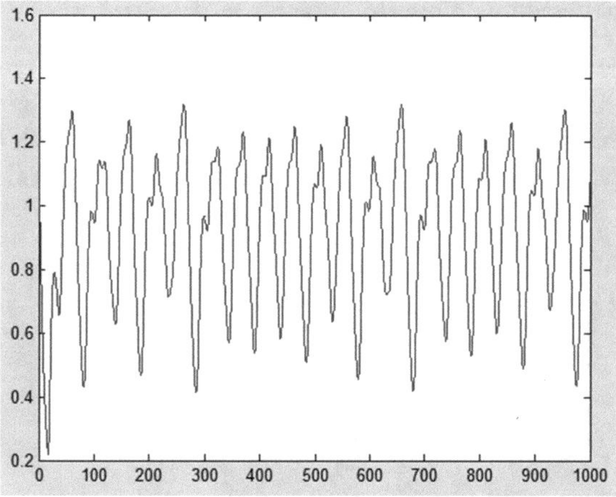

Fig. 2. Mackey-Glass Chaotic Sequence Diagram

The weight modification in BP neural network is formally taken as the state transfer equation, and the output of the neural network is used as the measurement update equation, which is a specific mathematical model such as (17) and (18).
The state transfer equation is (17).

$$w_{k+1} = w_k + u(k) \tag{17}$$

The measurement equation is (18).

$$y_k = h(w_k, u_k) + v(k) \tag{18}$$

The network weight w indicates that the two parameter vectors from time k are updated as shown in (19).

$$w_k = \begin{bmatrix} w_k(f) \\ w_k(g) \end{bmatrix} \tag{19}$$

The equation of state (17) represents the process of correction of neural network weight in the system prediction, the state of the system is set by the weight parameter w_k of the network, where $u(k)$ indicates the dynamic noise of the system.

The output of the Eq. (18), which is the output of the model of the state space, and $y(k)$ is a nonlinear function of weight vector w_k and input vector u_k, where $v(k)$ represents the measurement noise of the system.

Both the dynamic noise $u(k)$ and the measurement noise $v(k)$ are zero-mean-white noise whose variances are given by $E[u_i u_j^T] = Q_k$ and $E[v_i v_j^T] = R_k$. Respectively, the Q_k dynamic noise covariance and the R_k measured noise covariance are diagonal matrices.

4.2 Simulation Results of Chaotic Time Series Prediction

Using the extended Kalman filter structure and the mathematical model given above, the following 200 of one thousand data points are simulated by Matlab, and the following predicted fitting results were obtained. The predicted fitting was shown in Fig. 3, the predicted value error was shown in Fig. 4, and the predicted mean square error was shown in Fig. 5.

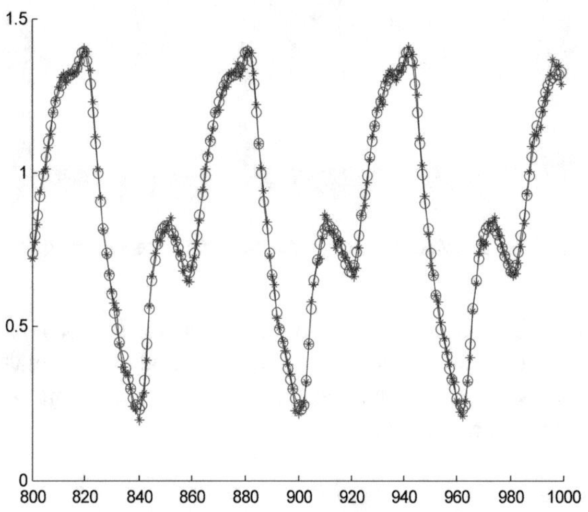

Fig. 3. The fitting results of Mackey-glass chaotic time series prediction

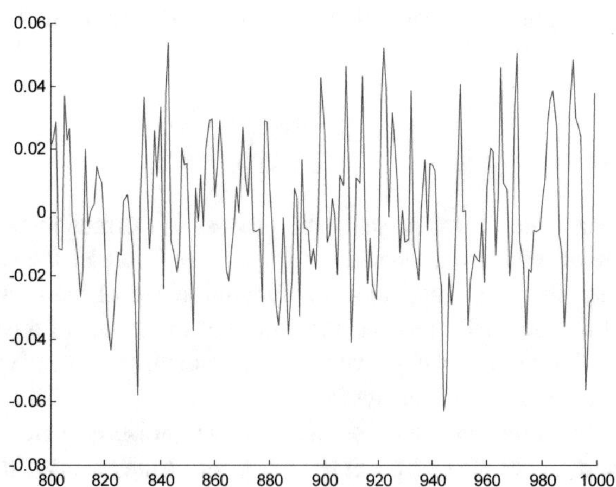

Fig. 4. Prediction error of Mackey-glass chaotic time series

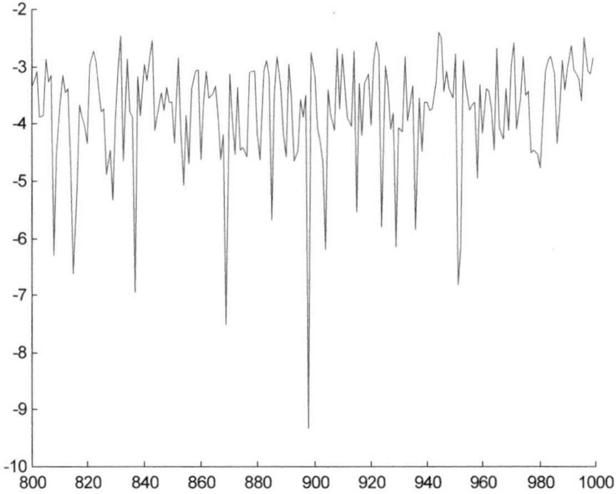

Fig. 5. The mean square error of Mackey-glass chaotic time series prediction

From the fitting curve of Fig. 3, the fitted value is in good to the actual value, the time series is very close, and the fitting error is quite small. The results show that using BP neural network training weight method combined with extended kalman filtering algorithm to predict mackey-glass time series has good effectiveness.

5 Conclusion

BP neural network model and its BP algorithm are described in the paper, the chaos and its chaotic time series model and its characteristics are briefly introduced, and a new chaotic time series prediction model is established based on BP neural network algorithm and EKF algorithm. Finally, Matlab was used to carry out the simulation experiment. The results show that the chaotic prediction model is effective, which provides an effective and feasible way for the prediction of a kind of time series with highly complex nonlinear dynamic relations. It is widely used in the fields of astronomy, hydrology, meteorology, automatic control, economy, etc., and the prediction of chaotic time series is also a hot topic in academic research.

Acknowledgement. The authors would like to thank the anonymous reviewers for their valuable comments. This work was supported by Initial Scientific Research Fund of FJUT (GY-Z12079), Pre-research Fund of FJUT (GY-Z13018), Fujian Provincial Education Department Youth Fund (JAT170367, JAT170369), Natural Science Foundation of Fujian Province (2018J01640) and China Scholarship Council (201709360002).

References

1. Zhang, H., Li, R.: Bernstein neural network chaotic sequence prediction based on phase space reconstruction. J. Syst. Simul. **28**(4), 880–889 (2016)
2. Zhang, S., Hu, Y., Bao, H.: Parameters determination method of phase-space reconstruction based on differential entropy ratio and RBF neural network. J. Electron. (China) (S1993-0615) **31**(1), 61–67 (2014)
3. Lee, C.M., Ko, C.N.: Neurocomputing 73, 449 (2009)
4. Ai, H., Shi, Y.: Study on prediction of haze based on BP neural network. Comput. Simul. **35** (1), 402–405 (2015)
5. Zhang, J., Tan, Z.: Prediction of the chaotic time sevies using hybrid method. Syst. Eng. Theor. Pract. **33**(3), 763–769 (2013)
6. Nie, Y., Wu, J.: An online time series prediction method and its application. J. Beijing Univ. Technol. **43**(3), 386–393 (2017)
7. Li, S., Luo, Y., Zhang, M.: Prediction method for chaotic time series of optimized BP neural network based on genetic algorithm. Comput. Eng. Appl. **47**(29), 52–55 (2011)
8. Zhang, H., Li, R.: Chaotic time sevies prediction of full-parameters continued fraction based on quantum particle sarm optimization algorithm. Control Decis. **31**(1), 52–58 (2016)
9. Hao, J., Tang, D.: Research of run off prediction based on generalized regregression neural network model. Water Resour. Power **34**(12), 49–52 (2016)

Machine Learning Techniques for Single-Line-to-Ground Fault Classification in Nonintrusive Fault Detection of Extra High-Voltage Transmission Network Systems

Hsueh-Hsien Chang[1(✉)] and Rui Zhang[2]

[1] Department of Computer and Communication, Jin Wen University of Science and Technology, New Taipei 23154, Taiwan
h.h.johnson.chang@gmail.com

[2] School of Management, Shanghai University of Engineering Science, Shanghai 201620, China

Abstract. This paper presents artificial intelligence (AI) approaches for fault classifications in non-intrusive single-line-to-ground fault (SLGF) detection of extra high voltage transmission network systems. The input features of the AI algorithms are extracted using the power-spectrum-based hyperbolic S-transform (PS-HST) for reducing the dimensions of the power signature inputs measured by using non-intrusive fault monitoring (NIFM) techniques. To enhance the identification accuracy, these features after pre-processing are given to AI algorithms for presenting and evaluating in this paper. Different machine learning techniques are then utilized to compare which classification algorithms are suitable to diagnose the SLGF for various power signatures in a NIFM system.

Keywords: Artificial Intelligence (AI)
Nonintrusive Fault Monitoring (NIFM) · Transmission network systems

1 Introduction

The power system faults which are mainly short circuit phenomena between the phases or phase and ground can lead to excessively high currents or over voltages which cause extensive damages to the devices.

Chen et al. [1] have proposed an adaptive phasor measurement unit (PMU) for transposed and un-transposed parallel transmission lines to estimate fault location with respect to different faults for a vertical configuration encountered in Taiwan. The data is sampled at sampling rate of 3.84 kHz. As a result, the performance of the proposed protection algorithm is almost independent of fault types, locations, resistance, and fault inception angles. However, the development of the scheme is based on the distributed line model and the PMUs at both ends of lines [2].

Da Silva et al. [3] have proposed a hybrid fault location algorithm for three-terminal transmission lines using wavelet transform (WT) to analyze the high frequency

P. Krömer et al. (Eds.): ECC 2018, AISC 891, pp. 109–116, 2019.
https://doi.org/10.1007/978-3-030-03766-6_12

components of the current and/or voltage signals for traveling waves from the fault point to the terminals. However, as concluded in [3], this method shows that it is influenced when subjects to a signal–noise rate (SNR) lower than 60 dB.

The non-intrusive monitoring techniques are a low-cost and easy application because it has only one set of voltage and current sensors installed at the electrical service entry (ESE) [4]. The authors [5] have proposed a particle swarm optimization (PSO) and genetic algorithm (GA) to optimize the training parameters of the back-propagation artificial neural network (BP-ANN) for improving the recognition accuracy in load identifications. Furthermore, the authors [6] have used wavelet multiresolution analysis (WMRA) technique and Parseval's theorem to identify load events in multiple load operations. Currently, the authors [7] have utilized a power-spectrum-based WT to identify transmission-line fault location for a centralized power distribution system of intelligent buildings. Figure 1 shows a representative non-intrusive fault monitoring (NIFM) system of extra high-voltage power transmission networks for a multiple generation power grid. As shown in the Fig. 1, the three-phase currents are measured at ESE and sent to the meter database management system (MDMS). These data are processed by the NIFM algorithms to identify the fault classes and types.

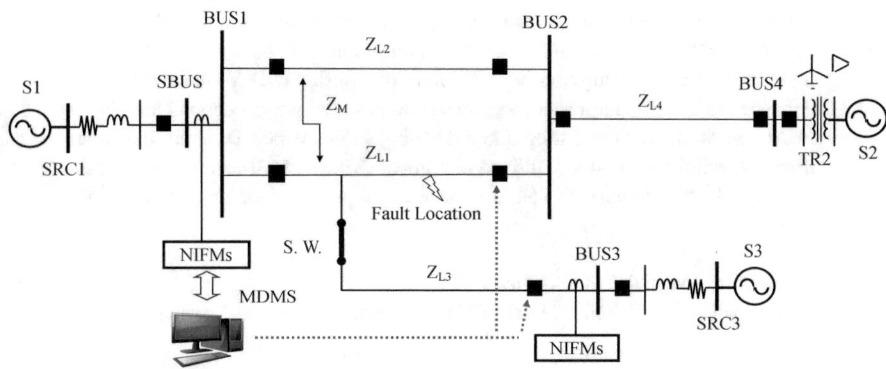

Fig. 1. Fault protection scheme on power transmission networks for a NIFM system.

In this paper, the accuracy rate and the computation requirement of the proposed various artificial intelligent (AI) methods, ex., BP-ANN, SVM, and k-NN, were verified in a simulated model system by using Electromagnetic Transient Program (EMTP). These results show a suitable fault classification approach is applied to develop a reliable nonintrusive fault classification system.

2 Proposed Methods

2.1 Power-Spectrum-Based Hyperbolic S-Transform

The complexity of the fault signatures is not directly processed by the AI algorithm. Therefore, some transformation techniques are needed to reduce the dimension of fault signatures.

In order to determine the discrete hyperbolic S-transform (HST), discretization of the signal *x(t)* is represented as follows:

$$X(m) = \frac{1}{N}\sum_{m=0}^{N-1} x(k)e^{-i2\pi nk} \tag{1}$$

$$HST[n,j] = \sum_{m=0}^{N-1} X(m+n)G(m,n)e^{i2\pi mj} \tag{2}$$

here,

$$G(m,n) = \frac{2|f|}{\sqrt{2\pi}(\gamma_f + \gamma_b)}\mathrm{e}^{\frac{-f^2V^2}{2n^2}} \tag{3}$$

$$V = \frac{(\gamma_f + \gamma_b)}{2\gamma_f\gamma_b}t + \frac{(\gamma_f - \gamma_b)}{2\gamma_f\gamma_b}\sqrt{t^2 + \lambda^2} \tag{4}$$

where N is total number of samples; γ_f and γ_b are the parameters of forward-taper and backward-taper, respectively; λ^2 is the positive curvature; and ζ is a translation factor to set the peak of the hyperbolic window at $(\tau - t) = 0$. In this paper, the γ_f, γ_b, and λ^2 are 0.2, 0.1, and 312.5, respectively.

The relationship of the power spectrum between the discrete signal $x[n]$ and each of the HSTCs can be built by Parseval's Theorem [8]. In this paper, the number of input neurons is selected from level 200 to level 600 at 25 intervals, yielding 17 scales of PS-HSTCs.

The power spectrum in Figs. 2 and 3 measured respectively on SBUS and BUS3 expresses that the three different fault types for the single-line-to-ground fault (SLGF) have different distributions of energy from level 200 to level 600. Consequently, in terms of the power spectrum in high levels (in high-frequency domain), different fault types can be identified using particular energy spectra as the features in the NIFM system.

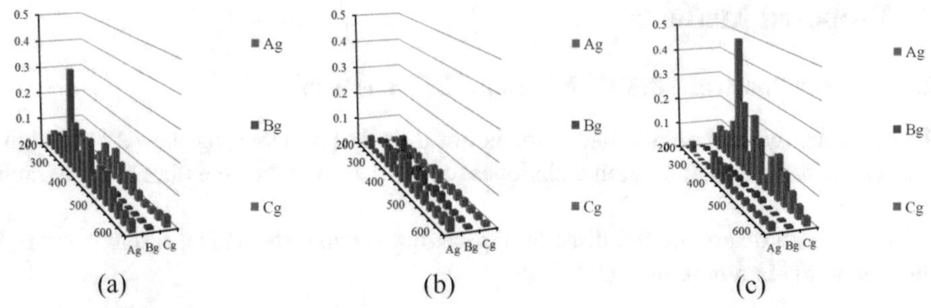

Fig. 2. Distributions of each level for the SLGF measured on SBUS. (a) Phase A. (b) Phase B. (c) Phase C.

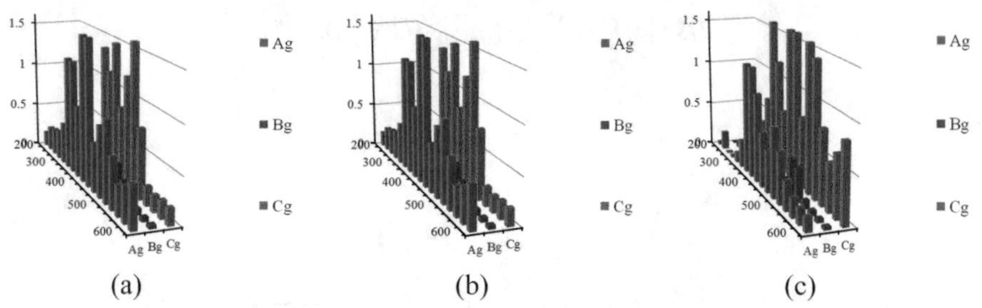

Fig. 3. Distributions of each level for the SLGF measured on BUS3. (a) Phase A. (b) Phase B. (c) Phase C.

2.2 BP-ANN

Most BP-ANN applications use gradient-descent training methods combined with learning via back propagation for single- or multilayer perceptron networks. These multi-layer perceptrons can be trained using analytical functions for applying a backward error-propagation algorithm to update interconnecting weights and thresholds [6].

A supervised MFNN is formed from one input layer, multiple hidden layers, and one output layer. The input, output, and hidden layers of the BP-ANN are as follows:

(a) Input layer: the selected PS-HSTCs information. In this paper, the number of input neurons is selected from level 200 to level 600 at 25 intervals, yielding 17 scales of PS-HSTCs for the fault event signals of each phase at ESE.
(b) Output layer: the number of output neurons is only one to show as a phase and ground indicator for connection status of fault type. In the case study, Ag, Bg, and Cg are represented as 1, 2, and 3, respectively.
(c) Hidden layer: two hidden layers are used in this work to enhance the efficiency of disaggregation. The common number of neurons in a hidden layer is the square-rooted sum of the number of neurons in an input layer and that in an output layer. Therefore, the number of nodes in one hidden layer is 5.

2.3 SVM

The support vector machine (SVM) is one of the machine learning techniques that use a hyperplane to separate the attribute space, then maximizing the margin between the instances of different classes [9]. An SVM constructs a hyperplane or set of hyperplanes for classification in a high- or infinite- dimensional space. Normally, a good separation is achieved by the hyperplane that has the largest distance to the nearest training data points of any class, the larger the margin the lower the generalization error of the classifier. Parameter C is a regularization parameter which helps implement a penalty on the misclassifications that are performed while separating the classes. Thus helps in improving the accuracy of the output.

In general, the radial basis function (RBF) kernel is a reasonable choice. The RBF kernel nonlinearly maps samples into a higher dimensional space. The RBF kernel can handle the case when the relation between class labels and attributes is nonlinear.

The value of gamma (γ) may play an important role in the SVM model. Changing the value of gamma may change the accuracy of the resulting SVM model. So, it is a good practice to use cross-validation to find the optimal value of gamma. In this paper, the C-SVC and RBF are employed for the SVM type and kernel type, respectively. The parameters C and γ are selected to be 10 by the cross-validation for the accuracy of the SVM model.

2.4 k-NN

The k-Nearest Neighbors (k-NN) is a non-parametric method used for classification. The k-NN algorithm is among the simplest of all machine learning algorithms. In k-NN classification, the output is a class membership. An object is classified by a majority vote of its neighbors, with the object being assigned to the class most common among its k nearest neighbors.

The accuracy of the k-NN algorithm can be severely degraded by the presence of noisy or irrelevant features. The best choice of k depends upon the data; generally, larger values of k reduce effect of the noise on the classification, but make boundaries between classes less distinct [10]. In classification problems, it is helpful to choose k to be an odd number as this avoids tied votes. In this paper, k is set to be 3.

3 Simulations and Results

3.1 Environment

In this case study, the AI algorithm in the NIFM system identifies different fault types at secondary side of CCVT of two different power generation buses (SBUS and BUS3) in three-phase 230 kV multiple terminal electric power transmission system networks, as shown in Fig. 4. In this paper, the fault location is set to the node L1F5. Each example of the power feature includes a fault resistance variation from 0Ω to 200Ω at 10Ω intervals and an inception angle of voltage signal variation of each fault resistance from 0^0 to 360^0 at 10^0 intervals, yielding 777 examples of power features for each bus (SBUS or BUS 3) and (3×777) raw data for each power signature (I_a, I_b, I_c), given that three fault types are respectively Ag, Bg, and Cg for SLGF on phase A, B, and C. The full input dataset can create a matrix with a size of $(3 \times 777 \times 3)$ by (1×17) using the proposed PS-HST selection, which includes the training dataset and the test dataset. The full raw data set is randomly partitioned into two subsets with equal size of data. One set of the data is for training and the other is for testing. The AI simulation programs were carried out using HeuristicLab [11]. The program was run to classify faults on a PC equipped with a 3.10 GHz Intel Core i5- 4440 CPU.

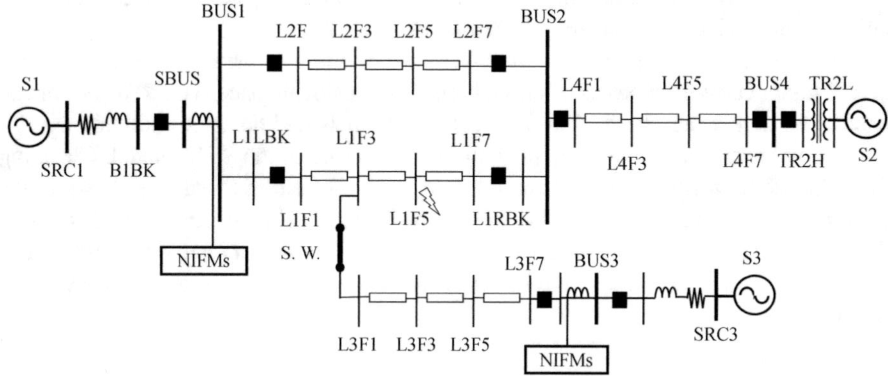

Fig. 4. Power transmission networks for a NIFM system.

3.2 Results

For three different fault classes measured on SBUS, the results for the training and test recognition accuracy of SLGF classification in average are respectively above 99.83% and 99.68% for the proposed AI algorithms from Table 1. Furthermore, the average value of the test recognition accuracy for the k-ANN is higher than that of the test recognition accuracy for other algorithms. Obviously, the average value of the test recognition accuracy is higher than that of the training recognition accuracy for the k-NN.

Table 1. Results of SLGF measured on SBUS.

	BP-ANN			SVM			k-NN		
	A	B	C	A	B	C	A	B	C
Recognition accuracy in training (%)	100	99.57	100	100	99.83	100	100	99.49	100
Average	99.86			99.94			99.83		
Recognition accuracy in test (%)	100	99.57	100	100	99.05	100	100	99.66	100
Average	99.86			99.68			99.89		
Execution time (s)	7.30	7.39	7.16	0.31	0.25	0.23	0.09	0.05	0.04
Average	7.28			0.27			0.06		

Regarding the average execution time, the time when the k-NN is used, is shorter than that of other methods. The average times of the BP-ANN and SVM are 7.28 s and 0.27 s, respectively.

Based on SLGF classes measured on BUS3 in Table 2, the average values of the training and test recognition accuracy are above 99.66% and 99.74%, respectively. In addition, the average value of the test recognition accuracy of BP-ANN is higher than that of the test recognition accuracy for other methods. Furthermore, the results for the training and test recognition accuracy in averages of Table 2 are 100% and 99.97%, respectively. Besides, the average value of the test recognition accuracy is higher than that of the training recognition accuracy for the k-NN.

Table 2. Results of SLGF measured on BUS3.

	BP-ANN			SVM			k-NN		
	A	B	C	A	B	C	A	B	C
Recognition accuracy in training (%)	100	100	100	100	100	100	98.97	100	100
Average	100			100			99.66		
Recognition accuracy in test (%)	100	99.99	99.91	99.83	99.48	99.91	98.48	99.91	99.91
Average	99.97			99.74			99.77		
Execution time (s)	8.50	5.94	6.11	0.33	0.29	0.26	0.04	0.05	0.04
Average	6.85			0.29			0.04		

In terms of the execution time, the k-NN is also smaller than other algorithms. The average times of the BP-ANN and SVM are 6.85 s and 0.29 s, respectively.

4 Conclusion

This paper presents PS-HST feature extraction method combined with various AI algorithms to enhance the recognition accuracy of fault classifications in extra high-voltage transmission networks using NIFM techniques.

The dimensions of data inputs for machine learning pre-processing can be effectively reduced by using feature extraction. To verify the validity of the proposed theoretical constructs, three popular AI algorithms in NIFM are compared in this paper for recognition accuracy and computation time. The obtained results show that the average values of the training and test recognition accuracy are respectively above 99.66% and 99.68%; however the algorithm is BP-ANN, SVM, or K-NN measured on SBUS or BUS3. Moreover, the average value of the test recognition accuracy is higher than that of the training recognition accuracy for the k-NN. In terms of the execution time, the BP-ANN is higher than other algorithms.

In future works, the proposed constructs will be completed for other fault types, ex., double-line-to-ground fault (DLGF) and balanced faults.

Acknowledgement. The authors would like to thank the Ministry of Science and Technology of the Taiwan, Republic of China, for financially supporting this research under Contract No. MOST 107-2221-E-228-001.

References

1. Chen, C.S., Liu, C.W., Jiang, J.A.: A new adaptive PMU based protection scheme for transposed/un-transposed parallel transmission lines. IEEE Trans. Power Deliv. **17**(2), 395–404 (2002)
2. Eissa, M.M., Masoud, M.E., Elanwar, M.M.M.: A novel back up wide area protection technique for power transmission grids using phasor measurement unit. IEEE Trans. Power Deliv. **25**(1), 270–278 (2010)
3. Da Silva, M., Oleskoviczb, M., Coury, D.V.: A hybrid fault locator for three-terminal lines based on wavelet transforms. Electr. Power Syst. Res. **78**, 1980–1988 (2008)
4. Hart, G.W.: Nonintrusive appliance load monitoring. Proc. IEEE **80**(12), 1870–1891 (1992)
5. Chang, H.H., Lin, L.S., Chen, N., Lee, W.J.: Particle-swarm-optimization-based nonintrusive demand monitoring and load identification in smart meters. IEEE Trans. Ind. Appl. **49**(5), 2229–2236 (2013)
6. Chang, H.H., Lian, K.L., Su, Y.C., Lee, W.J.: Power-spectrum based wavelet transform for nonintrusive demand monitoring and load identification. IEEE Trans. Ind. Appl. **50**(3), 2081–2089 (2014)
7. Chang, H.H.: Non-intrusive fault identification of power distribution systems in intelligent buildings based on power-spectrum-based wavelet transform. Energy Build. **127**, 930–941 (2016)
8. Chang, H.H., Linh, N.V., Lee, W.J.: A novel nonintrusive fault identification for power transmission networks using power-spectrum-based hyperbolic s-transform-part i: fault classification. In: IEEE 54th Annual Industrial and Commercial Power Systems (I&CPS), Niagara Falls, ON, Canada (2018)
9. https://docs.orange.biolab.si/3/visual-programming/widgets/model/svm.html
10. Everitt, B.S., Landau, S., Leese, M., Stahl, D.: Miscellaneous Clustering Methods, In Cluster Analysis, 5th edn. Wiley, Chichester (2011)
11. Wagner, S., et al.: Architecture and design of the heuristiclab optimization environment. In: Advanced Methods and Applications in Computational Intelligence, Topics in Intelligent Engineering and Informatics Series, pp. 197–261. Springer (2014)

Crime Prediction of Bicycle Theft Based on Online Search Data

Ning Ding[1(✉)], Yi-ming Zhai[1], Xiao-feng Hu[2], and Ming-yuan Ma[1]

[1] School of Criminal Investigation and Counterterrorism, People's Public Security University of China, Beijing, China
dingning_thu@126.com

[2] School of Information Technology and Network Security, People's Public Security University of China, Beijing, China

Abstract. In today's big data era, the development of the internet provides new ideas for analyzing and forecasting various types of criminal activities. The huge number of bicycle theft crime in China has become a social security problem that has to be solved urgently in our country. Can we use of internet search behavior to predict the trend of bicycle theft? Cluster analysis, correlation analysis and linear regression method are utilized to analyze the daily time series of bicycle theft in Beijing from 2012 to 2016 and the relationship between online search data, Taobao bicycle sales data and theft of bicycle theft. The results show that the crime of bicycle theft is mainly concentrated in the summer and autumn and the crime is the lowest near the New Year. What's more, in the morning of the day is the high incidence of such cases. The correlations between bicycle theft and the Baidu Index, Taobao bicycle sales data are significantly strong. The established multivariate linear regression model has a R-square of 0.804. The research provides an effective new idea for predicting the crime of bicycle theft and the tendency of other types of crime and provides the basis for the intelligence judgment and police dispatch.

Keywords: Baidu index · Bicycle theft · K-means · Correlation analysis
Multiple linear regression

1 Introduction

In the current intelligence-led modern policing work, the acquisition, analysis, judgment, and early warning of numerous intelligence information can provide important clues for all types of criminal offenses. Massive Internet resources provide abundant intelligence for policing work. Public security agencies should be good at extracting useful information from disorderly network data and correlating them with related criminal activities, so as to provide a powerful theory for the prediction and early warning of criminal activities. Stand by. Therefore, this article analyzes the association between open information and bicycle theft cases data obtained from the Internet in order to broaden the idea of Intelligence acquisition in modern public security work and promote the change from public security work to intelligence-led active policing.

P. Krömer et al. (Eds.): ECC 2018, AISC 891, pp. 117–128, 2019.
https://doi.org/10.1007/978-3-030-03766-6_13

With the continuous advancement of urbanization in China, the proportion of urban floating population continues to expand, and bicycle theft rate is growing as a unique social image of urban theft crime [1]. As a large bicycle country, China has a wide user base. The datum shows that China maintains the third place with 65 bicycles per 100 households [2]. In the study of the theft of bicycle crime, on the one hand, the factors influencing the theft of bicycle crime are complex and varied. For example, the actual number of bicycles and the appearance of new bicycles may affect the theft of bicycle crime. On the other hand, it is difficult to obtain the crime data needed for the study of the theft of bicycle crime and the current data on the number of bicycles. These are the problems that need to be solved to study the theft of bicycle crime.

When analyzing the factors affecting the theft of bicycle crime, the activity of the theft of bicycle crime can be indirectly reflected in two aspects: the sales of bicycles on Taobao and the web search data reflecting social attention. In recent years, with the rapid development of Internet technology, especially the emergence of smart phones in recent years, the Internet has become an inseparable part of people's lives, and online search engines have become the most common channel for users to obtain massive amounts of Internet information resources. A considerable amount of search data accumulated on the search engine server can objectively record the search behaviors and search needs of netizens. This cannot only reflect the daily attention of the majority of netizens to a certain degree, but also display the orientation of public opinions in front of hot events.

In order to explore the relationship between Internet search data, Taobao sales of bicycle data and crimes of theft of bicycles, this article first analyzed the level of bicycle theft cases in B cities in the five years from 2012 to 2016, as well as the months and times. The K-Means cluster analysis was conducted to explain the time trend of bicycle theft cases from the perspectives of Routine Activity Theory (RA Theory), traditional folklore and Rational Choice Theory. Finally, this paper establishes a multiple linear regression model by analyzing the relationship between the number of bicycle theft cases in B city and Taobao sales bicycle data and Baidu index data. This model can explain 80.4% of the variables. This study provides an effective new method for predicting the tendency of the crime of bicycle theft and provides the basis for the intelligence research and judgment and the police command and dispatch of this type of crime.

2 Related Literature

Regarding the theft of bicycle crime, there are a lot of studies at home and abroad. The domestic scholar Yu Dahong analyzes the four characteristics of the theft of bicycle crimes: (1) The diversity of means (2) The system of "theft, sale, purchase" is self-contained (3) The cost of committing the crime is low, and the success rate is high (4) The action is conspicuous [3]. In the study of this crime abroad, Levy et al. found that the number of bicycle thefts around the Washington subway station was positively correlated with the number of bicycles and potential perpetrators around each subway station based on bicycle census data and related crime data. The more business there are, the less likely there is a bicycle theft [4]. Chen and Lu used GIS and social network analysis methods to analyze the perpetrator data of electric bicycle thefts in Beijing from 2010 to 2012. It was found that electric bike burglars in Beijing are mostly foreign residents and most of the common criminal groups from the same town [5]. Ji et al.'s investigation of Nanjing Railway Station found that bicycle theft crime will affect people's choice of transportation mode when they are rushing to the train station. For example, some low-income workers who go to work by train will not ride a public bicycle to the train station, but workers who have experienced bicycle theft will be more willing to choose public bicycles as a means of transportation [6].

The authenticity, accuracy, and widespread use of web search data makes the analysis of social and economic behaviors based on web search data has gradually become a new hot spot for scholars in various fields [7, 8]. The research, of social science research abroad using search data first originated from epidemiological surveillance. The establishment of a relational model in 2004 by Johnson et al. demonstrated a strong correlation between influenza incidence and medical website search data [9]. Subsequently, Google's engineer Ginsberg and others published a paper on Nature several weeks before the outbreak of H1N1, introducing the "Google Flu Trends" (GFT), which not only successfully predicted the spread of H1N1 throughout the United States, but even Specific regions and states, but judged very timely [10]. Domestic research using online search data mainly focus on the analysis of economic trends and network public opinion control. For example, Xiang Yi et al. used Baidu Index to forecast stocks [11]; Liu Taoxiong discussed whether Internet search behavior is helpful to predict macroeconomics [12]; Chen Tao et al. analyzed and compared Google Trends in the event of an unexpected event. The characteristics of the time and space dimensions of Baidu index in the degree of public opinion on the Internet [13].

3 Temporal Pattern of Bicycle Theft Crime

The time element is the basic element in the crime scene and is the primary condition for analyzing the law of crime [1]. This article will use the time element as the primary analysis element of the bicycle theft situation. It can be seen from Fig. 1 that the monthly changes in bicycle theft crimes in each year are similar to the overall trend. In the January-February period, the number of crimes was low, and then it increased all the way to the peak after May, and started to decline from August. The number of crimes only showed a downward trend. The emergence of this law can be explained by the theory of daily activities proposed by Cohen and Felson [14], which attribute the conditions of crime occurrence to three points: (1) Potential and criminal capable offenders (2) The offender finds a suitable target or victim; (3) There is no protector who can protect the target or the victim. In the theory of daily activities, except for extreme weather that is not suitable for people to go out, high temperatures may increase people's activities and promote more social interactions. At the same time, they also increase the likelihood of finding a suitable target for potential offenders. The above reasons provide more opportunities for crime. According to the theory of daily activities, during the April-August period, the temperature of the climate gradually increased with time. People generally went out for more activities and the time to leave their homes increased. Therefore, the theft of bicycle cases was also high during this period. At the same time, folk customs are also an important factor influencing the amount of crime. From January to February, it is China's Lunar New Year. During the Spring Festival, people returning to their hometowns during the Lunar New Year is a tradition of the Chinese nation. At this time, the number of migrant workers remaining in the city of B has significantly decreased. Lowering the two-way reduction in the number of potential perpetrators and potential victims has also brought about a drastic reduction in the total amount of crime.

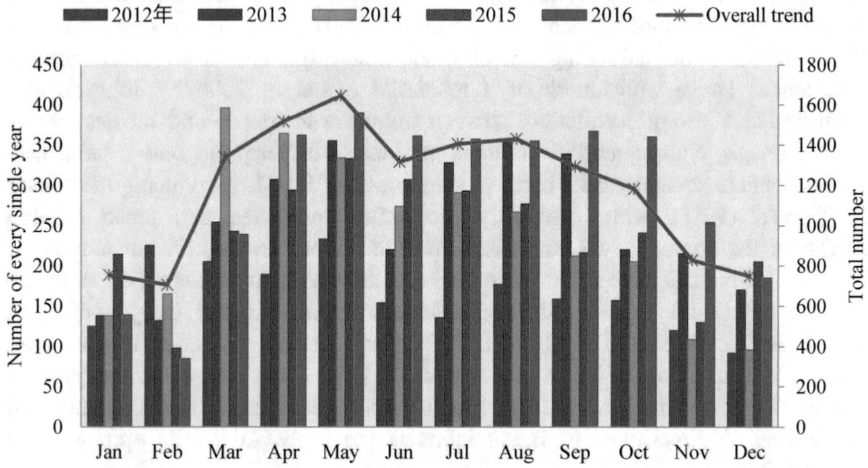

Fig. 1. Monthly and overall trend of bicycle theft cases from 2012 to 2016

From the overall trend of Fig. 2, we can see that bicycle theft crimes in City B maintained a very high number of cases during the day, especially in the morning, peaked at 8 am., and had the lowest number of incidents at 4 am. in the morning. From 4am to 8am in the morning, it is the period with the most significant increase in the day, followed by a decrease in volatility up to 18 points, and a significant decrease after 18 points. From 2012 to 2016, the distribution of different times during the day can be seen that there is a significant difference in the amount reported between day and night, in the morning and in the afternoon, so bicycle thefts show a clear time difference in the distribution of time. Analyzing the mean value of crimes represented by the dotted line in Fig. 2, we can find that the number of bicycle theft reports is higher than the average from 7 am to 20 pm, which is an important period of high incidence and is of great significance for policing work.

Crime pattern theory believes that criminal activities are most likely to occur in areas where potential offenders and potential victims are in the same space and time period [15]. For the theft of bicycle crimes, the analysis of time data can reveal the crime. The month in which the potential offenders and potential victims have the highest coincidence of activity time within one year, and the period of highest

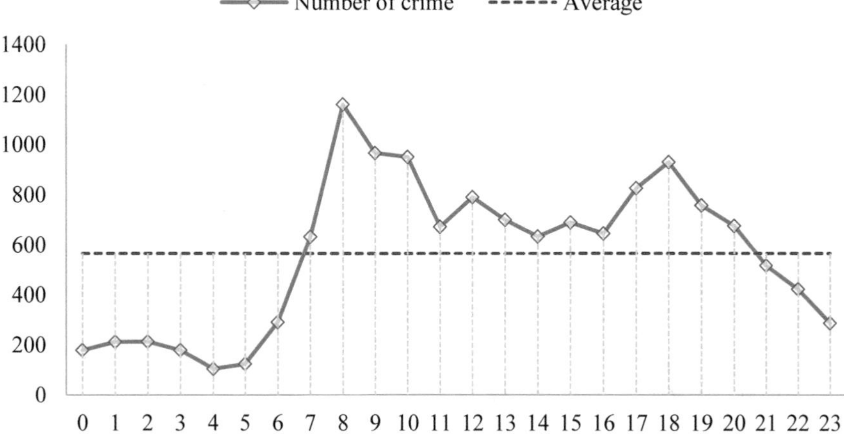

Fig. 2. Hourly distribution of bicycle theft cases from 2012 to 2016

coincidence in one day, and through this more regular prevention of such crimes, this paper introduces the statistics of cluster analysis. Methods to analyze this.

Cluster analysis, as a commonly used statistical analysis method, is widely used in various fields. For example, Li Shengzhu based on the crowdsourcing network platform influence point of view, based on the collection of the data of the major national crowdsourcing network platform website, the application of poly. Analytical methods were studied to sort out the main types of crowdsourcing web platforms [16]; Zhang Dan and He Yue used the social network of SNS websites as the research object, combined with the basic ideas of cluster analysis and social network analysis, and proposed social psychology. The theory-based framework for the analysis of SNS relationships and the use of experiments to verify the feasibility of the analytical framework [17]; Zhao Amin and Cao Guiquan, using 16 provincial government microblogs as research samples, using factor analysis and cluster analysis, To evaluate the influence of Weibo on government affairs and to compare empirical research, it was found that the level of influence of government microblog influence was not balanced; there was relatively less high-impact government microblog, and the influence of 16 microblogs on government affairs was a huge "pyramid" structure at the bottom. Distribution [18]. At the same time, the cluster analysis method is also a commonly used method for data mining and crime space-time analysis [19, 20]. The K-means clustering analysis method is widely used in various fields including criminology, such as Nath K-means cluster analysis method, because of its advantages such as simplicity, accuracy, and ease of operation. Analysis of crime patterns [21]. This section also uses this method to analyze the distribution of public transport plagiarism crimes at different times of the day and in different months of the year. Clustering results are calculated by SPSS 19.0 software. The general steps are:

Algorithm. 1. General steps of k-means clustering analysis method

Enter: hours set $D = \{x_1, x_2, \ldots, x_m\}$;
Clustering number: k.
Process:
randomly select k samples from D as the initial mean vector $\{\mu_1, \mu_2, \ldots, \mu_k\}$
repeat
make $C_i = \emptyset (1 \leq i \leq k)$
for $j = 1,2,\ldots,m$ **do**
calculate the distance between the sample and the mean vector: $d_{ji} = \left\| x_j - \mu_i \right\|_2$;
determined the cluster marker x_j after choosing the nearest mean vector: $\lambda_j = arg\ min_{i \in \{1,2,\ldots,k\}} d_{ji}$;
divide the sample x_j into the corresponding cluster: $C_{\lambda_j} = C_{\lambda_j} \cup \{x_j\}$;
end for
for i = 1,2,…,k **do**
calculate the new mean vector: $\mu_i' = \frac{1}{|C_i|} \sum_{x \in C_i} x$;
if $\mu_i' \neq \mu_i$ **then**
update the current mean vector μ_i to μ_i'
else
keep the current mean vector unchanged
end if
end for
until all the current mean vectors are not updated
Output: the clusters are divided into $C = \{C_1, C_2, \ldots, C_k\}$

As shown in Fig. 3, a cluster analysis of the total number of theft of bicycle crimes for a total of 60 months from 2012 to 2016 was conducted. The high-risk months for such crimes were found from March to October, and February was the low of such crimes. The month of January, November and December is the month of transition from low to high. During the K-means clustering of bicycle crimes at various times during the day, as shown in Fig. 4, a large number of criminal records with a time of 00:00 were found. After comparative analysis, it was found that such time was taken by the data collection department when the crime time was unknown. Is marked time, so the crime data of this time stamp is removed from the time of 0:00–0:59. Through the cluster analysis, it is found that the high-risk period for the theft of bicycle crimes within one day is 8:00–10:29 and 18:00–18:59, which is just the time for people to commute or go out to play. The peak, and people who choose to use bicycles as a means of transport, will also take the bicycle out of a safe area (such as a residential area) and enter a relatively unfamiliar place (such as a shopping mall) with a higher risk of loss. According to the crime model theory, this is the area where the potential theft of bicycle offenders coincides with the potential victims' space and time. The possibility of bicycle theft is most likely during this period of time. From the classification results, it can be seen that there are not many people who choose to steal bicycles at night. At this time, bicycles are generally parked in residential areas. At this time, the risk of stealing bicycles into residential areas is relatively high.

In order to better understand this psychology of bicycle theft, we introduce rational choice theory (Rational Choice Theory). In criminology, the rational choice theory adopts the utilitarian viewpoint, that is, people have the reasoning ability, attach importance to means, purpose and cost-effectiveness, and can make rational choices.

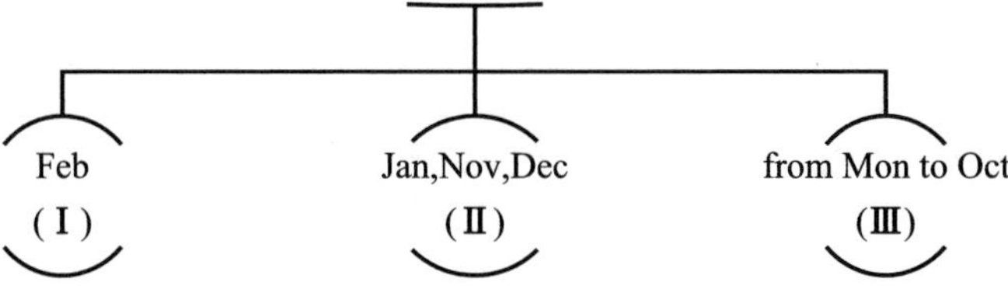

Fig. 3. K-means clustering of the theft of bicycle crimes in every month from 2012 to 2016

This method was designed by Cornish and Clarke to help understand situational crime prevention [22]. Suppose that crime is a purposeful act designed to meet the perpetrators' general needs in terms of money, status, sexuality, and pleasure. To meet these needs involves how to make (sometimes rather low-level) decisions and make choices, but the crime People are also limited by conditions such as personal ability and availability of information. Therefore, bike robbers will face many disadvantages such as the surveillance system, security personnel, and more comprehensive property protection facilities when they enter residential areas late at night. The proceeds from the theft of bicycles are generally not high, and the human rights of criminals will not be beneficial. I chose to sneak into residential areas for stealing late at night.

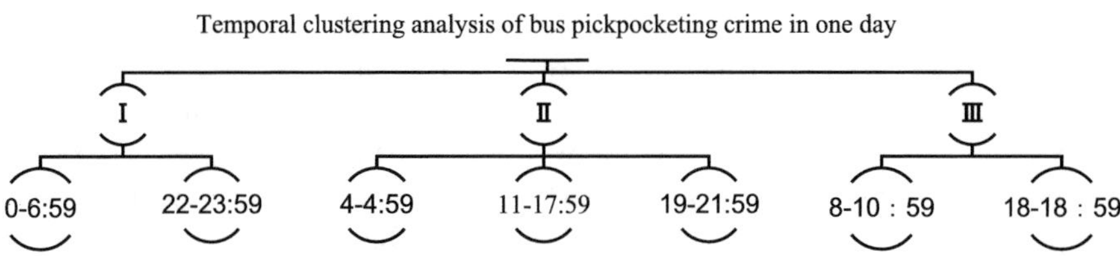

Fig. 4. K-means clustering of the theft of bicycle crimes in every hour from 2012 to 2016

4 The Correlation Between Related Factors and Bicycle Theft Crime

The overall number of bicycles in China is an important factor affecting the number of theft cases. However, due to its statistical difficulty, there is currently no accurate statistical data. Therefore, the data used by Taobao.com to sell bicycles include: volume, sales, and number of active babies. To reflect the current development of the bicycle market. This article analyzes the sales of Taobao bikes from August 2014 to September 2017. As shown in Fig. 5, bicycle sales reflect significant interannual changes, with high sales in spring and summer and low sales in autumn and winter. After the popularity of shared bicycles in 2017, the overall sales volume of bicycles was lower than that of previous years. However, it is worth noting that a social phenomenon is that with the increasing popularity of cycling bicycles, more and more citizens have become cycling enthusiasts, so people who purchase mountain bikes in stores are increasing. This article will use bicycles as the tipping point at a price of 1,000 yuan to distinguish professional mountain bikes selected by cycling enthusiasts from ordinary bicycle users. On the way, it can be seen that even after 2017, people's demand for mountain bikes is still growing steadily. Therefore, theft of bicycle crimes will not be significantly reduced due to the appearance of shared bicycles, and the types of theft bicycles will be more inclined to Higher-priced professional mountain bikes.

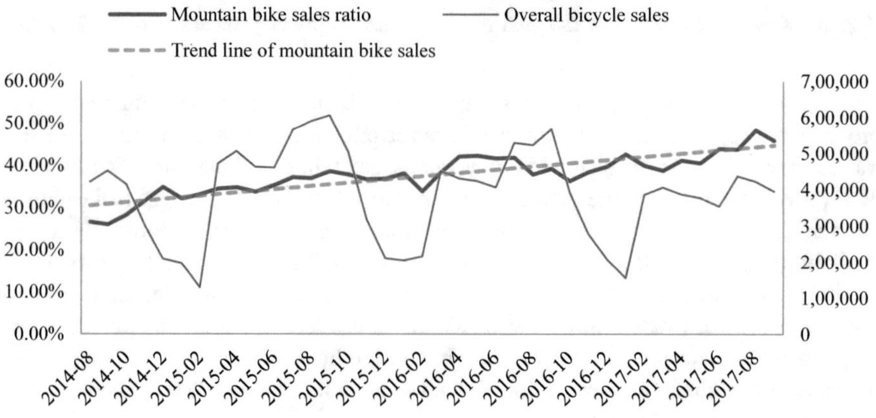

Fig. 5. Total bicycle sales on taobao.com and the proportion of mountain bike sales

In order to explore the relationship between Taobao bike sales data and the number of theft bicycle cases, this article uses SPSS 19.0 software to correlate the number of bicycles sold by Taobao in each month from August 2014 to September 2017 and the number of theft bicycle cases. In the correlation analysis of this paper, the results show that Taobao sales bicycle sales, trading volume and the number of active baby are positively correlated with the number of theft bicycle cases at the level of 0.01, this significant correlation is multi-linear afterwards. Regression analysis provides the basis (Table 1).

Table 1. Correlation analysis between taobao sales data of bicycle and cases data of bicycle theft

Project	Correlation	
	Pearson correlation	Significant (bilateral)
Sales Volume	0.686**	0.000
Trading Volume	0.673**	0.000
Active Commodity Number	0.498**	0.000

On the other hand, online search data can also reflect the current trend of theft of bicycle crimes. After analyzing the Baidu Index and bicycle crime data, it can be seen

Table 2. Correlation between the Baidu Index and the number of bicycle theft cases in Beijing from 2012 to 2016

Year	Daily data		Cumulative daily data	
	Pearson correlation	Significant (bilateral)	Pearson correlation	Significant (bilateral)
2012	0.512**	0.000	0.994**	0.000
2013	0.200**	0.000	0.999**	0.000
2014	0.356**	0.000	0.989**	0.000
2015	0.441**	0.000	0.999**	0.000
2016	0.351**	0.000	0.998**	0.000

**. Significant correlation at 0.01 (bilateral).

that for the search index, after a new thing emerges and quickly attracts public attention, its search index gradually increases. However, as time goes on, the popularity of the item gradually decreases, and the search index also decreases. However, the base of the index does not change much. Correspondingly, although the number of crimes will increase or decrease in different periods, the crime The quantity will increase or decrease to a certain extent and remain unchanged. Therefore, according to this feature, when analyzing the relationship between Baidu Index and the number of crimes, the cumulative value of the two types of data should also be analyzed as a variable.

From the test results in Table 2, we can see that the number of theft bicycle crimes in 2012–2016 is significantly related to the Baidu Index at 0.01. This correlation also reflects the current widespread social phenomenon: after bicycles are stolen, most people will first choose to search for solutions on the Internet, so when the Baidu index we observed has a higher and higher trend In the coming period, bicycle theft crime will also usher in a period of high incidence.

Table 3. Model summary of regression

Model summary	
R	0.897
R^2	0.804
Adjusted R^2	0.762
Std. Error of the Estimate	45.237

5 Multiple Linear Regression Analysis of the Number of Theft Cases

Regression analysis is widely used in various fields including crime analysis. In order to further examine the relationship between Taobao's sales of bicycle data and online search data theft and theft of bicycle cases, Taobao bicycle sales, volume, active baby The number of stolen bicycle crimes corresponds to the accumulated data of Baidu Index as independent variables X1, X2, X3…, the number of theft of bicycle crimes as the dependent variable Y to establish a multiple linear regression model, where y_t is the observation of Y and $\hat{y}_t$ is the estimate of y_t. $e_t = y_t - \hat{y}_t$ is called Residual. $\hat{\beta}$ should minimize the sum of residual squares $\sum_{t=1}^{T} e_t^2$.

$$Y = X_1\beta_1 + X_2\beta_2 + X_3\beta_3 + \ldots + \varepsilon \tag{1}$$

$$\hat{\beta} = (X^T X)^{-1} X^T Y \tag{2}$$

When using SPSS 19.0 to establish a multiple linear regression model, in addition to adding the sales data, sales volume, and number of active babes of Taobao's sales bicycles to the model, the more relevant Baidu index cumulative values in Table 3 are also added to the model. The fitted regression equation is:

The number of theft bicycle cases = −960.526 + 0.000000844 × sales - 0.001 × trading volume + 0.061 × Active commodity number + 0.0088 × cumulative Baidu index of Beijing, China - 0.001 × cumulative Baidu index of China.

According to the regression statistics, the R-squared of the regression model is 0.804, which has a good statistical significance (Tables 3 and 4).

Table 4. Multivariate linear regression analysis of the number of bicycle theft cases in City Beijing, China

Model	Unstandardized Coefficients		standardized Coefficients	t	Sig.
	B	Std. Error	Beta		
(Constant)	−960.526	201.654		−4.763	0.00
Sales Volume	8.44E−07	0	1.363	2.184	0.039
Trading Volume	−0.001	0.001	−1.241	−1.652	0.112
Active Commodity Number	0.061	0.028	0.542	2.136	0.044
Cumulative Baidu Index of Beijing, China	0.008	0.002	14.974	4.536	0.00
Cumulative Baidu Index of China	−0.001	0	−14.466	−4.378	0.00

6 Conclusions

This article analyzes the theft of bicycle cases based on web search data and Taobao sales data, and aims to provide new ideas for public security agencies to collect criminal information and broaden the channels for information acquisition. First of all, police officers should focus on patrolling streets and commercial areas where bicycles are parked in areas where the crime of bicycle theft is high, and reduce the chances of criminals committing crimes. In addition, the simple and fast way of obtaining network search data solves the problem of statistical data lagging behind. Police personnel in each area can monitor the relevant Baidu Index keywords and combine with other relevant network open data to keep abreast of new trends and new features that may occur in different types of crimes.

In the current era of big data, the introduction of big data ideas into the criminal analysis field is an important direction of policing work in the future [23, 24]. Future crime analysis should also include urbanization, migrant population, regional economic development, residents' focus, public opinion trends [25], and more. Data from different sources of social information were included to stimulate the unlimited potential of intelligence-led policing.

Acknowledgement. This work is supported by basic research program of People's Public Security University of China (No. 2018XKZTHY16) (No. 2016JKF01307) and National Key R&D Program of China (No. 2017YFC0803300).

References

1. Guo, F., Li, C., Zhao, Q.: Empirical analysis of bicycle theft scenarios — a case study of H in second-tier cities. J. GANSU Police Vocat. Coll. **12**(02), 48–53 (2014)
2. Huaon: 2017-2022 China Bicycle Industry Market Research and Investment Analysis Report. Beijing Aikaidete Consulting Co., Ltd., Beijing, China (2017)
3. Yu, D.: Current status and governance of bicycle theft activities. Journal Of Jiangxi Public Security College (02), 84–87 (2008)
4. Levy, J.M., Irvin-Erickson, Y., La Vigne, N.: A case study of bicycle theft on the Washington DC Metrorail system using a Routine Activities and Crime Pattern theory framework. Secur. J., 1–21 (2017)
5. Chen, P., Lu, Y.: Exploring co-offending networks by considering geographic background: an investigation of electric bicycle thefts in Beijing. Prof. Geogr. **70**(1), 73–83 (2017). https://doi.org/10.1080/00330124.2017.1325753
6. Ji, Y., Fan, Y., Ermagun, A., Cao, X., Wang, W., Das, K.: Public bicycle as a feeder mode to rail transit in China: the role of gender, age, income, trip purpose, and bicycle theft experience. Int. J. Sustain. Transp. **11**(4), 308–317 (2017)
7. Wang, Y.: A study of internet attention of public culture service system–taking baidu index as example. J. Mod. Inf. **37**(01), 37–40 (2017)
8. Li, Y., Wen, R., Yang, L.: The relationship between the online search data and the automobile sales–based on the keywords by text mining. J. Mod. Inf. **36**(08), 131–136 (2016)
9. Johnson, H.A., Wagner, M.M., Hogan, W.R., Chapman, W.W., Olszewski, R.T., Dowling, J.N., Barnas, G.: Analysis of Web access logs for surveillance of influenza., 20041202-1206 (2004)
10. Ginsberg, J., Mohebbi, M.H., Patel, R.S., Brammer, L., Smolinski, M.S., Brilliant, L.: Detecting influenza epidemics using search engine query data. Nature **457**(7232), 1012–1014 (2009)
11. Xiang, Y.: Could Baidu Index Forecast the Trend of Stock? Master, Southwestern University of Finance and Economics (2016)
12. Liu, T., Xu, X.: Can internet search behavior help to forecast the macro economy? Econ. Res. J. (12), 68–83 (2015)

13. Chen, T., Lin, J.: Comparative analysis of temporal - spatial evolution of online public opinion based on search engine attention–cases of google trends and Baidu Index. J. Intell. **32**(03), 7–10 (2013)
14. Cohen, L.E., Felson, M.: Social change and crime rate trends: a routine activity approach. Am. Sociol. Rev. **44**(4), 588–608 (1979)
15. Santos, R.B.: Crime analysis with crime mapping. Sage Publications (2016)
16. Li, S.: Cluster analysis on the influence evaluation of crowdsourcing network platform in China. J. Intell. **36**(08), 144–149 (2017)
17. Zhang, D., He, Y.: Study on SNS network based on cluster analysis. J. Intell. **31**(05), 62–65 (2012)
18. Cao, G., Zhao, A.: Positive study on evaluation and comparison of government affairs micro - blog influence: based on factor analysis and cluster analysis. J. Intell. **33**(03), 107–112 (2014)
19. Murray, A.T., Estivill-Castro, V.: Cluster discovery techniques for exploratory spatial data analysis. Int. J. Geogr. Inf. Sci. **12**(5), 431–443 (1998)
20. Murray, A.T., McGuffog, I., Western, J.S., Mullins, P.: Exploratory spatial data analysis techniques for examining urban crime: Implications for evaluating treatment. Br. J. Criminol. **41**(2), 309–329 (2001)
21. Nath, S.V.: Crime pattern detection using data mining, pp. 41–44. IEEE (2006)
22. Cornish, D.B., Clarke, R.V.: Understanding crime displacement: an application of rational choice theory. Criminology **25**(4), 933–948 (1987). https://doi.org/10.1111/j.1745-9125.1987.tb00826.x
23. Peng, Z.: Big data: the magic for intelligence - led policing to come true. J. Intell. **34**(05), 1–6 (2015)
24. Zhang, L.: The paradigm change of public security intelligence studies from the perspective of "Big Data". J. Intell. **34**(07), 9–12 (2015)
25. Su, Y., Wu, H., Chen, Y., Hu, W.: Using CCLM to promote the accuracy of intelligent sentiment analysis classifier for chinese social media service. J. Netw. Intell. **3**(2), 113–125 (2018)

Parametric Method for Improving Stability of Electric Power Systems

Ling-ling Lv[1(✉)], Meng-qi Han[1], and Linlin Tang[2]

[1] Institute of Electric Power, North China University of Water Resources and Electric Power, Zhengzhou 450011, People's Republic of China
lingling_lv@163.com

[2] Harbin Institute of Technology, Shenzhen 518055, People's Republic of China

Abstract. In this paper, the parametric method is used to bring the robust controller into the excitation system and adjust the required motor damping according to the actual situation. Poles assignment technique is used to design robust controller, so that the states of the closed-loop can rapidly go back to the desired position. The robustness of the system is also enhanced in the presence of disturbance and uncertainty. Simulation results show that the proposed design method ensures that the system operates safely at the rated power, and greatly reduces equipment damage due to the overload.

Keywords: Excitation system · Poles assignment · Robustness

1 Introduction

Stability of power systems is very important in present age, because stability is the guarantee of faultless systems operation. With the increasing growth in the power capacity, China's power grid has been expanding, the power grid construction has been continuously strengthened and gained a rapid development, the power transmission capacity has increased year by year. Therefore, the number of China's electricity user is also very large. It is particularly important to make the power systems operate safely.

A variety of malfunctions occur in the process of the power system operating. Although the timely troubleshooting will greatly reduce the possibility of damage to the equipment, the interference during the malfunction occurrence cannot be excluded correctly, timely and effectively. For instance, when the system overshoot caused is higher than the rated value, it will lead to power system collapse, the motor burned and other serious conditions. Various uncertainties in the power system can damage the electrical equipment, for prevent this occurrence, the motor signal should be properly controlled.

In recent years, Power System Stabilizer (PSS) [1] has been applied in power systems. Low frequency oscillations are observed when large power systems are interconnected by relatively weak tie-lines. PSS are incorporated in the excitation system of the generators to enhance damping of these low frequency oscillations. With the increasing power of the large scale of the system and the system structure and operation mode of increasingly complicated, the power system requires that the generator excitation controller has higher reliability, stability, economy and flexibility. For example, literature [2] proposed a novel approach to tune proportional-integral (PI),

P. Krömer et al. (Eds.): ECC 2018, AISC 891, pp. 129–137, 2019.
https://doi.org/10.1007/978-3-030-03766-6_14

proportional-integral-derivative (PID) and lead-lag PSS for a single machine infinite bus (SMIB) network by using backtracking search algorithm (BSA) to damp out low-frequency oscillation. The efficacy of BSA tuned PSS has been investigated by comparing the simulation results with fixed gain conventional PSS. Over time, Ting-Chia Ou has put forward a design of NIDC (Novel Intelligent Damping Controller) [3]. The proposed NIDC consists of a PID linear controller, an adaptive critic network and a proposed functional link-based novel recurrent fuzzy neural network (FLNRFNN). Test results show that the proposed controller can achieve better damping characteristics and effectively stabilize the network under unstable conditions.

Targeting on the model characteristics and characteristics of synchronous motor excitation system, a closed-loop feedback control system is established in this paper, the deviation of multiple output variables including the terminal voltage is taken as the optimal feedback input signal and put into the parameter controller, to eliminate the fluctuation of the excitation voltage, to achieve the effect of intelligent control and save large amounts of manpower and material resources, which will be described in detail in the following.

2 Design Principle of Robust Parametric Method

Parametric design method applied in this paper is firstly put forward in reference [4], and it is characterized of introducing the concept of arbitrary parameter matrix to build the mathematical formulation of the control laws to be designed. By this approach, all the requested control laws can be explicitly provided with complete freedom degree of design. Thus, combined with objective functions reflecting system performances, these control laws can be further optimized to achieve desired system performance, such as robustness. In the following, parametric design method will be stated in detail.

2.1 Parametric Expression of Control Laws

The considered linear invariant system is shown as

$$\dot{x} = Ax + Bu \tag{1}$$

Where, $x \in \mathbb{R}^n$ and $u \in \mathbb{R}^r$ are the state vector and are also input vector of the system, respectively; $A \in \mathbb{R}^{n \times n}$, $B \in \mathbb{R}^{n \times r}$ are the coefficient matrix of the system.

If (A, B) is controllable, select the state feedback control law:

$$u = Kx, K \in \mathbb{R}^{r \times n} \tag{2}$$

Then a closed-loop control system with Eqs. (1) and (2) can be given by

$$\dot{x} = A_c x, A_c = A + BK \tag{3}$$

Based on this, we can design controllers by poles assignment technique. In the following, the idea of solving parametric state feedback

For the sake of find the matrix K and make the closed-loop system obtain the ideal eigenvalue, set matrix $F \in \mathbb{R}^{n\times n}$ as the diagonalization matrix of the desired eigenvalue set for A_c, and V as the eigenvector matrix A_c, thus

$$(A+BK)V = VF \tag{4}$$

If V is reversible, so

$$K = WV^{-1} \tag{5}$$

According to literature [4], Eq. (4) can be transformed into the form of Matrix equation Sylvester

$$AV + BW = VF \tag{6}$$

Where,

$$W = XV \tag{7}$$

Since (A, B) is controllable, the following polynomial decomposition exists

$$(zI - A)^{-1}B = N(z)D^{-1}(z) \tag{8}$$

Where, $N(z) \in \mathbb{R}^{n\times r}$, $D(z) \in \mathbb{R}^{r\times r}$ are the right coprime matrix polynomial of Z. Let

$$D(Z) = [d_{ij}(z)]_{r\times r}, N(Z) = [n_{ij}(z)]_{n\times r}, w = \max\{w_1, w_2\}$$

Where,

$$w_1 = \max_{i,j\in\overline{1,r}}\{{}^{\circ}\left(d_{ij}(z)\right)\}, w_2 = \max_{i\in\overline{1,n}, j=\overline{1,r}}\{{}^{\circ}\left(n_{ij}(z)\right)\}$$

Then, $N(z)$ and $D(z)$ can be rewritten as

$$\begin{cases} N(z) = \sum_{i=0}^{w} N_i z^i, N_i \in \mathbb{C}^{n\times r} \\ D(z) = \sum_{i=0}^{w} D_i z^i, D_i \in \mathbb{C}^{r\times r} \end{cases} \tag{9}$$

It has been known previously that Matrix (A, B) is controllable, $N(s) \in \mathbb{R}^{n\times r}[z]$, $D(s) \in \mathbb{R}^{r\times r}[z]$ are the right coprime polynomials, meets the right coprime decomposition. If $(N(s), D(s))$ has expression (9), for preset matrix $F \in \mathbb{R}^{n\times n}$, the solution of the Sylvester matrix equation is express as

$$\begin{cases} V(Z) = N_0Z + N_1ZF + \cdots + N_wZF^w \\ W(Z) = W_0Z + W_1ZF + \cdots + W_wZF^w \end{cases} \tag{10}$$

Where, $Z \in \mathbb{R}^{r \times n}$ is an arbitrary parameter matrix and represents the degree of freedom of $(V(Z), W(Z))$.

After obtaining Matrix $V(Z), W(Z)$, then cite Eq. (5), the feedback gain value K can be obtained according to Eq. (5).

2.2 Improvement of the Objective Function

In order to make each performance of the system achieve the desired value, it is necessary to optimize the objective function of the system, so that the system has a very good robustness. Refer to literature [5] for the specific calculation process. By applying additional conditions on Feedback gain matrix K and Matrix V, the obtained free parameter Z can be used to make the system get some of the expected performance.

If there is the following disturbance in the closed-loop system

$$A + BK \to A + BK + \Delta(\varepsilon) \tag{11}$$

Where, $\Delta(\varepsilon) \in \mathbb{R}^{n \times n}$ indicates possible disturbance in the closed-loop system. According to literature [6], small gain of controller means robustness. On the other hand, [7] the smaller the value of $\mathcal{K}_F(V)$, the higher the sensitivity, and the better the robustness is. Therefore, the objective function can be given as

$$J(Z) \triangleq \alpha(\|K\|_F^2 + \mathcal{K}_F(V)) \tag{12}$$

Where, α is the weight coefficient and means the proportion of constraint on a variety of targets.

Once the minimum value of J is reached in the calculation process, take the corresponding parameter matrix $Z \in \mathbb{Z}$ as the optimal decision matrix Z_{opt}. Then, substituting Z_{opt} into Eqs. (9) and (10), we can calculate the optimal matrix V_{opt} and W_{opt}, and further calculate the optimal feedback gain K_{opt}.

2.3 Poles Assignment Based on the Damping Ratio

In the excitation system, whether the system's recovery capability is fast mainly depends on its motor damping. Therefore, in the study process of the power system failure, it is not enough to just consider the robustness of the system, the poles assignment of the damping ratio must be matched for restore the stability of the power system after the failure in time.

The rotor motion equation of the generator after Laplace's transformation can be formulated as:

$$T_J \frac{s^2}{w_0} \Delta\delta + D \frac{s}{w_0} \Delta\delta + K_1 \Delta\delta = 0 \tag{13}$$

Which is equivalent to

$$s^2 + 2\xi_n w_n s + w_m^2 = 0 \tag{14}$$

A pair of conjugate characteristic roots of the system without the controller are

$$s_{1,2} = -\xi_n w_n \pm j w_n \sqrt{1 - \xi_n^2} \tag{15}$$

Where, ξ_n is the system damping ratio, and w_n is the system's mechanical oscillation frequency with no damping. As the system damping value D is not large, the obtained damping ratio ξ_n will be very small, which cannot effectively achieve the desired value. Therefore, we usually add the excitation controller to the system, and change the damping ratio by increasing the damping coefficient, so that obtain the desired value.

The generator rotor equation of motion after adding the controller to the system is:

$$T_J \frac{s^2}{w_0} \Delta\delta + (D + D_E) \frac{s}{w_0} \Delta\delta + K_1 \Delta\delta = 0 \tag{16}$$

The conjugate characteristic roots with the controller are

$$s_{1,2}^{'} = -\xi_n^{'} w_n \pm j w_n^{'} \sqrt{1 - \xi_n^{'2}} \tag{17}$$

The overshoot formula for the second-order system is as follows

$$\sigma\% = e^{-\frac{\pi\xi}{\sqrt{1-\xi^2}}} \times 100\% \tag{18}$$

When the overshoot is set to 0.1, the damping ratio $\xi = 0.5912$, thereby the closed-loop poles reach the specified position by increasing the additional damping ratio, thus to maintain the stability of the system.

3 Application of Robust Control in Excitation Control

Excitation system control is a hot topic in today's power system research. The continuous extension of control theory, the excitation control technology will gain the in-depth development. In this section, mathematical model of the excitation control is introduced and simulation of excitation system by parametric controller is presented.

3.1 Mathematical Model of the Excitation System

The application of the excitation system can maintain the generator voltage or a certain voltage in the grid constant, meanwhile control the reactive power distribution of the generators running side by side. When the load of the generator changes, the excitation system can make the machine terminal voltage constant by adjusting the strength of the magnetic field, make reasonable allocation of parallel operation between the reactive power distribution units and improve system stability at the same time.

In this paper, the angular velocity and electromagnetic power have been added to provide feedback based on the terminal voltage as the feedback input, thus to ensure the overall stability of the power system operation.

Here, the equilibrium point linearization equation is borrowed from [8]:

$$\begin{cases} \Delta P_e = Q_E \Delta\delta + R_E \Delta E_q \\ \Delta P_e = Q_E^{'} \Delta\delta + R_E^{'} \Delta E_q^{'} \\ \Delta P_e = Q_V \Delta\delta + R_V \Delta V_t \end{cases} \tag{19}$$

In the formula,

$$Q_E = \frac{E_q V_s}{X_{d\Sigma}} \cos\delta, R_E = \frac{V_s}{X_{d\Sigma}} \sin\delta, Q_E^{'} = \frac{E_q^{'} V_s}{X_{d\Sigma}^{'}} \cos\delta + \frac{V_s^2 \left(X_{d\Sigma}^{'} - X_{d\Sigma}\right)}{2X_{d\Sigma}^{'} X_{d\Sigma}},$$

$$R_E = \frac{V_s}{X_{d\Sigma}^{\prime}} \sin\delta, Q_V = S_E + \frac{R_v X_s X_d E_q V_s \sin\delta}{X_{d\Sigma} P}, R_V = \frac{R_E X_{d\Sigma}}{\left(X_s^2 E_q + X_s X_d V_s \cos\delta\right) P},$$

$$X_{d\Sigma} = X_d + X_T + \frac{1}{2} X_L, X_{d\Sigma}^{'} = X_d^{'} + X_T + \frac{1}{2} X_L, X_s = X_T + \frac{1}{2} X_L,$$

$$T_d^{\prime} = T_{d0} \frac{X_{d\Sigma}^{\prime}}{X_{d\Sigma}}, P = (E_q^2 X_s^2 + V_s^2 X_d^2 + 2X_s X_d E_q V_s \cos\delta)^{\frac{1}{2}}.$$

Without considering the dynamic process of the excitation system and ignoring the damping, the system linearization equation is:

$$\dot{X} = AX + BU \tag{20}$$

Where

$$A = \begin{bmatrix} \frac{Q_E - Q_V}{T_d^{'} Q_V} & Q_E^{\prime} & -\frac{R_V Q_E}{T_d^{'} Q_V} \\ -\frac{w_0}{H} & -\frac{D}{H} & 0 \\ \frac{Q_E - Q_V}{T_d^{'} R_V Q_V} & \frac{Q_E^{'} - Q_V}{R_V} & -\frac{Q_E}{T_d^{'} Q_V} \end{bmatrix} \tag{21}$$

$$B = \begin{bmatrix} \frac{R_E^{'}}{T_{d0}} & 0 & \frac{R_E^{'}}{T_{d0} R_V} \end{bmatrix} \tag{22}$$

$$X = [\,\Delta P_e \quad \Delta w \quad \Delta V_t\,]^T \tag{23}$$

$$U = \Delta E_t \tag{24}$$

3.2 Simulation and Design of Excitation Controller for a Single Machine Infinite System

In the single machine infinite system, set the simulation parameters as follows:

$$X_d = 3.534, X_d^{'} = 0.318, H = 8.0, D = 5.0, X_T = 0.1, X_L = 1.46, T_{d0} = 10.0, U_s = 1.$$

When the system's initial operating point $\delta = 70^{\circ}$, it can be obtained combining the Eqs. (21–24). After judging the system is completely controllable, find its poles value by system damping. When the controller is not added to the system, its the conjugate characteristic roots is $-0.4205 \pm 5.2058i$.

From Eq. (15) it can be calculated that $w_n = 5.2228$. For the desired damping ratio $\xi = 0.5912$, by Eq. (17), the expected conjugate eigenvalues of the system should be: $-3.0877 \pm 4.2123i$. Without loss of generality, set another pole as -7.3381.

So,

$$s_1 = -3.0877 + 4.2123i, s_2 = -3.0877 - 4.2123i, s_3 = -7.3381.$$

Set the robust control system with the damping controller as Type I excitation system, meanwhile set the quadratic optimal control system as Type II excitation system.

Utilizing the parametric robust design Algorithm provided in Sect. 2 and literature [8], the feedback gains of Type I and Type II systems can be obtained as:

$$K_{\text{I}} = \begin{bmatrix} -62.8423 & 11.9988 & -150.2212 \end{bmatrix}, K_{\text{II}} = \begin{bmatrix} -55.6 & 5.1 & -20.8 \end{bmatrix}.$$

Take them into a single machine infinite system separately, observe and analyze their respective control effects through simulation experiment comparison. Let that the operating point of the system $\delta = 70^{\circ}$, and the line disturbance is added to the system between 0.1 s and 0.4 s, then the system returns to normal running. The simulation resolves of the corresponding variables are shown in Fig. 1.

As can be seen from the figure above that, Type I controller has stronger anti-interference ability than Type II controller, and the recovery time is shorter. As experiment results show that, proposed parametric robust design algorithm, with smaller overshoot and less settling time.

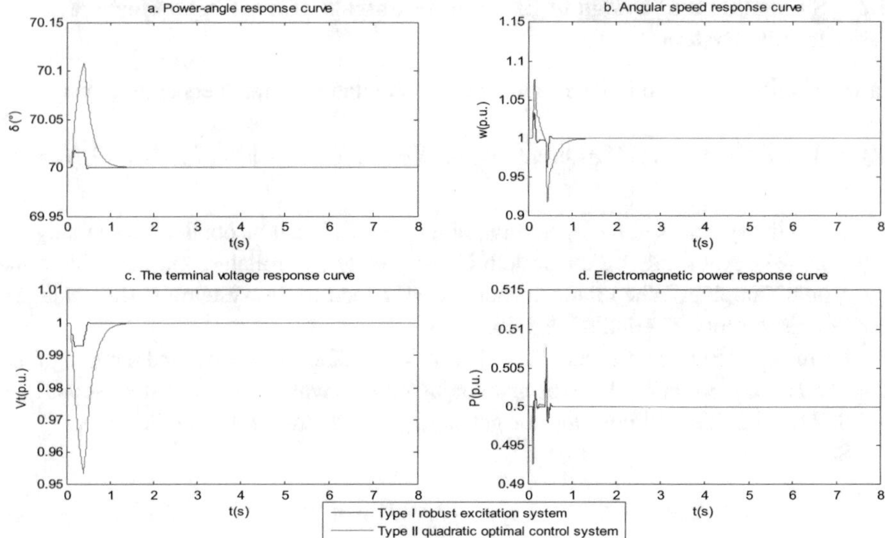

Fig. 1. Dynamic response curve under simulation experiments

4 Conclusions

In this paper, the parametric method is applied to bring the robust controller into the excitation system. The simulation results by parametric method is compared with that by the quadratic optimal control method. It proves that the parametric method can effectively improve the robustness and the adaptive adjustment of the system, and the closed loop system with this parametric robust controller has good dynamic performance and improved transient stability.

Acknowledgements. This work is supported by the Programs of National Natural Science Foundation of China (Nos. 11501200, U1604148, 61402149), Innovative Talents of Higher Learning Institutions of Henan (No. 17HASTIT023), China Postdoctoral Science Foundation (No. 2016M592285).

References

1. Naresh, G., Raju, M.R., Narasimham, S.V.L.: Enhancement of Power System Stability employing cat swarm optimization based PSS. In: 2015 International Conference on Electrical, Electronics, Signals, Communication and Optimization (EESCO), Visakhapatnam, India, pp. 1–6. IEEE (2015)
2. Shafiullah, M., Rana, M.J., Coelho, L.S., et al.: Power system stability enhancement by designing optimal PSS employing backtracking search algorithm. In: 2017 6th International Conference on Clean Electrical Power (ICCEP), Santa Margherita Ligure, Italy , pp. 712–719. IEEE (2017)

3. Ou, T.C., Lu, K.H., Huang, C.J.: Improvement of transient stability in a hybrid power multi-system using a designed NIDC (Novel Intelligent Damping Controller). Energies **10**(4), 488 (2017)
4. Zhou, B., Duan, G.R.: A new solution to the generalized Sylvester matrix equation AV-EVF = BW. Syst. Control Lett. **55**(3), 193–198 (2006)
5. Lv, L.-L.: Pole Assignment and Observers Design for Linear Discrete-Time Periodic Systems. Harbin Institute of Technology (2010)
6. Varga, A.: Robust and minimum norm pole assignment with periodic state feedback. IEEE Trans. Autom. Control **45**(5), 1017–1022 (2000)
7. Lv, L., Duan, G., Zhou, B.: Parametric pole assignment and robust pole assignment for discrete-time linear periodic systems. SIAM J. Control Optim. **48**(6), 3975–3996 (2010)
8. Lu, Q., Wang, Z., Han,Y.: Optimal Control of Power Transmission System. Science Press, pp. 158–183 (1982)

Application of Emergency Communication Technology in Marine Seismic Detection

Ying Ma[1,3], Nannan Wu[2], Wenbin Zheng[1(✉)], Jianxing Li[1,3], Lisang Liu[1], and Kan Luo[1,3]

[1] Fujian University of Technology, Fuzhou 350118, China
may@fjut.edu.cn, 87747318@qq.com
[2] Earthquake Administration of Fujian Province, Fuzhou 350001, China
[3] Research and Development Center for Industrial Automation Technology of Fujian Province, Fujian University of Technology, Fuzhou 350118, China

Abstract. The use of emergency communication technology in earthquake emergency communications was studied in this paper. Based on the existing emergency communication technology of Fujian Seismological Bureau, the paper puts forward a solution to meet the interconnection, voice interworking and data sharing between maritime emergency vessel and shore command. The program is implemented in the real-time monitoring project of the marine geophysical platform in the exploration of the deep structure of the crust in the western part of the Taiwan Strait, Which examines the feasibility of the program. The system has made a try in the field of marine seismic observation and communication, which has filled the gap of maritime earthquake emergency information interaction ability.

Keywords: Emergency communication · Real-time monitoring
Marine exploration · Marine earthquake · Satellite communications

1 Introduction

In recent years, the deep structure detection of the western part of the Taiwan Straits has been carried out by the seismological research institutes of Taiwan and Fujian. The exploration work was organized by the Seismological Bureau of Fujian province. Relevant experts from China Earthquake Administration came to guide research work during the period of exploration. Experts from the Institute of Geophysics, China Earthquake Administration, geophysics exploration center of China Earthquake Administration and Jiangxi Seismological Bureau participated in the survey. Taiwan Ocean University also took part in the exploration work. The purpose of the survey is to get information about the underground three dimensional crustal structure in Fujian province and its offshore area by means of detection work. It will provide a scientific basis for future research on the prediction of strong earthquake trend, formulation of strategic decision for earthquake prevention and disaster reduction, planning and utilization of land and resources, and construction of major projects [1]. The detection work is the first time to detect the deep crustal structure of the western Taiwan Strait on

P. Krömer et al. (Eds.): ECC 2018, AISC 891, pp. 138–145, 2019.
https://doi.org/10.1007/978-3-030-03766-6_15

the sea. The project team carried out the test and application of the air gun source system of the geophysical platform on the Yanping 2 scientific research ship.

2 Mission Target

Throughout the sea detection process, the earthquake site emergency communications technology personnel need to use temporary construction of the communication system to ensure the smooth flow of land and sea communications, and to achieve the maritime detection process of real-time image transmission. The system enables the shore command to keep abreast of maritime detection and to make timely decisions. Emergency communications technology in the marine seismic exploration work is still less. The previous marine earthquake detection work is through the shortwave radio station to achieve voice communication, not related to shipwreck video and data real-time communication. Emergency communication technology is a means and method for ensuring the emergency rescue and necessary communication. In the event of natural or man-made emergencies, we should make full use of various communication resources to achieve and ensure the smooth flow of emergency information transmission path. There are no precedents to apply this special communication mechanism to marine seismic detection. This is the use of emergency communications technology and marine seismic research work combined with a bold attempt [2].

In the process of marine detection, the task of emergency communication is to record the whole sea detection process in real time. Detailed audio-visual and related data will be provided for future development of detection technology. At the same time, the provisional command department, which is set up at the Xiamen seismic survey and Research Center, will be able to get all the information and audio data in real time so as to prepare the seismologists at any time to carry out the operation safety assessment according to the real-time picture, and to conduct the command and dispatch immediately through the communication system. At the same time, in order to be able to analyze the test data in a timely manner, the marine communication system is required to return the real-time latitude and longitude and time data from each source to the temporary command. Therefore, the marine communication system of each detection is required to record and transmit all audio and video signals of the geophysical platform during the marine operation at the same time, so as to realize the two-way real-time communication of audio and video, and also carry out the transmission of data files by the source of the seismic source. Since the scientific research vessel is temporarily rented, the required communication system also needs to be modular dismantling along with the marine geophysical exploration platform.

3 Demand Analysis

In order to achieve real-time audio video and graphic data transmission between sea and land, it is necessary to establish a mobile broadband communication system that can meet the requirements. The main problems and analysis are as follows:

(1) Offshore operating area not only no operator mobile phone signal coverage, but also no public 4G wireless network access signal. So the land and sea data transmission can only be completed through the satellite communication system. Satellite communication has a wide range of coverage, stable and reliable, free from ground conditions, flexible mobility and so on. It can provide large-span, large-scale, long-distance mobile communications services. Its technical characteristics are very consistent with the requirements of emergency communications systems. Satellite communication has an irreplaceable position and role in maritime emergency communications [3].
(2) The offshore platform is always in a voyage bump with the fixed satellite service (FSS), and the moved satellite service (MSS) services need to be used to achieve data communication. Because the frequency of FSS is C, Ku, Ka and other frequency bands, although it has the characteristics of high transmission bandwidth and high transmission rate, it can only be used for satellite fixed station, and it is not suitable for non positioning real-time communication on board. Although MSS can realize the service of data information transmission by mobile communication, the frequency of its use is L, S frequency band, the transmission bandwidth is small, the transmission rate is low [1], and the interaction effect of streaming media information is not ideal. According to the analysis of the quality requirements of maritime transmission information, the Satcom on the move satellite communication system is selected. Mobile communication is designed to meet the needs of users to transmit broadband video information through dynamic satellite. It uses the Ku band of fixed orbit satellites to transmit wideband video information by mobile communication, which combines the advantages of FSS and MSS. Through the "Satcom on the move" system, the ship can track the platform of satellite in real time and transmit the multimedia information such as voice, data and image in real time, which can meet the needs of multimedia communication under the condition of mobile platform.
(3) There are many typhoons along the coast of Fujian in summer. The equipment system needs waterproof and moistureproof measures. The satellite antenna for outdoor work must be reliable and fixed so as not to be affected by wind and rain and damage the equipment [2].
(4) The layout of audio and video acquisition equipment should be able to effectively capture the desired picture and audio signals. Therefore, the location of the signal collection point should be arranged according to the test point, and the influence of ambient noise and light on the signal acquisition should be considered, and the convenience and reliability of the cable laying should be taken into account.

4 Mission Solution

After the construction of the first phase equipment of the Fujian earthquake rescue team and the construction of the digital earthquake observation network of China Earthquake Administration, the Seismological Bureau of Fujian province now has the following emergency communication equipment. Including 2 devices can realize remote mobile video transmission: portable video conferencing terminal Tandberg Tactical MXP, 2 sets of on-board equipment of maritime satellite communications ThraneThrane EXPLORER 527, 2 portable equipment of maritime satellite communications ThraneThrane EXPLORE 700, and 2 sets of portable maritime satellite communication equipment ThraneThrane EXPLORE 500. These devices are only EXPLORER 527 can be counted as Satcom system, it consists of two parts of the terminal and tracking antenna, can provide 128 kbit/s exclusive bandwidth and maximum bandwidth of 464 kbit/s shared bandwidth of IP data services. Its omni-directional satellite antenna can automatically track the satellite, can also communicate in the drive, installation and removal is also more convenient, so be selected as our core communications equipment at sea. But because the EXPLORER 527 is BGAN business vehicle-mounted equipment. It is designed and manufactured as a special equipment for land vehicles. Whether it can be used with the ship for offshore operations platform has no relevant records and cases for reference. So we decided to do the full test before the task in order to grasp its performance characteristics [5–7].

As the test results will be affected by sea climatic conditions, hydrological conditions and submarine terrain conditions and other factors. And the effect will change with the sailing of the operating vessel. So it is particularly important to have real-time communications between the operating vessels and between the operating vessels on the shore. In addition to the “Yanping 2” expedition ship’s geophysical platform needs to carry out communication security, there are OBS (submarine seismograph) delivery ship needs to carry out communication transmission. The difference is that the expedition ship needs to communicate with the shore temporary command by real-time video. And the OBS shipments only need to transfer video recordings. But the two ships have to meet the needs of real-time voice communications. The overall communication connection diagram shown in Fig. 1.

Before the formal implementation of the transmission mission, our communications security personnel on the land using the vehicle for simulation testing. After that, we will correct the transmission requirements in real time according to the test results and the work of the ship survey platform. And then we in accordance with the actual situation of the test boat to monitor the transport platform to build the situation. Through the “Yanping 2” scientific survey ship field survey, we design the main equipment layout shown in Fig. 2. Maritime satellite antenna fixed on the roof of the gun array ship. The stern platform is a safe control area. This area does not allow people to enter a lot of time. So in order to be able to monitor the safety of sea and onboard equipment, we need to install two surveillance cameras. According to monitoring requirements analysis, the camera should have remote control rotation, optical zoom and night shooting function, in order to observe the scene from all angles to observe the field operations. Therefore, we installed in the gun array ship with No. 1 and No. 2 camera. The position of the No. 1 camera is at the stern, it is mainly used to observe the gun array equipment into the sea process and into the sea after the blasting scene, and can record the gun array was the process of towing the work of the state. The No. 2 camera is used to record the working conditions in the gun array. No. 3 camera is mainly used to record the work of the pressurized cabin, the contents of the shooting only need to show in the gun control room. Both the camera and the maritime satellite are powered by the gun control cabin. The computer output of the

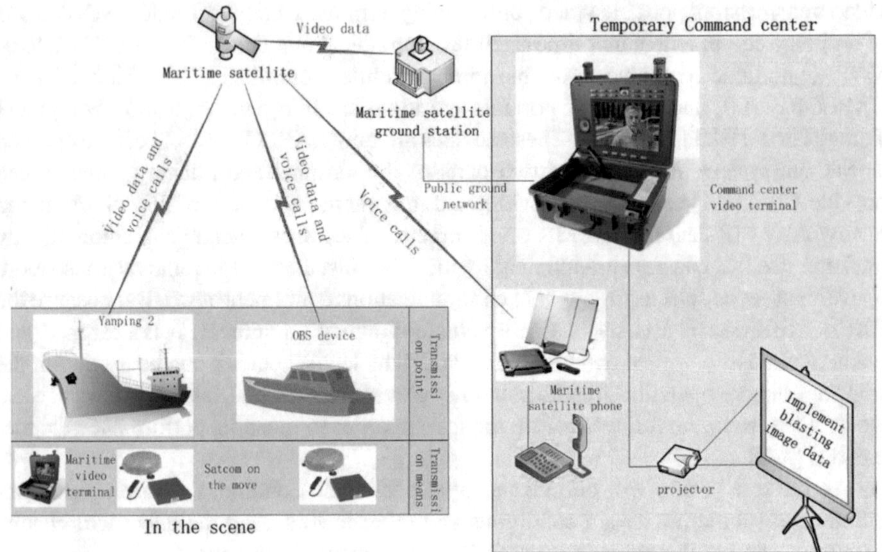

Fig. 1. Communication network diagram

navigation signal and the gun control system display signal through the VGA-AV converter to convert the AV signal. They are sent to the AV matrix along with the video signal of camera number 1, for operator scheduling (The actual implementation is shown in Fig. 4). The selected signal source is placed on the large screen monitor in the conference room and transmitted via the maritime satellite to the shore temporary command. One of the signals of Navigate or monitor was send to the cockpit as needed. The maritime satellite phone in the conference room can communicate with the headquarters while transmitting real-time images.

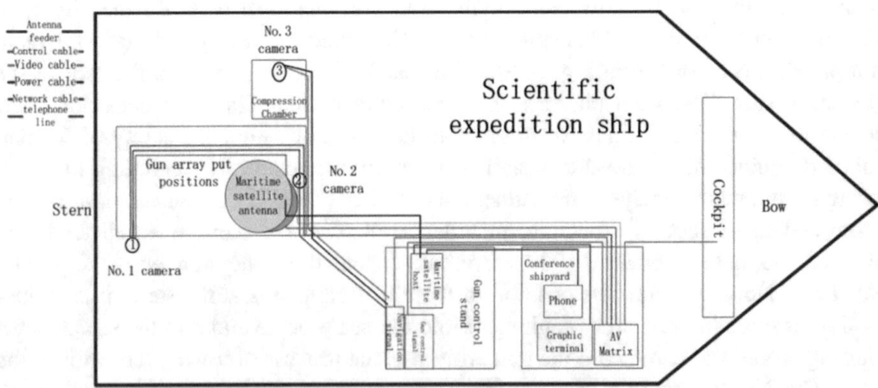

Fig. 2. Main equipment wiring diagram

5 Task Implementation

Before the mission was implemented, all equipment was debugged on the shore by multi state. The system is tested by the air gun source system on Yanping 2 scientific research vessel. During the experiment, the availability of the offshore transmission channel was tested (the transmission path of the data stream is shown in Fig. 3). Before the implementation of the official sea survey, two hard disk recorders were added. One is placed in the command post, and the other is in the conference module. In order to record all the images taken by the cameras as a technical file, we will leave them for future inspection. After the commissioning of the shore, the whole system arrived in Xiamen 1 weeks in advance. Due to the limitation of the length of antenna cable and the working environment, we have arranged the position of satellite antenna and host computer according to local conditions. Subsequently, maritime satellite communication systems were installed on the temporary headquarters of Xiamen Seismic Survey and Research Center, the “Yanping 2” scientific research vessel and the OBS launch vessel. A video conference system was set up at the Xiamen earthquake investigation and research center and the Yanping 2 research ship. Under the precondition of good sea condition, no wind wave and cloud influence, and with the efforts of the field communication support personnel to deal with the difficult problems flexibly according to local conditions, the actual measurement has reached the predetermined target. Overseas in Xiamen, we have achieved the goal of sending back to the command headquarters real-time the working scenes of “Yanping No. 2” scientific research vessel, such as trial voyage test, gun array hanging test, gun array towing test, gun array recovery test, air gun source excitation test, safety inspection and so on (the actual implementation effect is shown in the right picture of Fig. 4.) The OBS delivery ship also made good use of our video communication system. The whole system guarantees the smooth communication among the headquarters of Xiamen Survey Center, the “Yanping 2” scientific research vessel and the OBS launching vessel, and ensures the

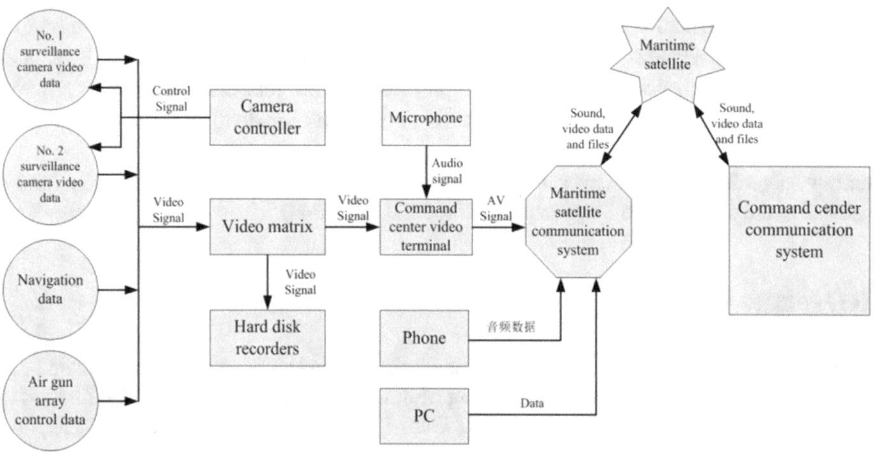

Fig. 3. Data transmission route

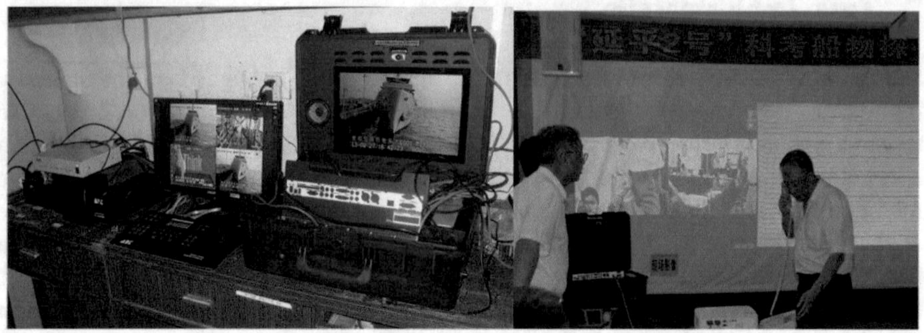

Fig. 4. The actual monitoring transmission renderings

smooth delivery of leadership orders and the smooth reporting of the offshore operation platform. The equipment was relatively stable and successfully completed the maritime communication support task.

6 Conclusion

Through the design and implementation of communication transmission in the “deep crustal structure exploration in the western Taiwan Strait”, the application of Fujian earthquake emergency communication system in marine seismic exploration has been realized for the first time. The successful completion of the task is due to careful and meticulous programming and multiple targeted simulation exercises. It is proved that the mobile communication equipment of the maritime satellite based on the land environment design can meet the communication and transmission needs of the “Yanping 2” scientific research vessel under the sea condition of little wind and good weather conditions. The temporary marine emergency communication system and monitoring system are running steadily, and the system is in normal coordination. During the implementation of the task, technicians have accumulated experience in the operation of the system module disassembly and operation, which has laid a good foundation for us to do a good job in the marine seismic field communication support, and has made a good reserve of experience and means for dealing with emergencies.

Acknowledgement. In this paper, the research was supported by Scientific Research Fund of Fujian Provincial Education Departmen (JAT160339, JA15343).

References

1. Ma, Y., Wu, N.: Research on seismic site emergency rescue traffic path analysis system based on public image information. J. Inf. Hiding Multimed. Signal Process. **9**(3), 577–585 (2018)
2. Wu, N., Hong, H., Huang, H., Huang, S., Guo, J., Wang, Q.: Implementation of integrated modification to in-situ emergency communication command vehicle in Fujian Province. South China J. Seism. **32**(2), 87–91 (2012). (in Chinese)

3. Debruin, J.: Establishing and maintaining high-bandwidth satellite links during vehicle motion. IEEE Control. Mag. **28**(1), 93–101 (2008)
4. Shang, J., Li, L., Liu, C., et al.: Exploration and experiment of low information rate satellite communication system in S-band. Telecommun. Eng. **56**(1), 54–59 (2016)
5. Wang, H.: Changes in satellite mobile communication system. Satell. Netw. **12**, 36–42 (2013). (in Chinese)
6. Lv, Z., Liang, P., Chen, Z.: Development status and trends of satellite mobile communications. Satell. Appl. **2016**(01), 48–55 (2016)
7. Andrew, D.S., Paul, S.: Utilizing the Globalstar network for satellite communications in low earth orbit. In: 54th AIAA Aerospace Sciences Meeting, pp. 1–8 (2016)

Electrical Energy Prediction with Regression-Oriented Models

Tao Zhang[1,2], Lyuchao Liao[1,2], Hongtu Lai[1], Jierui Liu[1,2(✉)], Fumin Zou[1,2], and Qiqin Cai[1,2]

[1] Fujian Key Lab for Automotive Electronics and Electric Drive, Fujian University of Technology, Fuzhou 350118, Fujian, China
TaoZ0721@163.com, fjachao@gmail.com, jierui1204@gmail.com, htlai@fjut.edu.cn

[2] Fujian Provincial Big Data Research Institute of Intelligent Transportation, Fujian University of Technology, Fuzhou 350118, Fujian, China

Abstract. Electrical energy consumption analysis is critical to saving energy, and therefore more and more attention has been paid to make a consumption prediction. However, there are so many factors in the building that affect the energy consumption of electrical appliances that it's hard to get an efficient method. To address these problems, the traditional linear regression model, SVM-based model, Random Forest (RF) and XGBoost algorithm were employed to explore the relationship between factors and consumption. The experimental results show that XGBoost is an efficient method to explore correlation pattern and to make a consumption prediction; removal of lighting factor show a more reasonable result to the prediction accuracy; and factor of temperature shows a more significant for consumption prediction than of humidity. This finding would be benefit to energy consumption modelling and improving prediction accuracy.

Keywords: Electrical energy consumption · Electrical data mining
Regression prediction model · Electric quantity prediction

1 Introduction

With the rapid growth of population and the continuous improvement of people's living standards, the building area and building energy intensity have been increasing year by year. Relevant information shows that China's building energy consumption has exceeded one third of the country's total energy consumption, ranking first in energy consumption, and building energy efficiency has become crucial. Therefore, the analysis of energy use in buildings has become the subject of numerous studies [1]. In an UK study on residential buildings [2], power consumption in TV and consumer electronics operating in standby increased by 10.2%. Therefore, limited energy data analysis can help us understand and quantify the relationship between different variables.

Constructing energy consumption prediction is difficult because the energy consumption behavior of buildings is complex, and the influencing factors are uncertain,

P. Krömer et al. (Eds.): ECC 2018, AISC 891, pp. 146–154, 2019.
https://doi.org/10.1007/978-3-030-03766-6_16

leading to frequent fluctuations in demand [3]. These fluctuations are due to non-linear factors such as building construction characteristics, occupant behavior, climatic conditions and subsystem components etc. Currently in the architectural design and energy simulation, the static fixed hypothesis is usually used to briefly describe the behavior of the building user, for example, the assumption that the user arrives at the building at 8:00 on weekdays and leaves at 17:00. These assumptions are mainly based on data provided by relevant standards and norms or personal experience of designers [4]. However, there are few literatures to conduct a more comprehensive analysis and comparison of this model and application. Since most of the lighting fixtures used in this study are LEDs, light energy consumption accounts for 1% to 4% of building energy consumption. In contrast, electrical energy consumption accounts for 70% to 79% of building consumption [5]; therefore, a predictive analysis method of building energy consumption was proposed for the perspective of electrical energy consumption.

In order to comprehensively analyze the impact of various factors on electrical energy consumption, four regression-oriented methods were employed to model energy consumption dataset and then to make a consumption prediction in this study.

The rest of the paper is organized as follows. Section 2 briefly reviewed the existing related works; Sect. 3 introduced experimental method and results; and finally, the work of this paper is summarized.

2 Related Works

Li et al. [6] obtained linear equations between the annual cooling load, thermal load, total load and influencing factors of residential buildings through orthogonal experiment, software simulation and linear regression. The comparison between the simulation software and the prediction equations were used to prove the reliability of the equation. Using this prediction equation, the main factors affecting the cold and heat load of residential buildings could be find out from various influencing factors. In this way, it's possible at a certain extent to explore the main factors to make energy-saving design and renovation of residential buildings, and ignore secondary factors and reduce unnecessary workload; however, there remains a need for efficient method to explore latent patterns.

To improve the prediction accuracy of energy consumption in college buildings, combining the advantages of traditional gray prediction model and neural network prediction model, a radical basis function (RBF) neural network energy consumption prediction algorithm was established. Lin et al. used this method to synthesize the advantages of the gray system theory and the advantages of neural network self-learning and self-organization [7]; and the case study showed that compared with the traditional grey theory and the RBF neural network prediction model, the relative error between the prediction and actual values of the combined model was reduced by 5.4% on average, which provided a decision basis for building energy conservation assessment and design.

Besides, an energy consumption model was also presented to utilize the data of energy factors, such as occupant schedules, operations and equipment, especially on tenants in buildings [8]; in this case, a random forest algorithm was employed to analyze the ranking of the importance of variables, and Gaussian process regression models was used to verify the energy consumption results of individuals, offices and retail tenants in commercial buildings. This method was mainly to determine the impact of energy use factors on the energy consumption of each tenant (office and retailer).

There are many related papers studying on analyzing energy data and trying to determine main factor, but more studies are needed to effectually get correlation between factors. In this study, a various factors analysis method was proposed to identify combinations of variables and to estimate energy consumption of house appliances.

3 Experimental Method and Results

3.1 Dataset Description

The experiments employ Public data coming from [9]. The energy metering of the data was done with an M-BUS energy counter, collecting energy consumption information every 10 min. Energy information was collected via an energy monitoring system connected to the Internet and reported via email every 12 h. Since most lighting fixtures are LEDs, light energy consumption accounts for 1% to 4% of the total, compared to 70% to 79% of electrical energy consumption per month [5].

Although there were no weather stations outside the home, weather data from the nearest airport weather station (Chièvres Airport, Belgium) was combined with this study to assess its impact on projected energy consumption. The downloaded weather data [5] is at hourly intervals and all variables is listed in Table 1.

The data including the temperature and humidity taken by the wireless sensor was obtained at an average frequency of 10 min, and the time span of the data set was 137 days (4–5 months). Figure 1 shows an overview of the energy consumption of the data set, which is the energy consumption waveform for the entire period, and the energy consumption curve shows a high degree of variability.

3.2 Data Preprocessing

Cross-validation was used to divide the data set into 75% training set (14, 801, 26) and 25% test set (4934, 26) for separated verification. Part of data from appliance energy consumption, including T1, RH1, T2 and RH2, were selected to be analyzed; and the results are shown in Fig. 2(1).

The latent correlation pattern was explored at first. As shown in Fig. 2(1), T1 is at a positive correlation (0.06) with Appliances energy consumption, with the scatter plot tending to normal distribution; and T1 is at a negative correlation (−0.02) with Light energy consumption, with the scatter plot also tending to a normal distribution. Similarly, RH1 is at a positive correlation (0.11) with Appliances energy consumption.

Secondly, the data quality was evaluated and cleaned. the scatter plot of relation between time and electrical energy consumption is shown in Fig. 2(2). As can be seen, there are many outliers in the experimental dataset; furtherly, the number of

Table 1. Data variables and description

Data variables	Units	Number of features
Appliances energy consumption	Wh	1
Light energy consumption	Wh	2
T1, Temperature in kitchen area	°C	3
RH1, Humidity in kitchen area	%	4
T2, Temperature in living room area	°C	5
RH2, Humidity in living room area	%	6
T3, Temperature in laundry room area	°C	7
RH3, Humidity in laundry room area	%	8
T4, Temperature in office room	°C	9
RH4, Humidity in office room	%	10
T5, Temperature in bathroom	°C	11
RH5, Humidity in bathroom	%	12
T6, Temperature outside the building (north side)	°C	13
RH6, Humidity outside the building (north side)	%	14
T7, Temperature in ironing room	°C	15
RH7, Humidity in ironing room	%	16
T8, Temperature in teenager room 2	°C	17
RH8, Humidity in teenager room 2	%	18
T9, Temperature in parents room	°C	19
RH9, Humidity in parents room	%	20
To, Temperature outside (from Chièvres weather station)	°C	21
Pressure (from Chièvres weather station)	mm Hg	22
RHo, Humidity outside (from Chièvres weather station)	%	23
Windspeed (from Chièvres weather station)	m/s	24
Visibility (from Chièvres weather station)	km	25
Tdewpoint (from Chièvres weather station)	°C	26

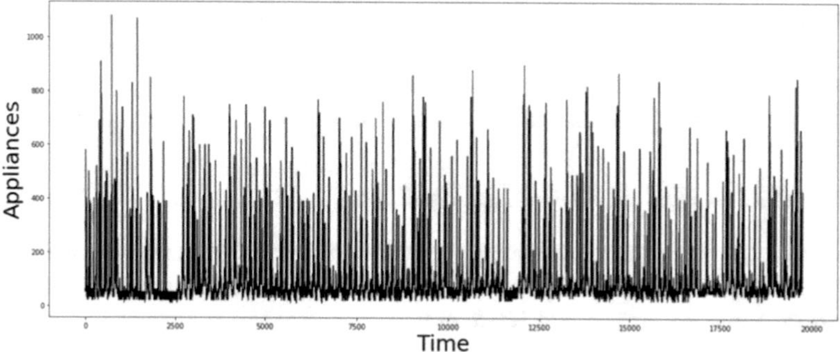

Fig. 1. Overview curve of appliances energy consumption

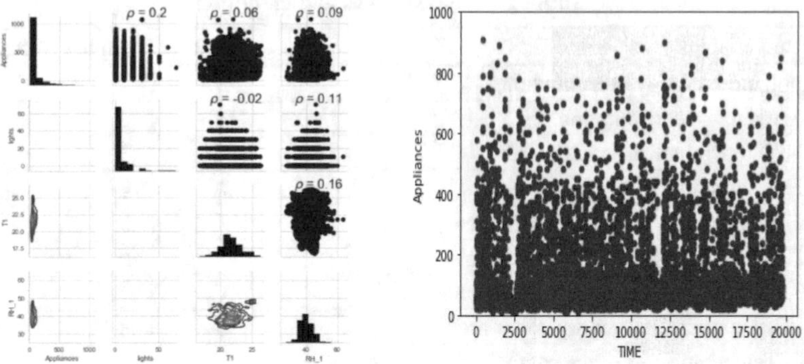

(1) Correlation analysis for part of factors (2) Outlier analysis with scatter of utilizing time

Fig. 2. Overview analysis of electrical energy consumption dataset

occurrences of each value of the electrical energy consumption data set is shown in Table 2, showing that many data are far from the mean value and have rare occurrences. These outliers should be deleted to clean experimental dataset.

Table 2. Electrical energy consumption elements

Energy consumption	50	60	40	...	900	860	1070
Quantity	4368	3282	2019	...	1	1	1

In order to furtherly analyze the correlation degree of each variable of the electrical energy consumption, the top 10 factors with the most relevant correlations are shown in Fig. 3(1). The heat map of each variable was plotted to visually determine the most correlated variable to electrical energy consumption.

The experimental results show that none of variables is far from correlation between each other. To furtherly determine the extent to which each variable affects the energy consumption of the appliance, the random forest feature importance assessment is used to estimate the average impurity attenuation with 10,000 decision trees; from the results shown in Fig. 3(2), it is found that each variable has an impact on the energy consumption of the appliance and should be selected as an influencing factor.

3.3 Regression Models for Energy Prediction

To get a better performance of energy prediction, several efficient algorithms were chosen to fit the energy consumption dataset, including Linear Regression, SVM-based Regression, Random Forest, and XGBoost.

A linear regression model with multi-variable was established to fit energy consumption at first. This linear regression model employs all available predictors. Let the random variable y change with m independent variables $x_1, x_2, \ldots, x_m$ and then have the following linear relationship:

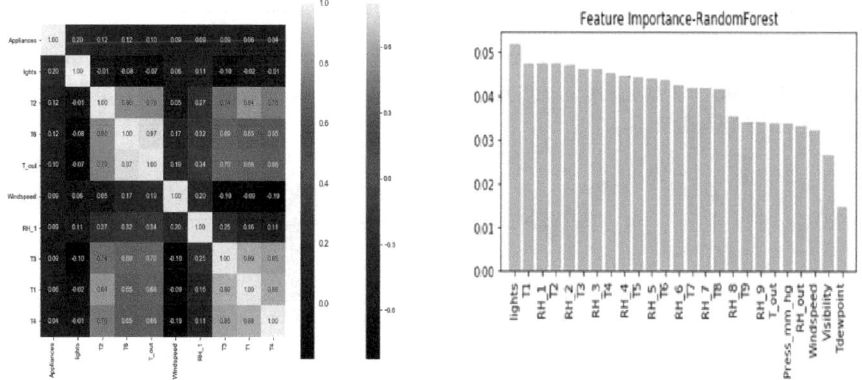

(1).Heat map of correlated factors (2). Factor importance assessment with RF

Fig. 3. Factor analysis of electrical energy consumption dataset

$$y = \beta_0 + \beta_1 x_1 + \cdots + \beta_m x_m + \in \quad (1)$$

Among them, the regression coefficient $\beta_0 \beta_1 \ldots \beta_m$ are m + 1 parameters estimated and $\in$ is a random variable (remaining parameter).

Secondly, a SVM-based regression prediction was employed. The basic idea is using a limited training data to establish a continuous functional relationship between input and output, and make the error smaller between the regression predicted value and the output value when the regression function is as smooth as possible [10]. For the training sample $T = \{(x_1, y_1), (x_2, y_2), \ldots, (x_n, y_n)\}$. Assume that the functional relationship between the input x and y is

$$y = WTx + b \quad (2)$$

Where W is the weight coefficient vector and b is the bias term.

Thirdly, Random Forest [11], being a tree-based model, was employed. In this model, each tree is constructed using random samples of selected predictors. The idea is to correlate trees and improve predictions. Generally, the more trees mean the more accuracy, but the experiments show R-square (R^2) seem to be non-change and the inefficient after increasing above 250 trees, so the tree number was set to 250 in this study.

Lastly, XGBoost algorithm was employed to explore this dataset, which is a large-scale parallel algorithm developed on the basis of the Gradient Boosting Decision Tree (GBRT). Compared with the traditional GBRT, the algorithm can perform parallel computing with multi-core CPU; so it gets more than 10 times performance than the other same-type algorithms. In addition, the traditional GBRT only utilizes the first-order derivative of the Taylor expansion, but XGBoost performs the second-order derivative expansion on the target error function, so it gets an improved accuracy than others.

3.4 Model Evaluation and Results

General indicators were employed to evaluate the prediction accuracy, including Root mean squared error (RMSE), R-squared/R^2, mean absolute error (MAE) and Median absolute error (MedAE).

$$RMSE = \sqrt{\frac{\sum_{i=1}^{n}(Y_i - \hat{Y}_i^2)^2}{n}} \tag{3}$$

$$R^2 = 1 - \frac{\sum_{i=1}^{n}(Y_i - \hat{Y}_i^2)^2}{\sum_{i=1}^{n}(Y_i - \bar{Y}_i)^2} \tag{4}$$

$$\mathrm{MAE} = \frac{\sum_{i=1}^{n}|Y_i - \hat{Y}_i|}{n} \tag{5}$$

$$\mathrm{MedAE} = \mathrm{median}(|\mathrm{y}_1 - \hat{\mathrm{y}}_1|, \ldots, |\mathrm{y}_\mathrm{n} - \hat{\mathrm{y}}_\mathrm{n}|) \tag{6}$$

Where Y_i is the actual measured value (energy consumption), $\hat{Y}_i$ is the predicted value, and n is the measured number.

The prediction results with different models was shown as receiver operating characteristic curve (ROC). As can be seen in Fig. 4, the ROC curves of four models show consumption prediction at different extent. Especially with XGBoost model, the dataset was fitted of most reasonable.

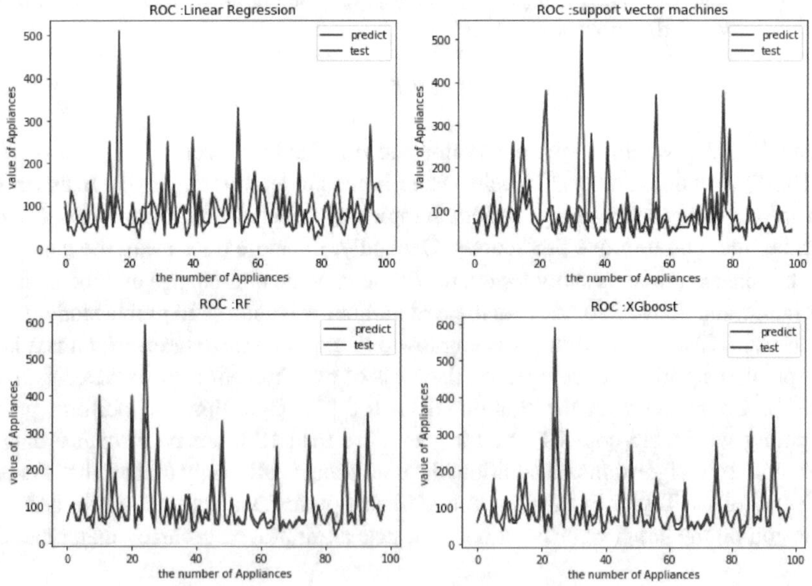

Fig. 4. ROC curves with different models

Table 3. Prediction experiments with different models

Model	RMSE	R^2	MAE	MedAE	Time (s)
LM	82.04	0.184	47.4	28.95	0.642
SVM	86.62	0.12	38.37	12.11	29.44
RF	65.9	0.462	27.16	10	32.02
XGBoost	**59.69**	**0.545**	**26.67**	**9.80**	**7.60**

Furtherly, the indicators, including RMSE, MAE and MedAE, were introduced to evaluate fitting accuracy and computing performance. The results in Table 3 show that the energy consumption data are fitted considerably. Especially, XGBoost has lower RMSE, MAE and MedAE than others, and it also has higher R^2, indicating that XGBoost could be a reasonable model to predict energy consumption.

Table 4. XGBoost prediction with different factors

Factors	RMSE	R^2	MAE	MedAE	Time (s)
Removal of lighting	57.92	0.571	25.66	9.97	7.55
Temperature only	58.83	0.558	26.77	10.1	3.39
Humidity only	62.36	0.503	28.35	10.71	4.40

In order to explore the key factors to predict energy consumption, experiments of XGBoost with differential inputting parameters were done. The results in Table 4 show that removal of lighting parameter would take a better prediction accuracy. Meanwhile, comparing to the humidity parameter, the temperature owns a preferable action to predict energy consumption.

Specifically, it could be known from Table 4 that the R^2 of XGBoost model was increased to 0.571 after removing the light energy consumption parameter, and was reduced to 0.503 only with the parameter of humidity.

4 Conclusion

Energy consumption analysis has caught more and more attention in recent years, and an increasing number of studies have worked on energy consumption analysis, but there are so many affect factors that it's hard to efficiently predict energy consumption. To address these problems, the traditional linear regression model, SVM-based model, Random Forest (RF) and XGBoost algorithm were employed to explore the relationship between factors and consumption. The experimental results show that XGBoost is an efficient method to explore correlation pattern and to make a consumption prediction; removal of lighting factor show a more reasonable result to the prediction accuracy; and factor of temperature shows a more significant for consumption prediction than of humidity.

Our results show a feasible method to fit historical data of energy consumption and to make a prediction. To improve prediction accuracy, further studying should be taken to analyze energy consumption with more factors in the future, and time-oriented activity modeling should be taken to understand latent energy-consumption pattern from humanity behavior.

Acknowledgment. This work was supported in part by Projects of National Science Foundation of China (No.41471333); project NGII20170625 of CERNET Innovation Project; project 2017A13025 of Science and Technology Development Center, Ministry of Education, China; project 2018Y3001 of Fujian Provincial Department of Science and Technology; projects of Fujian Provincial Department of Education (JA14209, JA15325).

References

1. Barbato, A., Capone, A., Rodolfi, M., et al.: Forecasting the usage of household appliances through power meter sensors for demand management in the smart grid. In: 2011 IEEE International Conference on Smart Grid Communications (SmartGridComm), pp. 404–409. IEEE (2011)
2. Firth, S., Lomas, K., Wright, A., et al.: Identifying trends in the use of domestic appliances from household electricity consumption measurements. Energy Build. **40**(5), 926–936 (2008)
3. Fan, L.J.: Energy consumption forecasting and energy saving analysis of urban buildings based on multiple linear regression model. Nat. Sci. J. Xiangtan Univ., 123–126 (2016)
4. Zhou, Y.P., Yu, Z., Li, J., Huang, Y.J., Zhang, G.Q.: Review of measuring methods and prediction models of building occupant behavior. HV & AC, pp. 11–18 (2017)
5. Candanedo, L.M., Feldheim, V., Deramaix, D.: Data driven prediction models of energy use of appliances in a low-energy house. Energy Build. **140**, 81–97 (2017)
6. Li, A.Q., Bai, X.L.: Study on predication of energy consumption in residential buildings. Build. Sci. **8**, 006 (2007)
7. Chao, Z., Siming, L., Qiaoling, X.: College building energy consumption prediction based on GM-RBF neural network. J. Nanjing Univ. Sci. Technol. **38**, 48–53 (2014)
8. Yoon, Y.R., Moon, H.J.: Energy consumption model with energy use factors of tenants in commercial buildings using Gaussian process regression. Energy Build. **168**, 215–224 (2018)
9. UCI: https://archive.ics.uci.edu/ml/datasets/Appliances+energy+prediction. Accessed 31 July 2018
10. Zhou, F., Zhang, L.M., Qin, W.W., Wu, X.G., Lin, J.Y.: Energy consumption forecasting model and abnormal diagnosis of large public buildings based on support vector machine. J. Civ. Eng. Manag. **6**, 014 (2017)
11. James, G., Witten, D., Hastie, T., Tibshirani, R.: An Introduction to Statistical Learning, vol. 112. Springer, New York (2013)

A Strategy of Deploying Constant Number Relay Node for Wireless Sensor Network

Lingping Kong(✉) and Václav Snášel

Faculty of Electrical Engineering and Computer Science,
VSB-Technical University of Ostrava, Ostrava, Czech Republic
konglingping2007@163.com

Abstract. The wide application of wireless sensor network facilitates a great variety of utilizations including remote monitoring, air condition evaluating, tracking and targeting, and so on. However, the performance of a wireless network is constraint by low-power and low-capacity. Hence, long-distance communication is not available for a homogeneous network, which hinders the growth of wireless network. This work proposed a strategy for deploying Relay nodes into the network based on directional shuffle frog leaping algorithm. The Relay node is an advanced node which is more powerful than common node and it can decrease the workload of inner nodes and improve the transmission situation of outer nodes. The experiment simulates other two algorithms as the comparison tests. The performance is good as evidenced by the experiment results of common nodes coverage, connectivity of Relay nodes and the fitness value.

Keywords: Relay node · Network connectivity
Directional shuffle frog leaping algorithm

1 Introduction

A wireless sensor network is a system which consists of many low-cost and low-power sensor nodes. Those sensor nodes are equipped with devices that can sense, calculate and transit packets [1]. Due to the power constraint, long-distance of communication by a sensor is not afforded [2]. Many researchers are dedicated to improve the situation and to prolong the network lifetime. Placing a small number of Relay nodes is one of the important approaches [3–5]. The Relay nodes have more power and cost much than a common sensor node, A Relay node is also called an advanced node due to its more powerful ability than a common node and it can decrease the workload of inner nodes and improve the transmission situation of outer nodes. And the related works about Relay nodes deployment usually are classified into two groups, a constant number or a minimum number [6–8]. There are many ways to accomplish the Relay nodes arrangement, swarm intelligence algorithm is one of the widely used patterns [9]. The author Hashim [10] uses artificial bee colony algorithm to optimize the Relay nodes' positions

P. Krömer et al. (Eds.): ECC 2018, AISC 891, pp. 155–161, 2019.
https://doi.org/10.1007/978-3-030-03766-6_17

for lifetime maximization while it uses the minimum spanning tree protocol to set up the network backbone in the first phase.

The rest of paper is organized as follows: Sect. 2 discusses the related works of optimization algorithms, especially the shuffle frog leaping algorithm variants. Section 3 presents the detail process of applying the directional shuffle frog leaping algorithm for placing Relay nodes. Section 4 gives the results of comparison experiments. Section 5 concludes the paper.

2 Related Works

Shuffle frog leaping algorithm (SFLA) is proposed by Eusuff [11], this algorithm combines the dividing group local search and information exchange global search together for finding the optimization solution. Its ability of adaptation to the dynamic environment makes this algorithm an important memetic algorithm [12,13]. SFLA is a branch of swarm optimization and meta-heuristic algorithm, in order to amplify its search ability, there are many improved version of SFLA have been introduced [14], such as Augmented shuffled frog leaping algorithm (ASFLA) [15], Antipredator adaptation shuffle frog leaping algorithm (AASFLA) [16], Directional shuffle frog leaping algorithm (DFSLA) [17], cognitive behavior shuffle frog leaping algorithm (CB-FSLA) [18], and so on.

ASFLA is proposed by Kaur. The operation of SFLA only optimize the worst individual of a group may result in a local optimum, so the author improves the updating solution by adding the best individual of a group movement scheme, this step could play a role in jumping out of a local best location for more solution space. AASFLA is proposed by Anandamurugan and successfully applies it into wireless sensor network for choosing the cluster heads, this algorithm can avoid the local searching by importing the idea of antipredator capabilities. CB-SFLA is introduced by Zhang, the algorithm enhances the performance of SFL algorithm in solving the optimization problems by adding the cognitive behavior factor. The factor is associated with the comparison results of current error value to the best after one individual moving repeat multivariate space. The author's idea is based on the Thorndikes Law of Effect [19], in which it stated that reinforce a random behavior becomes more probable in the future. The improvements of this algorithm are very clear with the problem scale increasing. DFSLA is proposed for improving the low rate of convergence, this algorithm includes grouping updating and global information exchange two operation modes. In this work, the author states that if one individual moves and gets better in one direction, then this individual may become better in a big probability after continuously moving in that direction. Except for that, the worst individual also uses some similarities of best ones from all of the groups to decide the movement, this is called advantages sharing.

3 Deploying Relay Nodes Process

The simulated wireless sensor network is a 200×100 units square area, and there are 100 sensor nodes randomly and uniquely deployed in the field. The sensing radius for each sensor node is the same, 20 units length. The DFLSA is originally designed for finding global minima of Standard Test function (Rosenbrock, Rastrigin and so on). Here we make a slight modification for putting this algorithm into WSN utilization, as placing a number of 20 Relay nodes constructing a two-layer network. The aim of this work is for decreasing the inner nodes' workload and improving the scalability and robustness of the network, at the same time not to increase the computing complexity of DSFLA. The way to accomplish the goal is to maximize the connecting sensor nodes with the Relay node and to minimize the distance between them. The process detail is summarized in the following steps:

Step 0: Initial setting. Define the population, each individual stands for a solution to the problem. In this case, each individual is an array with a pre-defined number of Relay node locations. Each location of the Relay node is composed of two-dimensional coordinates $\{x, y\}$ (as the network is 2D). The population number is set at n, the group number is m, and importantly $n \times m$ is an integer. So there are $\frac{n}{m}$ individuals in one group. The population $P = \{P_1, P_2, ..., P_n\}$, one individual $P_i = [x, y], i \in [1, ..., n]$, the dividing groups are marked as $G = \{G_1, G_2, ..., G_m\}$.

Step 1: Initialization. Randomly initial the positions of Relay nodes for all the individuals. The position of Relay node is constraint to the network boundary during the whole process.

Step 2: Evaluation and Sorting. For evaluating the good or bad of the population, we design a fitness function (See Eq. 1). Each individual with a higher fitness value is better than one with lower fitness value. Based on the fitness value then sorting the population in a decreased order. The first one individual in this sorting list is the global best, marked as g_{best}.

Step 3: Dividing groups. Partition the population into m sub-population as $G_j, j \in [1, ..., m]$ groups. The first one in sorting list goes to group one, the second one goes to group two, as this way, the m_{th} goes to G_m, do the same dividing process till each individual belongs to a certain group.

Step 4: Updating. There are two modes of evolving procedure, *global shuffle* and *information exchange exploration*. *Information exchange exploration* is for updating the location of one individual, and after several times this operation, the population needs *global shuffle* process, which means that we need to evaluate the population and to rank the population again.

Information exchange exploration mode proceeds by updating the worst individual in one group based on defined equations (see Eqs. 3 and 4).

Global shuffle mode first collects the new-generated individuals and evaluates them using the fitness function. Then sorting the population based on fitness value and compare the first one in new sorting list to the g_{best}, g_{best} always store the best fitness value one. Subsequently, re-group the population as Step 3.

Step 5: Termination condition. The optimization process will stop when it either runs the threshold number of iteration times or the g_{best} individual is satisfied with the needs.

Fitness function: A good designed evaluating function could speed up the evolving of a population. The coverage of Relay nodes and the distance between common nodes and Relay nodes are important factors to the network. So the fitness function F_n will be composed of two elements, *avg* and *domi*. *avg* is the summation of the average distance between Relay nodes to its connected common nodes. *domi* calculates the non-reduplicate number of dominated common nodes by the Relay nodes. In the Eq. 1, n stands for the total number of sensor nodes in the network.

$$F_n = n - domi - avg \tag{1}$$

Information exchange exploration: The *information exchange exploration* mode works by updating the worst individual in each group. Re-set the moving threshold value as the biggest steps that one individual can move in one direction. Suppose in one moment, the current group G_k is under updating process, $G_k = [G_{k,1}, G_{k,2}, ..., G_{k,n/m}]$, $G_{k,1}$ is the first individual of group G_k and also the best one. $G_{k,n/m}$ is the last element of group G_k and also the worst one. $G\{k, 1\}$ and $G\{k, \frac{n}{m}\}$ could be showed as $G_{k,1} = [k|x_1, k|y_1]$ and $G_{k,\frac{n}{m}} = [k|x_{\frac{n}{m}}, k|y_{\frac{n}{m}}]$ respectively. The detail process includes the following steps:

Step 0. Set one moving step counter $MS = 0$. The process starts from defining a 2D direction array, as $f = [f_1, f_2]$, and the value of f is produced by the following Eq. 2.

$$f_i = \begin{cases} -1, & if \quad k|i_{\frac{n}{m}} - k|i_1 \leq 0, \quad i = \{x, y\} \\ 1, & \quad Otherwise \end{cases} \tag{2}$$

Step 1. Compare the MS value to the moving threshold value. If MS is smaller, continue to move $G_{k,n/m}$ along the f direction based on the following Eq. 3, *rand* is random number with $rand \in [0 - 1]$, then mark $MS = MS + 1$. Otherwise, go to Step 3.

$$\begin{aligned} k|x_{\frac{n}{m}} &= k|x_{\frac{n}{m}} + f_i \left| k|x_{[\frac{n}{m}} - k|x_1 \right| \times rand \\ k|y_{\frac{n}{m}} &= k|y_{\frac{n}{m}} + f_i \left| k|y_{[\frac{n}{m}} - k|y_1 \right| \times rand \end{aligned} \tag{3}$$

Step 2. This step is to judge whether $G_{k,\frac{n}{m}}$ gets better or not. If it does, then go back into the previous operation. There are two different ways for this individual if it does not get better. One, the MS value is not one, then it stops updating, jump out its *information exchange exploration*. Two, move $G_{k,\frac{n}{m}}$ based on the following equation.

$$\begin{aligned} k|x_{\frac{n}{m}} &= (\textstyle\sum_{j=1}^{m} k|x_1) \div m \\ k|y_{\frac{n}{m}} &= (\textstyle\sum_{j=1}^{m} k|y_1) \div m \end{aligned} \tag{4}$$

Step 3. If the updating iteration times are over, then stop the process. Otherwise start again from step 1.

4 Experimental Results

The constant number of Relay nodes will be deployed in a 200 × 100 units square network. The network is composed of 100 sensor nodes with 20 units length communication radius. The experiment population number is 32. And the moving step threshold is 2, which means one individual could be updated twice in a certain direction at most. There are three different optimization algorithms that are applied to this network for placing the Relay nodes, they are SFLA, ASFLA, and DSFLA, and each algorithm runs 100 iteration operations. The Performance is good as evidenced by the experiment results of common nodes coverage, connectivity of Relay nodes and the fitness value.

Figure 1 is the simulated network with Relay nodes in three methods. Figure 1(a) is the result of SFLA. Figure 1(b) is the result of ASFLA, the last one is DSFLA showed as Fig. 1(c).

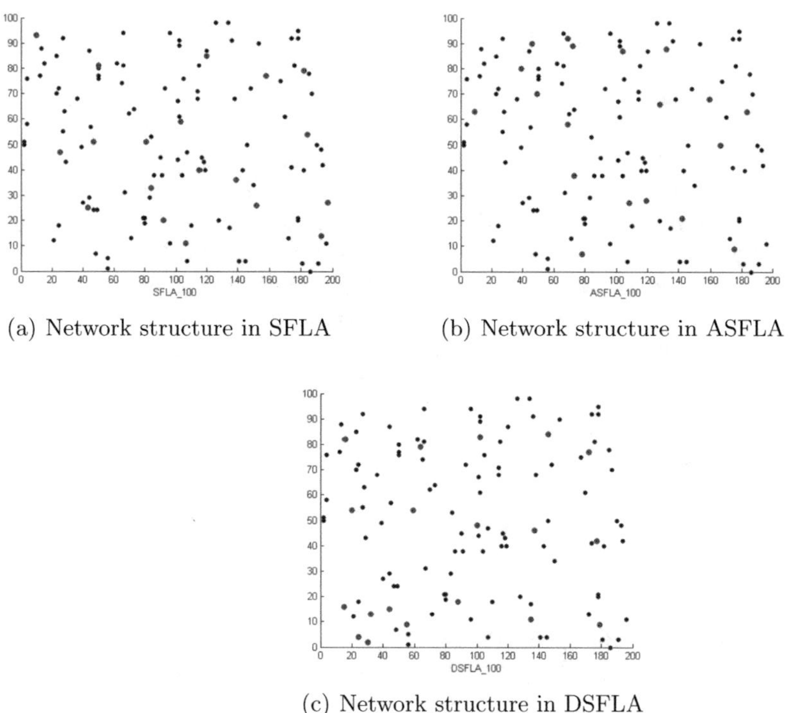

(a) Network structure in SFLA

(b) Network structure in ASFLA

(c) Network structure in DSFLA

Fig. 1. Simulated network structure figure

Table 1 is the final results with fitness values (FV), coverage number (COV) and connective number (CNV) of three simulated algorithms. The coverage number calculates the number of non-repeatable common nodes that are dominated by Relay nodes. The connective number means how many Relay nodes have the connecting ways with others. From the table, it tells there are 95% sensor nodes could be linked with Relay nodes directly in the DFSLA method, the other two methods only covers 84% of sensor nodes. In the same while, there is one Relay node left which is unconnected to any other one in AFSLA and SFLA methods. However, DFSLA method connects all the Relay nodes.

Table 1. Comparison results in three algorithms

Values	Methods		
	ASFLA	SFLA	DSFLA
FV	71.7908	72.6482	85.4519
COV	84	84	95
CNV	19	19	20

5 Conclusion

In this paper, we introduce a way to deploy a constant number of Relay nodes in a wireless sensor network based on directional shuffle frog leaping algorithm. A fitness function is designed for this algorithm which accelerates the convergence speed and guides simulations towards optimal solutions. In the meanwhile, there is no complicated unceasing checking validity and correcting steps adding to the algorithm. The experimental results show that our scheme has a good performance in fitness value, coverage common node number, and the Relay nodes placed in the network connected to each other.

References

1. Dapeng, W., Jing, H., Honggang, W., Chonggang, W., Ruyan, W.: A hierarchical packet forwarding mechanism for energy harvesting wireless sensor networks. IEEE Commun. Mag. **53**(8), 92–98 (2015)
2. Lloyd, E.L., Guoliang, X.: Relay node placement in wireless sensor networks. IEEE Trans. Comput. **56**(1), 134–138 (2007)
3. Yung, F.H., Chung-Hsin, H.: Energy efficiency of dynamically distributed clustering routing for naturally scattering wireless sensor networks. J. Netw. Intell. **3**(1), 50–57 (2018)
4. Younis, M., Kemal, A.: Strategies and techniques for node placement in wireless sensor networks: a survey. Ad Hoc Netw. **6**(4), 621–655 (2008)
5. Liquan, Z., Nan, C.: An effective clustering routing protocol for heterogeneous wireless sensor networks. J. Inf. Hiding Multimed. Signal Process. **8**(3), 723–733 (2017)
6. Senel, F., Mohamed, F.Y., Kemal, A.: Bio-inspired relay node placement heuristics for repairing damaged wireless sensor networks. IEEE Trans. Veh. Technol. **60**(4), 1835–1848 (2011)

7. Gwo-Jiun, H., Tun-Yu, C., Hsin-Te, W.: The adaptive node-selection mechanism scheme in solar-powered wireless sensor networks. J. Netw. Intell. **3**(1), 58–73 (2018)
8. Dejun, Y., Satyajayant, M., Xi, F., Guoliang, X., Junshan, Z.: Two-tiered constrained relay node placement in wireless sensor networks: computational complexity and efficient approximations. IEEE Trans. Mob. Comput. **11**(8), 1399–1411 (2012)
9. Chin-Shiuh, S., Van-Oanh, S., Tsair-Fwu, L., Quang-Duy, L., Yuh-Chung, L., Trong-The, N.: Node localization in WSN using heuristic optimization approaches. J. Netw. Intell. **2**(3), 275–286 (2017)
10. Hashim, A., Babajide, O.A., Mohamed, A.A.: Optimal placement of relay nodes in wireless sensor network using artificial bee colony algorithm. J. Netw. Comput. Appl. **64**, 239–248 (2016)
11. Eusuff, M., Kevin, L., Fayzul, P.: Shuffled frog-leaping algorithm: a memetic metaheuristic for discrete optimization. Eng. Optim. **38**(2), 129–154 (2006)
12. Rahimi-Vahed, A., Ali, H.M.: Solving a bi-criteria permutation flow-shop problem using shuffled frog-leaping algorithm. Soft Comput. **12**(5), 435–452 (2008)
13. Fang, C., Ling, W.: An effective shuffled frog-leaping algorithm for resource-constrained project scheduling problem. Comput. Oper. Res. **39**(5), 890–901 (2012)
14. Elbeltagi, E., Tarek, H., Donald, G.: A modified shuffled frog-leaping optimization algorithm: applications to project management. Struct. Infrastruct. Eng. **3**(1), 53–60 (2007)
15. Kaur, P., Shikha, M.: Resource provisioning and work flow scheduling in clouds using augmented Shuffled Frog Leaping Algorithm. J. Parallel Distrib. Comput. **101**, 41–50 (2017)
16. Anandamurugan, S., Abirami, T.: Antipredator adaptation shuffled frog leap algorithm to improve network life time in wireless sensor network. Wirel. Pers. Commun. **94**, 1–12 (2017)
17. Lingping, K., Jeng-Shyang, P., Shu-Chuan, C., John, F.R.: Directional shuffled frog leaping algorithm. In: International Conference on Smart Vehicular Technology, Transportation, Communication and Applications, pp. 257–264. Springer (2017)
18. Xuncai, Z., Xuemei, H., Guangzhao, C., Yanfeng, W., Ying, N.: An improved shuffled frog leaping algorithm with cognitive behavior. In: Intelligent Control and Automation, pp. 6197–6202. IEEE (2008)
19. Catania, A. C.: Thorndike's legacy: learning, selection, and the law of effect. J. Exp. Anal. Behav. **72**(3), 425–428 (1999)

Noise-Robust Speech Recognition Based on LPMCC Feature and RBF Neural Network

Hou Xuemei[1(✉)] and Li Xiaolir[2]

[1] School of Information Engineering, Chang'an University, Xi'an 710054, People's Republic of China
houxuemei@xupt.edu.cn
[2] College of Automation, Xi'an University of Posts and Telecommunications, Xi'an 710121, China

Abstract. To solve the problem that recognition rates of speech recognition systems decrease in the noisy environment presently, the Linear Predictive Mel cepstrum coefficient (LPMCC) is used as feature parameter and uses character possessing LPMCC and RBF neural network which have optimal approach capability and the fast training speed, adopts clustering algorithm and entire-supervised algorithm and realizes a noise-robust speech recognition system based on RBF neural net-work. The hidden layer training of clustering algorithm used K-means clustering algorithm and output layer learning used linear least mean square. The adjustment of the entire parameters of entire-supervised algorithm is based on grads decline method. It is a kind of supervised learning algorithm and can choose excellent parameters. Experiments show that entire-supervised algorithm have higher recognition rates in different SNRs than clustering algorithm.

Keywords: Speech recognition · RBF neural network · LPCMCC
Clustering algorithm · Entire-supervised algorithm

1 Introduction

To obtain a close result in a noise environment and a net sound environment is one of practical Speech Recognition problems. Speech Recognition in the realization process usually involves a number of factors needing to consider. Because of the randomness of the speech signal, as well as the human auditory very shallow understanding of the mechanism, the current noise environment in the speech recognition system can not meet all practicality, and Speech Recognition practical research has been the focus of the industry.

In this paper, we make a combination of the Mel frequency which is conformed to the human auditory characteristics and LP cepstrum coefficients, forming a LP Mel cepstrum (Linear Predictive Mel Cepstral Coefficients, LPMCC). Then we use LP Mel cepstrum as the speech feature parameter, RBF neural network model as recognition network, and respectively use clustering algorithm and the entire-supervised algorithm, getting a recognition rate in different SNR and vocabulary, in visual C++ platform using two algorithms realizing isolation word speech recognition system which is based

P. Krömer et al. (Eds.): ECC 2018, AISC 891, pp. 162–168, 2019.
https://doi.org/10.1007/978-3-030-03766-6_18

on the RBF neural network. The experimental results show that this method has strong anti-noise performance and identify effective.

LP cepstrum coefficient (Linear Predictive Cepstrum Coefficients, LPCC) is the most commonly used feature parameter. LPCC is a cepstrum coefficient based on actual frequency scale, but frequency of voice that human hear and actual frequency are not a linear proportional. According to the experiment results that the feature parameters based on extracted from human auditory model have better robustness than other parameters [1]. Mel frequency band division is an engineering simulation of human auditory characteristic. Besides the perception on high and low vowels, there is loudness perception in human auditory perception. The perception of loudness is related to the speech frequency width. Mel frequency scale nonlinearly maps (warping) voice frequency to a new frequency scale, which can richly reflect the nonlinear perception characteristic of frequency and amplitude to human auditory, frequency analysis and spectrum synthesis characteristic human shows when hearing complex voice. According to the experimental results of the perceptions of human hearings to frequency and amplitude, if we extract speech feature in this scale, the feature make more correspondence with human auditory characteristic [2, 3]. So normal LPC is made nonlinear changes further by means of Mel scale according to auditory characteristic, and LPMCC is obtained. LPMCC Algorithm, which considers channel excitation and human hearing, has an efficiency and feasibility.

2 RBF Neural Network Training Algorithm

2.1 Clustering Algorithm

I. Hidden layer training

We use unsupervised training to complete the study of the hidden layer and adopt k-means clustering algorithm. That is, concentrating the square of the distance from each sample point to the center of the cluster, summing them, and minimize. The algorithm is as follows:

(1) Initialize $C_j, j = 1, 2, \cdots N$ the center of the cluster; generally we set C_j as the first sample of input, and then we set ε the stopping threshold.
(2) Cycle beginning.
(3) Make all samples be on the principle of minimum distance clustering. Videlicet, That the principle according to $C_{j^*} = \min\left\|x_i - c_j\right\|$, returning x_i to Θ_j, Θ is a assembly of cluster, Θ_j is No. J cluster.
(4) Calculate the sample-averaging of the cluster center

$$C_j = \frac{1}{M_j}\sum_{x_i \in \theta_j} x_i \ \ (i = 1, 2 \cdots, K) \tag{1}$$

M_j is the number of sample collection.
(5) Calculate average distortion and relative distortion.

Average distortion:

$$D^{(n)} = \frac{1}{m}\sum_{r=1}^{m} \min d(X_r, C_j) \tag{2}$$

Xr is training sequences, $r = 1,2,\ldots,m$
Relative distortion:

$$\widetilde{D}^{(n)} = \left|\frac{D^{(n-1)} - D^{(n)}}{D^{(n)}}\right| \tag{3}$$

(6) The end of judgment

While $\widetilde{D}^{(n)} \leq \varepsilon$, end the cycle. Conversely, return to (2).

After the sample clustering, we can calculate the normalized parameter of the gauss kernel, gauss radius is σ_j^2 [4]. The parameter is the measure of the scope which is the input data of each node.

$$\sigma_j^2 = \frac{1}{M_j}\sum_{x_i \in \theta_j} (x_i - C_{1,j})^T (x_i - C_{1,j}) \tag{4}$$

II. The output layer training Hidden layer training

The study of output layer is tutor type. And it use linear least square method (Least Mean Square, abbr. as LMS). This method doesn't need iterative calculation. Its convergence speed is very fast. The purpose of LMS is that making the expected output of network and mean square error of the actual output to be a minimum [5]. Videlicet, meet $\|Y - W\Phi\|^2$ to be a minimum, thus we can find the estimator w_{ij} of $\hat{w}_{ij}$. Here, Y is the output vector, W is the weight matrix from the hidden layer to the output layer, Φ is the output vector of hidden layer. According to differential method, we can this formula:

$$W = (\Phi^T \Phi)^{-1} \Phi^T Y \tag{5}$$

Thus, the value of mean-square deviation can be a minimum. Generally, to prevent abnormal status of the matrix Φ, we often express W as this:

$$W = (\Phi^T \Phi + \eta\|\Phi^T \Phi\|)^{-1} \Phi^T Y \tag{6}$$

Here, we usually suppose η to be a positive number reaching to 0. Then we can get estimated value of the parameter w_{ij} [6].

2.2 Entire-Supervised Training Algorithm Clustering Algorithm

The basic thinking of entire-supervised algorithm: In the network all the parameters adjustment is a supervised learning process, in order to reach the purpose that making the performance index to be a minimum.

The performance index of RBF neural network:

$$E_i = \frac{1}{2}(y_i - \overset{\wedge}{y_i})^2 \qquad i \leq 1, 2, \cdots, N \tag{7}$$

$\overset{\wedge}{y_i}$ is corresponding to the expected output value of the No, i input vector, y_i is actual output value of the No and i input vector. N is sample number. If we make a combination of the seeking parameter (the centre of RBF network is $C = [c_1, c_2, \cdots, c_h]_{p \times h}$, Width is $\sigma = [\sigma_1, \sigma_2, \cdots \sigma_h]_{h \times 1}$ and the Link weight vector is $W = [w_{11}, \cdots, w_{ij}, \cdots, w_{ho}]_{h \times o}$, forming a ensemble is $Z = \{W, C, \sigma\}$, and using the performance index as a optimal objective function

$$\min_Z E_i = \frac{1}{2}(y_i - \overset{\wedge}{y_i})^2 \tag{8}$$

To adjust parameters, the learning process of RBF network can be seen as a process that seeking the non-blinding minimum of multi-variable function [7]. That is, learning of the entire network is a supervised learning process. Especially, learning of the centre is also a supervised learning process. Thereby, it prevents the problem in which through the conventional algorithm unsupervised learning caused hidden layer node centre sensitivity to initial value [8].

In this paper, we use the error-correction algorithm which is based on the gradient descent. Specific algorithm steps are as follows:

I. Initialization. Arbitrarily set the value of w_i, C_i and σ_i. Preset permissible error and learning rate η_1, η_2, η_3,

II. Recycle until reaching the permissible error and appointed repetitions.

(1) Calculate $e_j, j = 1, 2, \cdots N$

$$e_j = d_j - f(X_j) = d_j - \sum_{i=1}^{M} w_i \bullet G(X_j, C_i) \tag{9}$$

(2) Calculate the changes of the output unit weight

$$\frac{\partial E(n)}{\partial w_i(n)} = -\frac{1}{N}\sum_{j=1}^{N} e_j \exp\left(\frac{-\|X_j - C_i\|^2}{2\sigma_i^2}\right) \tag{10}$$

Change the weights:

$$w_i(n+1) = w_i(n) - \eta_1 \frac{\partial E(n)}{\partial w_i(n)} \tag{11}$$

(3) Calculate the changes of the hidden unit centre

$$\frac{\partial E(n)}{\partial T_i(n)} = -\frac{w_i}{N\sigma_i^2}\sum_{j=1}^{N} e_j \exp\left(\frac{-\|X_j - C_i\|^2}{2\sigma_i^2}\right) \bullet (X_j - C_i) \tag{12}$$

Change the centre:

$$C_i(n+1) = C_i(n) - \eta_2 \frac{\partial E(n)}{\partial C_i(n)} \tag{13}$$

(4) Calculate the changes of the function width

$$\frac{\partial E(n)}{\partial \sigma_i(n)} = -\frac{w_i}{N\sigma_i^3}\sum_{j=1}^{N} e_j \exp\left(\frac{-\|X_j - C_i\|^2}{2\sigma_i^2}\right) \bullet \left(\|X_j - C_i\|^2\right) \tag{14}$$

Change the width:

$$\sigma_i(n+1) = \sigma_i(n) - \eta_3 \frac{\partial E(n)}{\partial \sigma_i(n)} \tag{15}$$

(5) Calculate the error

$$E = \frac{1}{2N}\sum_{j=1}^{N} e_j^2 \tag{16}$$

3 Experimental Results

3.1 Speech Data

In this experiment, we directly make the speech data file which is obtained by systematic sampling as processing object and the experimental speech sample as isolated word. And set speech signal sample rate equal to 11.025 kHz, set frame length N equal to 256. In this experiment we used 10 word, 20 word, 30 word, 40 word and 50 word. They are 9 person's pronunciation under different SNR (Clean dB, 15 dB, 20 dB, 25 dB, 30 dB). Every person and every word pronounces 3 times. And we use it as training database. And we use another 7 persons' pronunciation to take speech recognition, in order to get the results of RBF neural network speech recognition under different SNR and vocabulary.

3.2 Network Training

I. Experiment 1: clustering training algorithm

In 10-words noiseless environment, 270 * 1024 feature vectors which are used for training generate a code book. And its clustering dimension is 1024, clustering size is 10. According to the nearest neighbor rule, let all the training characteristics into 10 clusters. Calculate each clustering centre and relative distortion. If distortion measure is less than the pre-set threshold (in the experiment,), the income clustering centre is the hidden node function centre. According to formula (7), we can get the function radius j. And according to the known output layer information (the words classification number), we can calculate the connection weight which is from the hidden layer to the output layer by using LMS.

II. Experiment 2: entire-supervised training algorithm

In 10-words noiseless environment, we use 10-words noiseless speech training network. Each characteristic training file is corresponding to a words- classification number. We use gradient descent algorithm. And according to the words-classification number, we constantly modify the network weights until the pre-set error precision is met. In this experiment, set network learning rate equal to 0.001. Set error precision equal to 10–5. Set the biggest learning times equal to 1000.

3.3 Network Recognition

After establishing RBF neural network model, we can make an identified test to the words enter network of the test suite, respectively. If we enter a 1024-dimensional feature vector word and calculate it through the hidden layer and output layer, we can get the classification number of each word. Then compare this classification number and it of the input feature vector. If they are equal, the recognition is right. Contrarily, it is wrong. Lastly, we can get the final recognition rate after calculating the ratio of the right recognition number and the number of awaiting recognition word.

3.4 Results and Conclusions

Table 1 is the experimental results of the above two training method in different SNR and vocabulary.

Firstly, from Table 1 we can see that RBF neural network that is used in speech recognition has got a better recognition rate. And with the increase in vocabulary recognition rate will rise. The reason is that with the increase in vocabulary the number of training hidden node will increase, and then network training will be more fully, and the robustness of the system will be enhanced. So the recognition rate rises. Secondly, comparing the training results of this two training methods, we can see that recognition rate of the entire-supervised training algorithm is obviously higher than it of the conventional clustering algorithm. It fully demonstrates that the entire-supervised training algorithm plays a more important part in the performance of RBF network, and

Table 1. The recognition rates of clustering and entire-supervised training method (%)

Words amount	Training methods	SNR(dB)				
		15 dB	20 dB	25 dB	30 dB	Clean dB
10	Clustering training	84.62	85.21	85.36	85.53	86.22
	Entire-supervised training	86.46	87.23	89.35	89.26	91.38
20	Clustering training	85.17	85.89	86.79	87.24	88.57
	Entire-supervised training	88.74	89.35	90.14	91.36	92.54
30	Clustering training	86.82	88.23	89.16	89.89	91.18
	Entire-supervised training	89.36	90.26	91.67	91.95	93..05
40	Clustering training	88.37	89.37	89.96	90.26	92.13
	Entire-supervised training	90.88	91.12	92.78	92.78	93.75
50	Clustering training	90.56	92.37	92.54	93.11	94.12
	Entire-supervised training	91.26	92.87	92.33	93.56	94.21

it makes RBF network possesses stronger classification ability. It also has the disadvantages which makes the training speed slower. That could be improved in the future study.

References

1. Hou, X., Zhang, X.: A speech recognition method of isolated words based on modified LP cepstrum. J. Taiyuan Univ. Technol. **506–510**, 37 (2006)
2. Yan, T., Yun, X., Jin, F., Zhu, Q.: RBF neural networks and their application to output–based objective speech quality assessment. Acta Electronica Sinica **1282–1285**, 32 (2004)
3. Guo, J.J., Luh, P.B.: Selecting input factors for clusters of Gaussian radial basis function network to improve market clearing price prediction. IEEE Trans. Power Syst. **665–672**, 18 (2003)
4. Jingjiao, L., Jie, S., Li, Z., Tianshun, Y.: Hybrid model of hidden markov models network model in speech recognition. J. Northeast. Univ. (Nat. Sci.), **144–147**, 120 (2006). http://www.springer.com/lncs. Accessed 21 Nov 2016
5. Shi, X., Gu, M., Wang, T., He, Z.: Sequential cluster method and its application on neural network based speech recognition. J. Circuits Syst. **99–103**, 5 (2000). http://www.springer.com/lncs. Accessed 21 Nov 2016
6. Hoshimi, M., Niyada, K.: Method and apparatus for speech recognition. J. Acoust. Soc. Am. **109**(3), 864 (2018)
7. Parthasarathy, S., Rose, R.C.: System and method for mobile automatic speech recognition. J. Acoust. Soc. Am. **126**(6), 3373 (2018)
8. López-Espejo, I., Peinado, A.M., Gomez, A.M., et al.: Dual-channel spectral weighting for robust speech recognition in mobile devices. Digit. Signal Process. **75**, 13 (2018)

Research on Web Service Selection Based on User Preference

Maoying Wu and Qin Lu(✉)

Qilu University of Technology, Jinan, China
786622186@qq.com

Abstract. At the present stage, weight is often used to express the user preference to QoS (Quality of Service). Due to the user's subjective judgment and the fuzziness of preference description, weight calculated through the traditional weighting method is difficult to express the user preference correctly. To solve the fuzziness of QoS attribute preference description and improve the correctness of service selection, the improved order relation analysis method (G1-method) by fuzzy number is adopted to represent the subjective weight of the user firstly; and the entropy weight method is adopted to determine the objective weight of the QoS attribute; finally, the objective weight is used to revise the subjective weight to calculate the comprehensive weight. Based on the user preference, the service is selected by improving the TOPSIS method with COSINE similarity. According to the experiment, the uncertainty of user preference description is effectively solved, the accuracy of service selection is improved through the improved TOPSIS method, and the selected service is more in line with the user requirement.

Keywords: Service selection · User preference · QoS attributes
TOPSIS method

1 Introduction

With a large same or similar Web services deployed on the Internet, QoS becomes the key to differentiate these Web services. However, different users are often have different need for QoS attributes. Some users think that price is more important, while others think that response time is more important. So the user preference needs to be considered during service selection. Weight is a scale reflecting the significance level of a criterion. The higher the weight of a criterion, the higher the significance level of the criterion, and the more it affects the outcome of decision-making [1]. However, in real life, it is easier for people to give "likes", "dislikes" and "prices are more important than response time" for QoS attributes. It is difficult to accurately give the weight of each attribute and to express a preference with a quantitative value. To satisfy the user demands under each situation, service that satisfy the preference demand of most users shall be selected; therefore, the QoS attribute weighting is studied in this paper.

In existing research, the method of determining weights is divided into two types: 1. subjective weighting, 2. objective weighting. The subjective weighting method weights each attribute according to the user's preference. The common methods of

P. Krömer et al. (Eds.): ECC 2018, AISC 891, pp. 169–179, 2019.
https://doi.org/10.1007/978-3-030-03766-6_19

subjective weighting include AHP (analytic hierarchy process) [2], expert investigation method (Delphi method) [3] and G1 method [4]. The objective weighting method determines weight by comparing the information content and changes of different QoS attribute values. The common methods of objective weighting include Gini coefficient method [5] and entropy weight method [6].

Currently, the service selection based on service preference has some research results. The cloud model is proposed by Professor Li Deyi, academician of Chinese Academy of Engineering. It is an uncertain transformation model to deal with qualitative concepts and quantitative descriptions. In [7], the author uses the cloud model to determine the user's subjective information. Cloud model can well express the uncertain description for the user preference by transforming the qualitative into the quantitative. However, it is too subjective and completely from the perspective of user, the correlation existing between the QoS attributes is ignored. In [8], it expresses the subjective uncertainty of user through the numerical features of the cloud model, the comprehensive weight is calculated by combining with the objective weight. However, the method of obtaining subjective weight in this paper is similar to the AHP method. It is necessary to compare the QoS significances pairwise, hence the consistency can't be guaranteed. In [9], the entropy weight method is adopted to calculate the objective weight of each QoS attribute while the subjective preference of user is considered less frequently. In [10, 11], the uncertainty is solved through the fuzzy logic, the uncertain demands of user are determined through the fuzzy AHP method. The AHP method is widely used due to its advantages of simplicity, practicality, and structure. The AHP method can transform qualitative problems into quantitative problems. The user's weighting the attribute is actually the process of transforming the qualitative into quantitative. However, the AHP method requires the consistency of the judgment matrix. In practice, as the order increases, it is often difficult to guarantee the consistency of the matrix. Although the fuzzy analytic hierarchy process solves the uncertainty problem caused by subjective factors, it still needs to check the consistency of the judgment matrix.

Based on the above analysis, a weighting method integrating the triangular fuzzy G1 method and entropy weight method is proposed to determine the weight of QoS attribute for the user request; the accuracy of TOPSIS method is improved with the COSINE similarity. The G1 method determines the attribute weight by the user's ordering the significance level of attribute, and require no consistency check. For the user, it is always much easier to provide the attribute sequencing than to give the attribute weight. When the values of the same attribute between Web services have little difference, although the attribute is very important, it is actually not comparable and should give smaller weight. When the values of the same attribute between different Web services have larger difference, it should give higher weight. The entropy weight method is used to reflect the attribute information amount by calculating the entropy value of each attribute to determine the weight of the attribute according to the amount of information. It is obviously reasonable to solve the objective weight through the entropy weight method. Finally, the comprehensive weight value after combined weights can express users' preference for attributes accurately.

2 Determination of Weight

2.1 G1 Method and Steps

The G1 method, a subjective weighting method, is proposed by Liu Yajun. It is an improvement of the analytic hierarchy process, avoiding the shortcomings of the traditional analytic hierarchy process, and making it simpler to use. The principle of G1 method is to order the indexes according to the user preference, judge the significance level of the two adjacent indexes and calculate the weight of each index according to the significance level. The advantage of this method is that it requires no consistency check. For the Web Service with several QoS attributes, the AHP method will cause the inconsistency due to the excess judgment orders. For the user, it will be much easier to order the significance level of attributes. Therefore, it is appropriate to obtain the subjective weight of the user through the G1 method.

The Specific Steps of the G1 Method are as Follows

(1) If there are n QoS attributes $\{C_1, C_2, \ldots, C_n\}$. The user sorts the attributes according to their preference for the attributes. If the attribute Ci is more important than the attribute $C_{j.}$, we denoted it as $C_i > C_j$.
(2) The user selects a favorite attribute from the n attribute as C_1.
(3) The user selects the most important attribute from the remaining n-1 attributes as C_2.
(4) In turn, a preference sequence $C_1 > C_2 > \ldots\ldots\ldots C_n$ is formed.
(5) Referring to Table 1, the ratio of importance of adjacent attributes in preference sequences is given, denoted as r_k. r_k represents the ratio of importance of C_{k-1} to C_k, k = 2, 3, 4,…, n.$\sum_{k=2}^{n}\prod_{i=k}^{n} r_i$

Table 1. The ratio of the importance of the two elements

r_k	Relative importance
1.0	Equally important
1.2	A little important
1.4	Obviously important
1.6	Strong important
1.8	Extremely important

The weights are calculated by the following Eqs. (1) and (2)

$$W_n = 1/\left(1 + \sum_{k=2}^{n}\prod_{i=k}^{n} r_i\right) \tag{1}$$

$$W_{n-1} = r_k W_n \tag{2}$$

W_n represents the weight of the nth attribute in the preference sequence, and r_k represents the ratio of importance degree.

2.2 Fuzzy Number Improvement G1 Method

In real life, due to the diversity of QoS attributes and the ambiguity of people's perception of QoS attributes, it is often difficult to give accurate preference degree after comparing two attributes. Therefore, people often use fuzzy numbers to represent uncertain information. The interval number is common fuzzy number. But sometimes the interval may be too large, and it is easy to cause errors after interval operation. The triangular fuzzy number [12] can not only keep the variable as interval, but also give the intermediate value with the highest probability of value. The triangular fuzzy number can solve the problem that the object can not provide accurate measurement, but used to use natural language to evaluate them. So this article uses triangular fuzzy numbers to represent the importance evaluation given by the user.

Definition 2.1: If $a = [\bar{a}, a, \underline{a}]$, $0 < \bar{a} < a < \underline{a}$, It is called a triangular fuzzy number. $a = [\bar{a}, a, \underline{a}]$ and $b = [\bar{b}, b, \underline{b}]$ are two arbitrary triangular fuzzy numbers, m is an arbitrary positive real number. The Operation rule of trigonometric fuzzy numbers is as follows:

$$a + b = [\bar{a}, a, \underline{a}] + [\bar{b}, b, \underline{b}] = [\bar{a} + \bar{b},\ a + b, \underline{a} + \underline{b}] \tag{3}$$

$$a \times b = [\bar{a}, a, \underline{a}] \times [\bar{b}, b, \underline{b}] = [\bar{a} \times \bar{b},\ a \times b, \underline{a} \times \underline{b}] \tag{4}$$

$$m \times a = [\mathrm{m}\bar{a}, ma,\ m\underline{a}] \tag{5}$$

$$a^{-1} = [1/\bar{a}, 1/a, 1/\underline{a}] \tag{6}$$

This paper refers to Table 1 to convert the comparison language set into a triangular fuzzy number, as shown in Table 2. We denoted it as $R_k = [\overline{R_k}, R_k, \underline{R_k}]$, Among $\overline{R_k}$, R_k, $\underline{R_k}$ respectively indicates the most conservative result, the most likely result, and the most optimistic result. The user gives a preference sequence of attributes $C_1 > C_2 > \ldots\ldots C_n$. Then, referring to Table 2, the ratio R_k of importance degrees of adjacent attributes is given, then the weight α_j of each attribute is obtained according to Eqs. (1) and (2).

Table 2. Comparison level and corresponding triangular fuzzy number

Relative importance	R_k
Equally important	(1.0, 1.0, 1.1)
A little important	(1.1, 1.2, 1.3)
Obviously important	(1.3, 1.4, 1.5)
Strong important	(1.5, 1.6, 1.7)
Extremely important	(1.7, 1.8, 1.8)

Definition 2.2: R is a real number set, $a > 0$, $b \in R$, $i \in [-1, 1]$. Then a + bi is the contact number, a is a certain number, b is an uncertain number, and i is an uncertainty variable. The contact number is a mathematical tool provided in the set pair analysis; the set pair analysis theory can be used to connect the certainty quantity and uncertainty

quantity. The triangular fuzzy number is the combination of certainty and uncertainty. The median value is certain value, while the upper and lower two values are uncertain values. So reference to [13], we can convert between the triangular fuzzy number and the contact number, and defuzzificate triangular fuzzy number by the contact number. If there is a triangular fuzzy number a = $[\bar{a}, a, \underline{a}]$, it is converted to the contact number $1 - a + (\underline{a} - \bar{a})$. The range of i is: $\left[\frac{\bar{a}-a}{\underline{a}-\bar{a}}, \frac{a-a}{\underline{a}-\bar{a}}\right]$. In the decision model of contact number, The value of $i = \frac{a}{a+\bar{a}-\underline{a}}$.

2.3 Objective QoS Weight

In the above section, although the improved G1 method can greatly improve the randomness of users' subjective weighting. The independent subjective weighting fails to reflect the relation between the QoS attribute values, so in order to make the weights more scientific, we correct the subjective weights through objective weighting. The concept of entropy comes from thermodynamics. Later, the introduction of information theory, it is widely used in various fields. Entropy method is an objective weighting method. The principle is to calculate the information entropy of the index. The smaller the attribute entropy, the larger the difference in the attribute values of the different candidate services with the same attribute, we weight the attribute with higher weight. The larger the entropy. The smaller the difference in the attribute values of the different candidate services with the same attribute, we weight the attribute with lower weight. Specific steps are as follows:

The first step: We process the QoS attribute values according to the following Equation. The QoS attributes include two types, one is cost type and one type is benefit type. The smaller the value of the cost type, the more it is preferred by the requester, such as response time, price. Benefit-type attributes, the higher the value, the more favored by the requester, such as reliability, stability. We normalize the attributes into two types, the Eqs. (7) and (8) is as follows:

$$Q_j = \frac{q_{\max} - q_{ij}}{q_{\max} - q_{\min}} \quad q_{\max} \neq q_{\min} \tag{7}$$

$$Q_j = \frac{q_{ij} - q_{\min}}{q_{\max} - q_{\min}} \quad q_{\max} \neq q_{\min} \tag{8}$$

The second step: we give the information entropy solution Eq. (9)

$$S_j = M\sum_{i=1}^{n} P_{ij} \cdot \ln p_{ij} \tag{9}$$

M is constant value, $M = (-\ln n)^{-1}$, S_j is the information entropy value of the jth attribute, and n is the number of candidate services. P_{ij} is the proportion of the i service under the jth QoS attribute. $P_{ij} = C_{ij} / (\sum_{i=1}^{n} C_{ij}))$, C_{ij} is the normalization matrix.

The third step: calculating the weight of each attribute according to the entropy value, the Eq. (10) is follow as

$$\beta_j = (1 - S_j) / \sum_{j=1}^{m} 1 - S_j \tag{10}$$

m is the number of attributes and S_j is the entropy of the jth attribute.

2.4 Comprehensive QoS Weight Calculation

Although subjective weighting method can well express users' preference for attributes, but ignores the inherent association of QoS values. The objective weighting method considers the relationship between values but ignores the subjective preference of users. Therefore, in order to make the weight take into account the correlation of QoS attributes, it can also consider the user's subjective preference. The combined weighting method is adopted in this paper to take the subjective weighting and objective weighting into comprehensive consideration. The Eq. (11) of combination weight is as follows:

$$W_j - \lambda\alpha_j + (1 - \lambda)\beta_j \quad \sum_{1}^{j} W_j = 1 \;\; j = 1, 2, 3, \cdots, n. \tag{11}$$

λ is Confidence degree, λ = [0,1]. The proportion of subjective weight in synthetic weight is adjusted by λ. The Confidence degree indicate user's confidence to the subjective weight he specifies and the attention to the weight value. The greater the user's confidence, the higher the value of confidence degree. When λ equals 1, the weight value is completely determined by the user. On the contrary, when λ equals 0, the weight value is completely determined by object weight. In general, λ = 0.5.

3 The Improved TOPSIS to Select Services

TOPSIS (Technique for Order Preference by Similarity to an Ideal Solution) method is common in the multi-objective decision [13], this method is widely used in service selection and multi-objective decision. The principle of TOPSIS method is to find the best and worst values for each attribute in the n alternative objects to constitute the positive ideal solution and negative ideal solution. Each attribute of the alterative service is taken as each point in the space to calculate the Euclidean distance between each point and the positive ideal solution and the negative solution for sequencing. The traditional TOPSIS method has the following problems: 1. Because the index weight fails to express the user demand completely, the TOPSIS result will be influenced. 2. The Euclidean distance method is used to calculate the distance from the QoS attribute value of alterative service to the ideal solution. When the Euclidean distances of the two different evaluations to the best and worst solutions are the same, there always be errors because of the comparison cannot be made according to the close degree. 3. The traditional TOPSIS method has the problem of inverted order; when the attribute is increased or reduced, there will be errors.

The first one, in this paper, the triangular fuzzy G1 method is adopted to calculate the subjective weight, the objective weight is calculated through the entropy weight method. The subjective weights are synthesized to calculate the comprehensive weight that can better express the user demand. For the second problem, [14] adopts the Mahalanobis distance to improve the TOPSIS method; however, the application of Mahalanobis distance requires the number of service attribute to be smaller than the number of candidate service, otherwise, there will be covariance matrix without inverse matrix. Obviously, it won't be established all the time, so that it will be strongly constrained in practical use. In this paper, the COSINE similarity method improvement TOPSIS is adopted to calculate the distance from the attribute value to the ideal solution. COSINE similarity is a method to calculate the similarity between the two high dimensional vectors. The method is first to map the attribute data into the vector space and then measure the similarity between the two vectors by measuring the space angle cosine. The included angle between the two individual vectors is between 0° and 180°, the larger the included angle, the lower the similarity. Set the two vectors A = [a1,a2,…,an] and B = [b1,b2,..,bn]. Then the similarity between vectors A and B is Eq. (12):

$$sim = \frac{A \cdot B}{\|A\| \cdot \|B\|} = \frac{\sum_{i=1}^{n} \left(A_i - \overline{A}\right) \times \left(B_i - \overline{B}\right)}{\sqrt{\sum_{i=1}^{n} \left(A_i - \overline{A}\right)^2} \times \sqrt{\sum_{i=1}^{n} \left(B_i - \overline{B}\right)^2}} \tag{12}$$

Among them, $\overline{A}$ and $\bar{B}$ are the average of vector A and vector B, which are used to correct the problem that the COSINE similarity method is not sensitive to values. COSINE distance between two vectors is Eq. (13):

$$D = 1 - \text{sim}(A, B) \tag{13}$$

The specific process of the improved TOPSIS is as follows

(1) To normalize the QoS attribute matrix according to Eqs. 7 and 8.
(2) To establish the weighted matrix through the comprehensive weight of all attributes.
(3) To determine the best, the worst ideal solution, the positive ideal solution is the maximum V_{ij} of the weighted matrix, the negative ideal solution is the minimum v_i of the weighted matrix. Set the positive ideal vector and negative ideal vector respectively as A and B.

$$\begin{aligned} &\mathrm{A} = (\mathrm{A}_1,\ \mathrm{A}_2,\ \cdots,\ \mathrm{A}_n),\ \mathrm{B} = (\mathrm{B}_1,\ \mathrm{B}_2,\ \cdots,\ \mathrm{B}_n), \\ &\mathrm{A_j} = \max\{\mathrm{v_{ij}}\}, \mathrm{b_j} = \min\{\mathrm{v_{ij}}\}, (\mathrm{i} = 1,\ 2,\ \cdots,\ \mathrm{m}; \mathrm{j} = 1, 2, \cdots, \mathrm{n}) \end{aligned} \tag{14}$$

(4) We calculate the COSINE similarity and distance between candidate services and positive and negative ideal solutions based on Eqs. 12 and 13.
(5) We calculate the fit degree of each schema based on distance.

$$C_i = \frac{d^-}{d_i^+ + d_i^-} \tag{15}$$

d_i^+ is the distance from vector to positive ideal vector, and d_i^- is the distance from vector to negative ideal solution vector.

4 Experimental Analysis

4.1 Simulation Experiment

This section illustrates the method of this article and verificate its feasibility through simulation experiments. There are 6 candidate services that meet the user's functional requirements and have different QoS. Among them, there are 4 QoS attributes of price, response time, reliability and accuracy. As shown in Table 3.

Table 3. Candidate service group

Candidate service	QoS Attribute			
	Price	Response time	Reliability	Accuracy
ws1	12	34.7893	78.4566	45.6723
ws2	22	22.1274	89.3785	87.3469
ws3	23	12.2141	86.6794	23.4532
ws4	11	13.5325	9.1233	63.4821
ws5	26	35.1428	25.4521	37.4531
ws6	45	3.2326	73.6742	43.8790

The price and response time are cost-type attributes that can be normalized by Eq. 7; reliability and accuracy are benefit-type attributes, that van be normalized by Eq. 8. The normalized matrix F_{ij} is as follows:

$$F_{ij} = \begin{bmatrix} 0.9706 & 0.0111 & 0.8639 & 0.3478 \\ 0.6765 & 0.4079 & 1.0000 & 1.0000 \\ 0.6471 & 0.7185 & 0.9664 & 0.0000 \\ 1.0000 & 0.6772 & 0.0000 & 0.6265 \\ 0.5588 & 0.0000 & 0.2035 & 0.2191 \\ 0.0000 & 1.0000 & 0.8043 & 0.3197 \end{bmatrix}$$

We calculate the subjective weights. First, the user sorts the four attributes into price > response time > reliability > accuracy according to their own preferences. Then refer to Table 2 to compare the importance of the two adjacent indicators. The result of the comparison is $R_2 = (1.3,1.4,1.5)$, $R_3 = (1.5,1.6,1.7)$, $R_4 = (1.3,1.4,1.5)$.

Then we calculate weights according to Eqs. 1 and 2, the weights of price, response time, reliability and accuracy are (0.37,0.40,0.43), (0.29,0.29,0.29), (0.17,0.18,0.19), (0.11,0.13,0.15), The triangular fuzzy weight is defuzzificated by contact number to get the weight α = (0.34,0.29,0.21,0.16).

We obtain the objective weight through entropy weight method in the 2.3 section. According to the normalized matrix, the data in our matrix are substituted into Eqs. 9, and 10 to obtain the objective weight β = (0.29,0.19, 0.38, 0.14). Finally, we integrate the objective weight and the subjective weight to obtain the combined weight W_j according to Eq. 11. W_j = (0.32, 0.24, 0.29, 0.15). In this paper, $\lambda = 0.5$. In practice, the user can change the comprehensive weight by adjusting the λ according to his own needs. We get the decision matrix R_{ij} by considering the weight.

$$R_{ij} = \begin{bmatrix} 0.311 & 0.003 & 0.250 & 0.052 \\ 0.216 & 0.098 & 0.290 & 0.150 \\ 0.207 & 0.172 & 0.280 & 0.000 \\ 0.320 & 0.163 & 0.000 & 0.094 \\ 0.179 & 0.000 & 0.059 & 0.033 \\ 0.000 & 0.240 & 0.233 & 0.048 \end{bmatrix}$$

According to Eq. 14, we determine the positive ideal solution R + and the negative ideal solution R-.

$$\begin{aligned} \mathrm{R}^{+} &= (0.320\ 0.240\ 0.290\ 0.150) \\ \mathrm{R}^{-} &= (0.000\ 0.000\ 0.000\ 0.000) \end{aligned}$$

We calculate the cosine distance from each scheme to the positive ideal solution R+ and the negative ideal solution R- according to Eqs. 12 and 13.

$$\begin{aligned} \mathrm{DR}^{+} &= (0.11,\ 0.19\ ,0.24,\ 0.02,\ 0.17,\ 0.16) \\ \mathrm{DR}^{-} &= (0.9,\ 0.91,\ 0.91\ ,0.89,\ 0.91,\ 0.9). \end{aligned}$$

If the distance from a service to a positive ideal solution is closer, the distance from the negative ideal solution is farther, the service will be more in line with the user demands. According to Eq. 15, we obtain the fitting degree Ci of each service, Ci = (0.89, 0.82, 0.80, 0.98, 0.84, 0.85). We can know that W4 > W1 > W6 > W5 > W2 > W3.

4.2 Experimental Analysis

The results of this experiment are compared with those obtained by the literature [9] and the literature [10], as shown in Table 4.

In this paper, the user preference to attributes is ordered as price > response time > reliability > accuracy. According to the comparison, document [9] adopting the

Table 4. Comparison with results

Method	Result of service sorting	Weight
The method of this paper	W4 > W1 > W6 > W5 > W2 > W3	0.32 0.24 0.29 0.15
The method in [8]	W1 > W2 > W3 > W6 > W4 > W5	0.29 0.19 0.38 0.14
The method in [9]	W1 > W2 > W5 > W3 > W4 > W6	0.34 0.18 0.27 0.21

entropy weight method has larger differences in no matter the weight or result with the user expectations; document [10] adopting the combination based on the fuzzy AHP method and the principal component analysis method has certain difference in the result of weight and the user expectations. The experiment is more in line with the user demands in no matter the attribute weight or the service sequencing result. Based on the fuzzy G1 method and entropy weight method, not only the subjective demands of use can be satisfied, but also the subjective weight can be revised in the objective method. The weight will be more scientific and reasonable, the selected service will be more in line with the user request.

This paper compares the improved TOPSIS of COSINE similarity and the traditional TOPSIS used in [10] from the selection service accuracy. According to the Fig. 1, it is obvious that the accuracy of the improved TOPSIS method through COSINE similarity is obviously improved, which avoid the error that has same distance from object to the positive ideal solution and the negative ideal solution resulting in not be judged.

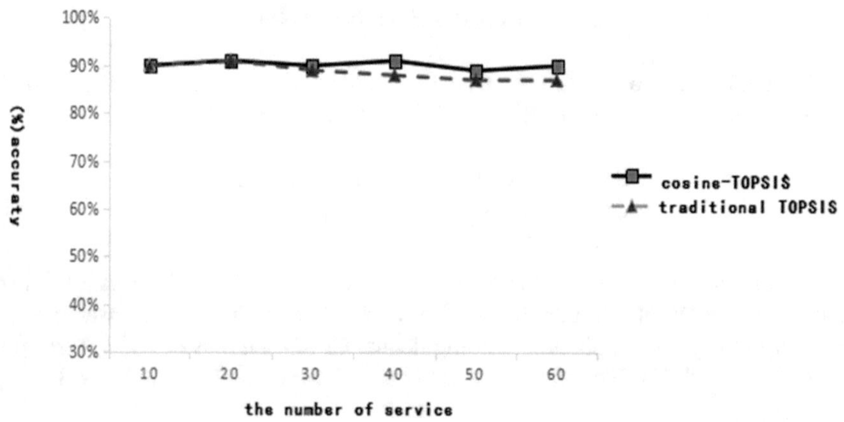

Fig. 1. Comparison of accuracy.

5 Conclusion

The traditional user-weighted method is difficult to express the user preference with the weight accurately. A triangular fuzzy improved G1 method proposed in this paper can express the user preference better according to the features of triangular fuzzy numbers. Also, the entropy weight method is adopted to revise the subjectivity caused by G1 method and make the weight values more objective and scientific through the comprehensive weighting. In this paper, the traditional TOPSIS has been improved to enhance the accuracy of service decision and make the selected service more in line with the user demands.

Acknowledgments. This work was supported by Key Research and Development Plan Project of Shandong Province, China (No. 2017GGX201001).

References

1. Song, J.: Research and Application of Multi-attribute Decision-making Algorithm. North China Electric Power University, Beijing (2015)
2. Ding, X., Zhang, D.: K-means algorithm based on AHP and CRITIC integrated weighting. J. Comput. Syst. **25**(7), 182–186 (2016). 888/j.cnki.csa.005267
3. Chen, Y.: Expert investigation method. Prediction **4**, 63–64 (1983)
4. Wang, X., Guo, Y.: Consistency analysis of judgment matrix based on G1 method. CMS **14**(3), 65–70 (2006). https://doi.org/10.16381/j.cnki.issn1003-207x.2006.03.012
5. Li, G., Cheng, Y., Dong, L.: Research on objective weighting method of gini coefficient. Manag. Rev. **26**(1), 12–22 (2014). https://doi.org/10.14120/j.cnki.cn11-5057/f.2014.01.004
6. Zhang, L., Dong, C., Yu, Y.: A service selection method supporting mixed QoS attributes. J. Comput. Appl. Softw. **33**(9), 15–19 (2016). https://doi.org/10.3969/j.issn.1008-0775.2016.07.004
7. Xie, H., Li, S., Sun, Y.: Research on DEMATEL method for solving attribute weights based on cloud model. Comput. Eng. Appl. **7**, 17–25 (2018). https://doi.org/10.3778/j.issn.1002-8331.1610-0169
8. Fan, Z., Li, N., Hao, B.: A method for uncertain weight calculation of QoS attributes of web services. Softw. Eng. **19**(7), 14–17 (2016). https://doi.org/10.3969/j.issn.1008-0775.2016.07.004
9. Sun, R., Zhang, B., Liu, T.: A web service quality evaluation ranking method using improved entropy weight TOPSIS. J. Chin. Comput. Syst. **38**(6), 1221–1226 (2017)
10. Sun, X., Niu, J., Gong, Q.: A web service selection strategy based on combined weighting method. J. Comput. Appl. **34**(8), 2408–2411 (2017)
11. Duan, J.: Research on Web Service Composition Method Based on User Preference. Southwest University, Chongqin (2014)
12. Zhang, S.: Several fuzzy multi-attribute decision making methods and their applications. Xidian University of Electronic Technology, Xian (2012)
13. Chen, Y., Yan, H., Guo, C.: QoS quantization algorithm for web services based on triangular fuzzy numbers. Microprocessor **37**(4), 38–42 (2016). https://doi.org/10.3969/j.issn.1002-2279.2016.04.011
14. Wang, W.: A TOPSIS improved evaluation method based on weighted mahalanobis distance and its application. J. Chongqing Technol. Bus. Univ. **35**(1), 40–44 (2018). https://doi.org/10.16055/j.issn.1672-058x.2018.0001.007

Discovery of Key Production Nodes in Multi-objective Job Shop Based on Entropy Weight Fuzzy Comprehensive Evaluation

Jiarong Han[1], Xuesong Jiang[1(✉)], Xiumei Wei[1], and Jian Wang[2]

[1] Qilu University of Technology (Shandong Academy of Sciences), Jinan, China
974262807@qq.com

[2] Shandong College of Information Technology, Jinan, China

Abstract. The multi-objective Job Shop complex network model based on data information is a new idea to solve the transformation of multi-objective shop scheduling problem in recent years. Finding key nodes on the complex networks model is the focus of this paper. The existing key nodes recognition method ignores the overall characteristics of the network, is susceptible to subjective factors, and does not apply to data based complex networks model. According to the characteristics of subjective and objective weighting, the entropy weight method in fuzzy mathematics is applied to the method of analytic hierarchy process (AHP). The next step is to establish a key nodes recognition method suitable for new model–Entropy weight fuzzy comprehensive evaluation method. To some extent, this method has made up for the lack of subjectivity and index capability of the method of analytic hierarchy process. Finally, the simulation results show that the method can effectively mine the key nodes in the model, and prove the rationality and effectiveness of the method.

Keywords: Entropy weight fuzzy comprehensive evaluation
Intelligent manufacturing · Industrial big data
Multi-objective job shop problems · Complex networks
Discovery of key nodes

1 Introduction

With the continuous development of intelligent manufacturing, in the face of more complex production and more and more data accumulation, the model optimization of the traditional process industry is limited by the development. Researchers are trying to find a breakthrough, using complex networks modeling and producing data as model nodes, which is just one of the new ideas for the transformation of production scheduling problems in multi-objective workshops. The model starts from the data characteristics of multi-objective manufacturing, analyzes the data generated in the production process, and fully considers the role of data in production. Literature 1 [1] put forward an application research framework for complex networks theory for discrete manufacturing processes. This paper summarizes the product assembly, effectiveness evaluation and other two aspects which are closely related to the manufacturing process, and summarizes and analyzes the current achievements. Literature 2 [2] based on the combination of the

P. Krömer et al. (Eds.): ECC 2018, AISC 891, pp. 180–190, 2019.
https://doi.org/10.1007/978-3-030-03766-6_20

complexity of manufacturing system and the complex networks, takes Job-shop as the research object, carries out the network bottleneck analysis and the network propagation characteristics of the complex production process of the manufacturing system, and analyzes the influence of the change of the network topology on the whole operation of the workshop manufacturing execution system. Literature 3 [3] analyses the basic characteristics of industrial data, the quality of data and the quality control methods of large industrial data, and puts forward some preliminary suggestions on the direction of the quality control of industrial large data. However, because the research on the application of complex networks to the production of multi-objective job shop is still in its infancy, and most of them are industry oriented to discrete modeling, so there is no more literature for reference.

After building complex networks model, finding key nodes and optimizing key nodes is the next priority. There are few key nodes in complex networks, but it has great influence on the whole network, and can determine the structure and function of the network [4, 5]. In the existing research, the main method is to sort the network nodes according to a certain evaluation index, and some of the nodes in the front rank are key nodes [6]. The evaluation indexes of network nodes mainly include degree, betweenness, clustering coefficient, eigenvector and so on. Different indexes reflect the importance of evaluation nodes from different aspects [7]. The researchers compared the scientific key nodes to evaluate model, but these model are subjective, and have no objective analysis of the contribution rate of various evaluation indicators, and lack of deep mining of the internal relations of the indicators [8, 9]. In view of the shortcomings of analytic hierarchy process method in identifying the objectivity of the method and the importance of quantifying the importance of nodes, a fuzzy comprehensive evaluation method based on entropy weight based on AHP is proposed in this paper. The method uses analytic hierarchy process method to empower different indexes, get node importance, excavate the key nodes of network according to the rank order, and introduce the Entropy weight method to correct the result, so as to realize the combination of static and dynamic, subjective and objective empowerment. Finally, a simulation experiment is carried out to prove the validity and rationality of the method. It can be applied to complex networks model in multi-objective job shop.

2 Multi-objective Job Shop Complex Networks Model

Based on the data based multi-objective job shop network real process entity, this paper uses $G = (N, E, W)$ to represent the complex networks model, where N represents a set of nodes, E is a set of edges, W represents a set of weights, and this is a network with a directed weighting. The construction of model needs several important links, such as data processing, establishing edges and determining weights.

2.1 Data Preprocessing

The data in production is directly derived from the sensors distributed in the production units of the workshop, such as temperature sensors, pressure sensors, speed sensors and so on. According to the set parameters, the sensor returns to the current data to the server every other time. Due to the different data types and sizes, data pretreatment is needed:

I. A series of data transmitted by the sensor is used as a time sequence of data, the first data of the sequence is numbered 0, and a certain sequence value at one time is "null" (the sensor does not respond to the request or the failure of the data), and all the sequences are rewritten to "null" at this time.
II. The next step is to generate logical sequence based on the data sequence: the first data at the beginning of processing is 0 moments, corresponding to the logical sequence value of 0. Starting from second numbers, if the data sequence is increased compared with the previous data, the value generated by the logical sequence is 1, unchanged at 0, and reduced to −1. If null data is encountered, the data will not generate logical sequence items with the next data, and the above rules will continue to be executed from the third point.

2.2 The Establishment of Complex Networks Model

In data based complex networks model, model node *R* is no longer a specific production link, but a collection of data points. The relationship between data nodes and processes is shown in Fig. 1 From the graph we can see that a process entity contains multiple data.

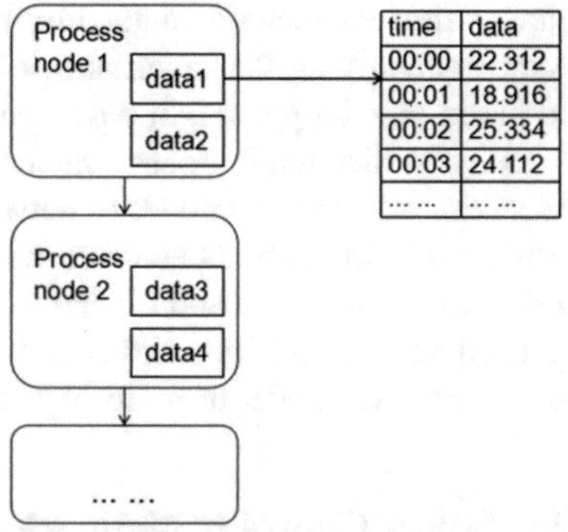

Fig. 1. Process and node representation

In the process of data preprocessing, data processing is used as logical sequence. Next, we use Apriori algorithm to mine the relationship between these logical sequences. There are two nodes in *A* and *B*. If *A*, *B* is the same as increasing or decreasing $A \rightarrow B$; *A* increases *B* decreases, or *A* decreases when *B* increases with $A \rightarrow -B$; there is no correlation between two nodes and $A \rightarrow \neg B$ indicates. The relationship between nodes *A* and *B* is only three $A \rightarrow B$, $A \rightarrow -B$, $A \rightarrow \neg B$. The logical representation and probability of all events are shown in Table 1.

Based on the above analysis, we can get the formula (1) of support for each event. When $\xi = 0$, P(others) = P(1,0) + P(0,1) + P(0,−1) + P(−1,0) + ξ, if the value of P

Table 1. The logical representation and probability representation of all events.

Logical representation of events	Logical representation of probability	probability
$A \to B$	$P(A \cup B)$	P(1,1) + P(−1,−1) + ξ
$A \to -B$	$P(A \cup -B)$	P(1,−1) + P(−1,1) + ξ
$A \to \neg B$	P(others)	P(1,0) + P(0,1) + P(0,−1) + P (−1,0) + ξ

(others) is greater than or equal to 30% (node node, there is no correlation between the points). The value of c(others) = 1/3(c(1,0) + c(0,1‖−1) + c(−1,0)) is calculated, and its value is greater than 44%. There is no correlation between *A* and *B*. If P (others) does not satisfy the minimum support or minimum confidence, then there may be an association between *A* and *B*, and there is a correlation between the value of P $(A \cup B)$ and $P(A \cup -B)$ when the value is greater than 40%. After determining the relationship between *A* and *B*, the corresponding confidence can be calculated by formula (2). If the minimum support of the two nodes is 40% and the minimum confidence is 60%, there is a strong association rule between the two nodes, and the edges between *A* and *B* two nodes are established. If the minimum support 40% is not satisfied or the minimum confidence 60% is not satisfied, there is no strong association rule between *A* and *B*, and no edge is established between *A* and *B*.

$$\begin{cases} s(AB) = s(1,1) + s(0,0) + s(-1,-1) \\ s(A-B) = s(1,-1) + s(0,0) + s(-1,1) \\ s(others) = s(1,0) + s(0,1) + s(0,-1) + s(-1,0) \end{cases} \quad (1)$$

$$\begin{cases} c(AB) = \frac{1}{3}(c(1,1) + c(0,0) + c(-1,-1)) \\ c(A-B) = \frac{1}{3}(c(1,-1) + c(0,0) + c(-1,1)) \end{cases} \quad (2)$$

In order to describe the relationship between data accurately, we need to add weights to each side in model. The weight set of edges can be expressed as $W = \{w_{ij} = f(n_i,n_j) \mid i, j \in (1,2,3, \ldots, n)\}$, where W_{ij} represents the weight from the node *i* to the edge of node *j*. If *A*, *B* two nodes are associated, there will be some functional relationship between the data generated by the upstream node and the data generated by the downstream node. The nodes n_i and n_j are two adjacent nodes in the node set *N*, and the direction of the side is directed from n_i to n_j, then there is a function relationship $n_j = f(n_i)$ between the values of n_i and n_j at a certain time, and the function expression can be obtained by the data sequence of node n_i and node n_j.Then the weight value between n_i and n_j can be represented as $W_{ij} = f'(n_i)$. If only two nodes are linear, the weight W_{ij} is constant.In the actual production process, the function relationship is not always linear, so in many cases, the value is changed. When we set up nodes, when the value of n_i is *x*, the value of node n_j is *y*, then the calculation formula of weight W_{ij} can be expressed as:

$$W_{ij} = \frac{\partial(f(n_i))}{\partial n_i} \tag{3}$$

When the relationship between two points is nonlinear, the weight obtained will be a function expression of function value changing with upstream node data. Weight reflects the upstream node's influence on downstream nodes. A weighted complex networks model based on data can be constructed, and the model diagram is described in the simulation experiment.

3 Entropy Weight Fuzzy Comprehensive Evaluation Method

Analytic hierarchy process method is an analytic method of hierarchical weight decision problem. A multi-objective decision problem is divided into a number of ordered low order levels, and the method of solving the judgment matrix eigenvector is solved. The weight of the lowest level to the top layer is obtained by weighting. Entropy weight method is an objective weighting method. It determines the weight of the index according to the size of the information load of each index. The greater the difference of the index, the greater the amount of information and the identification of the index. Using Entropy weight method to improve analytic hierarchy process method can make weight distribution more objective and accurate.

3.1 Complex Networks Key Nodes Evaluation Index

Based on the existing complex networks evaluation index research results, node degree, node betweenness, Clustering coefficient, and node eigenvector are selected to identify key nodes [10].

I. Degree: The node degree in the network is defined as all the number of edges directly connected to the node. It is indicated by k_i that the larger the node degree is, the more important the node is in the network.

II. Betweenness: The node betweenness through node r refers to the ratio of the number of paths passing through node r to the total number of shortest paths in all shortest paths in the network, and formula is (4). Among them, δ_{ij} is the shortest path number between nodes i and j, and $\delta_{ij}(r)$ is the number of r passing through the shortest path between nodes i and j.

$$CB(r) = \sum_{i,j \in N \neq j} \frac{\delta ij(r)}{\delta ij} \tag{4}$$

III. Clustering coefficient: The sum of the shortest distance between node i and the remaining nodes in the network is formula (5), where d_ij is the shortest distance between i and j.

$$Ci(\mathrm{i}) = \sum_{j \in N,\ \mathrm{i} \neq j} \mathrm{dij} \tag{5}$$

IV. Eigenvector: The eigenvector index of node i refers to the eigenvector of the maximum eigenvalue corresponding to the network adjacency matrix, formula is (6), and λ is the main eigenvalue of the network adjacency matrix, $e = (e_1,e_2,\ldots\ldots,e_n)^T$ is λ eigenvector, is 1, i and j have side connections; the value is 0 without side connections.

$$Ce(i) = \lambda \sum_{j \in N. i \neq j} \theta ijej \tag{6}$$

In order to sort an important network node (key nodes), the importance of the node (NIP) is determined by the above four indicators (7), of which v_1, v_2, v_3, v_4 are the weight coefficients, which determine the proportion of various evaluation indexes in the identification of key nodes.

$$NIP_{\mathrm{i}} = \mathrm{v}_1 \times D_{\mathrm{ei}} + \mathrm{v}_2 \times B_{\mathrm{ei}} + \mathrm{v}_3 \times C_{\mathrm{ei}} + \mathrm{v}_4 \times E_{\mathrm{ei}} \tag{7}$$

3.2 Determine the Weight of the Index

According to complex networks key nodes evaluation index, a hierarchical model is set up, as shown in Fig. 2. The relative importance of all the indexes relative to the final key nodes is judged, and the two indexes compare with each other is used to construct the judgment matrix $C = (c_{ij})_{m \times m}$, in which the value reference ratio scale method of c_{ij} (Table 2) is used, and m is the number of the evaluation indexes.

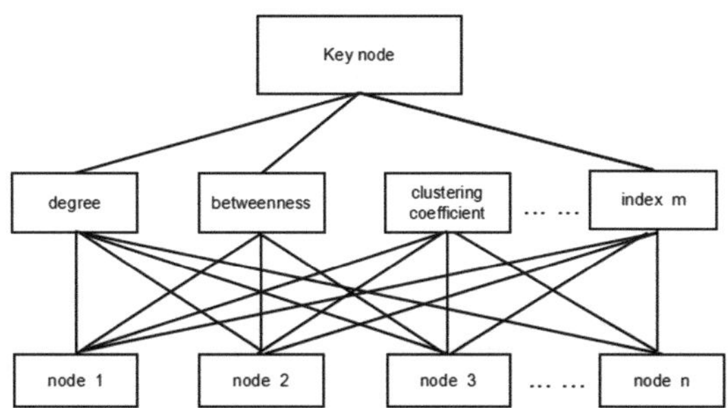

Fig. 2. Hierarchical hierarchical structure model

Table 2. Fundemental scale

c_{ij}	The extent of the index *i* is more important than the index *j*
1	i = j
3	i > j
5	i ≫j
7	i ≫>j
9	i ≫≫ j

Calculating the weight matrix *V* and formula of each index (8), where *x* represents the eigenvector matrix of matrix *C*, and *d* is the column where the largest eigenvalue.

$$VAHP - \mathrm{j} = \frac{x(\mathrm{j}, d)}{\sum_{i=1}^{m} x(j, d)} \tag{8}$$

3.3 Entropy Weight Method Correction

The number of network nodes is *n*, and the node set is *A*, $A = \{A_1, A_2, \ldots, A_n\}$, and the evaluation index is *m*. The set of indexes is set to *S*, $S = \{S_1, S_2, \ldots, S_m\}$,the initial decision matrix is $A' = (A'_{ij})_{n\times m}$. The matrix is standardized by formula (9), and the decision matrix $B = (B'_{ij})_{n\times m}$ is obtained. Using Entropy weight method, formula (10) calculates the difference coefficient of *j*(j = 1, 2, …, m), and $k = 1/ln\ m$. Finally, formula (11) is used to calculate the final weight, and the final weight is used to calculate the importance of each node, and the importance of each node is ordered. The part of the node is complex networks key nodes.

$$B_{ij} = \frac{A_{ij}}{\sum_{i=1}^{n} A_{ij}} \tag{9}$$

$$E_j = 1 + \mathrm{k}\sum_{i=1}^{n} \mathrm{B_{ij}}\ \ln\ \mathrm{B_{ij}} \tag{10}$$

$$\mathrm{V_j} = \mathrm{V_{AHP-j}}\ \times E_j, \mathrm{j} = 1, 2, \cdots, \mathrm{m} \tag{11}$$

4 Simulation Experiment Analysis

4.1 Complex Networks Model

This paper uses an example of a process for producing glass fibers in an alkali-free kiln process. Figure 3 shows a process for producing glass fibers in an alkali-free kiln process. The production process includes 12 different links, a total of 139 data sensor receiving points. Select a production line's data, and establish the data based complex networks model through the second chapter's method. Its visualization effect is shown in Fig. 4. As shown, the graph contains some isolated nodes, which contain some monitoring nodes, which are caused by node properties. In the model, the edge reflects the existence of association, and the weight of the edges represents the correlation between the data. The nodes degree, node betweenness, Clustering coefficient and node eigenvector of different nodes are calculated by using complex networks related formula. The calculation results of evaluation index and importance degree are shown in Table 3.

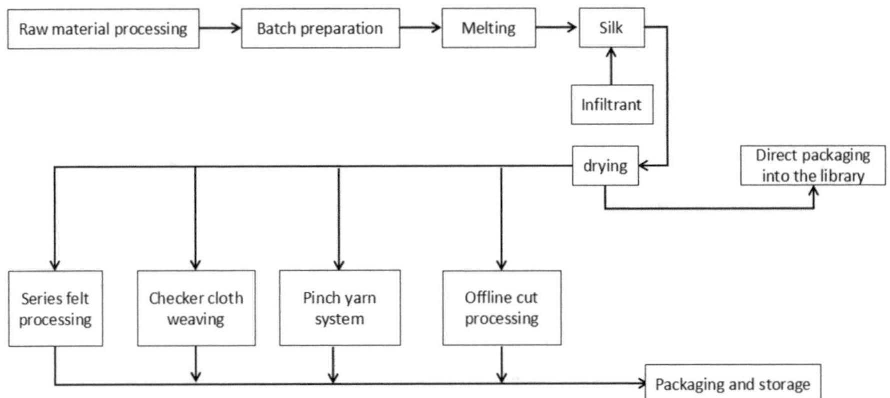

Fig. 3. Flow chart of tasks for the industrial process of glass fiber production

4.2 Simulation Experiment

According to the existing complex networks key nodes research results [10], the AHP algorithm is used to construct a judgment matrix, assuming that B_e is of the greatest importance, D_e and E_e are slightly larger and more important, and the importance of C_e is minimal (this assumption is determined by the nature of complex networks). The ratio of importance to them is shown in matrix C, where x_d is the eigenvector column corresponding to the maximum eigenvalue, and V is the weight ratio of the four evaluation index. The weight of key nodes can be recognized by four kinds of evaluation index.

$$C = \begin{bmatrix} C & De & Be & Ce & Ee \\ \text{De} & 1 & 1/3 & 3 & 1 \\ \text{Be} & 3 & 1 & 5 & 3 \\ \text{Ce} & 1/3 & 1/5 & 1 & 1/3 \\ \text{Ee} & 1 & 1/3 & 3 & 1 \end{bmatrix} \begin{matrix} xd & v \\ -0.2447 & 0.2003 \\ -0.8174 & 0.5114 \\ -0.1612 & 0.0880 \\ -0.3445 & 0.2003 \end{matrix}$$

The weights are corrected by Entropy weight method, and the final weights are assigned to $v_1 = 0.1242$, $v_2 = 0.5990$, $v_3 = 0.0077$, $v_4 = 0.2691$. The importance of

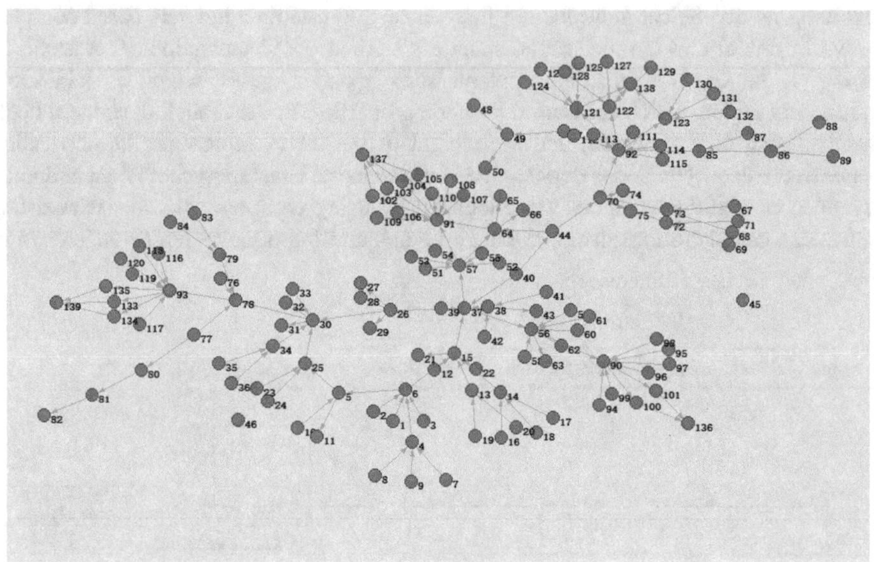

Fig. 4. Maximum connected subgraph of a complex network in a fiberglass job shop with an alkali-free kiln process.

Table 3. Complex networks evaluation index and importance degree

Node	D_e	B_e	C_e	E_e	NIP
R_1	0.002264	0.05342	0.00574	0.02648	0.04102
R_2	0.03658	0.01899	0.00324	0.03473	0.08547
……	……	…….	……	……	……
R_{139}	0.01393	0.22023	0.00113	0.1125	0.01843

each node is calculated according to the weight ratio, and the former part of the node is key nodes of the network, and Table 4 shows the first 10 nodes of the rank of importance. In order to verify the effectiveness of this method, the difference degree of importance of each node is calculated by means of mean square error, as shown by Table 5. From the results of the two tables, we can see that the method of key nodes identified by the AHP method modified by Entropy weight method and the original AHP algorithm is feasible and effective, but the difference degree between the node importance of this method is greater, and the difference of the single difference of the network is increased by 12.8% before the correction, so the recognition is higher. The results are more accurate.

Table 4. Node importance ranking results

	Node sorting
Analytic hierarchy process method	$R_{92} > R_{70} > R_{91} > R_{98} > R_{90} > R_{30} > R_{12} > R_{25} > R_{136} > R_{37}$
Entropy weight method correction	$R_{92} > R_{56} > R_{15} > R_{70} > R_{91} > R_{98} > R_{90} > R_{30} > R_{12} > R_{25}$

Table 5. Difference degree of identification before and after correction

AHP method	Entropy weight method correction	Percentage of promotion
0.002351	0.002652	12.8

5 Summary

The use of Entropy weight fuzzy comprehensive evaluation method to identify key nodes in the multi-objective job shop complex networks model based on data information has a more important application value. This method is based on the existing complex networks key nodes recognition research, and applies the analytic hierarchy process method and Entropy weight method idea to the key evaluation. This method improves the shortcomings of the existing recognition methods in the importance of objectivity and node quantification, and realizes the combination of static and dynamic empowerment, subjective and objective empowerment. The simulation experiment proves the effectiveness of the method, the recognition degree and the recognition result are more objective and accurate. It can be successfully applied to the data based complex networks model.

Acknowledgments. This work was supported by Key Research and Development Plan Project of Shandong Province, China (No. 2017GGX201001).

References

1. Zhang, F., Jiang, P.: A summary of the application research of complex network theory in the production process of discrete workshops. Ind. Eng. **19**(6), 1–8 (2016). https://doi.org/10.3969/j.issn.1007-7375.2016.06.001
2. Feng, H.: Research on Job-Shop Multi-bottleneck Recognition Method Based on Complex Network. Xinjiang University (2016)
3. Duan, C.: Discussion on data quality control of industrial big data under the background of intelligent manufacturing. Mech. Des. Manuf. Eng. **2**, 13–16 (2018). https://doi.org/10.3969/j.issn.2095-509X.2018.02.003
4. Callaway, D.S., Newman, M.E., Strogatz, S.H., et al.: Network robustness and fragility: percolation on random graphs. Phys. Rev. Lett. **85**(25), 5468 (2000)
5. Cohen, R., Erez, K., Ben-Avraham, D., et al.: Breakdown of the internet under intentional attack. Phys. Rev. Lett. **86**(16), 3682 (2001). https://doi.org/10.1103/PhysRevLett.86.3682

6. Xuan, Z., FengMing, Z., KeWu, L.: Finding vital node by node importance evaluation matrix in complex networks. J. Phys. **61**(5), 50201 (2012). https://doi.org/10.7498/aps.61.050201
7. Lü, L., Chen, D., Ren, X.L., et al.: Vital nodes identification in complex networks. Phys. Rep. **650**, 1–63 (2016). https://doi.org/10.1016/j.physrep.2016.06.007
8. Nan, H.E., DeYi, L.I., WenYan, G.A.N.: Mining vital nodes in complex networks. Comput. Sci. **34**(12), 1–5 (2007). https://doi.org/10.3969/j.issn.1002-137X.2007.12.001
9. Beijing: Mining vital nodes in complex networks. Computer Science (2007)
10. Zhang, X., Li, Y., Liu, G., et al.: Complex network node importance evaluation method based on node importance contribution. Complex Syst. Complex. Sci. **11**(3), 26–32 (2014)

A Data Fusion Algorithm Based on Clustering Evidence Theory

Yuchen Wang(✉) and Wenqing Wang

School of Automation, Xi'an University of Posts and Telecommunications, Xi'an, China
18291929956@163.com

Abstract. Based on the idea of clustering and evidence theory, a new measurement data fusion algorithm is proposed. Firstly, all the measured values are clustered into groups according to the hierarchical clustering method, the best clustering group is selected, and each group is given different weights. Secondly, the set consisting of each group of measured values is regarded as identification framework, then the measured values in the group are converted into the corresponding evidence, which is fused after the evidence is modified, and the fused evidence is regarded as the weight of each measured value. Finally, after the data is weighted and summed within the group, weighted summation between groups to obtain the fusion result. The validity of the method is verified by the data simulation.

Keywords: Data fusion · Hierarchical clustering · Evidence theory

1 Introduction

Data fusion is mostly applied to multi-sensor data fusion, that is, data analysis and processing are performed on multiple sensor measurements to obtain a more reliable data than a single sensor. It is also possible to analyze and process a batch of redundant data measured by the same sensor in a short period of time to obtain more accurate and reliable data within the allowed time range.

Many scholars all over the world have done extensive and in-depth research on data fusion. Reference [3] proposes the application of Bayesian estimation method in multi-source data fusion. Reference [4] is based on self-learning least squares weighted data fusion algorithm, and uses the state estimation characteristics and related historical information of Calman filter to improve the accuracy of the algorithm fusion. References [5–7] propose to estimate the similarity of each data based on confidence distance and improve the consistency of data fusion. Reference [8] uses fuzzy theory to calculate the mutual support degree of each node and carries out data fusion. The above algorithms all directly or indirectly assume that the measured value obeys the normal distribution of a certain parameter or knows the prior information of the measured value. However, in the actual measurement, due to the lack of measurement data, it can not accurately estimate the distribution function that the measured values obey, and the system error of the measurement device or the environmental error caused by the change of the measurement environment make the same batch of measurement data not

P. Krömer et al. (Eds.): ECC 2018, AISC 891, pp. 191–198, 2019.
https://doi.org/10.1007/978-3-030-03766-6_21

be distributed in a friendly manner. Reference [1] proposes an evidence-based theory to combine the evidences of measured values without hypothesis parameters, but fusion errors may occur for unfriendly distributions of measured values; Reference [2] proposes observation grouping of measurement data, and then used evidence theory for data fusion. However, its grouping is inaccurate, and there is the possibility of an error group, which affects the fusion of data.

In order to improve the accuracy of data fusion, this study proposes a new data fusion algorithm based on the idea of clustering and evidence theory. The algorithm firstly performs hierarchical clustering on the measured values, then uses evidence theory to perform data fusion on each type of data, and finally performs weighted summation on all types of data, making the data fusion more accurate and more reliable.

2 Data Fusion Algorithm

2.1 Clustering Algorithm

Selection of Clustering Algorithm. Cluster analysis originates from taxonomy and is an unsupervised classification method. It is essentially a collection of similarity elements, and the elements with large similarity are aggregated into one type. According to specific applications, we classify clustering algorithms into divide clustering, hierarchical clustering, and neural network clustering, etc.

Reference [9] proposes divide clustering algorithm needs to know the number of clusters in advance. Reference [10] points out that Neural network clustering algorithm generally has shortcomings such as slow convergence, long training time, and unexplained. Reference [11] shows Hierarchical clustering can display the input samples by genealogical map, and then select appropriate classification according to different thresholds, which is suitable for the requirements of this paper.

Basic Idea of Hierarchical Clustering. Hierarchical clustering algorithm uses the distance threshold as a criterion for determining the number of clusters. Each sample is self-contained, and then gradually merged according to the distance criterion to reduce the number of clusters until the requirement of classification is reached.

2.2 The Basic Idea of Evidence Theory

DS Evidence theory is a complete theory of dealing with uncertain problems, characterized by "interval estimation" rather than "point estimation" for uncertain information. Assume that Ω is an recognition framework consisting of hypothetical methods, which is a finite number of complete and mutual, 2^{Ω} being the power set of Z. The basic probability assignment on the recognition framework Z is mass function (basic trust assignment function), expressed as $2^{\Omega} \rightarrow [0, 1]$. It satisfies:

$$\begin{cases} m(\varphi) = 0; \\ \sum\limits_{A \subseteq \Omega} m(A) = 1; \end{cases} \tag{1}$$

where A that make $m(A) > 0$ is called a focal element.

When using evidence theory, the m trust function value of each unrelated evidence in recognition framework is calculated, and then the evidence combination is used to combine the evidence to get the value of synthetic trust function of the elements in recognition framework.

2.3 Algorithm Description

A set of velocity data of a uniform flow is measured by a velocity sensor in a short period of time. Firstly, the hierarchical clustering method in MATLAB is used to cluster measured data, the pedigree clustering graph is generated, and different classification results are obtained according to the different classification threshold. Then all measured values in a separate set of data after each classification are regarded as the recognition framework Ω, and each measured value is converted into various evidences, and the generated evidence is corrected and combined, the mass function of the measured values after the combination is the weight of the measured values, and weighted fusion of each measured value to obtain fusion value of this group of data. Finally, the data of each group is weighted and fused to the final fusion value according to different classification weights. The best fusion value is obtained by comparing each classification result.

The algorithm mainly involves the following problems: (1) How to use hierarchical clustering to group data and select thresholds after grouping. (2) After grouping, how to determine trust distribution of each measured value and how to correct it. (3) After obtaining various evidence, the selection of evidence combination rules has a great impact on the accuracy of fusion.

Hierarchical Clustering Grouping. Because of disturbances form environment and other factors, even uniform flow, the flow velocity measured in a very short time are not the same, but they are always distributed around some values. In this paper, the hierarchical clustering method is used to classify measured values and the minimum distance criterion is used to calculate them.

(1) The N measured values are self-contained and the Euclidean distance criterion is used to calculate the distances of various types. The $N * N$ dimensional distance matrix $D(n)$ is obtained.
(2) Find out the smallest elements in $D(n)$ and merge the corresponding two classes into one class.
(3) Calculate the distance of the new category after merging to obtain the distance matrix $D(n+1)$.
(4) Go to step (2) and repeat calculation and consolidation until all samples are grouped together. According to the output clustering graph, different distance thresholds are selected. Each time the measured values can be divided into k groups. The data of each group is L_j, and exists $\sum\limits_{j=1}^{k} L_j = N$.

Trust Allocation Process. After grouping all measured values N into clusters, the grouped measured values are converted into evidence and trust distribution is performed. If divided into j groups, one group has L_j, then there is $\sum_{j=1}^{k} L_j = N$.

Assuming that the average value of the distance between the measured value S_n and L_j measured values is d_n, then $d_n = \sum_{m=1}^{L_j} |S_n - S_m| / L_j$, where $|S_n - S_m|$ is the distance from the measured value S_n to S_m. At this time, the average distance between the L_j measured values is $\overline{d}$, then $\overline{d} = \sum_{n=1}^{L_j} d_n / L_j$. If $d_n \geq \overline{d}$, then S_n belongs to the large deviation measurement set. Otherwise, it belongs to the small deviation measurement set.

For any measured value S_n, there exists $\Delta \geq 0$, which makes the true value $S_0^{(j)}$ falls within the neighborhood of the measured value S_n, where $\Gamma_n = [S_n - \Delta, S_n + \Delta]$, Δ is the degree of deviation between the measured value and the true value. If S_n belongs to the large deviation measurement set, it is relatively far from the true value, where Δ takes the difference between the maximum measured value and the minimum measured value $d_{\max}$. If S_n belongs to the small deviation measurement set, it is relatively close to the distance from the true value, where Δ takes the average distance $\overline{d}$ between each measured value.

For L_j measurements, assuming that there are Z measurements in the neighborhood of S_n, the Z measurements are considered to be close to the true value $S_0^{(j)}$ with the same probability, so the same trust function is assigned. The basic trust function of the initial evidence e_n obtained by S_n is:

$$m_n(X_Z) = 1/Z(\forall X_Z \in \Gamma_n) \tag{2}$$

After the L_j initial evidence $e_i(i = 1, 2 \cdots L_j)$ is generated by the L_j measured values, the initial evidence may contain both large deviation and small deviation measurement sets, which affect the fusion precision, so L_j initial evidence should be modified:

(1) If S_{k1} and S_{k2} belong to the small deviation measurement set, the ratio of their basic trust assignments is:

$$m_i(S_{k1})/m_i(S_{k2}) = d_{k2}/d_{k1} \tag{3}$$

(2) If S_k belongs to a small deviation measurement set and S_r belongs to a large deviation measurement set, the ratio of their basic trust assignments is:

$$m_i(S_k)/m_i(S_r) = d_{\max}/d_k \tag{4}$$

A set of correction coefficient $\{\omega_r\}$, $r = 1, 2 \cdots, L_j$ is generated by formula (3) and (4), and L_j evidence is corrected by using correction coefficient. There exists:

$$\hat{m}_i(S_n) = \omega_n * m_i(S_n) \Big/ \sum_{r=1}^{L_j} \omega_r * m_i(S_r) \tag{5}$$

Evidence Combination. Because the evidence theory is applicable to the situation of independent evidence, the initial evidence generated by formula (1–5) may produce higher conflicts, and the combination of Dempster rules may be unreasonably combined. Therefore, this paper will prorate the probability of supporting evidence conflict to each measured value $S_n (n = 1, 2 \cdots, L_j)$ by proportion, and combination formula is:

$$m(S_n) = \prod_{i=1}^{L_j} m_i'(S_n) + c * \overline{m_i'}(S_n) \tag{6}$$

where, c is the conflict factor, $\overline{m_i'}(S_n)$ is the average basic trust distribution of S_n in all the evidence, and there are

$$c = 1 - \sum_{n=1}^{L_j} \prod_{i=1}^{L_j} m_i'(S_n) \tag{7}$$

$$\overline{m_i'}(S_n) = \sum_{n=1}^{L_j} m_i'(S_n) / L_j \tag{8}$$

In the synthetic evidence, $m(S_n)$ is the weight obtained by S_n, then the fusion result of the L_j measured values is:

$$S_0^{(j)} = \sum_{n=1}^{L_j} (S_n * m(S_n)) \tag{9}$$

The fusion result of the group j data is obtained by formula (2–9), and the same fusion method is used to obtain the fusion result $S_0^{(j)} (j = 1, 2 \cdots, k)$ of all the k groups. The number of each group is regarded as a weight coefficient, and weighted fusion is performed on the k group data. The fusion result is:

$$S_0 = \sum_{j=1}^{k} (L_j \cdot S_0^{(j)}) / N \tag{10}$$

3 Experimental Simulation

In order to verify this algorithm, the experimental simulation of the algorithm is given in this paper. In the simulation, a set of velocity data of a uniform flow is selected, and the reference value of the uniform flow is 25.0 dm/s.

Test data is:

S = [21.70 21.85 22.02 22.31 22.45 22.48 22.70 22.75 22.86 22.95 23.00 23.05 23.10 23.25 23.42 23.5 23.65 23.9 24.06 24.15 24.23 24.80 24.85 24.93 25.02 25.09 25.10 25.14 25.15 25.19 25.25 25.32 25.46 25.53 25.55 25.60 25.65 25.75 25.80 25.83 25.89 25.90 26.35 26.38 26.45 26.68 26.70 26.85 27.00 27.10 27.15 27.35 27.72 27.80 27.89 27.93 28.00 28.12 28.36 28.5];

There are 60 measurements in total. The hierarchical clustering method is used to classify the measured data. The simulation results are shown in Fig. 1.

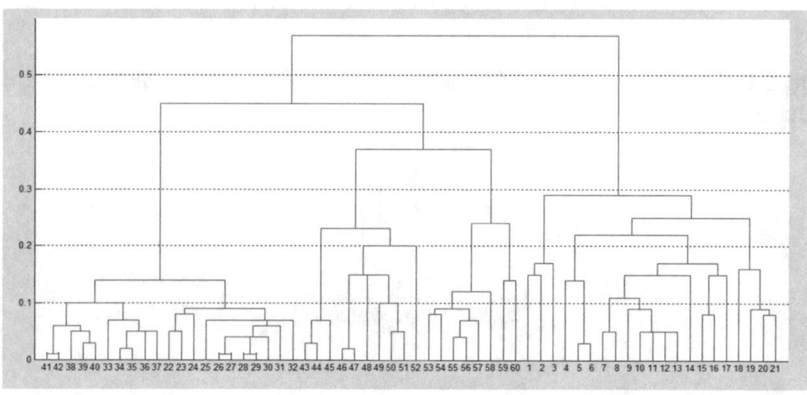

Fig. 1. Pedigree clustering

In Fig. 1, the abscissa is the respective measured value, and the ordinate is the classification threshold T. In order to select the optimal fusion scheme, the threshold T takes different values respectively, and the corresponding clustering is shown in Table 1.

Table 1. Clustering results

Threshold	Result	Group
$0.29 < T < 0.38$	{{41,42,38,39,40,33,34,35,36,37,22,23,24,25,26,27,28,29,30,31,32}, {43,44,45,46,47,48,49,50,51,52}, {53,54,55,56,57,58,59,60}, {1,2,3,4,5,6,7,8,9,10,11,12,13,14,15,16,17,18,19,20,21}}	four
$0.38 < T < 0.45$	{{41,42,38,39,40,33,34,35,36,37,22,23,24,25,26,27,28,29,30,31,32}, {43,44,45,46,47,48,49,50,51,52,53,54,55,56,57,58,59,60}, {1,2,3,4,5,6,7,8,9,10,11,12,13,14,15,16,17,18,19,20,21}}	three
$0.45 < T < 0.56$	{{41,42,38,39,40,33,34,35,36,37,22,23,24,25,26,27,28,29, 30,31,32,43,44,45,46,47,48,49,50,51,52,53,54,55,56,57,58,59,60}, {1,2,3,4,5,6,7,8,9,10,11,12,13,14,15,16,17,18,19,20,21}}	two

from the above table, we can see that data fusion is performed on different clustering results and the optimal fusion result will be selected.

The algorithm is used to fuse the data of each group, and take the third group in four categories {27.72, 27.80, 27.89, 27.93, 28.00, 28.12, 28.36, 28.5} as an example. As shown in Table 2.

Table 2. Group 3 evidence of trust function

Evidence	Basic Trust Distribution							
	$m_3(S_{53})$	$m_3(S_{54})$	$m_3(S_{55})$	$m_3(S_{56})$	$m_3(S_{57})$	$m_3(S_{58})$	$m_3(S_{59})$	$m_3(S_{60})$
e_{53}	0.0486	0.1459	0.1765	0.1851	0.1851	0.1615	0.0486	0.0486
e_{54}	0.0656	0.1969	0.2381	0.2497	0.2497	0	0	0
e_{55}	0.0539	0.1617	0.1955	0.2050	0.2050	0.1789	0	0
e_{56}	0.0539	0.1617	0.1955	0.2050	0.2050	0.1789	0	0
e_{57}	0.0539	0.1617	0.1955	0.2050	0.2050	0.1789	0	0
e_{58}	0	0	0.2332	0.2446	0.2446	0.2134	0.0643	0
e_{59}	0.0486	0.1459	0.1765	0.1851	0.1851	0.1615	0.0486	0.0486
e_{60}	0.0486	0.1459	0.1765	0.1851	0.1851	0.1615	0.0486	0.0486
Synthetic Evidence	0.0467	0.1400	0.1984	0.2081	0.2081	0.1543	0.0263	0.0182
S_0^3	27.9597							

from the above table, we can see that the third group of data fusion result is 27.9597.

Similarly, after the clustering into different groups, the fusion results of each group are obtained as shown in Table 3.

Table 3. Data fusion results of each group

Result	Group1	Group2	Group3	Group4	Fusion results	Deviation
4	25.3426	26.8381	27.9597	23.0085	25.1239	0.1239
3	25.3426	27.3442	23.0085		25.1261	0.1261
2	25.9615	23.0085			24.9280	0.072
1	25.2663				25.2663	0.2663

from the above table, we know that the result is the best when the data are divided into two groups.

Comparing results: Using the hierarchical clustering grouping, the optimal grouping can be more intuitively selected to obtain an optimal fusion result, and after the clustering grouping, the fusion is closer to the true value than the direct fusion, and the accuracy is higher.

4 Conclusion

In this paper, we proposes a new data fusion algorithm based on the idea of hierarchical clustering and evidence theory, which does not need to obtain the prior knowledge of the data, and does not require the data to obey the true distribution of the same parameter. When there are more data, the data are hierarchically clustered and the optimal grouping is selected according to the need to get the best fusion result. The next step is to simplify the grouping process and make the algorithm more concise.

Acknowledgements. This work is supported by Shaanxi Provincial Education Department industrialization project (16JF024) and is the key project in the field of industry (2018ZDXM-GY-039).

References

1. Xiong, Y., Ping, Z.: Data fusion algorithm inspired by evidence theory. J. Huazhong Univ. Sci. Technol. (Nat. Sci. Edn.), **39**(10), 50–54 (2011). https://doi.org/10.13245/j.hust.2011.10.007
2. Wang, W., Yang, Y., Yang, C.: A data fusion algorithm based on evidence theory. Control Decis. 1427–1430 (2013). https://doi.org/10.13195/j.kzyjc.2013.09.027
3. Sun, Z.: Bayesian estimation method for multi-source data fusion. J. Qilu Univ. Technol. (Nat. Sci. Edn.) 73–76 (2018). https://doi.org/10.16442/j.cnki.qlgydxxb.2018.01.016
4. Si, Y., Yang, X., Chen, Y., et al.: Multisensor weighted data fusion algorithm based on global state estimation. Infrared Technol. **36**(5), 360–364 (2014)
5. He, L., Zhang, C., Jiang, P.: A new method of conflict evidence fusion based on confidence distance. Appl. Res. Comput. **31**(10), 3041–3043 (2014)
6. Liang, X., Liu, X.: Improved consistency data fusion algorithm based on multicast tree. J. Huazhong Univ. Sci. Technol. (Nat. Sci. Edn.), **45**(3), 374–379 (2011). https://doi.org/10.19603/j.cnki.1000-1190.2011.03.007
7. Wang, H., Deng, J., Wang, L., et al.: Improved consistency data fusion algorithm and its application. J. China Univ. Min. Technol. **38**(4), 590–594 (2009)
8. Dou, G., Wan, R., Zhang, X.: Data fusion algorithm based on fuzzy theory for wireless sensor networks. Microelectron. Comput. **29**(9), 133–136 (2012). https://doi.org/10.19304/j.cnki.issn1000-7180.2012.09.033
9. Wang, S., Dai, F., Liang, B., et al.: A path based partition clustering algorithm. Inf. Control **40**(1), 141–144 (2011)
10. Feng, L.: Improved BP neural network algorithm and its application. Comput. Simul. **27**(12), 172–175 (2010)
11. Chen, X., Lou, P.H.: Application of improved hierarchical clustering algorithm in document analysis. J. Numer. Methods Comput. Appl. **30**(4), 277–287 (2009)

Low-Illumination Color Image Enhancement Using Intuitionistic Fuzzy Sets

Xiumei Cai, Jinlu Ma(✉), Chengmao Wu, and Yongbo Ma

Xi'an University of Posts and Telecommunications, Xi'an 710121, China
404628426@qq.com

Abstract. Because low illumination color image has the features of low brightness, poor contrast and dark color, and the enhancement effect of traditional image enhancement algorithm is very limited. A low illumination image enhancement algorithm based on fuzzy set theory is proposed, by transformed the RGB image into HSV space, and the brightness component V of the image is used to enhance the image in fuzzy plane. The experimental results show that this method is better than the traditional enhancement according to fuzzy set and the operation efficiency is higher, which can realize the clearness processing of low illumination image effectively.

Keywords: Intuitionistic fuzzy sets · Low illumination image
Contrast enhancement

1 Introduction

Images taken at night or under insufficient light have problems of low grayscale, low contrast and blurred edges. It is difficult to extract the effective information from the original image because the human eye has poor resolution to the low illumination image and even cannot distinguish some local details. Therefore, the enhancement of the low-illumination color image [1, 2] to obtain the color image suitable for the human eyes observation can improve the image quality significantly and observe more details.

Image enhancement methods are generally divided into frequency domain and transform domain. The frequency domain method [3] is based on the modification of the image's Fourier transform; the spatial domain method [4] directly processes the grayscale of the image pixels. The gray-scale transformation method is a common algorithm in space domain method, that is, through gray mapping function originally narrow grayscale range wider, and makes the processed image contrast enhancement. However, in low-light images with low contrast, the change of gray range has more significant effect on image quality, and the details become difficult to distinguish. In principle, both have their shortcomings. First, the gray level of the image is reduced, and some details disappear. Second, the noise that is not visible in the underexposed area of the image will appear. These traditional image enhancement technologies largely do not consider the fuzziness of the image, but simply change the contrast of the whole image or suppress the noise, thus, often suppress the noise and weaken the detail of the image.

P. Krömer et al. (Eds.): ECC 2018, AISC 891, pp. 199–209, 2019.
https://doi.org/10.1007/978-3-030-03766-6_22

In order to solve the above problems, a new low illumination color image enhancement algorithm is proposed, which converts the color space of the image from the RGB space into the HSV space [5], to maintains the color consistency; then transforms the image from the space domain using the membership function to the fuzzy domain, and enhance the image on the fuzzy feature plane to increase the contrast of the image and improve the deficiency of the traditional image.

2 HSV Color Space Model

At present, most of the image processing uses the RGB model for image enhancement. The RGB model obtains various colors by weighting the three color components of red (R), green (G), and blue (B). However, it is susceptible to the effects of illumination changes, and there is a high correlation between the three primary color components. Changing the color information of a channel often affects the information of other channels. Therefore, the image color distortion will be caused by the direct nonlinear processing of the color components of the image. The HSV model is a color model created by hue (H), saturation (S), and value (V) based on the visual characteristics of the color. HSV space is not only more suitable for the description of human color sense than RGB space, but also can separate various components effectively, making chromaticity and color saturation and brightness approximately orthogonal, which brings great convenient for subsequent true color image enhancement.

In the process of illumination compensation, the RGB image is converted to HSV space, and enhance the value component while the hue and saturation are kept unchanged, and then inverse transfer the value component with the hue and saturation components to generate new images. The transformation expression from RGB space to HSV space is as follows:

$$H = \begin{cases} 0, \; S = 0 \\ 60 \times \dfrac{G-B}{S \times V}, \; V = R \& G \geq B \\ 60 \times \dfrac{2+(B-R)}{S \times V}, V = G \\ 60 \times \dfrac{4+(R-B)}{S \times V}, V = B \\ 60 \times \dfrac{6+(G-B)}{S \times V}, V = R \& G < B \end{cases} \tag{1}$$

$$S = \frac{\max(R,G,B) - \min(R,G,B)}{\max(R,G,B)} \tag{2}$$

$$V = \max(R,G,B) \tag{3}$$

Where, R, G and B are respectively normalized RGB space values. The component H in the range (0, 360], both of the S and V components are in the range of [0, 1).

If $i = H/60$, $f = H/60 - i$, Where i is the divisor divisible by 60, and f is the remainder divisible by 60. Then:

$$P = V(1 - S) \tag{4}$$

$$Q = V(1 - Sf) \tag{5}$$

$$T = V[1 - S(1 - f)] \tag{6}$$

In the range of (0, 360], the transformation expression from HSV space to RGB space is as follows:

$$\begin{aligned}
&\text{when } i = 0, R = V, G = T, B = P \\
&\text{when } i = 1, R = Q, G = V, B = P \\
&\text{when } i = 2, R = P, G = Q, B = T \\
&\text{when } i = 3, R = P, G = Q, B = V \\
&\text{when } i = 4, R = T, G = P, B = V \\
&\text{when } i = 5, R = V, G = P, B = Q
\end{aligned}$$

3 Intuitionistic Fuzzy Set Enhancement Algorithm

3.1 Intuitionistic Fuzzy Set Algorithm

Intuitionistic fuzzy Sets [6, 7] (IFSs) is a generalization concept of fuzzy set theory. On the basis of fuzzy set theory, intuitionistic fuzzy set theory provides a solid mathematical basis to deal with the hesitation of uncertain information. Intuitionistic fuzzy sets are better able to react in human-like behavior than traditional fuzzy sets [8, 9]. For a finite set U, its intuitionistic fuzzy set can be expressed as:

$$A = \{(u, \mu(u), v(u), \pi(u)) | u \in U\}$$

where $\mu(u) + v(u) + \pi(u) = 1$, the functions $\mu(u)$, $v(u)$ and $\pi(u)$ denote the membership degree, nonmembership degree, and hesitation degree. $\mu(u)$ can be constructed using REFs [10], logarithmic functions, exponential functions, S and Z functions, or others [11].

There are three phases involved into the mostly fuzzy image processing approaches:

(1) Fuzzification Ψ, viz., the input data U (histograms, gray levels, features, etc.) is converted into a membership plane.
(2) Operation Γ, viz., some valid operator is applied in the membership plane for modified the membership value.

(3) Defuzzification Φ, viz., inverse transformed from the fuzzy domain to the spatial domain to complete the decoding and output data *X*(histograms, gray levels, features, etc.)

U is given by the following processing chain:

$$X = \Phi(\Gamma(\Psi(U))) \tag{7}$$

The block diagram of the intuitionistic fuzzy set enhancement algorithm is shown in Fig. 1. Since a foreground/background area has correlations in both spatial and frequency domain, it is necessary to divide an image into foreground and background areas for image enhancement. An original color image $I(u)$ is separated into the foreground area $I_1(u)$ and background area $I_2(u)$ though a threshold. After the fuzzy transformation and normalization, the filtered image $D(u)$ is combined with the original image by using the fusions #1, #2 and #3 to acquire an enhanced image.

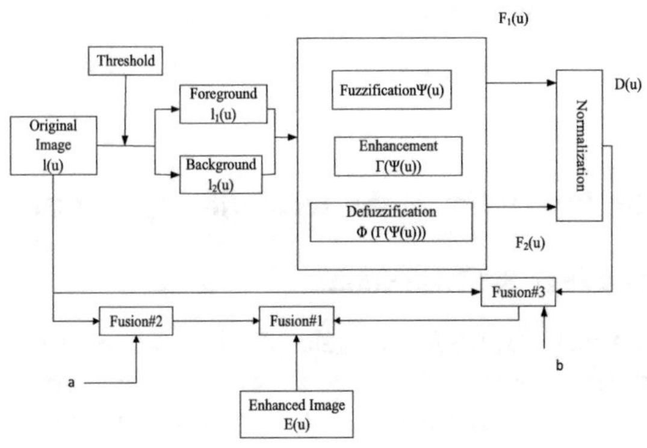

Fig. 1. The block diagram of the intuitionistic fuzzy set enhancement algorithm

3.2 Selection of Threshold

Global thresholding methods are easy to implement and also computationally less involved, such as the Otsu, minimum error, and Parzen window estimate methods [12]. But each way has pros and cons. In this section, we adopt an iterative strategy to automatically divide the image into the foreground and background areas. The selection of foreground or background does not require an input, for example, the size of the image, the gray scale, or the image features, etc.

For an input image, the following iterative strategy is used to find a global threshold:

(1) Initialize the global threshold T,

$$T = 0.5(I_{\max} + I_{\min}) \tag{8}$$

where $I_{\max}$ and $I_{\min}$ denote the maximum and minimum gray values of the image.

(2) Segment the picture using T. This engenders two groups of pixels: $I_1(u) = \{i | i \geq T, i = 0, 1, \ldots, 255\}$, $I_1(u)$ composing of all pixels with gray values $\geq T$, $I_2(u) = \{i | i < T, i = 0, 1, \ldots, 255\}$, $I_2(u)$ composing of all pixels with gray values $< T$.
(3) Calculate the average gray values m1 and m2 for the pixels in $I_1(u)$ and $I_2(u)$, respectively.
(4) Compute a new threshold value: $T' = 0.5(m_1 + m_2)$, If $|T - T'| > \lambda$, let $T = T'$, and repeat Steps (2) through (4). Or else, achieve a final segmentation threshold T'. Where, λ is a predefined constant.

3.3 Fuzzy Domain Image Processing

The intuitionistic fuzzy generator in the foreground and background area of the fuzzy plane transforms each pixel from the grayscale plane to the membership plane.

Use REFs to construct membership function of IFSs. If functions θ_1 and θ_2 are two automorphisms in a unit interval, an *REF* can be defined as:

$$REF : [0,1] \times [0,1] \rightarrow [0,1] \tag{9}$$

$$REF(x,y) = \theta_1^{-1}(1 - |\theta_2(x) - \theta_2(y)|) \tag{10}$$

Let $\theta_2(x) = x^2, 0 \leq x \leq 1$. Hence, a REF is defined as:

$$REF(x,y) = \theta_1^{-1}(1 - |x^2 - y^2|) \tag{11}$$

Subsequently, let $\theta_1(x) = \log((e-1)x + 1)$. Through inverse function, we get the inverse function of θ_1:

$$\theta_1^{-1}(x) = 0.582(e^x - 1) \tag{12}$$

where $e = \exp(1)$, $1/(e-1) \approx 0.582$, Therefore, the *REF* becomes:

$$REF(x,y) = 0.582[g(x) - 1] \tag{13}$$

where, $g(x) = \exp(1-(x+y)|x-y|)$. After that, we utilize the gray value x_u at point u in an image block (viz., the foreground or background area) to replace the variable x, and the average gray value of the block (viz., m_i $(i = 1,2)$) to substitute the variable y in (16). Then, the fuzzification for the foreground area can be expressed as:

$$\mu_i(u) = 0.582[g(u)-1] \tag{14}$$

where $g(u) = \exp(1-(x_u+m_i)|x_u-m_i|)$, $\mu_i(u)$ $(i = 1,2)$ denote the membership functions of the foreground and background areas. The fuzzification function is considered as the belongingness of a pixel to the image block. When traversing the entire image; the pixel plane is converted into the membership plane according to (14).

If $\mu(u)$ represent the membership function of each pixel in image I, which $v(u)$ represents its non membership function, there is a certain degree of hesitation $\pi(u)$ in allocating the subordinate value of each pixel. which,

$$v(u) = (1-\mu(u))/1+\lambda\mu(u) \tag{15}$$

$$\pi(u) = 1-\mu(u)-\frac{1-\mu(u)}{1+\lambda\mu(u)} \tag{16}$$

In this case, the range of the membership value is $\{\mu(u), \mu(u)+\pi(u)\}$.

After the fuzzification, a proper enhancement operation is necessary to enlarge the belongingness of the points whose gray levels are close to the average gray value of an image block and lessen the belongingness of those points whose gray levels are far from that average. The relationship between the enhancement and the fuzzification function is shown in Fig. 2.

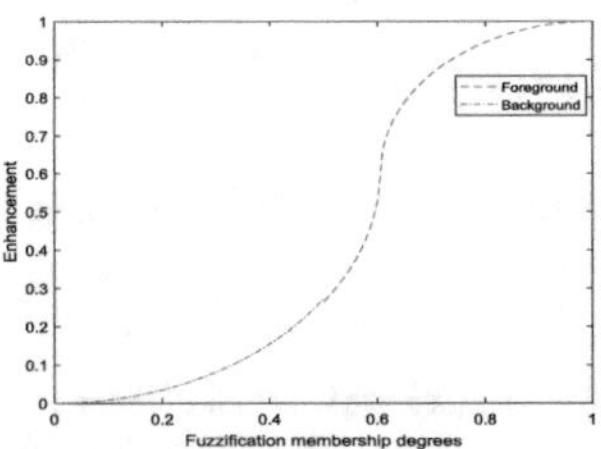

Fig. 2. Relationship between the enhancement and the fuzzification membership function

It can be seen from Fig. 2 that the fuzzy function of foreground and background can be well transitioned after some fuzzy enhancement. The function of fuzzy enhancement is as follows:

$$\text{If } (\mu_i)_{\min} \le \mu_i(u) \le (\mu_i)_{\max},\ \mu'(u) = \mu_i(th) - \sqrt{\mu_i^2(th) - \mu_i^2(u)} \tag{17}$$

$$\text{If } \mu_i(th) < \mu_i(u) \leq (\mu_i)_{\max},\ \mu'(u) = \mu_i(th) + \sqrt{\mu_i(th)^2 - \mu_i(u)^2} \tag{18}$$

where $\mu_i(th) = 0.582[g(th) - 1]$, $g(th) = \exp(1 - (th + m_i)|th - m_i|)$, $\mu_i'(u)$ $(i = 1, 2)$ are the enhancement membership degrees; *th* is the threshold according to the iterative strategy as mentioned in 2.2. $(\mu_i)_{\min}$ and $(\mu_i)_{\max}$ respectively, denote the minimum and maximum membership degrees of the foreground and background area, m_i $(i = 1, 2)$ denote the average gray values of the foreground and background areas, respectively.

Meanwhile, the fuzzy enhanced nonmembership function and the hesitation function can be determined by the formula (15) and the formula (16).

For the new fuzzy feature plane, the enhanced image can be obtained from the fuzzy domain to the gray space through an inverse transform.

$$\text{If } m_i \leq x_u \leq x_{\max}, F_i(u) = \sqrt{m_i^2 - \log(1 + 1.718\mu_i'(u))} \tag{19}$$

$$\text{If } th \leq x_u \leq m_i, F_i(u) = \sqrt{m_i^2 + \log(1 + 1.718\mu_i'(u))} \tag{20}$$

For different applications, fused #1, #2, and #3 as shown in Fig. 1 can be chosen as the arithmetic or logic operations, or parameterized logarithmic image processing (PLIP) [14]. Since arithmetic operations show better performance than PLIP [15], we adopt the arithmetic addition and multiplication in this paper. Then the enhanced image is:

$$E(u) = a \cdot I(u) + b(D(u) \otimes I(u)) \tag{21}$$

$$D(u) = \Phi(\Gamma(\Psi(I(u)))) \tag{22}$$

where a and b are the scaling factors; $I(u)$ denotes the original image; functions Φ, Γ, and Ψ denote the fuzzification, fuzzy enhancement, and defuzzification operations implemented orderly on $I(u)$; and $\otimes$ is the dot-product operation.

4 Intuitionistic Fuzzy Set Enhancement Algorithm Based on HSV Color Model

This paper combines the fuzzy set theory (based on IFSs) with the HSV color model and proposes a new fuzzy enhancement scheme, which is called the low illumination color image enhancement algorithm based on intuitionistic fuzzy sets. The scheme excels at highlighting details, such as the edge details of dim images. Transform color images from RGB space with closely related color components to HSV space to maintain color constancy. The image is transformed from the spatial domain into the fuzzy domain by using the luminance transform of the V component in the HSV model as a variable, and then enhanced the image on the feature plane to increase the contrast of the image. The algorithm steps are as follows:

Step 1: Input the image to be processed, transform the image into HSV color model, and extract the value component V
Step 2: Select the appropriate global thresholds to separate the foreground and background of the image
Step 3: Use the formula (14) to calculate the fuzzy feature plane of the image
Step 4: Enhance the fuzzy plane by using formula (17) and (18) to obtain a new fuzzy feature plane
Step 5: Using the formula (19) and (20) to do the inverse transformation to the new Fuzzy feature plane, transform the image from the fuzzy domain to the gray space, output the fuzzy enhancement variable and re-assign the V component again.
Step 6: Combining the original image and the fuzzy image by a combination of arithmetic addition and multiplication using formula (21), and outputting the fuzzy enhanced image.

5 Experimental Results and Analysis

Four low-light images of different scenes and types were selected in the experiment, and the RGB model-based fuzzy algorithm, HSV model enhancement algorithm, the IFS [13] algorithm and the HSV model-based intuitionistic fuzzy enhancement algorithm were used to compared through experimental results. The following 4 groups of images are the original image, and the different four algorithms processing effect results. As shown in Figs. 3, 4, 5 and 6.

(a) Original (b) FS-RGB (c) HSV (d) IFS (e) IFS-HSV

Fig. 3. The enhancement of the image ‘dusk’

Through the original image of Fig. 3, 4, 5 and 6 and the effect diagram processed by the four algorithms, it can be intuitively seen that the image processed by the algorithm has better enhancement effect, and the running time, the contrast between original image and each enhanced algorithm, the information entropy and the similarity between the original image and each enhanced algorithm (structural similarity index measurement system, SSIM) have given a objective evaluation. As shown in Tables 1, 2, 3 and 4.

Table 1 shows that the color image enhancement algorithm based on intuitionistic fuzzy sets has more advantages in time running. The statistical data in Tables 2 and 3 show that although the algorithms based on fuzzy set algorithm and HSV model have improved the contrast, the information entropy is affected. The algorithm based on intuitionistic fuzzy sets can solve this problem well. Table 4 compares the similarity between the four algorithms and the original image. It can be found that the image processed by the intuitionistic fuzzy set algorithm with color model added is less similar than other algorithms, that is, the enhancement effect is most obvious.

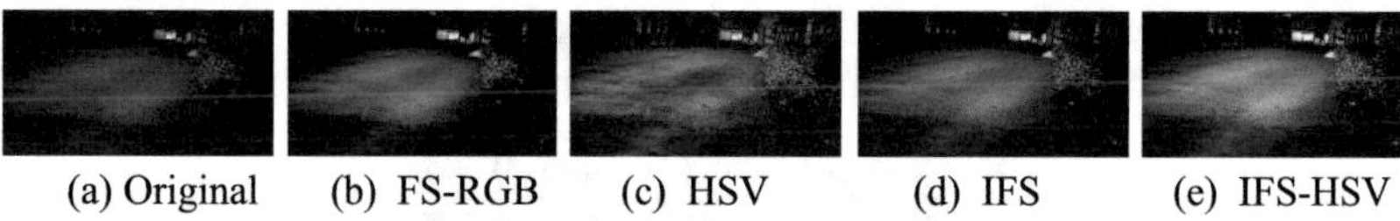

(a) Original (b) FS-RGB (c) HSV (d) IFS (e) IFS-HSV

Fig. 4. The enhancement of the image ‘road’

(a) Original (b) FS-RGB (c) HSV (d) IFS (e) IFS-HSV

Fig. 5. The enhancement of the image ‘scenic’

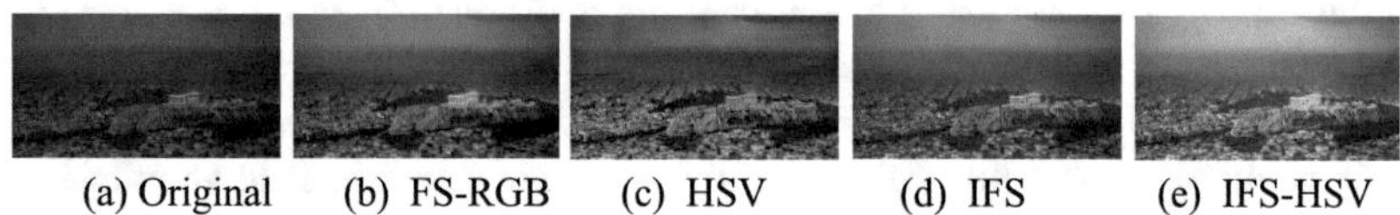

(a) Original (b) FS-RGB (c) HSV (d) IFS (e) IFS-HSV

Fig. 6. The enhancement of the image ‘city’

Table 1. The comparison of the running time (units)

Image	FS	HSV	IFS	IFS-HSV
dusk	2.235710	2.121260	1.702215	1.688734
road	0.641642	0.598399	0.523345	0.518440
scenic	1.377615	1.248350	1.230965	1.201130
city	2.488920	2.115735	1.782043	1.549208

Table 2. The comparison of the contrast

Image	Original	FS	HSV	IFS	IFS-HSV
dusk	0.0004	0.0005	0.0011	0.0015	0.0019
road	0.0002	0.0005	0.0009	0.0020	0.0025
scenic	0.0012	0.0014	0.0025	0.0029	0.0033
city	0.0011	0.0013	0.0021	0.0025	0.0030

Table 3. The comparison of the information entropy

Image	Original	FS	HSV	IFS	IFS-HSV
dusk	7.1340	6.4077	6.7420	7.0392	7.8763
road	6.0994	6. 5819	6.6378	7,3275	7.8642
scenic	8.9013	9.1368	9.8520	9.9902	10.3433
city	8.1257	8.2690	9.1744	9.7881	9.9899

Table 4. The comparison of the SSIM

Image	FS	HSV	IFS	IFS-HSV
dusk	0.9984	0.8783	0.7426	0.6953
road	0.9822	0.8875	0.7569	0.7122
scenic	0.9896	0.7993	0.6932	0.6599
city	0.9855	0.8988	0.8212	0.7077

6 Conclusion

For night, and the problem of the lower brightness and contrast of low illumination image, based on the color retention of HSV color model, an intuitionistic fuzzy set algorithm is used to enhance the brightness of the low illumination image, and the intuitionistic fuzzy set is more efficient than the traditional fuzzy set algorithm to deal with the uncertain region in the image because of its nonmembership degree and hesitation degree. The experimental results show that the proposed algorithm reduces the computational time while enhancing the contrast of image. It is an efficient low-illumination image enhancement algorithm.

Acknowledgements. This work was supported by the Department of Education Shaanxi Province (16JK1712), Shaanxi Provincial Natural Science Foundation of China (2016JM8034, 2017JM6107), and the National Natural Science Foundation of China (61671377, 51709228).

References

1. Du, Y., Wu, G., Tang, G.: Auto-encoder based clustering algorithms for intuitionistic fuzzy sets. In: International Conference on Intelligent Systems and Knowledge Engineering, pp. 1–6 (2017). https://doi.org/10.1109/iske.2017.8258819
2. Lee, S.L., Tseng, C.C.: Color image enhancement using histogram equalization method without changing hue and saturation. In: IEEE International Conference on Consumer Electronics – Taiwan, pp. 305–306. IEEE (2017). https://doi.org/10.1109/icce-china.2017.7991117
3. Bhairannawar, S., Patil, A., Janmane, A., et al.: Color image enhancement using Laplacian filter and contrast limited adaptive histogram equalization. In: IEEE International Conference on Innovations in Power and Advanced Computing Technologies, vol. 8(27), pp. 32–34 (2018). https://doi.org/10.1109/ipact.2017.8244991
4. Huang, K., Wang, Q., Wu, Z.: Natural color image enhancement and evaluation algorithm based on human visual system. Comput. Vis. Image Underst. **103**(1), 52–63 (2006). https://doi.org/10.1016/j.cviu.2006.02.007
5. Pal, S.K., King, R.A.: Image enhancement using smoothing with fuzzy sets. IEEE Trans. Syst. Man Cybern. **11**(7), 494–501 (1981). https://doi.org/10.1109/tsmc.1981.4308726
6. Hung, W.L., Yang, M.S.: Similarity measures of intuitionistic fuzzy sets based on Hausdorff distance. Pattern Recogn. Lett. **25**(14), 1603–1611 (2004). https://doi.org/10.1016/j,patrec.2004.06.006

7. Hung, W.L., Yang, M.S.: Similarity measures of intuitionistic fuzzy sets based on LP metric. Int. J. Approximate Reasoning **46**(1), 120–136 (2007). https://doi.org/10.1016/j,ijar.2006.10.002
8. Chaira, T.: Intuitionistic fuzzy segmentation of medical images. IEEE Trans. Bio. Eng. **57** (6), 1430–1436 (2010). https://doi.org/10.1109/tbme.2010.2041000
9. Atanassov, K.T.: Intuitionistic fuzzy set. Fuzzy Set. Syst. **20**(1), 87–96 (1986). https://doi.org/10.1016/s0165-0114(86)80034-3
10. Bustunce, H.: Restricted equivalence functions. Fuzzy Sets Syst. **157**(17), 2333–2346 (2006). https://doi.org/10.1016/j.fss.2006.03.018
11. Ananthi, V.P., Balasubramaniam, P., Lim, C.P.: Segmentation of gray scale image based on intuitionistic fuzzy sets constructed from several membership functions. Pattern Recogn. **47** (12), 3870–3880 (2014). https://doi.org/10.1016/j.patcog.2014.07.003
12. Wang, S., Chung, F., Xiong, F.: A novel image thresholding method based on Parzen window estimate. Pattern Recogn. **41**(1), 117–129 (2008). https://doi.org/10.1016/j.patcog.2007.03.029
13. Deng, H., Deng, W., Sun, X., et al.: Mammogram enhancement using intuitionistic fuzzy sets. IEEE Trans. Biomed. Eng. **PP**(99), 1 (2016). https://doi.org/10.1109/tbme.2016.2624306
14. Panetta, K., Agaian, S., Zhou, Y., et al.: Parameterized logarithmic framework for image enhancement. IEEE Trans. Syst. Man Cybern. Part B Cybern. Publ. IEEE Syst. Man Cybern. Soc. **41**(2), 460–473 (2011). https://doi.org/10.1109/tsmcb.2010.2058847
15. Panetta, K., Zhou, Y., Agaian, S., et al.: Nonlinear unsharp masking for mammogram enhancement. IEEE Trans. Inf. Technol. Biomed. Publ. IEEE Eng. Med. Biol. Soc. **15**(6), 918–928 (2011). https://doi.org/10.1109/titb.2011.2164259

Design of IoT Platform Based on MQTT Protocol

Shan Zhang[1(✉)], Heng Zhao[1], Xin Lu[1], and Junsuo Qu[2]

[1] School of Communication and Engineering, Xi'an University of Post and Telecommunications, Xi'an 710121, China
494905890@qq.com

[2] School of Automation, International Research Center for Wireless Communication and Information Processing of Shaanxi Province, Xi'an, China

Abstract. Against to the campus custom requirements, this paper designs the platform of IoT of campus application. In the platform design, the technical architecture is divided into five aspects, sensing layer, access layer, storage layer, service layer and application layer. The functional architecture is divided into nine independent functional modules, standardized interfaces and separated between UI and Server which are developed by SSH framework. The platform focus on designing modular, universal, read-write separation and load balancing, improving high-impact access for IoT application. The test results show that the platform meets the expected design goals. Through the actual scene application, the platform achieves the expected effects.

Keywords: IoT · MQTT · SSH · Load balancing

1 Introduction

With the rapid development of the IoT industry, "Internet of Everything" is gradually becoming a reality, thus changing all aspects of our life. Whether it is improving productivity, reducing production costs and management costs, or improving the standard of living and quality of residents, the IoT will achieve significant success in the market [1–3]. Now the global market has increased dramatically, and accessible devices have experienced explosive growth. By 2020, it is estimated that the access volume of IoT devices worldwide will reach 50 billion [4–6]. The IoT devices are designed to connect with other devices and use Internet protocols to transfer information. At present, Xively is one of the best IoT platforms. It provides an open API interface and socket communication based on the MQTT protocol [7].

2 Basic Concepts of the MQTT Protocol

Message Queuing Telemetry Transport (MQTT) is a protocol developed by IBM that is used in IoT for data transmission [8, 9], which is designed for restricted devices and low bandwidth, high latency or unreliable networks. In MQTT publisher and subscriber (or clients) do not need to know each other's identity, when a user posts a message

P. Krömer et al. (Eds.): ECC 2018, AISC 891, pp. 210–216, 2019.
https://doi.org/10.1007/978-3-030-03766-6_23

M to a particular topic T, all users subscribed to topic T will receive message M. Similar to Hyper Text Transfer Protocol (HTTP), MQTT relies on Internet Protocol (IP) and Transmission Control Protocol (TCP) as its underlying layer. Compared to HTTP, MQTT is designed as a protocol with lower overhead.

MQTT provides three levels of quality of service (QoS). Level 0 means that the message is transmitted at most once. Level 1 means that each message is transmitted at least once. Level 2 means that each message is transmitted exactly once, requiring a four-way handshake mechanism to ensure that the message is transmitted exactly once.

3 Platform Technology Architecture

The IoT platform based on the MQTT protocol adopts hierarchy systems, which divided into a sensing layer, an access layer, a storage layer, a service layer, and an application layer from bottom to top. The servers and databases in the entire platform are deployed on the Ali ECS cloud server, and the platform architecture can be flexibly adjusted relying on the extension of the service. The platform architecture diagram is shown in Fig. 1.

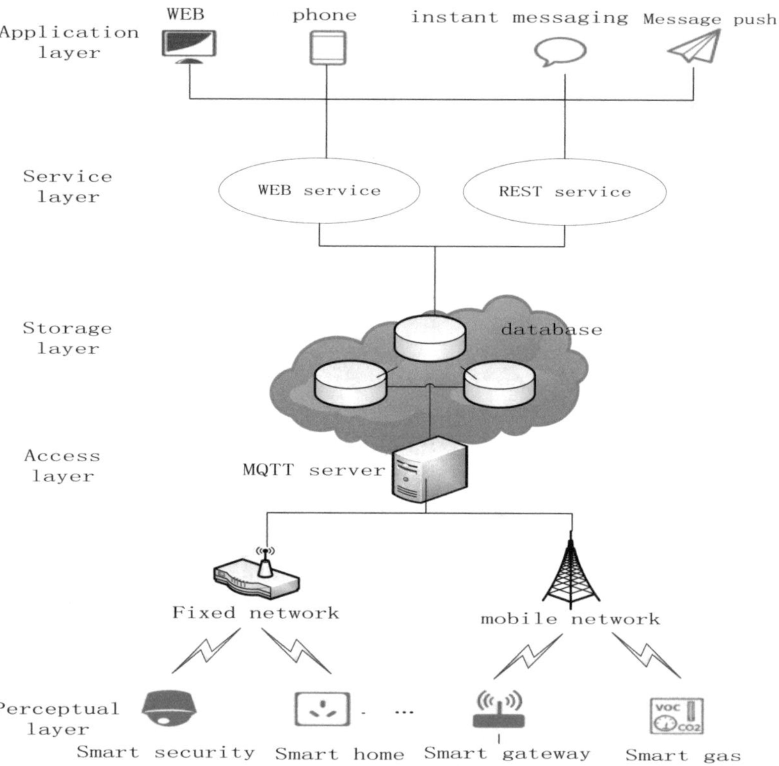

Fig. 1. Platform architecture diagram

The first layer is the sensing layer, which is mainly responsible for the collection of information. It is the most basic and core part of the IoT. The sensing layer identifies devices for operation and acquisition of relevant information through sensors such as readers, sensors, cameras, GPS locators, and RFID radio tags.

The second layer is the access layer, which is mainly responsible for the authorization and management of smart device connections in the sensing layer. The access layer accepts connection requests from devices on Cellular Mobile network (such as 2G, 3G, 4G, etc.) or fixed network (such as ADSL, FTTX, etc.) and authorizes devices in the server white list.

The third layer is the storage layer, which is mainly responsible for storing the sensor data uploaded by the smart device and the service data of the web platform. So as to cope with the demand expansion of the late business, the storage layer adopts a database cluster architecture combining read-write separation and high-availability load balancing technology to improve the database concurrency.

The fourth layer is the service layer, which is mainly responsible for providing WEB services and RESTful API interfaces for the application layer. WEB services are developed with the SSH open source framework to ensure that the platform has good scalability, maintainability and low coupling.

The fifth layer is the application layer, which is mainly responsible for providing users with WEB and mobile phone application services.

4 Platform Functional Architecture

The platform is mainly composed of a registration login module, a product management module, a data stream management module, a device management module, a data display module, an online debugging module, a product data module, a device trend module, and a data statistics module. The functional structure diagram is shown in Fig. 2.

5 Database High Concurrency Design

Most IoT platforms and software are multi-threaded, and concurrency errors are also difficult to detect which have caused serious accidents, including a blackout that left tens of millions of people without electricity [10]. So the MySQL database uses a combination of two technologies, read-write separation and high-availability load balancing, on high concurrency. The database architecture is shown in Fig. 3. The read and write separation uses the better performance of Mycat middleware. In the high-availability load balancing implementation, it adopts a combination of the haproxy, keepalive and mysql. It mainly takes into account that haproxy has its own mysql detection script without additional code. In the database group, the master server replication strategy is used the database servers (mysql-01 and mysql-02) responsible for write operations, ensuring that both servers can be used as real-time business databases and provide services at the same time; The database server (mysql-03 and mysql-04) of the operation is respectively used with the database server (mysql-01 and mysql-02) responsible for

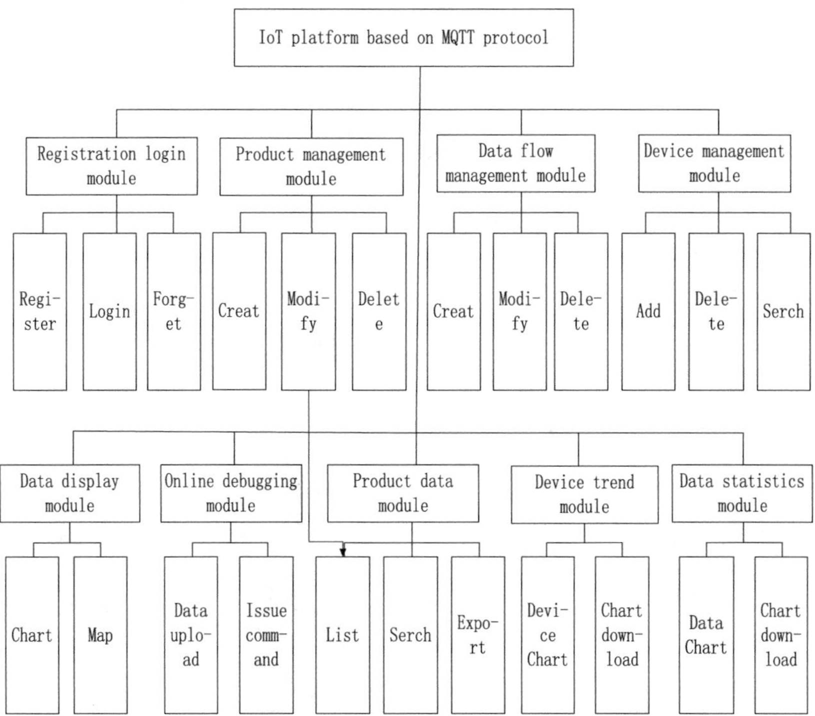

Fig. 2. Platform functional architecture

the write operation to adopt the master-slave replication strategy to ensure the data consistency of the master-slave database.

6 Platform Testing and Analysis

For the IoT platform, the number of concurrent access devices and the number of concurrent Web services connections were tested. The scheme is shown in Table 1.

The relevant statistics are shown in Table 2, where Samples (number of samples) is 5000, Average (average response time) is 783 ms/time, Error% (request failure rate) is 0.00%, and Throughput (requests per second) is 462.5. From the failure rate of the request in the statistics, it is obvious that the server can withstand simultaneous access of 5,000 devices. Simultaneously, you can also view the related connection statistics and the IP address and port number of the connected host by accessing the Connectors tab on the Apollo management page, as shown in Fig. 4. Among them, the Current Connected is 5000, which also indicates that the server can accept connection requests from 5000 devices at the same time.

Under the thread group of the "HTTP Request Thread", add a graphical result, view the result tree, and aggregate the report and other listeners and run the test plan. The result of the HTTP request is shown in Table 3. The Samples is 10000, the Average is

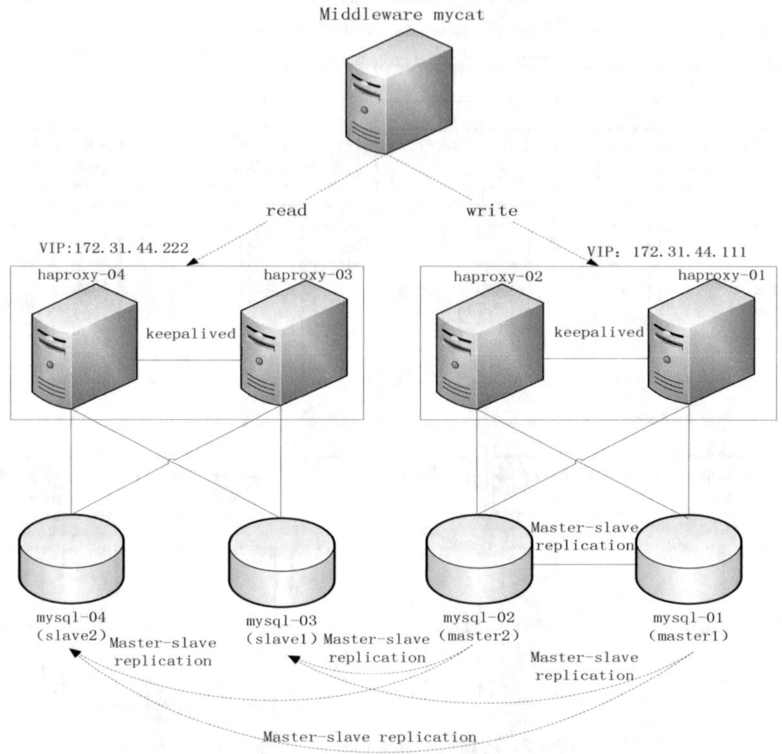

Fig. 3. Real-time change line chart

Table 1. Platform performance test plan

Number	Test items	Test program
1	Device access concurrency	At the same time, 5000 devices are connected, and the platform can respond normally
2	Web service concurrent connections	At the same time, 10,000 users access the Web service, and the platform can respond normally

Table 2. MQTT access statistics

Label	Samples	Average	Min	Max	Error%	Throughput
MQTT	5000	783	7	6553	0.00%	462.5/sec

3992 ms/time, the Error% is 0.00%, and the Throughput is 677.9. replying on the request failure rate, the platform can normally respond to 10,000 users accessing the Web service at the same time.

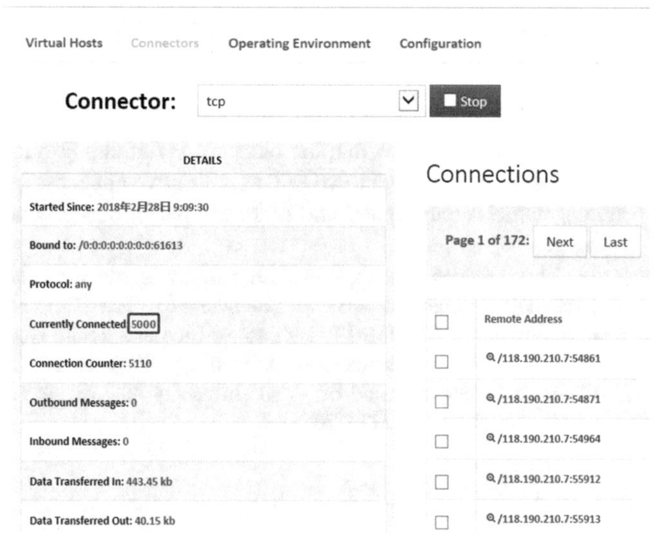

Virtual Hosts Connectors Operating Environment Configuration

Connector: tcp Stop

DETAILS

Started Since: 2018年2月28日 9:09:30

Bound to: /0:0:0:0:0:0:0:0:61613

Protocol: any

Currently Connected 5000

Connection Counter: 5110

Outbound Messages: 0

Inbound Messages: 0

Data Transferred In: 443.45 kb

Data Transferred Out: 40.15 kb

Connections

Page 1 of 172: Next Last

Remote Address

/118.190.210.7:54861

/118.190.210.7:54871

/118.190.210.7:54964

/118.190.210.7:55912

/118.190.210.7:55913

Fig. 4. Connection statistics details page

Table 3. HTTP request statistic

Label	Samples	Average	Min	Max	Error%	Throughput
HTTP	10000	3992	2	12909	0.00%	677.9/sec

For a single database server and a high concurrent server architecture, adjust the configuration and test the number of concurrent connections. The results are shown in Table 4.

Table 4. Concurrent test results

Configuration \ Success rate \ Concurrent number	50	500	5000	50000
Single database	100%	100%	100%	14.21%
High concurrency architecture	100%	100%	100%	100%

7 Conclusion

By testing the number of concurrent accesses and the number of concurrent connections of Web services, the platform was tested and verified from both functional and performance aspects. The results show that the platform is running normally and meets the expected design goals. It shows that MQTT as a lightweight message publishing/subscribing protocol, which concurrency and real-time performance is well.

Acknowledgements. This research was supported in part by grants from the International Cooperation and Exchange Program of Shaanxi Province (2018KW-026), Natural Science Foundation of Shaanxi Province (2018JM6120), Xi'an Science and Technology Plan Project (201805040YD18CG24(6)), Major Science and Technology Projects of XianYang City (2017k01-25-12), Graduate Innovation Fund of Xi'an University of Posts & Telecommunications (CXJJ2017012, CXJJ2017028, CXJJ2017056).

References

1. Zhang, X.: Talking about the application and development trend of domestic IoT. Software **33**(10), 116–117 (2012). https://doi.org/10.3969/j.issn.1003-6970.2012.10.037
2. Liu, Y.: Review of research on IoT technology. Value Eng. **22**, 226–227 (2013). https://doi.org/10.3969/j.issn.1006-4311.2013.22.125
3. He, W.: Key technologies and applications of the IoT. Inf. Comput. (20), 167–168 (2017). https://doi.org/10.3969/j.issn.1003-9767.2017.20.064
4. Wang, Q.: Thoughts on the development of IoT and the construction of China's IoT. Heilongjiang Sci. Technol. Inf. **38**(21), 164 (2014). https://doi.org/10.3969/j.issn.1673-1328.2014.21.154
5. Wang, F.: The exploration and future of china mobile IoT platform. Commun. World **30**, 36–37 (2017). https://doi.org/10.3969/j.issn.1009-1564.2017.30.022
6. Ding, Z.: Analysis on the development and construction of IoT in China. Inf. Syst. Eng. **5**, 123–124 (2017). https://doi.org/10.3969/j.issn.1001-2362.2017.05.087
7. Gao, N.: Application of IBM message middleware WebSphere MQ. Comput. Knowl. Technol. (Certification Exam) **6**(31), 8877–8879 (2010). https://doi.org/10.3969/j.issn.1009-3044.2010.31.083
8. Ren, X.: Message push server based on MQTT protocol. Comput. Syst. Appl. **23**(3), 77–82 (2014). https://doi.org/10.3969/j.issn.1003-3254.2014.03.012
9. Yao, D.: Research and implementation of IoT communication system based on MQTT protocol. Inf. Commun. (3), 33–35 (2016). https://doi.org/10.3969/j.issn.1673-1131.2016.03.014
10. Fan, X.: A review of software security research. Comput. Sci. **38**(5), 8–13 (2011). https://doi.org/10.3969/j.issn.1002-137X.2011.05.002

A Design of Smart Library Energy Consumption Monitoring and Management System Based on IoT

Chun-Jie Yang[1(✉)], Hong-Bo Kang[2], Li Zhang[2], and Ru-Yue Zhang[2]

[1] Xi'an University of Technology, Xi'an 710048, China
yyccjj007@163.com

[2] College of Automation, Xi'an University of Posts and Telecommunications, Xi'an 710121, China
khb2000@163.com

Abstract. Aiming at the energy consumption management of the library, a scheme of library monitoring system based on IoT is proposed. The library monitoring system is designed with embedded technology and LoRa technology, including the perception layer, network layer and application layer. The system consists of the underlying LoRa nodes, sink nodes, servers and monitoring terminals to achieve the library's internal power monitoring and management. In this paper, the system network topology, LoRa node hardware and software design, communication protocol design are introduced detailedly. Experiment tests indicate that the scheme has the long communication distance, low power consumption, easy networking and effective performance. The design provides a great solution for smart library energy consumption management and energy saving.

Keywords: Smart library · Internet of things technology
Energy consumption monitoring · LoRa

1 Introduction

The Internet of things (IoT) is the third wave of world industry following computer and Internet, The smart Library Based on the Internet of things will become a new model of the future library. The university library has many characteristics, such as long opening time, strong fluidity, more functional rooms, whose energy consumption is obviously higher than other public buildings, and the use of electricity is unreasonable. Therefore, the library environment and energy consumption management based on the advanced IoT technology is of great practical value. Low carbon, energy saving and green development will be a main direction for the future development of Smart Library [1, 2].

The paper proposes a solution of Library monitoring system based on LoRa communication technology. As a new kind of wireless communication technology of LPWAN, LoRa solves the balance between transmission distance and low power in traditional wireless sensor network. LoRa uses advanced spread spectrum modulation technology and encoding/decoding scheme, which increases the link budget and better anti-interference performance, and has better stability for deep fading and Doppler

P. Krömer et al. (Eds.): ECC 2018, AISC 891, pp. 217–224, 2019.
https://doi.org/10.1007/978-3-030-03766-6_24

frequency shift [3, 4]. In this paper, the LoRa nodes are used to collect the energy consumption and environmental parameters of the library, which are gathered to the gateway (sink nodes), and then uploaded to the remote server PC through the network. Finally, the monitoring of the energy (power) of the library is realized.

2 The Design of System

There are three layers in this system. The network topology is shown in Fig. 1.

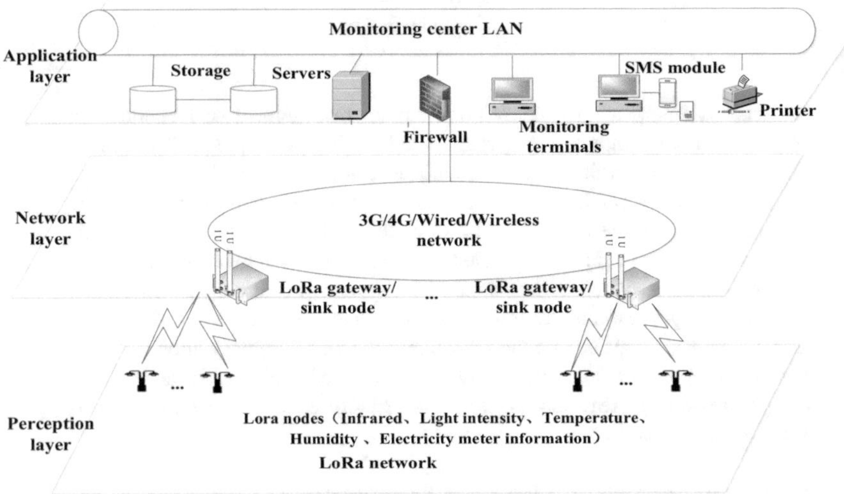

Fig. 1. The framework of energy consumption management system of smart library based on NB-IoT

(1) Perception layer. It is also called collection layer, the lowest level of the system, consisting of sensor nodes which collect the information of light intensity, infrared rays, temperature, humidity and electricity meters in the library, and convey information upward to network layer via LoRa network.
(2) Network layer. LoRa network architecture adopts typical star networks, dividing the controlled areas into different channels to reduce the mutual interference of signals. The sink node and acquisition nodes in the same areas are also in the same channel. The data from perception layer are passed to the LoRa sink nodes embedded with TCP/IP protocol, which will collect the data and send them to application layer via INTNET network [5–7].
(3) Application layer. There are server and application terminals in this layer. The data from network layer are received and integrated by the server which can store them in the data base. The proxy server is in charge of dealing the requests from clients and then distributes them to working servers evenly which is responsible

for the interaction details with clients. The application terminals contain PC monitor terminal and mobile intelligent terminal APP.

3 Design of System Hardware Platform

3.1 Terminal Monitoring Nodes Design

The structure of terminal nodes structure is shown in Fig. 2.

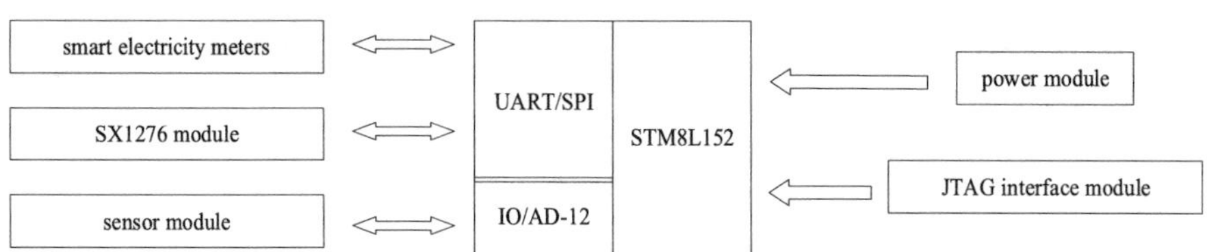

Fig. 2. Block diagram of terminal nodes composition

The terminal nodes complete data acquisition from the controlled objects in the field and transfer data upward to the corresponding sink node. STM8L152 produced by STMicroelectronics is used as MCU. The controller adopts special technology of ultra-low leakage, and the current is as low as 0.3μA in the lowest power mode. STM8L152 is very suitable for the data acquisition in the monitoring system.

(1) Sensor Circuit Design

The information collected in the library include smart electricity meters, infrared sensors, light intensity, temperature and humidity. The smart electricity meters are distributed in the metering terminals, which comply with all criteria of the standard about wired or wireless physical interfaces of the smart meters (Refer to industry standard DL/T 645-1997/2007 for implementation).The system adopts RS485 interface to collect smart electricity meters.

(2) LoRa Nodes Design

SX1276 produced by SEMTECH Company is the key chip of RF module in the system, which is a semi-duplex transceiver providing ultra-long range while maintaining low current consumption. The circuit including: analog switch circuit, transmitting circuit and receiving circuit.SX1276 requires mode switching when sending or receiving data [8].

3.2 Sink Nodes Design

The sink node is located at the network layer and connects the collect nodes and the server. It is a data gateway in the heterogeneous network and plays a key role in the system. The structure of the sink node is shown in Fig. 3. The node adopts

STM32F407 as the main controller, SX1276 as the LoRa receiver (Polling each terminal nodes in the same area), and ENC28J60 as the Ethernet driver which contains the communicating interface connected with MCU in SPI mode. Other circuits are not redundant.

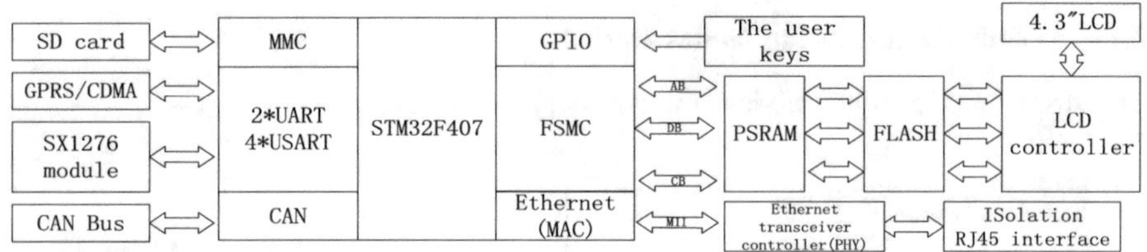

Fig. 3. Block diagram of sink node

TCP/IP network protocol is embedded in sink nodes which can realize data collection, protocol conversion, encapsulation and packaging of TCP/IP network protocol, data uploading, remote command delivering downwards, local storage debugging and other functions.

4 Design of System Software Platform

4.1 LoRa Communication Protocol Design

The system adopts LoRa private network protocol. According to the location, the LoRa nodes are partitioned by the system, and which have the same channel in the same region. After receiving the synchronous time instruction sent by the host (sink node), the node device synchronizes its own time. There are time information of a single time slice and the total number of time slices in the synchronization time. After synchronizing of system, each node can send data according to its device address number in the corresponding time slice, so that the data of all nodes will not conflict. The Synchronous clock information produced by the host is sent at regular intervals to ensure that all nodes have the same step, and the newly added nodes can quickly access the network [9, 10].

4.2 Software Design of Sink Node

Sink node software design includes FreeRTOS system design, network programming, data processing and server design. Software design adopts the top-down design method. FFreeRTOS has been transplanted to the sink node, which can create collection, local storage, network upload and other tasks. Collection task collects data of controlled objects; Local storage task (transplanted the FATFs open source file system) gets the valid data and stores them; Web upload task (transplanted Lwip embedded TCP/IP protocol stack) retrieves valid data and packages it into JSON format and uploads it to the remote server [11].

Network transmission is dominated by TCP/IP stack embedded in the MCU. Once the network transport service is started, it will take control of the ENC28J60 Ethernet controller, and all network data will be processed by the service. The network module is responsible for uploading the data by connecting the server.

4.3 Server Soft Design

The server consists of proxy server and working servers, whose processing is shown in Fig. 4. The server adopts TCP/IP protocol as communication protocol and adopts concurrent model of multithreading by which the Reactor event processing model is adopted. The general model of the server consists of a main thread Reactor + multiple worker threads (there are difference between the proxy server and the working server). The server is deployed typically in Linux environment, therefore, the system adopts the most efficient I/O multiplexing mechanism.

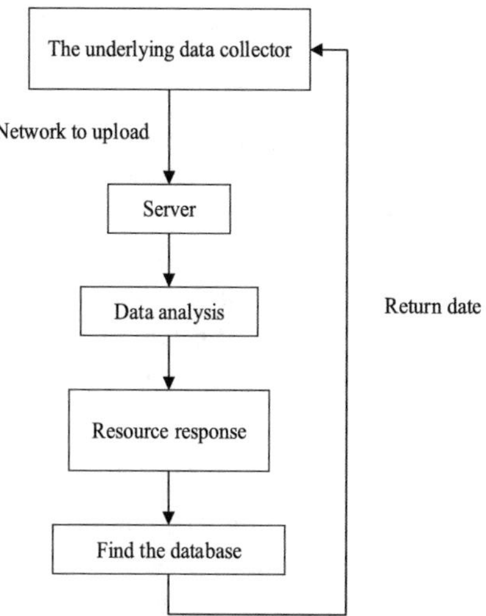

Fig. 4. Server processing process

In order to achieve load balancing, the proxy server uses the consistent hash algorithm implemented by the R-B tree and MD5 algorithm which is very efficient. The working server is added to the R-B tree with pre-configured weights. When the proxy server needs to assign a work server to a client, find a node in the R-B tree with the nearest large value of the client data created previously, and then completes the uniform distribution to the client connection. The database connection pool will allocate a certain number of database connections in advance, and every time the client wants to insert data into the database, it can directly use the already allocated connection. The interface of system front-end display are shown in Fig. 5.

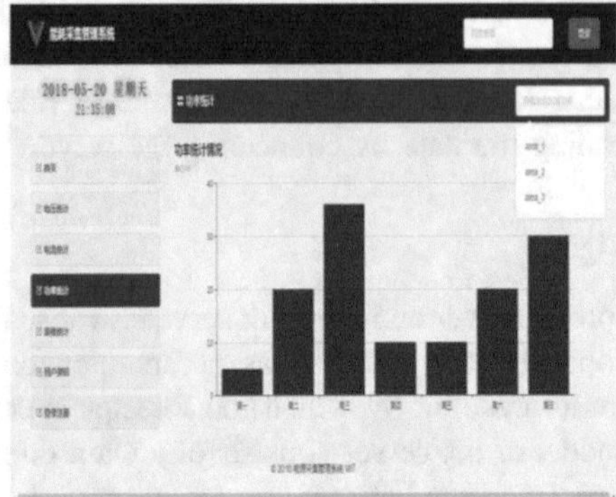

Fig. 5. The display interface of system front-end

5 System Testing

5.1 The Scheme of Testing

Taking a university library which has 4 floors as the test object, deploy the monitoring system. Each floor area is about 7500m2. According to the position of the lamps inside, the infrared module is arranged nearby to detect the personnel information; the photosensitive module, the temperature and humidity module are evenly distributed in every room; sink nodes are placed nearby work place of the staff on duty. The three-phase intelligent meters and LoRa sink nodes are installed in the distribution box of reading room to realize energy consumption monitoring and data transmission.

5.2 Test Results

According to the scheme, the RF center frequency of the LoRa nodes is 470 M and the transmitting power is 20dbm. The communication test results between the LoRa acquisition nodes and the sink nodes in the library is shown in Table 1, which shows that with the increase of the distance between the floor and the communication within the library, the data packet loss rate increases gradually, and the whole run is stable relatively.

Table 1. Library test results

Number	The test floor	Data communication packages	Packet loss rate
1	First floor	100	0%
2	Second floor	98	2*
3	Third floor	95	5%
4	Fourth floor	93	8%

For refine management, each function area of every floor is subdivided, for example, the reading room is divided into: seat area, bookshelf area and indoor background light area, the sensor nodes collect accurately the information of each sub-area. According to the data, the center makes real-time lighting planning and intelligent control of associated equipment. Comparison of energy consumption is shown in Fig. 6, the average energy consumption can be reduced by about 15%.

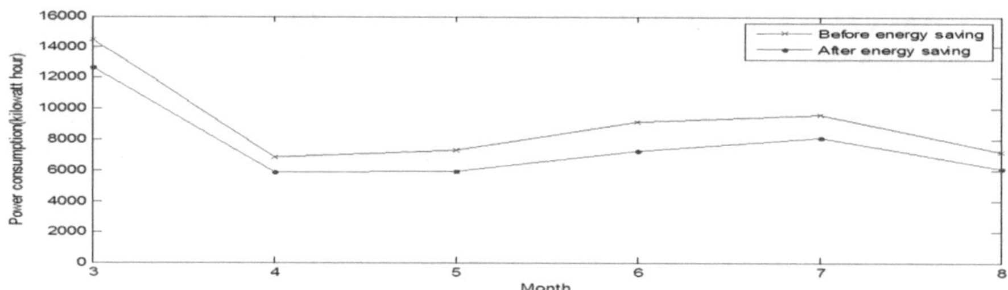

Fig. 6. Comparison of energy consumption between before and after energy saving

6 Conclusions

In this paper, a kind of energy consumption monitoring scheme of library is presented. LoRa is introduced into the communication network of the system, which can realize data interaction between the nodes in the bottom layer of LoRa and the Ethernet. The server adopts the design idea of multithreading, high concurrent and load balancing.

From the experiment results, the system has many advantages—satisfactory running state, long communication distance, convenient and fast networking, low cost, which can meet the requirements of smart library to achieve the purpose of controlling and regulating energy consumption. The system has higher practical value and better market application promise.

Acknowledgment. This work is supported by the Shaanixi Education Committee Project (14JK1669), Shaanxi Technology Committee Project (2018SJRG-G-03).

References

1. Wang, S.: On three main features of the smart library. J. Libr. Sci. China **6**, 22–28 (2016)
2. Huang, J., Xu, X., Li, J.: Application of intelligent lighting technology for energy-saving design in library buildings. Build. Energy Effic. **5**, 110–113 (2017)
3. Lewark, U.J., Antes, J., Walheim, J., et al.: Link budget analysis for future E-band gigabit satellite communication links. CEAS Space **4**(1), 41–46 (2013)
4. Min, H., Cheng, Z., Huang, L.: A design of wireless meter reading system based on RF. Comput. Meas. Control **2**, 639–642 (2014)
5. Aref, M., Sikora, A.: Free space range measurements with semtech loratm technology. In: 2nd International Symposium on Technology and Applications, DAACS—SWS, pp. 19–23, Fenburg (2014)

6. Wang, P., Wang, W.: Design of energy consumption data collector based on 3G. Comput. Meas. Control **12**, 4202–4206 (2015)
7. Martinez, B., Monton, M., Vilajosana, I., Prades, J.: The power of models. modeling power consumption for IoT devices. IEEE Sens. J. **15**(10), 5777–5789 (2015)
8. Lora family | wireless & RF ICs for ISM band applications | semtech. http://www.semtech.com/wireless-rf/lora.html. Accessed 21 Nov 2016
9. Zheng, T., Gidlund, M., Åkerberg, J.: Wirarb: a new MAC protocol for time critical industrial wireless sensor network applications. IEEE Sens. J. **16**(7), 2127–2139 (2016)
10. Marín, I., Arias, J., Arceredillo, E., Zuloaga, A., Losada, I., Mabe, J.: Ll-mac: a low latency MAC protocol for wireless self-organised networks Microprocess. Microsyst **32**(4), 197–209 (2008)
11. Hu, C., Zhao, Q.C., Feng, H.-R.: Research and implementation of a kind of intelligent data-collection system in the internet of things. Electron. Meas. Technol. **37**(6), 108–114 (2014)

A Random Measurement System of Water Consumption

Hong-Bo Kang[1(✉)], Hong-Ke Xu[1], and Chun-Jie Yang[2]

[1] Institute of Electrical and Control Engineering, Chang'an University, Xian 710064, China
khb2000@qq.com, xuhongke@chd.edu.cn

[2] College of Automation, Xian University of Posts and Telecommunications, Xian 710121, China
yyccjj007@163.com

Abstract. There is an "one switch" method in the water management in larger institutions, not focusing enough on internal control, and the essence is the lack of good monitoring method. The paper proposes a random measurement system of water consumption based on mix network of ZigBee and NB-IoT, consists of two components called detecting node and cooperative node. The system can realize the functions of detection and transmission of the parameters such as flow, velocity, time, frequency, failure etc., then data will be uploaded to the IoT platform. Experiment results show that system performance is effective and utility and can provide an effective assessment for strategy of using water.

Keywords: Water consumption · Random measurement · ZigBee NB-IoT

1 Introduction

The protection of water resources lies firstly in the rational allocation and the effective monitoring. The purpose of monitoring is detecting the occurrence of unreasonable water consumption, and the allocation of water resources is based on assessment of water consumption. Here's the situation: larger institutions have adopted the "one switch" extensive management, focusing on total consumption, not the process, ignoring the features of water consumption, so the random measurement means is acquired to meet the true needs of water.

Water consumption is measured mainly by water meter. Waterways belong to infrastructure, the distribution of water network is complex, so it is impossible for water meter to be installed to every branch, furthermore, water meter installation and maintenance costs are high, the statistics process is also hard, and the renovation of old device is impracticable.

P. Krömer et al. (Eds.): ECC 2018, AISC 891, pp. 225–231, 2019.
https://doi.org/10.1007/978-3-030-03766-6_25

2 The Composition and Principle of System

The essence of the problem is that water resources statistics lack effective means. In the paper, a random check system of water consumption based on Internet of things is proposed [1, 2] (Fig. 1).

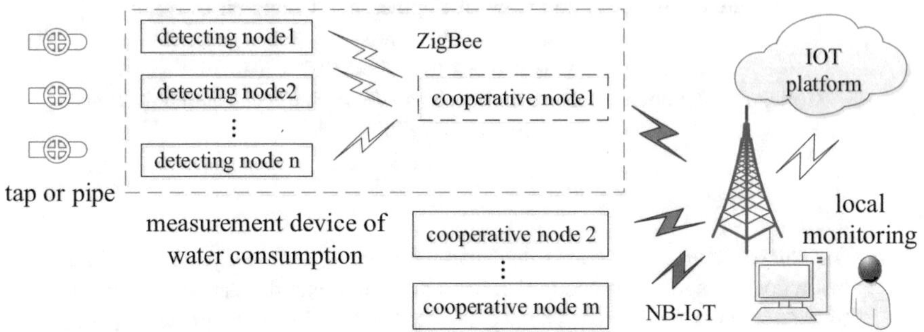

Fig. 1. Radom measurement system of water consumption

The main equipment of the system includes detecting node and communication node. A simple water consumption monitoring system will consist of one cooperative node and several detecting node. Detecting node collects the consumption parameters, such as speed and flow, after data collection is completed, information is sent to the cooperative node by its own data format, then the cooperative node will transmit data to the internet of things platform through NB-IoT connection. One cooperative node and several detecting nodes form a ZigBee network, cooperative node playing the role as an end-node of NB-IOT network that means the cooperative node runs like a protocol converter, converting ZigBee protocol to NB-IOT protocol [3].

The system structure can be divided into two layers, the primary function of the ZigBee layer is data collection, focusing on the smaller area. The NB-IoT layer can realize distance communication, focusing on wide areas interconnection. The water consumption monitoring device is the core part of the system, it can also exist separately when the NB-IoT network is turned off, all detecting data is kept locally with the ID and timestamp. On completion of the measurement, data can be copied, analyzed, and transmitted and so on [4].

3 Hardware Design

3.1 Detecting Node Design

The design can be divided into mechanical part and electrical part. It's important to note that detecting node is equipped with universal interface in mechanical design and can be easily connected to the end of any tap. When water flows through the detecting node, electricity can be generated by the pressure of water flowing and keep the detecting node itself working. If the water flow is cut off, the electricity will disappear after a time delay. The mechanical design will not be detailed here; the hardware structure of detecting node is shown in Figs. 2 and 3.

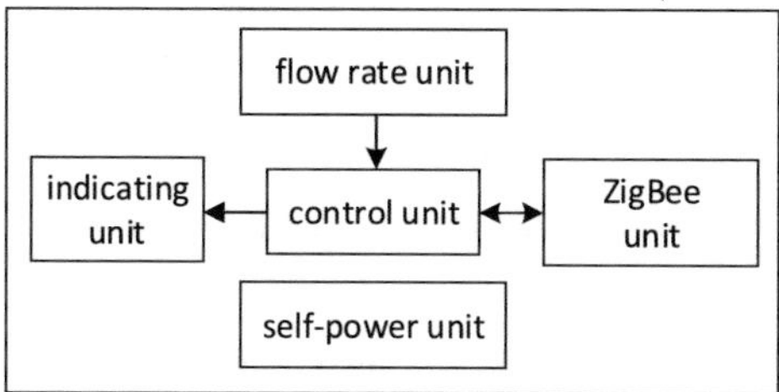

Fig. 2. Hardware of detecting node

The hardware of detecting node control part is made up of five units. The control unit finishes the core functions such as process control, data processing, real-time communication, it selects compact STC12C2052 as control core. The flow rate unit can detect the velocity, the theory goes that sampling the voltage self-powered which is proportional to flow velocity, when the value of voltage is got, then the flow velocity can be reverse deduced, furthermore, if it is integrated, the result will be amount of water flow; ZigBee unit works as a terminal node, the ZigBee unit select CC2530. Self-power unit has a small turbine which can generate power between 0 V and 20 V, its value always changes with the flow rate, the unit also is equipped with voltage stabilizer and micro battery. The Indicating unit includes tricolor LEDs which will changes in color and flicker frequency when the flow rate is changing, this unit can remind consumers to save water.

3.2 Cooperative Node

Cooperative node accesses the ZigBee network and NB-IoT network at same time. In ZigBee network, cooperative node plays the role like a coordinator, starting and maintaining the whole network, besides, all the detecting nodes data will be transmitted to the cooperative node. In NB-IoT network, cooperative node plays the role like a terminal node, it is the data flow origin of NB-IoT network, cooperative node communicates with IoT Platform using COAP protocol. The hardware structure of cooperative node is shown in Fig. 2, Other intelligent terminals can access the IoT platform using http/https protocol.

Cooperative node complete following functions such as display, data analysis, data framing, data storage, data remote transmission, it is made up of eight units. Control unit is control and communications center, it selects STM32F105MCU, communicating with ZigBee unit and NB-IoT unit through the serial ports. ZigBee unit still choose CC2530 as master chip. NB-IoT unit uses the BC95 module which works only a few minutes a day, and keeps deep sleeping during the rest time. Storage unit uses the T-flash card, data is stored as tables. Display unit can recycling present the detecting nodes data which select 2.74 inch e-ink screen, and connects MCU with SPI interface.

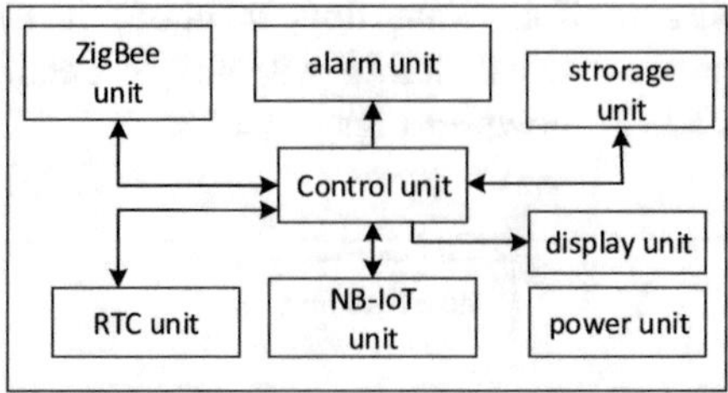

Fig. 3. Hardware of cooperative node

RTC unit can provides timestamp, the unit select PCF8563 chip supported by 32Kcrystal oscillator, its interface to MCU is IIC. Alarm unit use sound and light module, when the tap is not cut off after used, water supplies are damaged and so on, and the local and remote alarm will be triggered. Power unit can provide energy for other units, it is equipped with high capacity lithium-ion battery, being able to maintain the system working for a week.

4 Software Design

Software design can be divided into following three parts, detecting node module, cooperative node module and IoT platform. Detecting node module is design is relatively simple, when the tap is turned on, detecting nodes generate power and start to work, then the program module will be executed and the data detected will be transmitted 20 times per second. If the tap is turned off, the detecting node can keep working for about ten seconds, then the communication stops. The module and IoT platform would not be explained in the paper.

As to detecting node module, the program development is complex. The program module deals with the communication, when the data is received accurately, operation such as data parsing and data filtering are following, the data processed will be integrated to get flow, according to the change of flow. A sparse algorithm is put forward to improve efficiency of storage. The next step of data processing is data framing according to custom protocol specification, besides data processed such as flow and rate, additional information like type, time, ID, address etc. will be packed and stored. Furthermore, if data need to upload to IoT platform, data must be further processed and let it in accordance with JSON format before transmitting [5, 6].

Data obtained are discrete form through experiments, then the water flow different time periods can be calculated, the result multiply by cardinal number will be the total flow of a district, as is shown in Fig. 4. Above this, the flow, rate, and the peak can be easily calculated and stored to database. According to the statistical data, the distribution map of water consumption can be built and it can provide a reliable reference for allocation of water. When water scarcity happens, it is easy to take step such as valve shutoff, rate-limiting, volume limiting to assure working order, and improve the utilization of water resource.

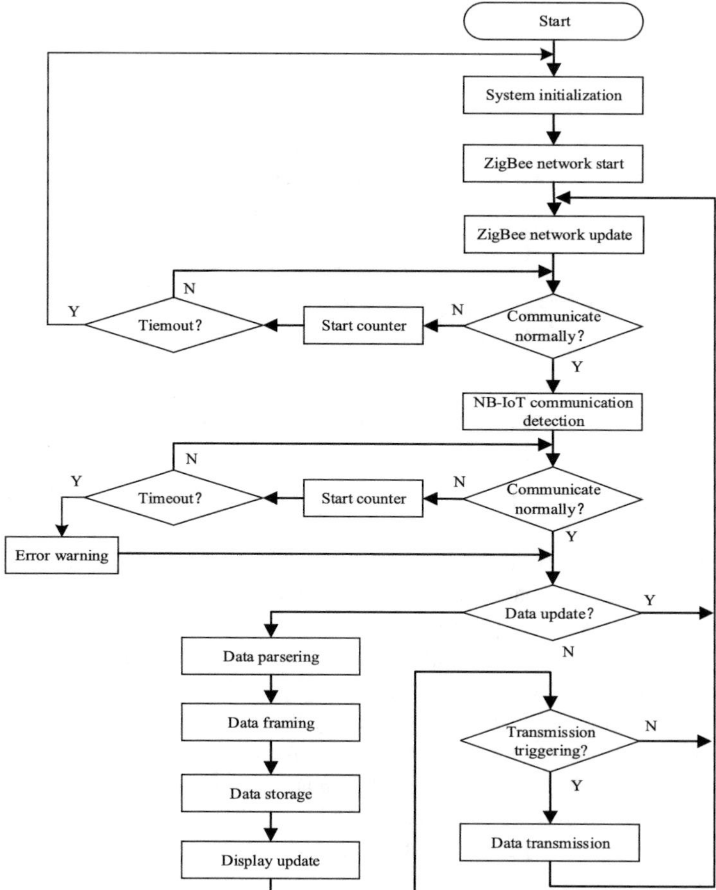

Fig. 4. Flow chart diagram of cooperative node

5 Results and Analysis

In order to verify the feasibility of the design, the system is stalled on first floor and second floor in laboratory building, including two cooperative nodes and six detecting nodes and divided into two groups. After handling the raw data, the result is shown in Fig. 5.

From Fig. 5, the data of two floors change regularly, the curve of the first floor is similar to the second one, and there will be a significant increase on the hour, and the value reaches its peak around ten o'clock, obviously, peak occurs during recess time and the biggest break is at ten o'clock, in line with the rules of water consumption. There is another phenomenon in the Fig. 5 that the water consumption of first floor is larger than the second's in average, because of larger population flow of first floor. The water consumption differs each year and each month, so the week's consumption

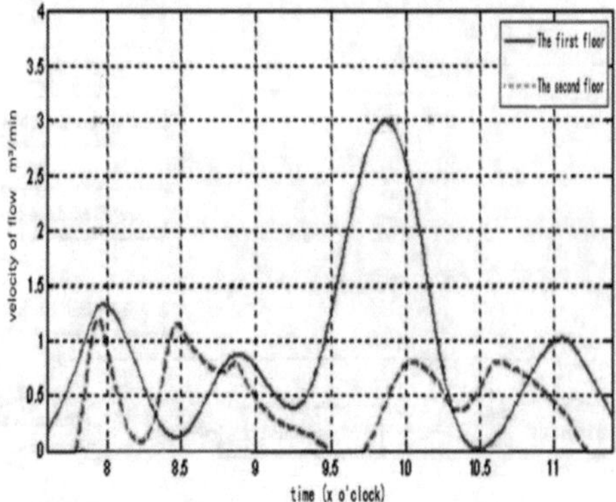

Fig. 5. Measurement of water velocity

would be minimal units of statistics, then the total flow of a building or one area can be easily calculated.

As is shown in Fig. 6, the experiment takes half a cubic meter as object of study, experiments reveals the relation between the measuring accuracy and velocity. Along with the increase of velocity, the error of volume is in decline, and the trend see a steep drop at velocity of 0.4 m/s to 0.6 m/s. That means measurement accuracy increases in proportion to velocity and may be relative accuracy when the velocity greater than 0.6 m/s. under normal conditions of operating, the velocity of flow would be kept at 0.9 m/s around, so the error can be negligible. At the beginning and ending time of operation of water valve, the error could be great, but this process is less than 0.5 s in duration, so the percentage of the error would be small, and the problem can be solved by compensation, the longer the water valve is open, the higher accuracy for the measurement.

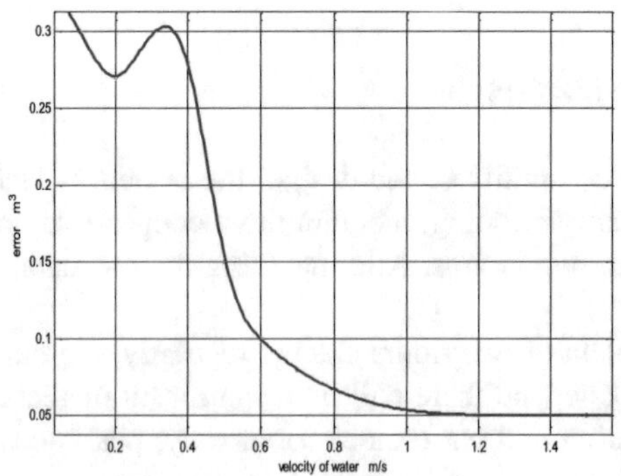

Fig. 6. Measurement of error

6 Conclusions

The system is prospective study which aims to fine management of water in large institutions, combining with the mainstream of ZigBee and NB-IoT technology, taking advantage of low power LAN and WAN. ZigBee network can effectively reduce the network energy consumption and improved robustness of the system, but the communication distance is short. with the help of public communication network, NB-IoT can realize remote communication, meeting the requirement of transmission of non-real-time data, and the penetrating of signal is high, the two technology making a good match with each other. The experiments have proved the system perform steadily, the result is accurate and the communication is reliable. The system can be extended and tailored flexibly to meet different needs and have offered the essential realization means for the measurement system of water consumption.

Acknowledgment. This work is supported by the Shaanixi Education Committee Project (14JK1669), Shaanixi Technology Committee Project (2018SJRG-G-03).

References

1. Huang, L.-C., Chang, H.-C., Chen, C.-C., Kuo, C.-C.: A ZigBee-based monitoring and protection system for building electrical safety. Energy Build. **43**(6), 1418–1426 (2011)
2. Atzori, L., Iera, A., Morabito, G.: The Internet of things: a survey. Comput. Netw. **54**(15), 2787–2805 (2010)
3. Ha, J.Y., Park, H.S., Choi, S., Kwon, W.H.: Enhanced hierarchical routing protocol for ZigBee mesh networks. IEEE Commun. Lett. **11**(12), 1028–1030 (2007)
4. Yao, M., Wu, P., Qu, W., Sun, Q.: Research on intelligent water meter based on NB-IoT. Telecom Eng. Tech. Stand. **6**, 32–35 (2018). https://doi.org/10.13992/j.cnki.tetas.2018.06.010
5. Carroll, R.J., Ruppert, D., Stefanski, L.A.: Measurement error in nonlinear models. Chapman and Hall, London (1995)
6. Hanfelt, J.J., Liang, K.Y.: Approximate likelihood for generalized linear errors-in-variables models. R. Statist. Soc. B. **59**(2), 627–637 (1997)

Design of the Intelligent Tobacco System Based on PLC and PID

Xiu-Zhen Zhang(✉)

School of Information Science and Engineering,
Fujian University of Technology, Fuzhou 350118, China
289126754@qq.com

Abstract. In order to further improve the quality and production of flue-cured tobacco, the system takes PLC as the control core, and the change of the real-time temperature and humidity in the flue-cured tobacco room can be detected by the wall-mounted temperature and humidity transmitter. The dynamic switching control methods of PID and Fuzzy PID are adopted, combining the advantages of both to drive circulation fan and blower to work accurately, and to control the air supply fan and the dehumidification fan precisely. The experimental results show that the intensive automatic flue-cured tobacco system based on PLC and PID reduces the fluctuation of temperature and humidity, improves the control precision of automatic flue-cured tobacco, reduces the labor intensity of workers, and has significant economic benefits.

Keywords: Intelligent tobacco · Temperature and humidity control
PLC · Fuzzy PID

1 Introduction

China is a major consumer of tobacco and tobacco leaf is an important part of agricultural products, and the annual average output accounts for 33.3% of the world [1]. The quality of tobacco leaf baking directly affects the production efficiency of tobacco [2]. Traditional manual work has high labor intensity, and the tobacco capacity is small [3]. The baking process is still based on human eyes and hands to make subjective judgment, and the process operation has strong subjectivity [4]. The different types of tobacco leaves yellow and dry in the baking process, in addition to the subjective judgment of the different technicians, there is a severe or backward phenomenon in the process of baking, which leads to the low quality of the tobacco, and the economic benefit is poor [5].

Based on the above situation, this paper puts forward the intensive automatic design of the flue-cured tobacco. The temperature and humidity control system, which is controlled by PLC, can monitor the change of temperature and humidity in the room in real time. The PID and fuzzy PID dynamic switching control algorithm is used to control the actuator operation according to the actual temperature and humidity setting value of the flue-cured tobacco process. Realize the control of blower, circulating fan, discharge fan, air supply blower, and electric furnace to automatically adjust the temperature and humidity in the tobacco room, Accurately judge the yellow and dry

P. Krömer et al. (Eds.): ECC 2018, AISC 891, pp. 232–240, 2019.
https://doi.org/10.1007/978-3-030-03766-6_26

status of tobacco leaves during the intensive baking process, improve the quality of tobacco leaves production, reduce the labor intensity of tobacco workers, and truly realize the intelligent control of tobacco baking [6, 7].

2 The Overall Composition of Intelligent Flue-Cured Tobacco System

The overall structure of the intensive automatic flue-cured tobacco system consists of temperature and humidity acquisition, A/D conversion module, PLC controller and various actuators. The overall structure module of the system is shown in Fig. 1. The threshold of temperature and humidity of tobacco leaves baked in the oven was collected in real time, and the parameter adjustment of stage time was specified. In manual mode, the output controls are controlled by their own buttons, which means that the output relay of PLC is connected, and the motor is controlled by KM suction; In the automatic mode, the data that is collected by the temperature and humidity transmitter, is converted to digital by A/D, and the Fuzzy PID dynamic switching control algorithm is used to accurately control the operation of each motor. When temperature and humidity are too high or too low in the flue-cured room, and there are sensor, motor and power supply failure, the system will alarm.

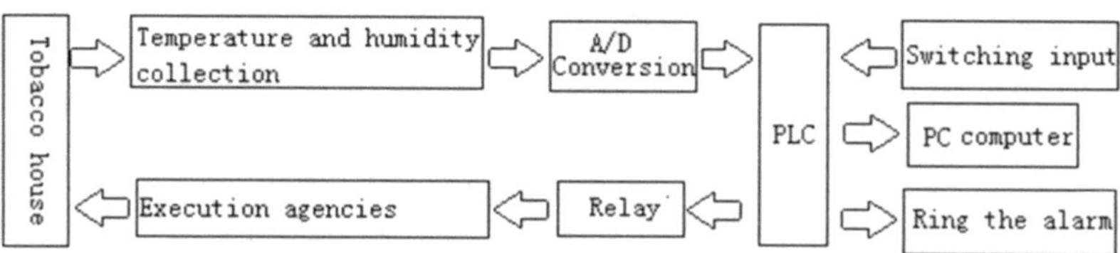

Fig. 1. The overall structure of the intelligent tobacco system

The electric control of the intensive automatic flue-cured room is supplied with 220 v power. The air blower and the circulating fan are respectively controlled by two relays, and the speed-controlled is carried out by the series resistance, and the work indicator light is provided. The positive and reverse rotation of the supplemental fan and the exhaust fan are controlled by a pair of relays. The electric furnace, the electromagnetic valve is controlled by a relay. Besides, there are instructions for operation, alarm and emergency stop.

The I/O connection diagram design of intensive automated tobacco system: it is input with manual and automatic control mode, which can save I/O points and reduce cost, without affecting the operation. The output adopts AC/DC control mode, the motor is controlled by AC contactor with AC power, and the DC power supply is used for indicator lamp.

3 Software Programming

3.1 Main Program Flow Chart

The overall design of the system flow chart is shown in Fig. 2. After system initialization, the manual and automatic control mode shall be judged. In the automatic mode: the current or voltage generated by the temperature and humidity transmitter is transmitted to the PLC analog module for data processing, and the Fuzzy control or the PID control algorithm program is selected by the processing result, and the PLC program is executed to determine whether the motor needs to be driven to control the work of the actuator. At the same time, it is judged whether the field value transmitted from the temperature and humidity transmitter has reached the task value, if the condition is met, the action will not be performed; If not, the action will be performed and the baking will be finished.

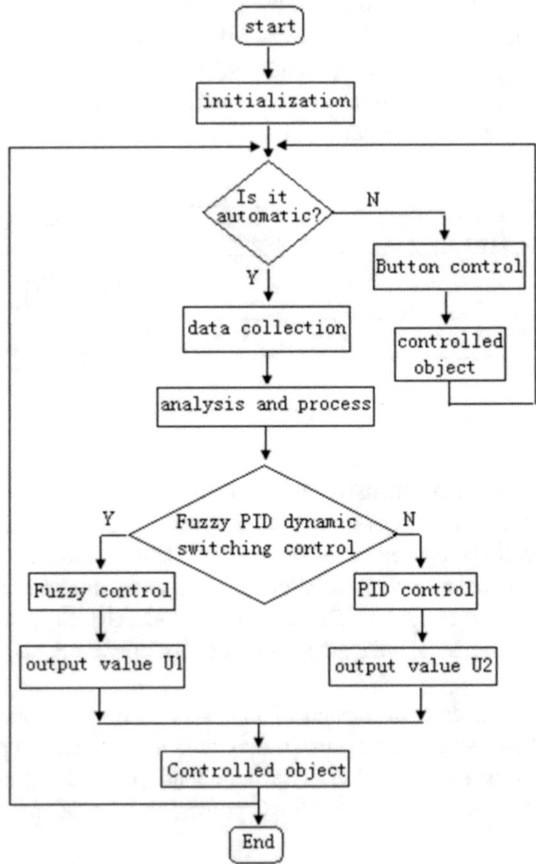

Fig. 2. Chart of main system flow

3.2 The Main Part of PLC Controlled of the System

Speed Regulation of Blower and Circulating Fan. The three processes of tobacco leaf yellowing, color fixing and drying are the process of flue-cured tobacco. In this process, the temperature increases continuously, which requires the blower to continuously enhance the gas, and the circulating fan provides the circulation air to ensure the uniform processing. The three processes of tobacco leaf yellowing, color fixing and drying are the process of flue-cured tobacco. In this process, the temperature increases continuously, which requires the blower to continuously enhance the gas, and the circulating fan provides the circulation air to ensure the uniform processing. The blower speed control has three gears, which are low speed, medium speed and high speed respectively, by controlling the relay (KM1 and KM2), the fan speed control shift is realized (0 0 stop, 0 1 low speed, 0 0 medium speed, 1 1 high speed).The same as by controlling the relays (KM3 and KM4), the speed change of the circulating fan can be realized.

Control of Dehumidification Fan and Air Supply Fan. The whole process of tobacco leaf baking, the change of humidity is relatively small, and the amplitude fluctuation is small, but the humidity control can't be ignored. When the humidity is higher than the set value, the PLC output drives the exhaust fan to work, and it is dehumidifying; On the contrary, it is humidifying. When the humidity meets the set value, the state remains. The operation is adjusted repeatedly so that the humidity meets the set point.

When the temperature in baking room is higher than the set point, the air supply fan is opened to accelerate the air circulation, and the temperature of the room is lowered; On the contrary, open a large one-step damper, until the damper is fully open. But it's still lower than the set value, open the blower, close the air supply fan, and the temperature of the room rises, and the blower will stop when the ambient temperature meets the set point.

3.3 Safety Work of PLC Controlled of the System

The safe work of the flue-cured tobacco room is very important, such as the safety of life, the safety of flue-cured tobacco and so on [8]. Based on this, safety alarm and emergency stop device are designed. If the temperature exceeds the upper limit of the maximum temperature, the system will give an alarm; The furnace is started and the access is not closed, which will also give an alarm, and the alarm light will be on.

The process of flue-cured tobacco is finished, the system will stop running automatically. The temperature drops to the allowable range, the access will be opened, then the staff can go to collect the tobacco; The stop button is pressed before the end of the work, the motor of the room will stop running. Similarly, the door will not open until the temperature drops to the allowable range. When an emergency occurs, press the emergency stop button, and the system will reset all operating actuators and handle the occurrence.

4 Implementation of Fuzzy PID Dynamic Switching Controller in Flue-Cured Tobacco System

4.1 Fuzzy PID Dynamic Switching Works

Based on the step response, whether the dynamic performance is good or not is mainly determined by the first two cycles of the system response [9]. To obtain an accurate conversion time, it is necessary to obtain the data of the deviation e. If the set value E_0 is smaller than the absolute value of the deviation E, the Fuzzy controller is selected, which not only can achieve a fast response, but also can control the overshoot amount; If the set value E_0 is greater than the E when the system response tends to be stationary, the PID (Proportional Integral Derivative) controller is selected [10]. It can effectively eliminate the static error and compensate the defect of Fuzzy controller, so that the system has better static performance and control precision. The flow chart of the switching control process is shown in Fig. 3.

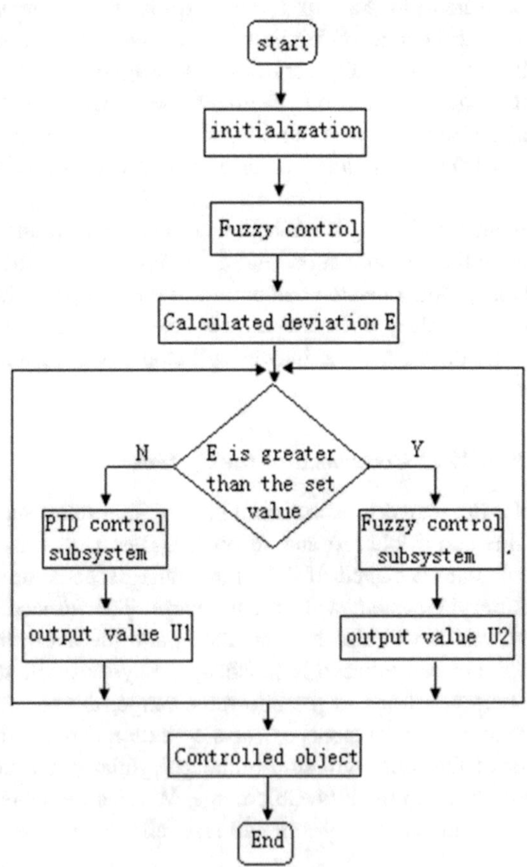

Fig. 3. Fuzzy PID switching control flow chart

4.2 Fuzzy Controller Programming

Design of the Fuzzy Controller The two-dimensional Fuzzy controller is used in this paper [11]. It can be use to generalize that control strategy of operator and establish the membership degree through the determination of Fuzzy language variable, and the Fuzzy language assignment table is obtained as shown in Tables 1, 2 and 3. The Fuzzy rule control table is set up by experience method, the Fuzzy control look-up table is generated, and stored in the register of PLC.

Table 1. Assignment table of deviation e

	−6	−5	−4	−3	−2	−1	0	1	2	3	4	5	6
PB										0.1	0.4	0.8	1.0
PM									0.2	0.7	1.0	0.7	0.2
PS								0.9	1.0	0.6	0.2		
0					0.1	0.6	1.0	0.6	0.1				
NS			0.2	0.6	1.0	0.9							
NM	0.2	0.7	1.0	0.7	0.2								
NB	1.0	0.8	0.4	0.1									

Table 2. Assignment table of deviation change rate ec

	−6	−5	−4	−3	−2	−1	0	1	2	3	4	5	6
PB										0.1	0.4	0.8	1.0
PM									0.2	0.7	1.0	0.7	0.2
PS								0.9	1.0	0.7	0.1		
0					0.1	0.5	1.0	0.5	0.1				
NS			0.1	0.7	1.0	0.9							
NM	0.2	0.7	1.0	0.7	0.2								
NB	1.0	0.8	0.4	0.1									

Table 3. Assignment table of output control quantity U

	−6	−5	−4	−3	−2	−1	0	1	2	3	4	5	6
PB											0.4	0.8	1.0
PM									0.2	0.7	1.0	0.3	
PS								0.9	0.9	0.7			
0					0.1	0.5	1.0	0.5	0.1				
NS				0.7	0.9	0.9							
NM		0.3	1.0	0.7	0.2								
NB	1.0	0.8	0.4										

The Flow Chart of Fuzzy Controller Programming. In the PLC, the digital quantity obtained in the sampling is converted into corresponding fuzzification domain element by quantification factor, and then according to the look-up table, the quantized value *U* of the corresponding output control amount is obtained, finally, it is multiplied by *Ku* to obtain the final actual output control amount. The program performs a cyclic scan method, and the programming flow is shown in Fig. 4.

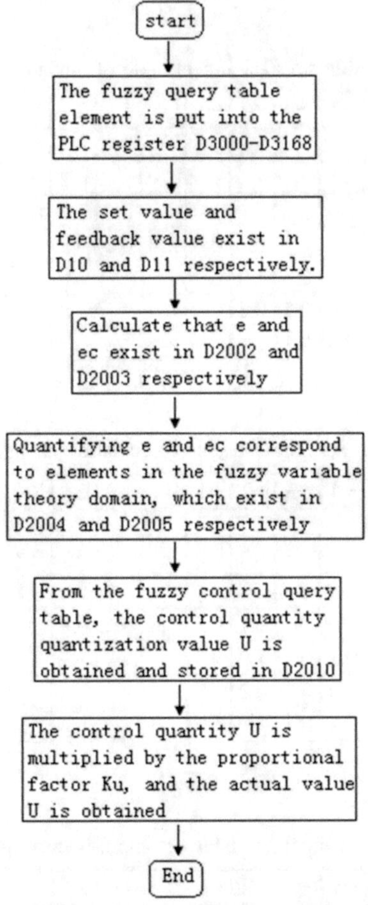

Fig. 4. Fuzzy controller flow chart

The input quantity is graded, such as the basic domain of *e* is [−11, 11], the *ec* is [−5.5, 5.5], and the control quantity *U* is [−11,11], by A/D conversion the digital quantities are respectively [−550,550], [−275,275], [−550,550].The quantization factors are $Ke = 0.55$, $Kec = 1.09$, and the scale factor is $Ku = 1.83$, and the fuzzification domain is [−6,6]. The quantization levels of deviation *e* and control quantity *u* are less than −550 corresponding to −6, and −550 ~ −450 corresponding to −5; etc., until they are greater than 550 corresponding to 6, that is, the corresponding fuzzy set theoretic field elements after quantization. The quantization level of *ec* is the same.

5 Testing and Analysis

The system is running in automatic mode, the temperature and humidity changes of the tobacco baking process is recorded, selected the record from flue-cured tobacco yellowing stage, make the curve and compare with the theoretical curve. In which that humidity change is small, and the temperature change is shown in Fig. 5, close to the test results, the error is within the allowable range, and the system is running well with the baking control requirements.

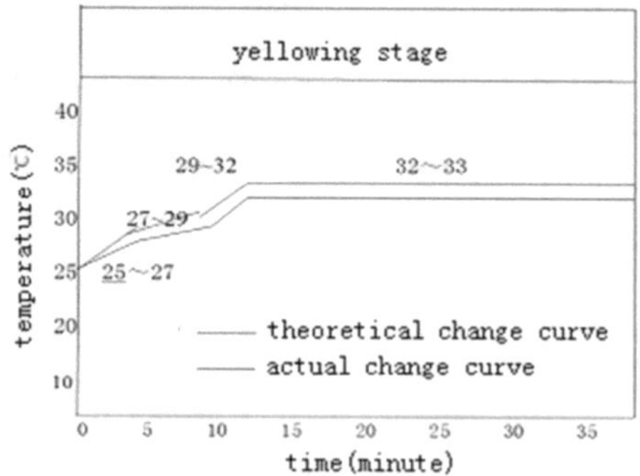

Fig. 5. Curve of theoretical and actual changes in temperature of the flue-cured tobacco

6 Conclusion

With PLC as the control core, the temperature and humidity of the flue-cured tobacco were collected and processed in real time. The Fuzzy PID dynamic switching controller was adopted, the advantages of the Fuzzy algorithm and PID control algorithm were combined to realize complementary advantages of the both. The experimental results show that the temperature and humidity can rapidly follow the set value of baking, improve the quality, have high accuracy, work stability and have practical application value.

References

1. Duan, S., Zhu, H.: Research progress in intelligent control technology of tobacco bulk curing. Acta Agriculturae Jiangxi **25**(2), 107–109 (2013)
2. Sun, J.: The design of temperature and humidity control system for intelligent tobacco baking hous. Instrum. Technol. **11**(2), 20–25 (2010)
3. Li, Z., Li, T.: MCU-based intelligent control system for flue-cured tobacco. Hubei Agric. Sci. **50**(2), 395–397 (2011)

4. Wang, G., Zhang, Z., Zhang, L.: Design of temperature and humidity control instrument for intelligent tobacco house based on STM32. J. Chin. Agric. Mech. **36**(2), 280–282 (2015)
5. Hu, J., Sun, J.: Research on multi-parameter and real-time detector of flue-cured tobacco house. J. Chin. Agric. Mech. **37**(1), 178–181 (2016)
6. Jia, F., Liu, G.: Comparison of different methods for estimating nitrogen concentration in flue-cured tobacco leaves based on hyperspectral reflectance. Field Crops Res. **150**(8), 108–114 (2013)
7. Zhou, S., He, Q., Wang, X.: An insight into the roles of exogenous potassium salts on the thermal degradation of flue-cured tobacco. J. Anal. Appl. Pyrolysis **123**(6), 385–394 (2016)
8. Zhou, S., Wang, X.: Quantitative evaluation of CO yields for the typical flue-cured tobacco under the heat-not-burn conditions using SSTF. Thermochim. Acta **208**(5), 7–13 (2015)
9. Zhao, X., Xu, L.: Constant Tension Winding System of Corn Directional Belt Making Machine Based on Self-adaptive Fuzzy-PID Control. Trans. Chin. Soc. Agric. Mach. **46**(3), 90–96 (2015)
10. Lu, Y.: Application Research of PLC Temperature Control System Based on Fuzzy Neural Network PID. Bull. Sci. Technol. **34**(1), 155–158 (2018)
11. Rubaai, A., Castro, M.J., Ofoli, A.R.: Design and implementation of parallel fuzzy PID controller for high-performance brushless motor drives an intetrated environment for rapid control prototyping. IEEE Trans. Ind. Appl. **44**(4), 1090–1098 (2008)

A Simple Image Encryption Algorithm Based on Logistic Map

Tsu-Yang Wu[1,2], King-Hang Wang[3], Chien-Ming Chen[4(✉)], Jimmy Ming-Tai Wu[5], and Jeng-Shyang Pan[1,2]

[1] Fujian Provincial Key Lab of Big Data Mining and Applications, Fujian University of Technology, Fuzhou 350118, China

[2] National Demonstration Center for Experimental Electronic Information and Electrical Technology Education, Fujian University of Technology, Fuzhou 350118, China

wutsuyang@gmail.com, jspan@cc.kuas.edu.tw

[3] Department of Computer Science and Engineering, Hong Kong University of Science and Technology, Clear Water Bay, Hong Kong

kevinw@cse.ust.hk

[4] Harbin Institute of Technology (Shenzhen), Shenzhen 518055, China

chienming.taiwan@gmail.com

[5] College of Computer Science and Engineering, Shandong University of Science and Technology, Qingdao 266590, China

wmt@wmt35.idv.tw

Abstract. Based on the better properties of chaotic maps, the studies of chaotic-based image encryption have received much attentions by researchers in recent years. In this paper, an image encryption algorithm based on logistic map with substitution approach is proposed. Meanwhile, we make the line chart, histogram, and pixel loss analysis to show the performance of our algorithm.

Keywords: Image · Encryption · Logistic map · Histogram analysis

1 Introduction

With the fast development of technologies [2,5,6], images are widely used in different areas such as medicine, military, social network, etc. How to protect the security of transmitted images via public channel becomes to be an important research issue. Encryption is a cryptographic primitive which provides confidentiality of data. However, the traditional symmetric key encryption algorithms such as DES and AES are designed to encrypt text. It leads researchers to design new algorithm to protect digital image for secrecy.

Chaotic maps [3,4] provide better properties: ergodicity, sensitivity, and pseudo randomness:

1. *Ergodicity.* Given a fixed domain, the chaotic map can traverse the whole corresponding range within a finite time;

P. Krömer et al. (Eds.): ECC 2018, AISC 891, pp. 241–247, 2019.
https://doi.org/10.1007/978-3-030-03766-6_27

2. *Sensitivity.* For an arbitrarily small perturbation or change, the result of chaotic map may return significantly different values;
3. *Pseudo randomness.* Coming from the ergodicity and the sensitivity properties.

Recently, some literatures [1,7–17] studied the possibility of applying chaotic maps on image encryption.

In this paper, we adopt substitution approach to propose an image encryption algorithm based on logistic map. In our algorithm, logistic map is used to generate random sequence. Based on the generated sequence, we can substitute the position of plain image to generate cipher image. In the implementation and analysis, we demonstrate the line chart analysis, histogram analysis, and pixel loss analysis of the proposed algorithm based on three gray images.

2 A Chaotic-Based Image Encryption Algorithm

In this section, we propose a image encryption algorithm based on logistic map. Firstly, we introduce logistic map.

2.1 Logistic Map

The logistic map [11] is defined by

$$x_{n+1} = \alpha \cdot x_n \cdot (1 - x_n), \tag{1}$$

where $x_n \in [0, 1]$ for $n = 0, 1, 2, \ldots$ and $\alpha \in [3.5699456, 4]$ is a control parameter to present the chaotic behavior. The relationship between α and chaotic behavior is depicted in Fig. 1.

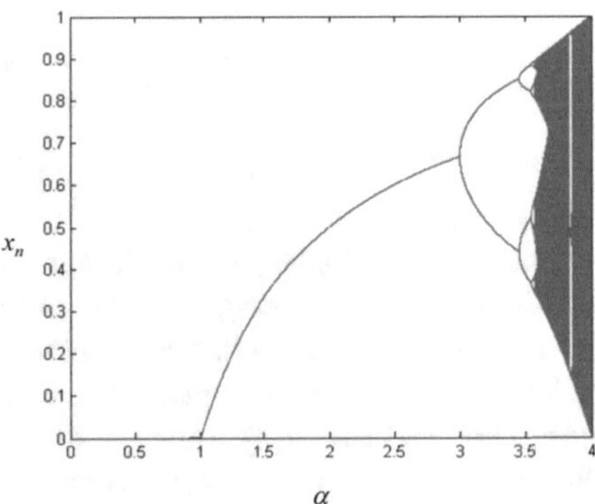

Fig. 1. The relationship between α and chaotic behavior

2.2 Detailed Algorithm

Our proposed chaotic-based image encryption algorithm consists of four steps: chaotic sequence generation, chaotic subsequence generation for column, chaotic subsequence generation for row, and image encryption. We assume that I is a gray image with size $m \times n$. The used notations are summarized in Table 1.

Table 1. Notation

Notations	Meanings
I	A gray image with size $m \times n$
α	A control parameter
Λ'	A subsequence $\Lambda' = \{\lambda'_i\}_{i=1}^m$, where $\lambda'_i \in [1, m]$
Γ'	A subsequence $\Gamma' = \{\gamma'_i\}_{i=1}^n$, where $\gamma'_i \in [1, n]$
C	A cipher image with same size $m \times n$

1. *Chaotic sequence generation.* This step is to generate an infinite chaotic sequence $\{x_i\}_{i=1}^{\infty}$ using the following iterations

$$x_{n+1} = \alpha \cdot x_n \cdot (1 - x_n) \tag{2}$$

with an initial value $x_1 \in [0, 1]$ and $\alpha \in [3.5699456, 4]$.
2. *Subsequence generation for column.* This step is to generate a subsequence Λ' with m elements which is used to encrypt I by executing the following procedures:
   (a) To select a sequence $A = \{x_i\}_{i=1}^{5m}$.
   (b) To randomly permute sequence A, we can find another sequence $A' = \{x'_i\}_{i=1}^{5m}$. Meanwhile, we record the position of each x'_i. For example, the position of x'_1 is 1.
   (c) To sort A' by ascendant, the resulted sequence is $A'' = \{x''_i\}_{i=1}^{5m}$.
   (d) To select previous m elements of A'', the resulted sequences is $A''' = \{x'''_i\}_{i=1}^{m}$.
   (e) To retrieve the position of each x'''_i, we can obtain a subsequence $\Lambda = \{\lambda_i\}_{i=1}^m$, where $\lambda_i \in [1, m]$.
   (f) To randomly permute sequence Λ, we can find another sequence $\Lambda' = \{\lambda'_i\}_{i=1}^m$, where $\lambda'_i \in [1, m]$.
3. *Subsequence generation for row.* This step is to generate a subsequence $\Gamma' = \{\gamma'_i\}_{i=1}^n$ which is used to encrypt I, where $\gamma_i \in [1, n]$. This step is similar the step of Subsequence generation for column.
4. *Encryption.* This step is to generate a ciphertext image C with the same size as the plain image I by

$$C(i, j) = I(\lambda'_i, \gamma'_j), \tag{3}$$

for $i \in \{1, 2, \ldots, m\}$ and $j \in \{1, 2, \ldots, n\}$.

3 Implementation and Analysis

We have implemented our proposed algorithm with MATLAB on a Windows 7 32-bits desktop machine, running against some sampled gray images: Lena (300×300), Scenery (400×400), and Landscape (650×434).

3.1 Line Chart Analysis

The line chart of a plain image shows the statistically distribution of the pixels. Figures 2 and 3 show three sets of the line chart of a plain image and the corresponding cipher image. As we can see the statistical information has been destroyed after encryption.

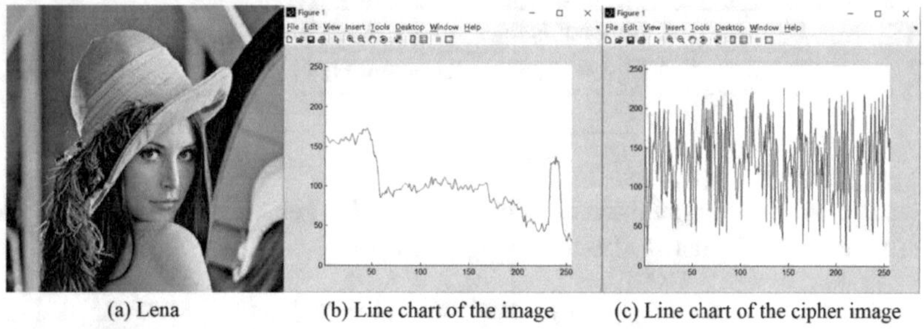

(a) Lena (b) Line chart of the image (c) Line chart of the cipher image

Fig. 2. Line chart analysis of Lena

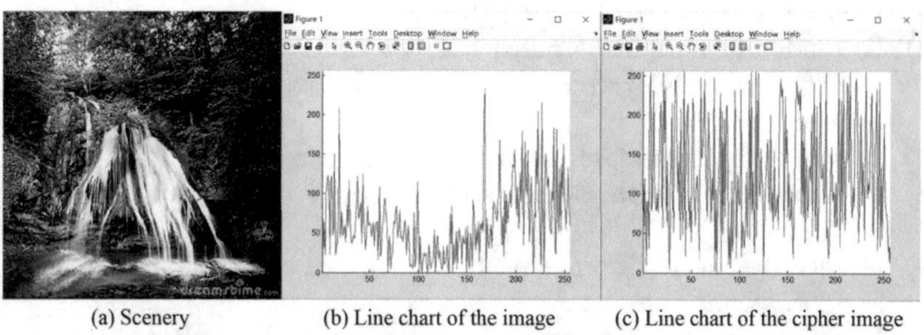

(a) Scenery (b) Line chart of the image (c) Line chart of the cipher image

Fig. 3. Line chart analysis of Scenery

3.2 Histogram Analysis

The histogram of a plain image shows the statistically distribution of the pixels. Figures 4 and 5 show three sets of the 3D histogram of a plain image and the corresponding cipher image. As we can see the statistical information has been destroyed after encryption.

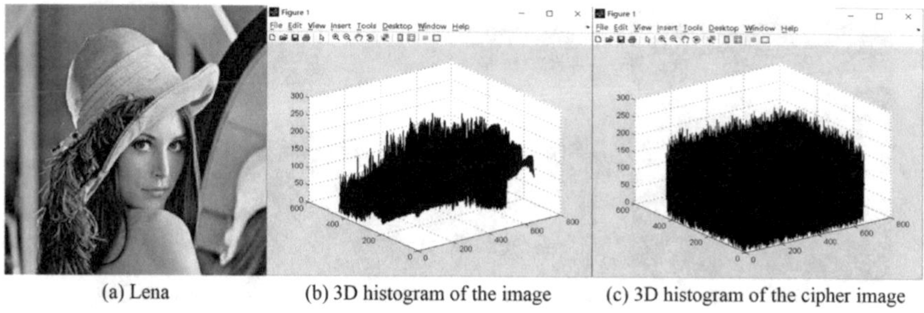

(a) Lena (b) 3D histogram of the image (c) 3D histogram of the cipher image

Fig. 4. 3D histogram analysis of Lena

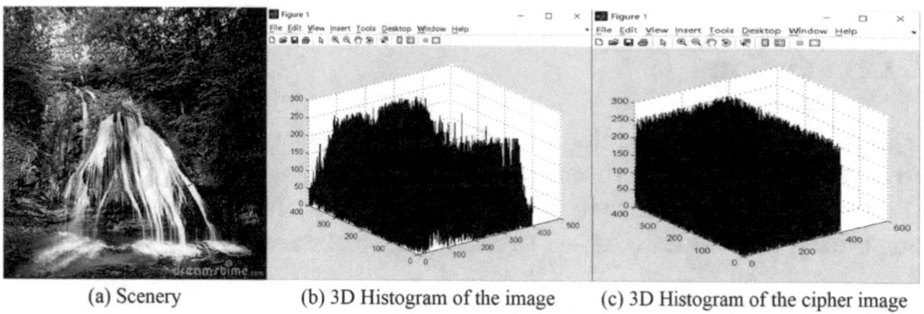

(a) Scenery (b) 3D Histogram of the image (c) 3D Histogram of the cipher image

Fig. 5. 3D histogram analysis of Scenery

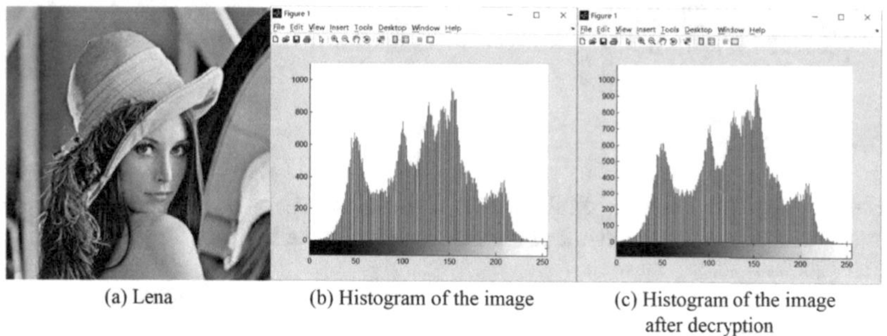

(a) Lena (b) Histogram of the image (c) Histogram of the image after decryption

Fig. 6. Pixel loss analysis of Scenery

3.3 Pixel Loss Analysis

In this subsection, we demonstrate that our image encryption algorithm cannot lead pixel loss problem after decryption in Figs. 6 and 7. We also use histogram to show this fact.

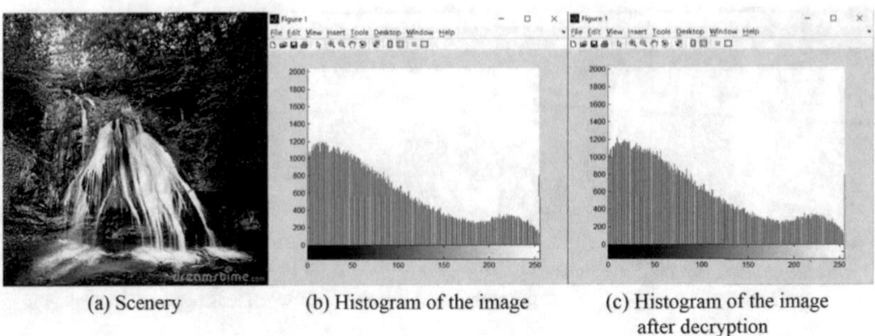

(a) Scenery (b) Histogram of the image (c) Histogram of the image after decryption

Fig. 7. Pixel loss analysis of Lena

4 Conclusion

We have proposed a chaotic-based image encryption algorithm. Meanwhile, Three analysis are made to evaluate the performance of our algorithm. Due to limited space, we will make correlation analysis, NPCR (number of pixels change rate), and UACI (Unified average changing intensity) in the future.

Acknowledgments. The work of Tsu-Yang Wu was supported in part by the Science and Technology Development Center, Ministry of Education, China under Grant no. 2017A13025 and the Natural Science Foundation of Fujian Province under Grant no. 2018J01636. The work of Chien-Ming Chen was supported in part by Shenzhen Technical Project (JCYJ20170307151750788) and in part by Shenzhen Technical Project (KQJSCX20170327161755).

References

1. Behnia, S., Akhshani, A., Mahmodi, H., Akhavan, A.: A novel algorithm for image encryption based on mixture of chaotic maps. Chaos Solitons Fractals **35**(2), 408–419 (2008)
2. Chang, F.C., Huang, H.C.: A survey on intelligent sensor network and its applications. J. Netw. Intell. **1**(1), 1–15 (2016)
3. Chen, C.M., Wang, K.H., Wu, T.Y., Wang, E.K.: On the security of a three-party authenticated key agreement protocol based on chaotic maps. Data Sci. Pattern Recognit. **1**(2), 1–10 (2017)
4. Chen, C.M., Xu, L., Wu, T.Y., Li, C.R.: On the security of a chaotic maps-based three-party authenticated key agreement protocol. J. Netw. Intell. **1**(2), 61–66 (2016)
5. Chen, X., Peng, X., Li, J.B., Yu, P.: Overview of deep kernel learning based techniques and applications. J. Netw. Intell. **1**(3), 83–98 (2016)
6. Fournier-Viger, P., Lin, J.C.W., Kiran, R.U., Koh, Y.S., Thomas, R.: A survey of sequential pattern mining. Data Sci. Pattern Recognit. **1**(1), 54–77 (2017)
7. Gao, T., Chen, Z.: A new image encryption algorithm based on hyper-chaos. Phys. Lett. A **372**(4), 394–400 (2008)

8. Huang, X.: Image encryption algorithm using chaotic chebyshev generator. Nonlinear Dyn. **67**(4), 2411–2417 (2012)
9. Hussain, I., Shah, T., Gondal, M.A.: An efficient image encryption algorithm based on S8 S-box transformation and NCA map. Opt. Commun. **285**(24), 4887–4890 (2012)
10. Liu, H., Wang, X.: Color image encryption based on one-time keys and robust chaotic maps. Comput. Math. Appl. **59**(10), 3320–3327 (2010)
11. Wang, X., Guo, K.: A new image alternate encryption algorithm based on chaotic map. Nonlinear Dyn. **76**(4), 1943–1950 (2014)
12. Wang, X., Liu, L., Zhang, Y.: A novel chaotic block image encryption algorithm based on dynamic random growth technique. Opt. Lasers Eng. **66**, 10–18 (2015)
13. Wu, T.Y., Fan, X., Wang, K.H., Pan, J.S., Chen, C.M., Wu, J.M.T.: Security analysis and improvement of an image encryption scheme based on chaotic tent map. J. Inf. Hiding Multimed. Signal Process. **9**(4), 1050–1057 (2018)
14. Xu, L., Li, Z., Li, J., Hua, W.: A novel bit-level image encryption algorithm based on chaotic maps. Opt. Lasers Eng. **78**, 17–25 (2016)
15. Ye, G., Huang, X.: A feedback chaotic image encryption scheme based on both bit-level and pixel-level. J. Vib. Control. **22**(5), 1171–1180 (2016)
16. Zhao, J., Wang, S., Chang, Y., Li, X.: A novel image encryption scheme based on an improper fractional-order chaotic system. Nonlinear Dyn. **80**(4), 1721–1729 (2015)
17. Zhu, H., Zhang, X., Yu, H., Zhao, C., Zhu, Z.: An image encryption algorithm based on compound homogeneous hyper-chaotic system. Nonlinear Dyn. **89**(1), 61–79 (2017)

Design and Analysis of Solar Balance Cars

Shan-Wen Zheng[1], Yi-Jui Chiu[2(✉)], and Xing-Die Chen[2]

[1] School of Material Science and Engineering, Xiamen University of Technology, No. 600, Ligong Rd, Xiamen 361024, Fujian Province, China
[2] School of Mechanical and Automotive Engineering, Xiamen University of Technology, No. 600, Ligong Rd, Xiamen 361024, Fujian Province, China
chiuyijui@xmut.edu.cn

Abstract. With the development of modern technology and the improvement of people's living standards, transportations such as automobiles, motorcycles, and sharing bicycles have emerged. However, motor vehicles cause many problems. It carries out the structural design of the solar balance cars and draws its structural model with UG software. The software ANSYS was used to analyze the mechanics of the solar balance cars. The designed indicators met the requirements for use.

Keywords: Solar energy · Balance cars · Structural design ANSYS finite element analysis

1 Introduction

Recently, with the development and wide usage of science technology, the number of cars has boomed and the road has become more narrow. So the emergence of balance car is inevitable, with its smart structure and operation, bringing people more convenient traveling mode. But at present, balance cars are only electric type, which is a little simplex and is different from the green commuting. Environmental pollution caused by social progress enables lots of people to attach more importance to the utilization of new energy, including solar energy, a kind of energy that the benefits outweigh the drawbacks. Considering balance car a kind of short-distance traveling tool, solar energy can offer it adequate energy for using. Therefore, in the future, solar balance car will be recognized by more people, and it is worthy being researched and discussed.

In contrast to other new energy, solar energy is a type of high availability and ideal renewable energy. Researchers are in this field, such as Scribner [1] measured the wheel uniformity and calculated radial force variation in order to eliminate vibration and testify it balanced when rolling. Lipomi et al. [2] made the fabrication of collapsible and portable devices because the intrinsic flexibility of thin-film materials imparted mechanical resilience. Koberle et al. [3] provided cost-supply curves for CPS and PV, basing on geoexplicit information on solar radiation and land cover type, exploring individual potential and interdependencies. Martin and Li [4] summarized the current progress in fuels' generation directly driven by solar energy, discussed the fundamental mechanisms and gave proposals for future research. Carneiro et al. [5] focused to explore a reentrant model by analyzing the variation in the PR which was

P. Krömer et al. (Eds.): ECC 2018, AISC 891, pp. 248–255, 2019.
https://doi.org/10.1007/978-3-030-03766-6_28

with the change of the function of geometrical and basic material parameters. Casella et al. [6] found some new applications of organic Rankine cycle system, such as concentrating solar energy, automotive heat recovery and so on.

The intention of this paper is to design a solar balance car used in peoples' daily life. On the basis of meeting stiffness, strength requirements and the feasibility analysis of several programs, the final program is determined, and vibration analysis is performed.

2 Theories Analysis

2.1 Theories Research

In this paper, the safety factor method is used to determine whether the stress meets the strength requirements. Whether the maximum stress is taken by the car or whether it is within the allowable stress range of the material. In static analysis, the allowable stress of car is

$$[\sigma] = \frac{\sigma_b}{n} \tag{1}$$

where σ_b is the ultimate strength, n is the safety factor.

Because the structure of solar balance car was complex, we should simplify the structure first. This paper exports the equations of the system.

$$[M]\{\ddot{X}\} + [K]\{X\} = 0 \tag{2}$$

Defined the position vector {X} as [Δ]{u}, where [Δ] was the modal matrix of the system. Equation (2) could be changed as follow:

$$[I]\{\ddot{u}\} + [A]\{u\} = 0 \tag{3}$$

In which:

$$[\Delta]^T[M][\Delta] = [I] = \begin{bmatrix} 1 & 0 & \cdots & 0 \\ 0 & 1 & \cdots & \vdots \\ \vdots & 0 & \ddots & 0 \\ 0 & 0 & 0 & 1 \end{bmatrix} \tag{4}$$

$$[\Delta]^T[K][\Delta] = [A] = \begin{bmatrix} \bar{\omega}_1^2 & 0 & \cdots & 0 \\ 0 & \bar{\omega}_2^2 & \cdots & \vdots \\ \vdots & 0 & \ddots & 0 \\ 0 & \cdots & 0 & \bar{\omega}_n^2 \end{bmatrix} \tag{5}$$

The natural frequency of the mistuned system was expressed as follow:

$$\bar{\omega}_n = \frac{\omega_n}{\sqrt{\frac{EI}{\rho AL^4}}}, n = 1, 2, 3 \cdots\cdots \tag{6}$$

2.2 Parameters Proposing

In the early stage of designing the balance cars body, it's the selection of structure that is essential. Because there is no frame, unitary construction body has small quality, low height and sensitive behavior. But is should bear all force. Owing to the weak ability that solar panels convert sunlight into electricity, the body structure need to be designed more lightweight to enhance the workpiece ratio of cars.

On the basis of existing balance cars, I refer to the parameters of Xiaomi No.9 balance car to design the main body parameters of solar energy balance car as Table 1:

Table 1. Design of solar energy balance cars

Width	60 cm	Height	60 cm
Tire diameter	28 cm	Height From The Ground	10 cm
Net weight	$\leq$ 12 kg	Load	$\leq$ 100 kg
Maximum gradeability	$\geq$ 10°	Average Speed	15 km/h
Lifetime	10000 h		

3 Finite Analysis

3.1 Model Import

As the car shown in the Fig. 1, it is known that the main frame is magnesium aluminum alloy with a density of 1800 kg/m^3, an elastic modulus of 40 GPa and poisson ratio of 0.28. Bearing made of materials for 45 steel, which is relatively great density of

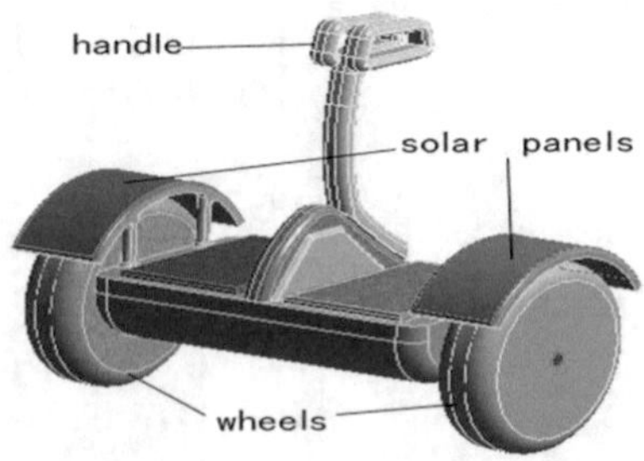

Fig. 1. Model import of solar balance car

7850 kg/m^3, the elastic modulus of 200 GPa and poisson ratio is 0.3. Now under the load of 100 kg of gravity, the deformation and stress distribution of the whole solar balance cars will be analyzed.

3.2 Contacts Installing

All contacts of the assembly are set as binding contacts (Fig. 2).

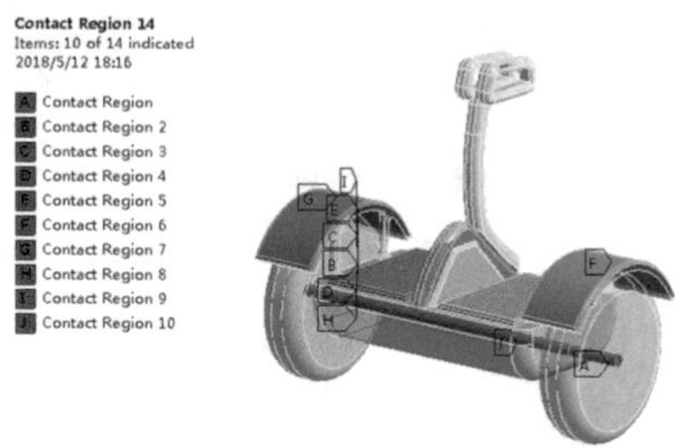

Fig. 2. Contacts installing

3.3 Mesh

In finite analysis, mesh is important, which can not only save the computation time of solution, but improve calculation accuracy. The model contains 79591 elements and 198830 nodes (Fig. 3).

Fig. 3. Mesh divides

4 Statistic Analysis

4.1 Strength Analysis on Both Sides of Load

Load and constraint: as shown in the Fig. 4, fixed constraint is applied to the wheel, and 500 N load is applied to the left and right pedals respectively, whose direction is vertically downward. Results solving show in the Figs. 5 and 6.

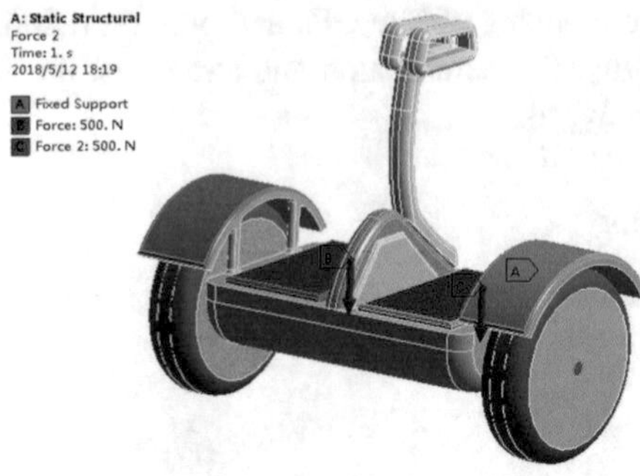

Fig. 4. Load and constraint in solar balance car

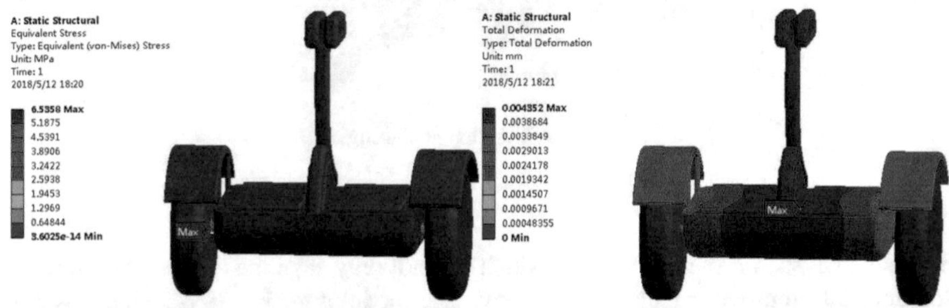

Fig. 5. Overall stress distribution

Fig. 6. Overall deformation distribution

According to the calculation results, the maximum stress of the model is located at the both ends of the bearing, which is 6.5 MPa, much smaller than 355 MPa, the strength of 45 steel. So it meets the strength requirements. The maximum deformation is 0.004 mm that is less than most balance cars', so it is negligible.

4.2 Strength Analysis on One Side of Load

Now it's another research about strength when people lifting one foot.

Load and constraint: as shown in the Fig. 7, fixed constraint is applied to the wheel, and 1000 N load is applied to one pedal, whose direction is vertically downward. Results solving show in the Figs. 8 and 9.

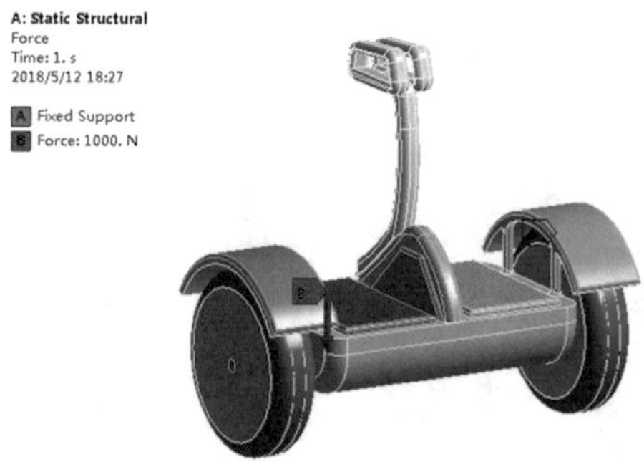

Fig. 7. Load and constraint in solar balance car

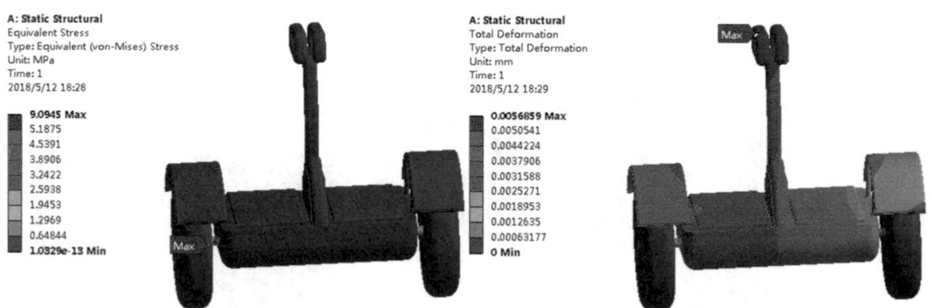

Fig. 8. Overall stress distribution

Fig. 9. Overall deformation distribution

5 Dynamic Analysis

Because solar balance car is under the pressure from people's quality, and it's susceptible to bumpy roads and lots of collision. Therefore, dynamic analysis is important as well. Table 2 shows the first six natural frequencies (Hz) of solar balance car. Figure 10 shows the first six mode shapes.

Table 2. The first six natural frequency

Mode	Frequency (Hz)	Maximum deformation (mm)	Maximum deformation position
1	68.504	31.757	The whole upper part of the handle
2	83.589	48.672	The outer edge of the left solar panel
3	83.656	48.682	The outer edge of the right solar panel
4	103.31	49.667	The middle upper part of the handle
5	154.90	37.881	The front upper edge of the handle
6	189.61	38.836	The outside corner of the two solar panels

(a) 1st (b) 2nd

(c) 3rd (d) 4th

(e) 5th (f) 6th

Fig. 10. The first six modes of the structure

By calculating, the natural frequency of the first mode is 68 Hz. And the car's critical speed is 4080RPM. Therefore, to keep away from the natural frequency of the system as far as possible, the working speed should be reasonably designed so as to avoid resonance.

6 Conclusion

As global energy has got into shortages and many other severe problems, solar energy plays an important role in human's society. This paper designs a solar balance car. The author uses ANSYS finite element analysis software to analyze the structure. Statistic analysis and modal analysis are used to ensure that the design of the car meets the requirement. The application of solar balance car can save much energy, decrease traffic congestion and make people's life more convenient.

Acknowledgement. This study is sustained by Fujian recommended the National College Students' innovation and entrepreneurship training program No.420.

References

1. Scribner. The case for on-car wheel balancing and wheel-to-hub indexing. Brake Front End **1**, 36–38 (2015)
2. Lipomi, J., Tee, C.-K., Vosgueritchian, M., et al.: Stretchable organic solar cells. Adv. Mater. **23**(15), 1771–1775 (2011)
3. Koberle, C., Gernaat, E.H.J., van Vuuran, P., et al.: Assessing current and future techno-economic potential of concentrated solar power and photovoltaic electricity generation. Energy **89**(9), 739–756 (2015)
4. Martin, D., Li, K.: Conversion of solar energy to fuels by inorganic heterogeneous systems. Chin. J. Catal. **32**(6), 879–890 (2011)
5. Carneiro, V.H., Puga, H., Meireles, J.: Analysis of the geometrical dependence of auxetic behavior in reentrant structures by finite elements. Acta Mech. Sin. **32**(2), 295–300 (2016)
6. Casella, F., Mathijssen, T., Colonna, P., et al.: Dynamic modeling of organic rankine cycle power systems. J. Eng. Gas Turbines Power Trans. ASME, **135**(4), 042310–1-042310-12 (2013)

Design and Analysis of Greenhouse Automated Guided Vehicle

Xiao-Yun Li, Yi-Jui Chiu(✉), and Han Mu

School of Mechanical and Automotive Engineering, Xiamen University of Technology, No. 600, Ligong Rd, Xiamen 361024, Fujian Province, China
chiuyijui@xmut.edu.cn

Abstract. In order to realize agricultural automation and large-scale production, a greenhouse automated guided vehicle is designed. It is equipped with a objective table and retractor device, using a track-type transmitting motion. The automatic guided vehicle is imported into the ANSYS through the graphical data conversion. The finite element model of the structure is generated by grid division. The finite element analysis of the structure is carried out. The strength and stiffness characteristics of the structure are calculated in the static analysis. The frequency of the vehicle calculated by modal analysis. The results show that the stiffness and strength meets the requirements of using, and the resonance damage can be avoided by avoiding the natural frequency of the vehicle. The greenhouse automatic guided vehicle can quickly transport materials to designated locations.

Keywords: ANSYS · Greenhouse · Automatic guided · Modal analysis

1 Introduction

As the greenhouse agriculture develops, only large-scale agricultural production and automated production can meet the requirements of the times. In the traditional agricultural goods handling, it mainly achieves the transportation of goods on the road. But in the greenhouse, it mostly relies on human resources to transport the goods, which requires more manpower and time. The automatic guided vehicle can transport the required materials to a fixed location, which facilitates the handling process and saves labor while improving labor efficiency. Its application is conducive to improve the economic efficiency of large-scale production.

Automatic greenhouse transport trucks mainly include the research on automatic handling technology and the design of vehicle and control systems. Researches are in this field, such as Jorgensen et al. [1] defined the requirements and scope of scripts which is used for controlling the automated guided vehicle in agricultural planting. The purpose is to achieve unmanned field operations that run through and cover entire fields. Martínez-Barberá and Herrero-Pérez [2] introduced a navigation system for flexible AGVs used for warehouse stock operations and frequent ground changes. Ni et al. [3] designed SCADA five-DOF handling robot structure according to carrying functional requirements and work characteristics of the SCARA robots. José et al. [4] based on an optimality property, proposed new mixed integer linear programming

P. Krömer et al. (Eds.): ECC 2018, AISC 891, pp. 256–263, 2019.
https://doi.org/10.1007/978-3-030-03766-6_29

(MILP) formulations for three versions of the problem and found that the proposed GA procedure can yield optimal or near optimal solutions in reasonable time. Bechtsis et al. [5] proposed a Sustainable Supply Chain Cube (S2C2) based on the design and planning of the AGV (Automatic Guided Vehicle) system and the modern supply chain (SC). An et al. [6] introduced a new type of robot navigation scheme: SLAM, which can build the environment map in a totally strange environment, and at the same time, locate its own position so as to achieve autonomous navigation function.

The intention of this paper is to design a automatic transfer vehicle used in the greenhouse. On the basis of meeting stiffness, strength requirements and the feasibility analysis of several programs, the final program is determined, and vibration analysis is performed.

2 Theories Analysis

2.1 Theories Research

In this paper, the safety factor method is used to determine whether the stress meets the strength requirements. Whether the maximum stress is taken by the vehicle or whether it is within the allowable stress range of the material. In static analysis, the allowable stress of vehicle is

$$[\sigma] = \frac{\sigma_b}{n} \tag{1}$$

where σ_b is the ultimate strength, n is the safety factor.

Because the structure of greenhouse automatic guided vehicle was complex, we should simplify the structure first. This paper exports the equations of the system.

$$[M]\{\ddot{X}\} + [K]\{X\} = 0 \tag{2}$$

Defined the position vector {X} as [Δ]{u}, where [Δ] was the modal matrix of the system. Equation (2) could be changed as follow:

$$[I]\{\ddot{u}\} + [A]\{u\} = 0 \tag{3}$$

In which:

$$[\Delta]^T[M][\Delta] = [I] = \begin{bmatrix} 1 & 0 & \cdots & 0 \\ 0 & 1 & \cdots & \vdots \\ \vdots & 0 & \ddots & 0 \\ 0 & 0 & 0 & 1 \end{bmatrix} \tag{4}$$

$$[\Delta]^T[K][\Delta] = [A] = \begin{bmatrix} \bar{\omega}_1^2 & 0 & \cdots & 0 \\ 0 & \bar{\omega}_2^2 & \cdots & \vdots \\ \vdots & 0 & \ddots & 0 \\ 0 & \cdots & 0 & \bar{\omega}_n^2 \end{bmatrix} \tag{5}$$

The natural frequency of the mistuned system was expressed as follow:

$$\bar{\omega}_n = \frac{\omega_n}{\sqrt{\frac{EI}{\rho A L^4}}}, n = 1, 2, 3 \cdots\cdots \tag{6}$$

2.2 The Establishment of a Modal

We use tracked automatic handling vehicle in greenhouse. The system includes a body, a objective table, retractor device, tracks and wheels. Three-dimensional model of the automatic guided vehicle is shown in Fig. 1. Through the calculation of the track transmission force, relevant parameters of the tracks can be obtained. The parameters of the automatic guided vehicle are shown in the following Table 1.

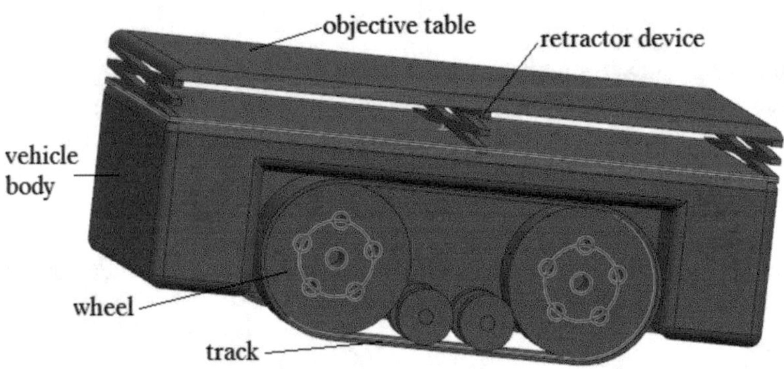

Fig. 1. Three-dimensional model of the automatic guided vehicle

3 Finite Analysis

The 3D geometry model is built by ANSYS software. The rubber is taken as the material for the tracks. The density of rubber is 1.10 g/cm^3, and its material constant C10, C01 and incompressibility parameter D1 are 8.0×10^7 Pa 5.3×10^7 Pa and 0.0001 Pa respectively. The 40Gr is taken as the material for the vehicle wheels and the aluminum alloy is taken as the material for the other structures. Their specific parameters are shown in the following Table 2. After the mesh is divided, the finite

Table 1. The parameters of the automatic guided vehicle

Mass	1575.4 kg	Wheel diameter	0.3 m
Vehicle-body dimension	1.5 m*1 m* 0.4 m	Coefficient of rolling resistance	0.1
Maximum laden mass	300 kg	Maximum gradeability	30°
Track width	0.17 m	Engine power	60 kw

element model is shown in Fig. 2. Set the physical field to mechanical. The grid is meshed according to the size of different parts. Before the simulation analysis, we need to check the grid quality. The model contains 40609 elements and 112554 nodes.

Table 2. Structure material parameter

	Density	Shear Modulus	Young Modulus	Poisson ratio
40Gr	7.90 g/cm^3	4.62×10^{12} Pa	1.2×10^{13} Pa	0.3
Aluminum alloy	2.77 g/cm^3	2.67×10^{10} Pa	7.1×10^{10} Pa	0.33

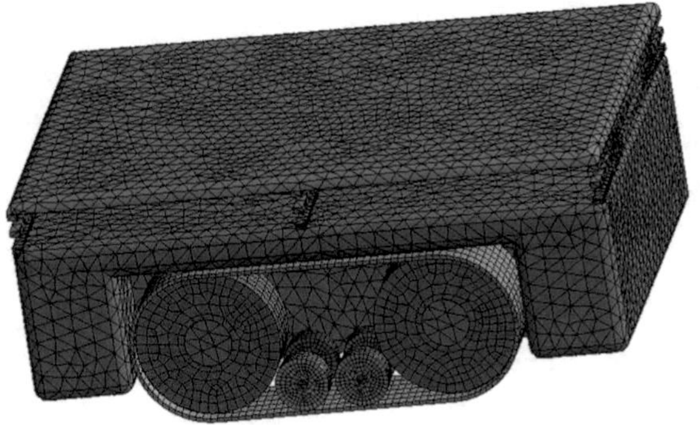

Fig. 2. Mesh divides

The loading of greenhouse automatic guided vehicle is shown in Fig. 3. The fixed support is provided at the tracks. The maximum laden mass of the automatic handling vehicle is 300 kg, so the force applied to the objective table is 2940 N, and the area of the objective table is 1.5 m*1.0 m. By calculating the stress applied to this surface is 0.00784 MPa.

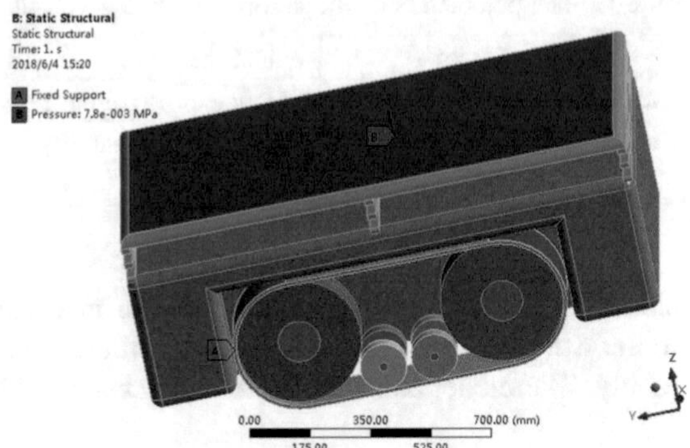

Fig. 3. The loading of greenhouse automatic guided vehicle

3.1 Static Analysis

By solving the ANSYS processor, the stress distribution and the deformation of the frame can be obtained, as shown in Figs. 4 and 5.

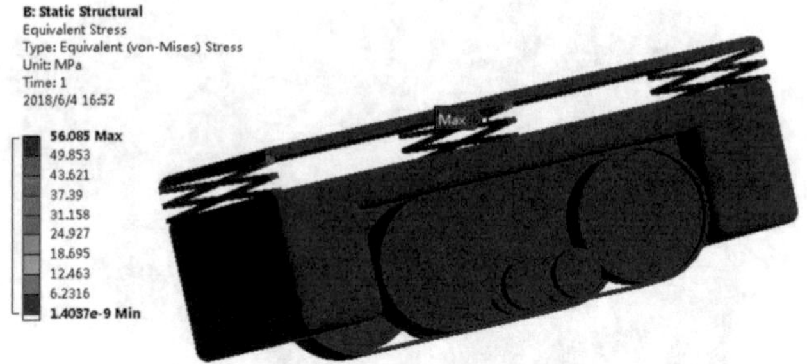

Fig. 4. The stress distribution of the vehicle

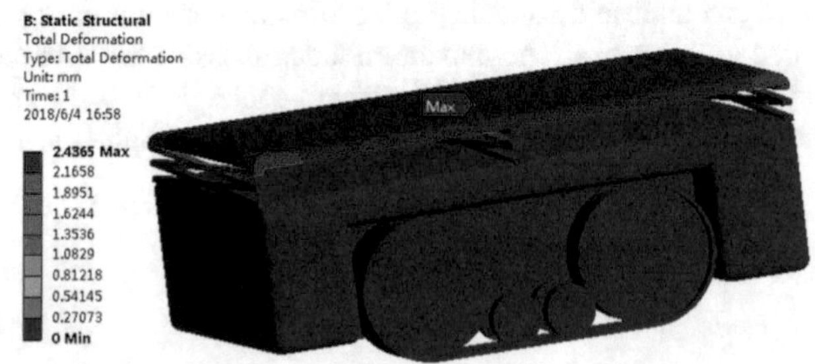

Fig. 5. The total deformation of the vehicle

According to the calculation results, the maximum stress of the model is located in the retractor device in the middle, which is 56.085 MPa. The aluminum alloy is a plastic material and the safety factor is [n] = 1.5 ~ 2. Here we take 2. The tensile strength of the material is 170 ~ 230 Mpa. Its allowable stress is calculated as $[\sigma] = 85\text{–}153$ Mpa. The maximum stress of the model is less than the allowable stress. So it meets the strength requirements. The maximum deformation is located on the top side of the axially moving plate, and the maximum total deformation is 2.4365 mm.

3.2 Dynamic Analysis

Because the greenhouse automatic guided vehicle is under pressure from the vegetables or fruits, we do pre-stress modal analysis to analyze the natural frequency and mode shape of the pre-stressed structure. Table 3 shows the first six natural Frequencies (Hz) of automatic guided vehicle. Figure 6 shows the first six mode shapes of automatic guided vehicle.

Table 3. The first six modes of the automatic guided vehicle

Mode	Frequency (Hz)	Maximum deformation (mm)	Maximum deformation position
1	14.099	4.9073	The lateral position of the objective table
2	14.678	2.9225	The objective table
3	17.536	3.0299	The objective table
4	20.959	4.8168	The vertical position of the objective table
5	26.215	1.2671	The upper part of the track
6	51.758	3.8432	The retractor device

The maximum deflection of the first six modes of a pre-stress vehicle mainly occurs at the objective table, which is mostly caused by the pressure of the material. Therefore, in the practical application process of the handling vehicle, the influence of pre-stressing should be considered. And to keep away from the natural frequency of the system as far as possible, the working speed should be reasonably designed so as to avoid resonance.

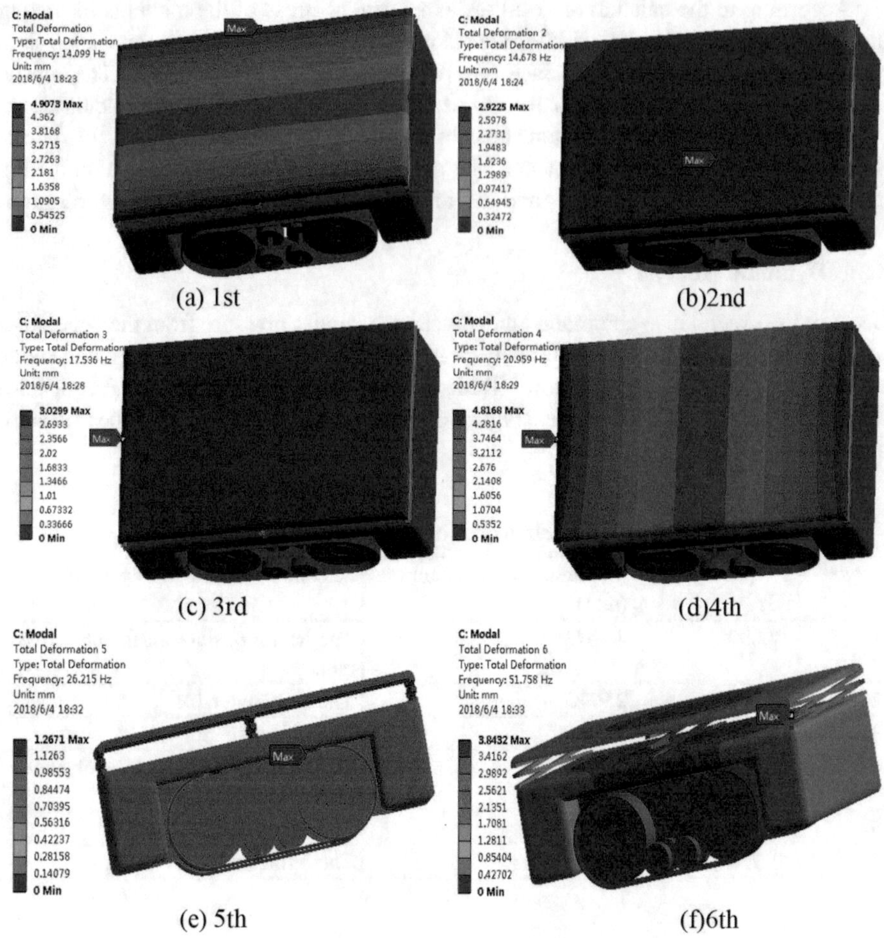

Fig. 6. The first six modes of the structure.

4 Conclusion

As automated technology enters agricultural production, automatic handling technology plays an important role in the greenhouse. This paper designs a greenhouse automatic guided vehicle based on transmitting motion. The author uses ANSYS finite element analysis software to analyze the structure. Static analysis and modal analysis are used to ensure that the design of the vehicle meets the requirements. The application of automatic greenhouse guided vehicle can increase agricultural productivity and reduce labor costs.

Acknowledgements. This study is sustained by Graduate Technology Innovation Project of Xiamen University of Technology No. 40316076, and Fujian recommended the National College Students' innovation and entrepreneurship training program No.420.

References

1. Jorgensen, R.N., Norremark, M., Sorensen, C.G., Nils, A.A.: Utilising scripting language for unmanned and automated guided vehicles operating within row crops. Sci. Direct **62**, 190–203 (2008)
2. Martínez-Barberá, H., Herrero-Pérez, D.: Autonomous navigation of an automated guided vehicle in industrial environments. Robot. Comput. Integr. Manuf. **26**, 296–311 (2010)
3. Ni, W., Wang, W.L., Zhao, X.H.: Structural design of SCARA handling robot with five degrees of freedom. Equip. Mach. **8**, 36 (2017)
4. José, A.V., Subramanian, P., Abraham, M.: Finding optimal dwell points for automated guided vehicles in general guide-path layouts. Int. J. Prod. Econ. **170**, 856–861 (2015)
5. Bechtsis, D., Tsolakis, N., Vlachos, D.: Sustainable supply chain management in the digitalization era: the impact of automated guided vehicle. J. Clean. Prod. **142**, 3970–3984 (2017)
6. An, F., Chen, Q., Zha, Y.F., Tao, W.Y.: Mobile robot designed with autonomous navigation system. J. Phys. Conf. Ser. **910**(1), 162–168 (2017)

Constructed Link Prediction Model by Relation Pattern on the Social Network

Jimmy Ming-Tai Wu[1], Meng-Hsiun Tsai[2(✉)], Tu-Wei Li[2], and Hsien-Chung Huang[3]

[1] College of Computer Science and Engineering, Shandong University of Science and Technology, Qindao, Shandong, China

[2] Department of Management Information Systems, National Chung Hsing University, Taichung, Taiwan
mht@nchu.edu.tw

[3] Office of Physical Education and Sport, National Chung Hsing University, Taichung, Taiwan

Abstract. For the link prediction problem, it commonly estimates the similarity by different similarity metrics or machine learning prediction model. However, this paper proposes an algorithm, which is called Relation Pattern Deep Learning Classification (RPDLC) algorithm, based on two neighbor-based similarity metrics and convolution neural network. First, the RPDLC extracts the features for two nodes in a pair, which is calculated with neighbor-based metric and influence nodes. Second, the RPDLC combines the features of nodes to be a heat map for evaluating the similarity of the node's relation pattern. Third, the RPDLC constructs the prediction model for predicting missing relationship by using convolution neural network architecture. In consequence, the contribution of this paper is purposed a novel approach for link prediction problem, which is used convolution neural network and features by relation pattern to construct a prediction model.

Keywords: Link prediction problem · Convolution neural network
Relation pattern · Social network

1 Introduction

While the online social networks, such as Facebook, Twitter, Youtube, etc., are explosive growth, how to find missing relationship has become a crucial challenge that is attracting the attention of academic and industry researcher to investigating social network structures. It has many applications, such as the recommendation of friends in a social network [1,2], a recommender system can help users find software tools that match their interest for Github [3] and further recommends friends on different types of social network, including Twitter and location-based Foursquare [4].

P. Krömer et al. (Eds.): ECC 2018, AISC 891, pp. 264–271, 2019.
https://doi.org/10.1007/978-3-030-03766-6_30

In recent years, the theorists and researchers of the social network proposed some algorithms of link prediction problem. For example, Ehsan and Maseud introduced a new unsupervised structural link prediction algorithm based on ant colony optimization [5]; Liang and Shuai introduced an ensemble approach based on neighbor-based metrics [6] to decomposing traditional link prediction problems into subproblems, and reduce the size of data size without drop down the accuracy; Leskovec et al. proposed a logistic regression based on degrees of the nodes and tried to obtain high accuracy on directed social network [7]. Those methods used machine learning framework to achieve the successful results in the experiments on link prediction problem.

As mentioned in the above, most of the previous works on link prediction using machine learning algorithms based on similarity metrics just focus on the information between two nodes and ignored the pattern of relationship in the whole graph. Further, due to the memory size and the algorithm limitations, traditional link prediction methods cannot handle a large-scale social network and lose a log of information in a big data environment. Therefore, this paper suggests a deep learning framework based on a new feature extraction method to improve the performance of the link prediction issue for a real-life social network.

To deal with large-scale social networks, a novel algorithm is thus proposed in this paper, it applies a deep learning model and a new feature extraction method to improve the accuracy of prediction. The feature extraction of this research, which called pattern of relationship, based on traditional neighbor-based metrics according to the related edges which are between the target node and the more important nodes for this target node in the social network. And it further lets measure of similarity between two nodes be a heat map. In addition, the proposed approach used convolution neural network, which is a branch of deep learning technique, to recognize the consistency between the patterns of relationship on two nodes.

2 Related Work

In this section, it introduces the techniques of link prediction, including the topology-based metrics, social theory based metrics and the prediction approaches based on machine learning algorithms.

2.1 Topology-Based Metrics

Most similarity metrics focus on the topological information on the social network. These indexes stem from the evaluation that two nodes have contact with similar nodes or paths. Liben-Nowell and Kleinberg discussed the metrics based on topology in early period [8], and after their work, that leads to many topology-based metrics be proposed.

2.2 Social Theory Based Metrics

Moreover, the academics in the link prediction field of the social network not only use topology-based metrics, but also use a lot of state-of-the-art social theories, including node centrality [9], structural balance [7,10], community [11] and closure, to improve the algorithms for predicting missing link or solving other problems in a social network.

2.3 Machine Learning Algorithms for Link Prediction

In the other side, there are many approaches proposed for link prediction problem by machine learning algorithms based on different similarity metrics, including neighbor-based, path-based, random walk based and external information from social network theories in past few years. It can be defined as a classification problem with two classes: existent links and non-existent links. And academics in the social network field built many supervised classification learning models to solve link prediction problem, e.g. decision tree [12,13], support vector machines [13,14], logistic regression, näive Bayes [12,13], random forests [12], Support Vector Regression (SVR) [15], restricted Boltzmann machine [16], and so on.

3 The Proposed Method: RPDLC Algorithm

This study employed a link prediction algorithm, which is called Relation Pattern Deep Learning Classification (RPDLC), based on the deep learning framework and pattern of relationship to detect the missing edges on the social network. The process consisted of extracting feature and training deep learning classifier tasks.

3.1 The Neighbor-Based Metrics Were Used in Algorithm

The two neighbor-based metrics, including Sorensen Index (SI) and Hub Depressed (HD), was used in the proposed algorithm. Because these have several characteristics, including simplicity, lower time complexity and wide-spread for using.

The neighbor-based metrics use neighbors of each node and common neighbors of two nodes to compute the similarity. Assume that N_x is a node in the social network, $\Gamma(N_x)$ is the set of neighbors based on x and $|\Gamma(N_x)|$ is the amount of $\Gamma(N_x)$. In the previous works, there are 11 kinds of metrics introduced to scale the similarity between two nodes in social networks.

Sørensen Index. Sørensen Index(SI) considers the size of the common neighbors, and focuses on the lower degrees of nodes that would have higher relationship likelihood [17]. More specifically, the higher value of SI is because both two nodes have low degree. The metric is defined as Eq. 1.

$$SI(N_x, N_y) = \frac{|\Gamma(N_x) \bigcap \Gamma(N_y)|}{|\Gamma(N_x) + \Gamma(N_y)|} \tag{1}$$

Hub Depressed. Hub Depressed is used to similar as HP, but it is considered by the nodes with higher degree [18]. More specifically, the higher value of HP is because of the higher degree node in a pair. The metric is defined as Eq. 2

$$HD(N_x, N_y) = \frac{|\Gamma(N_x) \bigcap \Gamma(N_y)|}{max(|\Gamma(N_x), \Gamma(N_y)|)} \tag{2}$$

3.2 The Pseudo Code of RPDLC Algorithm

The algorithm in this study was proposed for predicting existent and non-existent relationship, which is combined with neighbor-based metrics, the notion of relationship pattern and convolution neural network architecture. The pseudo code and description are provided in this section.

Algorithm 1. The RPDLC Algorithm with Shortest-Path and Influence Nodes

Input: S, the social network dataset (S_p, the positive samples that have existent edge, S_n, the negative samples that do not have existent edge); N, all nodes of dataset; δ, the threshold value of $N_{influence}$; $\bar{L}$, the threshold value of shortest path between two nodes;

Output: *Accuracy*; *AUC*

```
set δ value, 1≤ δ ≤ N;
set N_influence is the set of important nodes;
set n = the size of N;
set N_d = the set of CentralityDegree(N);
set M = the similarity metric;

while the size of N_influence < δ do
    N_influence append the node N_h, which owes the highest value in N_d;
    N_d pop out the node N_h;
end while

for a = 0; a ≤ n; a++ do
    for i = 0; i ≤ n; i++ do
        P_a = extract the features (N_a, N_i) by M;
    end for
    for b = 0; b ≤ n; b++ do
        if shortest path between N_a and N_b < L̄ then
            for i = 0; i ≤ N.length; do
                P_b = extract the features (N_b, N_i) by M;
            end for
        end if
    end for
    normalize all features with z-score normalization;
    heatmap H = P_a + P_b;
end for

while Accuracy and AUC not good enough do
    train convolution neural network prediction model
end while
return AUC, Accuracy;
```

In the Algorithm 1, it generates the neighbor-based features by HD or SI metric between N_a and $N_i \in N$, which is in the Eq. 3, whereas is identical with N_b.

$$\sum_{i=1}^{N} \textit{Neighbor-based metric}(N_a, N_i) \quad (3)$$

Afterwards, the algorithm combines feature vector of N_a and N_b and used z-score normalization, which is in the Eq. 4. To reshape to a new matrix $Features\,(m, len)$, and len as the all numbers of N in the social network. Then, let $Feature\,(m, len)$ be the heat map of relationship pattern between two nodes with all other nodes in the set of N. Finally, the prediction model used a convolution neural network framework to classify and predict the missing link.

$$Z_i = \frac{X_i - \mu}{\sigma} \quad (4)$$

where μ is mean and σ is standard deviation.

As the social network becomes larger as the number of nodes is extremely growing, that leads to an inestimable computation in the process of extracting feature with the set of whole nodes in the dataset.

For this reason, there is an approach with degree centrality for decreasing the number of parameter in each sample. The degree centrality of the node, which is one of the most wide-spread measures for counting the number of edges to a node.

The deep learning framework of the proposed method uses the concepts to build CNN architecture for predicting missing link. First, the framework builds the layers as the AlexNet [19] that stacks the convolution layers with only 3×3 kernels and max-pooling layers, further stacks the full-connected layers to be the last layers of the framework, and uses SeLU to be the activation function for converting output value between layers. Second, the framework uses dropout to avoid overfitting for improving the generalization ability of prediction model [22]. At last, the framework employs Adam optimizer [23] to dynamically adjust the learning rate for controlling the training speed.

4 Experiment

In this section, we present the setup and result of our experiments. The experiment's objective is to predict the missing link in three social network datasets with different types.

4.1 Experimental Setup

Each original dataset is randomly divided into 10 equal size subsamples. Of the 10 subsamples, 2 subsamples were retained as the test data for the testing model, and the other 8 subsamples are used to be training data for the training model.

To estimate the link prediction performance of the proposed algorithm and other baseline algorithms, the experiments use a standard metric, the Area Under receiver operating characteristic Curve (AUC) [27], to measure the accuracy of link prediction models. The AUC metric can be interpreted as the probability that a randomly chosen missing edge is given a higher score than a randomly chosen non-existent edge [18]. Among n independent comparison. There are numbers of $n^{'}$ when the missing edge has a higher score than the non-existent edge and numbers of $n^{''}$, which contain the set by the score is equal between missing and non-existent edge. Define the AUC as Eq. 5.

$$AUC = \frac{n^{'} + 0.5n^{''}}{n} \tag{5}$$

4.2 Comparisons with Other Algorithms

There were considered with five baseline and link prediction algorithms, including CN [28], AA [29], RWR [30], FL [31] and PNC [32], and that were discussed in section of related work. And the environment for training CNN models is based on Tensorflow 1.5.0 and GTX 1080Ti GPU.

As shown in Table 1, the RPDLC algorithm with SI metric achieved the highest performance on both Jazz and NetScience datasets, and the RPDLC algorithm with HD achieved the highest performance on Facebook dataset.

Table 1. AUC value of different algorithms in three datasets.

	RPDLC (SI)	RPDLC (HD)	PNC	FL	WR	AA	CN
Jazz	0.9999	0.9549	0.9665	0.9422	0.7077	0.8437	0.8334
NetScience	0.9853	0.5532	0.9843	0.8956	0.9490	0.7659	0.6894
Facebook	0.7709	0.9680	0.9603	0.9160	0.9389	0.7603	0.7442

5 Conclusions

To summarize, the present study is preliminary research on link prediction problem based on a convolution neural network model with a novel feature extraction approach, which is using influence node and neighbor-based metrics. A primary contribution is that proposed a link prediction algorithm with deep learning can obtain high performance for predicting missing relationship. Despite the RPDLC algorithm had high performance, it still had some defects. The RPDLC algorithm uses only neighbor-based metrics and restricts the number of metrics for extracting feature, and not considers other similarity metrics. In the future work, much more also needs to be attempted if the RPDLC algorithm uses other similarity-based metrics, such as path-based metrics, random walk based metrics and social theory based metrics, and the combination by different set of metrics, such as set by fewer metrics to increase the efficiency or set by more metrics to get more feature with different characteristic.

References

1. Barbieri, N., Bonchi, F., Manco, G.: Who to follow and why: link prediction with explanations. In: The 20th ACM SIGKDD International Conference on Knowledge Discovery and Data Mining, pp. 1266–1275. ACM (2014)
2. Tang, J., Chang, S., Aggarwal, C., Liu, H.: Negative link prediction in social media. In: The Eighth ACM International Conference on Web Search and Data Mining, pp. 87–96. ACM (2015)
3. Zhou, J., Kwan, C.: Missing link prediction in social networks. In: International Symposium on Neural Networks, pp. 346–354. Springer (2018)
4. Hristova, D., Noulas, A., Brown, C., Musolesi, M., Mascolo, C.: A multilayer approach to multiplexity and link prediction in online geo-social networks. EPJ Data Sci. **5**(1), 24 (2016)
5. Sherkat, E., Rahgozar, M., Asadpour, M.: Structural link prediction based on ant colony approach in social networks. Phys. A Stat. Mech. Appl. **419**, 80–94 (2015)
6. Duan, L., Ma, S., Aggarwal, C., Ma, T., Huai, J.: An ensemble approach to link prediction. IEEE Trans. Knowl. Data Eng. **29**(11), 2402–2416 (2017)
7. Leskovec, J., Huttenlocher, D., Kleinberg, J.: Predicting positive and negative links in online social networks. In: The 19th International Conference on World Wide Web, pp. 641–650. ACM (2010)
8. Liben-Nowell, D., Kleinberg, J.: The link-prediction problem for social networks. J. Assoc. Inf. Sci. Technol. **58**(7), 1019–1031 (2007)
9. Liu, H., Hu, Z., Haddadi, H., Tian, H.: Hidden link prediction based on node centrality and weak ties. EPL (Europhys. Lett.) **101**(1), 18004 (2013)
10. Cartwright, D., Harary, F.: Structural balance: a generalization of heider's theory. Psychol. Rev. **63**(5), 277 (1956)
11. Pirouz, M., Zhan, J., Tayeb, S.: An optimized approach for community detection and ranking. J. Big Data **3**(1), 22 (2016)
12. Scellato, S., Noulas, A., Mascolo, C.: Exploiting place features in link prediction on location-based social networks. In: The 17th ACM SIGKDD International Conference on Knowledge Discovery and Data Mining, pp. 1046–1054. ACM (2011)
13. De Sá, H.R., Prudêncio, R.B.: Supervised link prediction in weighted networks. In: The 2011 International Joint Conference on Neural Networks (IJCNN), pp. 2281–2288. IEEE (2011)
14. Li, X., Chen, H.: Recommendation as link prediction in bipartite graphs: a graph kernel-based machine learning approach. Decis. Supp. Syst. **54**(2), 880–890 (2013)
15. Hua, T.-D., Nguyen-Thi, A.-T., Nguyen, T.-A.H.: Link prediction in weighted network based on reliable routes by machine learning approach. In: 2017 4th NAFOSTED Conference on Information and Computer Science, pp. 236–241. IEEE (2017)
16. Yu, X., Chu, T.: Dynamic link prediction using restricted Boltzmann machine. In: 2017 Chinese Automation Congress (CAC), pp. 4089–4092. IEEE (2017)
17. Sørensen, T.: A method of establishing groups of equal amplitude in plant sociology based on similarity of species and its application to analyses of the vegetation on danish commons. Biol. Skr. **5**, 1–34 (1948)
18. Zhou, T., Lü, L., Zhang, Y.-C.: Predicting missing links via local information. Euro. Phys. J. B **71**(4), 623–630 (2009)
19. Krizhevsky, A., Sutskever, I., Hinton, G.E.: ImageNet classification with deep convolutional neural networks. In: Advances in Neural Information Processing Systems, pp. 1097–1105 (2012)

20. Simonyan, K., Zisserman, A.: Very deep convolutional networks for large-scale image recognition. arXiv preprint arXiv:1409.1556 (2014)
21. Klambauer, G., Unterthiner, T., Mayr, A., Hochreiter, S.: Self-normalizing neural networks. CoRR, abs/1706.02515 (2017)
22. Srivastava, N., Hinton, G., Krizhevsky, A., Sutskever, I., Salakhutdinov, R.: Dropout: a simple way to prevent neural networks from overfitting. J. Mach. Learn. Res. **15**(1), 1929–1958 (2014)
23. Kingma, D.P., Ba, J.: Adam: a method for stochastic optimization. arXiv preprint arXiv:1412.6980 (2014)
24. Gleiser, P.M., Danon, L.: Community structure in jazz. Adv. Complex Syst. **6**(4), 565–573 (2003)
25. Newman, M.E.J.: Finding community structure in networks using the eigenvectors of matrices. Phys. Rev. E **74** (2006). arXiv: physics/0605087
26. McAuley, J., Leskovec, J.: Learning to discover social circles in ego networks. In: NIPS, p. 9 (2012)
27. Hanley, J.A., McNeil, B.J.: The meaning and use of the area under a receiver operating characteristic (ROC) curve. Radiology **143**(1), 29–36 (1982)
28. Chen, J., Geyer, W., Dugan, C., Muller, M., Guy, I.: Make new friends, but keep the old: recommending people on social networking sites. In: The SIGCHI Conference on Human Factors in Computing Systems, pp. 201–210. ACM (2009)
29. Adamic, L.A., Adar, E.: Friends and neighbors on the web. Soc. Netw. **25**(3), 211–230 (2003)
30. Pan, J.-Y., Yang, H.-J., Faloutsos, C., Duygulu, P.: Automatic multimedia cross-modal correlation discovery. In: The Tenth ACM SIGKDD International Conference on Knowledge Discovery and Data Mining, pp. 653–658. ACM (2004)
31. Papadimitriou, A., Symeonidis, P., Manolopoulos, Y.: Fast and accurate link prediction in social networking systems. J. Syst. Softw. **85**(9), 2119–2132 (2012)
32. Yu, C., Zhao, X., An, L., Lin, X.: Similarity-based link prediction in social networks: a path and node combined approach. J. Inf. Sci. **43**(5), 683–695 (2017)

Digital Simulation and Intelligence Computing

A Tangible Jigsaw Puzzle Prototype for Attention-Deficit Hyperactivity Disorder Children

Lihua Fan[1(✉)], Shuangsheng Yu[2], Nan Wang[3], Chun Yu[1], and Yuanchun Shi[1]

[1] Department of Computer Science and Technology, Tsinghua University, Beijing 100084, China
air8218@163.com, chunyu@tsinghua.edu.cn, shiyc@mail.tsinghua.edu.cn

[2] Department of Electronic Engineering, Tsinghua University, Beijing 100084, China
wss15@mails.tsinghua.edu.cn

[3] Academy of Arts and Design, Tsinghua University, Beijing 100084, China
wn17@mails.tsinghua.edu.cn

Abstract. In our country, it is estimated that nearly 20 million children have Attention-deficit hyperactivity disorder (ADHD). There are three main methods of treating children with ADHD including EEG biofeedback, sensory integration training, and behavioral therapy. The problem is that only the EEG biofeedback process in three methods has the data that can be recorded and evaluated. In the field of human-computer interaction, researchers pay attentions to how to improve children's attention by digital children's toys with TUI (Tangible User Interface). This paper's motivation was to record and evaluate the attention of ADHD children in physical interactive process. We integrated the idea of TUI, the principles of sensory integration training and behavioral therapy. We designed and implemented a physical interactive digital platform with TUI- a tangible jigsaw puzzle prototype. There were two games were implemented in this platform, namely number sorting and catching mice.

Keywords: A tangible jigsaw puzzle · ADHD children
Physical interactive digital game

1 Introduction

ADHD is a very common mental disorder in children. Children with ADHD are characterized by hyperactivity, impulsivity, and attention deficit. The reason is that the physiological defects of brain development cause the attention to be ineffective and continuously concentrated on the corresponding sensory channels. Children with ADHD have generally poor academic performance and are difficult to get along with others at home and at school.

For children with ADHD, current methods for improving and treating children's attention include drug therapy, EEG biofeedback, sensory integration training, and

P. Krömer et al. (Eds.): ECC 2018, AISC 891, pp. 275–286, 2019.
https://doi.org/10.1007/978-3-030-03766-6_31

behavioral therapy (drug therapy is beyond the scope of this topic and will not be discussed later). The problem is that only the EEG biofeedback process has the data of brain waves can be measured, recorded, and evaluated. In the process of sensory integration training and cognitive-behavioral therapy, there are no records and evaluation of behavioral data, such as action data. Also, the assessment of treatment effect is based on a post-treatment scale assessment [7].

This paper's motivation was to record and evaluate the attention of ADHD children in physical interactive process. We designed and implemented a physical interactive digital game with TUI. The essence of TUI is that digital information is embedded in physical entities. In the field of human-computer interaction, there are more and more researches pay attentions to how to improve children's attention by digital children's toys with TUI. This game would integrate the idea of TUI, the principles of sensory integration training and behavioral therapy. The children could use other sensations such as tactile, spatial, and proprioceptive in the process of interacting with physical digital game.

We have prototyped a smart tangible puzzle system, which include (1) an Edison-based tangible puzzle which extends SPI interface to support color screen, (2) an audio processing algorithm to detect "blow", (3) an tracking application to capture location of puzzles, and (4) two initial games based on the platform.

This article would be elaborated as follows: (1) the related works; (2) the design scheme of the physical interactive puzzle game; (3) the realization of the prototype system; (4) the realization of two games: the number sorting and catching mice.

2 Related Works

We firstly introduce the basic principle of EEG biofeedback training, Sensory Integrative training and Cognitive-behavioral therapy, as shown in Fig. 1, and the research progress and the advantages and disadvantages of these treatments. Secondly, we introduce the study of the TUI in the field of the human-computer interaction. Lastly, we propose our analysis and ideas.

Fig. 1. Sensory integration and EEG biofeedback.

Through the mind control game, EEG biofeedback training is to train the brain, change the brain electrical waveform, and then achieve the purpose of improving the functional state of the brain. In the past decade, it has been used as the main means of treating children with ADHD in major hospitals and commercial institutions in China. This training method has been proven to have a good therapeutic effect [4, 6, 12, 19]. The shortcoming is that the training process is boring, the fatigue is obvious, and the sustained effect is short. Besides, because only having children's visual to participate, this method is not conducive to the comprehensive development of children's mind and body. Also, in the field of human-computer interaction, EEG biofeedback is integrated in children' digital agent to enhance the attention of children. The principle of it is: when the value of the attention of children drop to a threshold by EEG biofeedback, the digital agent adjusts the value of the attention of child by voice, video, the movement of robot etc. [2, 15, 18].

Sensory Integrative training stimulates different sensory information, promotes the development of brain nerve cells again, and restores the patient's ability of integration. Training methods include balance seesaw, cylinder, slide, balance table, trampoline, skateboard, puzzle, building blocks, etc. [5]. Also, it is suggested that Ping-Pong, bicycle, drawing, and jigsaw puzzle can be trained at home and at school [11]. Studies have shown that the sensory integration disorder rate in children with ADHD is 81.39% [21]. After the children having sensory integration training, their attention and behavioral characteristics have been significantly improved, and the efficiency is close to 70% [10]. The disadvantage is that training requires the participation of professionals, which can only be carried out in hospitals, and it takes a long time, is expensive, and is difficult to sustain. In addition, researchers designed mobile phone software for sensory integration training, which mainly used to improve hand-eye coordination and reasoning ability [20].

The jigsaw puzzle is one important way in sensory integration training, which is widely used to train children' attention. Also, it has been proved that can give a more comprehensive training for the cognition of children, which trains memory, perception, observation, and the capacity to grasp.

Cognitive-behavioral therapy trains ADHD children through a series of cognitive-related game tasks, including selection of graphics, finding differences, matching and classification (animals, tools, locations, and colors), storytelling, logical alignment, etc. [13]. A study [9] validated that the game could enhance the cognitive and cognitive enhancement could affect the severity of ADHD symptoms. The results of the relevant evaluation scales show that the severity of ADHD symptoms can be effectively improved and can last for three months. In addition, Mackie et al. also studied the relationship between cognitive control and attention function [14], and concluded that attention function includes cognitive control. The disadvantage of cognitive-behavioral therapy is effective results require long-term cooperation by parents and teachers.

In the field of human-computer interaction, MIT Media Laboratory's researchers have developed a series of digital children's toys and teaching aids with TUI such as quantity, cardinality, speed, ratio, probability, accumulation, feedback, etc. Digital MiMs [22], which help children to learn abstract concepts better. Puchi Planet [1], developed by the Media Design Institute of Keio University in Japan, is a TUI toy designed for long-term hospitalized children. Towards Utopia [3] is a TUI geographic knowledge learning system designed for children aged 7–10. Tangibility impacts reading and spelling acquisition of young children that having dyslexia [8]. Tangible blocks are used to teach programming games [16]. Patients can use a tangible device to record their pains [17].

In physical interaction, children can use other sensations other than sight and hearing, such as tactile, spatial, and proprioceptive. Compared with interacting with computer, physical interaction can avoid the disadvantages of physical and mental development problems caused by computer operations – only watching and listening. Besides, the idea of physical interaction in improving children's attention is consistent with sensory integration training to some extent. The difference is that physical entities in physical interactions have digital information and technology, while physical entities in sensory integration training are not related to digital information and technology.

Thus, we developed a platform of physical entities - Tangible Jigsaw Puzzle, which not only could include the function of traditional Jigsaw Puzzle, but also could include some other cognitive-related games with digital information and technology.

3 Design

Our research problem is the recording and evaluation of the attention of ADHD children by Tangible Jigsaw Puzzle game. Firstly, we will select some basic cognitive tasks as the content of the attention training. We then will find the relational model between the basic cognitive tasks and the multi-model information which is based on the data of the EEG biofeedback and behavioral characteristics. EEG biofeedback can be recorded the attention of ADHD children in the game and control the game schedule. In view of the result of the relational model, we establish a model to evaluate the degree of the attention. In the end, we apply the relation between the different cognitive tasks and the attention to design of the game. Then, the result will be used for designing of our game in order to record, analyze and improve the ADHD children' attentions.

We established the platform of Tangible Jigsaw Puzzle game. Unlike traditional jigsaw puzzle, we presented a new interactive way using multi-sensory channel. We show the ideas of how children interact with the puzzle pieces in Fig. 2.

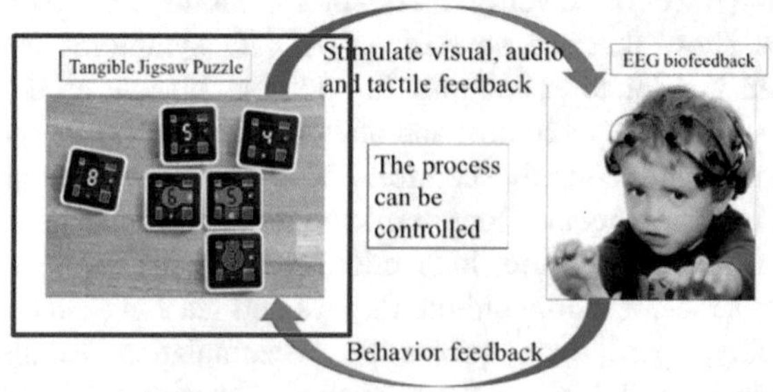

Fig. 2. Ideas about interaction.

This platform consists of several intelligent puzzle blocks. Each puzzle block is itself a small computer with wireless communications capability, which can provide visual (display), auditory (sound) and tactile (vibration) feedback. Meanwhile, each puzzle block supports behavior feedback including voice input and gesture input. Children can blow and pick up the puzzle block, which is defined as a new interactive way to control the puzzle pieces. In this paper, due to the complex of our research, we have initially implemented this system and gave the idea of EEG biofeedback.

4 System Implementation

Hardware: Intel NUC5i5RYK mini intelligent computer for the server, Intel Edison for the clients, Logitech C270 Webcam. Software: Visual studio 2010 C++ for the server, Intel IoTDev Kit built Eclipse C++ for the Edison.

4.1 The Design of the System

The system includes a server with a camera and three clients, as shown in Fig. 3. The communication mechanism between the server and client is sockets. The server's work contains three steps: a. the server receives the signal from clients; b. based on the received signals and three two-dimensional codes that a camera has scanned real-time, the server decides whether the game is correct; c. the server sends this signal to clients.

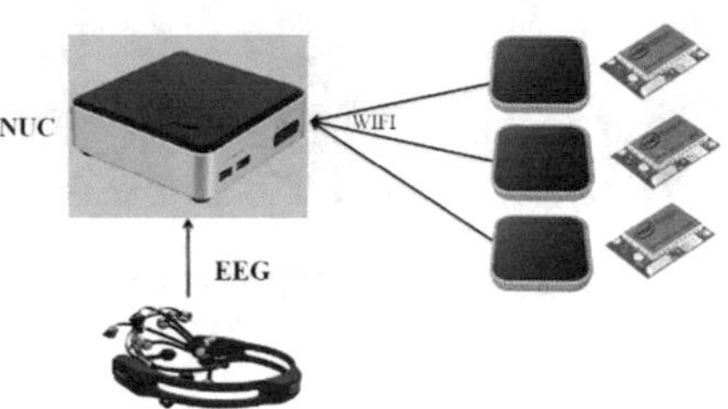

Fig. 3. Components of the system.

Each puzzle block is a client using an Intel Edison with some sensors and interfaces, which includes two types of the work: **a.** every client sends the signal to the host and receives the signal from the server; **b.** clients perceive users' behavior and interact with users. Brain-computer devices will control the progress of the game as a system component.

4.2 The Prototype System

Our prototype system includes the display module, the tracking module and the sensor module, as shown in Fig. 4. The display module includes the function of the image's display. The tracking module recognizes the location of the puzzle blocks. The sensor

Fig. 4. The prototype system.

module is used to the interaction. In the following sections, we will introduce every module.

The Display Module. The display module is the core of the system, which is used to display game-related images, as shown in Fig. 5. The left figure shows the connection between the screen and the Edison module, and the right figure is the effect of display screen.

Fig. 5. The display module.

The principle of the display: (1) The TFT screen with SPI bus interface; (2) Displaying program is divided into three parts: the SPI bus driver (initialization, communications transmission), the screen driver (initialization screen, image display), image processing algorithm library (curve drawing, filling), SPI bus driver uses libmraa; (3) The screen driver chip is ST7735S and the driver code initialize the screen, dot display, line display and position settings. The image display algorithm transplants Adafruit-GFX library.

The Tracking Module. The tracking module is used to track and identify the specific location of each puzzle block in the game. And the principle is the two-dimensional code recognition technology. We define that each puzzle block has a unique two-dimensional code number. At first, we paste the appropriate size of two-dimensional codes in the rear of each puzzle block in order to the camera can scan these codes of three blocks in real-time. Then the server can track and recognize the location of each block of the puzzle, and the relative position between the puzzle blocks.

As shown in Fig. 6, the left is the location that the two-dimensional code pasted. And the right is a display on the server about tracking and recognizing status in real-time. Especially, yellow border on each block is the identified area of the two-dimensional code and figures are predefined numbers of two-dimensional code. “False” blow in the right diagram represents that the numerical order currently is wrong.

Fig. 6. The tracking module.

The Sensor Module. We define that each puzzle block is a sensor module, which includes a 1.4-in. screen, a MPU6050 sensor, a sound sensor and a sensor button, as shown in Fig. 7. MPU6050 sensor is used to detect the action of the user. The sound sensor is used to detect the voice of the user. The button sensor is used to detect the signal that the user sends to the host.

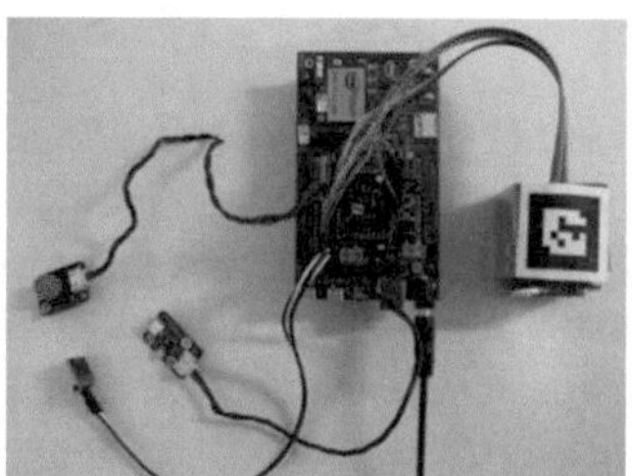

Fig. 7. The sensor module.

Image Storage. Each puzzle block stores images having the same content and the same number, which contains two types of the image. One is the image of number sorting and catching mice. Other is the image of smiley face and sad face. We define a code for each image in order to the implement of the system. According to the corresponding specification of selected screen, the image format that we use is bmp with 128 * 128 pixels.

5 The Platform of the Tangible Jigsaw Puzzle

Unlike traditional Jigsaw Puzzle, the platform of the tangible jigsaw puzzle has enhanced the game by information technology. We implemented two games called the number sorting and catching mice on this platform, as shown in Figs. 8-1 and 8-2.

Fig. 8.1. The number sorting.

Fig. 8.2. Catching mice.

5.1 Switch the Game

We give a definition that different images represent different games and that blowing to the puzzle block can change the image on the puzzle block. Thus, the user blowing one time to the puzzle block indicates one changing of the image. Consequently, we define a threshold of the sound sensor. When the value collected from the user's blowing is greater than the threshold, it means the completion of the blow.

5.2 The Number Sorting

We define that each puzzle block represents a number. When the game starts, you will see a number showed on each puzzle block. What the user should do is to pick up puzzle blocks and then exchange their order, until the blocks is in descending or ascending order.

The Logic of the Number Sorting

1. Switching the image: changing the current image into the image of number sorting by blowing the puzzle block.
2. Sending the signal: After sorting numbers, the user presses the button on each puzzle block. Meanwhile, every block sends the signal to the server through Wi-Fi. The content of the signal is the combinational code of the image's number and the block's number.

3. Matching: The server has saved the correct combination of the two-dimensional code and the image's number. In matching process, the server compare the combination which the client sends and the camera scans with the combination stored in the server to determine whether the sorting is correct or not.
4. Feedback: If the matching is successful, the feedback signal is one; on the contrary, the feedback signal is zero.
5. The show of the result: when the feedback signal that each puzzle block received is 1, the smiling face will display on each screen. When the feedback signal is 0, the screen will display a sad face on each screen.

5.3 Catching Mice

We define that each puzzle block represents a hole. When the game starts, a hole with a mouse or without a mouse will be randomly displayed on blocks. What the user should do is to pick up the block with a mouse as quick as possible. We give three kinds of the image.

The Logic of Catching Mice

1. Switching the image: changing the current image into the image of catching mice by blowing to the puzzle block.
2. Catching mice: Puzzle blocks randomly display the image with the mouse, which the user has picked up as quickly as possible.
3. Detecting and recording: The motion sensor detects whether the puzzle block is picked up or not in a certain time. At the same time, scanned two-dimensional code is recorded the code of the picked block.
4. The show of the result: By the data of the motion sensor and two-dimensional code, the server compute times that the mouse is caught.

In this project, the goal of the research is to establish a basic relationship between cognitive tasks and the attention. We implemented the number sorting and catching mice, which were the content of basic cognitive tasks. Number sorting can train ADHD children' ability of looking carefully and the permutations and combinations. Catching mice can train ADHD children' ability of looking carefully and controlling the reaction.

5.4 EEG Biofeedback

We set up a theoretical relationship between the basic cognitive task and the attention in this paper.

EEG engagement (E) measured by EEG equipment is formula (1). The activity level of α, β and θ waves of EEG waveform can be obtained through the lead of EEG equipment. A certain moment can reflect the concentration of an individual in a task. The greater the value is, the higher the concentration is. However, it is necessary to determine whether there is a certain degree of concentration in a period of time.

$$\mathrm{E} = \frac{\alpha}{\beta + \theta} \tag{1}$$

It is assumed that when a selected cognitive task C is executed, its attention is CA (T) in time T. In time T, the average level of EEG concentration is Ce. The relationship model between EEG data and attention is defined as (2).

$$\mathrm{CA(T)} = \mathrm{Ce} \tag{2}$$

Since the user experiment does not begin at present, we cannot give accurate results of the relationship between attention and cognitive tasks. Thus, this result will be given in the future study.

6 Discussion

In our prototype system, only three blocks are used. But in user experiment, the number of the puzzle block we will use is three to nine. The more blocks are used, the greater difficulty the game is.

In the present study, we try to use a variety of sensors, such as touch sensors, light sensors, temperature sensors, sound sensors, buttons, sensors and motion sensors. Problems we encountered were how sensors, user experience and the design could work very well together. First, different sensors have different characteristics, and the sensitivity, accuracy and range are limited. In the second, the content of the game to some extent determines the interactive way. And the game's interaction is determined by different sensors. Lastly, we have to consider whether the user experience is enjoyable or not and whether the interactive way of the game is natural or not. To address these issues, we also need long-term research and exploration.

7 Further Research Direction

Due to the current display screen's refresh rate is too slow, we will increase the refresh rate to meet the needs of the game. The appearance of the puzzle block-the screen and the processor are connected by lines now, which is too bulky and inconvenient to operate. To achieve a better user experience, we will design a special circuit board and package.

8 Conclusion

This paper integrates the idea of TUI, the principles of sensory integration training and cognitive-behavioral therapy to propose a novel physical interactive platform. The study includes two breakthroughs, which are the implementation of the display of the image and the design of the tangible jigsaw puzzle. As far as we knew that there was no interface of the image display on the development tools, we gave a solution which could meet our requirement in our game.

Compared with the traditional jigsaw puzzle, the design of tangible jigsaw puzzle is novel. Especially, each puzzle block can sense the user's voice and movements. We created and implemented a new interactive action, in which blowing to the block can switch the image and blocks can sense the user picking up or down. What's more, the game will provide the ADHD children with monitor (images), speaker (sounds) and vibrator (vibrations), which can stimulate their audio, visual and tactile senses. In the future, we will continue this study step by step.

Acknowledgments. This work is supported by the National Key Research and Development Plan under Grant No. 2016YFB1001402, the Natural Science Foundation of China under Grant No. 61572276 and 61672314, Tsinghua University Research Funding No. 20151080408, National Social Science Fund of China under Grant No. 15CG147.

References

1. Akabane, S., Leu, J., Iwadate, H., Choi, J.W., Chang, C.C., Nakayama, S., et al.: Puchi planet: a tangible interface design for hospitalized children. In: CHI 2011 Extended Abstracts on Human Factors in Computing Systems, pp. 1345–1350. ACM (2011). https://doi.org/10.1145/1979742.1979772
2. Andujar, M., Gilbert, J.E.: Let's learn! Enhancing user's engagement levels through passive brain-computer interfaces. In: CHI 2013, pp. 703–70. ACM (2013). https://doi.org/10.1145/2468356.246848
3. Antle, A.N., Wise, A.F., Nielsen, K.: Towards Utopia: designing tangibles for learning. In: International Conference on Interaction Design and Children, pp. 11–20. ACM (2011)
4. Bakhshayesh, A.R., Hänsch, S., Wyschkon, A., et al.: Neurofeedback in ADHD: a single-blind randomized controlled trial. Eur. Child Adolesc. Psychiatry **20**(9), 481–491 (2011). https://doi.org/10.1007/s00787-011-0208-y
5. Wang, C.: Risk Factors of ADHD Children with Different Degree Sensory Integration Dysfunction and Curative Effects of Sensory Integration Training, Doctoral dissertation (2005)
6. Zhou, D.: Electroencephalical biofeedback therapy for 134 cases of hyperactivity-type attention deficit hyperactivity disorder. Chin. Pediatr. Integr. Tradit. West Med. **5**(3), 261–262 (2013)
7. Mahone, E.M., Schneider, H.E.: Assessment of attention in preschooler. Neuropsychol. Rev. **22**(4), 361–383 (2012)
8. Fan, M., Antle, A.N., Hoskyn, M., et al.: Why tangibility matters: a design case study of at-risk children learning to read and spell. In: CHI Conference on Human Factors in Computing Systems 2017, pp. 1805–1816. ACM (2017). https://doi.org/10.1145/3025453.3026048
9. Halperin, J.M., Marks, D.J., Bedard, A.C., et al.: Training executive, attention, and motor skills: a proof-of-concept study in preschool children With ADHD. J. Atten. Disord. **17**(8), 711–721 (2013). https://doi.org/10.1177/1087054711435681
10. Cheng, H., et al.: Influence of sensory integration disorder on cognitive function in children with ADHD. Cap. Med. **11**(12), 19–20 (2004)
11. Cheng, H., et al.: Application of sensory integration to the treatment of ADHD children. Nanfang J. Nurs. **10**(2), 17–18 (2003)
12. Li, J.: To analyze the effect of EEG biofeedback on 48 children with ADHD. Med. J. Chin. People's Health **20**(21), 2478–2479 (2008)

13. Lauth, G.W., Schlottke, P.F.: Child Attention Training Manual. Sichuan University Press (2013)
14. Mackie, M.A., Dam, N.T.V., Fan, J.: Cognitive control and attentional functions. Brain Cogn. **82**(3), 301–312 (2013). https://doi.org/10.1016/j.bandc.2013.05.004
15. Marchesi, M.: BRAVO: a brain virtual operator for education exploiting brain-computer interfaces. In: CHI 2013 Extended Abstracts on Human Factors in Computing Systems, pp. 3091–3094. ACM (2013). https://doi.org/10.1145/2468356.2479618
16. Melcer, E.F., Isbister, K.: Bots & (Main) frames: exploring the impact of tangible blocks and collaborative play in an educational programming game. In: CHI Conference on Human Factors in Computing Systems (2018). https://doi.org/10.1145/3173574.3173840
17. Price, B.A., Kelly, R., Mehta, V., Mccormick, C., Ahmed, H., Pearce, O.: Feel my pain: design and evaluation of painpad, a tangible device for supporting inpatient self-logging of pain. In: CHI Conference, pp. 1–13 (2018). https://doi.org/10.1145/3173574.3173743
18. Szafir, D., Mutlu, B.: Pay attention! Designing Adaptive agents that monitor and improve user engagement. In: ACM Conference on Human Factors in Computing Systems, pp. 1580–1586. ACM (2012). https://doi.org/10.1145/2207676.2207679
19. Liu, T., Wang, J., Chen, Y., Song, M.: Neurofeedback treatment experimental study for adhd by using the brain-computer interface neurofeedback system. In: World Congress on Medical Physics and Biomedical Engineering. IFMBE Proceedings, vol. 39, pp. 1537–1540 (2013). https://doi.org/10.1007/978-3-642-29305-4_404
20. Fang, W.-P., Pen, S.-Y.: A mobile phone base sensory integration training software. In: 2011 Fifth International Conference on Genetic and Evolutionary Computing, pp. 276–278 (2011). https://doi.org/10.1109/icgec.2011.68
21. Yan, C.: The intervence study of sensory integration training to ADHD children. Psychol. Res. **1**(6), 28–31 (2008)
22. Zuckerman, O., Arida, S., Resnick, M.: Extending tangible interfaces for education: digital montessori-inspired manipulatives. In: SIGCHI Conference on Human Factors in Computing Systems, pp. 859–868. ACM (2005). https://doi.org/10.1145/1054972.1055093

Harmony Search with Teaching-Learning Strategy for 0-1 Optimization Problem

Longquan Yong(✉)

School of Mathematics and Computer Science,
Shaanxi University of Technology, Hanzhong 723001, China
yonglongquan@126.com

Abstract. 0-1 optimization problem plays an important role in operational research. In this paper, we use a recently proposed algorithm named harmony search with teaching-learning (HSTL) strategy which derived from Teaching-Learning-Based Optimization (TLBO) for solving. Four strategies (Harmony memory consideration, teaching-learning strategy, local pitch adjusting and random mutation) are employed to improve the performance of HS algorithm. Numerical results demonstrated very good computational performance.

Keywords: 0-1 optimization problem · Operational research · Harmony search
Teaching-learning-based optimization

1 Introduction

0-1 optimization problem, $\min f(x) = x^T Ax, \quad \text{s.t. } x \in \{0, 1\}^n$, where $A \in R^{n\times n}$, play an important role in discrete mathematics, operational research and computer science. For example, network optimization, assignment problem, and Knapsack problem, all these problems have 0-1 variables, and the aim of these problems is to solve many 0-1 optimization problems [1–3]. Several methods were developed to solve such problems that can be classified into two classes. Exact methods, like branch and bound, continuous method, can give the exact solutions [4–6]. However, in the worst case, the computation time increases exponentially with the size of the problems, especially while for problems with a high number of variables ($n > 200$) [7]. The second class contains metaheuristic methods which give sub-optimal solutions but in reasonable times compared to exact methods [8]. Metaheuristic has been proven to be an effective way to solve complex engineering problems. Metaheuristic are designed to tackle complex optimization problems where other optimization methods have difficulty to solve.

Harmony Search (HS) is a new metaheuristic algorithm and it is based on natural musical performance process that arises when a musician examines for a better state of harmony. The music harmony is a combination of sounds that have aesthetics satisfaction. Harmony in nature is a special loud sound between several sound waves that have different frequencies. The musical performances are seeking to find a nice harmony (perfect state) determined as standard aesthetic, such as the optimization process looks for an optimal solution to an objective function. In music improvisation, each musician

P. Krömer et al. (Eds.): ECC 2018, AISC 891, pp. 287–296, 2019.
https://doi.org/10.1007/978-3-030-03766-6_32

performs any launch in the margin possible, creating the set of harmony vectors. If all launches are a good harmony, this experience is stored in the memory of each musician, and so the ability to create a good harmony is increased next time [9, 10].

The HS algorithm has powerful exploration ability in a reasonable time but is not good at performing a local search. In order to improve the performance of the harmony search method, several variants of HS have been proposed [11–16]. These variants have some improvement on continuous optimization problems. However, their effectiveness in dealing with discrete problems is still unsatisfactory. Especially for a high-dimensional discrete optimization problem, HS is apt to appear premature convergence and stagnation behavior.

Teaching-Learning-Based Optimization (TLBO) algorithm is a new nature-inspired algorithm; it mimics the teaching process of teacher and learning process among learners in a class. TLBO shows a better performance with less computational effort for large scale problems [17, 18]. In addition to, TLBO needs very few parameters. In the TLBO method, teacher phase relying on the best solution found so far usually has the fast convergence speed and the well ability of exploitation; it is more suitable for improving accuracy of the global optimal solution. Learner phase relying on other learners usually has the slow convergence speed; however it bears stronger exploration capability for solving multimodal problems.

To overcome the inherent weaknesses of HS, we propose a novel harmony search algorithm based on Teaching-Learning (HSTL). In the HSTL method, an improved Teaching-Learning strategy is employed to enhance the performance of dealing with discrete problems by HS method.

The remainder of the paper is organized as follows. Section 2 introduces the HS algorithm with 0-1 Variable. The teaching-learning-based optimization (TLBO) algorithm and the proposed algorithm (HSTL) are introduced in Sect. 3. Experimental results are discussed in Sect. 4. Finally, Sect. 5 concludes this paper.

2 Harmony Search Algorithm with 0-1 Variable

The steps harmony search algorithm with 0-1 variables are as follows:

Step 1. Initialize the problem and algorithm parameters. Optimization problem is specified as follows:

$$\text{Minimize} f(x) \text{ subject to } x_i \in \{0,1\}, i = 1,2,\cdots,D,$$

where $f(x)$ is an objective function; x is the set of each decision variable x_i; D is the number of decision variables, X_i is the set of the possible range of values for each decision variable, $X_i : x_i^L \leq X_i \leq x_i^U$, here $x_i^L = 0, x_i^U = 1$. The HS algorithm parameters are also specified in this step. These are the harmony memory size (HMS), or the number of solution vectors in the harmony memory, i.e. population size; harmony memory considering rate (HMCR); pitch adjusting rate (PAR); and the number of improvisations (Tmax), or stopping criterion.

Step 2. Initialize the harmony memory. The HM matrix is filled with as many randomly generated solution vectors.

$$x_i^j = x_i^L + (x_i^U - x_i^L) \times rand, j = 1, 2, \cdots, \text{HMS}.$$

Every variable is replaced by the nearest integer, that is $x_i^j = \text{round}(x_i^j)$. For example, let $x = (0.8, 0.3, 0.9)$. Then $\text{round}(x) = (1, 0, 1)$.

Step 3. Improvise a new harmony. Generating a new harmony is called 'improvisation'. A new harmony vector, $x^{new} = (x_1^{new}, x_2^{new}, \cdots, x_D^{new})$, is generated based on three rules: (1) memory consideration, (2) pitch adjustment and (3) random selection. The procedure works as Fig. 1.

```
For each i ∈ [1,2,···,D] do
  If rand < HMCR
      x_i^new = x_i^j (j = 1,2,···,HMS)          //memory consideration
      If rand < PAR
          x_i^new = x_i^new ± r × BW(i)          //pitch adjustment
          x_i^new = min[max(x_i^new, x_i^L), x_i^U]  //truncation processing
      End
  Else
    x_i^new = x_i^L + rand × (x_i^U − x_i^L)     //random selection
  End
  x_i^new = round(x_i^new)
End
```

Fig. 1. Generating a new harmony by classical HS algorithm

$x_i^{new}(i = 1, 2, \cdots, D)$ is the ith component of x^{new}, and $x_i^j(j = 1, 2, \cdots, \text{HMS})$ is the ith component of the jth candidate solution vector in HM. Both r and rand are uniformly generated random number in the region of [0, 1], and BW is an arbitrary distance bandwidth.

Step 4. Update harmony memory. If the new harmony vector $x^{new} = (x_1^{new}, x_2^{new}, \cdots, x_D^{new})$ is better than the worst harmony in the HM, judged in terms of the objective function value, the new harmony is included in the HM and the existing worst harmony is excluded from the HM.

Step 5. **Check stopping criterion.** If the stopping criterion (Tmax) is satisfied, computation is terminated. Otherwise, Steps 3 and 4 are repeated.

3 HSTL Algorithm

3.1 The TLBO Algorithm

Teaching-Learning-Based Optimization (TLBO) algorithm is a new nature-inspired algorithm. It mimics the teaching process of teacher and learning process among learners in a class. In the TLBO method, it's the task of teacher to try to increase mean knowledge of all learners of the class in the subject taught by him or her depending on his or her capability. Learners make efforts to increase their knowledge by interaction among themselves. A learner is considered as a solution or a vector, different design variables of vector will be analogous to different subjects offered to learners and the learners' result is analogous to the 'fitness' as in other population-based optimization techniques. The teacher is considered as the best solution obtained so far. The process of working of TLBO is divided into two phases, 'Teacher Phase' and 'Learner Phase'.

(1) **Teacher Phase.** Assume there are 'D' number of subjects (i.e. design variables), 'NP' number of learners (i.e. population size), x_i^{best} is the best learner (i.e. teacher) in subject i ($i = 1, 2, \cdots, D$). The works of teaching as follows:

$$x_i^{j,new} = x_i^{j,old} + rand \times \left(x_i^{best} - T_F \times Mean_i\right),\ Mean_i = \frac{1}{NP}\sum_{j=1}^{NP} x_i^j,$$
$$j = 1, 2, \cdots, NP,\ i = 1, 2, \cdots, D,$$

where $x_i^{j,old}$ denotes the result of the jth ($j = 1, 2, \cdots, NP$) learner before learning the ith ($i = 1, 2, \cdots, D$) subject, $x_i^{j,new}$ is the result of the jth learner after learning the ith subject. T_F is the teaching factor which decides the value of $Mean_i$ to be changed. The value of T_F is generated randomly with probability as T_F = round (1 + $rand$). When the leaner x^j finished his or her learning from teacher, update the x^j by $x^{j,new}$ if $x^{j,new}$ is better than $x^{j,old}$.

(2) **Learner Phase.** Another important approach to increase knowledge for a learner is to interact with other learners. Learning method is expressed as Fig. 2.

3.2 The HSTL Algorithm

In the classical HS algorithm, a new harmony is generated in step 3 of Fig. 1. After the selecting operation in the step 4, the population variance may increase or decrease. With a high population variance, the diversity and exploration power will increase, in the same time the convergence and the exploitation power will decrease accordingly. Conversely, with a low population variance, the convergence and the exploitation power will increase; the diversity and the exploration power will decrease. So it is

---

For each learner x^j $j = 1, 2, \cdots, NP$

Randomly select another learner x^k $(j \neq k)$

If x^j is superior to x^k then

$$x^{j,new} = x^{j,old} + rand \times (x^j - x^k)$$

Else

$$x^{j,new} = x^{j,old} + rand \times (x^k - x^j)$$

End

End

If $x^{j,new}$ is superior to $x^{j,old}$ **then**

$x^j = x^{j,new}$

End

Fig. 2. The procedure of learner phase

significant how to keep balance between the convergence and the diversity. Classical HS algorithm loses its ability easily at later evolution process, because of improvising new harmony from HM with a high HMCR and local adjusting with PAR. And HM diversity decreases gradually from the early iteration to the last. However, in HS algorithm, a low HMCR employed will increase the probability (1-HMCR) of random select in search space; the exploration power will enhance, but the local search ability and the exploitation accuracy can't be improved by single pitch adjusting strategy.

To overcome the inherent weaknesses of HS, we develop a novel harmony search algorithm combining teaching-learning strategy (HSTL). The improvisation of new target harmony in HSTL algorithm is shown as Fig. 3.

4 Computational Results

In this section we solve some 0-1 optimization problems in order to illustrate the efficiency of the HSTL method. All the experiments were performed on MatlabR2009a system with Intel(R) Core(TM) 4 × 3.3 GHz and 2 GB RAM. To efficiently balance the exploration and exploitation power of the HSTL algorithm, HMCR, BW and TLP are dynamically adapted to a suitable range with increase of generations. Let HMS = 50, $HMCR_{max} = 0.95$, $HMCR_{min} = 0.65$, $TLP_{max} = 0.55$, $TLP_{min} = 0.15$, $BW_{max} = 0.5$, $BW_{min} = 0.1$, $T_{max} = 50n$.

$x^{new} = x^{worst}$ // select x^{worst} as optimization target vector

For ***i*****=1 to D do**

If *rand*<HMCR //(a) Harmony memory consideration

$$x_i^{new} = x_i^j \ (j = 1, 2, \cdots, \text{HMS})$$

Elseif *rand*<TLP //(b) Teaching-Learning strategy

If *rand* <0.5 // Teaching

$$x_i^{new} = x_i^{best} + (2 * rand - 0.5) * \left| x_i^{best} - x_i^{worst} \right|$$

Else // Learning

Randomly select r_1 and r_2 from {1,2,…,HMS}

If x^{r_1} *is better than* x^{r_2}

$$x_i^{new} = x_i^{new} + rand \times (x_i^{r_1} - x_i^{r_2})$$

Else

$$x_i^{new} = x_i^{new} + rand \times (x_i^{r_2} - x_i^{r_1})$$

end

End

$$x_i^{new} = \min(\max(x_i^{new}, x_i^L), x_l^U)$$

Elseif rand<PAR // (c)Local pitch adjusting strategy

$$x_i^{new} = x_i^{new} \pm rand \times \text{BW}(i)$$

$$x_i^{new} = \min(\max(x_i^{new}, x_i^L), x_l^U)$$

Elseif rand<p_{m} // (d)Random mutation operator

$$x_i^{new} = x_i^L + (x_i^U - x_i^L) \times rand$$

End

$$x_i^{new} = round(x_i^{new})$$

End

Fig. 3. The improvisation process of new harmony in HSTL algorithm

4.1 Problems

Problem 1. Where the matrix A is given by

$$a_{ii} = 4n, \quad a_{i,i+1} = a_{i+1,i} = 1, \quad a_{ij} = 0, \quad i = 1,\ 2,\ \cdots,\ n.$$

Here $A \in R^{n\times n}$ is a dominant tridiagonal matrix, and the unique solution is $x^* = (0,\ 0,\ \cdots,\ 0)^T \in R^n$.

Problem 2. Where the matrix A is given by

$$a_{ii} = (-1)^{i-1}(2n), \quad \text{rand('state',0)},\ a_{ij} = \text{rand}, \quad i = 1,\ 2,\ \cdots,\ n.$$

Here $A \in R^{n\times n}$ is a dominant tridiagonal matrix, and the unique solution is $x_i^* = 0$, if i is an odd number; $x_i^* = 1$, if i is an even number [19].

Problem 3. Where the matrix A is given by

$$a_{ii} = -10, \quad a_{i,i+1} = a_{i+1,i} = 1, \quad a_{ij} = 0, \quad i = 1,\ 2,\ \cdots,\ n.$$

Here $A \in R^{n\times n}$ is a dominant tridiagonal matrix, and the unique solution is $x^* = (1,\ 1,\ \cdots,\ 1)^T \in R^n$.

4.2 Results

In order to eliminate the influence of random number,10 independent runs of HSTL were carried out and the best, the mean, the worst fitness values, the standard deviation (Std), and success rate (SR, numbers of finding best fitness value divide 10) were recorded in Tables 1, 2 and 3.

Table 1. The Results for 10 Runs on problem 1

n	Best	Mean	Worst	Std	SR
20	0	48	80	41.31182	0.4
30	0	60	120	63.24555	0.5
40	0	48	160	77.28734	0.7
50	0	0	0	0.00000	1
60	0	0	0	0.00000	1
100	0	0	0	0.00000	1
200	0	0	0	0.00000	1
300	0	0	0	0.00000	1
400	0	0	0	0.00000	1
500	0	0	0	0.00000	1
1000	0	0	0	0.00000	1

Table 2. The Results for 10 Runs on problem 2

n	Best	Mean	Worst	Std	SR
20	−314.48836	−309.90883	−291.59069	9.65451	0.8
30	−700.47285	−690.45394	−667.07647	16.13198	0.7
40	−1238.95087	−1225.78235	−1195.05578	21.20334	0.7
50	−1929.92243	−1924.48305	−1875.52863	17.20083	0.9
60	−2773.38752	−2773.38752	−2773.38752	0.00000	1
100	−7662.18325	−7662.18325	−7662.18325	0.00000	1
200	−30593.72008	−30593.72008	−30593.72008	0.00000	1
300	−68764.61048	−68764.61048	−68764.61048	0.00000	1
400	−122184.85445	−122184.85445	−122184.85445	0.00000	1
500	−190854.45200	−190854.45200	−190854.45200	0.00000	1
1000	−762942.74336	−762627.02204	−760837.30115	710.42316	0.8

Table 3. The Results for 10 Runs on problem 3

n	Best	Mean	Worst	Std	SR
20	−162	−158.4	−156	3.09839	0.4
30	−242	−238.4	−236	3.09839	0.4
40	−322	−321.4	−316	1.89737	0.9
50	−402	−402	−402	0.00000	1
60	−482	−482	−482	0.00000	1
100	−802	−802	−802	0.00000	1
200	−1602	−1602	−1602	0.00000	1
300	−2402	−2402	−2402	0.00000	1
400	−3202	−3202	−3202	0.00000	1
500	−4002	−4002	−4002	0.00000	1
1000	−8002	−8001.4	−7996	1.89737	0.9

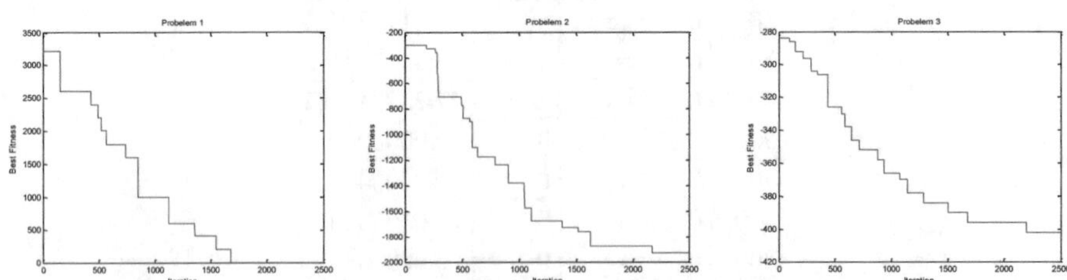

Fig. 4. The convergence of the best fitness for all problems with n = 50

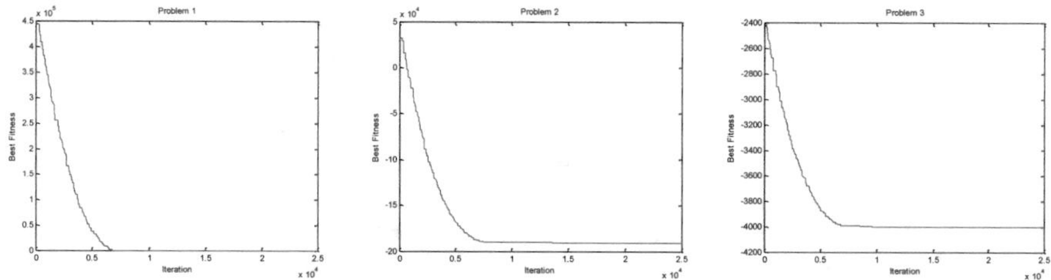

Fig. 5. The convergence of the best fitness for all problems with $n = 500$

Figures 4 and 5 show the convergence of the best fitness in the population for all problems with $n = 50$ and $n = 500$.

5 Conclusion

We have given HSTL algorithm for solving the 0-1 optimization problems. Primary results show that HSTL algorithm has fast convergence speed from Figs. 4 and 5. Future works will also focus on studying the applications of HSTL algorithm for solving engineering optimization problems, such as assignment problem, DNA computing, feeder automation planning, load identification.

Acknowledgment. This work is supported by Project of Youth Star in Science and Technology of Shaanxi Province (2016KJXX-95)

References

1. Dantzig, G.B., Fulkerson, D.R., Johnson, S.M.: Solution of a large-scale traveling salesman problem. Oper. Res. **2**, 393–410 (1954)
2. Gomory, R.E.: Outline of an algorithm for integer solutions to linear programs. Bull. Am. Math. Soc. **64**, 275–278 (1958)
3. Nemhauser, G.L., Wolsey, L.A.: Integer and Combinatorial Optimization. Wiley, New York (1988)
4. Barnhart, C., Johnson, E.L., Nemhauser, G.L., et al.: Branch-and-price: column generation forsolving huge integer programs. Oper. Res. **48**, 316–329 (1998)
5. Wolsey, L.A.: Integer Programming. Wiley, New York (1998)
6. Jiinger, M., Liebling, T., Naddef, D., et al.: 50 Years of Integer Programming 1958-2008: From the Early Years to the State-of-the-Art. Springer, Berlin (2010)
7. Li, D., Sun, X.L.: Nonlinear Integer Programming. Springer, New York (2006)
8. Jourdan, L., Basseur, M., Talbi, E.G.: Hybridizing exact methods and metaheuristics: a taxonomy. Eur. J. Oper. Res. **199**(3), 620–629 (2009)
9. Zong, W.G., Kim, J.H., Loganathan, G.V.: A new heuristic optimization algorithm: harmony search. Simul. Trans. Soc. Model. Simul. Int. **76**(2), 60–68 (2001)
10. Zhao, X., Liu, Z., Hao, J., et al.: Semi-self-adaptive harmony search algorithm. Nat. Comput. **16**(4), 1–18 (2017)

11. Yong, L., Liu, S., Zhang, J., Feng, Q.: Theoretical and empirical analyses of an improved harmony search algorithm based on differential mutation operator. J. Appl. Math. **2012**, Article ID 147950
12. Tuo, S., Yong, L., Zhou, T.: An improved harmony search based on teaching-learning strategy for unconstrained optimization problems. Math. Probl. Eng. **2013**, Article ID 413565, 29 pages. https://doi.org/10.1155/2013/413565
13. Tuo, S., Yong, L., Deng, F.: A novel harmony search algorithm based on teaching-learning strategies for 0-1 knapsack problems. Sci. World J. **2014**, Article ID 637412, 19 pages. https://doi.org/10.1155/2014/637412
14. Tuo, S., Zhang, J., Yong, L., Yuan, X.: A harmony search algorithm for high-dimensional multimodal optimization problems. Digit. Signal Process. Rev. J. **46**(11), 151–163 (2015)
15. Tuo, S., Yong, L., et al.: HSTLBO: a hybrid algorithm based on Harmony Search and Teaching-Learning-Based Optimization for complex high-dimensional optimization problems. PLoS ONE **12**, e0175114 (2017)
16. Wang, L., Hu, H., Liu, R., et al.: An improved differential harmony search algorithm for function optimization problems. Soft. Comput. **4**, 1–26 (2018)
17. Rao, R.V., Savsani, V.J., Vakharia, D.P.: Teaching-learning-based optimization: an optimization method for continuous non-linear large scale problems. Inf. Sci. **183**(1), 1–15 (2012)
18. Rao, R.V., Savsani, V.J., Balic, J.: Teaching-learning-based optimization algorithm for unconstrained and constrained real-parameter optimization problems. Eng. Optim. **44**(12), 1447–1462 (2012)
19. Pardalos, P.M.: Construction of test problems in quadratic bivalent programming. ACM Trans. Math. Softw. **17**(1), 74–87 (1991)

Grid-Connected Power Converters with Synthetic Inertia for Grid Frequency Stabilization

Weiyi Zhang(✉), Lin Cheng, and Youming Wang

Xi'an University of Posts & Telecommunications, Xi'an 710121, China
zhangweiyi@xupt.edu.cn

Abstract. Grid-connected power converters with synthetic inertia have been experiencing a fast development in recent years. This technology is promising in renewable power generation since it contributes to the grid frequency stabilization, like how a synchronous machine does in a traditional power system. This paper proposes optional active power control strategies for grid-connected power converters to let them have inertial response during big load changes and grid contingencies. By giving mathematical expressions, the control parameters are clearly related to the power loop dynamics, which guides the control parameter tuning. The local stability is also investigated. A preliminary simulated and experimental verification is given to support the control strategy.

Keywords: DC-AC power converter · Grid-connected power converter
Power converter control · Synthetic inertia

1 Introduction

Most Distributed Generation (DG) systems are based on renewable energy sources. Nowadays, these systems are required to participate in grid regulation and offer supporting services to improve the grid operation stability. Recent and incoming connection requirements for grid-connected power converters are more demanding regarding grid-supporting. Therefore, the modern and future renewable-based power generation systems should have droop characteristics as a complement of maximum power point tracking (MPPT) algorithms. These droop functionalities are fully feasible from renewable energy sources that work with a capacity reserve or in parallel with any energy storage devices. In addition, providing novel ancillary services, like synthetic inertia, by grid-connected converters, have been increasingly studied to improve the frequency stability and contribute to the inertial response. Providing synthetic inertia gives rise to a different converter control paradigm compared to the traditional style, where the converter dynamics are mainly characterized by the phase-locked loop (PLL) [1] and the power control loop that defines the operating point [2].

The idea of specifying grid-connected converters with inertia and droop characteristics is well accepted because of the successful operation of the traditional power system, which relies on the electromechanical characteristics of the numerous synchronous generators. In the past years, the converter control based on the emulation of

P. Krömer et al. (Eds.): ECC 2018, AISC 891, pp. 297–303, 2019.
https://doi.org/10.1007/978-3-030-03766-6_33

the synchronous generator electromechanical characteristics, mostly the emulation of the swing equation, has been studied intensively since its first publication [3], where the virtual synchronous machine is proposed. Relevant studies have been conducted from different perspectives like inertia emulation characteristic [4], PLL-less control [5], providing virtual impedance [6], adaptive inertial response [7], primary frequency and voltage regulation [8], as well as stability analysis [9], where the impact of the droop and inertia parameters on the system dynamics are analyzed.

All the aforementioned works have given characteristic analyses or implementation proposals in different aspects of synchronous generator emulation control, whereas the transient analysis and experimental verification in presence of grid frequency variations are not thoroughly shown. This paper introduces the synchronous power control (SPC) strategy for grid-connected converters to provide inertia emulation and primary frequency control. Particularly, compared with existing studies, the transient response of the converters in presence of grid frequency changes is studied analytically and validated in experiments in this paper.

2 Grid-Connected Power Converters Based on the SPC

The SPC endows grid connected voltage source converters (VSC) with virtual electromechanical characteristics, as an emulation and enhancement of synchronous generators.

For common synchronous machines, considering a mainly inductive output impedance and a synchronized condition (a small value of δ), (1) and (2) can be simplified as:

$$P = \frac{EV}{X}\delta = P_{\max}\delta, \tag{1}$$

$$Q = \frac{V(E-V)}{X}. \tag{2}$$

As shown in (1) and (2), synchronous machines regulate the active and reactive powers by adjusting the load-angle and the magnitude of the electromotive force through the governor and the exciter, respectively. Similarly, the SPC controls the active and reactive powers by adjusting its inner voltage phase-angle and magnitude, respectively, similar to a synchronous machine, rather than the conventional in-phase and in-quadrature current control performed in the decoupled rotating $(d-q)$ reference frame.

3 Synchronous Power Loop Control

The mathematical model of the active PLC of the SPC is shown in Fig. 1.

The synchronous angular speed ω is adjusted according to the error in the converter's power control, which will further modify the load-angle δ to regulate the

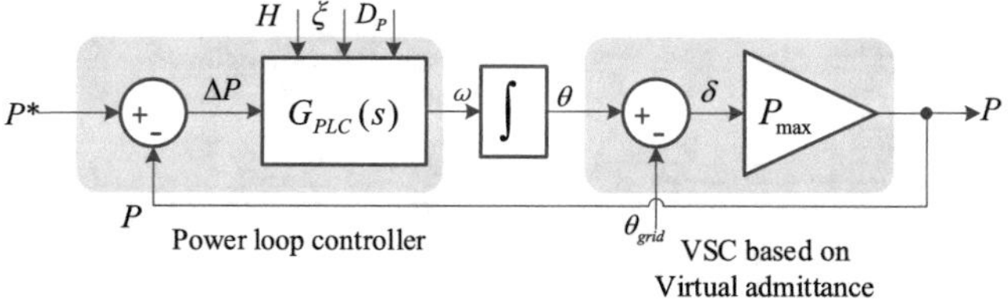

Fig. 1. Mathematical model of the SPC's active power control.

generated active power. In this way, even though the grid voltage phase-angle θ_{grid} is unknown and can be variable in a realistic operation, ω can always be adjusted to eliminate the power control error, and meanwhile maintains the synchronization with the grid frequency ω_g. $G_{PLC}(s)$ represents the transfer function between the active power control error. ΔP is the difference between the power reference (P^*) and the power injected in the point of common coupling (PCC), P, and P_{max} is the gain between δ and P, which is defined in (1).

The design of the SPC's PLC is discussed in this section based on the above modeling of the active PLC.

3.1 Controller Based on the First-Order Torque Equation

The synchronous machine first-order torque equation (1st - OTE) for small signal variation of the rotor angular frequency, ω, around the rated rotor angular frequency, ω_s, can be expressed in terms of power as:

$$P_m - P_e = \omega_s(Js + D)\omega. \tag{3}$$

Based on the 1st - OTE, the PLC in Fig. 2, $G_{PLC}(s)$, could be obtained as shown in

$$G_{PLC}(s) = \frac{1}{\omega_s(Js + D)}, \tag{4}$$

which considers both the inertia, J, and the damping, D, terms. This transfer function is referred as the Mechanical PLC (MPL) in this paper. This active power loop control strategy is typical and is derived from the swing equation.

According to (4), the resulting second-order closed-loop transfer function would have the following form:

$$\frac{P}{P^*}(s) = \frac{\omega_n^2}{s^2 + 2\xi\omega_n s + \omega_n^2}, \tag{5}$$

where:

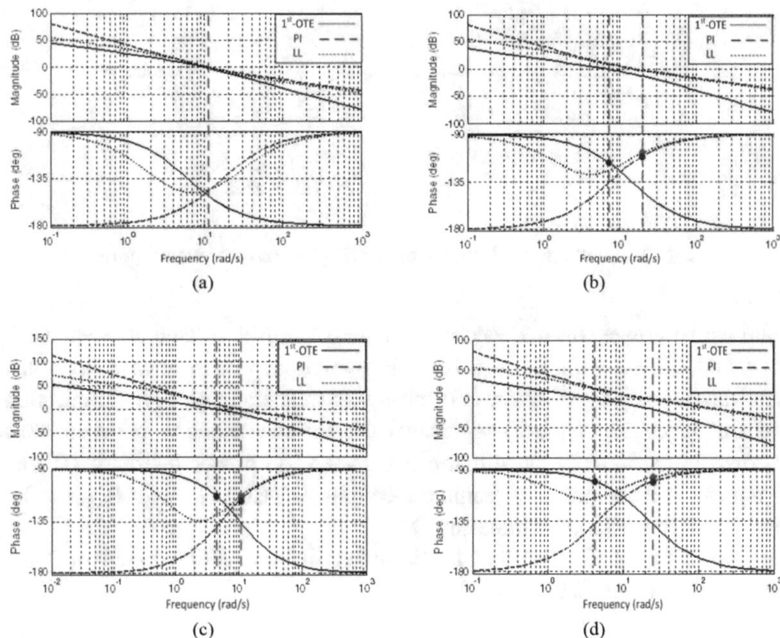

Fig. 2. Bode plot based on different power loop controllers and different parameter values: (a) H = 5 s, ζ = 0.3, (b) H = 5 s, ζ= 0.7, (c) H = 10 s, ζ= 0.7, (d) H = 5 s, ζ= 1.1.

$$\xi = \frac{D}{2}\sqrt{\frac{\omega_s}{JP_{\max}}}, \tag{6}$$

$$\omega_n = \sqrt{\frac{P_{\max}}{J\omega_s}}, \tag{7}$$

being P_{max} as defined in (1).

The MPL controller gains, J and D, should be set according to:

$$J = \frac{2HS_N}{\omega_s^2}, \tag{8}$$

$$D = \frac{2\xi}{\omega_s}\sqrt{\frac{2HS_N P_{\max}}{\omega_s}}. \tag{9}$$

3.2 PI Power Loop Controller

The commonly used PI controller can also be used as an alternative to implement the PLC. The PI-based PLC makes the output regulated power equal to the reference value in steady state, even if there are variations in the grid frequency. The PI controller used for the power loop has the following form:

$$G_{PLC}(s) = K_X + \frac{K_H}{s}. \tag{10}$$

Using it as the power loop controller block in Fig. 2, the resulting closed-loop transfer function can be written as:

$$\frac{\partial P}{\partial P_{ref}}(s) = \frac{2\xi\omega_n s + \omega_n^2}{s^2 + 2\xi\omega_n s + \omega_n^2}, \tag{11}$$

where the damping coefficient and natural frequency are respectively given by:

$$\xi = \frac{K_X}{2}\sqrt{\frac{P_{\max}}{K_H}}, \tag{12}$$

$$\omega_n = \sqrt{P_{\max} K_H}. \tag{13}$$

Therefore, the PI-based PLC gain, K_H, should be set according to:

$$K_H = \frac{\omega_n^2}{P_{\max}}. \tag{14}$$

The natural frequency, ω_n, in this case, can be translated to the moment of inertia, J, by equating the ω_n in (7) and (13). Then (14) changes to:

$$K_H = \frac{1}{J\omega_s}, \tag{15}$$

4 Stability and Dynamics Based on Different Power Loop Controllers

The stability of the active PLC and the power control dynamics are analyzed in this section, considering three different types of PLC mentioned in the last section.

4.1 Stability Based on Different Power Loop Controllers

The system based on the 1st - OTE, (5), is a standard second-order system, which is known that its stability is determined by the closed-loop poles, accordingly, by ξ. Comparing (5) and (11), all the three types of PLC lead to the same closed-loop poles, expressed by ωn and ξ. Therefore, the stability is mainly determined by the value of ξ, no matter which PLC is used. However, since the PI and the lead-lag (LL) controllers introduce additional zero to the system, the phase-frequency characteristics can be different based on different PLCs, which may lead to different phase margins. Figure 2 shows the Bode plot of the systems based on different PLCs.

Figure 2(a), (b), (c) and (d) all show that the phase margin resulted from different PLCs are close to each other. The only observed difference among different PLCs is the crossover frequency, which mainly reflect the dynamic characteristics.

4.2 Dynamics Based on Different Power Loop Controllers

Figure 3 is plotted in order to demonstrate the relationship between the settling time and the inertia constant.

It is known that the settling time of a standard second-order system is inversely proportional to the natural frequency ω_n, therefore, the settling time of the system based on the 1st - OTE is proportional to the square root of the inertia constant H,

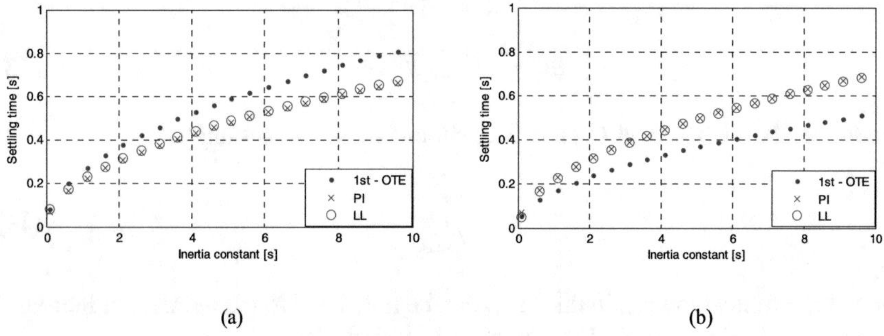

Fig. 3. Relationship between the inertia constant and the closed-loop step response settling time: (a) $\zeta = 0.7$, (b) $\zeta = 0.8$.

according to (7) and (8). Figure 3 demonstrates that the systems based on the PI or LL controllers follow a similar characteristic, as an emulation of the synchronous generator dynamics.

Figure 3(a) shows that when $\zeta = 0.7$, the settling time of the system based on the PI or lead-lag controller is smaller than the one based on the 1st - OTE. Figure 3(b) shows that when $\zeta = 0.8$, the situation is the opposite. Comparing Fig. 3(a) with (b), it is found that the settling time of the systems based on the PI or lead-lag controllers is not relatively affected by the change of the damping factor, while the settling time of the system based on the 1st - OTE is more affected.

5 Conclusion

This paper presented three different PLC strategies for grid-connected converters based on the Synchronous Power Controller, to provide inertia emulation and primary frequency control features to power converters linked to renewable energy sources.

The PLC was designed to provide damping, inertia emulation and *P-f* droop characteristics with considerations of stability and dynamics. The frequency support characteristics of the controlled converter were particularly analyzed and validated in this paper. The analytical relation between the grid frequency deviation and the active power change was derived based on the accurate modeling of the active PLC. The experimental tests, done in a 10 kW regenerative source test bed, equipped with a frequency programmable voltage ac-source, have endorsed the analytical analysis. In this regard, the inertial and droop characteristics were clearly shown under frequency sweeps and the converter is controlled with different sets of parameters. The simulation and experimental results validate the performance of the three models of PLC presented in this paper. Therefore, the inertia constant, damping factor and droop gain can be accurately given for achieving good grid-interaction dynamics and also complying with the current TSO requirements.

References

1. Teodorescu, R., Liserre, M., Rodriguez, P.: Grid Converters for Photovoltaic and Wind Power Systems. Wiley, Chichester (2011)
2. Akagi, H., Hirokazu, E., Aredes, M.: Instantaneous Power Theory and Applications to Power Conditioning. Wiley, Hoboken (2007)
3. Beck, H.P., Hesse, R.: Virtual synchronous machine. In: Proceedings of EPQU, pp. 1–6 (2007)
4. Van Wesenbeeck, M.P.N., Haan, S.W.H.D, Varela, P., Visscher, K.: Grid tied converter with virtual kinetic storage. In: IEEE Bucharest PowerTech, pp. 1–7 (2009)
5. Zhang, L., Harnefors, L., Nee, H.P.: Power-synchronization control of grid-connected voltage-source converters. IEEE Trans. Power Syst. **25**(2), 809–820 (2010)
6. Rodriguez, P., Candela, I., Citro, C., Rocabert, J., Luna, A.: Control of grid-connected power converters based on a virtual admittance control loop. In: Proceedings of EPE, pp. 1–10 (2003)
7. Torres, M.A., Lopes, L.A., Moran, L.A., Espinoza, J.R.: Self-tuning virtual synchronous machine: a control strategy for energy storage systems to support dynamic frequency control. IEEE Trans. Energy Convers. **29**(4), 833–840 (2014)
8. Vrana, T.K., Hille, C.: A novel control method for dispersed converters providing dynamic frequency response. Electr. Eng. **93**(4), 217–226 (2011)
9. D'Arco, S., Suul, J.A., Fosso, O.B.: Control system tuning and stability analysis of virtual synchronous machines. In: IEEE Energy Conversion Congress and Exposition (ECCE), pp. 2664–2671 (2013)

A BPSO-Based Tensor Feature Selection and Parameter Optimization Algorithm for Linear Support Higher-Order Tensor Machine

Qi Yue[1(✉)], Jian-dong Shen[1], Ji Yao[1], and Weixiao Zhan[2]

[1] Xi'an University of Posts and Telecommunications, Xi'an 710121, China
yueqi6@163.com

[2] China Academy of Information and Communications Technology, Beijing, China

Abstract. Feature selection is one of the key problems in the field of pattern recognition, computer vision, and image processing. With the continuous development of machine learning, the feature dimension of the object is becoming higher and higher, which leads to the problem of dimension disaster and over fitting. Tensor as a powerful expression of high dimensional data, can be a good solution to the above problems. Considering the much redundancy information in the tensor data and the model parameter largely affects the performance of linear support higher-order tensor machine (SHTM), a BPSO-based tensor feature selection and parameter optimization algorithm for SHTM is proposed. The algorithm can obtain better generalized accuracy by searching for the optimal model parameter and feature subset simultaneously. Experiments on USF gait recognition tensor set show that compared with the ordinary tensor classification algorithm and GA-TFS algorithm, this algorithm can shorten the time of large-scale data classification, reduce about 22.06% time-consuming, and improve the classification accuracy in a certain extent.

Keywords: Tensor feature selection · Parameter optimization
Support higher-order tensor machine · Tensor rank-one decomposition

1 Introduction

With the continuous development of information technology, the size of data continues to increase. Analysis of high dimensional data, such as hyperspectral image and video image, is becoming an urgent problem to solve [1–3]. At present, the high dimensional data feature is generally represented as a vector mode, so that the characteristic dimension of the expression is generally higher, and the data structure information is lost. Ideally, the higher dimension of feature leads to the better effect of recognition accuracy. But in reality, it is difficult to provide sufficient samples, the increase of feature dimension can only lead to the "dimension disaster" and "over fitting" problem [4–6], which severely reduce the recognition accuracy and computing speed of the algorithm.

P. Krömer et al. (Eds.): ECC 2018, AISC 891, pp. 304–311, 2019.
https://doi.org/10.1007/978-3-030-03766-6_34

Tensor as a powerful expression of high dimensional data, can be a good solution to the above problems. Many real-world image and video data, such as gray level face images and color video sequences, are more naturally represented as tensors [7, 8]. To construct learning models and design fast algorithms for tensor data become a research hot topic.

Along with the development of the tensor decomposition theory and the multi linear subspace learning theory, many researchers have proposed a variety of tensor learning algorithms with tensor as the input data. Tao proposed a supervised tensor learning framework with tensor as input data, which extend SVM and Fisher discriminant analysis into tensor form [9]; Wu proposed a seed space embedding dimension reduction method [10]; Liu proposed a local maximum interval classifier for image and video classification [11]; Suykens proposed nonlinear tensor learning model, which can deal with the tensor data of the non-square matrix after the expansion of the matrix [12]. Signoretto proposed a kernel model based on the cumulative amount for the classification of multi-channel signals [13]. Hao presented a linear support high order tensor machine model (SHVM) for tensor model representation and classification [14, 15], which avoid the shortcomings of time-consuming and local optimum problems in supervised tensor learning framework by combined with the advantages of support vector machine model and tensor rank one decomposition. The tensor rank one CP decomposition can better reflect the tensor data structure information and the inherent correlation, especially for higher order tensor. However, the classification results of the tensor learning framework of Hao is sensitive to the parameters. And the redundant information of tensor also greatly reduces the classification performance of SHVM.

In this paper, considering the problem of tensor feature redundancy information and classifier parameter affect classifier performance, a tensor feature selection method based on Filter-Wrapper mixture model is proposed. In the algorithm, the feature of low dimension is obtained by CP tensor decomposition, and the binary particle swarm algorithm is used to fuse the Filter and Wrapper model. The algorithm can effectively improve the classification performance, by searching the optimal classifier parameters and feature subset simultaneously. Experiments on USF gait recognition tensor sets show that, compared with the ordinary tensor classification algorithm and GA-TFS algorithm, this algorithm can shorten the time of large-scale data classification, reduce about 22.06% time-consuming, and improve the classification accuracy in a certain extent.

The rest of paper is organized as follows. In Sect. 1, the theory of tensor algebra and tensor rank one decomposition are brief introduced. In Sect. 2, the proposed algorithm is discussed. The experimental results and analyses are presented in Sect. 3. Section 4 gives conclusion of the paper.

2 The Proposed Optimization Algorithm

Feature selection is one of the key problems in the field of pattern recognition, computer vision, and image processing. Based on different evaluation criteria, feature selection can fall into the filter model and the wrapper model. Filter model is fast and efficient, and it has strong universality and robustness. It is suitable for processing large scale data sets and online data. Wrapper model has better classification performance, but its robustness and universality are poor. In this study, we fuse the advantages of filter model and wrapper model based on Binary Particle Swarm Optimization (BPSO).

Considering that SHVM is vulnerable to parameter setting, therefore, we use BPSO to select the optimal feature subset and model parameter in the SHTM simultaneously. The particle code, redundant coefficient, fitness function and Algorithm steps are described in details as follows.

2.1 Particle Code

Given the N order tensor $\mathcal{X} \in R^{I_1 \times I_2 \times \cdots \times I_N}$, which contain $I_1 \times I_2 \times \cdots \times I_N$ elements. We use Alternating Least Square Method to compute rank-one decomposition, the results can be obtained as follow:

$$\mathcal{X} \approx \sum_{r=1}^{R} \mathbf{x}_r^{(1)} \circ \mathbf{x}_r^{(2)} \circ \cdots \mathbf{x}_r^{(N)} \tag{1}$$

From formula (2), it can be deduced that the tensor have N modes, and each mode can be decomposed into R dimensional vector, as $\mathbf{x}_1^{(n)}, \mathbf{x}_2^{(n)}, \cdots, \mathbf{x}_R^{(n)} \in R^{I_n}$, $n = 1, 2, \cdots N$. Each row vector is related, as a feature, the vector can either choose or not at the same time. Therefore, it can be deduced that the size of N order tensor feature set is $\sum_{n=1}^{N} I_n$.

For each mode decomposition vector, use binary to indicate whether the feature is selected or not. It can be deduced that N segment binary code is required to represent each mode feature selected state. The length of each segment of the binary code is $I_1, I_2, \cdots, I_N$. Because the SHVM is sensitive to the parameters, the binary representation of the parameter is added in the particle.

Figure 1 shows the binary chromosome representation of our design method. In Fig. 1, Bc represents the binary value of SHVM penalty parameter, BF_1 to BF_N represents the feature mask of the tensor mode space. The length of each mask is from I_1 to I_N. The value "1" represents the feature is selected, the value "0" represents the feature is not selected.

Bc	BF_1	$\cdots$	BF_i	$\cdots$	BF_N

Fig. 1. Particle code design.

2.2 Redundant Coefficient

According to the concept of mutual information, a new correlation redundancy analysis method is proposed in reference [16], as shown in formula (2). In formula (2), N is s the feature quantization level and C_c is Class space dimension.

$$J(f_j) = \frac{I(C,f_j)}{\log(C_c)} - \frac{1}{|S_{i-1}|}\sum_{f_i \in S_{i-1}} \frac{I(f_i,f_j)}{\log(N)} \tag{2}$$

2.3 Fitness Function

The fitness function is the direction of the particle swarm search. In order to obtain better classification results with fewer features, the classification accuracy and feature dimension need to be considered simultaneously. Therefore, the fitness function can be expressed as formula (3). In the formula, a_1 and a_2 is constant, which is used to adjust the accuracy rate and the ratio between the characteristic dimension.

$$F = a_1 Accuracy + a_2 \frac{1}{Feature\_\dim} \tag{3}$$

The redundant coefficient is defined as shown in the formula (3). In the formula f_j is alternative feature for select, x_{ij} is the j component for the i particle.

$$\beta_{ij} = J(f_j) * x_{ij} \tag{4}$$

2.4 The Proposed Algorithm

Input tensor training set $\mathcal{S} = \{\mathcal{X}_i \in R^{I_1 \times I_2 \times \cdots \times I_N}, y_i \in \{-1, 1\}, i = 1, 2, \cdots L\}$, each of these categories corresponds to the training set elements is $p_j, j = 1, 2$, and satisfied $p_1 + p_2 = L$. Set the number of particle swarm N, the minimum and maximum flight velocity of the particles $V_{\min}$ and $V_{\max}$, the maximum number of iterations T, the dimension of the feature dimension D, the number of selected features d, the fitness function threshold value T_h, the feature subset $S_i^k = \emptyset$.

Output: the optimal generalized accuracy, the optimal model parameter and the corresponding feature subset.

1. Apply ALS to conduct the tensor rank-one decompositions for all of the tensors.
2. Randomly generated initial population.
3. Detect whether to meet the convergence conditions, to meet the go to step 7, or to continue.
4. Update the position of particle. The calculation of mutual information between features and categories $I(f_i, C)$. Select the maximum mutual information of corresponding feature added feature subset S_i^k, and let $S_i^k = S_i^k + \{f_i\}$.
5. Test the maximum number of iterations, if less than it, then $k = k + 1$.

5.1. Detection feature subset is feature number less than the specified number of features, feature selection is redundant if it is less than the maximum coefficient of feature subset, until the number of specified number of features.
5.2. Calculate the feature subset fitness, update p_i, p_g, set up $S_i^k = \varnothing$, and go to 5.1.

6. Detect whether cross-border the maximum number of iterations or fitness function threshold. If the iteration is reached, output the corresponding feature subset.

3 Experiment and Analysis

3.1 Experiment Data Sets and Parameter Setting

In this section, we evaluate the performance of the proposed algorithm on three tensor databases (USFGait17_32x22x10, USFGait17_64x44x20 and USFGait17_128x88x20). The linear SHTM is used as baseline in order to comparison.

The sets of data standardization pretreatment, the distribution in the interval range of [0, 1]. And is divided into six parts and make sure that each sample has a similar distribution. Select five of the data sets parts as training samples, the remaining one as a test sample. All of the tensor is decomposed into rank- one tensor by alternating least square method. The training time does not include the time of tensor decomposition. The evaluation criterions are classification accuracy and operation time.

The SHVM is used to measure the classification performance of Wrapper model, the SHVM kernel function is Gauss radial basis function. Penalty parameter uses 16-bit binary code. Due to the tensor CP decomposition length selection method is not yet mature, the paper adopts the step search method to search the optimal decomposition length in the range of [3, 10].

BPSO algorithm parameters are set as follows: The weight coefficient is adjusted by the linear adjustment strategy proposed by the reference [10]. The number of particles $N = 100$, the tuning parameters in BPSO algorithm is $c_1 = c_2 = 2$, Particle maximum and minimum flying speed is $V_{\min} = -6$, $V_{\max} = 6$ respectively, the maximum number of iterations is $T = 2000$, the fitness function threshold value is $T_h = 0.9$.

3.2 Experimental Result Analysis

In order to verify the validity of the feature selection, the experiments conducted to compare the performance of proposed algorithm, GA-SHTM algorithm and SHTM algorithm. The results shown in Table 1, the experimental results contrast diagram as shown in Fig. 2(a), the training time comparison diagram as shown in Fig. 2(b).

From Table 1 and Fig. 2(a), we can see that the proposed algorithm is better than the non- feature selection algorithm in all data sets, and slightly higher than the GA based wrapper optimization algorithm. Figure 2(b) shows that the training time of proposed algorithm is significantly lower than that of the SHTM algorithm which did not use feature selection method. Compared with the genetic algorithm-based optimization algorithm, the proposed algorithm can save 11.2% of the time in the USF Gait1 data sets, can save 30% of the time in USF Gait2 data sets, and can save 25% of

Table 1. Comparison of the results of BFS-SHTM and SHTM on USF datasets

Datasets	Algorithm	Generalized accuracy (%)	Training time (s)	R	C
USF Gait1	BFS-SHTM	81.21	4.32	8	258
	SHTM	78.48	10.94	8	1
USF Gait2	BFS-SHTM	83.15	8.28	8	32
	SHTM	82.00	19.71	8	1
USF Gait3	BFS-SHTM	83.36	5.42	4	512
	SHTM	82.49	26.76	7	1

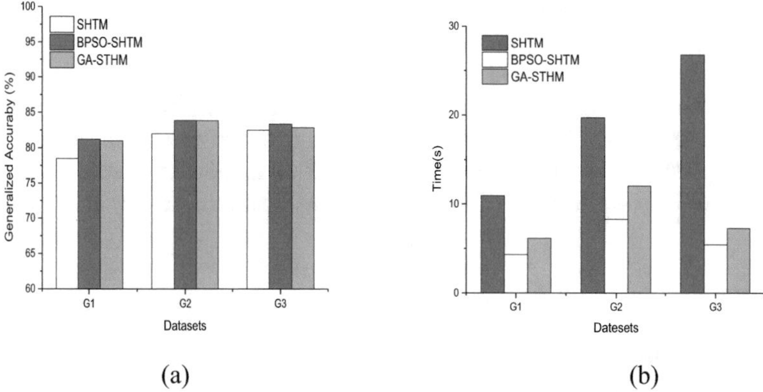

Fig. 2. (a) Average recognition rate comparison. (b) Average recognition time comparison

the time in the USF Gait3 data sets. So, it can be seen that the larger size of dataset, the lower time complexity of proposed algorithm. The saving time on USF Gait3 data set is less than USF Gait2 is because of the different rank, which resulting in the final saving time is not increased by the size of the data set (Table 2).

Table 2. Feature subsets obtained by the proposed algorithm on USF datasets.

Datasets	Mode	Selected feature subsets
USF Gait1	1	1110110001100111010000101100111
	2	0101111111010011010011
	3	1110011011
USF Gait2	1	1111110011010000101101101000100011101110111001100010110001011101
	2	110001000111010110100000011111101101000 11
	3	0011000010110101 1011
USF Gait3	1	1000101101111110101110000011000110011110011011111010000001001001101101100001 1001101011010010001000010101110000010000100111001011
	2	1110000100110001101000000110001010101100011111011011100000001100101 11000101011000101 0010
	3	1110011011101010001 1

Generally speaking, the algorithm combines the advantages of Filter model and Wrapper model. The Filter model is used to provide a better optimization of the Wrapper model, so it can save a lot of the calculation time. Compared with the optimization algorithm based on genetic algorithm, the proposed algorithm can obtain a more compact feature subset with lower feature dimension due to the feature redundancy coefficient.

4 Conclusion

In this paper, we study the feature selection and parameter optimization algorithm based on tensor data representation. A new Filter-Wrapper hybrid optimization algorithm is proposed. In the algorithm, the CP tensor decomposition is used to get low dimensional tensor features, and the mutual information feature redundancy factor is used to measure the redundancy between features. Experiments on USF gait recognition tensor sets shows that, compared with the ordinary tensor classification algorithm and GA-TFS algorithm, the proposed algorithm can significantly reduce the classification time of large-scale data and improve the classification accuracy in the same time.

References

1. Negi, P.S., Labate, D.: 3-D discrete shearlet transform and video processing. IEEE Trans. Image Process. **21**(6), 2944–2954 (2012). A Publication of the IEEE Signal Processing Society
2. Yan, T., Jones, B.E.: Color image processing and applications. Meas. Sci. Technol. **2**, 222 (2001)
3. Gavrila, D.M.: The visual analysis of human movement: a survey. Comput. Vis. Image Underst. **73**(1), 82–98 (1999)
4. Lu, H., Plataniotis, K.N., Venetsanopoulos, A.N.: MPCA: multilinear principal component analysis of tensor objects. IEEE Trans. Neural Netw. **19**(1), 18–39 (2008)
5. Yan, S., Xu, D., Yang, Q., et al.: Multilinear discriminant analysis for face recognition. IEEE Trans. Image Process. **16**(1), 212–220 (2007). A Publication of the IEEE Signal Processing Society
6. Lu, H., Plataniotis, K.N., Venetsanopoulos, A.N.: Uncorrelated multilinear discriminant analysis with regularization and aggregation for tensor object recognition. IEEE Trans. Neural Netw. **20**(1), 103–123 (2009)
7. Wang, H., Ahuja, N.: Compact representation of multidimensional data using tensor rank-one decomposition. In: International Conference on Pattern Recognition, vol. 1, pp. 44–47. IEEE (2004)
8. Geng, X., Smith-Miles, K., Zhou, Z.H., et al.: Face image modeling by multilinear subspace analysis with missing values. IEEE Trans. Syst. Man Cybern. Part B Cybern. **41**(3), 881–892 (2011). A Publication of the IEEE Systems Man & Cybernetics Society
9. Tao, D., Li, X., Hu, W., et al.: Supervised tensor learning. In: Proceedings of the IEEE International Conference on Data Mining, pp. 450–457 (2005)
10. Fei, W., Yanan, L., Yueting, Z.: Tensor-based transductive learning for multimodality video semantic concept detection. IEEE Trans. Multimed. **11**, 868–878 (2009)

11. Liu, Y., Liu, Y., Chan, K.C.C.: Tensor-based locally maximum margin classifier for image and video classification. Comput. Vis. Image Underst. **115**(115), 300–309 (2011)
12. Signoretto, M., Lathauwer, L.D., Suykens, J.A.K.: A kernel-based framework to tensorial data analysis. Neural Netw. Off. J. Int. Neural Netw. Soc. **24**(8), 861–874 (2011)
13. Signoretto, M., Olivetti, E., De Lathauwer, L., et al.: Classification of multichannel signals with cumulant-based kernels. IEEE Trans. Signal Process. **60**(5), 2304–2314 (2012)
14. Hao, Z., He, L., Chen, B., et al.: A linear support higher-order tensor machine for classification. IEEE Trans. Image Process. **22**(7), 2911–2920 (2013). A Publication of the IEEE Signal Processing Society
15. Savicky, P., Vomlel, J.: Exploiting tensor rank-one decomposition in probabilistic inference. Kybernetika **43**(5), 747–764 (2006)
16. Vinh, L.T., Lee, S., Park, Y.T., et al.: A novel feature selection method based on normalized mutual information. Appl. Intell. **37**(1), 100–120 (2012)

The Terrain Virtual Simulation Model of Fujian Province Based on Geographic Information Virtual Simulation Technology

Miaohua Jiang[1], Hui Li[1], Kaiwen Zheng[1], Xianru Fan[1], and Fuquan Zhang[2(✉)]

[1] Department of Geography, Ocean College, Minjiang University, Fuzhou 350108, China
jmhpr@outlook.com, 3123696349@qq.com, 1455460062@qq.com, 1430667453@qq.com

[2] Fujian Provincial Key Laboratory of Information Processing and Intelligent Control, Minjiang University, Fuzhou 350108, China
8528750@qq.com

Abstract. With the development of information technology, virtual simulation technology has been widely used in various disciplines. The application of virtual simulation technology based on geographic information is also advancing in the development of the times. Geographical information system is an important information carrier of realistic simulation, and three-dimensional (3D) map visualization provides an important foundation platform for mining, surveying and other industry simulation. In this study, we simulated terrain of Fujian province as an example, on the base of ASTER GDEM V2 data and using virtual simulation technology of geographical information. The results showed that: Compared with two-dimensional (2D) model map, 3D terrain model map has more expression of spatial geographical information. Comparing the MODIS model diagram with the partial model diagram of LC8, the overall effect of LC8 is better than that of the MODIS model diagram. The studies are intended to provide references for terrain simulation and development based on terrain simulation.

Keywords: Virtual simulation technology · Terrain · Three-dimensional

1 Introduction

The earth is the carrier of human activity [1]. Since ancient times, people have tried various methods to describe the surface characteristics of the earth in order to understand the natural world. Initially, people used pictographic symbols to describe the earth. With the progress of the times and the development of human civilization, people gradually realize that the fluctuation of the ground has a profound influence on the

M. Jiang and H. Li—Contributed equally to this paper.

P. Krömer et al. (Eds.): ECC 2018, AISC 891, pp. 312–318, 2019.
https://doi.org/10.1007/978-3-030-03766-6_35

natural environment such as temperature and vegetation. For this reason, the expression of the characteristics of the earth's surface has become a matter of increasing concern. With the emergence of romanticism, the natural landscape is one of its main forms of expression. Therefore, landscape painting has become the mainstream of this period. For example Perspective Map, Halo Map, Sticky Area Map, Landscape Map and so on.

Although the characteristics of the surface of the earth can be displayed using two-dimensional expressions, since the development of computer technology and photogrammetry, people are increasingly not satisfied with the display of two-dimensional planes. Three-dimensional model can avoid some drawbacks of two-dimensional model, define the key manufacturing information directly on the three-dimensional model, greatly improving the efficiency of the project and so on [2].

The fluctuation of the surface is closely related to the elevation of the surface of the earth. The degree of fitting and fidelity of the surface of the earth are closely related to the accuracy of the elevation. Now, taking Fujian province as an example, using the method of elevation data combined with virtual simulation technology, shows the three-dimensional terrain in Fujian province.

Fujian is located in the southeast of China and on the coast of the East China Sea. It is between 23°33′ to 28°20′ north latitude and 115°50′ to 120°40′ east longitude. Fujian Province is dominated by hilly terrain. The Western and central mountains form the skeleton of Fujian's topography. Both mountain belts are northeast-southwest, parallel to the coast. In Fujian, the peaks and peaks are towering, the hills are continuous, and the valleys and basins are interspersed. And mountains and hills account for more than 80% of the total area of the province. So in Fujian Province, the mountainous is 80% area, the water is 10% area, the field is 10% area. The northwest of Fujian province has higher topography and greater undulation, while the southeast of Fujian is lower and more flat. The terrain of Fujian province is presented with 3D virtual reality and realistic visual effect, so as to improve people's intuitive cognition of Fujian province's natural landform, and provide references for terrain simulation and topography based simulation and development. At the same time, the combination of electronic simulation technology and geographic information technology also reflects one of the main trends of geographical information development.

2 Materials and Methods

The elevation data in Fujian Province uses the "ASTER GDEM V2" data from http://www.gscloud.cn/.

ASTER GDEM data was jointly developed by METI of Japan and NASA of the United States and distributed free of charge to the public [3]. The data are based on detailed observations of the new generation of NASA Earth observation satellites TERRA [4, 5]. This digital elevation model contains 1.3 million stereoscopic images collected by advanced on-board thermal emission and anti-radiation (ASTER). Due to the exception in the local area of the original ASTER GDEM V1 data, the digital elevation data products processed by ASTER GDEM V1 also have data anomalies in individual regions. The ASTER GDEM V2 version uses an advanced algorithm to improve the V1 version of the GDEM image, improving the spatial resolution accuracy and elevation accuracy of the data. The algorithm reprocessed 1,500,000 images, of which 250,000 images were newly acquired after the release of the V1 GDEM data.

ASTER mapping data cover all land areas between 83° N and 83° S, covering 99% of the Earth's surface. The data is a digital elevation data product with a global spatial resolution of 30 m [4].

Surface texture used MODIS covering Fujian Province from http://www.gscloud.cn/, Landsat8 covering a part of Fujian from http://www.gscloud.cn/, and other random color bands.

MODIS is a new generation of optical remote sensing instruments in the world. It has 36 discrete spectral bands and a wide spectrum, covering from 0.4 microns (visible light) to 14.4 microns (thermal infrared). MODIS Multi-Band data can simultaneously provide information on land surface conditions, cloud boundaries, cloud properties, ocean water color, phytoplankton, biogeography, chemistry, atmospheric water vapor, aerosols, surface temperature, cloud top temperature, atmospheric temperature, ozone and cloud top height, etc. The true color synthesis using the red, green, and blue bands of the MODIS data was obtained, that is, the true color remote sensing image of Fujian Province was obtained. MODIS data has three resolutions of 250 m, 500 m, and 1000 m, and the data is usually hdf file. The band information in the MODIS data used for true color synthesis is shown in the following Table 1.

Table 1. MODIS real color synthesis

Band name	Wavelength	Spatial resolution
Red	0.62–0.67 μm	250 m
Green	0.54–0.57 μm	500 m
Blue	0.46–0.48 μm	500 m

The Landsat8 satellite carries two sensors, the OLI Land Imager and the TIRS Infrared Sensor. OLI includes all the bands of the ETM+ sensor. In order to avoid atmospheric absorption characteristics, OLI has readjusted the band. The larger adjustment is OLI Band 5 (0.845–0.885 μm), excluding the water vapor absorption feature at 0.825 μm; OLI panchromatic band Band8 band range narrow, this method can better distinguish vegetation and vegetation-free characteristics on panchromatic images. In addition, there are two additional bands: the blue band (band 1:0.433–0.453 μm) The main application of coastal zone observations, short-wave infrared band (band 9:1.360–1.390 μm) Including strong water vapor absorption characteristics can be used for cloud detection; The bands of near-red band 5 and short-wave red band 9 and MODIS are similar. Landsat8 data has three resolutions of 15 m, 30 m, and 100 m, and the data is usually tiff file. The band information in the Landsat8 data used for true color synthesis is shown in the following Table 2.

Three-dimensional virtual simulation refers to a realistic virtual environment that is generated using computer technology and has multiple perceptions such as vision, hearing, touch, and taste [6]. The techniques used in three-dimensional simulation include three-dimensional modeling technology and stereoscopic synthesis display technology etc.

Table 2. Landsat8 real color synthesis.

Band name	Wavelength	Spatial resolution
Red	0.64–0.67 μm	30 m
Green	0.53–0.59 μm	30 m
Blue	0.45–0.51 μm	30 m

VR technology is a high-speed development of contemporary information technology and the product of integration with other technologies. This simulation has the most basic characteristics. There are immersion, interaction and imagination. VR is a science that integrates people and information. The purpose is to express information through virtual experience. It is combined with multiple disciplines. Including artificial intelligence, cybernetics, computer graphics, database, human-computer interface technology, sensing technology, electronics, robots, real-time computing technology, multimedia and telepresence technology. Data Gloves (DG), Data Clothes (DS), Data Helmet (HID) and other devices make it easy for users to manipulate the virtual environment. The virtual world created by virtual reality technology must have three elements: dialogue with the virtual world, self-discipline of the virtual world, performance of the virtual world and the sense of presence [7].

Rendering with color strips enhances autonomy. You can try different color strips for rendering comparisons and highlight the corresponding features.

The virtual simulation design process of Fujian terrain can be divided into three steps. First, the downloaded ASTER GDEM V2 data is seamlessly spliced and all the DEM data covering Fujian Province is spliced into a whole. Using the vector data of Fujian province, the data of DEM is cropped to obtain the complete data of DEM in Fujian province.

Second, the data covering Fujian Province of MODIS L1B standard products were downloaded, and the resolution of band 3 and 4 (500 m) in MOD02 HKM range was increased to 250 m. The 250 m resolution band 3 and 4 were synthesized with band 1 (250 m) in MOD02QKM radiance. The synthesized image uses band 1 as red, band 4 as green, and band 3 as blue for RGB synthesis.

For LC8 data, the relevant images are downloaded and the downloaded remote sensing images are seamlessly spliced. Afterwards, it is cropped according to the vector boundary. It is best to use band 4 as red, band 3 as green, and band 2 as blue for RGB synthesis.

Finally, based on the MODIS RGB synthesis image and the LC8 RGB synthesis image, the 3D stereoscopic display effect is used for display processing. Then, the three-dimensional graphics and VR technology are combined to form a virtual simulation model of Fujian Province.

3 Experimental Analysis and Conclusions

Using related software and corresponding data processing, Fujian terrain simulation model is realized. On the basis of realizing Fujian terrain simulation model, simple line can be added to facilitate the visual analysis and understanding of Fujian terrain simulation model.

It is easy to conclude that the 2D rendering model is not realistic enough compared to the 3D rendering simulation model. The 3D rendering model highlights the ups and downs of the terrain and realizes the transformation process of the terrain model from a 2-dimensional plane to a 3-dimensional stereo, giving people a more realistic visual experience (Fig. 1).

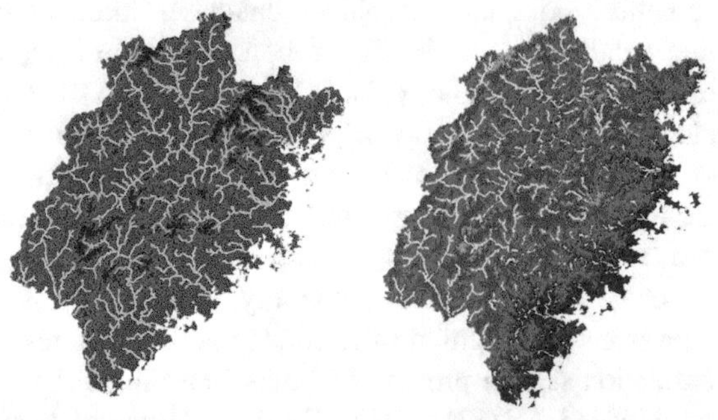

Fig. 1. 2D rendering model and 3D rendering simulation model.

The 3D model based on MODIS data is very different from the 3D model based on LC8. The 3D model based on LC8 is closer to the real world, while the color difference of the 3D model based on MODIS does not change much (Fig. 2).

Fig. 2. The 3D model based on MODIS and the 3D model based on LC8.

Compared with LC8 data, MODIS data has lower resolution and more cloud cover. For this reason, the true color synthesis effect of MODIS is not as good as LC8, resulting in a clear difference between the two.

On the basis of 3D visualization model, adding VR technology makes virtual reality integrate vision, hearing, and touch into a whole. By combining 3D visualization analysis with VR technology, the virtual simulation model of Fujian terrain has good visualization effect and fidelity.

4 Summary

The most important parts of the virtual terrain simulation experiment in Fujian province are the processing of DEM data, the combination of 3D stereo effect display and VR technology. The emphasis of fidelity is on the laying of textures. The use of MODIS data needs to be improved by cloud removal and atmospheric correction, making true color images more consistent with the texture characteristics of the object. When using LC8 data, we should pay attention to the selection of cloud quantity should be less than 10%, which can enhance the fidelity of the experiment.

In recent years, with the development of information technology, the application of electronic virtual simulation technology in various fields has been more and more extensive, and its application in geographical information is not widely exception [7, 8]. Geographical information system is an important information carrier of realistic simulation, among which 3D map visualization provides an important basic platform for mining, surveying and other industry simulation. This study discusses the process and matters needing attention of 3D virtual terrain simulation in Fujian province. It provides a reference for the study of virtual simulation terrain.

Acknowledgements. This study was supported by the National Natural Science Foundation of China (Grant No. 31470501) and the Program for New Century Excellent Talents in Fujian Province University (2015).

References

1. Wu, X., Tian, H., Zhou, S., et al.: Impact of global change on transmission of human infectious diseases. J. Sci. China (Earth Sci.) **57**, 189–203(2014)
2. Nie, G.P., Ren, G.J.: The application of virtual reality technology in teaching of industrial design–outline of the project for the human-computer interactive simulation of the upper limb operations, Hong Kong, China (2013)
3. Guo, X.Y., Zhang, H.Y., Zhang, Z.X., et al.: Comparison between ASTER-GDEM and SRTM3 data quality precision. J. Remote. Sens. Technol. Appl., 334–339 (2011)
4. Skidmore, W.: Thirty meter telescope detailed science case: 2015. J. Res. Astron. Astrophys. **15**, 1945–2140 (2015). Warren Skidmore on behalf of the TMT International Science Development Teams &TMT Science Advisory Committee
5. Liu, L., Wan, W.X., Chen, Y.D., et al.: Recent progresses on ionospheric climatology investigations. In: 12th Academic Annual Conference of the Institute of Geology and Geophysics, Chinese Academy of Sciences, Beijing, China (2013)
6. Xu, T.X.: Virtual Research and Implementation of Three-dimensional City Simulation system. Hubei University (2011)

7. Zhang, L.M., He, B.Y., Zhang, Y.F.: Virtual reality and three-dimensional visual simulation technology and its application in geographic information system. J. Xinjiang Univ. (Nat. Sci.), 41–45 (2003)
8. Fang, S, Wang, X.Y.: Design of geographic information virtual simulation experiment. J. Mod. Educ., 168–170 (2018)

A Bidirectional Recommendation Method Research Based on Feature Transfer Learning

Yu Mao[1], Xiaozhong Fan[1], Fuquan Zhang[1,2](✉), Sifan Zhang[1], Ke Niu[3], and Hui Yang[1]

[1] School of Computer Science and Technology, Beijing Institute of Technology, Beijing, China
{3120110367, fxz}@bit.edu.cn, 8528750qq.com, 451540776@qq.com, zlj-1943@163.com

[2] Fujian Provincial Key Laboratory of Information Processing and Intelligent Control (Minjiang University), Fuzhou 350121, China

[3] Computer School, Beijing Information Science and Technology University, Beijing, China
niuke@bistu.edu.cn

Abstract. In recommendation systems, data cold start is always an important problem to be solved. In this paper, aiming at problems such as few users, sparse evaluation data and difficulty of model start-up, a new bidirectional recommendation method based on feature transfer learning is proposed in the field of recommendation systems with two-way evaluation data. Based on the limited domain features, in order to transfer more useful information, we build a feature similarity based bridge between the target domain and the training field. First, we obtain the bidirectional recommendation matrix in the training field. Then, the feature space of users and items is vectorized to calculate the similarity between the target domain and the training domain. Finally, the feature transfer learning model is constructed to transfer the target domain, and the objective bidirectional recommendation matrix is obtained. The experimental results show that the method proposed in this paper can solve the data cold start problem in some bidirectional recommendation fields, and has achieved better results compared with the traditional recommendation method.

Keywords: Bidirectional recommend · Transfer learning Recommender system

1 Introduction

With the advent of the era of big data, people gradually come into the age of data overload from the age of data lack, and it is becoming more and more difficult for users to get the content they need from mass data. In this context, the recommendation system emerges as the times require. Recommendation system is an effective means to help users solve information overload by discovering items of interest to users [1]. The classic recommender system is composed of users, items and users' ratings to items, the key task is to predict the unknown score data users may give, and then recommend the items that users are interested in. However, traditional methods of recommendation

P. Krömer et al. (Eds.): ECC 2018, AISC 891, pp. 319–327, 2019.
https://doi.org/10.1007/978-3-030-03766-6_36

(such as collaborative filtering and matrix decomposition) depend only on the user's rating data on items, so there are some problems such as cold start and data sparsity.

The cold start problem [2, 3] is how to recommend for new users or new items without historical data. The cold start problem is a classic problem that has been widely concerned in collaborative filtering recommendation algorithm, the problem seriously affects the quality of the recommendation, and is detrimental to the long term development of the system. At present, some solutions are put forward for the cold start problem, for example, random recommendation method, average value method, mode method, information entropy method, etc. [4, 5]. Although these methods have achieved certain effect in solving the data cold start problem, in the actual situation, the amount of knowledge that users can provide and feedback is limited in a single field, so this problem has become an important bottleneck to improve the quality of recommendation system. Therefore, the research of cross domain recommendation algorithm is of great significance for solving practical problems in real life.

The transfer learning method is a new research direction in recent years to solve the cold start of data [6]. The target of this method is to apply the knowledge which has been studied in a field to a new field. In a number of related fields, there may be some common characteristic attributes between recommendation scores. Based on this idea, the accuracy of recommendation system can be effectively improved by transfer the relevant knowledge among different fields. In real life, there are bidirectional score data for users and items in some fields, not only the evaluation result of the users to the items, but also the feedback information of the items to the users. The bidirectional connection between these users and items can help the recommendation system to analyze more accurately, however, we often get a lot of evaluation data in a certain field, but we encounter data sparsity in another area that we need to predict. When the traditional recommendation model is used to matrix the target domain evaluation data, the matrix will be very sparse, especially in the phase of computing similarity, due to inadequate training data characteristics, many useful bidirectional evaluation features are ignored because they are not captured by models, therefore, the traditional recommendation method will lead to poor recommendation results.

Aiming at the problems of sparse data, too few features and inefficient use of bidirectional data in some fields, this paper proposes a new bidirectional recommendation model based on feature transfer learning, the method uses the idea of transfer learning to learn the related feature between fields, combines bidirectional evaluation data in the field, and effectively solve the data cold start problem in the target domain in the recommendation algorithm by using the related training field.

The structure of this article is as follows: The Sect. 2 introduces the research status of recommendation algorithm based on transfer learning; The Sect. 3 describes the bidirectional recommendation model; The Sect. 4 introduces a bidirectional recommendation method based on feature transfer learning; The Sect. 5 is the experiment and the result analysis; The Sect. 6 is the conclusion.

2 Related Works

At present, transfer learning can be divided into four types according to the content of learning: instance-based transfer, feature-based transfer, parameter-based transfer and knowledge-based transfer. Instance-based transfer [7] is mainly using the re-weighted weight of instance and sampling the key points to perform secondary analysis on the data in the training data-set, Dai et al. [8], reassigns weight of data, increase the weight of good data, reduce the value of bad data, so as to find out the importance of data. Feature based transfer [9] is mainly to discover the association knowledge between training field and target field, then describe these knowledge in the form of feature representation. Raina et al. [10], Lee et al. [11] used the algorithm of feature representing transfer thoughts, to learn association features by using feature construction method, and complete the marking task. Parameter-based transfer [12, 13] is based on the assumption that multiple field models have shared parameters, and coding field knowledge by parameters to carry out field transfer. Lawrence et al. [14] proposed a algorithm based on normal stochastic process to perform a parameter based multitask transfer learning task. Transfer based on relevance knowledge is to establish a mapping relationship between training field and target field, thus perform transfer learning by using the relationship of knowledge.

All of these studies are about transfer learning in target fields using external related data, and achieved good results. However, these methods require high correlation for training data, and are used to deal with a small number of classification problems, and it is difficult to achieve good results in the field recommendation with sparse data.

In addition, with the wide application of the transfer learning algorithm, some scholars have applied the transfer learning model to the recommendation system. Among them, Tang [15] proposed a method that apply the transfer learning method combing with the improved collaborative filtering algorithm to the vertical electricity supplier recommendation system, using the website order data and clicking data in the field of vertical e-commerce as the input data, meanwhile, the user's purchase behavior and click behavior are considered to improve the similarity between users. Ke [16] proposed a method for estimating the equilibrium parameter approximately by using the characteristic subspace distance, combining the collaborative filtering recommendation algorithm, meanwhile, feature transfer is carried out on the basis of the matrix decomposition model. Li et al. [17] proposed a method that construct codebook relationship between fields, using the class level relationship between user class and item class to represent the score pattern, then extend the codebook and transfer to restructure, so as to achieve the recommendation of target field. Pan et al. [18] use the user click rate and other auxiliary field data to build a complex matrix of user and item integration, and then combine the principled matrix factorization method to perform the transfer learning of the target field. These recommendation methods combining transfer learning have achieved some good results, but most of them are based on the analysis of one-way evaluation training data, and do not make full use of the characteristics of project attributes, ignoring the feedback information of items to users.

To sum up, the above methods have made some good progress in the recommendation system, but these methods do not specializes in the study of the feedback information for the users. In some fields, there is a bidirectional evaluation system in the target field, but the data are too sparse, and there are a lot of bidirectional evaluation data in the training field associated with it. In our proposal, by learning bidirectional data features in the training field, we build a bidirectional feature model for field transfer, so as to perform accurately recommendation in the target field.

3 Bidirectional Recommendation Model

In this section we will present the model of the bidirectional recommendation system. In some areas (e.g. campus recruitment), there will be both the user's evaluation data and the feedback information from the items to the users, thus forming a bidirectional data structure, the specific model framework is shown in Fig. 1.

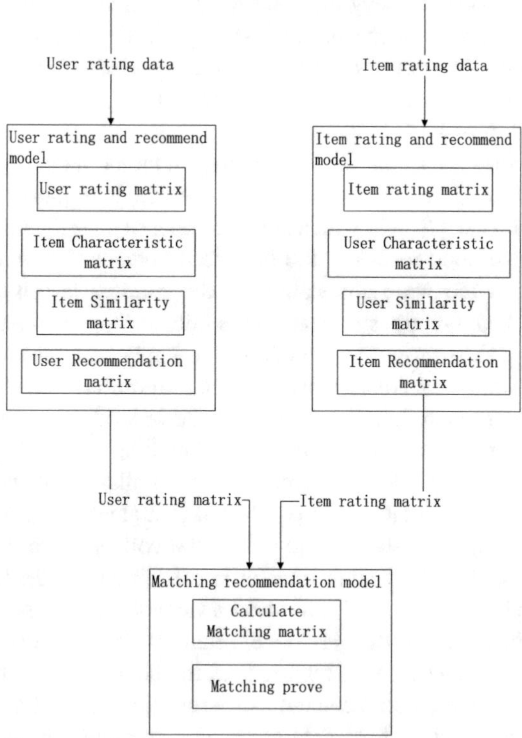

Fig. 1. The bidirectional recommendation model

In order to effectively utilize the bidirectional information of user and items, we propose bidirectional recommendation algorithm for bidirectional evaluation data in the field, and get the result of bidirectional recommendation matrix. The concrete steps are as follows:

Step 1: Matrixing the users' evaluation matrix of the items to generate user evaluation matrix E_u.
Step 2: Matrixing the items' evaluation matrix of the users to generate item evaluation matrix E_p.
Step 3: Generating item feature matrix T_p, this matrix is produced by calculating transpose the E_u matrix.
Step 4: Generating user feature matrix T_u, this matrix is produced by calculating transpose the E_p matrix.
Step 5: Compute the similarity among each user by calculating Manhattan distance of the E_u row vectors.
Step 6: Construct item similarity matrix Dp by using the result of step 5.
Step 7: Compute the similarity among each items by calculating Manhattan distance of the E_p row vectors.
Step 8: Construct user similarity matrix D_u by using the result of step 7.
Step 9: Compute the user recommendation matrix (R_{up}) based on (3.1).

$$U_i = \sum\nolimits_{m=0}^{M} simi_{\mathrm{Im}} \times pref_{\mathrm{Im}} \Big/ \sum\nolimits_{m=0}^{\mathrm{M}} simi_{\mathrm{Im}} \tag{3.1}$$

The U_i is the value of user rating item, the $simi_{\mathrm{Im}}$ is the similarity matrix of U_i to item I_m.

Step 10: Compute the item recommendation matrix (R_{pu}) based on (3.2).

$$P_i = \sum\nolimits_{n=0}^{N} simi_{Un} \times pref_{Un} \Big/ \sum\nolimits_{n=0}^{\mathrm{N}} simi_{Un} \tag{3.2}$$

Step 11: Combine the R_{up} and the R_{pu} to obtain the result matrix (RD_{up}) by using parameter δ and μ, based on (3.3).

$$RD_{up} = \left(R_{up} \times \delta + R_{pu} \times \mu\right)/(n+m) \tag{3.3}$$

The δ is related to importance of each value of user rating item, the μ is related to importance of each value of item rating user.

4 Bidirectional Recommendation Method Based on Feature Transfer

In real life, there are strong correlations between certain fields, which can usually be grouped in a large field, for example, movies and music can be summed up in the field of art, recruitment and registration can be summed up in the field of education. The feature of users and items in these areas have high similarity, and there is a hypothesis in the recommendation system that items with the same or similar attributes usually have similar behavior. Therefore, we can cluster users and items feature to get corresponding user class and item feature classes, the relationship between these classes can be shared in a certain range of fields, so as to achieve the effect of transfer learning.

Similarly, in the field of bidirectional data, we can also generate the relationship between the item class and the user attribute class, and then combine the previous section with the bidirectional recommendation model to make a precise recommendation.

Ideally, if the same user class and item feature class in the two fields are exactly the same in the same set, then we only need to generate a set of class sets. However, in reality, we usually use only one set of central points to determine a set of categories. In this paper, we use the orthogonal non negative matrix triangulation decomposition algorithm proposed in [Ding 2006] to generate the set of clusters. First, according to the formula (4.1), we construct the user-item feature matrix U and the item feature-item feature matrix V.

$$\min_{U \geq 0, V \geq 0, S \geq 0} \left\| X_{aux} - UST^{T} \right\|_{F}^{2} \tag{4.1}$$

Among them, X_{aux} is n × m user-item feature matrix, and U is n × k nonnegative orthogonal matrix, V is a nonnegative orthogonal matrix of m × l, and S is a non-negative orthogonal matrix of k × l, $\|\cdot\|F$ denotes the Frobenius norm, k is the number of user categories, and l is the number of item feature classes.

Secondly, through matrix calculus the matrix U and V to obtain the auxiliary matrix B_{uvv}, the detailed method is refer to the codebook construction algorithm in [Bin Li]. In the same way, through the item-user matrix X and the item feature- item feature matrix Y to acquire the reverse auxiliary matrix B_{xyy}. Then, combine the auxiliary matrix B_{uvv} and B_{xyy}, calculate the bidirectional auxiliary matrix B_{uvxy}.

Finally, combining with the train domain data, expand and reconstitute the auxiliary matrix B_{uvxy}, then fill the target domain matrix with user-item feature similarity value to obtain the finally Z_A.

5 Experiment

5.1 Experimental Corpus

In this section, we conducted several tests to evaluate the effect of our bidirectional recommendation model based on transfer learning. The test datasets we used in the experiments is a group general rating data from campus recruitment field and the field of students choosing tutors. Among them, take the field of campus recruitment as a training field, including 200 thousand student users, 100 thousand recruitment information, 500 thousand student evaluation data and 50 thousand recruitment company's feedback to students. Take the field of students choosing tutors as target area, which includes 1000 student users, 200 tutors, 10 thousand student evaluation scores and 100 tutors feedback information. The rating value for this dataset is ranged from 1 to 5, the more larger the number, the more user likes item. The rating value is counted by combine click-through rate and collection rate.

5.2 Experimental Design

The experiment use 5-fold cross validation method to evaluate the result. The idea is: choose 80% to be train data set, and the rest of 20% date to be test data set. We will describe the accuracy of the predictions of our technique through Root Mean Square Error (RMSE) and Mean Absolute Error (MAE).

$$RMSE = \sqrt{\frac{\sum_{(u,i)|R_{test}} \left(r_{u,i} - \widehat{r}_{u,i}\right)^2}{|R_{test}|}} \tag{5.1}$$

$$MAE = \frac{\sum_{(u,i)|R_{test}} \left|r_{u,i} - \widehat{r}_{u,i}\right|}{|R_{test}|} \tag{5.2}$$

The $\widehat{r}_{u,i}$ is the prediction score, the $r_{u,i}$ is the actual score.

5.3 Analysis of Experimental Results

In order to prove effect of our method, we compare the result for our approach and other techniques: Based on user collaborative filtering model, probabilistic matrix factorization, and collaborative filtering method based on transfer learning. From the experimental results in Table 1, it can be seen that the RMSE and MAE could be effectively improved by bidirectional transfer learning approach. Our result is even lower than the other recommendation algorithms.

Table 1. Experiment comparison

Technique	Relative standard	
	RMSE	MAE
UCFM	1.291	1.876
PMF	1.028	1.379
TLCF	0.976	1.132
Our approach	0.853	1.033

It can be seen from the experimental results above, user based collaborative filtering model perform poor recommendation results, the reason is mainly because the method only calculates user similarity based on user preferences, however, in some areas where data are sparse, there is a problem of data cold start. It is difficult to get enough user evaluation data. The method of probabilistic matrix factorization is better than that of user based collaborative filtering method, the reason is that the use of matrix decomposition technique avoids the sparsity of data to some certain extent, but it does not effectively use a large number of auxiliary data in the related fields. Although collaborative filtering based on transfer learning uses data in related fields, it does not take into account the existence of bidirectional data in some fields, so this method is less effective than recommendation method based on bidirectional transfer learning.

6 Conclusion

In this paper, a bidirectional recommendation method based on transfer learning is proposed, this method effectively uses the bidirectional data characteristics in some related fields and fills the target matrix with the orthogonal non negative matrix tri-angular decomposition algorithm. The experiment proves the effectiveness of the method. In the future work, we will further study how to get more data feature information, so as to further improve the accuracy of recommendation.

This research was supported by the State Key Laboratory of Digital Publishing Technology.

References

1. Bobadilla, J., Ortega, F., Hernando, A., Gutiérrez, A.: Recommender systems survey. Knowl. Based Syst. **46**, 109–132 (2013)
2. Adomavicius, G., Tuzhilin, A.: Toward the next generation of recommender systems: a survey of the state-of-the-art and possible extensions. IEEE Trans. Knowl. Data Eng. **17**(6), 734–749 (2005)
3. Yu, K., Schwaighofer, A., Tresp, V., Xu, X., Kriegel, H.P.: Probabilistic memory-based collaborative filtering. IEEE Trans. Knowl. Data Eng. **16**(1), 56–69 (2004)
4. Sun, M., Li, F., Lee, J., et al.: Learning multiple-question decision trees for cold-start recommendation, pp. 445–454. ACM (2013)
5. Zhou, K., Yang, S.H., Zha, H.: Functional matrix factorizations for cold-start recommendation. In: International ACM SIGIR Conference on Research and Development in Information Retrieval, pp. 315–324. ACM (2011)
6. Pan, S.J., Yang, Q.: A survey on transfer learning. IEEE Trans. Knowl. Data Eng. **22**(10), 1345–1359 (2010)
7. Bickel, S., Scheffer, T.: Discriminative learning for differing training and test distributions. In: Proceedings of the Twenty-Fourth International Conference on Machine Learning, DBLP, pp. 81–88 (2007)
8. Dai, W., Yang, Q., Xue, G.R., et al.: Boosting for transfer learning. In: International Conference on Machine Learning, pp. 193–200. ACM (2007)
9. Quattoni, A., Collins, M., Darrell, T.: Transfer learning for image classification with sparse prototype representations. In: IEEE Conference on Computer Vision and Pattern Recognition, CVPR 2008, pp. 1–8. IEEE (2008)
10. Raina, R., Ng, A.Y., Koller, D.: Constructing informative priors using transfer learning. In: International Conference on Machine learning, pp. 713–720 (2006)
11. Lee, H., Battle, A., Raina, R., et al.: Efficient sparse coding algorithms. In: International Conference on Neural Information Processing Systems, pp. 801–808. MIT Press (2006)
12. Stark, M., Goesele, M., Schiele, B.: A shape-based object class model for knowledge transfer. In: IEEE International Conference on Computer Vision, pp. 373–380. IEEE (2009)

13. Bonilla, E.V., Chai, K.M.A., Williams, C.K.I.: Multi-task Gaussian process prediction. In: Conference on Neural Information Processing Systems, Vancouver, British Columbia, Canada, December. DBLP (2008)
14. Lawrence, N.D., Platt, J.C.: Learning to learn with the informative vector machine. In: International Conference on Machine Learning, p. 65. ACM (2004)
15. Tang, J.: Submitted in partial fulfillment of the requirements for the degree of Master of Engineering. Nanjing University (2013)
16. Ke, L.: A survey of collaborative filtering based on transfer learning. Huaqiao University (2014)
17. Li, B., Yang, Q., Xue, X.: Can movies and books collaborate? Cross-domain collaborative filtering for sparsity reduction. In: IJCAI 2009, Proceedings of the International Joint Conference on Artificial Intelligence, Pasadena, California, USA, pp. 2052–2057, DBLP (2009)
18. Pan, W., Xiang, E.W., Liu, N.N., et al.: Transfer learning in collaborative filtering for sparsity reduction. In: Twenty-Sixth AAAI Conference on Artificial Intelligence, pp. 662–668. AAAI Press (2010)

FDT-MAC: A New Multi-channel MAC Protocol Based on Fuzzy Decision Tree for Wireless Sensor Network

Hui Yang[1], Linlin Ci[3], Fuquan Zhang[1,2](✉), Minghua Yang[3], Yu Mao[1], and Ke Niu[4]

[1] School of Computer Science and Technology, Beijing Institute of Technology, Beijing, China
zlj-1943@163.com, maoyu_bit@163.com, 8528750@qq.com

[2] Fujian Provincial Key Laboratory of Information Processing and Intelligent Control, Minjiang University, Fuzhou 350121, China

[3] Rocket Military Equipment Research Institute, Beijing, China
cilinlin@263.net, yang-mh@foxmail.com

[4] Computer School, Beijing Information Science and Technology University, Beijing, China
niuke@bistu.edu.cn

Abstract. The current popular multi-channel Medium Access Control (MAC) layer protocol adopts a strategy of separating control and data channels, in order to improve channel utilization and packet forwarding success rate. However, the randomness of the data access and the high mobility of the nodes will lead to switch channels too frequently, which will still increase the packet transmission latency. Therefore, this paper proposes a fuzzy decision tree for MAC (FDT-MAC) protocol, which constructs the fuzzy decision tree of the nodes sub-tree. FDT-MAC utilizes fuzzy reasoning method to construct the membership function of the node collision factor (NCF) and node relative direction (NRD) factor. So that the node can quickly select the appropriate channel according to the real-time status (Output value) of each channel, reduce the collision probability of the packet, and improve the delivery probability (90%) and throughput of the channel (750 bps).

Keywords: Multi-channel · MAC protocol · Fuzzy decision tree

1 Introduction

Nowdays, researchers have proposed several protocols that exploit the multi-channels available in order to increase throughput of channels. These protocols are designed for wireless sensor networks in orthogonal channels, the different devices can transmit in parallel on distinct channels with the MAC protocol, this parallelism increases the throughput and can potentially reduce the channel delay, provided that the channel access time is not excessive. However, how devices agree on the channel to be used for transmission and how they resolve potential contention for a channel, these choices affect the delivery probability and throughput characteristics of the protocol.

P. Krömer et al. (Eds.): ECC 2018, AISC 891, pp. 328–336, 2019.
https://doi.org/10.1007/978-3-030-03766-6_37

In this paper, we proposes a fuzzy decision tree for MAC (FDT-MAC) protocol, which compared with several EM-MAC, MC-MAC and APDM protocol. In Sect. 2, we introduces the related research work. Section 3 introduces the fuzzy theory and some basic conception and then proposes our method FDT-MAC. Section 4 discusses experiment results and compares the performance of the protocols.

2 Related Works

Many researchers aimed to provide reliable and real-time delivery probability and channel throughput under multi-channel environments.

DW-MAC [3], MMSN [5], and TM-MAC [4] are all multi-channel protocols of statically configured mobile wireless sensor networks. This protocol robs different channels by specifying neighbor nodes within two hops. Demand Wakeup MAC (DW-MAC) [3] automatically wakeup the channel according to demand, increase the effective channel capacity and the traffic load in one operating cycle, DW-MAC achieved low delivery delays under a wide range of traffic loads including unicast and broadcast traffic. In the case of unicast high channel load, DW-MAC reduces the channel transmission latency by 70%. In broadcast mode, DW-MAC can reduce the average channel delay by 50%. MM-MAC uses a non-uniform compensation algorithm to achieve a roughly balanced channel load, thereby reducing channel congestion time. Such protocols belong to the node-level multi-channel MAC protocol. Simulation and experimental results show that the network performance of the multi-channel protocol based on node granularity is greatly improved compared to the single-channel MAC protocol.

Tang et al. proposed a prediction mechanism for fast convergence of multi-channel MAC protocols: EM-MAC protocol [6]. The protocol stipulates that after the initialization of the receiving node, the Pseudo Random Function (PRF) is used to independently select its working time and channel scheduling. The protocol distributes network load balancing to different channels, reducing the probability of data collision in the channel (Data Collision Rate, DCR).

In 2017, Song, C proposed the Adaptive Multi-Priority Distributed Multichannel (APDM) [10], which assumes that packet generation has different priorities baseing on Poisson distribution, using Markov chain analysis model to optimizes the packet transmission probability, and the APDM protocol can ensure the security priority transmission of data packets and reduce the transmission delay of data packets. Ramanuja et al. [7] proposed a MAC protocol based on ad hoc network cells, which specifies the connected network elements as the prototype of the coarse-grained channel allocation mechanism. WU et al. [9] proposed a multi-channel MAC protocol that uses a coarse-grained channel allocation strategy to alleviate the problem of a small number of available channels.

3 FDT-MAC Protocol

3.1 Basic Parameter

In cybernetics, Fuzzy Decision refers to the mathematical theory and method of making decisions in a fuzzy environment [13]. The basic idea is: we assume that the value of the parameter under consideration is clear, and the predetermined decision strategy cannot be changed when the value has a small change [13].

Definition 1. Node Relative Direction (NRD). For the nodes in the same sub-tree in the channel, we specify that the relative angle of the relative motion is between 0 and 180°; the larger the angle of the direction, the larger the NRD value. The two nodes are facing each other (0°) and the NRD value is 0 because the node will encounter the node for a period.

Definition 2. Node Collision Factor (NCF). The NCF is the number of all nodes in the one-hop range, which means the same tree as the node and the next hop of the transmission path to the base station. When a node becomes the next hop of the data transmission path, the data collision probability of the channel will positively correlated with the number of nodes as the next hop. In order to simplify the protocol design, we use the number of nodes to describe the degree of possible collision.

3.2 Fuzzy Membership Function

We calculate the angle in order to obtain the NRD values, and then store the values in the node membership function, NRD divided into 5 levels, each level corresponds to the fuzzy set of NRD values on the domain [0, 180].

Very High (VH): indicates that the angle of NRD of the direction of motion of the node is greater than 135° less than 180°, and the rest is 0.

High (H): indicates that the angle of the NRD of the direction of motion of the node is greater than 90° and less than 180°, and the rest is zero.

- Low (L): indicates that the angle is greater than 0° and less than 90°, and the rest is 0.
- Very Low (VL): indicates that the node direction angle is less than 45°, and the rest is 0.

Analogously, NCF represents the number of neighbor nodes in the one-hop range. In this paper, the maximum number of neighbors for a node is set to 9. The NCF is divided into three levels using a simple trapezoidal function.

High (H): indicates that the number of neighbors of the node is greater than 6 and less than 9 (the maximum number of nodes predefined by the channel).

Medium (M): indicates the three cases where the number of neighbors of the node is between 0 and 9.

Low (L): indicates that the number of neighbors of the node is less than 3, and the rest is 0.

3.3 Fuzzy Rules

The fuzzy control rule is a key component of the fuzzy system,and the standard rule establishes the form of the statement as "if A, then B", where A and B are the values defined by the fuzzy set on the domain. Since there are two fuzzy parameters established. The fuzzy system adopts the classic Mamdani model, and we build 15 fuzzy rules, in which each rule is a fuzzy command, the two parameters are respectively NCF and NRD, and the control rules are shown as Table 1.

Table 1. Fuzzy Rules

NO.	IF NCF	And NRD	So Output is
1	H	VH	VH
2	H	H	VH
3	H	M	H
4	H	L	M
5	H	VL	L
6	M	VH	VH
7	M	H	H
8	M	M	M
9	M	L	L
10	M	VL	VL
11	L	VH	H
12	L	H	M
13	L	M	L
14	L	L	VL
15	L	VL	VL

3.4 Defuzzification

For output domain, we selected their arithmetic mean value if there are more than two maximum membership values corresponding to the Output value

$$x_0 = \frac{1}{N}\sum_{i=1}^{N} x_i, x_i = \max(\mu_x(x)) \tag{1}$$

$\mu_x(x)$ presents the membership function, and N is the total number of output values of the same maximum membership.

The interface for constructing the membership function in the multi-channel decision tree is constructed by using the Matlab 7.0 fuzzy inference plug-in. Figure 1 (b) depicts the input-output relationship diagram. It can be seen that as the NCF factor (0° to 180°) and the NRD factor (0 to 9) continue to increase, the corresponding defuzzified outputs increase, eventually reaching a relatively stable value.

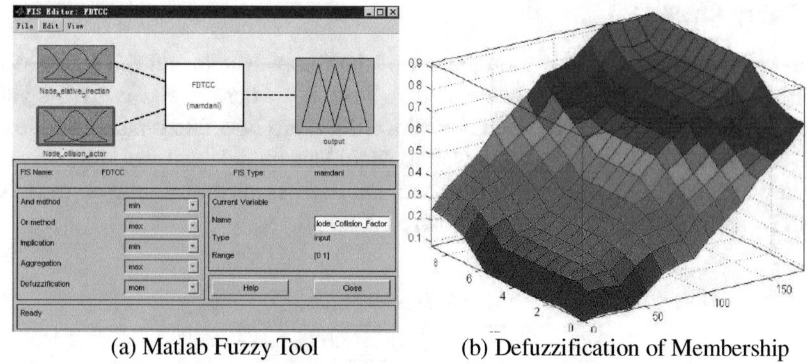

(a) Matlab Fuzzy Tool (b) Defuzzification of Membership

Fig. 1. **(a)** Matlab fuzzy tool. **(b)** Defuzzification of Membership

3.5 FDT-MAC Protocol Description

Each node dynamically maintains Node State Table (NST), and the NST stores Output defuzzification values, which updated automatically according to Table 1.

FDT-MAC adopts the breadth-first strategy of Fuzzy Decision Factor (FDF) to obtain the shortest path from each node to the base station node, mainly solving the problem of the node at the "cross-coincidence" node of the sub-tree, we can refer each leaf node should be classified into the sub-tree according to its output value (Fig. 2).

According to FDT-MAC, the balance of the sub-tree size in the graph needs to be maintained. So we can assign node 5 (shown by the light gray arrow) to the black subtree on the left side of the node, then ensuring that it is traversed in the first order, and that nodes 2, 8, and 9 are still available, the NCF is increased from 2 to 3; the intra-tree collision factor of the selected node 8 remains at 2; and the selection of node 9 increases additionally the depth of the tree. Therefore, the node 5 will select the node 8 in the subtree as its father node (Fig. 3).

4 Experiment and Analysis

4.1 Experiment Setting

We uses GloMoSim as simulation platform designed for wireless networks by UCLA, which is a free Qualnet version, and it supports high-level languages such as PARSEC, C and C++ etc.

The experimental area size is 200 m * 200 m, the dynamic period T of the network link is set to 100 min; the node communication radius is from 10 m to 35 m; according to the usual interference/communication ratio model, the node interference radius is set to 1.5 times the communication radius.

//each $node_n$ except sink node, quadruple $\{int(n), col(n), l_n, b_n\}$;
while $(n \le M-1)$
{
while $(b_n > 1)$ //total number of nodes in the entire channel range be M
{
// $b_n > 1$ presents subtree T_j, select $node_n$ belongs to the subtree
$|V_i| = |\{v|(v \in T_i) \cap (l_v \le l_n)\}|$;
$|V_j| = |\{v|(v \in T_j) \cap (l_v \le l_n)\}|$;
if $(|V_i| > |V_j|)$
{
$V_j = V_j \cup \{n\}$; // $node_n$ belongs to subtree j;
}
else if $(|V_i| < |V_j|)$
{
$V_i = V_i \cup \{n\}$; // $node_n$ belongs to subtree i;
}
else
{
if $\left(\sum_{p \in \{V_i \cup \{n\}\}} int(p) \ge \sum_{p \in \{V_j \cup \{n\}\}} int(p)\right)$
$V_j = V_j \cup \{n\}$; // $node_n$ belongs to subtree j;
else
$V_i = V_i \cup \{n\}$; // $node_n$ belongs to subtree i;
}
select $node_n$ belongs to subtree and parent node
{
//For the neighbor node set, CF is the standard selection father node;
}
else// $|TS_n| > 1$ exist more than two father node
{
$col(p) = min\{col(i) \mid i \in S_n - SN_n\}$;
//Select father node p;
}}

Fig. 2. FDT-MAC protocol description

4.2 Experiment Analysis

This section tests the performance of FDT-MAC and then compares with the current popular algorithms.

In first experiment, Fig. 4(a) shows that the delivery probability rate of FDT-MAC$_{NRD}$ and FDT-MAC$_{NCF}$ protocol under a single factor, which has maintained a certain increase in the initial stage. but the value is slower than EM-MAC, MC-MAC

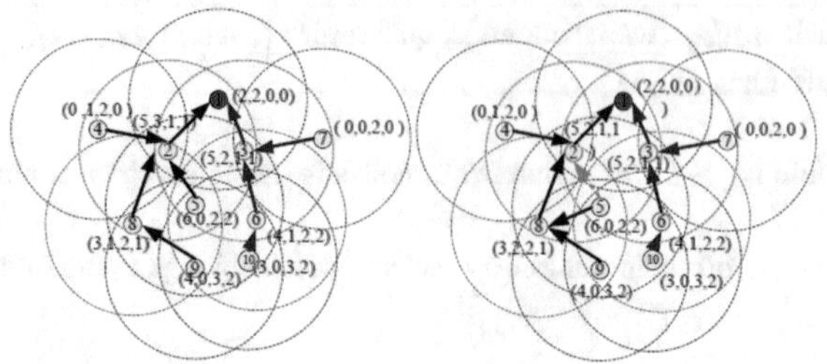

Fig. 3. **(a)** Select subtree without NCF. **(b)** Select subtree with NCF

and APDM. The Max value is only about 50%. The reason can be known that single parameter cannot solve the problem that the node may collide or the inconsistent direction, and frequently detach and enter the neighboring nodes, the probability channel ensures that the packet delivery success rate reaches about 90% quickly.

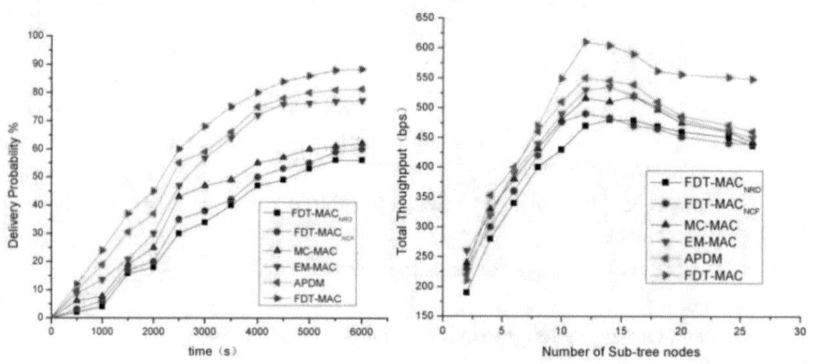

Fig. 4. **(a)** Delivery probability rate. **(b)** Total thoughput

Figure 4(b) shows the FDT-MAC protocol performs fine segmentation based on the fuzzy NCF and NRD. Under the condition of ensuring the maximum throughput of the channel, the same The sub-tree can accommodate 12 nodes, comparing with APDM (550 bps), FDT-MAC protocol has a maximum throughput of 600 bps. When the number of nodes reaches 26, The throughput of MC-MAC, EM-MAC and APDM protocols decline rapidly, while FDT-MAC protocol only declined slightly (550 bps). It is worth noting that in the case where the number of nodes in the same sub-tree is less than 5, the throughput of FDT-MAC is roughly equivalent to that of MC-MAC and EM-MAC protocols.

The second experiment is set at the non-random motion mode, it focus on the high probability of multi-channel FDT-MAC protocol packets in a stable network topology. It can be seen that FDT-MAC can adapt to the compared node formation movement model, compared with in the first experiment (Fig. 4(a)), FDT-MAC has basically reached more than 85% delivery success rate at 2500 s, which indicates that the FDT-

MAC protocol can configure the channel faster when the node group motion has certain regularity to keep the data successfully transmitted (Fig. 5(a)).

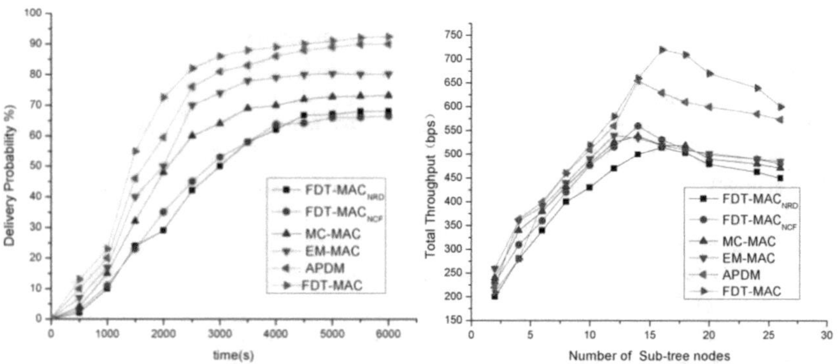

Fig. 5. **(a)** Delivery probability rate. **(b)** Total thoughput

Figure 5(b) shows that the total channel throughput for the number of different nodes in the sub-tree (up to 26 nodes). It can be seen from the experimental results that the FDT-MAC protocol performs fine segmentation based on the fuzzy node collision factor and the node relative direction factor network. The data collision in the generated sub-tree is lower than APDM. Under this condition by ensuring the maximum throughput of the channel, it can accommodate 16 nodes in the same sub-tree, which results in maximum throughout (750 bps).

5 Conclusions

This paper proposes a fuzzy decision tree for MAC(FDT-MAC) protocol for the multi-channel MAC of wireless networks. We construct the fuzzy decision tree of the nodes sub-tree, and utilize fuzzy reasoning method to construct the membership function of the NCF and NRD. These experiment results show that the nodes employed FDT-MAC can quickly select the appropriate channel according to the real-time channel status (output value), reduce the collision probability, and improve the delivery probability and throughput of channel.

Acknowledgement. This research is supported by the natural science foundation of China (grant no. 61063042), the State Key Laboratory of Digital Publishing Technology.

References

1. Jian, Q., Gong, Z.H., Zhu, P.D.: Overview of MAC protocols in wireless sensor networks. J. Softw. **19**(2), 389–403 (2008)
2. Zhou, G., Huang, C.D., Yan, T., et al.: MMSN: multi-frequency media access control for wireless sensor networks. In: Proceedings of the 25th IEEE International Conference on Computer Communications, pp. 1–13 (2006)
3. Sun, Y., Du, S., Gurewitz, O., Johnson, D.B.: DW-MAC: a low latency, energy efficient demand-wakeup mac protocol for wireless sensor networks. In: Proceedings of the 9th ACM International Symposium on Mobile ad Hoc Networking and Computing, pp. 53–62. ACM (2008)
4. Zhang, J.B., Huang, C.D., Sang, H., et al.: TM-MAC: an energy efficient multi-channel MAC protocol for ad hoc networks. In: Proceedings of the 2012 IEEE International Conference on Communications, pp. 3554–3561 (2012)
5. Chen, X., Han, P., He, Q.S., et al.: A multi-channel MAC protocol for wireless sensor networks. In: Proceedings of the 6th IEEE International Conference on Computer and Information Technology, pp. 224–228 (2015)
6. Tang, L., Sun, Y.J., Omer, G., David, B.: EM-MAC: a dynamic multichannel energy-efficient MAC protocol for wireless sensor networks. In: Proceedings of the 12th ACM International Symposium on Mobile Ad Hoc Networking and Computing, pp. 16–19 (2011)
7. Ramanuja, V., Sandeep, K.: Component based channel assignment in single radio, multi-channel ad hoc networks. In: Proceedings of the 12th Annual International Conference on Mobile Computing and Networking, pp. 378–389 (2006)
8. Hieu, K.L., Dan, H., Tark, A.: A control theory approach to throughput optimization in multi-channel collection sensor networks. In: Proceedings of the 6th international conference on Information processing in sensor networks, pp. 31–40 (2007)
9. Wu, Y.F., John, A., He, T., et al.: Realistic and efficient multi-channel communications in wireless sensor networks. In: Proceedings of the 27th IEEE International Conference on Computer Communications, pp. 1193–1201 (2008)
10. Song, C., Tan, G., Yu, C., Ding, N., Zhang, F.: APDM: an adaptive multi-priority distributed multichannel MAC protocol for vehicular ad hoc networks in unsaturated conditions. Comput. Commun. **104**, 119–133 (2017)
11. Zhou, G., He, T., Stankovic, A., et al.: RID: radio interference detection in wireless sensor networks. In: Proceedings of the 24th Annual Joint Conference of the IEEE Computer and Communications Societies, vol. 2, pp. 891–901, March 2014
12. Zheng, G.Q., Wei, Y., Estrin, D.: An cross-layer energy-efficient MAC protocol for wireless sensor networks. In: Proceedings of the Annual Joint Conference of the IEEE Computer and Communications Societies, vol. 3, pp. 1567–1576 (2009)
13. Zadeh, L.A.: Fuzzy Sets. Information and Control, pp. 338–353 (1965)

A Chunk-Based Multi-strategy Machine Translation Method

Yiou Wang[1] and Fuquan Zhang[2,3(✉)]

[1] Beijing Institute of Science and Technology Information, Beijing 100044, People's Republic of China
wangyiou90@163.com

[2] Digital Performance and Simulation Technology Lab, School of Computer Science and Technology, Beijing Institute of Technology, Beijing 100081, People's Republic of China
8528750@qq.com

[3] Fujian Provincial Key Laboratory of Information Processing and Intelligent Control, Minjiang University, Fuzhou 350121, People's Republic of China

Abstract. In this paper, a chunk-based multi-strategy machine translation method is proposed. Firstly, an English-Chinese bilingual tree-bank is constructed. Then, a translation strategy based on the chunk that combines statistics and rules is used in the translation stage. Through hierarchical sub-chunks, the input sentence is divided into a set of chunk sequence. Each chunk searches the corresponding instance in the corpus. Translation is completed by recursive refinement from chunks to words. Conditional Random Fields model is used to divide chunks. An experimental English-Chinese translation system is deployed, and experimental results show that the system performs better than the Systran system.

Keywords: Machine translation · Chunks parsing · Grammar induction Conditional random fields

1 Introduction

Nowadays, machine translation [1] is a trending topic in natural language processing, which is widely used in smart reading, interpreting [2], information spreading, etc. Example-Based Machine Translation (EBMT) [3, 4] and Statistical Machine Translation (SMT) [5, 6] are two typical translation methods based on corpus. The basic idea of EBMT is to match the input of sample language and the chunks in the database. It requires parallel corpus as data source, which provides bilingual information for translation engine. EBMT is based on text extracting and chunk combination (or other type of text chunk). In EBMT, the basic processing unit is a chunk. However, the main idea of SMT is to conduct statistical analysis on large amount of parallel corpus. SMT is designed to overcome the disadvantage of rule-based machine translation, which has strong dependency on specific languages.

However, both EBMT and SMT require a huge bilingual corpus database that need include all the possible combination of words and phrases, which is hard to achieve.

P. Krömer et al. (Eds.): ECC 2018, AISC 891, pp. 337–342, 2019.
https://doi.org/10.1007/978-3-030-03766-6_38

Also, the extraction algorithm to eliminate ambiguity efficiently in large-scale corpus is remained unknown. To improve the performance of these two machine translation methods, we propose a new Chunk-based Multi-strategy Machine Translation (CMMT) method. CMMT that uses chunk as the basic unit of translation is a combination of rule-based and statistics-based translation method.

2 The Theory of CMMT

Now, the theory of CMMT is illustrated.

Firstly, the Treebank [7] is constructed and it extracts grammar rules from parallel bilingual corpus database. There are two types of treebanks. One is annotation chunk Treebank [8], and the other one is annotation dependent Treebank [9, 10]. To construct a Chinese Treebank, we need to conduct grammar structural annotation for each sentence. This can be achieved through either manually annotating by linguists, or semi-automatic annotation by parsers. In the latter case, linguists usually have to check and correct the results, and the labor intensity depends on the level of correctness of the annotation. In CMMT, we use Penn Treebank. After executing all the grammar structure annotation for the entire context, each sentence is divided into chunks, and each chunk is made up of a group of words or smaller chunks. By processing and comparing all the combinations of the chunk sequence, we observe that the sequences frequently show up. Therefore, the sequences should be extracted and recorded to be used for the next step. Because the order of some observed sequences is very likely to change after translation (in the target language), we need to target the supplemental words of the bilingual corpus database and execute the targeted process, which means that we record every translated sequences for the observed sequences for future use.

Secondly, a finite automata that could be tracked and used by the translation engine is designed according to the obtained grammar rule. Each sequence found previously is shown as a path automatically, which connects the initial state and final state of the finite automata (and also possibly go thought an intermediate state). Different POS tags can be viewed as transitive tag inside the automata. Each transitive tag is combined with the translation sequence obtained through the final calibration process (in the target language). The input word by word conducted by the automata (when used for translation) is a sentence with POS tagging. Each chunk changes the current state of the automata. Therefore, the input sentence traverse the automata and appoint the translation mode.

Finally, the input sentence is assigned as several chunks after slicing and chunking in the stage of translation. Each chunk is searched in the corpus database. If the chuck exists, then the chunk is no longer divided. Otherwise, this chunk is divided into smaller sub-chunks. The worst case is that the chunk is neither found in the corpus database nor able to be divided into sub-chunks. In this case, a list of simple words will be obtained as last, where the translation is completely based on rule. In other cases, both rule-based and instance-based methods are used at the same time.

The structure of CMMT is illustrated in Fig. 1.

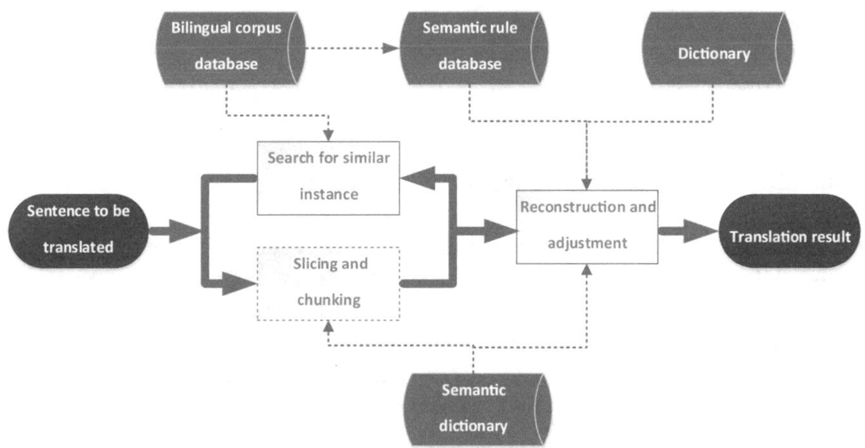

Fig. 1. The fundamental structure of CMMT

3 Chunking Model in CMMT

In this article, we conduct chunking on the input English sentence using the improved CRFs model. Conditional random fields (CRFs) [11] is a class of statistical modeling method, applied in sequence labelling and chunking. Any feature of the input sequence could be easily included into the model. The chunking steps are explained as below:

Input sentence

$S = \langle W, T \rangle$, where $W = w_1, w_2, \ldots, w_n$ is the word sequence of the sentence and $T = t_1, t_2, \ldots, t_n$ is the POS tagging sequence for each word. Then, we get the chunk description sequence as follows:

$EH = \{EG, WS\}$, where $EG = \{eg_{ij}\}$, which represents the components from the i-th word to the j-th word, and $WS = ws_1, ws_2, \ldots, ws_n$, which is the chunk sequence marked with sentence boundary detection information, where $ws_i = \langle w_i, t_i, bp_i \rangle$, and the value of bp_i is 0,1 or 2, which represents the word located in the middle of the component, left boundary and right boundary respectively. The conditional probability of state sequence with input sequence W is calculated as Formula (1).

$$P_{bp}(T/W) = \frac{1}{Z_w} e^{\left\{\sum_{i=1}^{n} \sum_{k} [bp_k f_k(t_{i-1}, t_i, W, i)]\right\}} \tag{1}$$

where Z_w is the normalization factor that ensures the sum of all probabilities of state sequence as 1, $f_k(t_{i-1}, t_i, W, i)$ is a Binary Characteristic Function, bp_k is the weight of $f_k(t_{i-1}, t_i, W, i)$. Then, The Most likely tagged sequence of the input sequence $W = w_1, w_2, \ldots, w_n$ is:

$$T^* = \arg_T \max P_{bp}(T/W) \tag{2}$$

Besides, the component boundary information of chunks could be predicted automatically. Considering the input sequence W_{ij}, T_{ij}, proper component boundary tagged sequence $BP_{ij} = bp_i, bp_{i+1}, \ldots, bp_j$ is selected to maximize $\mathrm{P}(BP_{ij}/T, W)$.

4 Evaluation of the Translation Results

In the first part of the experiment, we use a bilingual corpus database of 100 pairs of sentences. To obtain more reliable results, we divided the corpus into 5 groups, with each group consists of 20 sentences. Bilingual evaluation understudy (BLEU) is used to evaluate results. We calculate the BLEU score of the 5 parts respectively using 4-gram [12]. Therefore, for each system, we obtain 5 BLEU measurement samples. The average and the standard deviation of BLEU are shown in Table 1.

Table 1. BLEU score of 5 parts of corpus database (average and standard deviation)

Criteria	CMMT
Average	0.564
Standard deviation	0.018

In this experiment, we use two modes to evaluate the BLEU score. The first mode considers the BLEU calculated on the words of n-gram, whereas the second mode make use of the POS tags of words instead of words themselves. The results and average score of the two modes are shown in Table 2.

Table 2. BLEU score of the two modes

Modes	CMMT
BLEU (Based on Word)	0.564
BLEU (Based on POS tags)	0.895
Average	0.729

In the second part of the experiment, CMMT system is compared with some other popular English-Chinese translation systems, including Systran, Transman of IBM and manual translation. Fluency and Adequacy are calculated to evaluate translation quality. Experimental results (shown in Table 3) demonstrate that CMMT-based method achieves almost the same level of quality as current widely-used commercial machine translation system, and even surpass them to some extent. Comparing with Systran system, CMMT system reaches 68.3% in fluency, whereas only 56.9% by Systran, and we achieves an increase of nearly 12%.

Table 3. Comparison results of CMMT and other systems

Different systems	Fluency	Adequacy
CMMT	0.683	0.791
Systran	0.569	0.784
Transman	0.837	0.849
Manual	0.813	0.820

5 Conclusion

In this article, we propose a chunk-based multi-strategy machine translation method, CMMT. We implemented our system and achieve desirable results in the experiment. Experiment results show that our method performs well in both fluency and adequacy. The proposed method has the advantages from both rules and statistics. However, comparing to some specific practical translation methods, our method has not yet shown obvious advantages. The next step of our work includes improving the instance corpus database, increasing the post-processing quality of the translated text, and selecting better model features.

Acknowledgment. The authors are very grateful to Special Projects for Reform and Development of Beijing Institute of Science and Technology Information (2018) (Information rapid processing capacity building with applied artificial intelligence and big data technology) for the supports and assistance.

References

1. Babhulgaonkar, A.R., Bharad, S.V.: Statistical machine translation. In: 1st International Conference on Intelligent Systems and Information Management, pp. 62–67. Institute of Electrical and Electronics Engineers Inc. (2017)
2. Gong, H.: The role of speech recognition and machine translation in interpreting. Study Lang. Arts Sports **5**, 383–385 (2018)
3. Semmar, N., Laib, M.: Building multiword expressions bilingual lexicons for domain adaptation of an example-based machine translation system. In: 11th International Conference on Recent Advances in Natural Language Processing, pp. 661–669. Association for Computational Linguistics (2017)
4. Chua, C.C., Lim, T.Y., Soon, L.: Meaning preservation in example-based machine translation with structural semantics. Expert Syst. Appl. **78**, 242–258 (2017)
5. Mahata, S.K., Das, D., Bandyopadhyay, S.: MTIL2017: machine translation using recurrent neural network on statistical machine translation. J. Intell. Syst. (2018)
6. Wang, X., Lu, Z., Tu, Z., et al.: Neural machine translation advised by statistical machine translation. In: 31st AAAI Conference on Artificial Intelligence, pp. 3330–3336. AAAI press (2017)
7. Sun, L., Jin, Y., Du, L., Sun, Y.: Automatic extraction of bilingual term lexicon from parallel corpora. J. Chin. Inform. Process. **14**(6), 33–39 (2000)

8. Branco, A., Carvalheiro, C., Costa, F., et al.: DeepBankPT and companion Portuguese treebanks in a multilingual collection of treebanks aligned with the penn Treebank. In: 11th International Conference on Computational Processing of Portuguese, pp. 207–213. Springer (2014)
9. Badmaeva, E., Tyers, F.M.: A dependency treebank for Buryat. In: 17th International Conference on Intelligent Text Processing and Computational Linguistics, pp. 397–408. Springer (2018)
10. Bielinskiene, A., Boizou, L., Kovalevskaite, J., Rimkute, E.: Lithuanian dependency treebank ALKSNIS. In: 7th International Conference on Human Language Technologies - The Baltic Perspective, pp. 107–114. IOS Press (2016)
11. Song, D., Liu, W., Zhou, T., et al.: Efficient robust conditional random fields. IEEE Trans. Image Process. **24**(10), 3124–3136 (2015)
12. BLEU-WIKIPEDIA. https://en.wikipedia.org/wiki/BLEU. Accessed 12 Feb 2018

Customer Churn Warning with Machine Learning

Zuotian Wen[1], Jiali Yan[1(✉)], Liya Zhou[1], Yanxun Liu[1], Kebin Zhu[1], Zhu Guo[1], Yan Li[1], and Fuquan Zhang[2]

[1] Bank of China, Beijing 100091, China
516064077@qq.com, yanjiali_bit@163.com, 56239536@qq.com, 378039475@qq.com, moon_dr626@sina.com, 654558352@qq.com, 584081737@qq.com

[2] Fujian Provincial Key Laboratory of Information Processing and Intelligent Control, Minjing University, Fuzhou 350121, China
zfq@bit.edu.cn

Abstract. Customer churn refers to the phenomenon of suspension of cooperation between customers and enterprises due to the implementation of various marketing methods. The customer churn warning refers to revealing the customer churn pattern hidden behind the data by analyzing the payment behavior, business behavior and basic attributes of the customer within a certain period of time, predicting the probability of the customer's loss in the future and the possible reasons, and then guide the company to carry out customer retention work. After the forecast, the system can list the possible lost customers. And then the marketers can conduct precise marketing and improve marketing success rate. In this paper, we present a algorithm named *Customer Churn Warning* (CCW) to alert customers to churn.

Keywords: CCW · Machine learning · Customer churn warning Customer retention · Intelligent computing

1 Introduction

Today, with the increasingly mature marketing methods, our customers are still a very unstable group, because their market interests drive the levers to favor people, love and reason. When the marketing of the company does not meet the interests of the customer, there will be customer changes. Changes in customers often mean changes and adjustments in a market that can be a fatal blow to a local (regional) market. Customer churn and customer entry are common customer changes in the market. As we all know, the cost of developing a new customer is more than three times the cost of retaining an old customer. So how to improve customer loyalty, how to effectively carry out customer turnover warning, how to effectively retain customers is a problem that modern enterprise marketers have been discussing.

The reasons for customer churn can be broadly classified into two categories: active churn and passive churn. Active loss means that the customer takes the initiative to leave the business relationship with the company. For example, the customer's living

P. Krömer et al. (Eds.): ECC 2018, AISC 891, pp. 343–350, 2019.
https://doi.org/10.1007/978-3-030-03766-6_39

environment changes, the customer no longer needs the products currently purchased, the customer is insured by other companies, etc. Passive loss means that the customer involuntarily separates from the company. Usually caused by changes in the customer's economic situation or ability to pay. Through customer churn early warning analysis, companies can identify the most likely customers who are losing and the most likely causes of these customers, so that they can selectively and specifically take customer retention measures.

The customer churn warning refers to revealing the customer churn pattern hidden behind the data by analyzing the payment behavior, business behavior and basic attributes of the customer within a certain period of time, predicting the probability of the customer's loss in the future and the possible reasons, and then guide the company to carry out customer retention work. After the forecast, the system can list the possible lost customers. And then the marketers can conduct precise marketing and improve marketing success rate.

In this paper, we present a method named CCW using logistic regression to predict the possible loss of customers. The overall process is shown in Fig. 1. The CCW method is divided into 4 modules, the first one is the requirement analysis module, in which the analysis of business needs and the modeling data needed to be analyzed according to the business requirements. The second module is the model training module, which mainly focuses on model learning. Specifically, in this module, we need to do data extraction, data feature analysis, data set preparation, model construction, model evaluation, model deployment. The third module is the loss warning module. The main task of this module is to give the list of potential customers. The last module is the customer stratified retention module. In this module, we need to subdivide the cause of customer loss, formulate a retention strategy, and evaluate the cost of retention and the potential value of the customer. If the retention cost is greater than the customer's potential value, the online retention method is adopted; otherwise, the online and offline linkage retention method is adopted.

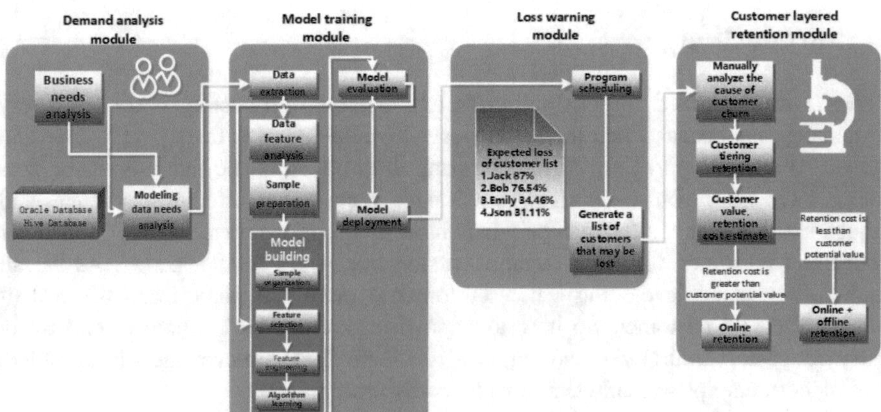

Fig. 1. The overview of our system

2 Related Work

The establishment of the customer churn early warning model mainly uses decision trees, neural networks, clustering algorithms, and regression algorithms. [2, 3] using decision tree algorithm for customer churn early warning modeling. [4–8] using Neural Networks for customer churn early warning modeling. [9, 10] using Support Vector Machines (SVM) for customer churn early warning modeling.

The decision tree is easy to understand and explain, but his accuracy is often not the highest. Neural networks tend to achieve better results, but his interpretation is low, and it is often difficult for business people to understand the results. Support vector machines are difficult to implement for large-scale training samples. In summary, we choose logistic regression for the interpretability and accuracy of the model.

3 Our Method

In this section, we briefly introduce logistic regression and propose our algorithm.

Logistic Regression is a regression problem for dealing with dependent variables as categorical variables. Commonly, it is a two-category or binomial distribution problem. It can also deal with multi-classification problems. It actually belongs to a classification method.

The relationship between the probability of the two-category problem and the independent variable is often an S-shaped curve, as shown in the Fig. 2, using the Sigmoid function. The Sigmoid function is:

$$f(x) = \frac{1}{1 + e^{x}}$$

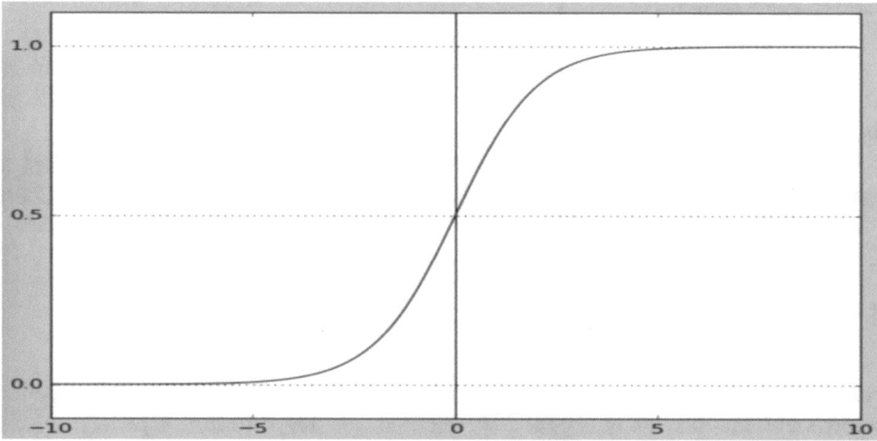

Fig. 2. Sigmoid function graph

The domain of the function is the whole real number, the value range is between [0, 1], and the result of the x-axis at 0 is 0.5. When the value of x is large enough, it can be regarded as a problem of 0 or 1 type. If it is greater than 0.5, it can be regarded as a type 1 problem, and if it is a type 0 problem, and just 0.5, it can be classified into a class 0 or a class 1. For a 0–1 type variable, the probability distribution formula for y = 1 is defined as follows:

$$P(y = 1) = p$$

The probability distribution formula for y = 0 is defined as follows:

$$P(y = 0) = 1 - p$$

The formula for the expected value of discrete random variables is as follows:

$$E(y) = 1 * p + 0 * (1 - p) = p$$

The linear model is used for analysis, and the formula is transformed as follows:

$$p(y = 1|x) = \theta_0 + \theta_1 x_1 + \theta_2 x_2 + \cdots + \theta_n x_n$$

In practical applications, the probability p and the dependent variable are often nonlinear. To solve this kind of problem, we introduce the logit transformation, which makes the linear relationship between logit(p) and the independent variable. The logistic regression model is defined as follows:

$$\mathrm{logit}(p) = \ln\left(\frac{p}{1-p}\right) = \theta_0 + \theta_1 x_1 + \theta_2 x_2 + \ldots + \theta_n x_n$$

By derivation, the probability p transform is as follows, which is consistent with the Sigmoid function, and also reflects the nonlinear relationship between the probability p and the dependent variable.

In this paper, we proposed CCW algorithm that is based on the improvement of logistic regression algorithm. The customer churn warning is a two-category problem. The conclusion we need to draw includes two possibilities, that is, the customer may lose and the customer will not lose.

In our method, we use logistic regression as the underlying algorithm to adjust to specific business scenarios. On this basis, propose our method– Customer Churn Warning (CCW).

4 The Experiment

Sample Selection. Our selection rules are shown in Fig. 3. Our observation period consists of 6 months, the base period is 3 months, and the performance period is also 3 months. The meaning of the time corresponding to these three periods is also explained below the figure.

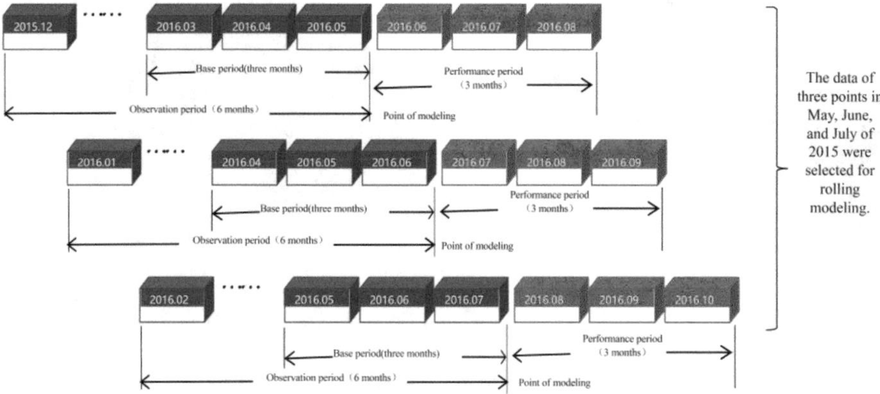

Fig. 3. Sample selection rules

Observation period: Evaluate whether customers are likely to lose during the performance period based on changes in customer behavior during this period.

Base period: the first three months of the modeling time is the base period, and the benchmark of the target customer when the loss is defined.

Performance period: three months after the modeling time is the performance period, and the performance of the target customer when the loss is defined.

Data Selection. The data information in our data set includes the customer's asset information, the customer's basic attributes, the customer's label information (including contributions, asset preferences, car signs, etc.), customer information and other data to be processed; Product data include product maturity, expected annuality, and so on. In addition, we hope to introduce external data, such as external credit data, loan data and consumption data. The purpose of introducing external data is to build a more complete customer portrait, hoping that the system will have a better understanding of customers.

5 Result Analysis

We collected one year of business data for data modeling and came to the following conclusions as show in Figs. 4 and 5.

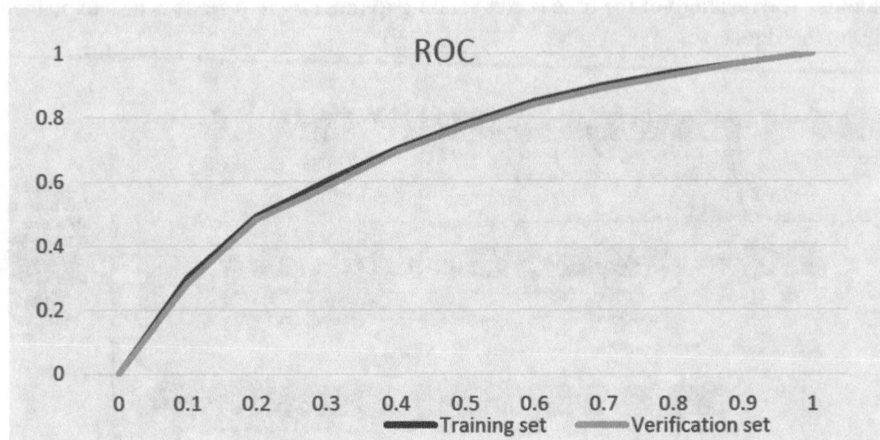

Fig. 4. ROC diagram (The difference between the training set and the verification set ROC curve is small, and the evaluation within the model sample is stable.)

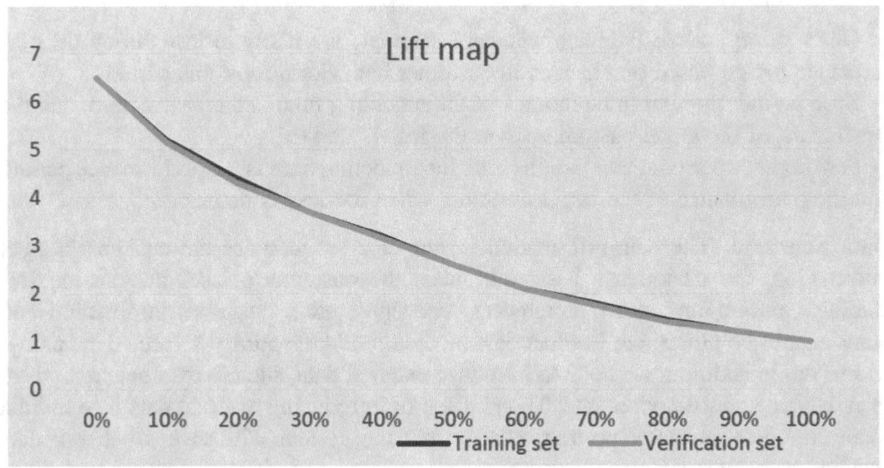

Fig. 5. Lift map (The difference between the cumulative lift of the training set and the verification set is small, and the evaluation within the model sample is stable.)

The possible reasons for customer churn are as shown in Fig. 6.

It can be seen that the use of logistic regression to deal with customer churn early warnings can obtain a more accurate list of possible lost customers, providing follow-up marketing personnel with the ability to save customers and improve marketing success rate.

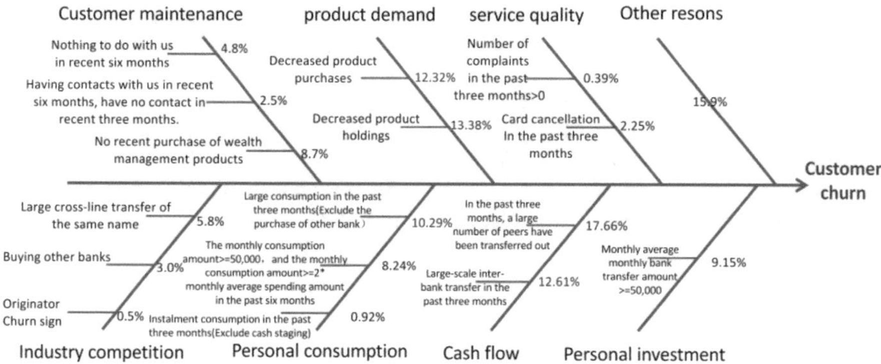

Fig. 6. The possible reasons for customer churn.

6 Conclusion

Establishing a loss warning model can help enterprises to better manage customer relationships, do customer care for high-risk customers, and do their best to retain them, which can strengthen the ability of enterprises to resist customer risk. Enterprises can set up a CCW-based loss warning system, select different coverage rates according to the cost budget, and can perform real-time scoring prediction for customers. Once the predicted loss probability exceeds the set threshold, the early warning system can issue a warning. Tell the company to focus on that customer.

Acknowledgement. The customer churn early warning model received the best creative solutions in the Bank of China ("Technology Leading" Innovation Forum) and has been included in key implementation projects. Thanks to the teammates who contributed to the competition, as well as the leading colleagues who planned to organize the competition, and the leaders who valued the project.

References

1. Zhong, J.: Research on customer churn prediction model for telecom enterprises. Xi'an University of Science and Technology (2014)
2. Yang, X.: Research and implementation of mobile communication user loss early warning model based on decision tree. Ocean University of China (2014)
3. Du, X., Wang, Z.: Decision tree based security customer churn model. Comput. Appl. Softw. (2009)
4. Liang, L., Weng, F., Ding, Y.: Application research of neural network in customer churn model. Commer. Res. **2**, 55–57 (2007)
5. Shao, S.: Analysis and prediction of insurance company's customer loss based on BP neural network. Lanzhou University (2016)
6. Lin, R., Chi, X.: Analysis model of bank customer churn based on artificial neural network. Comput. Knowl. Technol. **08**(3), 665–667 (2012)

7. Tian, L., Qiu, H., Zheng, L.: Modeling and implementation of telecom customer churn prediction based on neural network. J. Comput. Appl. **27**(9), 2294–2297 (2007)
8. Luo, B., Shao, P., Luo, J., et al.: Research on customer churn based on rough set theory-neural network-bee colony algorithm integration. Chin. J. Manag. **8**(2), 265 (2011)
9. Xia, G., Jin, Y.: Customer churn prediction model based on support vector machine. Syst. Eng. Theory Pract. **28**(1), 71–77 (2008)
10. Wang, G., Guo, Y.: Application research of support vector machine in telecom customer churn prediction. Comput. Simul. **28**(4), 115–118 (2011)

Quantum Identity-Authentication Scheme Based on Randomly Selected Third-Party

Xiaobo Zheng[1,2], Fuquan Zhang[3,4], and Zhiwen Zhao[1,5](✉)

[1] College of Information and Technology, Beijing Normal University, Beijing 100875, China
zhengxb@bistu.edu.cn, zhaozw126@126.com
[2] Information and Network Management Center, Beijing Information Science and Technology University, Beijing 100101, China
[3] School of Computer Science and Technology, Beijing Institute of Technology, Beijing, China
8528750@qq.com
[4] Fujian Provincial Key Laboratory of Information Processing and Intelligent Control, (Minjiang University), Fuzhou 350121, China
[5] Beijing Normal University, Zhuhai, China

Abstract. In our article we give a scheme to complete identity-authentication by quantum fingerprinting. Our scheme can overcome the vulnerability of one certification authority and tolerate a controlled error. The cost of our scheme is $\mathrm{k}\log_2(n)$ for n bit classical identity information while the complexity of quantum circuit in our scheme is also $\mathrm{k}\log_2(n)$.

Keywords: Quantum identity-authentication · Quantum fingerprinting

1 Introduction

The authentication service is concerned with assuring that a communication is authentic. At the time of connection initiation, the authentication service assures that the two entities are authentic (that is the entity that it claims to be). At the time, the service must assure that the connection is not interfered with in such a way that a third party can masquerade as one of the two legitimate parties for the purpose of unauthorized transmission or reception.

Kerberos [1] and X.509 [2] are two of the most widely used authentication services. Kerberos is an authentication service developed as part of Project Athena at Mit. An authentication server (AS) in Kerberos knows the passwords of all users and stores these in a centralized database. In addition, the AS shares a unique secret key with each user. If AS is attacked by a third party, all users can be fraud because their passwords are leaked. X.509 was initially issued in 1988. X.509 is based on the use of public-key cryptography and digital signatures. The heart of the X.509 scheme is the public-key certificate associated with each user. Their certificates are assumed to be created by some trusted certification authority (CA). The CA signs the Certificate with its secret

P. Krömer et al. (Eds.): ECC 2018, AISC 891, pp. 351–358, 2019.
https://doi.org/10.1007/978-3-030-03766-6_40

key. If the corresponding public key is known to a user, then that user can verify that a certificate signed by the CA is valid. If CA is attacked by a third party, all user's certificates are not valid because the certificate can be faked. When all user don't know whether AS or CA is attacked, the situation is even worse.

Huge user information was lost in Many network information security events because there is only one AS or CA. So We give a authentication scheme to use more than one AS or CA. Our scheme use the fragment of user information. If one AS or CA was attacked, attacker can't get integrity user information. At the same time, he can't fraud other by those information. Quantum communication can overcome the vulnerability of classical communication. Specially quantum communication can give unconditional security of communication channel [3, 4]. Many algorithms use entangle quantum states to assist classical communication. Entangle quantum states have some limitations in current technology. So our scheme select a more practical technology of quantum fingerprinting to complete identity-authentication.

2 Identity-Authentication Scheme Based on Randomly Selected Third-Party

2.1 A Subsection Sample

Identity authentication system generally involved three participants, a prover (Alice), a verifier (Bob) and an attacker. Some message on identity information must be exchanged from Alice to Bob. In some cases a trust-able certification authority (CA) has to be involved for judging arguments between Alice and Bob. An attacker wants to give some fraud message which makes Bob to believe a mock Alice. There are simple model of identity authentication (Fig. 1).

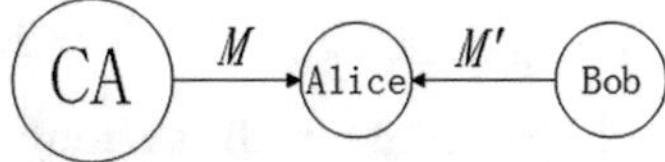

Fig. 1. A simple model

In this model Alice want to verify the identity of Bob. CA has the identity information of Bob. Alice asks Bob and CA to send Alices identity information. If Alice receive the same information from Bob and CA, she believes Bob. If not, she doest believe Bob. If this IA Model exchanges information by quantum, it is a QIA model. There are two ways to attack for this model. One way the attacker can create a message to send the Alice. Alice believe the message from Bob. The other way the attacker can create a message to send Alice. Alice believe the message from the trust-able certification (CA). We note the maximal success probabilities for two ways by P_b and P_c, respectively [5]. The probability of the model can be attacked successfully is

$$P_a = max\{P_b, P_c\} \tag{1}$$

So we want to reduce P_a by our scheme.

2.2 Randomly Selected Third-Party Scheme

With regard to the discussion below, please note that in the communication system served by our scheme, suppose Alice intends to confirm the Bobs identity and there are already K pieces of Bobs identity information which have been distributed to K CA centers, and each CA center has 1/K of Bobs identity information.

From Fig. 2 Alice wants to verify the identity of Bob. She gets m_i and $m_i^{'}$ from CA_i and Bob respectively. If m_i and $m_i^{'}$ are equal, Alice will accept Bob. Alice can repeat this process to reduce the P_a. Even if a attacker knows all identity information of Bob he still can't fraud Alice if he don't know what is m_i. But if a attacker can eavesdrop any of random CA or Bob he will have some probability to fraud Alice. So we give a quantum scheme to solve this problem.

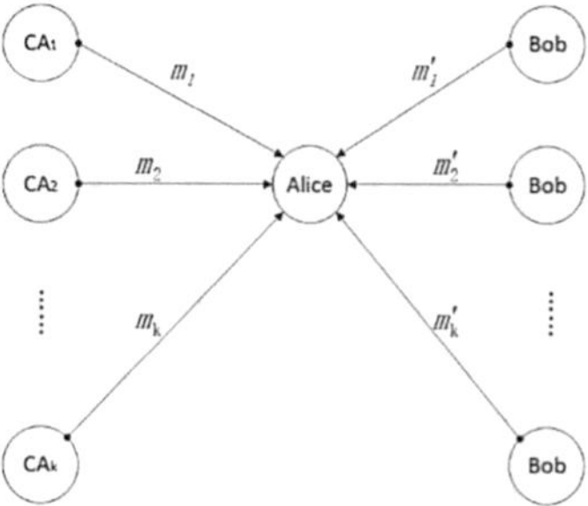

Fig. 2. Random CA authentication

3 Quantum Identity-Authentication Scheme

Our scheme uses a quantum fingerprinting [6] to compare the equality of m_i and $m_i^{'}$ from CA_i and Bob respectively. In [6] they repeat Fig. 3 k times to reduce an error probability. This repeating method can be restricted with the same quantum states and the same CA. Our scheme extend the repeating method to use any quantum states and random CA.

3.1 Quantum Fingerprinting

For each $m \in \{0, 1\}^n$, it can be defined a quantum state

$$|CA_m\rangle = \frac{1}{\sqrt{l}} \sum\nolimits_{j-1}^{l} |j\rangle |E_j(m)\rangle \tag{2}$$

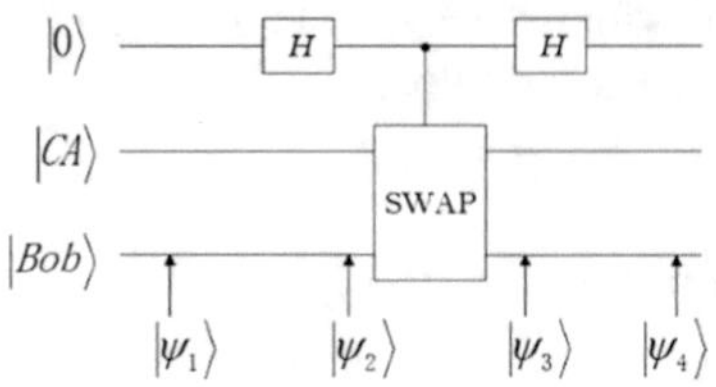

Fig. 3. Quantum one-side error probability measure

$\mathrm{E}: \{0,1\}^n \rightarrow \{0,1\}^l$ is an error correcting code with m $\in \{0,1\}^n$. $\mathrm{E}_j(m)$ denote the jth bit of $\mathrm{E}(m)$.$|Bob_{m'}\rangle$ has a same define.

$$\left|Bob_{m'}\right\rangle = \frac{1}{\sqrt{l}} \sum\nolimits_{j-1}^{l} |j\rangle \left|E_j(m')\right\rangle \tag{3}$$

In our scheme Bob and CA send quantum fingerprinting $|Bob_{m'}\rangle$ and $|CA_m\rangle$ to Alice. Alice must make a distinction between identical or un-identical of two fingerprinting. In fact noise is a great bane of information processing systems. Our scheme can tolerate a small enough δ difference when two fingerprinting is identical. For distinguishing our scheme use a technology which is named the controlled-SWAP in Fig. 3.

3.2 Quantum Circuit of One-Sided Error Probability

In Fig. 3 H is the Hadamard transform, which maps$|x\rangle = \frac{1}{\sqrt{2}}(|0\rangle + (-1)^x|1\rangle), x \in \{0,1\}$, SWAP is the operation $|CA\rangle|Bob \rightarrow\rangle Bob|CA\rangle$, c-SWAP is controlled SWAP (controlled by the first qubit. If the first qubit is $|1\rangle, |CA\rangle|Bob\rangle$ will swap to be $|Bob|CA\rangle$. Tracing through the execution of this circuit in Fig. 3, the following quantum states illustrate the process.

$$\Psi_1 = |0\rangle|CA\rangle|Bob\rangle$$

$$\Psi_2 = (H \otimes I)\Psi_1 = \frac{1}{\sqrt{2}}|0|CA\rangle|Bob\rangle + \frac{1}{\sqrt{2}}|1\rangle|CA\rangle|Bob\rangle$$

$$\Psi_3 = (c - SWAP)(H \otimes I)\Psi_1 = \frac{1}{\sqrt{2}}|0|CA\rangle|Bob\rangle + \frac{1}{\sqrt{2}}|1\rangle|Bob\rangle|CA\rangle$$

$$\Psi_4 = (H \otimes I)(c - SWAP)(H \otimes I)\Psi_1$$

$$= \frac{1}{\sqrt{2}}\left[\frac{1}{\sqrt{2}}(|0\rangle + |1\rangle)\right]|CA\rangle|Bob\rangle + \frac{1}{\sqrt{2}}\left[\frac{1}{\sqrt{2}}(|0\rangle - |1\rangle)\right]|Bob\rangle|CA\rangle$$

$$\frac{1}{2}|0\rangle(|CA\rangle|Bob\rangle + |Bob\rangle|CA\rangle) + \frac{1}{2}|0\rangle(|CA\rangle|Bob\rangle - |Bob\rangle|CA\rangle)$$

If $|CA\rangle$ and $|Bob\rangle$ are equal measuring the first qubit will outcome $|0\rangle$ with probability 1 for $\frac{1}{2}\left(1 + |\langle CA|Bob\rangle|^2\right) = 1$.

3.3 Quantum Identity-Authentication Scheme Based on Randomly Selected K Third-Parties

Now we give a more complexing model in Fig. 4. The process in Fig. 4 can illustrate the quantum states in the below

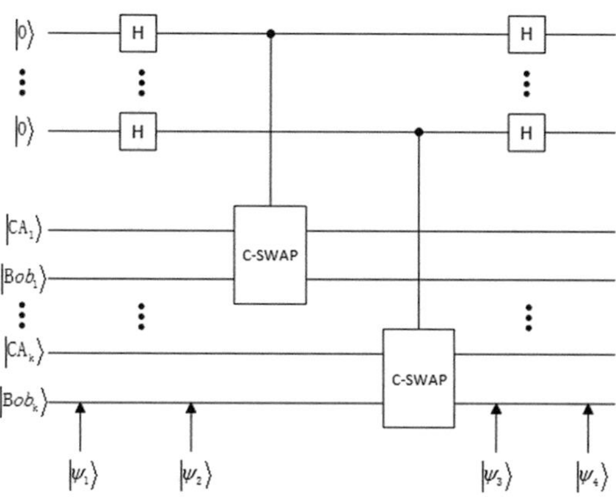

Fig. 4. k times of Quantum one-side error probability measure

Under the new model, we can recalculate the quantum state to get

$$\Psi_1 = |0\rangle^{\otimes k}\left|CA_1\right\rangle|Bob_1\rangle \cdots |CA_k\rangle|Bob_k\rangle$$

Now we define $\left|\varphi_i^0\right\rangle = |CA_i\rangle|Bob_i\rangle$ and $\left|\varphi_i^1\right\rangle = |Bob_i\rangle|CA_i\rangle$. so we aslo define

$$\left|\varphi_i^x\right\rangle = \left|\varphi_1^{x_1}\right\rangle\left|\varphi_2^{x_2}\right\rangle \cdots \left|\varphi_k^{x_k}\right\rangle \tag{4}$$

We can write

$$\begin{aligned}\Psi_1 &= |0\rangle^{\otimes k}\left|\varphi_1^0\right\rangle\left|\varphi_2^0\right\rangle \cdots \left|\varphi_k^0\right\rangle \\ &= |0\rangle^{\otimes k}\left|\varphi^0\right\rangle\end{aligned}$$

Where $\left|\varphi^0\right\rangle = \left|\varphi_1^0\right\rangle\left|\varphi_2^0\right\rangle \cdots \left|\varphi_k^0\right\rangle$.

$$\begin{aligned}\Psi_2 &= (H^{\otimes k} \otimes I)\Psi_1 \\ &= \sum_{x\in(0,1)^k} \frac{|x\rangle}{\sqrt{2^k}}\left|\varphi^0\right\rangle\end{aligned}$$

Then

$$\begin{aligned}\Psi_3 &= (c - SWAP)^{\otimes k}(H^{\otimes k} \otimes I)\Psi_1 \\ &= \sum_{x\in(0,1)^k} \frac{|x\rangle}{\sqrt{2^k}}|\varphi^x\rangle\end{aligned}$$

So

$$\begin{aligned}\Psi_4 &= (H^{\otimes k} \otimes I)(c - SWAP)^{\otimes k}(H^{\otimes k} \otimes I)\Psi_1 \\ &= \sum_{x\in(0,1)^k} \sum_{z\in(0,1)^k} \frac{(-1)^{x\cdot z}}{2^k}|z\rangle|\varphi^x\rangle \\ &= \sum_{z\in(0,1)^k} |z\rangle \sum_{x\in(0,1)^k} \frac{(-1)^{x\cdot z}}{2^k}|\varphi^x\rangle\end{aligned}$$

where $x \cdot z = x_1 \cdot z_1 + x_2 \cdot z_2 + \cdots + x_k \cdot z_k$. IF we measure the first k qubits of $|0\rangle$, the probability of $|z\rangle$ is

$$\begin{aligned}&\left\langle \sum_{x\in(0,1)^k} \frac{(-1)^{x\cdot z}}{2^k}|\varphi^x\rangle, \sum_{x\in(0,1)^k} \frac{(-1)^{x\cdot z}}{2^k}|\varphi^x\rangle \right\rangle \\ &= \sum_{x\in(0,1)^k} \sum_{x'\in(0,1)^k} \frac{(-1)^{(x+x')\cdot z}}{4^k} \langle\varphi^x|\varphi^x\rangle\end{aligned}$$

when $|z\rangle = |0\cdots0\rangle$ and all $|CA_i\rangle = |Bob_i\rangle$ we can get the probability of $|0\cdots0\rangle$ is 1.

4 QIAS Security

4.1 Security of Number of CA Center

The scheme in the research introduces random selected third-parties rather than a fix one, for avoiding a potential risk, i.e. fix one third-party may actively conspire with Alice to deceive other people. Particularly, if a third-party is needed to provide a proof about Alice, he can forswear himself. However, our scheme, it is nearly impossible for them to conspire with Alice. Moreover, if one thirdparty provides illusive identity information, and another third-party provides right information, also a conclusion can be drawn that the one third-party is possible to conspire with Alice. Hereby, more third-parties can restrict their behavior to some extent. In a word, the security and reliability of the thirdparty identification can be guaranteed by more random selected third-parties much better than fix one. In fact when the number of Distributed CA centers is small we can verify Alice information for all of CA centers. When the number of Distributer CA centers is enough large, we can repeat the verify process to randomly select more third-parties. QIAS which we give the quantum identity authentication has similar security and potential applications like [7] which they give a quantum digital signature.

4.2 Error Tolerate of QIAS

IF $|CA_i\rangle$ and $|Bob_i\rangle$ have some small difference for some kind of noise in quantum information outside system, QIAS scheme can control the error. IF $|\langle CA_i|Bob_i\rangle|^2 = \delta_i$, the probability for $|0\cdots0\rangle$ is

$$\sum_{x\in(0,1)^k}\sum_{x'\in(0,1)^k}\frac{1}{4^k}\left\langle\varphi^x\middle|\varphi^{x'}\right\rangle$$
$$=\frac{1}{2^k}(1+\sum_{i=1}^{k}\delta_i+\sum_{i,j,i\neq j}^{k}\delta_i\delta_j+\cdots+\delta_1\delta_2\cdots\delta_k)$$

Let $\delta = \min(\delta_1, \delta_2, \cdots, \delta_k)$, so

$$\begin{aligned}&\sum_{x\in(0,1)^k}\sum_{x'\in(0,1)^k}\tfrac{1}{4^k}\left\langle\varphi^x\middle|\varphi^{x'}\right\rangle\\ \geq\ & \tfrac{1}{2^k}(1+\sum_{i=1}^{k}\delta+\sum_{i,j,i\neq j}^{k}\delta^2+\cdots+\delta^k)\\ =\ & \tfrac{1}{2^k}(1+\delta)^k\end{aligned}$$

Now the low bounder of the probability for $|0\cdots0\rangle$ is $\frac{1}{2^k}(1+\delta)^k$. In fact [6, 8] give a repetition technique can reduce $1 - \delta$ to any $\varepsilon > 0$. So we can control δ to get enough confident probability of $|0\cdots0\rangle$.

4.3 The Complexity of Quantum Circuit

In classical computation we need compare $O(n)$ bits information to verify n bits Bob' identity for a simple model. if we change to k parties model we need compare $kO(n)$ bits information. There are too much users in many big application system. The cost of identity-authentication will be very high. The quantum fingerprint can reduce the cost significantly. In k parties model QIAS need $k(\log_2(n)+O(1))$ qubits to verify n bits Bob' identity information. In Fig. 3 we give a quantum circuit model to compare k parties quantum states. From [9] we can knew our quantum circuit belong to Simple Swapping Circuit. So the quantum circuit complexity is $O(k(\log_2(n))$. If someone improves our quantum circuit to mixed swapping circuit, the circuit complexity will be reduce to $O(\log_2 k(\log_2(n))$.

5 Conclusion

Traditional three-party communication models have been widely used. There are three parties Alice, Bob and a third party(or the referee) in this model. The vulnerability of this model is the third party. Our scheme can overcome the vulnerability and achieve identity authentication with the low cost.

References

1. Bryant, W.: Designing an authentication system: a dialogue in four scenes. Project Athena document, February 1988
2. I'Anson, C., Mitchell, C.: Security defects in CCITT recommendation X.509 the directory authentication framework. Computer communications Review, April 1990
3. Bennett, C.H., Brassard G.: An update on quantum crytography. Advances in Cryptology. In: Proceeding of Crypto 1984, Barbara, pp 475–480. Springer, Heidelberg (1985)
4. Mosca, M., Stebila, D., Ustaoglu, B.: Quantum Key Distribution in the Classical Authenticated Key Exchange Framework, Jun 2012. arXiv:1206.6150v1,27
5. Zeng, G.: Quantum Private Communication, 173 p. Higher Education Press, Beijing (2010)
6. Buhrman, H., Cleve, R., Watrous, J., de Wolf, R.: Quantum fingerprinting. Phys. Rev. Lett. **87**(16), 167902 (2001)
7. Gottesman, D., Chuang, I.: Quantum digital signatures. Technical Report, Cornell University Library, November 2001. arXiv:quant-ph/0105032
8. Ablayev, F.: Alexander Vasiliev, Quantum Hashing, Oct 2013. arXiv:1310.4922v1,18
9. Koca, C., Akan, O.B.: Quantum Memory Management Systems. ACM (2015). ISBN 9781-4503-3674-1/15/09

Application of R-FCN Algorithm in Machine Visual Solutions on Tensorflow Based

Yumeng Zhang[1], Yanchao Ma[1], and Fuquan Zhang[2,3(✉)]

[1] Beijing Information Science Technology University, Beijing, China
767641818@qq.com, 1141679423@qq.com

[2] School of Computer Science and Technology, Beijing Institute of Technology, Beijing, China
8528750@qq.com

[3] Fujian Provincial Key Laboratory of Information Processing and Intelligent Control, Minjiang University, Fuzhou 350121, China

Abstract. This paper presents a solution based on Tensorflow platform and R - FCN deep learning model about self-driving cars image processing. Through the Supervised learning of data sets, make them exercise the image segmentation and recognition of information, thus to self-driving cars driving decision-making support.

Keywords: Deep learning · Image processing · Machine vision Autonomous driving

1 Introduction

With the development of intelligentialize degree and the technology of the Internet of Things, more and more intelligent products have entered people's lives and have a great impact on people. Under such a background, the application of deep learning in the direction of image processing has gradually become a reality in the process of image processing, attracting many researchers to invest in this big topic for research. Since 1950, especially in the past ten years, companies and research institutions in the frontier of major computers have invested huge costs and made breakthroughs, showing its scientific value, application value and development prospects. It has a major impact on saving social resources, reducing traffic accident rates, and improving travel efficiency.

The main research in this paper is to apply the latest technology in the field of computer vision to process and train the image data obtained by the environment perception module to assist the computer in making behavior decisions.

P. Krömer et al. (Eds.): ECC 2018, AISC 891, pp. 359–366, 2019.
https://doi.org/10.1007/978-3-030-03766-6_41

2 Related Work

2.1 Tenserflow

TensorFlow [1] is Google's second-generation manual intelligence learning system based on DistBelief. Its name is derived from its operating principle. Tensor means an N-dimensional array, and Flow means a calculation based on a data flow graph. TensorFlow is a process in which tensors flow from one end of the flow graph to the other.

Since the release of Tensorflow in November 2015, it has been widely used and praised. In the comparison of open source frameworks for various deep learning, Tensorflow has an absolute advantage in Github's attention and user volume; The overall score of 88 in each dimension of the mainstream in-depth learning framework was also 16 points ahead of the second Caffe [2] (BVLC Institutional in-depth Learning Framework). It has been applied to machine learning systems, as well as to computer science related fields such as computer vision, language recognition, information retrieval, robotics, geographic information extraction, and natural language understanding.

2.2 Deep Learning

Deep learning [3] refers to a collection of algorithms that use various machine learning algorithms to solve various problems such as images and texts on a multi-layer neural network. Deep learning can be classified into neural networks from a large class, but there are many changes in the specific implementation. The core of deep learning is feature learning, which aims to obtain hierarchical feature information through a layered network, thus solving the important problem that requires manual design features. Deep learning is a framework that contains several important algorithms: Neural Networks, AutoEncoder, and Sparse Coding. For different problems (image, voice, text), different network models are needed to achieve better results.

3 Background

The Convolutional Neural Network (CNN) [4] has a wide range of applications in the field of computer vision. It is an algorithmic mathematical model that mimics the behavioral characteristics of animal neural networks and performs distributed parallel information processing. This kind of network relies on the complexity of the system to adjust the relationship between a large number of internal nodes to achieve the purpose of processing information.

The FCN [5] is a network structure that is improved on the basis of CNN. In the classic CNN classification, a softmax (output layer function) is usually connected at the end of the network structure. The FCN (Full Convolutional Neural Network) solves the problem of image segmentation of semantic level graphs. The FCN accepts input images of any size and restores the current image to the same size as the input image.

In 2016, Dai Jifeng [6] et al. proposed the concept of a position sensitive score map based on the network framework of target detection of regional complete convolutional networks. The structure of R-FCN is one of the FCNs. In order to include translation variance into the FCN, the researchers used the output of the FCN to design a set of position sensitive score maps. The location information of the object is included, and the top area is provided with a RoI Pooling layer to process the location information, and then there is no weight layer.

R-FCN can convert a general picture classification network into a network for target detection. It can be achieved 2.5 to 20 times faster than the faster rcnn [7]. It can achieve good image segmentation [8], which is very suitable for road image processing [9] - to quickly detect multiple different targets in the same image.

4 Experiment

4.1 Environmental Preparation

I manually built the environment configuration with Docker [10]. The detailed data is shown in the Table 1.

Table 1. Experimental context

Host configuration	GPU: GTX1080	Video Memory: 4G ~ 12G	Memory: 16G
Operating system version	Ubuntu 16.04 LTS		
Software and corresponding version	tensorflow0.11, python2.7		
Development tools	PyCharm2017/Vim/Docker		
Remote connection	Xshell-5.0, ssh		

4.2 Data Preparation

The data comes from a series of image data from open source community. The data was originally provided for the team to do research.

In this test, there were 10,000 pieces of data as a training set and 2000 pieces of data as a test set. All pictures are real road traffic scene pictures of 640*360 size. Each line in the label file (Label.idl) corresponds to a json string indicating the category and coordinates of the object to be identified in the chart. Category vehicles, bikers, pedestrians, and traffic lights. For example, Fig. 1.

4.3 Experimental Process

TensorFlow provides the TFRecord component to unify different raw data formats, using the storage format provided by TensorFlow, and to manage different attributes more efficiently. The data in the TFRecord file is stored in the format of Protocol Buffer, a tool provided by Google to process structured data.

Fig. 1. Experimental data sample

The code that converts local data to TFRcord format data, def _int64_feature defines the attribute that generates the integer type, and def _bytes_feature defines the attribute that generates the string type. Labels are the correct boundary values for each image in the training data and can be saved as an attribute in TFRcord. The image resolution of the Pixels training data can be used as an attribute in the example. Tostring converts the image matrix into a string for saving. Finally, all data is written to the Example Protocol Buffer data structure.

The method of reading the TFRcord format file. The TFRcord format file is read by creating a reader, and the queue is used to maintain a list of input files. There are two ways to read data from a file. Read only reads a sample at a time, and the read_up_to function reads multiple data at a time. Finally, you can enter data in a multithreaded manner.

Figure 2 shows a classic input data processing flow. When training a neural network, the data is subjected to a pre-processing stage before entering the network, and the data read in the pre-processing stage is from a file in the TFRcord format. There is a multi-threaded processing flow in TensorFlow that speeds up the process with a multi-threaded process. The multi-threaded processing of data in TensorFlow is implemented through queues.

After image data preprocessing, file format conversion, and input data strategy development, the next step is to concentrate entirely on the implementation of the model.

4.4 Core Algorithm Analysis

The R-FCN (ResNet-101) target detection network based on the TensorFlow platform is based on the deep learning classic network framework model in the paper. R-FCN (ResNet-101) basic skeleton is to remove the global average (global average) and classification fc layer in the implementation of ResNet-101 above, and add a 1024d 1*1 conv layer (convolution layer) to reduce Dimensions, then a conv layers of k*k*(C+1) channels are added to generate score maps.

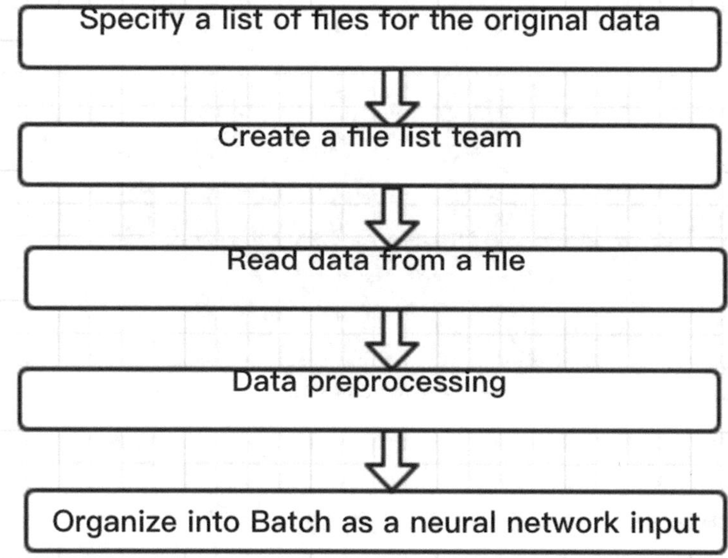

Fig. 2. Data processing flow chart

Similar to the fast rcnn network during training, its ground truth iou in forward propagation is a loss function greater than 0.5 divided into two parts as shown in Formula (1):

$$\mathrm{L}(\mathrm{s}, t_x, \mathrm{y}, \mathrm{w}, \mathrm{h}) = L_{cls}(s_{c^*}) + \lambda[c^* > 0]L_{reg}(t, t^*). \tag{1}$$

The following is a visual representation of the calculation process (Figs. 3 and 4).

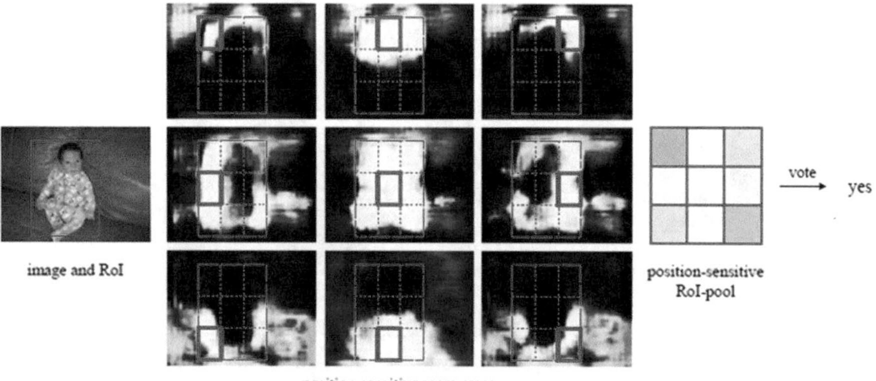

Fig. 3. Visualization of R-FCN (K*K = 3*3) character detection

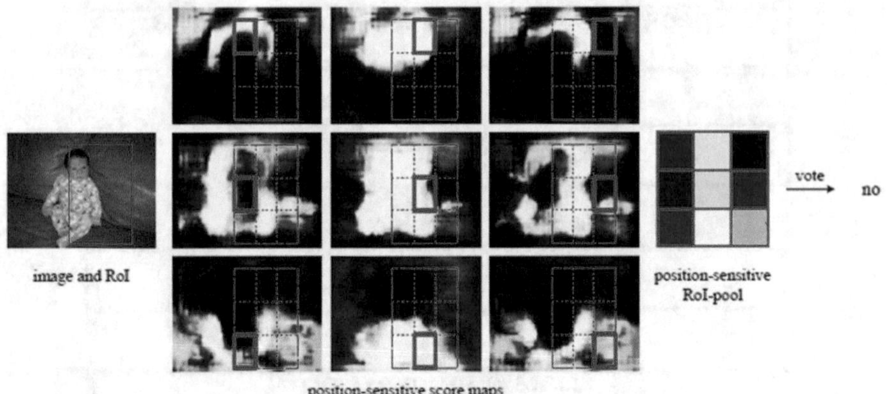

Fig. 4. Visualization When a region of interest does not truly cover the target

5 Experimental Results and Analysis

5.1 Evaluation Method of Experimental Results

The model evaluation method selected in this experiment is shown in Fig. 5, where IoU represents the accurate prediction probability of the model on the test set. The blue area in the formula indicates the predicted bounding box value area and the real bounding box. The intersection of the value regions is calculated according to the value of the blue region; the denominator in the formula is the union of the predicted bounding box value region and the true bounding box value region. The ratio of the numerator to the denominator calculates IoU. For example, with 0.5 as the segmentation, greater than 0.5 is an accurate prediction, and less than 0.5 is a misprediction. The mean precision of mAp (mean average precision) is the probability of describing the IoU accurate prediction test set, which is generally used as the basis for the accuracy of our reference model.

mAP： mean average precision. 平均精度

IoU：

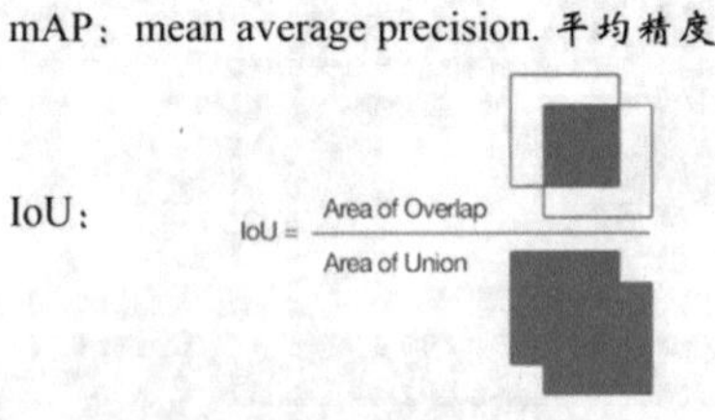

Fig. 5. Calculation formula of MAP

5.2 Experimental Results and Comparison

The three models were tested under the same conditions, and the results were recorded as follows. After comparison, we can see that the effect of the R-FCN model is the most ideal solution (Figs. 6, 7 and 8).

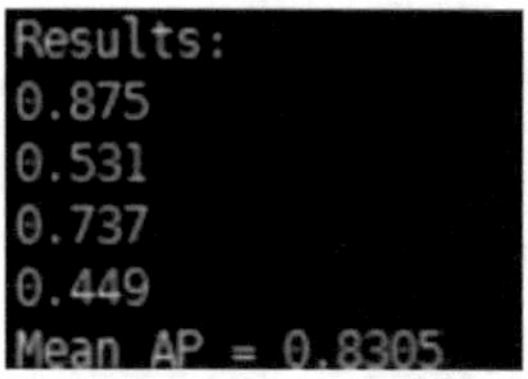

Fig. 6. mAp of fast rcnn (VGG_1024)

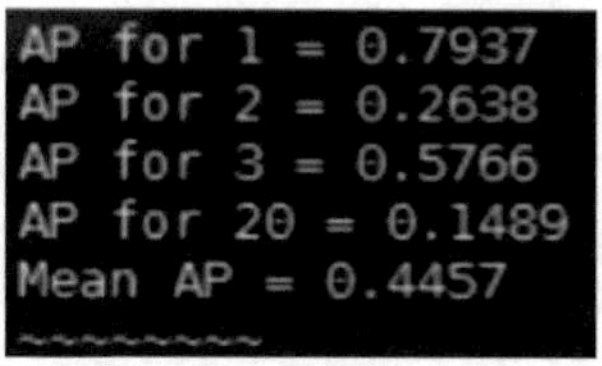

Fig. 7. mAp of faster (VGG_16)

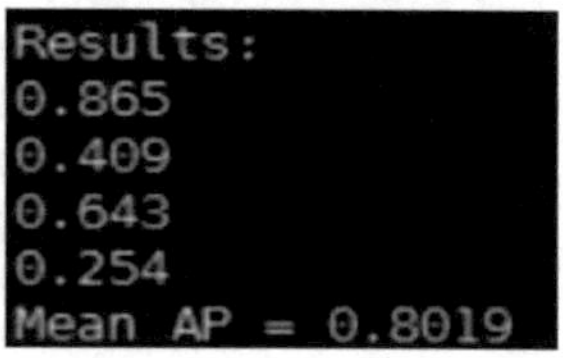

Fig. 8. mAp offfaster (VGG_16)

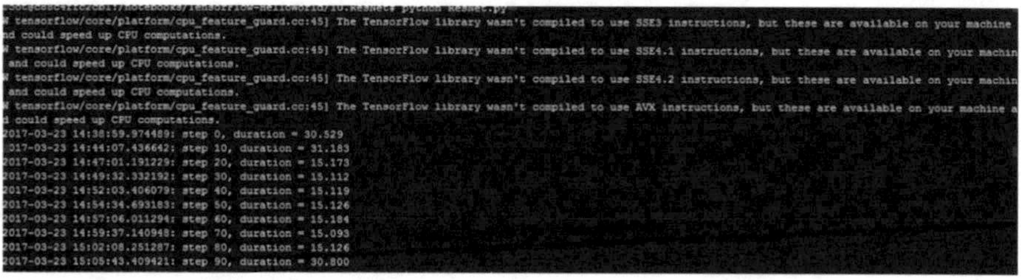

Fig. 9. Time consuming diagram of computation process

5.3 Performance Analysis

As can be seen in Fig. 9, ResNet-101 does a forward propagation, and each picture takes about 0.35 s. Therefore, such time-consuming online use is basically no problem, and it can be processed online in real time. When applied in a specific scenario, it is also necessary to consider the various hardware devices of the car in the entire platform.

6 Conclusion

This paper compares and studies the effect of R-FCN deep learning network on computer vision image processing based on Tensorflow platform. A more efficient and practical solution was proposed and proved. It provides reference value for research in other fields such as autonomous driving, and hopes to effectively reduce research and development costs and development difficulty. After further improvement, the accuracy of the model continues to increase is our next research direction.

Acknowledgement. Thanks to Beimen Shenzhou Special Vehicle Laboratory, School of Computer Science, Beijing Information Science and Technology University, School of Vehicle Engineering, Tsinghua University.

References

1. TensorFlow. https://en.wikipedia.org/wiki/TensorFlow
2. Nan, Y.: Research on Convolutional Neural Networks Based on Caffe Deep Learning Framework. Hebei Normal University, Hebei (2014)
3. Xiaochun, L.: Research on Classification and Recognition of Handwritten Images Based on Deep Learning. Donghua Institute of Measurement and Computing, Nanchang (2016)
4. Bouvrie, J.: Notes on Convolutional Neural Networks (2006)
5. Fully Convolutional Networks for Semantic Segmentation, CVPR 2015 best paper, key word: pixel level, fully supervised, CNN (2015)
6. Jifeng, D.: Object Detection via Region-Based Fully Convolutional Networks. People's Posts and Telecommunications Press (2016)
7. Girshick, R.: Fast R-CNN 2016 (2015)
8. Fu, R.: Research on Image Target Recognition Based on Deep Learning. National University of Defense Technology, Changsha (2014)
9. Guiying, Z.: Review of Research on Autopilot Algorithm Based on Computer Vision. Guizhou University, Guizhou (2016)
10. Docker. https://en.Wikipedia.org/wiki/Docker_(software)

Recent Advances on Information Science and Big Data Analytics

An Overview on Visualization of Ontology Alignment and Ontology Entity

Jie Chen[1,2], Xingsi Xue[1,2,3,4(✉)], Lili Huang[5], and Aihong Ren[6]

[1] College of Information Science and Engineering, Fujian University of Technology, Fuzhou 350118, Fujian, China
jack8375@gmail.com
[2] Intelligent Information Processing Research Center, Fujian University of Technology, Fuzhou 350118, Fujian, China
[3] Fujian Provincial Key Laboratory of Big Data Mining and Applications, Fuzhou 350118, Fujian, China
[4] Fujian Key Lab for Automotive Electronics and Electric Drive, Fujian University of Technology, Fuzhou 350118, Fujian, China
[5] College of Humanities and Law, Fujian Agriculture and Forestry University, Fuzhou 350002, Fujian, China
[6] School of Mathematics and Information Science, Baoji University of Arts and Sciences, Baoji 721013, Shaanxi, China

Abstract. Visualization of ontology alignment and ontology entity is becoming a method to help Solve the problem of semantic heterogeneity. This paper first introduces the basic concepts of ontology and ontology matching, and then illustrates the importance of ontology visualization technique. The existing partial visualization tools are discussed respectively in two aspects of visualization of ontology alignment and visualization entity on ontology. Finally, four researching directions on ontology visualization technique are pointed out for the future work.

Keywords: Ontology alignment · Ontology visualization
Ontology entity

1 Introduction

Ontology is an explicit specification of conceptualization [1]. The specification consists of a generic vocabulary and information structure for a domain. Ontology has been used in many fields to link information semantically in a standardized way. However, different tasks or different viewpoints lead to different conceptualization of ontological designers' interest in the same field. The subjectivity of ontology modeling leads to heterogeneous ontology [2], which is characterized by differences in terms and concepts. Examples of these differences include naming the same concepts with different words, naming different concepts with

P. Krömer et al. (Eds.): ECC 2018, AISC 891, pp. 369–380, 2019.
https://doi.org/10.1007/978-3-030-03766-6_42

the same words, and creating hierarchies for specific domain areas with different levels of detail, and so on. Ontology matching is the basic solution to the problem of ontology heterogeneity [2], which can identify the corresponding relationship between semantic related entities of ontology. Because of the increasing relevance of performing ontology matching, there are many fully automatic and semi-automatic ontology matching technologies. The performance of automated systems (in terms of precision and alignment callbacks) is limited because of the diminishing benefits of more advanced alignment techniques [3,6]. Therefore, the automatic generation of mappings should only be considered as the first step towards final alignment, and validation by one or more users is essential to ensure alignment quality [7]. If the user is required to participate in the ontology matching process, it is usually essential for the user to quickly grasp the content of the ontology. To achieve this, many ontology tools implement visual components.

This paper mainly introduces the Ontology matching visualization technology from the perspectives of the Ontology Alignment and Ontology Entity. The first contribution to this effort is to review the existing visualization tools and make comparisons. Its second contribution is to discuss the challenges in this area and to outline possible useful ways of addressing the identified challenges.

The remainder of the paper is organized as follows The remainder of the paper is organized as follows Sect. 2 presents the basic concepts in Visualization-technique-based ontology matching domain; Sects. 3 and 4 respectively overview visualization technique on ontology matching and visualization technique on ontology Sect. 5 shows four future research directions in Interactive ontology visualization techniques; and finally, Sect. 6 draws the conclusions.

2 Preliminaries

2.1 Ontology

Ontology is "an explicit specification of a conceptualization" [1], which defines a common vocabulary for the knowledge domain. Ontologies support the sharing of information structures, the reuse of domain knowledge, and the clarification of domain assumptions. Ontology provides a shared vocabulary, that is, those existing object types or concepts in a particular domain and their attributes and interrelationships; Ontology, in other words, is a special set of terms that are structured and more suitable for computer systems. In short, ontology is actually a formal representation of a set of concepts and their relations in a specific domain.

As a form of knowledge representation about the real world or one of its components, the current application fields of ontology include (but are not limited to): artificial intelligence, semantic web, software engineering, biomedical informatics, library science, and information architecture.

2.2 Interactive Ontology Matching

Ontology matching can establish semantic relations between the heterogeneous entities of two ontologies, and the alignment obtained is the basis of implementing the ontology inter-operation [4]. When the scale of the ontologies is large, it is impractical to match them manually in terms of both efficiency and effectiveness. Therefore, many ontology matching systems have been developed to match two heterogeneous ontologies without any human intervention. However, the performance of automatic matchers (in terms of precision and recall) is limited because of the bottleneck brought by the similarity measures [6]. This may be due to the complexity of the ontology alignment process and the specificity of each task, which makes no similarity measure distinguish all the semantically identical entities in various context. Therefore, the automatic generation of mappings should only be considered as the first step towards final matching, and validation by one or more users is essential to ensure alignment quality [7]. Interactive ontology matching is designed to enable users and automatic matchers to cooperate with each other within a reasonable time and generate high-quality ontology alignment. User Interface (UI) is one of the key components for implementing an effective interactive ontology matcher.

2.3 Graphical User Interface

User interface is an indispensable part of an interactive system to implement the human-machine interaction [8]. Since ontology is a complex knowledge base, the validation of ontology alignment is a task involving high memory loads. To verify each mapping, the structure and constraints of two ontologies need to be considered by the user, and other mappings and their logical results should also be kept in mind, which is impossible without the support of visual tools like UI. The purpose of ontology visualization is to help a user understand the detailed information inside an ontology. Given the complexity of ontologies and alignments, a critical aspect of visualizing them is not overwhelming the user [9]. People use working memory to understand things, but people's memory is limited. When there is too much information, people will be easily overwhelmed. In order to solve this problem, this limitation can be extended by grouping similar things, which is known as "chunking" [10] and it can be used to help a user promote the cognition process and reduce the memory load. In addition, another important aspect of ontology visualization is to provide a user with enough information to verify the correctness of each mapping [8], including vocabulary and structure information of ontology as well as other related potential mapping.

At present, visualization in interactive ontology matching is mainly divided into two categories: ontology alignment and ontology entity. The visualization ontology alignment is intended to visualize the matchings. The visualization on ontology entity visualizes the relationship between concepts and concepts within the same ontology.

3 Visualization of Ontology Alignment

3.1 Ontology Alignment

An ontology alignment is a set of correspondences between entities belonging to a pair of ontologies $O1$ and $O2$. Given two ontologies, a correspondence is a 5-tuple: $< i_d, e_1, e_2, r, n >$ [2,11], such that:

- i_d is an identifier for the given correspondence;
- e_1 and e_2 are entities, e.g., classes and properties of the first and second ontology, respectively;
- r is the semantic relation between e_1 and e_2 (for example, equivalence, more general, disjointness)
- n is a confidence measure number in the $[0, 1]$ range, which expresses how much the author or algorithm believes in the fact that the relation exists.

3.2 Visual Tools of Ontology Alignment

Visual tools on ontology alignment mainly visualize r and n in the tuple. The user modifies the generated ontology alignment by visualizing relationships and confidence measure number. The existing visualizing tools are introduced in details as follows:

COMA++. COMA++ [12] automatically generates a match between the meta-model and the target schema and draws a line between possible matches, as shown in the Fig. 1. Users can define their own lexical matches by interacting with the schema tree. When the mouse hovers over a potential correspondence line, the matching confidence is displayed, with values between 0 and 1. Each line displays a different color according to its confidence measure number (for example, if the number reaches 1, the color of the line is green; If the number is 0.5, the line is yellow).

COGZ. There is an interactive visualization plug-in called COGZ [13] in the protege. Like COMA++, COGZ uses a visual metaphor for matching communications. Candidate pairs are represented by red dotted lines, while verified pairs are represented by black solid lines [13]. The tool enables incremental search and filtering of source and target ontologies, and generation of corresponding. For example, when a user types a search term for the source ontology, after each click, the ontology tree representation is filtered to display only terms and hierarchies that match the search criteria. Other filtering capabilities allow users to focus on parts of the hierarchy or help hide unwanted information from the display.

COGZ uses prominent propagation to help users understand and navigate matching communications. When a user selects an ontology term, all matches except those related to the selected term are semi-transparent, and the associated

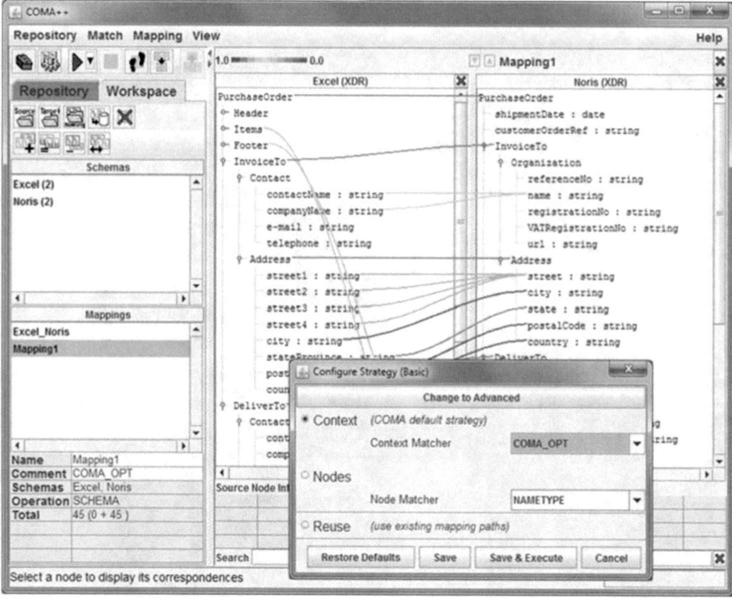

Fig. 1. COMA++ interface

matches are highlighted. To support large ontological navigation, the fish-eye magnification can be used. Fisheye zoom can produce distortion effect of the source tree and goal tree, so the selection of the term will be displayed in a normal font size, and other terms will be according to its relevance to the selected value gradually smaller.

AlViz. Similar to COGZ, AlViz [14] is a plug-in of Protege. AlViz was developed specifically to visualize ontology alignment. It applies multiple views through cluster diagram visualization and synchronous navigation in standard tree controls (see Fig. 2). The tool helps users understand the result of ontology matching by providing an overview of the ontology as a cluster. Cluster represents the abstraction of the original ontology graph. In addition, clusters are colored based on their similarity to the underlying concepts of other ontologies. For example, in Fig. 2, the four views of the tool visualize two ontologies named tourismA and tourismB. The nodes of the graphs and dots next to the list entries represent the similarity of the ontologies by color. The size of the nodes results from the number of clustered concepts. The graphs show that there *is_a* relationship among the concepts. Green indicates similar concepts available in both ontologies, whereas red nodes represent equal concepts [14]. The sliders to the right adjust the level of clustering. Different from the above two visualization tools, AlViz mainly highlights the similarity between concept clusters and aims to help users understand the results of ontology matching from a macro perspective.

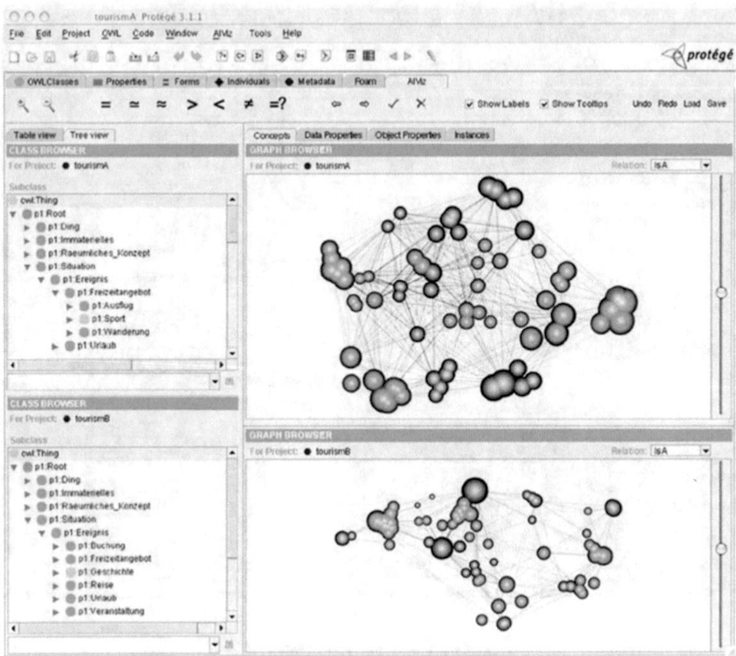

Fig. 2. Screenshot of AlViz plugin while matching two tourism ontologies

For a single alignment, which is represented in the standard tree control, it is not as straightforward as COMA++ and COGZ.

3.3 Discussion

According to the above description of Visual tools ontology alignment tools, they not only have the simple operation function of ontology entity visualization (as mentioned in the Sect. 4), but also have an editing function to help users modify the matching and make the matching more accurate. But there are some drawbacks to these tools. COMA++ has four problems: scalability issues, configuration effort, limited semantics of match mappings, limited accessibility. Although these problems were improved in COMA 3.0 [15], they were not complete enough. COGZ is a plugin for protege, which reduces some of the shortcomings of PROMPT [16] by adding filters to the candidate mapping list. Despite the improvements, there are still too many suggestions for users to choose from. AlViz adopts a multi-view approach, mainly to help users determine the location of most mappings, the type of alignment, the difference between the aligned ontology and the source ontology, and whether the choice of mapping directly or indirectly affects the ontology part that users care about. In other words, AlViz's role is to help users understand the two ontologies from a macro perspective, and to deal with the details. In general, the overall research direction of ontology alignment visualization tools is to visualize the semantic relation and confidence measure number between the target ontology and the source ontology, as well as some modification suggestions of the mapping, and thus to let users modify according to these contents to achieve the purpose of accurate matching. At present, the matching of small and medium-sized ontology has been gradually improved, while the matching of large ontology still has the disadvantages of too much information, messy view and being easy to get lost. Therefore, the field still needs to make significant progress in meeting the limitations of existing methods and meeting ontology matching requirements.

4 Visualization of Ontology Entity

4.1 Ontology Entity

In general, the entities inside an ontology consist of classes, attributes, and relationships that can formally describe a discourse domain. On this basis, an ontology can be defined as a triple $O = \{C, R, is\_a\}$ [17] where:

- $C = \{c_1, c_2, c_3, \cdots, c_n\}$ is the set of classes;
- $R = \{r_1, r_2, r_3, \cdots, r_m\}$ is the set of slots (properties) or binary roles/relations among classes;
- is_a, the inheritance relation.

4.2 Visual Tools on Ontology Entity

Ontology alignment visualization tool mainly visualizes ontology matching results, but cannot visualize the relationship between entity and entity in the same ontology. Therefore, such a problem will arise. The user does not know enough information to satisfy his opinion on correct modification. Therefore, the study of ontology entity visualization is also very important. The existing visualizing tools are introduced in details as follows:

OntoViz. OntoViz [18] is a visual plug-in for Protege. OntoViz uses a simple two-dimensional graphical visualization method to represent classes and relationships of ontologies. With OntoViz, one can visualize the property slots, inheritance, and role relationships for each class that the ontology contains. OntoViz provides configuration actions such as selecting which classes and instances to be included in the visualization, specifying colors for nodes and edges, and scaling. Figure 3 shows an OntoViz visualization example where one can see classes and their inheritance relationships, instances, and property slots. To the left of the view is a configuration panel where the user can choose which elements of a given ontology to display in the visualization.

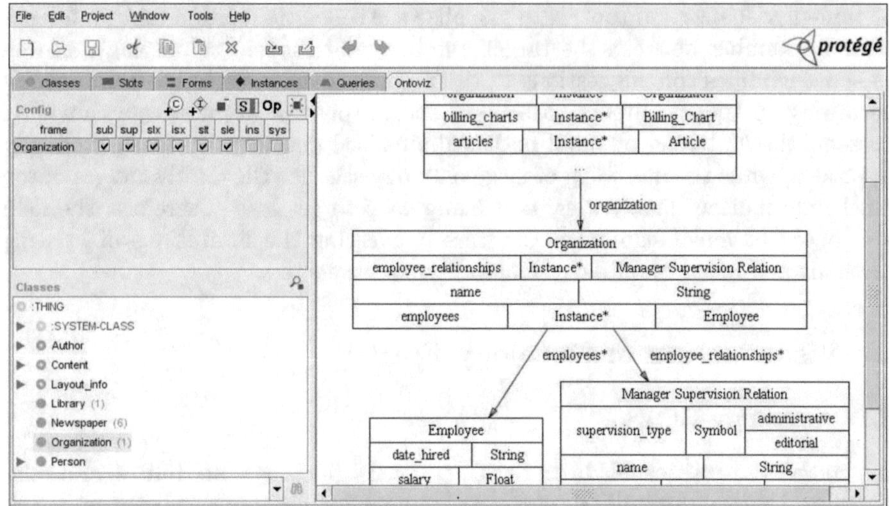

Fig. 3. An example of an OntoViz view

KC-Viz. KC-Viz [5] is an approach to visualizing entities. The importance of supporting tasks related to understanding an ontology's structure or a global model of the ontology is emphasized. In their work, an ontology is first pre-processed through a summarization algorithm to determine the nodes of most importance. This network of key concepts is then displayed in a node-link graph representation as shown in Fig. 4. This starting point of key concepts then support a middle-out approach to exploration. KC-Viz supports interactions such as zooming and history keeping to support the exploration tasks. In addition, each subtree that is hidden is indicated in a green arrow. Further, the sizes of a hidden subtree are displayed in brackets with two numbers, one indicating the number of immediate children, and the other indicating the number of total children. Although the authors have not yet conducted an evaluation on KC-Viz, it may lead to similar issues pertaining to node-link representations; that is, a very limited number of nodes can be displayed on the screen before the context or overview is lost through clutter or occlusion.

OntoViewer. OntoViewer [5] uses multiple and coordinated views and automatically analyzes the concepts, relationships, properties, and instances of the ontology. Figure 5 shows the OntoViewer overview, which has a well-defined region, which combines the visual and interaction patterns of information visualization technology (scaling, translation, selection, links, filtering, and rearranging). OntoViewer view is in the upper left corner to visual class hierarchy view, using a 2D hyperbolic tree, a focus + context technology, designed to reduce cognitive overload and users get lost in the process of interaction. This technique allows one to drag and drop classes, dynamically change the displayed hierarchy, and select. The view in the lower left corner shows the treeview, a simple and intuitive technique for listing the main aspects of an ontology's classes, relationships, and properties. The intermediate view represents another focus +

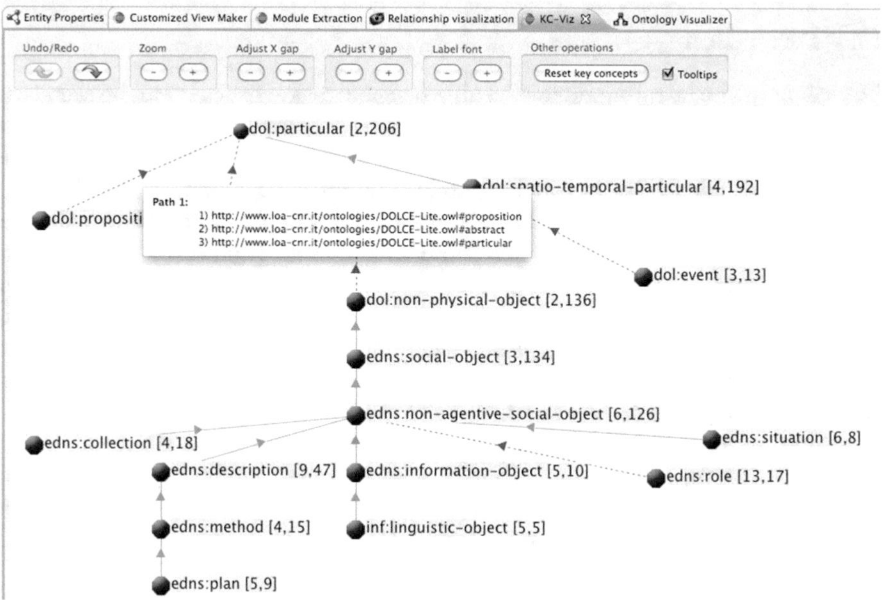

Fig. 4. KC-Viz ontology visualization

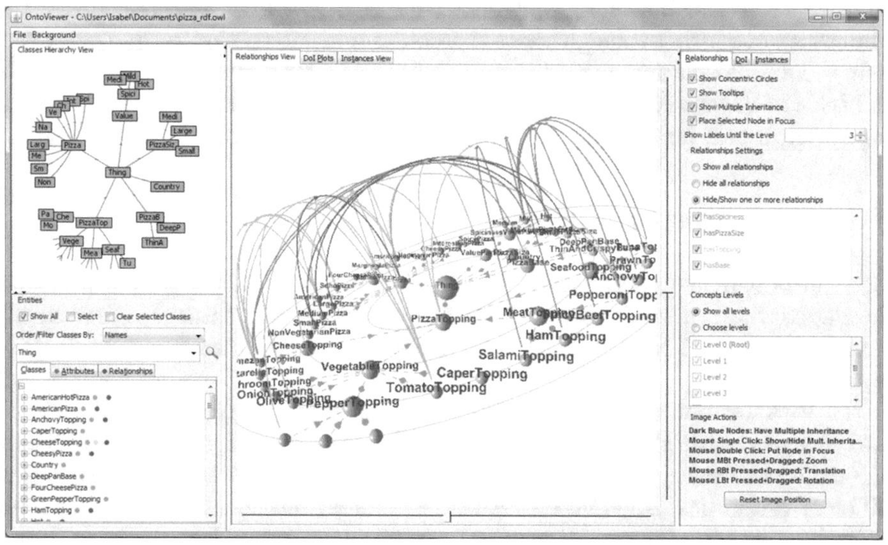

Fig. 5. OntoViewer overview

context visualization and hyperbolic tree: an extension of 2.5D radial tree, the class hierarchy in the XZ plane, according to the space (x, y plane), the relationship between the classes is colorful curve, avoid the overlap line. Interaction is achieved by scaling, translation, rotation, and selection.

4.3 Discussion

According to the introduction of the above visualization tools, it can be seen that ontology is visualized by using nodes to represent their entities and arcs to represent the relationships between entities. The operation functions of visual tools are mainly divided into: indent list, node link and tree, zoom, space fill, focus + context or distortion, and 3D information landscape. Different visual tools choose different operating functions and combine them organically, so that users can understand the information without getting lost in the ontology. There are some common problems with the four visualizations above. OntoViz is not suitable for visualizing large ontologies because the scope of visualization is limited to a few hundred entities. In addition, OntoViz does not allow one to browse multiple relationships. Although the authors of the KC - Viz have not yet conducted an evaluation on the KC - Viz, it may contribute to similar issues pertaining to the node - link representations; That is, a very limited number of nodes can be displayed on the screen before the context or overview is lost through clutter or occlusion. OntoViewer lacks search capabilities, and it can be difficult for users to find a concept.

In addition, when a large ontology are visualized, these visualization tools will be interfered and blocked due to the complexity of ontology structure. In addition, when zooming in, many visualizations are lost to the whole or the direction. In order to reduce information overload, many interaction and distortion techniques are studied. Although these distortion and interaction technologies can help reduce information overload on a smaller scale, they cannot fully scale to the full scale of the ontology (tens of thousands of nodes). While the OntoViewer has improved, it still feels a bit messy, especially when visualizing large ontologies.

5 Future Research Direction

Now, although many interactive ontology visualization techniques have been developed, there are still many defects. In this section, four future research directions are proposed for the defects of visualization tools introduced in Sects. 3 and 4.

A. Visualization of large ontologies. For ontology visualization tool, how to effectively visualize the information needed by users in a large ontology is one of the main challenges. A large ontology has hundreds of entities. Each entity has different properties. Different entities have different relationships. It can be seen that the amount of information of a large ontology is particularly large. Even if the ontology is visualized, users will feel pressured by too much information. Therefore, how to accurately visualize the information users need while hiding the information they don't need is very important. Multi-view visualization may solve this problem.

B. View layout. One of the main challenges for ontology visualization tools is how to browse the information presented in an orderly and clear manner. Most users don't seem to like messy and overly cluttered views, preferring visualizations that provide the possibility of browsing the information presented in an orderly and clear manner, even if in some cases it requires attention to specific parts of an ontology or hierarchy. This fact suggests that visualization should also leverage the semantic context of information, even user profiles, to guide and support hierarchical or ontological exploration.

C. Query and positioning function. Visualization should be combined with effective search tools or query mechanisms. For tasks related to locating a particular class or instance, browsing is not enough, especially for large ontologies.

D. Multi-user collaboration. When users are involved in editing ontology alignment, individual user suggestions may deviate, and the idea of multiple people online may be adopted to minimize human errors.

6 Conclusion

Visualization of ontology alignment and ontology entity is becoming a method to help solve the problem of semantic heterogeneity. This paper first introduces the basic concepts of ontology and ontology matching, and then illustrates the importance of ontology visualization technique. The existing partial visualization tools are discussed respectively from two aspects of visualization of ontology alignment and visualization entity on ontology. Finally, four researching directions on ontology visualization technique are pointed out for the future work.

Acknowledgments. This work is supported by the National Natural Science Foundation of China (No. 61503082), Natural Science Foundation of Fujian Province (No. 2016J05145), Scientific Research Foundation of Fujian University of Technology (Nos. GY-Z17162 and GY-Z15007) and Fujian Province Outstanding Young Scientific Researcher Training Project (No. GY-Z160149).

References

1. Gruber, T.R.: A translation approach to portable ontologies. Knowl. Acquis. **5**(2), 199–220 (1993)
2. Shvaiko, P., Euzenat, J.: Ontology matching: state of the art and future challenges. IEEE Trans. Knowl. Data Eng. **25**(1), 158–176 (2013)
3. Noy, N.: Algorithm and tool for automated ontology merging and alignment. In: Proceedings of AAAI, pp. 450–455 (2000)
4. Silva, I., Freitas, C.D.S., Santucci, G., Dal, C.M., et al.: Ontology Visualization: One Size Does Not Fit All (2012)
5. Silva, I.C.S.D., Freitas, C.M.D.S, Santucci,G.: An integrated approach for evaluating the visualization of intensional and extensional levels of ontologies. In: Beliv Workshop: Beyond Time and Errors - Novel Evaluation Methods for Visualization, pp. 470–486. ACM (2010)

6. Granitzer, M., Sabol, V., Onn, K.W.: Ontology alignment a survey with focus on visually supported semi-automatic techniques. Futur. Internet **2**(3), 238–258 (2010)
7. Meilicke, C., Shvaiko, P., Shvaiko, P.: Ontology alignment evaluation initiative: six years of experience. J. Data Semant. XV **6720**, 158–192 (2011)
8. Dragisic, Z., Ivanova, V., Lambrix, P.: User validation in ontology alignment. In: The Semantic Web-ISWC 2016. Springer, Cham (2016)
9. Shvaiko, P., Giunchiglia, F., Silva, P.P.D.: Web explanations for semantic heterogeneity discovery. Lect. Notes Comput. Sci. **3532**, 303–317 (2005)
10. Noy, N.F., Mortensen, J., Musen, M.A.: Mechanical Turk as an Ontology Engineer? Using Microtasks as a Component of an Ontology-engineering Workflow (2013)
11. Euzenat, J., Shvaiko, P.: Ontology Matching. Springer, New York (2007)
12. Aumueller, D., Do, H.H., Massmann, S.: Schema and ontology matching with COMA++. In: Proceedings of ACM SIGMOD International Conference on Management of Data 2005, pp. 906–908. ACM (2005)
13. Falconer, S.M., Noy, N.F., Storey, M.A.D.: Towards understanding the needs of cognitive support for ontology mapping. In: International Workshop on Ontology Matching, DBLP (2006)
14. Lanzenberger, M., Sampson, J.: AlViz - a tool for visual ontology alignment. In: Tenth International Conference on Information Visualization 2006, pp. 430–440. IEEE (2006)
15. Massmann, S., Raunich, S., Arnold, P.: Evolution of the COMA match system. In: International Conference on Ontology Matching 2011, pp. 49–60 (2011). CEUR-WS.org
16. Noy, N.F., Musen, M.A.: The PROMPT suite: interactive tools for ontology merging and mapping. Int. J. Hum.-Comput. Stud. **59**(6), 983–1024 (2003)
17. Amann, B., Fundulaki, I.: Integrating ontologies and thesauri to build RDF schemas. In: European Conference on Research and Advanced Technology for Digital Libraries 1999, pp. 234–253. Springer, Heidelberg (1999)
18. Singh, G., Prabhakar, T.V., Chatterjee, J.: OntoViz: visualizing ontologies and thesauri using layout algorithms. In: AFITA 2006: The Fifth International Conference of the Asian Federation for Information Technology in Agriculture, J.N. Tata Auditorium, Indian Institute of Science Campus, Bangalore, India, 9–11 November (2006)

An Improved Method of Cache Prefetching for Small Files in Ceph System

Ya Fan[1], Yong Wang[1], Miao Ye[2,3(✉)], Xiaoxia Lu[2], and YiMing Huan[2]

[1] School of Computer Science and Information Security, Guilin University of Electronic Technology, Guilin, Guangxi 541004, China

[2] School of Information and Communication, Guilin University of Electronic Technology, Guilin, Guangxi 541004, China

ym@mail.xidian.edu.cn

[3] Guangxi Colleges and Universities Key Laboratory of Cloud Computing and Complex Systems, Guilin University of Electronic Technology, Guilin, Guangxi 541004, China

Abstract. To improve the inefficiency of file access for massive small files in the distributed file system, prefetching the small files into the cache is the most conventional method to be adopted. As the number of cached files is proportional to time, there will be many redundant small files which occupy the cache space and haven't been read for a long time, the hit ratio will be decreased in this situation. To solve this defect, we proposed an improved LRU-W algorithm based on the file read times and the time interval of file read and designed a L2 cache optimization mechanism which can search file in the linked list with higher priority firstly and remove the files with lighter weight factor from the linked list with lower priority dynamically. The experiments and its result show that when prefetching massive small files, the proposed method in this paper can increase the hit ratio of cached files and improve the overall performance of Ceph file system.

Keywords: Massive small files · Ceph · LRU-W · L2 cache

1 Introduction

With the rapid development of cloud computing and big data, the global data is increasing exponentially, especially in the fields of e-commerce, social network and scientific computing, where more and more small files are produced. The high equipment cost and maintenance cost of traditional storage system make it difficult to meet the storage capacities and the file read requirements for massive small files. To overcome the lack of storage capacity, the distributed file system has been developed rapidly. But a distributed file system will always get performance bottlenecks, when faced with lots of read operations of massive small files. For now, most distributed file systems are built base on Linux and Unix systems, an operating system like them usually takes three times of disk I/O to read out a file. Due to the large amount of massive small files and the limited memory capacity of the hard device, it's hard to read

P. Krömer et al. (Eds.): ECC 2018, AISC 891, pp. 381–389, 2019.
https://doi.org/10.1007/978-3-030-03766-6_43

all the index node information into memory for one time, so we can't get all the file's location with one disk I/O. To improve the efficiency of massive file read operation in Ceph file system and the limitations of traditional LRU algorithm, an improved LRU_W cache algorithm has been proposed in this paper. This algorithm adds weight factor and visit times on the basis of LRU algorithm, and we also designed a L2 cache to increase the cache hit ratio. When a small file hasn't been read not for a long time, its weight will decline, if the weight is lower than the threshold, it will be removed from the cache, if the weight is higher than the priority threshold, it will be stored in the cache list with higher priority. The method in this paper can not only reduce the wasted capacity, but also ensure the small files with higher priority will be searched for firstly, which can improve the hit ratio in the cache.

2 Relative Work

For now, the most common cache algorithms are the LRU (Least Recently Used) algorithm [1], LFU [2] (Least Frequently Used) [3] algorithm FIFO (First In First Out) algorithm and the MRU [4] (Most Recently Used) algorithm. These cache algorithms are mainly implemented based on access time or access frequency, they all have lower complexity and can be implemented easily.

Perkowitz [5] et al. propose an improved LRU-Threshold cache algorithm to improve the problem about the time localization in the traditional LRU algorithm. Ding et al. [6] give an improved LFU cache algorithm to optimize the cache pollution problem in LFU cache algorithm. Niu [7] and others design an algorithm which include a strategy of data cache based on affecting factors and a method that removes files through the correlation of metadata, which can reshape the metadata and remove the data with lower affection factors dynamically to enhance the cache hit ratio. These improved algorithms described above can improve the cache hit ratio, but they just only focus on the access history of cached objects.

In order to improve the cache performance furtherly, Huang [8] and others add a cache predict module on the basis of cache replacement algorithm, which can infer that which file might be read next time by analyzing the history access information of the cached object, this strategy can improve the overall performance of the system by reducing the file access time in this system. When the cache capacity is insufficient, the cached file will be removed according to the original cache algorithm. Different cache strategies have different prediction methods. The most ordinary strategies of cache prefetching are as follows: prefetching strategy based on user access probability [9], prefetching strategy based on data mining [10] and prefetching strategy based on neural network [11].

3 Research Method

3.1 Method to Improve LRU Algorithm

Cache prefetching is to read out some files into cache based on its correlation before the user make a file read request, so that we can effectively reduce the response time to get target files. In this dissertation, the LRU algorithm is adopted to implement the cache prefetching function for small files. However, it shows in the experiment, when caching massive small files with the original LRU algorithm, the number of the cached file will increase with the quantity of the file read request, so there will be many small files that haven't been read for a long time but take up the cache space which will lead to the reduction of efficiency. Thus, in this paper, an improved LRU_W algorithm based on file access time and the time interval of file access is presented. When the file has been written to the cache, the algorithm will give it an initial weight factor Rw1, the value of weight factor will decay with the increase of the time interval but go up with the rising number of file read times. When the cached file has been read again, we need to recalculate the file's weight factor. the weight factor of the cache file is recalculated through formula (1).

$$Rw = Rw1 * e^{-(Nt-Nr)*t} + 1 \tag{1}$$

Where RW1 is the value of the last weight factor, Nt is the length of the LRU list, Nr is the number of file access, t is the time interval of file read, and 1 is the compensation factor. When the weight factor is less than the threshold, the file will be removed from the cache, so that we can increase the utilization of cache by removing some timeworn files which haven't been accessed for a long time.

3.2 L2 Cache Structure

We have designed a two-level cache strategy to optimize LRU_W algorithm, in this strategy, cached files will be given with different priorities according to the different value of weight factor, and the priority is proportional to the value of weight factor, the detailed implementation process is displayed in Fig. 1.

As the figure above shows, Q and Q1 respectively represent the primary cache and the secondary cache of the L2 cache structure, where the file priority in the secondary

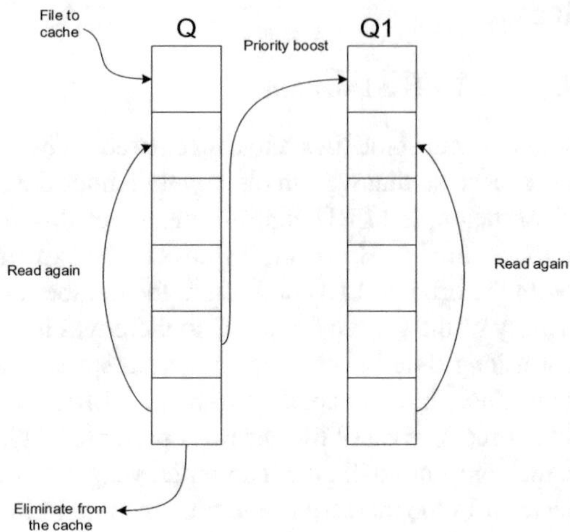

Fig. 1. L2 cache structure

cache is higher than that in the primary cache. The main implementation process of this caching strategy is as follows:

1. Put the recently added cached file into the first level of cache Q, record the arriving time and the access times, and also give this file an initial weight factor.
2. The value of weight factor is changing dynamically, if the weight factor of the file is bigger than the given threshold, the file will be put into the secondary cache Q1 which has the higher priority. Whenever user send a file access to the system, the fill will be first looked up in Q1, if it does not exist in Q1, then we look it up in Q, which ensures the files with higher weight factor could be looked up first.
3. The number of the cached file will go up with time, when the cache capacity is insufficient, the files with low weight factor will be removed from the lower-priority cache Q.
4. Due to the value of weight factor of cached file will Decay with time, when the value of the weight factor is lower than the threshold, the file will be eliminated from the cache, to avoid the waste of cache capacity.

The pseudocode of the L2 cache based on the improved LRU_W algorithm is shown in Table 1.

Table 1. Code for L2 cache strategy

Input: file access requests

Output : object files

```
cache(files) {
  Q<small file to cache> / / first-level cache
  Q1<small file to cache> / / secondary-level cache
  File access request arrived
  if file in Q1:
    Nr=Nr+1
    Rw=Rw1*e- (Nt-Nr)*t +1  / /the value of weight factor
when this file has been accessed again
    Return the target file
  elif file in Q:
    Nr=Nr+1
    Rw=Rw1*e- (Nt-Nr)*t+1
    Return the target file
  else:
    Read target file from the Ceph cluster
    Rw1=2 //initial value of weight factor
    Nr=1
  end if
  Rw=Rw1*e- (Nt-Nr)*t  / / After a certain time interval t,
 the value of weight factor
  if Rw>R:
    move file to Q1
  else:
    remove file from Q
  end if
 }
```

4 Results and Discussion

4.1 Experimental Environment

In order to test the hit ratio of the improved LRU_W cache algorithm and the efficiency of L2 cache strategy on the reading performance of the massive small files in Ceph file system, the following test environment was built: The Ceph cluster uses 5 machines, 1 monitoring node, 1 client node and 3 storage nodes, each storage node contains 2 OSD nodes, each file has three replicas, and the size of Ceph file block is set to 4 M. The hardware and software test environment are shown in Table 2.

Table 2. Software & hardware environment

Software & hardware information	
CPU	Intel(R) I3-3230 v2 @2.60 GHz
Memory	16 GB
operating system	Ubuntu 14.04
Ceph version	9.2.1
file system on OSD	XFS

4.2 Experimental Results and Analysis

In this paper, LRU_W algorithm is applied to the caching mechanism of the system. The test of cache optimization mechanism is mainly divided into three parts: relative hit ratio, absolute hit ratio and average read time.

(1) **The Relative Hit Ratio of LRU_W Algorithm**

We use the ratio of the number of the local cached files to the number of the stored files to test the relative hit ratio of LFU, LRU and LRU_W. The ratio of cached file is set to 10%, 20%, 30%, 40%, 30%, 55%, 60%, 65%, the test results are shown in Fig. 2.

From Fig. 2, we can see that the relative hit ratio of these three algorithms is proportional to the ratio of cached file, the LRU-W algorithm we have proposed apparently works well than the two other algorithms at a lower ratio of cached file, and it always works better than the other two algorithms.

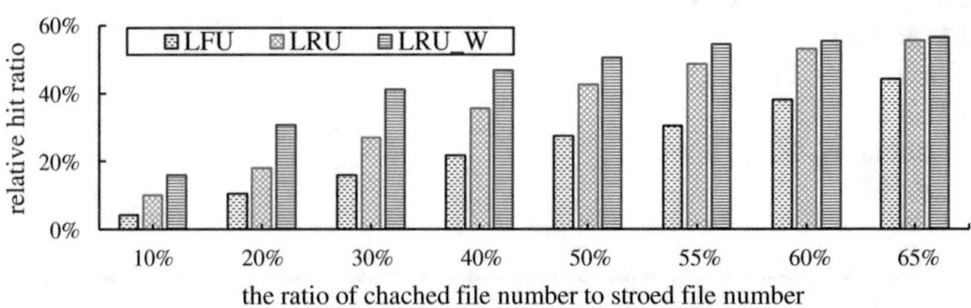

Fig. 2. The relative hit ratio of LRU_W, LRU, LFU

(2) **The Absolute Hit Ratio of LRU-W Algorithm**

We use the ratio of the number of the local cached files to the number of the stored files to test the absolute hit ratio of three different cache algorithms. The ratios are set to 10%, 20%, 30%, 40%, 30%, 55%, 60%, 65%, and the absolute hit ratios tested of LFU, LRU and LRU_W are shown in Fig. 3.

From Fig. 3, we can see that the absolute hit ratio of these three algorithms is rising with the increase of the cached file ratio, when the ratio is less than 40%, the absolute hit ratio of LRU_W algorithm is superior to the other two algorithms. When the ratio reaches 55%, the absolute hit ratio of these three cache algorithms is almost the same.

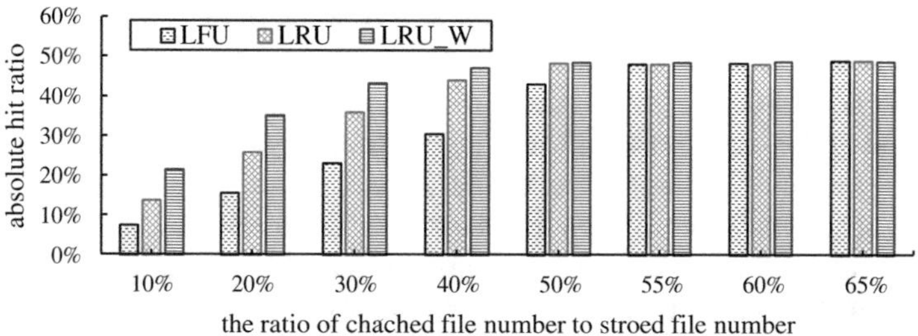

Fig. 3. The absolute hit ratio of LRU_W, LRU, LFU

(3) **System File Read Performance Test**

The main test in this module is to compare the search efficiency of small files while using cache optimization mechanism and not using cache optimization mechanism We have tested the average reading time of 1000, 3000, 6000 and 10000 small files respectively. Where, the file number of the caching mechanism is set to 2000, and the average time of small files access is shown in Fig. 4.

On Fig. 4, while a system without cache optimization mechanism, the average time of small files access increases greatly with the increasing number of read operations. The average file read time in the system with cache optimization mechanism is apparently shorter than the one in original storage system while reading the same number of small files.

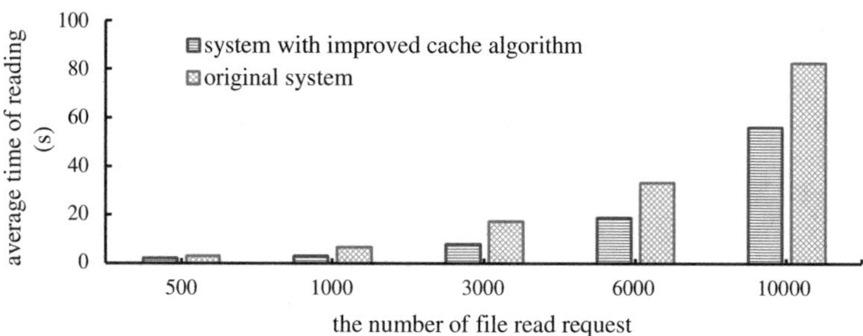

Fig. 4. The average time of reading in two patterns

5 Conclusions and Future Work

5.1 Conclusions

This paper presented an improved cache algorithm LRU-W on the basis of the LRU algorithm, which can improve the relative hit ratio and the absolute hit ratio in the cache, This algorithm is interval factor of file access and the times of file read factor are added to this algorithm and we also proposed a L2 cache strategy, which can ensure the file most likely to be read next time can be cached in a linked list with higher priority, so that it can be looked up firstly in a shorter linked list to reduce the hit time, on the other hand we have preprocessed the massive small files to combine these files with tight connection to each other, in order to read the files with high correlation to the target file into the cache. If the cache capacity is full, those files with lower priority will be eliminated from the cache to make sure there are enough space for the file with a high value of weight factor to store in the cache. The experimental results indicate that the LRU-W algorithm has a better performance than LRU algorithm and LRF algorithm when in face of massive file access of small files.

5.2 Future Work

Due to the weight factor is calculated dynamically and the data of cached file will be rewrite in real time, the device carries the cache system will always with high CPU load when faced with frequent file read requests, which will cause performance bottlenecks to the system, thus it's significant to optimize the elimination algorithm in the cache in the future work. What's more, the file preprocessing operation needs to take a while to calculate the relevance of small files on the basis of the historical record of access times in journal, which are about to store in Ceph file system, So that there is still work to be done in order to find a better way to combine the files which is highly correlated to each other.

Acknowledgement. This work is partly supported by the National Natural Science Foundation of China (Nos. 61662018, 61861013), the Project of Science and Technology of Guangxi (No. 1598019-2), Foundation of Guilin University of Technology (No. GUTQDJJ20172000019).

References

1. Beckmann, N., Sanchez, D.: Modeling cache performance beyond LRU. In: IEEE International Symposium on High PERFORMANCE Computer Architecture, pp. 225–236. IEEE Computer Society (2016)
2. Jaleel, A., Najafabadi, H.H., Subramaniam, S., et al.: CRUISE: cache replacement and utility-aware scheduling. ACM SIGARCH Comput. Arch. News **40**(1), 249–260 (2012)
3. Jung, D.Y., Lee, Y.S.: Cache replacement policy based on dynamic counter method. Adv. Sci. Lett. **19**(5), 1530–1534 (2012)
4. Jiang, B., Nain, P., Towsley, D.: LRU cache under stationary requests. In: ACM Sigmetrics Performance Evaluation Review, vol. 45(2) (2017)
5. Perkowitz, S.: A survey of Web cache replacement strategies. ACM Comput. Surv. **35**(4), 374–398 (2003)
6. Ding, J., Wang, Y., Wang, S., et al.: Design and implementation of high efficiency acquisition mechanism for broadcast audio material. In: IEEE/ACIS, International Conference on Computer and Information Science, pp. 667–670. IEEE (2017)
7. Niu, D.J., Cai, T., Zhan, Y.Z., et al.: Metadata caching subsystem for cloud storage. Appl. Mech. Mater. **214**, 584–590 (2012)
8. Huang, X.Y., Zhong, Y.Q.: Web cache replacement algorithm based on multi-markov chains prediction model. Microelectron. Comput. **5**, 123–125 (2014)
9. Wang, W.: Research on Web Cache and Prefetching Model Based on Access Path Mining. Southwest Jiaotong University (2014)

10. García, R., Verdú, E., Regueras, L.M., et al.: A neural network based intelligent system for tile prefetching in web map services. Expert. Syst. Appl. Int. J. **40**(10), 4096–4105 (2013)
11. Jing, C., Wang, M., An, P.C., et al.: 3D model prefetching system based on neural network. Comput. Appl. Softw. **32**(7), 182–185 (2015)

Solving Interval Bilevel Programming Based on Generalized Possibility Degree Formula

Aihong Ren[1(✉)] and Xingsi Xue[2,3]

[1] School of Mathematics and Information Science, Baoji University of Arts and Sciences, Baoji 721013, Shaanxi, China
raih2003@hotmail.com

[2] College of Information Science and Engineering, Fujian University of Technology, Fuzhou 350118, Fujian, China

[3] Intelligent Information Processing Research Center, Fujian University of Technology, Fuzhou 350118, Fujian, China

Abstract. This study proposes a method for dealing with interval bilevel programming. The generalized possibility degree formula is utilized to cope with interval inequality constraints involved in interval bilevel programming. Then several types of equivalent bilevel programming models for interval bilevel programming can be established according to several typical possibility degree formulas which are corresponding to different risk attitudes of decision makers. Finally, a computational example is provided to illustrate the proposed method.

Keywords: Interval number · Interval bilevel programming Generalized possibility degree formula

1 Introduction

Bilevel programming problem plays a significant role for its many successful applications such as supply chain planning, management, engineering design, etc. In real world situations, sometimes the parameters of the problems cannot be determined in a precise manner owing to various uncertain factors. In order to treat such situations, some studies utilize interval numbers to represent the uncertain parameters and develop the so-called interval bilevel programming. From the point of view of computational efficiency, interval bilevel programming is a simple and flexible tool for modelling uncertain bilevel programming compared with fuzzy or stochastic bilevel programming since only the lower and upper bounds of interval parameters are required.

In recent years, different solution approaches have been developed in the literature to tackle interval bilevel programming. For bilevel linear programming with interval coefficients, Abass [1] converted the problem into a deterministic bilevel optimization problem on the basis of the order relation as well as the

P. Krömer et al. (Eds.): ECC 2018, AISC 891, pp. 390–396, 2019.
https://doi.org/10.1007/978-3-030-03766-6_44

possibility degree of interval, and then employed the Kth best approach to solve the final model. Calvete and Galé [2] investigated bilevel programming with interval-valued objective functions and suggested two enumerative algorithms to work out the optimal value range. Subsequently, Nehi and Hamidi [3] extended the KBB algorithm in [2] and the RKBW algorithm proposed in this study to handle the general interval bilevel programming. In addition, both kinds of cutting plane methods were designed for interval bilevel programming in [4]. Ren et al. [5] combined normal variation of interval number, chance-constrained programming with the preference-based index to transform interval bilevel programming into a preference-based deterministic bilevel programming problem solved by estimation of distribution algorithm. Ren and Wang [6] proposed a method based on reliability-based possibility degree of interval to deal with the general interval bilevel programming.

In this study, we employ the generalized possibility degree formula proposed by Liu et al. [7] to deal with interval inequality constraints included in interval bilevel programming. In view of several typical possibility degree formulas, several equivalent bilevel programming models for interval bilevel programming can be built by considering decision makers' different attitudes. Finally, an illustrative example is given to show the applicability of the proposed method.

The remainder of this paper is organized as follows. Section 2 reviews some definitions used in this paper. In Sect. 3, by considering various risk attitudes of decision makers, we build several kinds of equivalent bilevel programming models by using several alternative possibility degree formulas. In Sect. 4, we employ a numerical example to illustrate the proposed method. Section 5 gives the conclusions.

2 Preliminaries

In this paper, an interval number $\hat{a}$ is denoted as $[\underline{a}, \overline{a}]$, where $\underline{a} \leq \overline{a}$. The center and the radius of interval $\hat{a}$ are defined as $m(\hat{a}) = \frac{\underline{a}+\overline{a}}{2}$ and $w(\hat{a}) = \frac{\overline{a}-\underline{a}}{2}$.

For any two intervals $\hat{a} = [\underline{a}, \overline{a}]$ and $\hat{b} = [\underline{b}, \overline{b}]$, the sum and product operations can be given as follows:

1. $\hat{a} + \hat{b} = [\underline{a} + \underline{b}, \overline{a} + \overline{b}]$;
2. $k\hat{a} = \begin{cases} [k\underline{a}, k\overline{a}],\ k \geq 0 \\ [k\overline{a}, k\underline{a}],\ k < 0 \end{cases}$.

Next, we recall a generalized possibility degree formula developed by Liu et al. [7].

Definition 1 [7]. *Let $\hat{a} = [\underline{a}, \overline{a}]$ and $\hat{b} = [\underline{b}, \overline{b}]$ with $\underline{a}, \underline{b} \geq 0$ be two interval numbers, then the possibility degree of $\hat{a} \geq \hat{b}$ is defined as:*

1. *If $\hat{a} \cap \hat{b} = \emptyset$, when $\overline{a} \leq \underline{b}$, then $P(\hat{a} \geq \hat{b}) = 0$; and when $\underline{a} \geq \overline{b}$, then $P(\hat{a} \geq \hat{b}) = 1$.*

2. If $\hat{a} \cap \hat{b} \neq \emptyset$, then

$$P(\hat{a} \geq \hat{b}) = \frac{\int_{\underline{b}}^{\overline{a}} f(x)dx}{\int_{\underline{b}}^{\overline{a}} f(x)dx + \int_{\underline{a}}^{\overline{b}} f(x)dx} \tag{1}$$

where the function $f(x)$ which is defined in $(0, +\infty)$ is the attitude function.

In the case of $\hat{a} \cap \hat{b} \neq \emptyset$, several representative possibility degree formulas are given through selecting distinct attitude functions corresponding to decision makers' various risk attitudes in [7]:

Case 1. If the decision maker is neutral, the function $f(x)$ should be taken as a constant. Taking $f(x) = c$ for example, by formula (1), one has

$$P(\hat{a} \geq \hat{b}) = \frac{\overline{a} - \underline{b}}{\overline{a} - \underline{a} + \overline{b} - \underline{b}}. \tag{2}$$

Case 2. If the decision maker becomes pessimistic, the function $f(x)$ may be considered to be monotonically decreasing. Taking $f(x) = \frac{1}{x}$ for example, one has

$$P(\hat{a} \geq \hat{b}) = \frac{ln\overline{a} - ln\underline{b}}{ln\overline{a} - ln\underline{a} + ln\overline{b} - ln\underline{b}}. \tag{3}$$

Case 3. If the decision maker becomes optimistic, the function $f(x)$ may be taken into account to be monotonically increasing. Taking $f(x) = \sqrt{x}$ for example, one has

$$P(\hat{a} \geq \hat{b}) = \frac{\overline{a}^{\frac{3}{2}} - \underline{b}^{\frac{3}{2}}}{\overline{a}^{\frac{3}{2}} - \underline{a}^{\frac{3}{2}} + \overline{b}^{\frac{3}{2}} - \underline{b}^{\frac{3}{2}}}. \tag{4}$$

3 Solution Method

Let $x \in R^p$ be the upper level variables and $y \in R^q$ be the lower level variables. Consider the following interval bilevel linear programming problem:

$$\begin{cases} \min\limits_{x} & [\underline{c}_1, \overline{c}_1]x + [\underline{d}_1, \overline{d}_1]y \\ & \text{where } y \text{ solves} \\ \min\limits_{y} & [\underline{c}_2, \overline{c}_2]x + [\underline{d}_2, \overline{d}_2]y \\ \text{s. t.} & [\underline{a}_s, \overline{a}_s]x + [\underline{b}_s, \overline{b}_s]y \geq [\underline{e}_s, \overline{e}_s], s = 1, 2, \cdots, m, \\ & a_t x + b_t y \geq e_t, t = m+1, m+2, \cdots, n, \\ & x \geq 0, y \geq 0, \end{cases} \tag{5}$$

where $[\underline{c}_l, \overline{c}_l] = ([\underline{c}_{l1}, \overline{c}_{l1}], [\underline{c}_{l2}, \overline{c}_{l2}], \cdots, [\underline{c}_{lp}, \overline{c}_{lp}])$, $[\underline{a}_s, \overline{a}_s] = ([\underline{a}_{s1}, \overline{a}_{s1}], [\underline{a}_{s2}, \overline{a}_{s2}], \cdots, [\underline{a}_{sp}, \overline{a}_{sp}])$, $l = 1, 2$, $s = 1, 2, \cdots, m$, are $p-$dimensional interval vectors;

$[\underline{d}_l, \overline{d}_l] = ([\underline{d}_{l1}, \overline{d}_{l1}], [\underline{d}_{l2}, \overline{d}_{l2}], \cdots, [\underline{d}_{lq}, \overline{d}_{lq}])$, $[\underline{b}_s, \overline{b}_s] = ([\underline{b}_{s1}, \overline{b}_{s1}], [\underline{b}_{s2}, \overline{b}_{s2}], \cdots, [\underline{b}_{sq}, \overline{b}_{sq}])$ are $q-$dimensional interval vectors; $[\underline{e}_s, \overline{e}_s]$ are interval numbers. a_t are $p-$dimensional crisp vectors, b_t are $q-$dimensional crisp vectors, and e_t are crisp numbers, $t = m+1, m+2, \cdots, n$.

In interval programming, a potential way to convert the interval inequality constraints into deterministic inequality forms is to make the interval constraints satisfied with a certain possibility degree level. In this paper, the generalized possibility degree formula introduced by Liu et al. [7] is applied to cope with the interval inequality constraints involved in problem (5). The advantage of this generalized possibility degree formula is that it is flexible to reflect various attitudes of decision makers by picking different attitude functions. In this sense, using formulas (2), (3) and (4), several crisp equivalent forms of the interval inequality constraints included in problem (5) are obtained under the consideration of several typical risk attitudes of decision makers, and then several kinds of equivalent bilevel programming models of problem (5) are built.

Based on the arithmetic operations among interval numbers, we have $[\underline{a}_s, \overline{a}_s]x + [\underline{b}_s, \overline{b}_s]y = [\underline{a}_s x + \underline{b}_s y, \overline{a}_s x + \overline{b}_s y], s = 1, 2, \cdots, m$. In this section, we only discuss the case of $\underline{a}_s x + \underline{b}_s y \geq 0$ and $\underline{e}_s \geq 0$, and the condition $[\underline{a}_s x + \underline{b}_s y, \overline{a}_s x + \overline{b}_s y] \cap [\underline{e}_s, \overline{e}_s] \neq \emptyset$.

For the case of $\underline{e}_s \leq \underline{a}_s x + \underline{b}_s y \leq \overline{a}_s x + \overline{b}_s y \leq \overline{e}_s$, if the upper and lower level decision makers hold neutral risk attitude, from formula (2), a crisp equivalent form of the s–th interval inequality constraint of the lower level programming problem is obtained as follows:

$$P([\underline{a}_s, \overline{a}_s]x + [\underline{b}_s, \overline{b}_s]y \geq [\underline{e}_s, \overline{e}_s]) = \frac{(\overline{a}_s x + \overline{b}_s y) - \underline{e}_s}{(\overline{a}_s x + \overline{b}_s y) - (\underline{a}_s x + \underline{b}_s y) + (\overline{e}_s - \underline{e}_s)} \geq \alpha_s,$$

where α_s, $s = 1, 2, \cdots, m$, represent the satisfactory degree given by decision makers.

Now the deterministic structure of the uncertain constraint region of problem (5) is generated. Obviously, minimizing the interval-valued objective functions at the upper and lower levels of problem (5) will be sufficient to minimize their center values.

Based upon the above discussions, the equivalent bilevel programming model of problem (5) can be built as follows:

$$\begin{cases} \min\limits_{x} & \frac{\underline{c}_1 x + \underline{d}_1 y + \overline{c}_1 x + \overline{d}_1 y}{2} \\ & \text{where } y \text{ solves} \\ \min\limits_{y} & \frac{\underline{c}_2 x + \underline{d}_2 y + \overline{c}_2 x + \overline{d}_2 y}{2} \\ \text{s.t.} & \underline{e}_s \leq \underline{a}_s x + \underline{b}_s y, \\ & \underline{a}_s x + \underline{b}_s y \leq \overline{a}_s x + \overline{b}_s y, \\ & \overline{a}_s x + \overline{b}_s y \leq \overline{e}_s, \\ & \frac{(\overline{a}_s x + \overline{b}_s y) - \underline{e}_s}{(\overline{a}_s x + \overline{b}_s y) - (\underline{a}_s x + \underline{b}_s y) + (\overline{e}_s - \underline{e}_s)} \geq \alpha_s, s = 1, 2, \cdots, m, \\ & a_t x + b_t y \geq e_t, t = m+1, m+2, \cdots, n, \\ & x \geq 0, y \geq 0. \end{cases} \tag{6}$$

Additionally, for pessimistic decision makers at the upper and lower levels, the following crisp equivalent form of the $s-$th interval inequality constraint is derived by formula (3):

$$P([\underline{a}_s,\overline{a}_s]x+[\underline{b}_s,\overline{b}_s]y\geq[\underline{e}_s,\overline{e}_s])=\frac{ln(\overline{a}_sx+\overline{b}_sy)-ln\underline{e}_s}{ln(\overline{a}_sx+\overline{b}_sy)-ln(\underline{a}_sx+\underline{b}_sy)+(ln\overline{e}_s-ln\underline{e}_s)}\geq\alpha_s.$$

Then we can establish another equivalent bilevel programming model of problem (5) as follows:

$$\begin{cases}\min\limits_{x} & \frac{\underline{c}_1x+\underline{d}_1y+\overline{c}_1x+\overline{d}_1y}{2}\\ & \text{where } y \text{ solves}\\ \min\limits_{y} & \frac{\underline{c}_2x+\underline{d}_2y+\overline{c}_2x+\overline{d}_2y}{2}\\ \text{s. t.} & \underline{e}_s\leq\underline{a}_sx+\underline{b}_sy,\\ & \underline{a}_sx+\underline{b}_sy\leq\overline{a}_sx+\overline{b}_sy,\\ & \overline{a}_sx+\overline{b}_sy\leq\overline{e}_s,\\ & \frac{ln(\overline{a}_sx+\overline{b}_sy)-ln\underline{e}_s}{ln(\overline{a}_sx+\overline{b}_sy)-ln(\underline{a}_sx+\underline{b}_sy)+(ln\overline{e}_s-ln\underline{e}_s)}\geq\alpha_s, s=1,2,\cdots,m,\\ & a_tx+b_ty\geq e_t, t=m+1,m+2,\cdots,n,\\ & x\geq0, y\geq0.\end{cases}\tag{7}$$

Similarly, for optimistic decision makers at the upper and lower levels, the following equivalent bilevel programming form of problem (5) can be constructed by substituting the interval inequality constraints with formula (4):

$$\begin{cases}\min\limits_{x} & \frac{\underline{c}_1x+\underline{d}_1y+\overline{c}_1x+\overline{d}_1y}{2}\\ & \text{where } y \text{ solves}\\ \min\limits_{y} & \frac{\underline{c}_2x+\underline{d}_2y+\overline{c}_2x+\overline{d}_2y}{2}\\ \text{s. t.} & \underline{e}_s\leq\underline{a}_sx+\underline{b}_sy,\\ & \underline{a}_sx+\underline{b}_sy\leq\overline{a}_sx+\overline{b}_sy,\\ & \overline{a}_sx+\overline{b}_sy\leq\overline{e}_s,\\ & \frac{(\overline{a}_sx+\overline{b}_sy)^{\frac{3}{2}}-\underline{e}_s^{\frac{3}{2}}}{(\overline{a}_sx+\overline{b}_sy)^{\frac{3}{2}}-(\underline{a}_sx+\underline{b}_sy)^{\frac{3}{2}}+(\overline{e}_s^{\frac{3}{2}}-\underline{e}_s^{\frac{3}{2}})}\geq\alpha_s, s=1,2,\cdots,m,\\ & a_tx+b_ty\geq e_t, t=m+1,m+2,\cdots,n,\\ & x\geq0, y\geq0.\end{cases}\tag{8}$$

For other situations such as $\underline{a}_sx+\underline{b}_sy\leq\underline{e}_s\leq\overline{a}_sx+\overline{b}_sy\leq\overline{e}_s$, some other equivalent bilevel programming forms of problem (5) can be also established by putting the corresponding constraint conditions into models (6), (7) and (8). Here we omit them. Clearly, crisp bilevel programming models (6), (7) and (8) can be solved by some existing efficient solution strategies.

4 Numerical Example

Consider the following numerical example to illustrate the proposed method:

$$\begin{cases} \max\limits_{x} & [1,2]x + [4,5]y \\ & \text{where } y \text{ solves} \\ \max\limits_{y} & [3,5]x + [6,8]y \\ \text{s. t.} & [3.5,4]x + [3.1,4]y \geq [10,15], \\ & y \leq 3, \\ & x \geq 0, y \geq 0. \end{cases} \tag{9}$$

For pessimistic decision makers at the upper and lower levels, from model (7), the equivalent bilevel programming model of this example can be obtained as follows:

$$\begin{cases} \min\limits_{x} & -[\frac{(x+4y)+(2x+5y)}{2}] \\ & \text{where } y \text{ solves} \\ \min\limits_{y} & -[\frac{(3x+6y)+(5x+8y)}{2}] \\ \text{s. t.} & 3.5x + 3.1y \geq 10, \\ & 4x + 4y \leq 15, \\ & \frac{ln(4x+4y)-ln10}{(ln(4x+4y)-ln(3.5x+3.1y))+(ln15-ln10)} \geq \alpha, \\ & y \leq 3, \\ & x \geq 0, y \geq 0. \end{cases} \tag{10}$$

Now we set the satisfactory degree $\alpha = 0.6$. By combining estimation of distribution algorithm [8] with some traditional technique, we solve model (10) and get the optimal solution as follows: $(x, y) = (0.75, 3.0)$. Putting this optimal solution into the interval inequality constraint involved in model (9), we have

$$\begin{aligned} & P(0.75 \cdot [3.5,4] + 3.0 \cdot [3.1,4] \geq [10,15]) \\ = & \frac{ln(4\times0.75+4\times3.0)-ln10}{(ln(4\times0.75+4\times3.0)-ln(3.5\times0.75+3.1\times3.0))+(ln15-ln10)} = 0.6387. \end{aligned}$$

Similarly, for optimistic decision makers at the two levels, according to model (8), problem (9) can be converted as:

$$\begin{cases} \min\limits_{x} & -[\frac{(x+4y)+(2x+5y)}{2}] \\ & \text{where } y \text{ solves} \\ \min\limits_{y} & -[\frac{(3x+6y)+(5x+8y)}{2}] \\ \text{s. t.} & 3.5x + 3.1y \geq 10, \\ & 4x + 4y \leq 15, \\ & \frac{(4x+4y)^{\frac{3}{2}}-10^{\frac{3}{2}}}{((4x+4y)^{\frac{3}{2}}-(3.5x+3.1y)^{\frac{3}{2}})+(15^{\frac{3}{2}}-10^{\frac{3}{2}})} \geq \alpha, \\ & y \leq 3, \\ & x \geq 0, y \geq 0. \end{cases} \tag{11}$$

Through solving model (11), we achieve the optimal solution as follows: $(x, y) = (0.75, 3.0)$. Then one has

$$P(0.75 \cdot [3.5, 4] + 3.0 \cdot [3.1, 4] \geq [10, 15])$$
$$= \frac{(4\times 0.75+4\times 3.0)^{\frac{3}{2}} - 10^{\frac{3}{2}}}{((4\times 0.75+4\times 3.0)^{\frac{3}{2}} - (3.5\times 0.75+3.1\times 3.0)^{\frac{3}{2}}) + (15^{\frac{3}{2}} - 10^{\frac{3}{2}})} = 0.6101.$$

Apparently, different possibility degrees of the interval inequality constraint included in model (9) are obtained with different possibility degree formulas. The possibility degree obtained from decision makers with pessimistic attitudes is 0.6387, while obtained from decision makers with optimistic attitudes is 0.6101. It is obvious that the former has higher possibility degree than the latter.

5 Conclusions

In this paper, we use the generalized possibility degree formula to reduce interval inequality constraints involved in interval bilevel programming into crisp equivalent forms. Then several equivalent bilevel programming forms for interval bilevel programming are constructed in consideration of decision makers' various attitudes. Finally, an illustrative example are carried out to exhibit the proposed method.

Acknowledgments. This work was supported by National Natural Science Foundation of China (No.61602010), Natural Science Basic Research Plan in Shaanxi Province of China (No.2017JQ6046) and Science Foundation of the Education Department of Shaanxi Province of China (No.17JK0047).

References

1. Abass, S.A.: An interval number programming approach for bilevel linear programming problem. Int. J. Manag. Sci. Eng. Manag. **5**(6), 461–464 (2010)
2. Calvete, H.I., Galé, C.: Linear bilevel programming with interval coefficients. J. Comput. Appl. Math. **236**(15), 3751–3762 (2012)
3. Nehi, H.M., Hamidi, F.: Upper and lower bounds for the optimal values of the interval bilevel linear programming problem. Appl. Math. Model. **39**(5–6), 1650–1664 (2015)
4. Ren, A.H., Wang, Y.P.: A cutting plane method for bilevel linear programming with interval coefficients. Ann. Oper. Res. **223**, 355–378 (2014)
5. Ren, A.H., Wang, Y.P., Xue, X.X.: A novel approach based on preference-based index for interval bilevel linear programming problem. J. Inequal. Appl. **2017**, 112 (2017). https://doi.org/10.1186/s13660-017-1384-1
6. Ren, A.H., Wang, Y.P.: An approach based on reliability-based possibility degree of interval for solving general interval bilevel linear programming problem. Soft Comput. 1–10 (2017). https://doi.org/10.1007/s00500-017-2811-4
7. Liu, F., Pan, L.H., Liu, Z.L., Peng, Y.N.: On possibility-degree formulae for ranking interval numbers. Soft Comput. **22**, 2557–2565 (2018)
8. Larranaga, P., Lozano, J.A.: Estimation of Distribution Algorithms: A New Tool for Evolutionary Computation. Kluwer Academic Publishers, Norwell (2002)

Two Algorithms with Logarithmic Regret for Online Portfolio Selection

Chia-Jung Lee(✉)

School of Big Data Management, Soochow University, Taipei, Taiwan
leecj2009@gmail.com

Abstract. In online portfolio selection, an online investor needs to distribute her wealth iteratively and hopes to maximize her final wealth. To model the behavior of the prices of assets in a financial market, we consider two measures of the price relative vectors: quadratic variability and deviation. There exist algorithms which achieve good performance in terms of these two measures. However, the theoretical guarantees depend on an additional parameter, which may not be available before the investor chooses her strategies. In this paper, the performances of the algorithms are tested using real stock market data to understand the influence of this additional parameter.

Keywords: Online portfolio selection · Regret
Quadratic variability · Deviation

1 Introduction

The *online portfolio selection* is an important problem in computational finance. An online investor has to allocate her wealth repeated over a set of assets in a finance market, and aims to maximize her final cumulative wealth, or to minimize the regret, which is difference between the performances of the best fixed strategy in hindsight to that of her strategies. A long line of research has worked on (e.g. [3], [6], [1], [4]), and one can find [7] as a nice survey.

In most of research, the theoretical regret bound is $O(\log T)$, where T is the number of trading periods. However, the prices of one asset at different moments may be dependent. In such a scenario, one can expect to obtain a smaller theoretical regret bound [5], [2], while to perform these algorithms, one needs a parameter which is related to the prices of the assets during the trading periods, and the theoretical regret bounds are also dependent on this parameter. Since the dependence on this parameter does not appear in the regret bound of $O(\log T)$, we would like to test the relation between the extra parameter and the performances of the algorithms in [2] and [5] using real stock market data.

2 The Problem Model

For the sake of simplicity, let $[n]$ denote the set $\{1, 2, \cdots, n\}$ for a positive integer n. For a vector $x \in \mathbb{R}^n$ and an index $i \in [n]$, x_i is the i'th element of x.

P. Krömer et al. (Eds.): ECC 2018, AISC 891, pp. 397–402, 2019.
https://doi.org/10.1007/978-3-030-03766-6_45

In the *universal portfolio selection model*, an online investor iteratively allocates her wealth over n assets in a financial market for T trading periods, and aims to maximize her final wealth. In each round $t \in [T]$, the online investor has to decide a distribution of her wealth, called the *portfolio vector*, $x_t \in \Delta_n$, where $\Delta_n = \{x \in [0,1]^n : \sum_i x_i = 1\}$ before the closing prices of the assets reveal. After that, she can receive a *price relative vector* $p_t \in \mathbb{R}^n$, where $p_{t,i}$ is the ratio of the closing price of the i-th asset in the trading period t to the last closing price. Hence, in the t-th trading period, the wealth of the investor will increase by a factor of $\langle p_t, x_t \rangle$, where $\langle x, y \rangle = \sum_{i \in [n]} x_i y_i$ is the inner product of two vectors $x, y \in \mathbb{R}^n$, and the final wealth is $W_0 \prod_t \langle p_t, x_t \rangle$ for some initial wealth W_0. One measure for the performance of a sequence of the portfolio vectors $x_1, x_2, \cdots, x_T$ is the *exponential growth rate*, which is

$$\log \prod_t \langle p_t, x_t \rangle = \sum_t \log \langle p_t, x_t \rangle.$$

A common benchmark strategy is the *constant rebalanced portfolio(CRP)*, which rebalances the portfolio vector in every period to a fixed distribution, and the *regret* of the online investor is defined as

$$\max_{x^* \in \Delta_n} \sum_t \log \langle p_t, x^* \rangle - \sum_t \log \langle p_t, x_t \rangle,$$

the difference between the exponential growth rate of the best CRP and that of her portfolio vectors.

One related problem is the *online convex optimization problem*, in which an online player iteratively makes decisions in T rounds. In each round $t \in [T]$, the player must choose a strategy $x_t \in \mathcal{X}$, for some convex feasible set $\mathcal{X} \subseteq \mathbb{R}^n$. She then obtains a convex loss function $f_t : \mathcal{X} \to \mathbb{R}$, and suffers a loss of $f_t(x_t)$. The goal of the online player is to minimize her total loss, which is $\sum_{t=1}^{T} f_t(x_t)$. The performance of her strategy is measured by the regret, which is the difference between the total loss she suffers and that of the best fixed strategy in hindsight, i.e.

$$\sum_t f_t(x_t) - \min_{x* \in \mathcal{X}} \sum_t f_t(x^*).$$

Note that the universal portfolio selection model can be seen as a special case of the online convex optimization problem with $\mathcal{X} = \Delta_n$, and $f_t(x) = -\log \langle p_t, x \rangle$.

3 Online Portfolio Selection Algorithms

In a real financial market, the fluctuating range of the prices for one asset is not volatile every day. To deal with this kind of scenario, two kinds of measures of the price relative vectors are proposed: the *quadratic variability* [5] and the *deviation* [2].

Hazan and Kale introduced the quadratic variability for a sequence of the price relative vectors $p_1, p_2, \cdots, p_T$, defined as

$$Q = \sum_{t=1}^{T} \|p_t - \mu\|^2 ,$$

where $\|\cdot\|$ is the L_2-norm and $\mu = \frac{1}{T}\sum_{t=1}^{T} p_t$ is the mean of the price relative vectors. Observe that if a sequence of the price relative vectors has a small quadratic variability, most of the price relative vectors center around their mean. They proposed Algorithm 1 [5], which is based on the "Follow-The-Regularized-Leader" (FTRL) scheme, and with a regret bound parametrized by the quadratic variability of the price relative vectors.

Algorithm 1. FASTER QUADRATIC-VARIATION UNIVERSAL ALGORITHM

1: **for** $t = 1$ to T **do**
2: Play $x_t = \arg\min_{x \in \Delta_n} \left(\sum_{\tau=1}^{t-1} \tilde{f}_\tau(x) + \frac{1}{2}\|x\|^2\right)$.
3: Receive the price relative vector p_t.
4: Let $\tilde{f}_t(x) = -\log\langle p_t, x_t\rangle - \frac{\langle p_t, x - x_t\rangle}{\langle p_t, x_t\rangle} + \frac{\delta\langle p_t, x - x_t\rangle^2}{8\langle p_t, x_t\rangle^2}$.
5: **end for**

Theorem 1 [5]. *For the portfolio selection problem over n assets, and a sequence of the price relative vectors $p_1, \cdots, p_T \in [\delta, 1]^n$ for some constant $\delta \in (0, 1)$, the regret of Algorithm 1 is bounded by $O\left((n/\delta^3) \cdot \log(Q + n)\right)$.*

Chiang et al. considered a more general measure, called deviation, which is defined as

$$D = \sum_{t=1}^{T} \|p_t - p_{t-1}\|^2 ,$$

where p_0 is the all-0 function. Note that deviation models the environment in which the difference of the individual vector to its previous one is usually small. Chiang et al. provided an algorithm for the more general online convex optimization problem. For the portfolio selection problem, the algorithm is given in Algorithm 2, and the regret bound parametrized by the deviation is shown.

Theorem 2 [2]. *For the portfolio selection problem over n assets, and a sequence of the price relative vectors $p_1, \cdots, p_T \in [\delta, 1]^n$ for some constant $\delta \in (0, 1)$, the regret of Algorithm 2 is bounded by $O\left((n/\delta^2) \cdot \log((n/\delta) D)\right)$.*

According to the definitions, both of the quadratic variability and the deviation can model a financial market, and therefore, there are two algorithms can attain a logarithmic regret bound. However, the additional parameter $\delta \leq \min_{t,i} p_{t,i}$ may not be obtained when the investor need to selects her strategies. One may wonder whether the performances of the algorithms will be affected by a very small value of δ. We would like to figure it out via some experiments.

Algorithm 2. DEVIATION UNIVERSAL ALGORITHM

Let $\beta = \delta^2/2$, and $\gamma = \sqrt{n}/\delta$.
Let $x_1 = \hat{x}_1 = (1/n, \cdots, 1/n)^\top$ be the uniform distribution over $[n]$.
for $t = 1$ to T **do**
  Play $\hat{x}_t$.
  Receive the price relative vector p_t.
  Compute $\ell_t = -\frac{p_t}{\langle p_t, x_t \rangle}$ and $H_{t+1} = I + \beta\gamma^2 I + \beta \sum_{\tau=1}^{t} \ell_\tau \ell_\tau^\top$.
  Let $y_{t+1} = x_t - H_t^{-1}\ell_t$, and update

$$x_{t+1} = \arg\min_{x \in \Delta_n} (x - y_{t+1})^\top H_t (x - y_{t+1}).$$

  Let $\hat{y}_{t+1} = x_{t+1} - H_{t+1}^{-1}\ell_t$, and update

$$\hat{x}_{t+1} = \arg\min_{x \in \Delta_n} (x - \hat{y}_{t+1})^\top H_{t+1} (x - \hat{y}_{t+1}).$$

end for

4 Experiments

In this section, the effect of the parameter δ on the above algorithms is examined with real stock market data.

Eight S&P500 stocks are randomly selected, and the historical stock prices from June, 2015 to May, 2018 are obtained from Yahoo! finance. After scaling the price relative vectors to fit the requirement of the algorithms, the quadratic variability of the generated vectors is about $Q = 0.56$ and the deviation is about $D = 1.13$. Moreover, the key parameter $\min_{t,i} p_{t,i}$ is 0.68. To understand the effect of δ on the algorithms, we then computed the performances, i.e the exponential growth rate, of Algorithm 1 and Algorithm 2 with parameters $\delta = \min_{t,i} p_{t,i} = 0.68$ and $\delta = 0.1$, respectively. The performance of Best CRP is also computed as a benchmark to obtain the regrets of the algorithms, which is the difference between the performance of Best CRP and that of the algorithm.

In Fig. 1, the differences for two algorithms as a function of the trading days is plotted. For Algorithm 1, the regret bound in Theorem 1 suggests that the performance with $\delta = \min_{t,i} p_{t,i}$ is better than that with $\eta = 0.1$. The experiment result approximately matched the theoretical bound, while the difference between them is small, less than 0.0018. A similar result occurred for Algorithm 2. Moreover, the difference between the performances with two different values of δ is slightly smaller than that of Algorithm 1, which implies that the performance of Algorithm 2 is stable as the chance of the parameter δ. In addition, the regrets of two algorithms with $\delta = 0.68$ are also compared in Fig. 2, and Algorithm 2 achieved a smaller regret. These experiment results showed that for online portfolio selection problem, the performance of Algorithm 2 is slightly better than that of Algorithm 1.

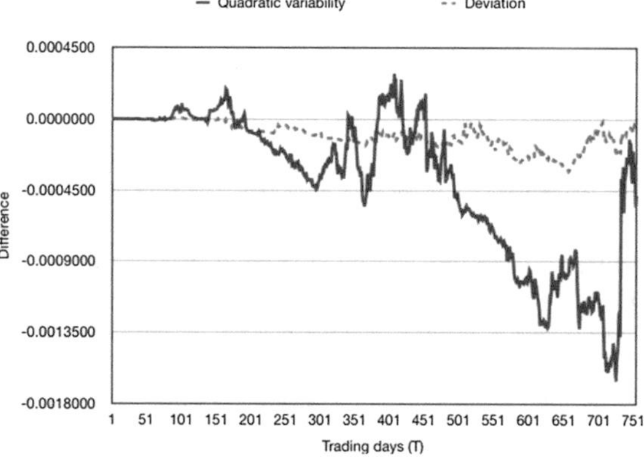

Fig. 1. The differences of the performances between the parameters $\eta = \min_{t,i} p_{t,i}$ to $\eta = 0.1$ for two algorithms.

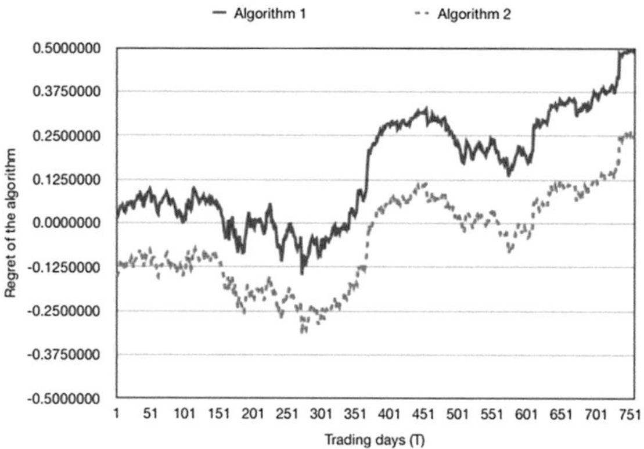

Fig. 2. The regrets of Algorithm 1 and Algorithm 2 with $\eta = 0.68$.

5 Conclusions

For online portfolio selection problem, two algorithms, which are designed for the scenario that the price relative vectors may have some correlation, are tested. Although these two algorithms have been shown to achieve logarithmic regrets, the regret bounds are dependent on a parameter δ, which is a lower bound on the components of the price relative vectors, and may not be available before the investor make decisions. To find out the effect of the parameter δ, the performances of these two algorithms are tested with different values of δ using real stock market data, and the experiment results showed that for both of these two algorithms, the regrets of different choices of the parameter δ are almost identical. It would be interesting to obtain an efficient algorithm with regret logarithmic in the quadratic variability or the deviation, and a more refined relation with δ.

References

1. Agarwal, A., Hazan, E., Kale, S., Schapire, R.E.: Algorithms for portfolio management based on the Newton method. In: ICML, pp. 9–16 (2006)
2. Chiang, C.K., Yang, T., Lee, C.J., Mahdavi, M., Lu, C.J., Jin, R., Zhu, S.: Online optimization with gradual variations. J. Mach. Learn. Res. Proceedings Track **23**, 6.1–6.20 (2012)
3. Cover, T.: Universal portfolios. Math. Financ. **1**, 1–19 (1991)
4. Das, P., Banerjee, A.: Meta optimization and its application to portfolio selection. In: Proceedings of International Conference on Knowledge Discovery and Data Mining (2011)
5. Hazan, E., Kale, S.: An online portfolio selection algorithm with regret logarithmic in price variation. Math. Financ. **25**(2), 288–310 (2015)
6. Helmbold, D.P., Schapire, R.E., Singer, Y., Warmuth, M.K.: On-line portfolio selection using multiplicative updates. In: ICML, pp. 243–251 (1996)
7. Li, B., Hoi, S.C.H.: Online portfolio selection: a survey. ACM Comput. Surv. **46**(3), 35 (2014)
8. Zinkevich, M.: Online convex programming and generalized infinitesimal gradient ascent. In: ICML, pp. 928–936 (2003)

Rank-Constrained Block Diagonal Representation for Subspace Clustering

Yifang Yang[1(✉)] and Zhang Jie[2]

[1] College of Science, Xi'an Shiyou University, Xi'an 710065, Shaanxi, People's Republic of China
yangyifang@xsyu.edu.cn

[2] School of Computer Science and Engineering, Yulin Normal University, Yulin 53700, Guangxi, People's Republic of China

Abstract. The affinity matrix is a key in designing different subspace clustering methods. Many existing methods obtain correct clustering by indirectly pursuing block-diagonal affinity matrix. In this paper, we propose a novel subspace clustering method, called rank-constrained block diagonal representation (RCBDR), for subspace clustering. RCBDR method benefits mostly from three aspects: (1) the block diagonal affinity matrix is directly pursued by inducing rank constraint to Laplacian regularizer; (2) RCBDR guarantees not only between-cluster sparsity because of its block diagonal property, but also preserves the within-cluster correlation by considering the Frobenius norm of coefficient matrix; (3) a simple and efficient solver for RCBDR is proposed. Experimental results on both synthetic and real-world data sets demonstrate the effectiveness of the proposed algorithm.

Keywords: Subspace clustering · Spectral clustering
Block diagonal representation

1 Introduction

During the past two decades, subspace clustering has been extensively studied. It arises in numerous applications in computer vision [1], image representation, and compression [2], motion segmentation [3], face clustering [4]. Recently, spectral methods based on the self-expressiveness model are used for constructing the affinity matrix between points. Sparse Subspace Clustering (SSC) [3] and Low-Rank Representation (LRR) [6,7] may be two most representative ones. The solutions obtained by SSC and LRR are block diagonal when the subspaces are independent. Beyond SSC and LRR, many other subspace clustering methods, For example, the Multi-Subspace Representation (MSR) [8] combines the idea of SSC and LRR, while the Least Squares Regression (LSR) [9] simply uses $\| Z \|^2$ and its effective is mainly due to its grouping effect for modeling the correlation structure of data. Afterwards, Correlation Adaptive Subspace Secmentation (CASS) [10] and Subspace Secmentation with Quadratic Programmimg

P. Krömer et al. (Eds.): ECC 2018, AISC 891, pp. 403–410, 2019.
https://doi.org/10.1007/978-3-030-03766-6_46

(SSQP) [11] are proposed, etc. These methods all pursue the block diagonal representation matrix by indirect methods. Different from these methods, Robust Subspace Segmentation with Block-diagonal Prior (BD-SSC and BD-LRR) [12] directly enforces the Z to be exactly k-block diagonal by considering SSC and LRR with a hard Laplacian constraint. It is verified to be effective in improving the clustering performance of SSC and LRR. However, the algorithm may not be stable as a result of stochastic sub-gradient descent solver.

In this paper, we propose a novel subspace clustering method, called rank-constrained block diagonal representation (RCBDR), for subspace clustering. Different from SSC, LRR, and LSR, RCBDR obtains block diagonal affinity matrix by directly inducing rank constraint to Laplacian regularizer. Compared with SSC and LRR, an efficient and simple solver is proposed. In particular, RCBDR guarantees between-cluster sparsity because of its block diagonal property, but also preserves the within-cluster correlation by considering the Frobenius norm of coefficient matrix.

The paper is organized as the following six sections. In Sect. 2, we briefly review SCC and LRR. In Sect. 3, we introduce the new method and its solution in detail. In Sect. 4, experiments on synthetic and real-world data sets are presented to demonstrate the effectiveness of the new method. Conclusions are made in Sect. 5.

2 Related Works

In this section, we briefly review the related work, such as sparse subspace clustering (SSC)and low-rank representation (LRR) before introducing our model.

2.1 SSC and LRR

SSC and LRR are two spectral clustering based methods, and they are the most effective approaches to subspace segmentation. The SSC model is formulated as follows.

$$\min \parallel X - XZ \parallel_F^2 + \lambda \parallel Z \parallel_0 \qquad s \cdot t \quad diag(Z) = 0 \tag{1}$$

where $\parallel Z \parallel_0$ is the l_0 norm and λ is used to balance the impact of the two terms. Since the above optimization problem is NP-hard, SSC can often be relaxed into following model under certain condition.

$$\min \parallel X - XZ \parallel_F^2 + \lambda \parallel Z \parallel_1 \qquad s \cdot t \quad diag(Z) = 0 \tag{2}$$

where $\parallel Z \parallel_1$ is the l_1 norm. It has been shown that when the subspaces are independent, The solution of Eq. (2) is block diagonal. However the solution may be too sparse, which causes each block may not be fully connected.

As for LRR, it imposes low-rank on Z. Due to the rank minimization problem is in general NP hard, thus it adopts nuclear norm to be a surrogate of the rank of Z. Then, the objective function of LRR model is written as following.

$$\min \parallel X - XZ \parallel_{2,1} + \lambda \parallel Z \parallel_* \qquad s \cdot t \quad diag(Z) = 0 \tag{3}$$

where $\| Z \|_*$ is the nuclear norm of Z, which is defined to be the sum of all singular values of Z. The solution of Eq. (3) is block diagonal when the subspaces from which the data are drawn are independent [9].

However, the block-diagonal structure obtained by these methods is fragile and will be destroyed when the signal noise ratio is small, the different subspaces are too close, or the subspaces are not independent. Hence the subspace segmentation performance may be degraded severely [12].

2.2 Block-Diagonal Structure in Subspace Clustering

The solutions of many existing subspace clustering methods obey the block diagonal property under certain subspace assumption(independent subspaces or orthogonal subspaces assumption [5], which is rather restrictive and does not apply to realistic data. Also, the block diagonal property of Z does not guarantee the correct clustering, since it may be too sparse, which causes each block may not be fully connected. Therefore, to get the correct clustering, we aim to obtain not only the block diagonal property of Z, but also expect each block is fully connected.

3 The Proposed Method

We observe that many existing methods all own the common block diagonal property by indirect methods, but they all suffer from heavy computational burdens when calculating Z. The work [12] enforces the Z to be exactly k-block diagonal by direct method, However, the algorithm may not be stable as a result of stochastic sub-gradient descent solver. For this reason, we seek to simultaneously obtain a block-diagonal affinity matrix and a solution which is computationally cheap.

Theorem 1. The multiplicity of the eigenvalue zero of the Laplacian matrix L_W is equal to the number of connected components in affinity matrix W.
Where W is affinity matrix, L_w is Laplacian matrix of W and $L_w = D - W$, D is a diagonal matrix whose element $D_{ii}(D_{ii} = \sum_{j=1}^{n} W_{ij})$ is the degree of the point x_i. Theorem 1 indicates that if the number of connected blocks in W is k, namely, clusters number of the data sets is k, then the $rank(L_w) = n - k$.

Theorem 2 (Ky Fans Theorem [13]). Let $L_Z \in \mathbf{R}^{n\times n}$ is real symmetry matrix, σ_i is the ith eigenvalue of L_Z, let $\sigma_1 \leq \sigma_2 \leq \cdots \sigma_n$, then

$$\sum_{i=1}^{k} \sigma_i(L_z) = \min_{F^TF=I} Tr(F^T L_Z F) \tag{4}$$

where $F \in \mathbf{R}^{n\times k}$ is the indicator matrix, which consist of k eigenvectors associated with the k smallest eigenvalues of L_Z. If the $\sum_{i=1}^{k} \sigma_i(L_Z) = 0$, then constraint $rank(L_Z) = (n - k)$ can be satisfied.

Hence, the objective function for recovering the block diagonal affinity matrix can be written as

$$\min_{Z,B,F} \frac{1}{2} \parallel X - XZ \parallel_F^2 + \frac{\beta}{2} \parallel Z - B \parallel_F^2 + 2\gamma Tr(F^T L_B F), \\ s,t \quad B \geq 0, diag(B) = 0, B^T = B \tag{5}$$

where the first term is used to regularize the feature reconstruction error and capture the global structure of data, and the second term makes the subproblems for updating Z and B strongly convex, while the third regularization term is used to pursue block diaconal affinity matrix so that the data sets can be classified correctly. The third Laplacian regularizer term guarantee between-cluster sparsity because of its block diagonal property, but it lacks the consideration of within-cluster correlation. Therefore, to enforce the within-cluster correlation, we introduce a Frobenius norm of Z into the objective function

$$\min_{Z,B,F} \frac{1}{2} \parallel X - XZ \parallel_F^2 + \frac{\alpha}{2} \parallel Z \parallel_F^2 + \frac{\beta}{2} \parallel Z - B \parallel_F^2 + 2\gamma Tr(F^T L_B F), \\ s,t \quad B \geq 0, diag(B) = 0, B^T = B \tag{6}$$

3.1 Solving the Optimization Problem

In this section, an alternative optimization algorithm is adopted to solve problem (8) with respect to three variables Z, B, and F.

A. Updating Z

When B and F are fixed, then Problem (6) becomes

$$\min_{Z} \frac{1}{2} \parallel X - XZ \parallel_F^2 + \frac{\alpha}{2} \parallel Z \parallel_F^2 \frac{\beta}{2} \parallel Z - B \parallel_F^2, \tag{7}$$

By setting the first-order derivative of the objective function in (7) with respect to Z to zero, the optimal solution of Z can be derived by

$$Z = (X^T X + \beta I)^{-1}(X^T X + \alpha I + \beta B) \tag{8}$$

B. Updating B

When Z and F are fixed, then Problem (6) becomes

$$\min_{B} \frac{\beta}{2} \parallel Z - B \parallel_F^2 + 2\gamma Tr(F^T L_B F), s,t \quad B \geq 0, diag(B) = 0, B^T = B \tag{9}$$

Problem (9) can be rewritten as

$$\min_{b_i \geq 0} \frac{\beta}{2} \sum_{i,j} (z_{ij} - b_{ij})^2 + 2\gamma \sum_{i,j} \parallel f_i - f_j \parallel_2^2 z_{ij} \tag{10}$$

Note that Problem (10) is independent for different i values, and we can solve the following problem for each individual i:

$$\min_{b_i \geq 0} \frac{\beta}{2} \sum_j (z_{ij} - b_{ij})^2 + 2\gamma \sum_j \parallel f_i - f_j \parallel_2^2 z_{ij} \tag{11}$$

Denote $\phi_{ij} =\parallel f_i - f_j \parallel_2^2$, and ϕ_i as a vector whose the jth element is ϕ_{ij}(and similarly for z_i and b_i). The problem (11) can be written in the vector form as

$$\min_{b_i \geq 0} \frac{\beta}{2} \parallel b_i - (\frac{2}{\beta} z_i - \frac{\gamma}{\beta} \phi_i) \parallel_2^2 \tag{12}$$

C. Updating F

When B and Z are fixed, then Problem (8) becomes

$$\min_B 2\gamma Tr(F^T L_B F) \tag{13}$$

From the Theorem 2, we can know that $F \in \mathbf{R}^{n \times k}$ consist of k eigenvectors associated with the k smallest eigenvalues of L_B.

4 Experiments

To evaluate the proposed RCBDR algorithm, we compare it with SSC [3], LRR [6,7], and LSR [9] on both synthetic and real data. These data sets contain quite different noise levels, thus are suitable for testing the influence of noise and corruption on the performance. The experiments are completed in Matlab R2016b and performed on a computer with Intel(R) Core(TM)i5-6500 CPU@3.20 GHz and Windows 7 Professional.

4.1 Synthetic Experiment

We give an intuitive example to illustrate the effectiveness of RCBDR. Synthetic data set are generated following the scheme in [8]. We construct $k = 5$ subspaces $\{S_i\}_{i=1}^5$ whose bases $\{U_i\}_{i=1}^5$ are computed by $U_{i+1} = TU_i, 1 \leq i \leq k$, where T is a random rotation matrix and $U_1 \in \mathbf{R}^{D \times m}$ is a random orthogonal matrix. We set $D = 30$ and $m = 5$. For each subspace, we sample$n_i = 50$ data vectors by $X_i = U_i Q_i, 1 \leq i \leq k$, with $Q_i \in \mathbf{R}^{m \times n_i}$ i.i.d. N(0,1) matrix. So we have $X \in \mathbf{R}^{D \times n}$, where $n = n_i \times k$.

The parameters for each method are tuned to achieve the best performance. Among of them, the parameters of RCBDR are $\alpha = 0.01$, $\beta = 2$ and $\gamma = 0.4$. Figure 1 shows that an illustrative comparison on the constructed affinity matrices for synthetic data set, of which 30% are randomly chosen and corrupted by adding Gaussian noise with a mean 0 and a variance 0.05^2. We can see that the block-diagonal affinity matrix constructed by SSC is very sparse, yet very dense by LRR and LSR. Our proposed RCBDR method is a compromise of SSC, LSR and LRR.

For comparing the robustness, we add dense Gaussian noise, with a mean 0 and a variance 0.3^2, further add noises uniformly distributed on [–0.5 0.5] to the clean data. Then, we test the performance of different algorithms on noisy synthetic data with an increasing percentage of corruptions. We repeat the experiment by 10 times. The related parameters setting is as follows: the parameter of SSC, LRR, and LSR is 0.2, 0.8, 0.3, respectively; the parameters of RCBDR are $\alpha = 1$, $\beta = 10$ and $\gamma = 0.5$. Figure 2 reports the mean accuracies of the four methods with respect to the percentage of corruptions. It shows that our proposed RCBDR outperforms other state-of-the-art methods with a clear margin.

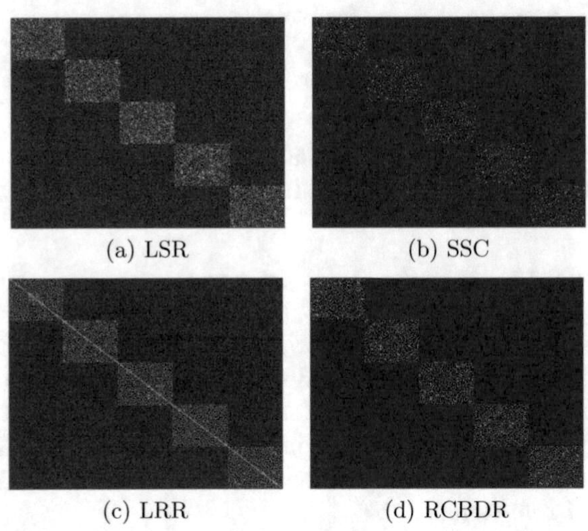

(a) LSR (b) SSC (c) LRR (d) RCBDR

Fig. 1. An illustrative comparison on the constructed sample affinity matrices for synthetic noisy samples from 5 subspaces. (a) SSC; (b) LRR; (c) BDR; (d) RCBDR.

4.2 Real Experiment on Extended Yale B Data Set

We test on the Extended Yale B database [14]. This dataset consists of 2,414 frontal face images of 38 subjects under 9 poses and 64 illumination conditions. For each subject, there are 64 images. Each cropped face image consists of 192×168 pixels. We then construct the data matrix X from subsets which consist of different numbers of subjects from the Extended Yale B database. In this experiment, we use the first k classes data($k = 5, 8, 10$), each class contains 64 images. Then the data are projected onto a 6^*k dimensional subspace for k classes clustering problem by PCA. Table 1 lists the segmentation accuracies of each method on the Extended Yale Database B. Table 1 shows the clustering result on the Extended Yale B database. For SSC, LRR and LSR, we cite the

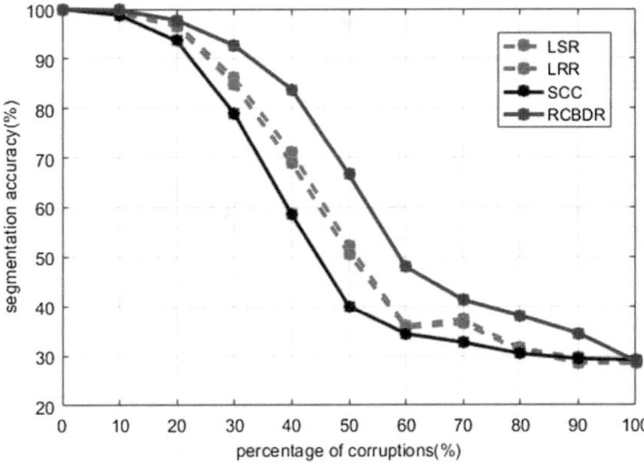

Fig. 2. Comparison on the synthetic data as the percentage of corruptions increases.

results reported in [10]. From Table 1, it can be found that RCBDR obtains the highest clustering accuracies on all these three clustering tasks. In particular, RCBDR gets accuracies of 94.31%, 87.35%, and 81.25% for face clustering with 5, 8, and 10 subjects, respectively. For face clustering with 8 and 10 subjects, both LRR and LSR perform much better than SSC, which can be attributed to the strong grouping effect of the two methods. However, both the two methods lack the ability of subset selection, and therefore may group some data points between clusters together. RCBDR not only preserves the grouping effect within cluster but also guarantee between-cluster sparsity because of its block diagonal property.

Table 1. The clustering accuracies (%) on the Extended Yale B database

	SSC	LRR	LSR	RCBDR
5 subjects	80.31	86.56	92.19	**94.31**
8 subjects	62.90	78.91	80.66	**87.35**
10 subjects	52.19	65.00	73.59	**81.25**

5 Conclusions

This paper aims at constructing the block diagonal affinity matrix by directly method. Based on rank constraint, we have proposed RCBDR method, which is a new spectral clustering with block diagonal property. Different from SSC, LRR, and LSR, RCBDR obtains block diagonal affinity matrix by directly inducing rank constraint to Laplacian regularizer. Compared with SSC and LRR, an efficient and simple solver is proposed. In particular, RCBDR guarantees between-cluster sparsity because of its block diagonal property, but also preserves the within-cluster correlation by considering the Frobenius norm of coefficient matrix. Tests on both the synthetic and the real data testify to the robustness of our robust RCBDR when compared with other state-of-the-art methods.

Acknowledgement. This work was supported by the Scientific Research Plan Projects of Shaanxi Education Department (No.17JK0610); the Doctoral Scientific Research Foundation of Shaanxi Province (0108-134010006).

References

1. Ma, Y., Yang, A.Y., Derksen, H., Fossum, R.: Estimation of subspace arrangements with applications in modeling and segmenting mixed data. SIAM Rev. **50**(3), 413–458 (2018)
2. Hong, W., Wright, J., Huang, K., Ma, Y.: Multi-scale hybrid linear models for lossy image representation. IEEETrans. Image Process. **15**(12), 3655–3671 (2006)
3. Elhamifar, E., Vidal, R.: Sparse subspace clustering: algorithm, theory, and applications. IEEE Trans. Pattern Anal. Mach. Intell. **35**(11), 2765–2781 (2013)
4. Ho, J., Yang, M.-H., Lim, J., Lee, K.-C., Kriegman, D.: Clustering appearances of objects under varying illumination conditions. In: IEEE Conference on Computer Vision and Pattern Recognition, pp. 11–18 (2003)
5. Lu, C., Feng, J., Lin, Z., Mei, T., Yan, S.: Subspace clustering by block diagonal representation. IEEE Trans. Pattern Anal. Mach. Intell. (2018)
6. Liu, G., Lin, Z., Yan, S., Sun, J., Yu, Y., Ma, Y.: Robust recovery of subspace structures by low-rank representation. IEEE Trans. Pattern Anal. Mach. Intell. **35**, 171–184 (2013)
7. Liu, G., Lin, Z., Yu, Y.: Robust subspace segmentation by low-rank representation. In: Frnkranz, J., Joachims, T. (eds.) Proceedings of the 27th International Conference on Machine Learning, ICML-10, Haifa, Israel, pp. 663–670 (2010)
8. Luo, D., Nie, F., Ding, C., Huang, H.: Multi-subspace representation and discovery. In: Joint European Conference Machine Learning and Knowledge Discovery in Databases, LNAI, vol. 6912, pp. 405–420 (2011)
9. Lu, C.-Y., Min, H., Zhao, Z.-Q., Zhu, L., Huang, D.-S., Yan, S.: Robust and efficient subspace segmentation via least squares regression. In: Proceedings of European Conference on Computer Vision (2012)
10. Lu, C., Lin, Z., Yan, S.: Correlation adaptive subspace segmentation by trace lasso. In: ICCV (2013)
11. Luo, D., Nie, F., Ding, C.H.Q., Huang, H.: Multi-subspace representation and discovery. In: ECML/PKDD, pp. 405–420 (2011)
12. Feng, J., Lin, Z., Xu, H., Yan, S.: Robust subspace segmentation with block-diagonal prior. In: Proceedings of IEEE Conference on Computer Vision and Pattern Recognition, pp. 3818–3825 (2014)
13. Fan, K.: On a theorem of wey concerning eigenvalues of linear transformations. Proc. Nat. Acad. Sci. USA **35**(11), 652–655 (1949)
14. Lee, K.-C., Ho, J., Kriegman, D.J.: Acquiring linear subspaces for face recognition under variable lighting. IEEE Trans. Pattern Recognit. Mach. Intell. **27**(5), 684–698 (2005)

A Method to Estimate the Number of Clusters Using Gravity

Hui Du(✉), Xiaoniu Wang, Mengyin Huang, and Xiaoli Wang

College of Computer Science and Engineering,
Northwest Normal University, Lanzhou 730070, China
51253474@qq.com

Abstract. The number of clusters is crucial to the correctness of the clustering. However, most available clustering algorithms have two main issues: (1) they need to specify the number of clusters by users; (2) they are easy to fall into local optimum because the selection of initial centers is random. To solve these problems, we propose a novel algorithm using gravity for auto determining the number of clusters, and this method can obtain the better initial centers. In the proposed algorithm, we firstly scatter some detectors on the data space uniformly and they can be moved according to the law of universal gravitation, and two detectors can be merged when the distance between them less than a given threshold. When all detectors no longer move, we take the number of detectors as the number of the clusters. Then, we utilize the finally obtained detectors as the initial center points. Finally, the experimental results show that the proposed method can automatically determine the number of clusters and generate better initial centers, thus the clustering accuracy is improved observably.

Keywords: Clustering · Number of clusters · Initial centers · Gravity Detector

1 Introduction

Cluster analysis is a tool for exploring the basic structure of a given data set, and is applied to a variety of engineering and scientific fields, such as medicine, sociology, biology, psychology, image processing and pattern recognition [1]. Clustering problems can be described as finding clusters in a data set, and within each cluster data have a high similarity, while have a low one between different clusters. K-means algorithm [2] is one of the most classical cluster methods, whose attractiveness lies in its simplicity. However, it has three main shortcomings. First, it is slow because it takes to complete the each iteration. Second, it has to specify the number of clusters k by users. Third, it empirically finds worse local optima when confined to poor initial centers. Many researchers propose solutions of these problems [3–8]. In order to solve the initialization problems of K-means algorithm, Grigorios designed the Min-Max K-means algorithm [13], which given weight values for each cluster according to the covariance of each cluster, then optimized the target of the algorithm. But this method needs to specify the number of clusters k. Kolesnikov et al. introduced a method for determining an optimal number of clusters based on parametric modeling of the

P. Krömer et al. (Eds.): ECC 2018, AISC 891, pp. 411–419, 2019.
https://doi.org/10.1007/978-3-030-03766-6_47

quantization error [12]. However, this method requires more computational time. Pelleg and Moore proposed X-means algorithm [9] for learning *k*, which tried over many values of *k* and obtained a scheme for each value. X-means uses the Bayesian Information Criterion (BIC) to score each model, and chooses the strategy with the highest BIC score for estimation of *k*. Hamerly and Elkan proposed G-means algorithm [10], which assumed the data in the same cluster obeyed the Gaussian distribution. The algorithm first gives a smaller number of clusters, and then uses statistical methods to estimate the data in the same cluster whether obey the Gaussian distribution. Those clusters which obey the Gaussian distribution will be does not split into two clusters. This procedure repeats until they get an ideal number of clusters. Fujita et al. proposed slope statistic non-parametric method to estimate the number of clusters [11]. This method can handle status when the dataset does not obey a mixture of Gaussian distribution, and when the number of parameters is large. These methods adopt the similar model, namely, given evaluation criteria, they try over many values of *k*in order to obtain the best result. Although these algorithms can obtain the true value of *k*, their computation costs are more expensive.

In this paper we propose a novel algorithm to automatically determine the number of clusters and initial centers for K-means algorithm. We scatter a certain amount of detectors on the data space uniformly, and use the law of universal gravitation for these detectors moving and merging to estimate the number of clusters. This method is useful and adoptable for many clustering algorithms other than K-means, but here we consider only the K-means algorithm for simplicity.

2 The Proposed Algorithm

We firstly scatter a certain amount of detectors in the data space by using uniform design method [14]. These detectors utilize the law of universal gravitation to move, specifically, larger mass detectors will attracted the smaller ones and smaller ones will move very near to larger ones. In our algorithm, we will combine two detectors into one when the distance between them is smaller than a threshold. We call it a stable state when all detectors do not move and cannot be merged. At this time, these detectors will be seen as initial centers and the number of them as the number of clusters.

2.1 Determine the Number of Clusters

The detail of the proposed algorithm is as follows.

First, put detectors. We use the uniform design to put a certain amount of detectors in the data space, e.g., red dots denote the detectors and blue stars represent the experimental data in Fig. 1(a). Since we use the GLP method [14], initial number of detectors is a prime number n. In experiments we let $n = 37$. We define a set $C(x)$ of data in the data space for the detector x, and call set $C(x)$ the class set of detector x. In the data space if and only if the degree of membership between data z and detector x is smaller than that between data z and any other detector, data z is in $C(x)$. In the other words, the set $C(x)$ is a cluster that the center is detector x.

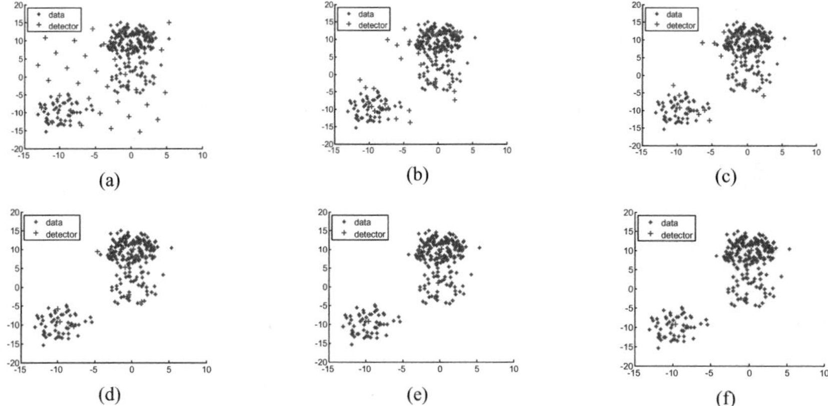

Fig. 1. Moving and merging of detectors

Second, move and merge. We define that the number of data points contained in the circle of radius E with one detector as the center is the mass of this detector. Thus the mass of any detector will be larger than or equal to 1. Based on the law of universal gravitation, the detector moving steps are as follows:

i. Compute the radius E according to the Eq. (1):

$$E = pS/(2\sqrt{n} - 1), \tag{1}$$

where p is a parameter that control the value of radius E, S is the farthest distance between different data points in data sets. According to the experimental results, the value of the parameter p is better to take the value between 0.8 and 2.0.

ii. Compute gravity between different detectors, and let $G = 1$ because of computing simply.
For each detector x, in its neighborhood (a circle of radius E with x as center), find out a detector y which has the largest gravity to x. We provide that the smaller mass detector will move towards the larger mass one, and suppose that the detector with smaller mass is y, then y will move towards the detector x with larger mass by Eq. (2) iteratively:

$$y(t+1) = (1-\lambda)y(t) + \lambda x(t), \;\; 0<\lambda<1, \tag{2}$$

where λ indicates the step size, and we take $\lambda = 1/3$ in our experiments, and t represents time.

iii. We will merge detector x and y into one when their distance is shorter than αE, namely, keep the larger mass detector, say x. After that, move the data of C(y) into $C(x)$ and delete the set $C(y)$ and the detector y. The parameter α is used to guide the speed of the merging. In our experiments, we set $\alpha = 0.25$.

iv. Repeat step (ii) to (iv) until all detectors no longer merging and moving. After that, these detectors are regarded as the initial centers and their number is the number of clusters. For example, the main process of detectors' moving and merging on synthetic datasets are shown in Fig. 1. This experiment result is that the number of clusters is 2, and it depends on the radius *E*. If *E* is too small, the detectors will not merge and move. Then, the number of clusters will be large. If *E* is too large, it will increase the computational consumption, and it will cause that the number of clusters becomes 1.

2.2 Complexity Analysis

The proposed method consists mainly of two steps. In the first step, we put detectors uniformly in data space and classify all data with time complexity $O(n)$, where n is the number of detectors. In second step, we move and merge the detectors until they do not meet the moving and merging conditions, with time complexity $O(TnN)$, where T is the number of iterations and N is the number of data ($n << N$). To sum up, the total time complexity of the proposed method is $O(TnN) + O(n) = O(TnN)$. Besides, the memory complexity of the proposed method is $O(nN)$.

3 Experiment Result and Analysis

3.1 Evaluation Metrics

We adopted three commonly used metrics to evaluate the performance of the algorithms, and these metrics are Davies Bouldin Index (*DBI*) [19] and Normalized Mutual Information (*NMI*) [17] and *Accuracy*. *DBI* computes the ratio of the degree of dispersion between different cluster data and the tightness between the same cluster data. *NMI* is used to directly contrast between the true labels and the class labels obtained from clustering algorithm. As the name suggests, *Accuracy* is used to calculate the proportion of clustering correctly.

3.2 Experiments

Experiment Environment and Datasets. The experiment environment contains: MATLAB R2010a, Intel(R) Xeon(R) CPU, 2.53 GHz, win7, 64bits operating system, 4G memory. In our experiments, the real-world data sets (UCI data sets [18], see Table 1) and two synthetic data sets S_1, S_2 have been used for clustering. Synthetic data S_1, S_2 are showed in Figs. 2 and 3.

Parameter. In the proposed method, the effect of the parameter p (in Eq. (4)) on the results is obviously. The parameter p controls the size of radius E, while the radius E is inversely proportional to the number of clusters. So when p becomes bigger, the number of clusters will be less. Experimental results on two synthetic data sets show the relationship between the parameter p and the number of clusters (see Figs. 2 and 3). Figures 2 and 3 show the results with the different value of p. Because we don't know the true number of clusters beforehand with using these synthetic data sets for experiments, we use DBI as a criterion to judge the clustering result. We highlight the best performances for each data set.

Table 1. A summary of real-world data sets (from UCI dataset)

Data set	Size	Dimensions	Classes
Iris	150	4	3
Wine	178	13	3
Yeast	1480	8	10
Zoo	101	16	7
Liver	345	6	2
Ecol	336	7	8
Sp	267	44	2
Pima	768	8	2
Breast	106	9	6
Hab	306	3	2
Hr	269	44	2
Io	351	34	2
Balance	625	4	3
Pendigittrain	7494	16	10

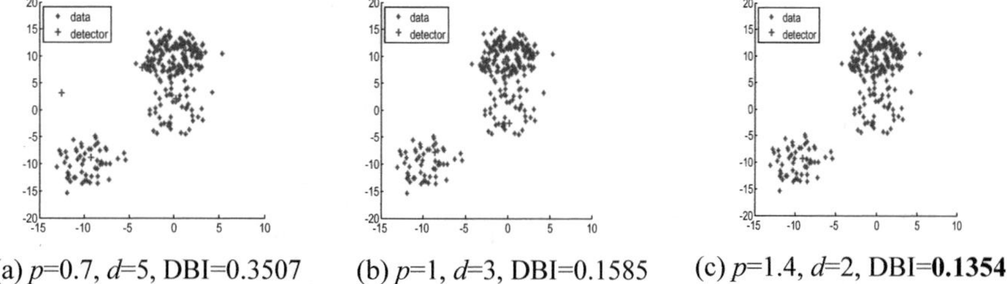

(a) p=0.7, d=5, DBI=0.3507 (b) p=1, d=3, DBI=0.1585 (c) p=1.4, d=2, DBI=**0.1354**

Fig. 2. The results with the different p for the data sets S_1

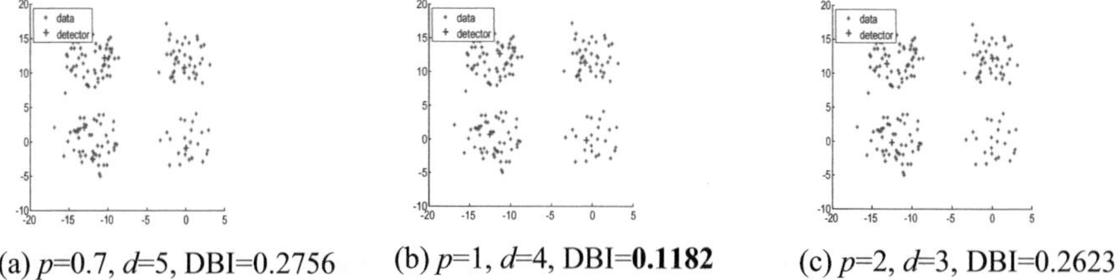

(a) p=0.7, d=5, DBI=0.2756 (b) p=1, d=4, DBI=**0.1182** (c) p=2, d=3, DBI=0.2623

Fig. 3. The results with the different p for the data sets S_2

In Fig. 2, we find the value of DBI is smallest when p = 1.4, and the number of clusters obtained is 2. In Fig. 3, the value of DBI is smallest when p = 1, and the number of clusters obtained is 4.

It is worth noticing that the number d of detectors is 5 in Fig. 2 (a), but the detector located upper left position will not be selected as the center when the initial

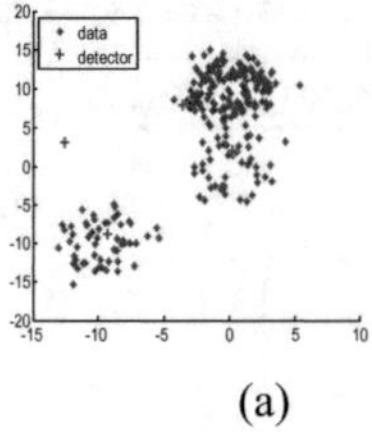

(a)

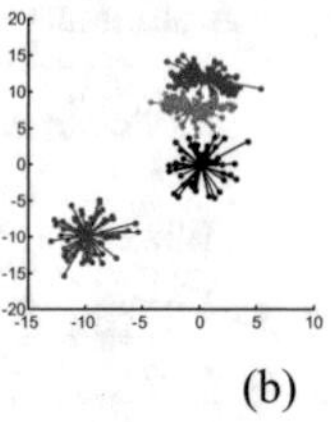

(b)

Fig. 4. (a) The result after moving and merging of detectors, and the number of detectors is 5. (b) Clustering result of the K-means algorithm on the synthetic data sets, and the detectors obtained from (a) are used to initial centers (the number is 5). But one detector doesn't be selected as the center at beginning, and the real number of clusters is 4 in the end.

distribution, so the real number of clusters is 4 (see Fig. 4). In our experiments, the number of clusters is the real number of clusters which is subtracted the number of the detectors that don't be selected as the initial centers.

Table 2 shows the value of p when get the true classes on real-world data sets. Experimental results show that when the parameter p in the value of 0.1 to 2.0, the proposed algorithm can get the true number of clusters.

Table 2. The value of p when get the true classes on real-world data sets

Data set	P	Classes
Iris	1.40	3
Wine	1.63	3
Yeast	1.38	10
Zoo	1.80	7
Liver	1.46	2
Ecol	1.50	8
Sp	2.00	2
Pima	1.60	2
Breast	1.70	6
Hab	1.35	2
Hr	2.00	2
Io	2.00	2
Balance	1.755	3
Pendigittest	1.565	10

Table 3. Comparison of NMI on UCI data sets (means NMI ± standard deviations)

Data set	Proposed method	K-means
Iris	**0.7419**	0.6639 ± 0.0612
Wine	**0.8347**	0.4249 ± 0.0015
Yeast	**0.3070**	0.2588 ± 0.0150
Zoo	**0.7828**	0.7090 ± 0.0031
Liver	**0.0005**	0.0003 ± 0.0001
Ecol	**0.6580**	0.5705 ± 0.0231
Sp	**0.0914**	0.0854 ± 0.0014
Pima	**0.0517**	0.0507 ± 0.0001
Breast	**0.5121**	0.3646 ± 0.0154
Hab	**0.0009**	0.0008 ± 0.0001
Hr	**0.0902**	0.0852 ± 0.0029
Io	0.0562	**0.0570 ± 0.0013**
Balance	**0.1170**	0.1103 ± 0.0403
Pendigittest	**0.6932**	0.6808 ± 0.0114

Results and Comparisons. From the above experiments, we can find that the result of clustering will be better if the data are spherically distributed around the center. In order to test the performance of our proposed algorithm in other circumstance, we do some experiments on real-world data because most real-world data are not spherical. Table 1 shows the summary of real-world data sets we used. The comparison of the *NMI* between the proposed method and K-means are showed in Table 3, in which we highlight the best performances for each data set. From Table 3 we can see that the experimental results of the proposed method are better than the K-means algorithm on all the data sets except the data set Io. Table 4 shows the comparison of Accuracy, and we find that the experimental results of the proposed method are better than the K-means on 9 data sets, while the experimental results of the K-means are better than proposed method on only 4 data sets.

Table 4. Comparison of Accuracy on UCI data sets (means Accuracy ± standard deviations)

Data set	Proposed method	K-means
Iris	**0.8867**	0.7011 ± 0.0315
Wine	**0.9494**	0.6113 ± 0.0586
Yeast	**0.5216**	0.3494 ± 0.0201
Zoo	**0.8020**	0.6912 ± 0.0074
Liver	**0.5565**	0.5431 ± 0.0057
Ecol	**0.6339**	0.4940 ± 0.0423
Sp	0.6217	**0.6356 ± 0.0564**
Pima	**0.6680**	0.6674 ± 0.0259
Breast	**0.4717**	0.3396 ± 0.0187
Hab	**0.5229**	0.5117 ± 0.0041
Hr	0.6245	**0.6364 ± 0.0329**
Io	0.6011	**0.6070 ± 0.0017**
Balance	0.5296	**0.6869 ± 0.0610**
Pendigittest	**0.6701**	0.6499 ± 0.0499

Applications in Texture Image Segmentation Clustering is one of mainly methods applied for image segmentation. We apply the proposed method for texture image segmentation to verify its effectiveness. In our experiments, we use two synthetic images constructed from Brodatz database [20] and one natural image selected from Berkeley database [21]. The segmentation results of experiments see Fig. 5. In this figure, image (a), (c) and (e) are original images, and their size are 150 × 150, 153 × 155 and 151 × 100 pixels respectively, and image (b), (d) and (f) are the segmentation results, where c denotes the number of clusters, and w represents the size of window, and p is the parameter in our algorithm (see Eq. (4)). The results of experiments show that the proposed method can be used for image segmentation.

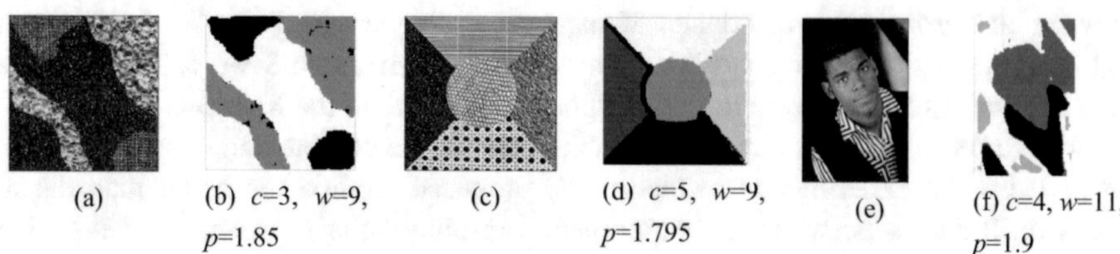

(a) (b) c=3, w=9, p=1.85 (c) (d) c=5, w=9, p=1.795 (e) (f) c=4, w=11, p=1.9

Fig. 5. (a), (c) and (e) are original images, (b), (d) and (g) are the results of segmentation respectively, where c denotes the number of clusters, and w represents the size of window, and p is the parameter in our algorithm (see Eq. (4)).

4 Conclusion

In this paper, we use the orthogonal design algorithms launch several detectors in the data space, then move and merge in accordance with the law of gravity. The number of detectors is the clustering number and the location of the probe is the initial center of the K-means algorithm when the detector is no longer moving. Experimental results show that when the parameter p in the value of 0.8 to 2.0, the proposed algorithm can get the ideal number of clusters. Due to our algorithm provides an ideal initial centers, it can get a good clustering result. The proposed algorithm framework automatically determines the number of clusters can also be used in other clustering algorithms which need to specify the clustering number.

Acknowledgment. This work is supported by the National Natural Science Foundation of China (No. 61472297 and No. 61402350 and No. 61662068).

References

1. Pena, J.M., Lozano, J.A., Larranaga, P.: An empirical comparison of four initialization methods for the K-means algorithm. Pattern Recognit. Lett. **20**, 1027–1040 (1999)
2. MacQueen, J.: Some methods for classification and analysis of multivariate observations. In: Proceedings of the 5th Berkeley Symposium on Mathematical Statistics and Probability, Berkeley, USA, pp. 281–297. University of California Press (1967)
3. Estivill, C.V., Yang, J.: Fast and robust general purpose clustering algorithms. Data Min. Knowl. Discov. **8**(2), 127–150 (2004)
4. Muchun, S.U., Chienhsing, C.H.O.U.: A modified version of the K-means algorithm with a distance based on cluster symmetry. IEEE Trans. Pattern Anal. Mach. Intell. **23**(6), 674–680 (2001)
5. Likas, A., Vlassis, M., Verbeek, J.: The global k-means clustering algorithm. Pattern Recognit. **36**, 451–461 (2003)
6. D'Urso, P., Giordani, P.: A robust fuzzy k-means clustering model for interval valued data. Comput. Stat. **21**(2), 251–269 (2006)
7. Chunsheng, H.U.A., Qian, C.H.E.N., et al.: RK-means clustering: K-means with reliability. IEICE Trans. Inf. Syst. **E91D**(1), 96–104 (2008)
8. Timmerman, M.E., Ceulemans, E., et al.: Subspace K-means clustering. Behav. Res. Methods **45**(4), 1011–1023 (2013)

9. Pelleg, D., Moore, A.: X-means: extending k-means with efficient estimation of the number of clusters. In: Proceedings of the 17th International Conference on Machine Learning, pp. 727–734 (2000)
10. Hamerly, G., Elkan, C.: Learning the k in k-means. In: Proceedings of the 17th Annual Conference on Neural Information Processing Systems, pp. 281–288 (2003)
11. Fujita, A., Takahashi, D.Y., Patriota, A.G.: A non-parametric method to estimate the number of clusters. Comput. Stat. Data Anal. **73**, 27–39 (2014)
12. Kolesnikov, A., Trichina, E., Kauranne, T.: Estimating the number of clusters in a numerical data set via quantization error modeling. Pattern Recognit. **48**(3), 941–952 (2015)
13. Tzortzis Likas, G.A.: The MinMax k-means clustering algorithm. Pattern Recognit. **47**, 2505–2516 (2014)
14. Fang, K.T., Shiu, W.C., Pan, J.X.: Uniform designs based on Latin squares. Stat. Sin. **9**(3), 905–912 (1999)
15. Fang, K.T., Wang, Y.: Number-Theoretic Methods in Statistics. Chapman and Hall, London (1994)
16. Zhang, L., Liang, Y., Jiang, J., Yu, R., Fang, K.T.: Uniform designs applied to nonlinear multivariate calibration by ANN. Anal. Chim. Acta **370**(1), 65–77 (1998)
17. Shang, F.H., Jiao, L.C.: Fast affinity propagation clustering: a multilevel approach. Pattern Recognit. **45**, 474–486 (2012)
18. http://archive.ics.uci.edu/ml/
19. Davies, D.L., Bouldin, D.W.: A cluster separation measure. IEEE Trans. Pattern Anal. Mach. Intel. (PAMI) **1**, 224–227 (1979)
20. http://www.ux.uis.no/~tranden/brodatz.html
21. http://www.eecs.berkeley.edu/Research/Projects/CS/vision/bsds/

Analysis of Taxi Drivers' Working Characteristics Based on GPS Trajectory Data

Jing You[1(✉)], Zhen-xian Lin[2], and Cheng-peng Xu[3]

[1] School of Communication and Information Engineering, Xi'an University of Posts and Telecommunications, Xi'an, China
youj5646@126.com

[2] School of Science, Xi'an University of Posts and Telecommunications, Xi'an, China
lzhx126@126.com

[3] Institute of Internet of Things and IT-Based Industrialization, Xi'an University of Posts and Telecommunications, Xi'an, China
xyfoal@xupt.edu.cn

Abstract. Using the large-scale taxi GPS trajectory data mining to analyze working characteristics of taxi drivers, it can provide reference help for those who want to work in taxis. This paper proposes a method for analyzing the working characteristics of taxi drivers based on Spark data processing platform. Firstly, the GPS trajectory data are cleaned and then imported into HDFS. Secondly, taxi drivers' work indicators and taxis' stop points are extracted. Then, the statistical method is applied to analyze the work indicators to obtain drivers' work feature, and the K-Means algorithm is used to cluster the stay points to get the drivers' three meals rest and then drivers' rest feature can be get. The results show that the Spark platform can quickly and accurately analyze the working characteristics of taxi drivers.

Keywords: Spark data processing platform · Taxi GPS trajectory data
Taxi drivers' work characteristics

1 Introduction

In recent years, vehicle GPS device have been rapidly popularized, and large-scale trajectory data generated by vehicle GPS device has become an important resource for data mining and research applications. By mining the GPS trajectory data of taxis, the working characteristics of the taxi driver group are analyzed, which is helpful to understand taxi industry situation and the working condition of taxi drivers, and providing reference help for those who want to work in taxis.

There are two aspects to the study of taxi drivers by using taxi GPS trajectory data: (1) When the taxi is in operation, some work analyzes the operational characteristics of the taxi [1–3], and the spatial and temporal distribution of the high-income taxis trajectory [4]; (2) When the taxi is not in operation, some work explores the residence of taxi drivers and analyze the law of work and rest [5, 6]. Although the above research has better analyzed the characteristics of a certain aspect of taxi drivers, it lacks an

P. Krömer et al. (Eds.): ECC 2018, AISC 891, pp. 420–430, 2019.
https://doi.org/10.1007/978-3-030-03766-6_48

analysis of the overall behavior (work and rest) of taxi drivers, and the efficiency is greatly reduced when faced with the processing of massive data.

This paper proposes a method for analyzing the working characteristics of taxi drivers, which is based on Spark data processing platform. The method extracts important work indicators and stay points, then analyzes the work indicators to get drivers' work feature, and uses K-Means algorithm to analyze drivers' rest feature. The experimental results show that the analysis method designed by Spark platform can quickly and accurately get the working characteristics of the taxi drivers. This method is also applicable to the analysis of the work characteristics of the courier and the delivery staff.

2 Spark Data Processing Platform and K-Means Algorithm

2.1 Spark Data Processing Platform

Spark is a big data memory computing framework similar to Hadoop MapReduce [7]. It can be used in different scenarios such as stream processing, iterative calculation, and batch processing [8]. Spark's operations are divided into two parts: transformation and action. The Resilient Distributed Dataset (RDD) is the core of Spark. The Spark data analysis process is shown in Fig. 1. Firstly, the Spark calculation engine calls the textfile operator to load the data in the HDFS (Distributed File System) into the cluster memory; secondly, the program creates a series of RDDs from the input data; thirdly, convert the RDD by calling the Spark operator; finally, the RDD is triggered to complete the calculation by calling saveAsTextFile action operator, and the final calculation result is stored in the HDFS [9]. Spark common operators are as follows: (1) map, map each data one-to-one to return a new RDD; (2) distinct, remove duplicate data from data; (3) filter, filter data as required; (4) sortBy, sort the data in the form of (key, value) by key value; (5) groupByKey, aggregate the value of the same key value for data in the form (key, value); (6) saveAsTextFile, save the data calculation results.

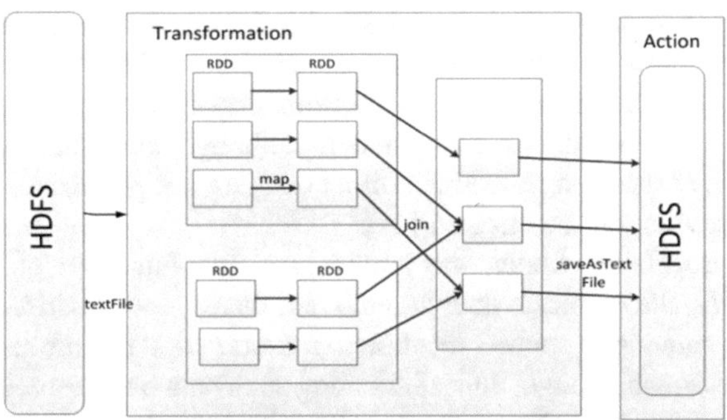

Fig. 1. Spark data analysis process

2.2 K-Means Algorithm

As an unsupervised learning clustering algorithm, K-Means divides similar data points by pre-set k values and initial centroid of each category, and the optimal clustering result is obtained by the iterative optimization of the divided mean values [10]. Given sample set $D = \{x_1, x_2, \ldots, x_m\}$, divide it into k clusters $C = \{C_1, C_2, \ldots, C_k\}$, the k-means algorithm minimizes the square error of the mean vector of the sample data and the cluster

$$E = \sum_{i}^{k} \sum_{x \in C_i} \|x - \mu_i\|_2^2 \tag{1}$$

3 Data Preprocessing

These data are the trajectory data of more than 14,000 taxis in Chengdu from August 18, 2014 to August 23, 2014 for 6 consecutive days, including 17 h from 6:00 to 24:00, and the sampling interval is about 30 s (some are 10 s), the daily data are about 57 million, the 6-day data are more than 300 million, and the data set is 13 GB. The taxi GPS data includes the taxi number, timestamp, longitude, latitude, and passenger status. Examples of GPS data are shown in Table 1. There are two types of passenger status, '1' means full load and '0' means empty cars.

Table 1. Example of taxi GPS data

Taxi number	Latitude	Longitude	Status	Timestamp
1	30.569583	104.068404	1	2014/8/18 10:13:34

Due to the measurement accuracy of the vehicles GPS and the influence of the high-rise buildings or trees in the city on the signal transmission, the GPS has a large error in the positioning, so that there are some abnormalities in the original data, including the latitude and longitude crossing; taxi drivers repeatedly repeat the meter, the vehicles are stuck in traffic or encounters a traffic light at the intersection, resulting in duplicate data. If these errors and abnormal data are not processed, the accuracy of the calculation results will be affected [11].

In data preprocessing, firstly, the latitude and longitude out of bounds data is rejected; secondly, the duplicate data is removed; finally, for the drift data, the data is characterized by moving to another location in a short period of time at a moving speed that exceeds a reasonable range, thus eliminating such data by the speed threshold. The speed of the car in the urban area is 60 km/h, thus the set speed threshold is V = 0.0166 km/s. Three consecutive points $P_i\{t_i, lat_i, lon_i\}$, $P_j\{t_j, lat_j, lon_j\}$, $P_k\{t_k, lat_k, lon_k\}$ in the trajectory data, calculate the speed between P_i and P_j, P_j and P_k respectively,

$$V_{ij} = \frac{dis\tan ce_{p_i p_j}}{t_j - t_i}, V_{jk} = \frac{dis\tan ce_{p_j p_k}}{t_k - t_j} \quad (2)$$

if $V_{ij} > V$ and $V_{jk} > V$, then the location point P_j is judged to be an abnormal point and is culled.

4 Design of Taxi Drivers' Work Characteristics Analysis Method

4.1 Taxi Drivers' Work Characteristics Analysis Method

Taxi drivers' daily working hours account for 41.6%–58.3% of their life, their movement track is random when working, and there is no fixed stopover points, but they will stay for a certain period of time during meals (breakfast, lunch, dinner). Combining the feature of taxi drivers' work and rest, this paper uses the Spark data processing platform to analyze the working characteristics of taxi drivers based on the GPS trajectory data of Chengdu taxis, as shown in Fig. 2. Firstly, taxi GPS trajectory are preprocessed with common operators such as map, distinct, filter, sortBy, groupByKey, etc. in Spark. Then some important work indicators and taxi stop points are extracted. Using the statistical method to analyze the work indicators to get taxi drivers' work feature, use K-Means clustering algorithm to cluster the stop points to get the taxi drivers' resting place cluster points, and then analyze the taxi drivers' rest feature.

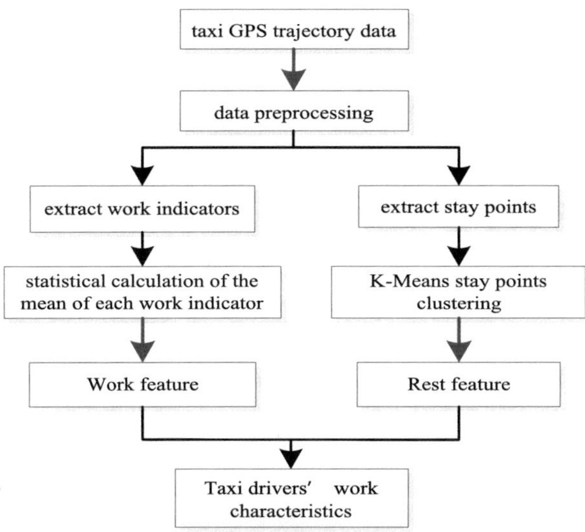

Fig. 2. Analysis method of taxi drivers' work characteristics

4.2 Extracting Work Indicators and Stay Points

The two algorithms of extracting important work indicators and extracting stay points are the core algorithms in the whole analysis method, both algorithms are based on the Spark platform.

Extracting Work Indicators. The distance and time difference of the interval are calculated by the latitude and longitude changes of adjacent time in the taxis' GPS trajectory data. we accumulate the interval distance to obtain single passenger delivery mileage, and calculate income according to the Chengdu pricing standard [12], then accumulate one day to get the taxi's mileage, passenger delivery mileage, passenger duration, order count, income, and finally calculate the loading rate and the single average mileage, pseudo code is as follows:

```
  1.Enter pre-processed taxi trajectory data
  2.Use map to read data by line, set key=taxi number,
values=GPS trajectory data
  3.Use groupByKey to group data by taxi number, and then
use map to process all data.
  4.While data is not null:
  5.  dist1+=get_distance(lat1,lon1,lat2,lon2) #driving
mileage
  6.  If status is 1:
  7.    dist2+=get_distance(lat1,lon1,lat2,lon2) #single
passenger mileage
  8.    time1+=time_diff(timestamp1,timestamp2) #passen-
ger duration
  9.    If whether the timestamp of two adjacent data is
the same hour period:
  10.       dist3+=get_distance(lat1,lon1,lat2,lon2)
#Hourly passenger mileage
  11.  elif state changed from 1 to 0:
  12.   dist4+=dist2 #passenger mileage
  13.   income+=income_sum(dist3) #income
  14.   count+=1 #order count
  15.   dist2=0
  16.rate=dist4/dist1 #load rate
  17.dist_ave=dist3/count #average mileage
```

Extracting Stay Points. The stay points are extracted by time threshold and distance threshold, the stay points are mainly used to analyze the feature of the taxi drivers at rest (breakfast, lunch, dinner). The minimum time threshold T = 10 min and the maximum distance threshold D = 0.1 km are set according to reference [13], the pseudo code is as follows:

```
  1.Enter pre-processed taxi trajectory data
  2.Use map to read data by line, set key=taxi number,
values=GPS trajectory data
  3.Use groupByKey to group data by taxi number, and then
use map to process all data.
  4.While data is not null:
  5.  if status is 0:
  6.    dist=get_distance(lat1,lon1,lat2,lon2)
  7.    if dist<D:
  8.      stay_time+=time_diff  #stay duration
  9.      add each GPS track data to the list of stay lo-
cations
  10.   else:
  11.     if stay_time>T:
  12.       k=1,calculate the cluster center of the stop
position list using K-Means
  13.       append GPS data of the center point and stay
duration to the stop list
  14.       clear the list of stay locations, stay_time=0
```

5 Experiment and Result Analysis

5.1 Experimental Environment

Hardware Environment. One host server is the master node, five virtual machines are slave nodes, Ubuntu14 operating system, the master node memory is 128G, and the slave nodes memory are 2G.

Software Platform. JDK7 version, Hadoop 2.6.5, Spark 1.6.3, python2.7. Spark supports Java, Scala, Python, and R language programming. Because the Python language is concise, it can be run directly without compiling and packaging. It is suitable for data analysis and display. Therefore, the code used in this experiment is based on Python language.

5.2 Experimental Results and Analysis

Analysis of the Feature of Taxi Drivers' Work. By the extraction algorithm of work indicators, hourly passenger delivery distance of each taxi is obtained. Then the sum of the hourly passenger delivery distance of all taxis per day is counted, and it is divided by the actual number of taxis per day, finally the hourly average passenger delivery distance for the week and Saturday is obtained, as shown in Fig. 3. According to Fig. 3, there are two low-peak passengers delivery distance in Chengdu during the week, from 12:00 to 13:00 and from 18:00 to 19:00, because these two time periods are lunch break and dinner break of taxi drivers, and 18:00 to 19:00 is the handover time of taxi drivers. On Saturday, the slow peak is 11:00 to 12:00, 13:00 to 15:00 and 17:00 to 19:00, and the peak is 20:00 to 23:00. Because the urban residents travel during the week is basically commuting time, and on Saturday is basically for dining and leisure, and August is summer, the evening activities are later than other seasons, and the evening travelers are more than other seasons. So taxi drivers' mileage change on Saturday is somewhat moderate compared to the week. It can also be concluded that the work intensity of taxi drivers on Saturday is lower than that in the week.

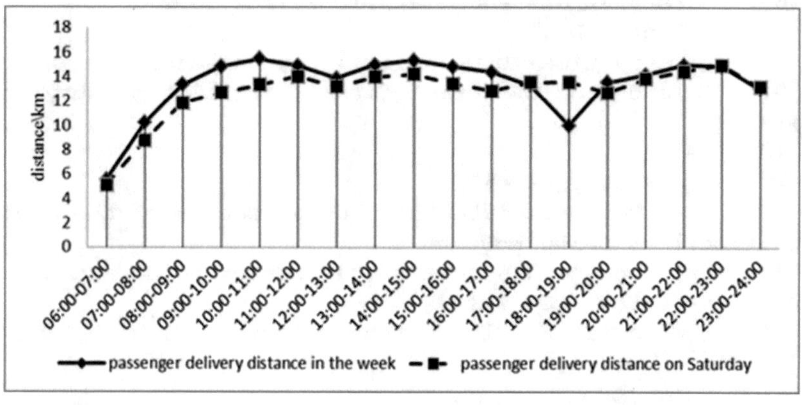

Fig. 3. Distribution of average passenger mileage per hour

The statistical method is used to calculate the mean value of each work index to obtain the results in Table 2. It can easily see that, except for driving mileage and single average passenger delivery mileage, taxi drivers' work indicators are higher than Saturday. The taxi driver in Chengdu travels an average of 348.63 km per day, of which the actual passenger mileage is 263.39 km, the orders is 37, the single average passenger delivery mileage is 6.84 km, and the income can reach 704.17¥. Taxi drivers can get a longer working time than the nine-to-five commuter, and the work intensity is higher than commuter. Because the state of urban residents gradually shifts from work in the week to leisure on Saturday, the working condition of taxi drivers are better than those on Saturday.

By counting taxi drivers' orders and income, the distribution of the orders and the income is obtained, as shown in Figs. 4 and 5. From Fig. 4, it can be clearly seen that taxi drivers' orders in Chengdu is mainly between 30 and 50 times, accounting for 79.8%, and the number of drivers with 50 or more orders is less, at 4.2%. As can be seen from Fig. 5, the daily income of taxi drivers is mainly distributed between 600¥ and 800¥, accounting for 64.2%, and only 1177 drivers can earn more than 800¥, accounting for 8.4%.

Table 2. Taxi drivers' work indicators

Date	Mileage (km)	Passenger mileage (km)	Loading rate	Passenger duration (h)	Single average mileage (km)	Orders	Income (¥)
2014/8/18	349.71	264.02	0.72	9.43	6.79	37	706.10
2014/8/19	351.06	263.71	0.72	9.38	6.88	37	704.64
2014/8/20	356.03	264.58	0.71	9.28	6.76	38	707.23
2014/8/21	344.78	262.61	0.73	9.66	6.87	36	702.58
2014/8/22	339.19	264.78	0.75	10.05	6.92	37	705.66
2014/8/23	351.02	260.61	0.70	8.91	6.83	35	698.83
average	348.63	263.39	0.72	9.45	6.84	37	704.17

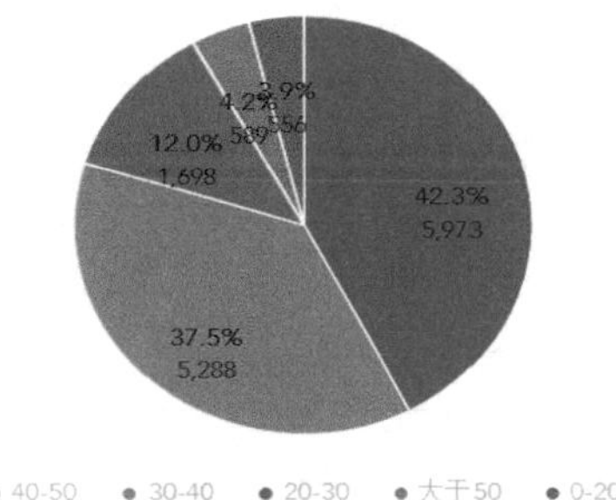

Fig. 4. Distribution of order count

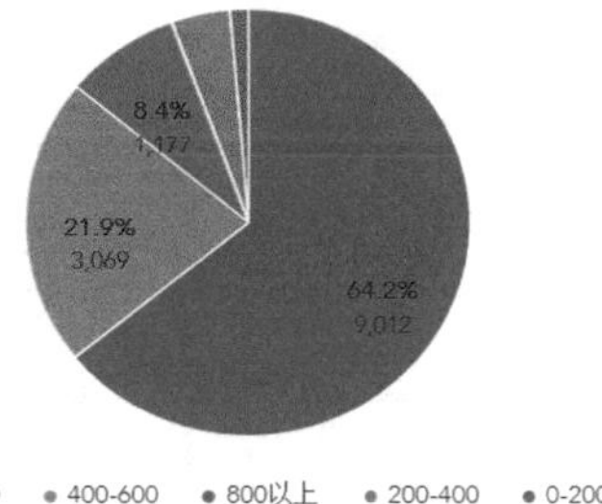

Fig. 5. Distribution of income statistic

Fig. 6. Distribution of cluster center points of three meals break

Analysis of the Feature of Taxi Drivers' Rest. According to during of stay and stay's time stamp, the stop points are divided into three categories: breakfast break, lunch break, dinner break. The duration of stay is 10 min to 30 min and the stay's timestamp is between 6 and 8 o'clock for breakfast break; the duration of stay is between 20 min and 40 min and the stay's timestamp is between 10 and 14 o'clock for lunch break; the duration of stay is between 20 min and 60 min and the stay's timestamp of 17 to 20 o'clock for dinner break. Then K-Means clustering algorithm is used to cluster the latitude and longitude of these three types of stay points. There are 8 cluster center points in each of the three meals, and the cluster center points are mapped to the map, as shown in Fig. 6. The points included in each cluster center of the three meals are counted to obtain the results as shown in Fig. 7.

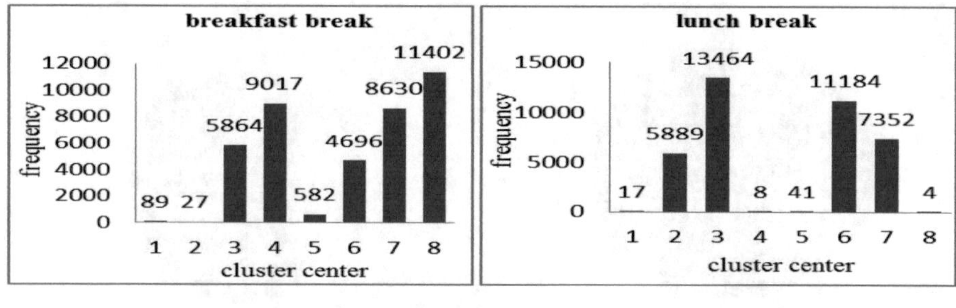

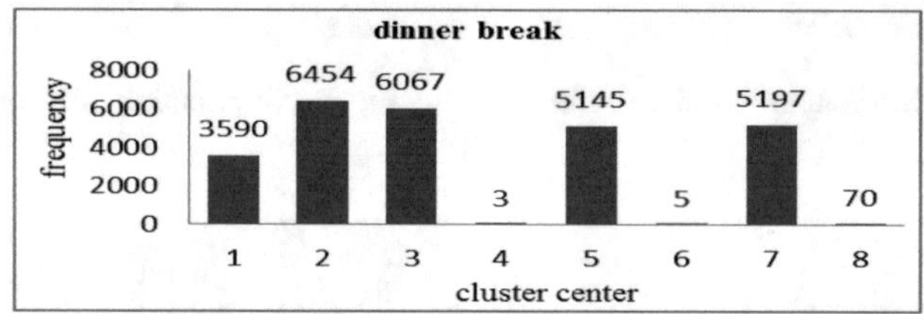

Fig. 7. Three meals break each cluster center point contains point statistics

In Fig. 6, orange stands for breakfast break, red stands for lunch break, blue stands for dinner break. From the picture, the taxi drivers' three meals are mainly within the Third Ring Road of Chengdu, and a few are outside. The rest of the breakfast and dinner areas are relatively concentrated, and the lunch break are relatively scattered. The three breaks in the Third Ring Road and Wenjiang District are relatively close, up to 369 m, and the farthest is only 2,230 m. From Fig. 7, it can get the Chengdu taxi driver's breakfast break mainly in the 4, 8 area, the lunch break is mainly in the 3,6 area, and the dinner break is mainly in the 2, 3 area. These areas are mainly located in the old central area of Chengdu, which is within the First Ring Road of Chengdu. Breakfast break in the 1, 2 area, lunch break 1, 4, 5, 8 area, dinner break 4, 6, 8 area contains fewer taxi drivers. It can be seen from Fig. 6 that these cluster center points are far from the old central area of Chengdu, and are all outside the Third Ring Road of Chengdu.

6 Conclusion and Discussion

Through analysis, we get the average time of taxi drivers in Chengdu to work more than 9 h a day, the work intensity is high, the daily mileage is more than 340 km, the actual passenger delivery mileage is more than 260 km, most drivers take orders 30 to 50 times, the income is 600¥ to 800¥, and the rest of the meal is basically in the old city center.

Using the Spark big data processing platform, we can quickly and accurately analyze the taxis' GPS trajectory data in Chengdu for 6 days. And the calculation of the extraction work indicators and the stop point can be completed in 40 min. The research method used is also applicable to the courier and the delivery staffs' characteristics. This article will further distinguish the categories of taxi drivers, analyze the differences in the work characteristics of different categories of taxi drivers, and study the factors affecting the income of taxi drivers of different categories.

Acknowledgements. This research was supported in part by Key Research and Application of Big Data Trading Market in Shaanxi Province (2016KTTSGY01-01).

References

1. Zhuang, L., Song, J., Duan, Z., et al.: Research on taxi operating characteristics based on floating car data mining. Urban Transp. **14**(1), 59–64 (2016). https://doi.org/10.13813/j.cn11-5141/u.2016.0109
2. Zhuang, L., Wei, J., He, Z., et al.: Taxi operation and management characteristics based on floating car data. J. Chongqing Jiaotong Univ. **33**(4), 122–127 (2014). https://doi.org/10.3969/j.issn.1674-0696.2014.04.25
3. Lv, Z., Jianping, W., Yao, S., et al.: FCD-based analysis of taxi operation characteristics: a case of Shanghai. Nat. Sci. East China Norm. Univ. **3**, 133–144 (2017). https://doi.org/10.3969/j.issn.1000-5641.2017.03.015
4. Liu, L., Andris, C., Ratti, C.: Uncovering cabdrivers' behavior patterns from their digital trace. Comput. Environ. Urban Syst. **34**, 541–548 (2010). https://doi.org/10.1016/j.compenvurbsys.2010.07.004
5. Zhang, J., Chou, P., Du, M.: Mining method of travel characteristics based on spatio-temporal trajectory data. Transp. Syst. Eng. Inf. **14**(6), 72–78 (2014). https://doi.org/10.16097/j.cnki.1009-6744.2014.06.011
6. Liu, H., Kan, Z., Sun, F., et al.: Taxis' Short-Term out-of-service behaviors detection using big trace data. Geomat.S Inf. Sci. Wuhan Univ. **41**(9), 1192–1198 (2016). https://doi.org/10.13203/j.whugis20150569
7. Xia, J., Wei, Z., Fu, K., et al.: Review of research and application on hadoop in cloud computing. Comput. Sci. **43**(11), 6–11+48 (2016). https://doi.org/10.1196/j.issn.1002-136X.2016.11.002
8. Karau, H.: Spark Fast Big Data Analysis, pp. 1–6. People Post Press (2015)
9. Duan, Z., Chen, Z., Chen, Z., et al.: Analysis of taxi passenger travel characteristics based on spark platform. Comput. Syst. Appl. **26**(3), 37–43 (2017). https://doi.org/10.15888/j.cnki.csa.005617
10. Zhou, Z.: Machine Learning, pp. 202–204. Tsinghua University Press, Beijing (2016)

11. Yang, Y., Yao, E., Pan, L., et al.: Research on taxi route choice behavior based on GPS data. J. Transp. Syst. Eng. Inf. Technol. **15**(1), 81–86 (2015). https://doi.org/10.16097/j.cnki.1009-6744.2015.01.013
12. Chengdu Taxi Network. Taxi rental standard in Chengdu downtown area. http://www.cdtaxi.cn/zujia.html,2018/06/11
13. Qi, X., Liang, W., Ma, Y.: Optimal path planning method based on taxi trajectory data. J. Comput. Appl. **37**(7): 2106–2113 (2017). https://doi.org/10.11772/j.issn.1001-9081.2017.07.2106

Research on Complex Event Detection Method Based on Syntax Tree

Wenjun Yang(✉) and Aizhang Guo

Qilu University of Technology (Shandong Academy of Sciences), Jinan, China
y_wjl114@163.com

Abstract. This paper focuses on the diversity of event flows and the limitation of memory. In this paper, an application tree structure is proposed to compress event storage, and a complex event detection method based on syntax tree is adopted. This method uses the strategy of constraint downshift and shared subsequence to achieve the goal of saving time and space. Constraint downshift prioritizes events with low pass rates and eliminates a large number of non-compliant events, thereby increasing efficiency. The shared subsequence is based on the existing matching results, and a new result sequence is constructed according to the query event pattern. In order to improve query efficiency and save storage space, nested queries are used to query complex events. The effectiveness of these methods was verified by experiments with these strategies, and the accuracy of the method was compared with the SASE method for complex event detection. Finally, summarize the paper and point out the next research direction.

Keywords: Syntax tree · Complex event detection · Shared subsequence Constraint down

1 Introduction

The existing process industry is relatively complex, which is characterized by uncertainty, coordination of multiple resources, large amount of data and information, and higher requirements for the coordination of management control [1]. Process enterprises improve their operational efficiency by coordinating and controlling the relationship among production planning, scheduling and production control. In fact, complex event detection is by collecting data from various production links, such as filtering, aggregation, rule constraint and so on, gradually aggregating the simple events into complex events, and refining out the abnormal situation and useful information in the process industry [2]. At the same time, complex event detection also faces the following problems:

(1) For massive real-time data, the pattern matching of complex event detection adopts a complete and complete match, the processing speed is low, and the computing resources and storage resources are greatly wasted.
(2) In the query processing of complex events, there is a lack of effective rules to describe the correlation and information between data. Although the query mode can be prefabricated according to historical data and domain experts, it is impossible to presuppose all query modes because of the complexity of real-time data.

P. Krömer et al. (Eds.): ECC 2018, AISC 891, pp. 431–440, 2019.
https://doi.org/10.1007/978-3-030-03766-6_49

To solve these problems, a complex event detection method based on syntax tree is proposed in this paper. The main contributions are as follows:

(1) By analyzing the similarity of test results, we propose to build shared common subsequences by using the detected result sequences to achieve incremental detection and shared detection.
(2) The user query rules are divided into multiple sub queries, so that sub queries can also form shared sequences and enhance the scalability of query statements.

2 Related Concepts and Description of the Problem

2.1 Related Concepts

Definition 1: Event model. In general, the model of events is a three tuple: (id, type, time), id represents the id of the detected object; type is the event type; time indicates the occurrence time of the event. This universal model can also be extended to E = (id, attributes, causality, start, end), start and end respectively indicate the start time and end time of the event, attributes = $\{a_1, a_2, \ldots, a_n\}$, $n \geq 0$, represents the collection of attributes of the event, causality = $\{c_1, c_2, \ldots, c_n\}$, $n \geq 0$, represents a set of causal events associated with the event.

Definition 2: Event operator [3]. Event operators should cover the patterns and semantics involved in process industry production as far as possible.

(1) Logic and, denoted by $X \wedge Y$, it means that both X and Y occur. $(X \wedge Y)$ represents the maximum interval between X and Y.
(2) Logic or, denoted by $X \vee Y$, means the occurrence of X or the occurrence of Y.
(3) Logic negation, denoted by $\neg X$, means that event X did not occur.
(4) SEQ Operator, denoted by $E = SEQ(E_1, E_2, \ldots E_n)$, indicates that events occur in order.
(5) Choice, denoted by X^Y, represents an event that conforms to the keyword Y, such as first event instance X^{First}, tail event instance X^{Last}, the nth event instance X^n, all event instances X^{All} and so on.

Definition 3: Attribute constraint. It generally contains temporal constraints and predicate constraints [4]. temporal constraint, as the name implies, describes the rules of events that should be met in time. A predicate constraint is a constraint on an event, such as E(e. id = "001")∧(e. type = "car")∧(a = "A")∧(t < 10 s), it can be used to indicate that the code is 001, the event category is car, and the dwell time at A is less than 10 s.

Definition 4: Complex event processing rules [5].

Event<event pattern>
If<conditions>
Do<action>

Every step of complex event detection must be processed according to this rule. This means that for complex event expressions that users want to query, if the events in the event flow satisfy the conditions, they trigger corresponding behaviors.

Definition 5: Tree structure. The query expression of complex events is represented by tree structure. For example, Build a query expression Q = SEQ(A, B, C).win:time (2 min) into a tree structure. According to the query expression Q, we need to query the ABC type events that are executed sequentially within two minutes. The entire construction process is shown in Fig. 1.

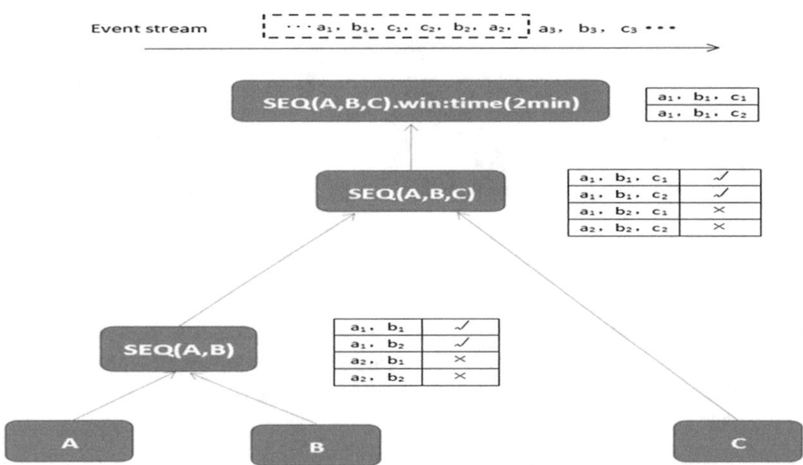

Fig. 1. Implementation of tree structure model

Definition 6: Candidate event instances. An instance that has occurred and has the same ID attribute as the current event instance [6].

Definition 7: Matching process. Use the tree structure shown in Fig. 1 to match the events involved in the user query with the candidate event instances, and then construct the result sequence.

2.2 Problem Description

Problem 1: In the process of complex event detection, tree structure detection will produce a large number of intermediate results. There are a large number of non conforming events in these intermediate results, which affect the efficiency of storage and invocation.

Problem 2: In the process of complex event detection, the query event pattern is parsed according to the user query requirements. There are often a lot of duplication in the result of a specific event flow and query pattern. For example, there are four types of ABCD events occurring in a time window, the query event pattern is SEQ(A, B, C, D), and meet the corresponding conditions of the corresponding event type. As shown in Fig. 2, according to the query conditions, the result sequence R1, R2, R3, R4 can be obtained. It can be seen from the result sequence that there is a duplicate part between the result sequences, that is, the shared subsequence, For example, the shared subsequences of R1 and R3 are (b_1, c_2, d_1), and the shared subsequences of R2 and R4 are (b_2, c_2, d_1).

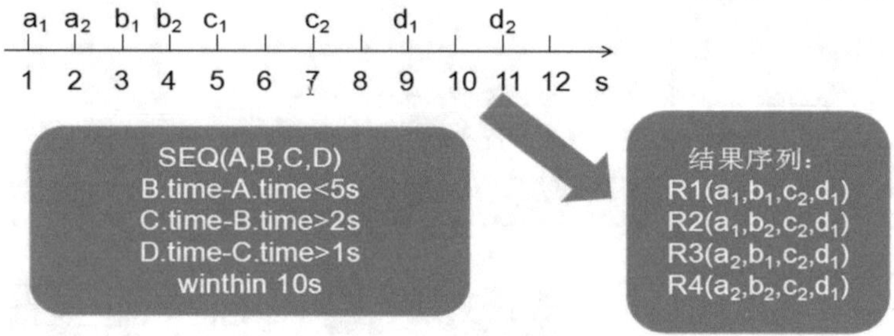

Fig. 2. Event pattern detection

Problem 3: In general, when a user query is used to detect complex events, it is necessary to parse the complex expression, extract the query parameters on the event, and write the parameters to the query statement to query the events that meet the conditions. When the query volume of users is very large, many query parameters are duplicated, which will lead to duplicate query. Storage and call will become a big problem. When the query volume of users is very large, many query parameters are duplicated, which will lead to duplicate query. Storage and call will become a big problem.

2.3 Method Description

Method 1: Constraint down. For problem 1, in order to improve the efficiency of detection, we can change the location of predicate constraint or time constraint to improve efficiency [7]. For example, the query mode is SEQ ((A, B), C) winthin 10 s, we change the constraint position of the query, that is, moving up and down in the tree structure, the query mode can be equivalent to SEQ ((A, B) winthin 10 s, C) winthin 10 s, which can filter out some insignificant intermediate results.

The whole tree structure is composed of multiple query constraints. First, we need to calculate the time consumed by various query constraints in event detection, which is expressed in T, and the unit is ms. Next, we take 3 query constraints as an example to calculate the time consumption (Table 1).

Table 1. Correlation variable definition

Variable meaning	Letter
Time consumption of comparison	r
Time consumption of acquiring attributes	s
The new generation of event time consumption	o
Probability of passing time constraint U1	P_{time}
Probability of passing type constraint U2	P_{type}
Probability of passing attribute constraint U3	P_{atti}

Assuming that there are n attribute constraints, the probability of passing is P_1, P_2, …, P_n, the probability that the data flow passes all the constraints is:

$$Patti = \prod_{i=1}^{n} P_i \tag{1}$$

In the matching process, the time consumption of attribute determination can be seen to be basically the same, so Tw = T1 = T2 = … Tn, the minimum of the time of consumption is:

$$T_U = T_{U1} + P_{time}T_{U2} + P_{time}P_{type}T_{U3} = n + 2s + Tw\left(1 + \sum_{i=2}^{n}\prod_{j=1}^{i-1} P_j\right) \tag{2}$$

The probability of a query condition is:

$$P = P_{time}P_{type} = 1 * 1/2 * \prod_{j=1}^{n} P_j \tag{3}$$

The query constraint that has three conditions are relatively simple and get T = 2n + j after taking in.

For a set of data streams, the time consumed for matching the most complex cases is:

$$T = n + 2s + Tw\left(1 + \sum_{i=2}^{n}\prod_{j=1}^{i-1} P_j\right) + n\prod_{j=1}^{n} P_j + \prod_{i=1}^{n} P_i/2 \tag{4}$$

Therefore, when there are multiple constraint conditions, the detection efficiency will be improved if the constraint conditions with low probability are passed first.

There is also a constraint downshift for testing with equivalent attributes [8], For example (A∧B) where [$attri_1$ $attri_2$], using the partition strategy for the selection of the candidate event instances, it can also improve detection efficiency and save time consumption.

Method 2: From the description of problem 2, a new sequence of results can be constructed by using a matched sequence of results, and a number of matching results can be output by a shared subsequence matching only once. When the result R1 is detected, the shared results are saved, and then when the detection of R3 is needed, the results are compared directly with the shared results to improve the detection efficiency.

There are two kinds of matching results, including semantic inclusion or semantic exclusion. For a complex event, the first step is to find out if there is a matching sequence that is exactly the same, and if not, to find sequences that are semantically inclusive or mutually exclusive.

For example, the detection of SEQ(A¬B) winthin 10 s indicates that the occurrence of B does not occur in A within 10 s, which is related to SEQ(A, B, …) winthin 10 s has semantic mutual exclusion.

Method 3: Nested query. Table 2, we can check the commodity in and out of the warehouse, query users are interested in different events, but the event attributes are fixed, so nested query can be adopted to achieve the expansion of query statements. For example, user A pays attention to goods with a commodity number of 007 in two minutes, and user B pays attention to goods with a commodity number of 123. Therefore, the combined storage of query statements constrained by this equivalent property can reduce memory and make it convenient to call. The nested query is:

```
Select*from pattern [every (a=commodity(id in [007, 123]
and action='in')->b=commodith(id=a. id and action!=a. action
and position=a. position)where time. winthin. (2 min))]
```

Table 2. The simple event of a commodity

Attribute name	Attribute value	Attribute meaning
id	[000–999]	Commodity number
action	[in, out]	Warehouse
position	[P1, P2, P3, P4]	Position
time	[00:00–23:59]	Time
weight	>0	Weight

3 Correlation Algorithm

The implementation of this algorithm is a syntactic tree structure model based on the improved matching tree basic model. Matching tree, the basic model, is based on the basic event as the leaf node of the matching tree, and the complex event intermediate node is obtained through constant constraints. When the node is reached and the user query event is iterated layer by layer, the complex event that satisfies the condition is considered to be detected. For the methods proposed in the previous section, we introduce the ideas of these strategies into the matching tree model. The algorithm is as follows:

Input : Query constraint Query(Q1, Q2, …, Qn), event stream Stream(id, attributes, causality, start, end), index structure Index.

Output : Syntactic tree Tree.

```
1. Prase(complex events)
2. User Query=(Q1, Q2, …, Qn)←Stream
3. Save relation(complex events)
4. Push_downconstraints()
5. for(i=1;i<=n;i++)
      if Index[i]. id==id
      Get Tree←Index[i]. pointer
      (id, attributes, causality, start, end)insert Tree
      tree=Tree(complex event)
6. t=Entical(tree)
     If t=tUContinment(tree)
     else tUExclude(tree)
7. if t is null
     Get candidate list()
8. if equivslence list exists
       Partition(Candidate instance)
9. if father node exists
     Report result
     Save result to Shared_subsequence()
```

First, the complex event is parsed (1), the event flow is processed according to the user's query statement (2), and the related complex events are stored (3), and the constraint is moved down (4). This part is to do the initialization to prepare for the next construction of the language tree structure. The syntax tree is constructed by the index structure (5). View Shared subsequence table whether to have the test results of the structure of the complex event. If there is such a test result, it is directly reference, If it does not exist, view the detection results that are included or mutually exclusive with the semantic structure of the complex event. The semantics include the parts that can use the same semantics. The semantic mutual exclusion can be used after conversion (6). If there is no shared subsequence and there is no semantic inclusion or semantic exclusion, it is necessary to extract relevant candidate event instances. If there is an equivalent test, then the candidate event instance is partitioned (7, 8). Determine whether the parent node of the candidate event instance exists, and if it exists, the result is feedback immediately, and the result is stored in the shared subsequence table for the subsequent detection is called (9).

4 Experimental Analysis

This section will verify the accuracy and validity of the mentioned optimization strategy and the complex event detection method based on the syntax tree through experiments. The experimental environment is built in the VMware virtual machine, the operating system of the host is Windows 7, the CPU is Intel i7 processor, the memory is 8G, the virtual machine environment CPU is dual core, the memory 1G, the operating system Ubuntu12. 04, Storm, Esper and other software are in the virtual machine.

In this experiment, in order to better verify the method proposed in this paper, we compare it with the traditional SASE system model algorithm [9]. lBy experiment, we compare the runtime of complex event detection algorithm based on syntax tree, and more importantly, the accuracy of complex event detection. Before doing the experiment, we first define 40 complex event patterns, in which group A sets up 10 complex event patterns that contain the common subsequence ((A∧B) (C∧D)), its purpose is to test the role of the shared result subsequence. In order to test the effectiveness of the constrained down-shift strategy, Group B sets 10 complex event detection modes that only change the position of the attribute constraint. According to the experience and the characteristics of data set, By changing the position of predicate constraint and time constraint, the probability events are first detected. The other 20 complex event detection modes were divided into two parts. Group C does not use the shared result subsequence method and the constrained downshift method. The complex event pattern cannot contain common subsequences. The two methods in group D are used, not only setting common subsequence, but also ensuring that the constraint condition is low. The following figure is the result of a contrast experiment.

From Fig. 3, we can see that the shared subsequence method and the constraint downshift strategy have obvious advantages in the complex event detection process, which greatly improves the detection efficiency. The result of the shared matching reduces the repeated matching in the complex event detection process, which saves time; The constraint downshift strategy takes priority to detect the event with low pass rate, which eliminates a large number of events at the beginning, reduces the workload of subsequent event detection, and saves time.

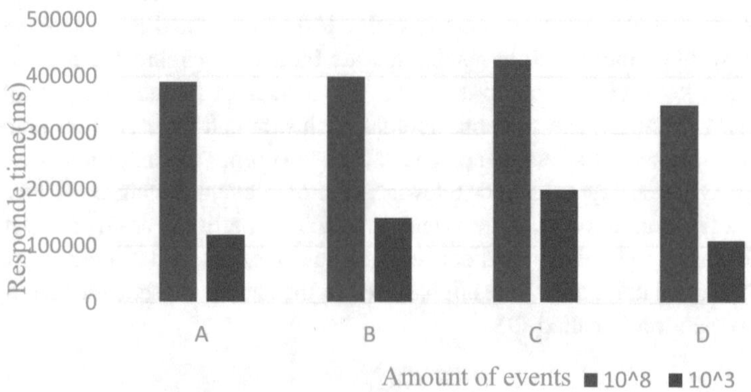

Fig. 3. The impact of different strategies on event detection

The experiment also compares the accuracy of the complex event detection, and compares the D method with the traditional SASE method and the A, B, and C group methods in this paper to make a more objective evaluation. The results of the experiment are as shown in Fig. 4.

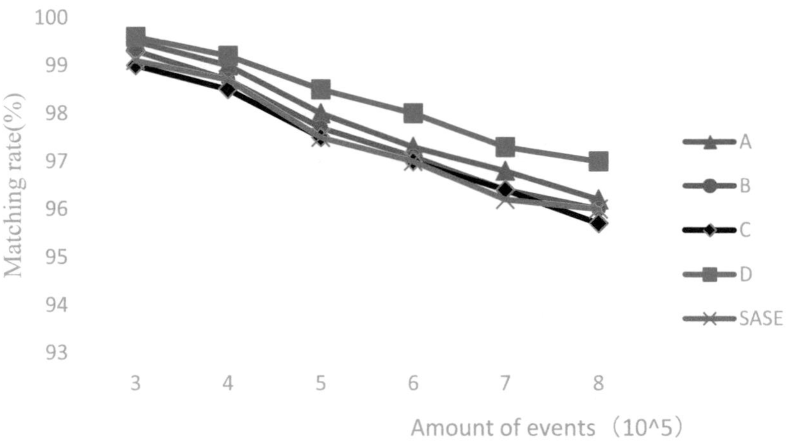

Fig. 4. Matching rate comparison diagram

The superiority of the D group method can be seen from Fig. 4. The strategy of adding common subsequence method and constraint downshift is not only short in matching time but also high in accuracy. Compared with the traditional SASE algorithm, the complex event detection method based on the syntax tree also has the advantage. The shared subsequence accelerates the detection efficiency, the detection and sharing are carried out simultaneously, and the content of the shared subsequence is constantly enriched, and the content of the detection is constantly comprehensive, thus the success rate of the event matching is raised.

5 Conclusion

In this paper, we apply the shared common subsequence and the constraint downshift strategy in the process of complex event detection, and apply the nested query strategy to the query process, and the validity of this idea is proved by the experiment. Although the method in this paper has shown good performance in the experiment, the real process industry environment is more complex, and the data and data types are more complex than the data in the experiment. At this stage, there is still no good way to excavate the information and the association between data, and the construction of many event patterns still needs to be completed based on historical data and artificial experience. In this experiment, the Eeper is a 4.11.0 version, and the next step is to improve the open source software according to the experimental requirements, and use a larger amount of events to carry out the experiment.

Acknowledgement. This work was supported by Key Research and Development Plan of Shandong Province (2017GGX201001).

References

1. Hu, C., Li, P.: Comparison of MES between productions of continuous industries and discrete industries. Control Instrum. Chem. Ind. **30**(5), 1–4 (2003)
2. Wang, F., Liu, S., Liu, P.: Complex RFID event processing. Int. J. Very Large Data Bases **18** (4), 913–931 (2009). https://doi.org/10.1007/s00778-009-0139-0
3. Dimitriadou, K., Papaemmanouil, O.: Explore-by-example: an automatic query steering framework for interactive data exploration. In: Proceedings of the 2014 ACM SIGMOD International Conference on Management of Data Snowbird, USA, pp. 517–528 (2014). https://doi.org/10.1145/2588555.2610523
4. Yi, H.: Research on reconfigurable manufacturing execution system for RFID-based real-time monitoring. Tsinghua University, (2011)
5. Liu, H.-L., Li, F.-F.: Processing nested query over event streams with uncertain timestamps. Chin. J. Comput., 123–134 (2016). https://doi.org/10.13190/j.jbupt.2017.02.008
6. Wang, Y., Mend, Y.: Method of complex events detection based on shared matching results. Appl. Res. Comput., 2338–2341 (2014). https://doi.org/10.3969/j.i55n.1001-3695.2014.08.023
7. Shahbaz, M., McMinn, P., Stevenson, M.: Automatic generation of valid and invalid test data for string validation routines using web searches and regular expressions. Sci. Comput. Program. **97**, 405–425 (2015). https://doi.org/10.1016/j.scico.2014.04.008
8. Wasserkrug, S., Gal, A.: Efficient processing of uncertain events in rule-based systems. IEEE Trans. Knowl. Data Eng. **24**(1), 45–58 (2012). https://doi.org/10.1109/TKDE.2010.204
9. Gyllstrom, D., Wu, E., Chae, H.J., et al.: SASE: complex event processing over streams. arXiv preprint arXiv:cs/0612128 (2006)

Theoretical Research on Early Warning Analysis of Students' Grades

Su-hua Zheng[1(✉)] and Xiao-qiang Xi[2]

[1] School of Science, Xi'an University of Posts and Telecommunications, Xi'an, China
1028743050@qq.com

[2] Institute of Internet of Things and IT-Based Industrialization, Xi'an University of Posts and Telecommunications, Xi'an, China
xxq@xupt.edu.cn

Abstract. Based on the basic theory of data mining, the classical association rule algorithm, Apriori algorithm is used to analyze the grade data of students majoring in computer science and technology and information and computing science of a university, which aims to find out the intrinsic links between the courses and put forward some meaningful early warning rules. Since a lot of rules that obtained by the Apriori algorithm do not conform to logic, effective rules need to be screened artificially according to the prior knowledge of courses sequence, which will waste a lot of time and effort. So SPADE algorithm based on sequential pattern mining is introduced to obtain early warning rules that base on time series. The results show that there is a strong correlation among professional core courses. The obtained rules can provide early warning for students, provide reference for teachers' teaching plans, and assist in the formulation of professional training programs.

Keywords: Data mining · Apriori algorithm · SPADE algorithm
Grade analysis · Curriculum link

1 Introduction

China is a big education country. According to the "2016 National Education Development Statistics Bulletin", the total number of higher education students in the country has reached 36.99 million, with more than 56,000 professional points. Each professional point has accumulated a large amount of valuable student grade data. Simply performing backups, queries, and statistics on these data does not effectively exploit the hidden value behind the data. Fully mining and using the internal correlation of these data is of great significance to improve the level of education. Data mining finds implicit rules by analyzing a large number of data. Common algorithms include decision tree algorithms, classification algorithms, association algorithms, neural network algorithms [1, 2] and so on. Data mining technology can be used to find the implicit association rules between the various courses for early warning of student performance, and provide the guidance for teacher's teaching plan arrangement and training program design.

P. Krömer et al. (Eds.): ECC 2018, AISC 891, pp. 441–449, 2019.
https://doi.org/10.1007/978-3-030-03766-6_50

Apriori algorithm is a classical algorithm in data mining and is widely used in various fields. Through the study of literature, the application of Apriori algorithm in student grade analysis can be summarized. It is mainly used to study the association between courses [3], the impact of student behavior and teacher behavior on student performance [4], and the various educational systems and decision analysis systems based on the creation of data warehouse and the application of association rules [5]. When mining association rules, Apriori algorithm is most often used with decision tree algorithm [6] and clustering analysis algorithm [7]. The Apriori algorithm is parallelized to improve its efficiency [8]. In some studies, the Apriori algorithm is optimized. The methods based on the compression matrix [9], the transaction marker [10], the data size division [11], the interest degree [12] and so on are introduced, to reduce the running time and improve the accuracy of the algorithm by reducing the number of scans on the transaction database and the invalid connections.

The Apriori algorithm is used to analyze the association rules of the core courses in the professional training program to find out the close links between the main courses, which can provide a very effective learning approach for mastering the professional knowledge. In addition, these rules can make the formulation of training programs more in line with the learning characteristics of students. However, these association rules cannot reflect the early warning relationship between courses in time series. There are a large number of non-logical rules, which need to be filtered by manual intervention. When the rule set is relatively large, it is very difficult to find out all the sequence-based association rules. According to the limitations of Apriori algorithm in mining sequence model, a new algorithm, Sequential Pattern Discovery Using Equivalence (SPADE) algorithm is introduced. Using SPADE algorithm, effective early warning rules between courses can be obtained directly, and the comprehensibility and practicability of the rule set can also be improved.

2 Fundamental Theories

2.1 Association Rules

There is some regularity between the values of two or more variables, which is called association [13]. Association rules look for correlations between different items that appear at the same time, aiming to find interesting associations or interconnections between item sets in a large amount of data. The classical association rule algorithms include Apriori algorithm, Fp-growth algorithm, etc. Now researchers are aiming at different research objectives, graph-based association rule mining algorithm, data stream-based association rule mining algorithm, sequence-based association rule mining algorithm are proposed [14], and make association rules better applied. Association rules are defined as: Given a data set D, $Item = \{i_1, i_2, \ldots i_m\}$ is a collection of all these items. $Tr = \{tr_1, tr_2, \ldots tr_k\}$ is used to represent a transaction set, Each transaction set is a non-empty subset of *Item*. The association rules in the data set are constrained by the degree of support and confidence. The support is the percentage of transactions in D that contain both X and Y, that is, the probability. Confidence is the percentage of Y, which is the conditional probability, if the transaction in D already contains X. The support of the rule $is \rightarrow it$ in D refers to the ratio of number of transactions containing *is* and *it* to the number of *Tr*. It is written as:

$$\text{Support}(is \rightarrow it) = |\{Tr : is \cup it \subseteq Tr, Tr \in D\}|/|D| \quad (1)$$

The confidence of the rule $is \rightarrow it$ in D refers to the ratio of the number of transactions containing *is* and *it* in *Tr* to the number of *Tr* that includes *is*. Referred to as:

$$\text{Confidence}(is \rightarrow it) = |\{Tr : is \cup it \subseteq Tr, Tr \in \text{D}\}|/|Tr : is \subseteq \text{Tr}, Tr \in \text{D}| \quad (2)$$

The minimum support and the minimum confidence are artificially given thresholds. Association rule mining is to find out all rules with all support and confidence greater than a given threshold in many rules [15].

2.2 Apriori Algorithm

As the most classical association rule algorithm, Apriori algorithm is widely used in various fields. By analyzing the relevance of data, the information obtained has important reference value in the decision-making process. The iterative method of layer-by-layer search is used to find out the relations of item sets in the database to form rules, and the process includes connection and pruning [16]. The Apriori algorithm is based on two core theories: the subset of frequent itemsets must be frequent itemsets; the super-set of infrequent itemsets must be infrequent itemsets [14]. The flow chart of the algorithm is as Fig. 1:

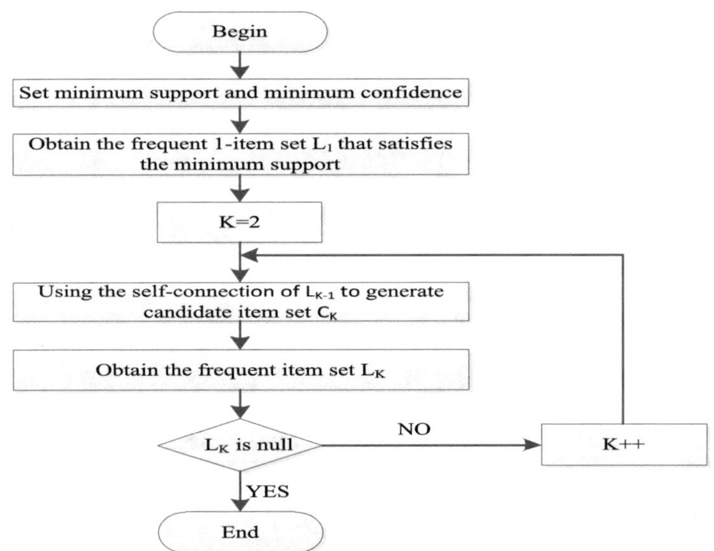

Fig. 1. The mining process of Apriori algorithm

2.3 SPADE Algorithm

Sequential pattern mining algorithm is an important branch of data mining, which can be used for biological sequence analysis, customer shopping analysis, website traffic analysis and so on.

The concept of sequence pattern was first proposed by Agrawal and Strikant [17]. That is, mining the rules that meet the minimum degree of support from the database of data sequences. Apriori-like algorithms are all based on the horizontal format algorithm, and an algorithm based on the vertical format was proposed by Zaki. SPADE is one of the classic algorithms [18]. The basic idea of the SPADE algorithm is to use the characteristics of the vertical representation of data. When generating frequent item sets, there is no need to scan the original database. Instead, the vertical data sequence of each item set is subjected to intersection operation. The item is frequently if the generated intersection is greater than the support degree [19]. The flow chart of the algorithm is as Fig. 2:

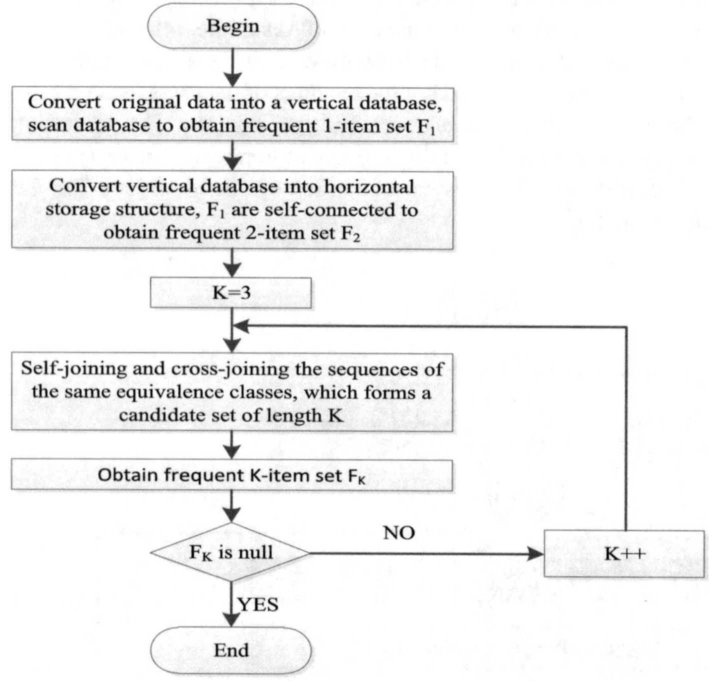

Fig. 2. The mining process of SPADE algorithm

The SPADE algorithm only scans the full database three times, so the operating efficiency is improved.

The Apriori algorithm and SPADE algorithm are applied to the exploration of curriculum relevance in our problem, and the algorithms are implemented in weka plaform and R language respectively to get meaningful curriculum association rules which can provide advice for students and teachers.

3 Application of Association Rules in the Relevance of Courses

3.1 Identify Data Mining Object and Goal

The data mining object is the grade data of students majoring in computer science and technology and information and computing science. The mining goal is to find a certain kind of association among the core courses of the specialty. This association will be used to provide the early warning for the students' grade, appropriate guidance for the arrangement of the teacher's teaching plan and the instructions of professional training program.

3.2 The Choice of Model

The model to use is determined according to the mining object and goal. The classic Apriori algorithm is used to find association rules in the data. Some rules that obtained by the Apriori algorithm have no practical meaning, such as mathematical modeling course was taught before the operating system, however, in the obtained rules, if the operating system is medium, and the good probability of mathematical modeling will be 83%, obviously not reasonable. The reasonable rules need to be picked up by human intervention in most of the literature according to the prior knowledge of courses sequence. This approach is feasible when database is small, while is impossible for large database and will consume a lot of time and energy. So SPADE algorithm is introduced, which is based on time series pattern mining. Given the sequence of courses, one can directly get links between successive courses without having to filter the rules. The Apriori algorithm is implemented on the weka platform, and the SPADE algorithm is implemented in the R language software.

3.3 Data Collection

The data in this paper is derived from the grade data of students in computer science and technology, information and computing science major of a university. From the professional training program, the core courses are selected as the data mining targets, including discrete mathematics, data structure, and computer composition principle, etc.

3.4 Data Preprocessing

Since there are some missing values or isolated points in the student grade data set, they need to be deleted or filled in according to the nature of the vacancy, which is called cleaning. Apriori algorithm and SPADE algorithm are one kind of algorithms used to mine the association rules of frequent set of Boolean, so the data set needs to be discretized to satisfy the data format of the algorithm. Subdivided discretization is used as: the grade meets or exceeds 85 for excellent, 75 to 84 for good, 60 to 74 for medium, less than 60 for fail.

3.5 The Application of Apriori Algorithm and Its Results

According to the preset minimum support and minimum confidence, the Apriori algorithm is used to mine association rules. Let the minimum support degree 0.01 and the minimum confidence degree 0.5. Here the concept of lift is introduced. For the rule that item set X deduces item set Y, the significance of the lift of the rule is to measure the independence of item set X and item set Y. its formula is:

$$\mathrm{Lift}(X \rightarrow Y) = \frac{Support(X \cup Y)}{Support(X) * Support(Y)} \tag{3}$$

When the lift is 1, the X and Y item sets are independent. When the lift is less than 1, it means that X and Y are mutually exclusive. At this case, regardless of how high the support and confidence of the rule that X item set deduces the Y item set, this association rule is considered as invalid and should be neglected. In general, when the lift is greater than 3, the strong association rule is regarded as valid. The obtained effective strong association rules are shown in Table 1.

Table 1. Partial strong association rules with a minimum support degree 0.01 and a minimum confidence degree 0.5

Pre rule	Post rule	Support	Confidence	Lift
DME[a]	DSE[b]	0.027	1.00	15.71
DME	CLDE[c]	0.018	0.67	4.31
DME	DPAE[d]	0.027	1.00	11.00
DME	OSE[e]	0.018	0.67	4.58
DME	CPE[f]	0.018	0.67	6.11
DME	CNE[g]	0.018	0.67	18.30

[a]Discrete mathematics excellent
[b]Data structure excellent
[c]Circuit and logic design excellent
[d]Database principle and application excellence
[e]Operating system excellent
[f]Compiling principle excellent
[g]Computer network excellent

It can be seen that discrete mathematics is the foundation of follow-up courses, such as operating system, computer composition principle, and computer network. Students with good discrete mathematics score will do better in subsequent core professional courses. Students should pay more attentions to this course. Teachers should improve the mathematical foundation of the students, and lay strong foundation for future professional courses. Data structure and database principle are also the courses that have strong correlations with other courses. The results show that there is a strong relationship among the core courses, and the training plan is at a reasonable level.

3.6 SPADE Algorithm Application and Results

Although the Apriori algorithm can find association rules between courses, a disadvantage of the algorithm is also apparent. If the rule that item set X deduces item set Y is one of strong association rules which satisfy minimum support and minimum confidence, it is also possible that Y deduces X is strong association rule (such as computer network excellence deduces discrete mathematics excellence). But it is clear that the second rule here is invalid, because the course of discrete mathematics was taught before the course of computer network. It will consume a lot of time and energy to filter out effective rules in the large number of produced rules by using manual intervention. The SPADE algorithm is one kind of algorithms that can get association rules based on time sequence. The data should be processed into a time-stamped format, as shown in Table 2.

Table 2. Partial data format table of SPADE algorithm

SID	EID	Event
1	10	Mathematical Analysis 1 fail Advanced Algebra excellent
1	20	Mathematical Analysis 2 medium Mathematical Logic and Graph Theory good
1	30	Ordinary Differential Equation good Operating System good
1	40	Mathematical Modeling medium Database Principles and Application medium
1	50	Probability Theory and Mathematical Statistics good
1	60	Operations Research and Optimization Algorithm medium
1	70	Coding Theory excellent

The cspade function of the aRuleSequence package in the R language is used to process the data. The association rules that based on the time series are shown in Table 3.

SPADE algorithm can obtain rules that are based on sequences, which can make the rules more clear and more alert awareness. If a student failed a mathematics analysis 1 in the first semester, he will need to pay more attention to mathematics analysis 2 in the next semester. Otherwise, the fail probability of mathematics analysis 2 will be very high. The teacher can judge the fail probability of the student in his or her subject according to the student's previous learning result, and pay more attention and guidance to the student, so the student can increase his ability to escape from fail condition. The study of basic mathematics is not only the foundation of further mathematics, but also the foundation of computer knowledge learning. For students, the basics of mathematics must be strengthened. The teacher should review the basic knowledge of mathematics before starting computer professional courses, which will increase the teaching results and cultivating quality obviously.

Table 3. Partial strong association rules obtained by SPADE algorithm with a minimum support 0.3

Pre-rule	Post-rule	Confidence
Mathematical Analysis 1 fail	Operating System fail	0.39
Operating System fail	Operational Research Optimization Algorithm fail	0.33
Mathematical Analysis 1 fail	Mathematical Analysis 2 fail	0.39
Mathematical Analysis 1 fail	Database Principle and Application fail	0.33
Advanced Algebra medium	High-level Language Programming medium	0.32
Mathematical Analysis 2 fail	Ordinary Differential Equation fail	0.33

4 Conclusion and Discussion

Using Apriori algorithm to analyze the grade of students in a college, the intrinsic links between the core courses are obtained. This result can provide some inspiration for teachers and students. SPADE algorithm is introduced to overcome the shortcoming of Apriori algorithm that cannot mine the rules with time order. The rules of warning significance were obtained for the follow-up learning of students and the follow-up teaching of teachers. Teachers must not only pay attention to the preparation of the pilot courses in the basic knowledge, but also to the connections between similar courses to help students to understand knowledge better. Students should fully understand the significance of the core curriculum in the knowledge system and lay a solid foundation for comprehensive application in the future. The information mining of student grade in colleges and universities is very valuable. We must make full use of these resources to improve the level of higher education.

References

1. Yao, S.L.: Application and research on correlation between colleges courses based on data mining. Bull. Sci. Technol. **28**(12), 232–234 (2012). https://doi.org/10.13774/j.cnki.kjtb.2012.12.018
2. Liu, Y.L.: Application of data mining to decision of college students' management. J. Chengdu Univ. Inf. Technol. **21**(3), 373–377 (2006). https://doi.org/10.3969/j.issn.1671-1742.2006.03.013
3. Zhu, D.F.: Applied research of association rules algorithm on Universities' Educational Management System. Dissertation for Master Degree, Zhejiang University of Technology (2013)
4. Tao, T.T.: An analysis of students' consumption and learning behavior based on campus card and cloud classroom data. Dissertation for Master Degree, Central China Normal University (2017)
5. Zhao, H.: The research and application of data mining technology in analysis for students' performance. Dissertation for Master Degree, Dalian Maritime University (2007)
6. Jiang, H.Y.: The application of Apriori association algorithm in student's results. J. Anshan Norm. Univ. **9**(2), 48–50 (2007). https://doi.org/10.3969/j.issn.1008-2441.2007.02.015

7. He, C., Song, J., Zhuo, T.: Curriculum association model and student performance prediction based on spectral clustering of frequent pattern. Appl. Res. Comput. **32**(10), 2930–2933 (2015). https://doi.org/10.3969/j.issn.1001-3695.2015.10.011
8. Hao, X.F., Tan, Y.S., Wang, J.Y.: Research and implementation of parallel Apriori algorithm on Hadoop platform. Comput. Mod. **2013**(3), 1–4 (2013). https://doi.org/10.3969/j.issn.1006-2475.2013.03.001
9. Li, Z.L.: Research and application of Apriori algorithm based on cluster and compression matrix. Dissertation for Master Degree, Suzhou University (2010)
10. Yang, C.Y.: Research and the application of Apriori algorithm in the analysis of student grade. Dissertation for Master Degree, Hunan University (2016)
11. Shao, X.K.: The Research on Apriori algorithm and the application in the undergraduate enrollment data mining. Dissertation for Master Degree, Beijing Jiaotong University (2016)
12. Dong, H.: Association rule mining based on the interestingness about vocational college courses. J. Jishou Univ. (Nat. Sci. Ed.) **33**(3), 41–46 (2012). https://doi.org/10.3969/j.issn.1007-2985.2012.03.011
13. Agrawal, R., Srikant, R.: Fast algorithms for mining association rules in large databases. In: International Conference on Very Large Databases, pp. 487–499. Morgan Kaufmann, San Francisco (1994)
14. Cui, Y., Bao, Z.Q.: Survey of association rule mining. Appl. Res. Comput. **33**(2), 330–334 (2016). https://doi.org/10.3969/j.issn.1001-3695.2016.02.002
15. Liu, J.Y., Jia, X.Y.: Multi-label classification algorithm based on association rule mining. J. Softw. **28**(11), 2865–2878 (2017). https://doi.org/10.13328/j.cnki.jos.005341
16. Zhao, H.Y., Li, X.J., Cai, L.C.: Overview of association rules Apriori mining algorithm. J. Sichuan Univ. Sci. Eng. (Nat. Sci. Ed.) **24**(01), 66–70 (2011). https://doi.org/10.3969/j.issn.1673-1549.2011.01.019
17. Srikant, R., Agrawal, R.: Mining sequential patterns: generalization sand performance improvements. In: Proceedings of the 5th International Conference on Extending Data Base Technology, pp. 3–7. Springer, London (1996)
18. Zaki. M.J.: SPADE: an efficient algorithm for mining frequent sequences. Mach. Learn. **42** (01), 31–60 (2001). https://doi.org/10.1023/a:1007652502315
19. Dang, Y.M.: Research on the sequential pattern mining algorithms. J. Jiangxi Norm. Univ. (Nat. Sci. Ed.) **33**(05), 604–607 (2009). https://doi.org/10.16357/j.cnki.issn1000-5862.2009.05.025

Study on the Prediction and Analysis of the Number of Enrollment

Xue Liu[1(✉)] and Xiao-qiang Xi[2]

[1] School of Communications and Information Engineering, Xi'an University of Posts and Telecommunications, Xi'an, China
1711311208@qq.com

[2] Institute of Internet of Things and IT-Based Industrialization, University of Posts and Telecommunications, Xi'an, China
xxq@xupt.edu.cn

Abstract. In the era of big data, the value of data has received unprecedented attention. Predictive analysis is an important direction of data application. Different data are suitable for different prediction models, and the prediction accuracy is different. In this paper, in order to accurately find a variety of data to adapt to the prediction model, the data of three different areas of high school enrollment in Shaanxi, Xi'an and China were predicted and analyzed. The results of polynomial fitting, grey model and grey prediction model based on wavelet transform are compared. After demonstration and analysis, the polynomial fitting is more effective to fill missing data. Grey prediction model and grey combination model based on wavelet transform are suitable for data prediction, and the grey combination model based on wavelet transform is more accurate than grey prediction model, and it is more suitable for the predictive analysis of enrollment number. The predicted results can provide reference and basis for the educational administrative departments to make decisions.

Keywords: Predictive analysis · Polynomial fitting · Grey model Wavelet transform

1 Introduction

In the era of big data, the value of data is constantly being valued. Among the various applications of data, predictive analysis refers to the study of the inherent laws of data, the establishment of approximate expressions, and the prediction of future development trends of data based on previous data, in order to facilitate people to prepare in advance. analysis, level evaluation and association analysis are three important research directions. This paper focuses on the research of prediction analysis. Predictive analysis refers to the study of the inherent laws of data, the establishment of approximate expressions, and the prediction of future development trends of data based on previous data, in order to facilitate people to prepare in advance. analysis refers to the study of the inherent laws of data, the establishment of approximate expressions, and the prediction of future development trends of data based on previous data, in order to facilitate people to prepare in advance.

P. Krömer et al. (Eds.): ECC 2018, AISC 891, pp. 450–459, 2019.
https://doi.org/10.1007/978-3-030-03766-6_51

At different stages of education, enrollment has always been a key process. Enrollment involves a wide range of enrollment, ranging from enrollment to national enrollment to small townships and counties. If we can know the number of students enrolled each year in advance, the early work of quota allocation, resource allocation and optimization can be carried out effectively.

In recent years, many education departments have taken the actual number of students enrolled as research objects. Hubei University of Engineering designs the grey GM (1,1) prediction of college enrollment to provide some reference for the decision-making of the relevant departments of the university [1]. The Public Safety Research Center of the Department of Engineering Physics of Tsinghua University adopts the grey system and neural network method, combines with the data of the number of enrollment in the general higher schools from 1970 to 2009, set up the common colleges and universities enrollment scale of the grey system GM (1,1) model and BP neural network model, in order to rationally formulate the school enrolment plan of general higher schools [2]. Qingdao University of Technology combines grey model and support vector regression machine to model the enrollment of colleges and universities through the comprehensive analysis of the enrollment situation over the years, and the suitable model is established to predict the future enrollment plan, which has a good guiding role in the work of school enrollment [3].Taking the Huanggang Vocational and Technical College as a case, the number of enrollment in the past years is taken as the research object, and the gray system GM(1,1) model is adopted. By establishing the different dimension number and the different time series model of the equal dimension, the enrollment scale of the 12-th five-plan period has been predicted reliably [4]. The College of Information Management of Chengdu University of Technology makes use of the grey prediction model to predict the number of freshmen enrolled during the annual enrollment period of colleges and universities in Sichuan Province, which can be used for reference in the employment trend in the future and let relevant departments make appropriate adjustments to future employment policies [5]. Grey system theory is used to analyze the enrollment and scale of primary school students and junior middle school by the Harbin Engineering University and the GM (1,1) model establishes to predict the long-term trend of high school students by analyzing census data. They concluded that there will be a shortage of students in higher education in a few years' later [6]. In order to further improve the gray model, a grey model based on wavelet is proposed, which can improve the results of prediction analysis.

2 Research Methods

The methods used in this paper include polynomial fitting, grey model and grey model based on wavelet transform. The fitted formula can be used for prediction and analysis. The polynomial fitting is mainly used to supplement the missing part of the data. The grey model needs to use the grey theory proposed by Professor Deng Julong in 1982 to predict the data through the grey model [7]. In order to improve the accuracy of the prediction, In this paper, the wavelet transform and grey prediction model are combined to predict the data. Through the above three methods, the number of national high school students, Shaanxi high school students and Xi'an high school students are forecasted and analyzed. It is convenient for the education department to make the corresponding decision based on the data.

2.1 Polynomial Fitting

Polynomial fitting is based on the principle of least square method, it can fit a given n discrete points, but the obtained curve does not require the given points, otherwise it will cause the phenomenon of over fitting and make the reflection inaccurate.

Given n discrete points (ti, yi) where i = 1, 2, ···n, one need to find the approximate curve $y_i = \varphi(t_i)$ and make the deviation, that is $\varphi(t_i) - y_i$, between each point on the approximate curve and the corresponding real point the minimum. Using the criterion of the least square method, the sum of square deviations $\sum_{i=1}^{n} (\varphi(x_i) - y_i)^2$ is minimized and the fitting curve will be obtained [8].

The following is the processing of polynomial fitting:

Let the fitting polynomial be

$$y(x) = a_m x^m + a_{m-1} x^{m-1} + \ldots + a_1 x + a_0, \tag{1}$$

a_i, i = 0, 1, ···, m is the undetermined coefficient. Calculate the sum of the distances from each point to the curve, that is, the sum of square errors:

$$R^2 = \sum_{i=1}^{n} [y_i - y(x)]^2 \tag{2}$$

In order to obtain the coefficient value a, the right side of the equation is respectively subjected to a_i partial derivation, and a system of equations can be obtained. The coefficient matrix of the system is a symmetric positive definite matrix, and there is a unique solution, and the polynomial Eq. (1) can be obtained by the solution. The calculated polynomial equation can be used for data prediction.

2.2 Grey Prediction Model

In order to ensure the feasibility of the model, the corresponding test bust be done for the existing data. The grey prediction model can only be established after the test has been qualified. Otherwise, the data must be transformed to meet the needs. The rating ratio of the data must fall within the required range [9].

Let the original sequence be

$$X^{(0)} = \left\{x^{(0)}(1), x^{(0)}(2), \cdots x^{(0)}(n)\right\} \tag{3}$$

The method of checking data is to calculate the ratio of the near series

$$\delta(k) = \frac{x^{(0)}(k-1)}{x^{(0)}(k)}, k = 2, 3, \cdots n, \tag{4}$$

and check whether they can fall to the interval $S = (e^{-2/(n+1)}, e^{2/(n+1)})$. The original data can be applied to the grey prediction model if all the order ratios meet the interval conditions, otherwise the data should be transformed appropriately, such as translation [10]. The grey prediction model is a scientific and quantitative forecasting of the data by processing the original data and establishing the grey model. It can calculate and predict the future through a small amount of existing data. GM (1,1) represents of the first order and one variable, G represent Grey and M represents Model.

The values for predicting the $n + 1$, $n + 2$, ⋯ moment are respectively $\hat{x}(n+1), \hat{x}(n+2), \cdots$.
Let the time series of the corresponding prediction model be

$$X^{(1)} = \left\{x^{(1)}(1), x^{(1)}(2), \cdots x^{(1)}(n)\right\} \tag{5}$$

which is the 1-AGO (first cumulative) of $X^{(0)}$ to eliminate the randomness and volatility of data. $x^{(1)}(m) = \sum_{i=1}^{n} x^{(0)}(i)$, $m = 1, 2, \cdots n$ and it corresponds the following recursive formula

$$\begin{cases} x^{(1)}(1) = x^{(0)}(1) \\ x^{(1)}(i) = x^{(0)}(i) + x^{(1)}(i-1) \end{cases} i = 1, 2, \cdots n \tag{6}$$

Using $X^{(1)}$ to calculate the parameters a, b in GM (1,1)

$$\hat{a} = [a, b]^T = \left(B^T B\right)^{-1} B^T Y_n \tag{7}$$

$$B = \begin{bmatrix} -\frac{1}{2}\left(X^{(1)}(1) + X^{(1)}(2)\right) & 1 \\ -\frac{1}{2}\left(X^{(1)}(2) + X^{(1)}(3)\right) & 1 \\ \cdots & \cdots \\ -\frac{1}{2}\left(X^{(1)}(n-1) + X^{(1)}(n)\right) & 1 \end{bmatrix} \tag{8}$$

$$Y_n = \left[X^{(0)}(2), X^{(0)}(3), \cdots, X^{(0)}(n)\right]^T \tag{9}$$

The model is as follows

$$\hat{x}^{(1)}(i+1) = \left(x^{(0)}(1) - \frac{b}{a}\right)e^{-ai} + \frac{b}{a} \tag{10}$$

The residual and posteriori error detection are carried out on the established model, and the prediction of the data can only be carried out after the model has passed the test [11].

(1) Residuals sequence

$$\begin{aligned} \varepsilon^{(0)} &= (\varepsilon(1), \varepsilon(2), \cdots, \varepsilon(n)) \\ \varepsilon(i) &= x^{(0)}(i) - \hat{x}^{(0)}(i), i = 1, 2, \cdots n \end{aligned} \tag{11}$$

Relative error sequence is

$$\Delta_k = \left(\frac{|\varepsilon(1)|}{x^{(0)}(1)}, \frac{|\varepsilon(2)|}{x^{(0)}(2)}, \ldots, \frac{|\varepsilon(n)|}{x^{(0)}(n)} \right). \tag{12}$$

The residual error is used to judge whether the model is good or bad. The bigger residual error corresponds to the worse model, the smaller residual corresponds to the accuracy of the model. When $k \leq n, \bar{\Delta} = \frac{1}{n} \sum_{k=1}^{n} \Delta_k$, is the mean relative error. Given α, When $\bar{\Delta} < \alpha$ and $\Delta_n < \alpha$, the model is called residual qualified [12] (Table 1).

Table 1. Precision grade table

Precision grade	Relative error α index critical point
First class	0.01
Second class	0.05
Third class	0.10
Forth class	0.20

(2) Posterior difference test
The post-test difference test is performed according to two indicators: accuracy test (post-test difference) and (small error probability). The variance of the original sequence and the residual sequence are s_1^2 and s_2^2, respectively.

$$c = \frac{s_2}{s_1}, \tag{15}$$

$$p = P\left\{ \left| \varepsilon^{(0)}(k) - \bar{\varepsilon}^{(0)} \right| < 0.6745 S_1 \right\}. \tag{16}$$

The calculation results of c and p will be taken to compare with the precision grade table to judge if the model is effective, detailed parameters can be seen from Table 2 [13].

Table 2. Small error probability and posterior error precision grade table

Precision grade	p-value	c-value
Good	$p > 0.95$	$c < 0.35$
Qualified	$0.8 \leq p < 0.95$	$0.35 < c < 0.50$
Barely qualified	$0.7 \leq p < 0.8$	$0.5 < c \leq 0.65$
Unqualified	$p \leq 0.7$	$c > 0.65$

2.3 Grey Prediction Model Based on Wavelet

Here the discrete data is regarded as a discrete signal $f(k)$, which is project into each positive subspace V_j and is called $f_j(k)$, and the projection will be calculated iteratively. The decomposition formula of the Mallat algorithm corresponding to the discrete wavelet transform is [14]:

$$\begin{cases} f_j(k) = \sum_l h_{l-2n} f_{j-1}(l) \\ c_{jk} = \sum_l g_{l-2n} f_{j-1}(l) \end{cases} \tag{17}$$

The corresponding Mallat reconstruction algorithm formula is

$$f(k) = \sum_{j=1}^{J} \sum_{n \in Z} c_{jk} \bar{g}_{k-2n} + \sum_{n \in Z} f_j(n) \bar{h}_{k-2n} \tag{18}$$

Wavelet transform is to decompose a signal $f(k)$ into detail signal c_{jk} (wavelet coefficient) with different scale and resolution and approximate signal $f_j(k)$ with very low scale and resolution. h_j and g_j are impulse responses of low pass filter H and high pass filter G respectively, $\bar{h}_j$ and $\bar{g}_j$ are impulse responses for reconstructing low and high pass filter respectively. Using the idea of filter, the signal can be divided into an approximate signal and a detail signal. Using the idea of filter, the signal can be divided into an approximate signal and a detail signal. The approximate signal belongs to the low frequency part and it is the real component and the fundamental part of the original signal. It represents the characteristics and the development law of the signal, and has a good resolution in the frequency domain. The detail signal belongs to the high frequency part and represents the nuances of the original signal and has good resolution in time domain [15].

After the original signal is decomposed by mallat algorithm, the approximate signal and the detail signal are obtained, and then the approximate signal is predicted according to the gray prediction model. The predicted data and detail signal are reconstructed by the mallat algorithm. That's what we're really predicting result.

3 Analysis on the Enrollment of Senior Middle School Students

Through the national statistical bulletin [16] of the national education development, the Shaanxi provincial education development statistical bulletin [17] and the Xi'an education development bulletin [18], the number of national high school enrollment in 2008–2015 years, the number of students in Shaanxi high school, and the number of students in the 2011–2017 years of Xi'an high school was collected. According to these three kinds of data, polynomial fitting, grey prediction model and wavelet based grey prediction model are carried out respectively. By comparing and analyzing the laws and differences, it is convenient for relevant departments to make arrangements and work in advance, and to formulate corresponding rules and regulations for management.

3.1 Processing Steps

1. Polynomial fitting
   Equation (1) is used to fit the N order data of the original data first, and then the square sum error is calculated by the Eq. (2). The minimum value is calculated and then the order and coefficient of the polynomial are determined, and so the final polynomial is obtained.
2. Grey model
   Before the gray model is established, the data needs to be tested by the formula (4). After the test, the parameter value is calculated by the formula (7) and brought into the formula (10), and the formula of the gray model is obtained.
3. Grey prediction model based on wavelet
   After the data is decomposed through the formula (17), the approximate signal is reconstructed by the formula (18). Finally, the approximate signal is predicted through the modeling step of the grey model and the model is set up.

All the above three modeling methods are realized by MATLAB programming. The three methods are verified and analyzed by the enrollment of senior high schools in China, Shaanxi and Xi'an.

3.2 The Comparison of the Three Models

Using the processing steps in Sect. 3.1, the corresponding results are shown through figures and tables in the following.

Figures 1, 2 and 3 are results of forecasting the number of senior middle school students in China, Shaanxi and Xi'an by polynomial fitting, grey model and combination model respectively. The prediction effect of polynomial fitting is the best, the polynomial tends to morbid when the fitting order is too high, so it is not convenient to use this method to predict. Comparing with the grey model, the combined model is closer to the original data, which proves that the combined model can meet the prediction requirements and achieve the optimization effect.

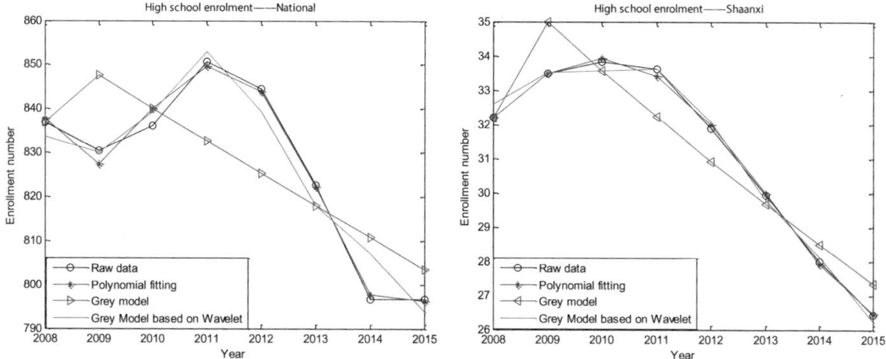

Fig. 1. National high school enrollment **Fig. 2.** Shaanxi senior high school enrollment

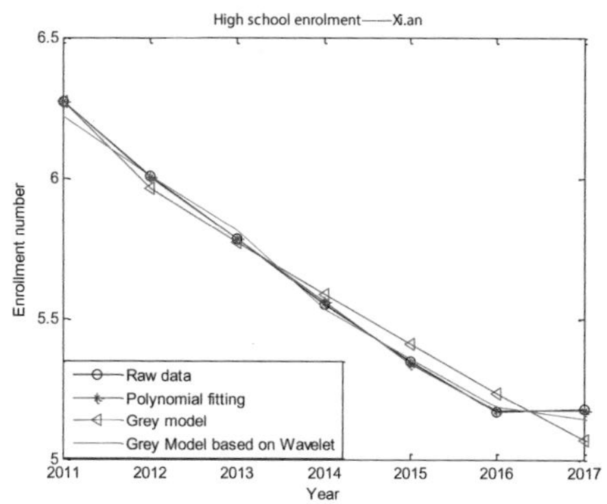

Fig. 3. Xi'an Senior High School enrollment

Table 3. Comparative tables of relative errors-National

Year	High school enrollment	Polynomial fitting relative error	Grey prediction relative error	Wavelet grey prediction relative error
2008	837.01	0.001	0.000	0.000
2009	830.34	0.004	0.021	0.022
2010	836.24	0.004	0.005	0.002
2011	850.78	0.001	0.021	0.023
2012	844.61	0.001	0.023	0.016
2013	822.70	0.001	0.006	0.000
2014	796.60	0.001	0.018	0.005
2015	796.61	0.001	0.009	0.013
	Sum	0.014	0.102	0.081
	Mean value	0.002	0.013	0.010

Table 4. Comparative tables of relative errors-Shaanxi

Year	High school enrollment	Polynomial fitting relative error	Grey prediction relative error	Wavelet grey prediction relative error
2008	32.20	0.0002	0.0000	0.0000
2009	33.49	0.0001	0.0446	0.0425
2010	33.84	0.0029	0.0079	0.0010
2011	33.61	0.0059	0.0414	0.0426
2012	31.88	0.0036	0.0301	0.0373
2013	29.94	0.0017	0.0089	0.0047
2014	27.98	0.0029	0.0177	0.0112
2015	26.43	0.0009	0.0341	0.0408
	Sum	0.0181	0.1847	0.1801
	Mean value	0.0023	0.0231	0.0225

Table 5. Comparative tables of relative errors-Xi'an

Year	High school enrollment	Polynomial fitting relative error	Grey prediction relative error	Wavelet grey prediction relative error
2011	6.27	0.000	0.000	0.000
2012	6.00	0.000	0.007	0.005
2013	5.78	0.000	0.002	0.006
2014	5.55	0.001	0.006	0.010
2015	5.35	0.002	0.011	0.009
2016	5.17	0.001	0.013	0.008
2017	5.18	0.000	0.021	0.016
	Sum	0.006	0.061	0.054
	Mean value	0.001	0.009	0.008

Through comparison of Tables 3, 4 and 5, the results of Figs. 1, 2 and 3 are more accurately confirmed. After three kinds of data are treated with three methods, the total and average relative error of polynomial fitting is minimal, and the total and average relative error of the combined model is less than that of the grey model. It is proved that the prediction effect of the combined model is more accurate than that of the grey model.

4 Conclusion

Using polynomial fitting, grey model and wavelet based grey combination model, this paper analyzes the enrollment number of senior high school students in China, Shaanxi province and Xi'an city, three administrative regions of different sizes. The fitting results are different for different methods. Polynomial fitting has the best fitting result and can be used to make up for the few missing or uncounted data. The enrollment of senior high school students satisfy the prediction conditions of grey model and a combination model, the latter has better fitting results and can be used to predict such data. Other province's enrollment data analysis also support this conclusion. The fitting result is valuable for the educational administration department. In fact, for the enrollment data, the best prediction is doing a good job of daily statistics. The further work of this paper is to develop a program to finish the fitting and prediction automatically.

References

1. Guo, H.R.: Prediction of college enrollment based on grey theory. J. Hubei Inst. Eng. **33**(6), 48–51 (2013)
2. He, C.H.: The prediction method of enrollment scale in colleges and universities. J. Tsinghua Univ. (2012). https://doi.org/10.16511/j.cnki.qhdxxb.2012.01.001
3. Zhu, Z. W.: Application of mathematical model in college enrollment prediction. Dissertation for Master Degree, Qingdao University of Technology (2010)
4. Zhou, Z.X.: Grey prediction of enrollment scale of Huanggang vocational college during the twelfth five-year plan period. J. Huanggang Voc. Techn. College **13**(04), 92–95 (2011). https://doi.org/10.3969/j.issn.1672-1047.2011.04.23
5. Luo, X.L.: Research on the application of GM (1,1) model in enrollment prediction of colleges and universities – a case study of colleges and universities in Sichuan Province. J. Guizhou Univ. (Nat. Sci. Edn.) **25**(04), 342–345 (2008)
6. Song, L.L.: Survival game of undergraduate education in colleges and universities based on shortage of students. Dissertation for Master Degree, Harbin Engineering University (2006)
7. Xiao, X., Miao, S.H.: Grey Prediction and Decision Making Method. Science Press, Beijing (2013)
8. Li, M.W., Sha, X.Y.: Improvement and application of grey prediction model based on GM (1,1). Comput. Eng. Appl. **52**(4), 25–26 (2016). https://doi.org/10.3778/j.issn.1002-8331.1506-0257
9. Liu, S.F., Xie, N.M.: Theory and Application of Grey System. Science Press, Beijing (2013)
10. He, M.F.: Population forecasting model based on grey system theory. South China University of Technology (2012)
11. Liu, S.F., Ceng, B., Liu, J.F., Xie, N.M.: Research on several basic forms and application scope of GM (1,1) model. Syst. Eng. Electr. Technol. **36**(3), 502–504 (2014)
12. Tong, M.Y.: Grey modeling method and its application in prediction. Dissertation for Doctor Degree, Chongqing University (2016)
13. Zhang, D.H., Jiang, S.F., Shi, K.Q.: The theoretical defect and improvement of the grey prediction formula. Syst. Theor. Pract. **122**(8), 140–142 (2002)
14. Zhao, J.P., Ding, J.L.: Prediction of road traffic accidents based on wavelet analysis and grey GM (1,1) model. Pract. Underst. Math. **45**(12), 119–124 (2015)
15. Zhang, H.B.: A theory based on wavelet analysis of the grey prediction method. Dissertation for Master Degree, Harbin University (2009)
16. Ministry of education: National Statistical Bulletin for the Development of Education in China, 2008–2015. http://www.moe.edu.cn/jyb_sjzl/sjzl_fztjgb/. Accessed 11 Mar 2018
17. Shaanxi provincial education department: Statistics Bulletin of Shaanxi Education Development, 2011–2017. http://www.xaedu.gov.cn/ptl/def/def/index_902_4657.html. Accessed 15 Mar 2018
18. Xi'an Municipal Bureau of Education: Statistics Bulletin on the Development of Education in Xi'an, 2011–2017. http://www.xaedu.gov.cn/ptl/def/def/index_902_4657.html. Accessed 21 Mar 2018

Information Processing and Data Mining

The Impact Factor Analysis on the Improved Cook-Torrance Bidirectional Reflectance Distribution Function of Rough Surfaces

Lin-li Sun[1(✉)] and Yanxia Liang[2]

[1] School of Automation, Xi'an University of Posts and Telecommunications, Chang'an West St., Chang'an District, Xi'an, China
rourouxiang@126.com

[2] Shaanxi Key Laboratory of Information Communication Network and Security, Xi'an University of Posts and Telecommunications, Xi'an, China

Abstract. The Cook-Torrance BRDF of the material is not energy balanced for the reflected radiance and the albedo converges to zero at grazing angles. This gap is filled by appropriate modifications of the Cook-Torrance BRDF model. The improved BRDF model is relevant with the surface roughness, the distribution of visible normal, the Fresnel factor and the geometrical attenuation factor. The model is applied to metallic surfaces with various values of root mean square and the geometrical attenuation factors as the incidence angle is increased. The improved model is analytic and suitable for Computer Graphics applications.

Keywords: BRDF model · Geometrical attenuation factor
Microsurface theory

1 Introduction

An important unsolved problem in the research field of computer vision is the evaluation of bidirectional reflectance distribution function (BRDF) on rough surfaces. In particular, the effect of roughness and the geometrical attenuation factor are quite critical, and the problem becomes significantly challenging.

The Cook-Torrance BRDF model [1] is widely applied in rendering, this kind of model is simpler than other known metallic models and its anisotropic form could provide good metallic impression. But the main problem of this model is that at grazing angles and at viewing directions below mirror direction, especially at grazing angles the reflected radiance significantly violates energy balance [2]. The Cook-Torrance BRDF model is not physically plausible.

The distribution of visible normals and geometrical attenuation factor influence the shadowing and masking effect, while the shadowing or masking occurs as a part of the incident ray or the outgoing ray is blocked by the neighboring topography. So the geometrical attenuation factor, gauged by the proportion of light that is not attenuated, demonstrates the effect of shadowing, masking, and shadowing-masking on rough

P. Krömer et al. (Eds.): ECC 2018, AISC 891, pp. 463–470, 2019.
https://doi.org/10.1007/978-3-030-03766-6_52

surfaces [3–5]. The distribution of visible normals and geometrical attenuation factor are important terms of BRDF, from physical models to geometrical models.

In this paper, we make some appropriate modifications of the Cook-Torrance BRDF model, and show the influence factors on the improved BRDF models.

2 The Improved BRDF Model and the Geometrical Attenuation Factor

2.1 Improved BRDF Model

If the material of the surface is rough conductor, the Cook-Torrance BRDF of the material is [6, 7]

$$f(\omega_i, \omega_m) = \frac{F(\omega_i, \omega_m) G_2(\omega_i, \omega_o) D(\omega_m)}{4|\omega_i \cdot \omega_g||\omega_o \cdot \omega_g|} \tag{1}$$

where ω_i is the direction of incident ray, ω_g is the direction of geometric normal, $\omega_o$is the direction of the outgoing ray. ω_m is the microfacet normal. $F(\omega_i, \omega_m)$ is the Fresnel factor which is 1 for a rough conductor. $G_2(\omega_i, \omega_o)$ is the geometrical attenuation factor. $D(\omega_m)$ represents the distribution of visible normals over the microsurface, it is the microfacet distribution function and it has some different types such as Beckmann, GGX and Gaussian [8].

This model is not energy balanced; the reflected radiance and the albedo converge to zero at grazing angles. We are using the $\max((\omega_o \cdot \omega_g), (\omega_o \cdot \omega_g))$ factor of Neumann's [2], which leads to a new BRDF model:

$$f(\omega_i, \omega_m) = \frac{F(\omega_i, \omega_m) G_2(\omega_i, \omega_o) D(\omega_m)}{4\max((\omega_o \cdot \omega_g), (\omega_o \cdot \omega_g))} \tag{2}$$

In this paper, we discuss Gaussian distribution for the microsurface and it is widely used in the optics literature. $G_2(\omega_i, \omega_o)$ is the geometrical attenuation factor, this factor depends on the distribution of visible normal on the microsurface. The widely used Smith geometrical attenuation factor is derived from the assumption that neglecting the correlation between height and slope, and the density of surface is Gaussian [9]. Nicodemus has defined the BRDF as the ratio of the total reflected flux of the observation direction ω_o to the incident flux, coming from the direction ω_i [10], however, we use the general form of the microfacet model for rough conductor is showed in Eq. (1).

The demarcation of the incident ray, denoting that the surface is between illuminated and shadow, occurs for a given observation direction. The surface is entirely illuminated when the incident angle is less than the critical point, while shadow occurs when the incident angle is bigger than the critical point. The illumination factor is put forward to illustrate the illumination on the surface of an object.

Note that, as the surface is without shadow, the BRDF is relevant with the incident and the outgoing directions and the distribution of visible normal, when the incident angle is less than the evaluated illumination factor. However, when the incident angle is larger than the demarcation angle, the geometrical attenuation factor must be taken into account. The illumination factor makes the decision whether to introduce the geometrical attenuation factor into the BRDF or not.

2.2 Blinn's Geometrical Attenuation Factor

The geometrical attenuation factor is first defined by Torrance and Sparrow representing the proportion of light that is not attenuated by the combination of microfacet shadowing upon incidence and by microfacet masking subsequent to reflection. The function of $G_2(\omega_i, \omega_o)$ is derived by explicit assumption that planar microfacets comprising rough surfaces arise from V-grooves, and his geometrical attenuation factor is simplified by adopting the same V-grooves assumption by BLinn [4]:

$$G(\theta_i, \theta_r, \varphi_r) = min\left(1, \frac{2cos\alpha cos\theta_r}{cos\beta}, \frac{2cos\alpha cos\theta_i}{cos\beta}\right) \tag{3}$$

where θ_i, θ_r is the tilt angle of incidence and reflection of ray respectively, φ_r is azimuthal angles of reflection ray, α is the polar angle from the mean surface normal to the microfacet normal. β is the incidence as measured from the microfacet (Fig. 1).

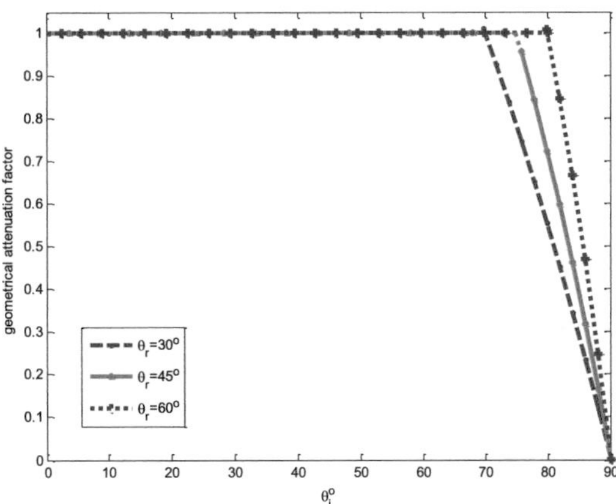

Fig. 1. Blinn's geometrical attenuation factor. The curves are broken lines. The incident angle is 30°, 45°, 60° respectively.

The BLinn's geometrical attenuation factor has a simple mathematical form, and it is widely used in 3D graphics simulation, physics rendering and computer graphics.

The BLinn's curve is a broken line. The inflection point occurs at θ_i = 70.1°, 73.3°, 80.6° when θ_i = 30°, 45°, 60° respectively. The inflection point is the critical point denoting the shadowing or masking is occurred when θ_i is larger than the demarcation angle.

2.3 Smith's Geometrical Attenuation Factor

Smith's geometrical attenuation factor is derived by the assumption that the distribution of height and slope on the rough surface is Gaussian. And for computational ease, Smith neglected the correlation between height and slope [5].

$$G(\theta_i) = \frac{1 - \frac{1}{2} erfc\left(\mu / \sqrt{2}\sigma\right)}{\Lambda(\mu) + 1} \tag{4}$$

where $\mu = cot\theta$, θ is the tilt of the incident ray, σ is the root mean square of height deviation. $\Lambda(\mu)$ is the Λ function, *erfc* is the error function complement. Equation (4) is simple and it is known to be the most physical realistic geometric-optics model (Fig. 2).

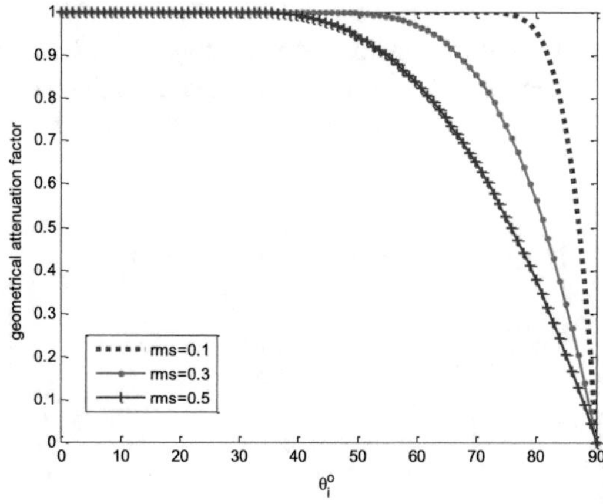

Fig. 2. Smith geometrical attenuation factor, as rms equals 0.1, 0.3, and 0.5. The curve is concentrated in the center as the surface roughness becomes larger.

We get the demarcation point: θ_i = 73.8°, 53.3°, 41.2° when rms = 0.1, 0.3, 0.5 respectively. When the rms value becomes larger, the curve is concentrated in the center. Smith's geometrical attenuation factor in Eq. (4) is of simple analytical form and as such may be useful approximation to the true shadowing functions in industry.

3 Influence on BRDFs

Light reflected by a surface is depended on the microscopic shape characteristic of the surface. The Blinn model is derived under the assumption that each specularly reflecting facet comprises one side of a symmetric V-groove cavity. All masking and shadowing effects take place within the cavities. The geometrical attenuation factor is relevant with the incident angle and the reflection angle.

The Smith model is defined based on the microgeometry that each surface point as being optically flat and it is significantly larger in scale than visible wavelengths. Each microsurface reflects light from incoming light direction to outgoing direction depends on the orientation of the normal on the microsurface. And the visibility of non-backfacing point on the microsurface depending on its height but not on its normal. Only those that incoming or the outgoing ray is not shadowed or masked can potentially contribute to the BRDF value.

In this section, we discuss the Smith geometrical attenuation factor acted on the improved BRDF model in greater detail and explain how geometrical attenuation factor and surface normal distribution are pertinent to the study of BRDF.

3.1 Distribution of Visible Normals

$D(\omega_m)$ represents the distribution of visible normals over the microsurface as we mentioned above in Eq. (1). We discussed the distribution of facets in geometrical attenuation factor. The Blinn model assumed the facets are long and symmetric V-groove cavity, and the Smith model assumed the facets are Gaussian distributed.

In this paper, we discuss Gaussian distribution of visible normals assumptions for the microsurface and they are widely used in computer graphics.

Figure 3 showed the Gaussian facet normal distribution when rms = 0.1, 0.3, 0.5 respectively. The curve is concentrated in the center as the surface roughness becomes larger.

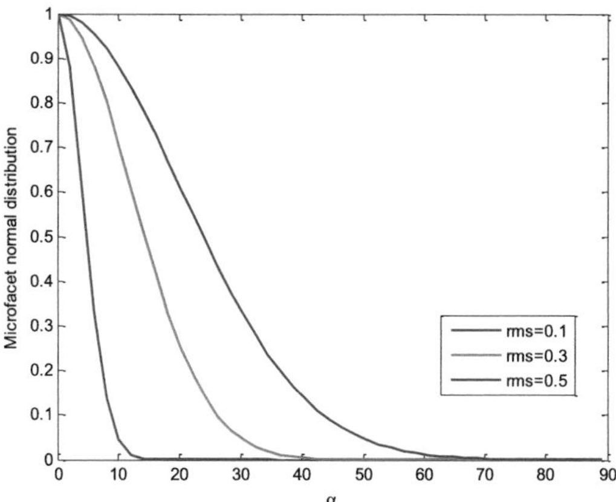

Fig. 3. The curve of Gaussian microfacet distribution function of visible facet normal when rms = 0.1, 0.3, 0.5 respectively.

3.2 Results and Analysis

Now we consider the impact factors of geometrical attenuation factor and the facet normal distribution function in the BRDF. The Fresnel term is 1 for rough conductor. When the incident angle is given, the geometrical attenuation factor is relevant with the roughness of the surface. And the facet normal distribution is Gaussian distributed, the outgoing angle is consequently be determined.

Figures 4, 5 and 6 show resulting curves from the analysis on the improved Cook-Torrance BRDF model. The roughness on the surface is evaluated by rms value, and it generated various Gaussian distribution of visible normal on microsurface. And the rms value affects the geometrical attenuation factor. When we increase the incident angle, the geometrical attenuation factor changes accordingly.

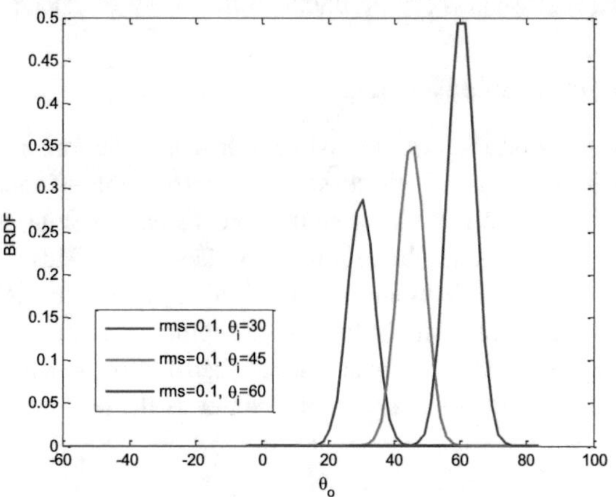

Fig. 4. The curve of improved Cook-Torrance BRDF model, when rms = 0.1, the incident direction is 30°, 45°, 60° respectively.

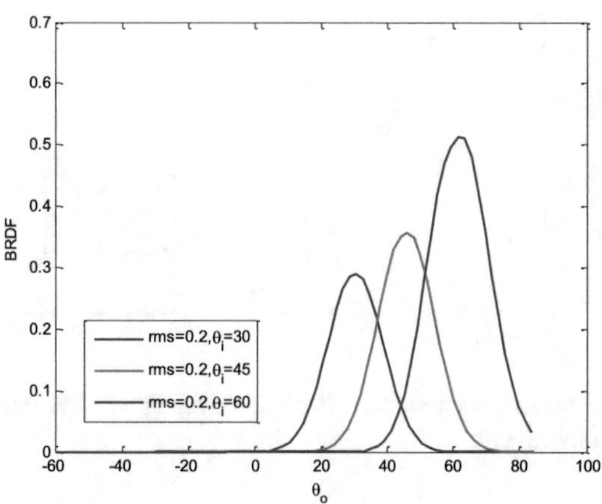

Fig. 5. The curve of improved Cook-Torrance BRDF model, when rms = 0.2, the incident direction is 30°, 45°, 60° respectively.

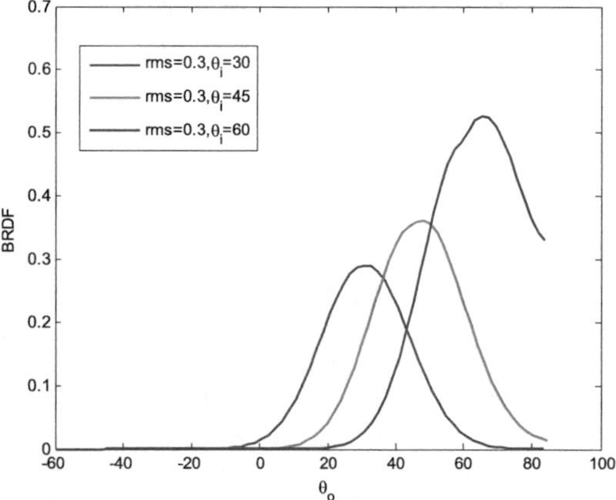

Fig. 6. The curve of improved Cook-Torrance BRDF model, when rms = 0.3, the incident direction is 30°, 45°, 60° respectively.

Table 1. Geometrical attenuation factor

rms value	$\theta_i = 30°$	$\theta_i = 45°$	$\theta_i = 60°$	$\theta_i = 75°$
rms = 0.1	1	1	1	0.9963
rms = 0.2	1	1	0.9981	0.8860
rms = 0.3	1	0.9996	0.9696	0.7357
rms = 0.4	1	0.9938	0.9085	0.6162
rms = 0.5	0.9998	0.9754	0.8365	0.5265

For a given rms value, the BRDF curve is related to the geometrical attenuation factor and changed by the incident angle. The geometrical attenuation factor is also related with the incident angle as well. As shown in Table 1, the value of geometrical attenuation factor becomes larger when the incident angle and the rms value increases.

The improved Cook-Torrance BRDF model corrects the defects of the original model that in grazing angles the reflected radiance can be unacceptable greater than the incoming radiance.

4 Conclusion

The Cook-Torrance BRDF model is not energy balanced at grazing angles, it is not physically plausible. We improved the model by modification on the term of denominator a little bit, and we take the microsurface theory and discussed the Gaussian distribution of visible normals assumptions on the microsurface, analyzed influence factor of roughness and geometrical attenuation factor when the incident angle increased. The representation successfully captures these very different reflectance characteristics.

Acknowledgement. This work was supported by the Department of Education Shaanxi Province, China, under Grant 2013JK1023, and Shaanxi STA International Cooperation and Exchanges Project (2017KW-011).

References

1. Cook, R., Torrance, K.: A reflectance model for computer graphics. Comput. Graph. **15**(3), 307–316 (1981). https://doi.org/10.1145/357290.357293
2. Neumann, L., Neumann, A.: Compact metallic reflectance models. Comput. Graph. Forum. **18**, 161–172 (1999). https://doi.org/10.1111/1467-8659.00337
3. Schlick: A customizable reflectance model for everyday rendering. In: Proceedings of Fourth Eurographics Workshop on Rendering, pp. 73—83 (1993)
4. Blinn, J.: Models of light reflection for computer synthesized pictures. Comput. Graph. SIGGRAPH, 192–198 (1977). https://doi.org/10.1145/563858.563893
5. Smith, B.: Geometrical shadowing of a random rough surface. IEEE Trans. Antennas Propag. **15**(5), 668–671 (1967). https://doi.org/10.1109/TAP.1967.1138991
6. Heitz, E., D'Eon, E., D'Eon, E., et al.: Multiple-scattering microfacet BSDFs with the Smith model. ACM Trans. Graph. **35**(4), 58, 1–14 (2016). https://doi.org/10.1145/2897824.2925943
7. Walter, B., Marschner, S.R., Li, H., et al.: Microfacet models for refraction through rough surfaces. In: Eurographics Symposium on Rendering Techniques, Grenoble, France. pp. 195–206 (2007). https://doi.org/10.2312/egwr/egsr07/195-206
8. Kurt, M.: An anisotropic BRDF model for fitting and Monte Carlo rendering. ACM Trans. Graph. (2010). https://doi.org/10.1145/1722991.1722996
9. Heitz, E., Dupuy, J., Hill, S., Neubelt, D.: Real-time polygonal-light shading with linearly transformed cosines. ACM Trans. Graph. **35**(4), 41, 1–8 (2016). https://doi.org/10.1145/2897824.2925895
10. Nicodemus, F.E., Richmond, J.C., Hsia, J.J., et al.: Geometrical considerations and nomenclature for reflectance. **160**, 94–145 (1977). https://doi.org/10.6028/nbs.mono.160

Analysis of Commuting Characteristics of Mobile Signaling Big Data Based on Spark

Cong Suo[1](✉), Zhen-xian Lin[2], and Cheng-peng Xu[3]

[1] School of Telecommunications and Information Engineering, Xi'an University of Posts and Telecommunications, Xi'an, China
suocong_xian@163.com

[2] School of Science, Xi'an University of Posts and Telecommunications, Xi'an, China
lzhxl26@126.com

[3] Institute of Internet of Things and IT-Based Industrialization, Xi'an University of Posts and Telecommunications, Xi'an, China
xyfoal@xupt.edu.cn

Abstract. The study of commuting mode is of great significance for reducing urban traffic pressure and constructing intelligent city. However the commonly used research methods are slow in computing when dealing with large-scale mobile signaling data. A method of parallel clustering and statistics using Spark is proposed. In this method, a large amount of data is cleaned on Hive and denoised. The user data is divided into different areas through the K-Means algorithm on the Spark, and then the spatial-temporal statistics are carried out in the different partition area. Finally, the location of the user's place of residence and work and the length of commuter distance and time are obtained, which can be used to divide users from the traditional nine-to-five and non-nine-to-five and provide an effective reference for urban planning and traffic congestion.

Keywords: Spark · Mobile signaling big data
Place of residence and workplace identification · Commuting distance
Commuting time

1 Introduction

Commuting is the process of travelling between a place of residence and a place of work. The studies of residents' commuting characteristics have become one of the hot issues at home and abroad [1–6]. At present, domestic studies on commuting characteristics are mainly concentrated in first-tier cities such as Beijing, Shanghai, Shenzhen and so on, which can alleviate urban pressure and construct a more rational urban layout. The data of research commuting mainly include questionnaire survey, taxi JPS positioning data, city card data and mobile phone signaling data [7–9].

From the perspective of data sources, the size and representativeness of the sample are the key issues to study the commuting efficiency and influencing factors of the whole city by means of questionnaire survey [10, 11]. Therefore, people began to use the card data and taxi JPS positioning data through statistical methods to study the spatial structure of different cities, occupation and housing balance, as well as

P. Krömer et al. (Eds.): ECC 2018, AISC 891, pp. 471–482, 2019.
https://doi.org/10.1007/978-3-030-03766-6_53

commuter travel. However, the collection of such data is related to the travel of users, and only includes commute by public transport mode. In contrast, with the advantages of a wide coverage and the continuity of space-time, the mobile signaling data can be better used in commuting research, such as the identification of residence and work places, the extraction of hot spot, urban commuter circle identification and so on [12–15]. The purpose of this paper is to study the commuting characteristics of residents in order to divide the crowd, including the place of residence and work, the commuting distance and the commuting time. However, multiple operations are required on a particular dataset during the analysis process. And with the development of mobile Internet, the scale of mobile signaling data is also expanding, so there are some problems such as slow computing speed and low efficiency.

In this paper, Hive is introduced into the preprocessing of raw data. A new method of calculating the commuter characteristics is constructed by combining parallel clustering algorithm and spatio-temporal statistics based on Spark.

2 Spark

Spark is an efficient distributed computing system for large-scale data processing. The memory-based iterative computing framework and Spark's core technology, resilient distributed dataset (RDD), make Spark especially suitable for applications where there are multiple operations on a specific dataset [16]. Currently, there are four main running modes of spark including local, standalone, yet another resource negotiator (YARN), mesos. The four main components of spark are shown in Fig. 1.

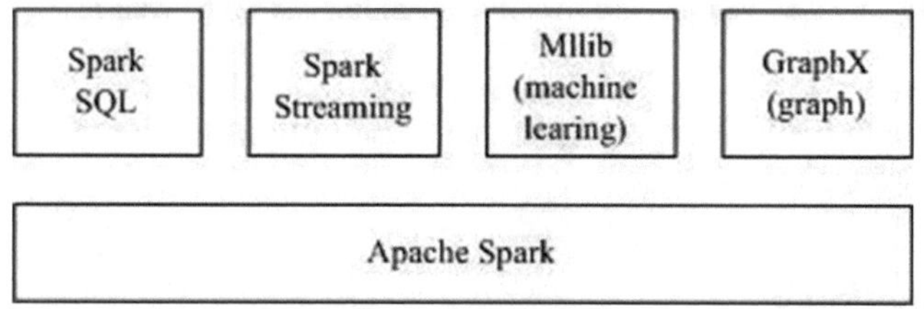

Fig. 1. The components of Spark

2.1 Composition of the Spark Architecture

The master-slave model is adopted in the distributed Spark cluster, where the master node runs the Master process and the slave node runs the Worker process. The Spark architecture is shown in Fig. 2.

Client submits the client application. Driver is the core component of Spark architecture. It provides the running environment of the program, starts the application program, creates SparkContext, and then converts the application program into a stage directed Acyclic Graph(DAG). Cluster Manager requests external services of resources through SparkContext. Executor is responsible for executing tasks submitted by TaskSchedule.

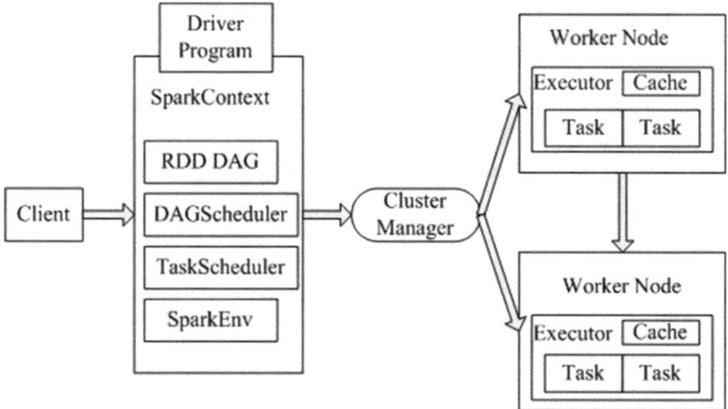

Fig. 2. Spark architecture composition

3 Data Preprocessing

3.1 Mobile Signaling Data

The mobile phone signaling is the communication data between the mobile phone user and the transmitting base station or the microstation. The data used in the experiment is provided by the China Unicom operator after the desensitization of the mobile phone. The composition of the original data is shown in Table 1 with six attributes, and the attribute meaning of the data is shown in Table 2.

Table 1. Mobile phone signaling data sample

User identification	Time stamp	Base station code	Cell code	Longitude	Latitude
LM2CcXMxTrgf + EEDOLIw	20170214074722	48415	25087	109.09	34.33
KStx6GZWe9Ft4YSi + yiA	20170213074753	48493	55902	110.17	37.75
ezqiRuo3aapXbAhyCA4w	20170305075231	47878	11753	108.81	34.59

Because there are many problems in the original mobile signaling data used in this paper, it is necessary to preprocess the mobile signaling data. The processing process is divided into data cleaning and data denoising.

3.2 Data Cleaning

The main way to clean data is shown in Fig. 3. Hive is used to clean data, create partition table according to time. Only four attributes of user data are extracted, which are user identification, timestamp, longitude and latitude. The processed data is stored in Hadoop distributed File system (HDFS).

Table 2. The attribute meaning of cell phone signaling data

Number	Attribute	The meaning of attribute
1	User identification	User unique Identification number after Mobile phone desensitization
2	Time stamp	Timing of signaling updates, accurate to seconds
3	Base station code	Location Area Code (LAC)
4	Cell code	Cell Identify Code (CID)
5	Longitude	Longitude at which the location is updated
6	Latitude	Latitude at which the location is updated

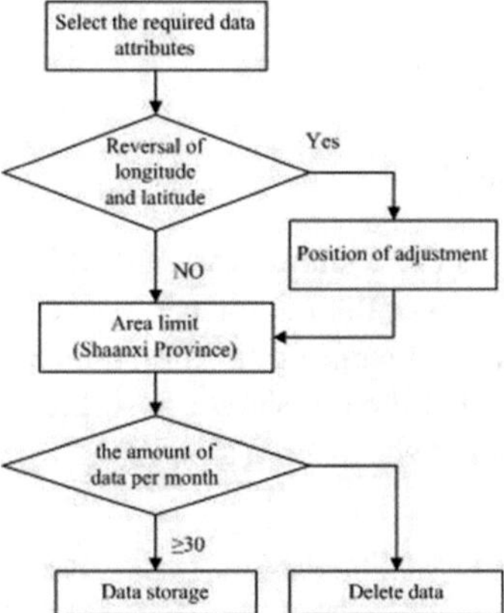

Fig. 3. Data cleaning

3.3 Data Denoising

There is a phenomenon of Mobile phone switching between two or more base stations in a short period of time in the cellular signaling data. The data corresponding to this phenomenon is called ping-pong switching data, and the processing principle is to take the speed between two points as the reference. Data denoising flow is shown in Fig. 4.

Given three adjacent points $A(lon1, lat1, t1)$, $B(lon2, lat2, t2)$, $C(lon3, lat3, t3)$, lon, lat, t represents longitude, latitude, and time respectively. v_1, v_2 represents the speed of AB, BC. v is the given threshold.

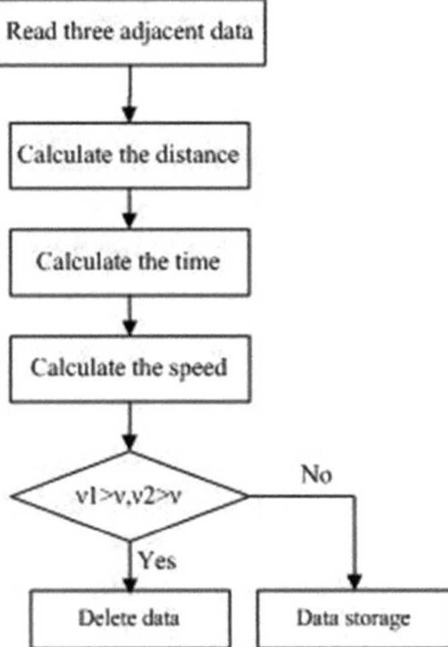

Fig. 4. Data denoising

4 Analysis of Commuting Characteristics

Identify the place of residence and work, calculate the commuting distance and time, and use these three dimensions to analyze the commuting characteristics of users.

4.1 K-Means Algorithm Under Spark

K-Means is a clustering algorithm based on distance partition. The K-Means cluster centers are calculated by iterative method. Finally, it is hoped that the square of the distance between each data point to the centroid of its category is minimized. The square error is generally used as the objective function.

$$\mathrm{E} = \sum_{i=1}^{k} \sum_{\mathbf{x} \in C_i} \|\mathbf{x} - \boldsymbol{\mu}_i\|_2^2 \tag{1}$$

$$\boldsymbol{\mu}_i = \frac{1}{|C_i|} \sum\nolimits_{\mathbf{x} \in C_i} \mathbf{x} \tag{2}$$

Where dataset $D = \{x_1, x_2, \ldots, x_m\}$, Partitioned cluster $C = \{C_1, C_2, \ldots C_k\}$.

The Spark platform implements K-Means algorithm through MLlib, runs multiple K-Means algorithms in parallel, and returns the cluster center of the best cluster. The size of K value in K-Means algorithm is related to the final clustering effect. Computing cost method is used to calculate the within set Sum of Squared error (WSSE), and the validity of clustering is measured according to WSSE to determine the final K value.

4.2 Identification of Place of Residence and Work

The user data is divided into regions by clustering algorithm, and the result is matched with the map. Then the data combination time of each region is analyzed. Set most likely residence time to 20: 00–8: 00 and work time to 9: 00–18: 00. Statistics of mobile phone signaling data in different regions and different time periods and the place of residence and the place of work are the most frequently appeared places respectively.

4.3 Commuting Distance

When the latitude and longitude of two points are known, the commute distance is calculated by using the haversine formula. In the actual calculation, there is a certain error between the location of the base station and the actual distance of the user, and the error range can be regarded as the same location within 1000 m. The haversine formula is as follows.

$$haver\sin(\frac{d}{R}) = haver\sin(\phi_2 - \phi_1) + \cos(\phi_1)\cos(\phi_2)haver\sin(\nabla\lambda) \tag{3}$$

$$haver\sin(\theta) = \sin^2(\theta/2) \tag{4}$$

R denotes the radius of the Earth, ϕ_1, ϕ_2 represent the latitude of two points, $\nabla\lambda$ represents the difference in longitude between two points, d is the distance sought.

4.4 Commuting Time

In the experiment, the signaling data particles of the mobile phone are coarse and have strong randomness, so the probability of the user producing the data in the commuting and commuting process is very small. In order to solve this problem, a method to calculate the average value of the time difference sequence is proposed in this paper.

Divide commuting into early commuting and late commuting. In the early commuting process, extract the time when the user last appeared at his place of residence and the earliest time he appeared at the work place in the morning, and during the late commuting process, extract the last time that the user appears in the work place and the earliest time appear in the place of residence every night, then calculate the corresponding time difference, the final calculation results will form a series of time differences composed of sequences. Setting the threshold of time difference, the unqualified calculation results are discarded from the sequence, and the average value of the final sequence is calculated as the last commuting time. With reference to the commuting time in real life, the threshold value selected in the experiment is set to be more than 10 min and less than 90 min.

4.5 Data Analysis

The process of data analysis is shown in Fig. 5 and divided into four parts including data preprocessing, K-Means clustering, spatio-temporal statistics, and result analysis.

The data is stored in HDFS and Spark reads the processed data converted to RDD from HDFS, and then uses SparkMLlib to cluster RDD with K-Means. The experiment is based on spatial location clustering, so it is necessary to extract the feature vectors related to spatial location consisting of the longitude and latitude, and the two attributes are transformed into a feature array for K-Means clustering.

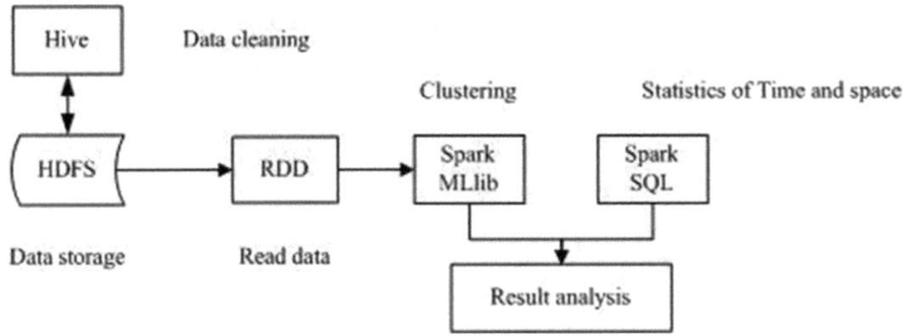

Fig. 5. Data analysis

Spatio-temporal statistics need to be calculated according to the results of clustering. The clustering results eventually form different regions, and then each region be analyzed. Using Spark SQL and statistical methods to combine time and space, calculate the amount of data in different time periods and different regions, and identify the occupation and residence of users according to the time period corresponding to residence and work place. Then the distance between the place of residence and work, that is the commuting distance of the user, is calculated by using the haversine formula. Finally, the commuting time of the user is calculated by the time difference sequence formed in the process of the early commuting and the late commuting.

The result analysis mainly matches the clustering results with the map through the visualization method, and then combines with the spatio-temporal analysis to obtain the commuting characteristics of users. Finally, the users are divided into different types of commuting according to the results of analysis.

5 Experiment and Result Analysis

5.1 Experimental Environment

Six computers are used in the experiment, including one master node and five slave nodes. The primary node is 32 cores, disk memory 1.67 T, memory 128 G. The slave node has 16 cores, 20 G of disk memory, and 2 G of memory. Operating system Ubuntu 14.01.1, Java version 1.7.0, Hadoop version 2.6.5, Scala version 2.11.8, Spark version 1.6.3. The cluster environment plan is shown in Table 3. Use the Scala language for application design.

5.2 Experimental Data

The experiment used data from Unicom operators for a total of 31 days from February 13 to March 15, 2017, with more than four million users and about 1.5 billion records the size of 40.4 G. In this paper, three different types of users are extracted for analysis, and the daily data of users are shown in Fig. 6. Users 1 and 3 have data on a daily basis, and user 2 has a total of 27 days of data.

Table 3. Node environment planning

Host name	Node	IP address
master-hadoop51	master	20.0.0.51
slave1-hadoop114	slaver	20.0.0.114
slave1-hadoop116	slaver	20.0.0.116
slave1-hadoop117	slaver	20.0.0.117
slave1-hadoop118	slaver	20.0.0.118
slave1-hadoop119	slaver	20.0.0.119

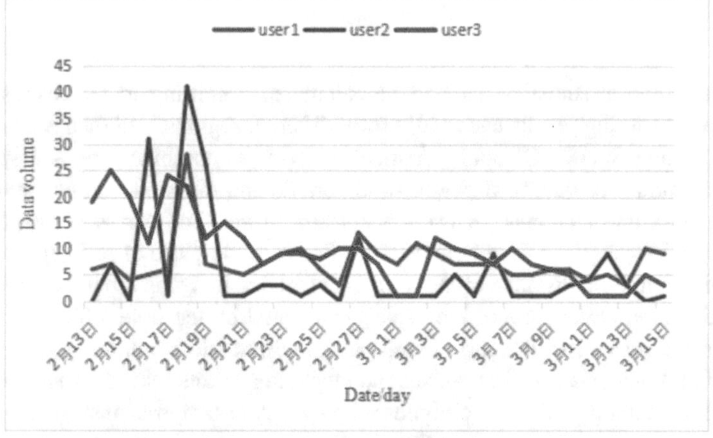

Fig. 6. The amount of data per day for the users

5.3 Analysis of Experimental Results

The data of each user is clustered with different K values, and then the final K value is determined according to the inflection point of calculated WSEE value. The relationship between the k value and the WSEE value is shown in Fig. 7. Finally, the K value of user 1 is 5, and the K value of user 2 and user 3 is 3. It can be seen from Fig. 7 that the WSSE value of user 2 is very small, so the data of user 2 is stable and the place of daily activity is relatively centralized. The value of WSSE of user 3 is the largest indicating that the data distribution of user 3 is relatively scattered, and the places where the user travel every day are relatively far away.

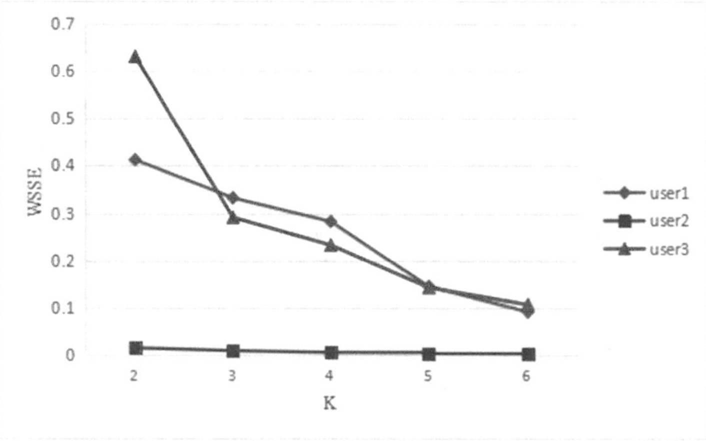

Fig. 7. K value corresponding to WSSE

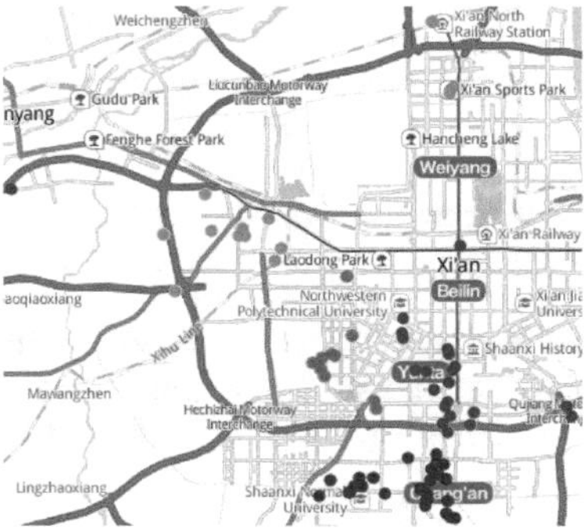

(a) User 1

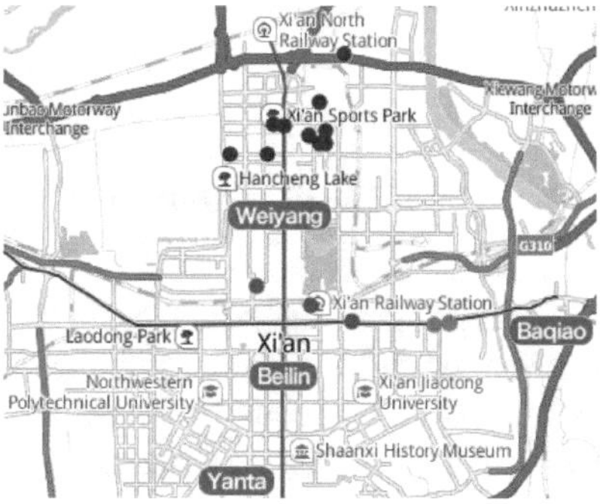

(b) User 2

(c) User 3

Fig. 8. Matching user data to map

After clustering, each user's data is divided into several distinct areas. The result is shown in Fig. 8 and different colors are used to represent different clusters.

The data for each area of the user is counted by time period, as shown in Fig. 9. As you can see from Fig. 9(a), user 1 has the most data in area 0 and is mostly concentrated between 19: 00 p.m. and 8: 00 a.m. So area 0 is the place where the user lives. Area 4's data are mostly concentrated between 9: 00 a.m. and 18: 00 p.m., so area 4 is used as the user's work area. The data in 0 and 4 regions are counted respectively, and the most frequent occurrences is that the place of residence and work of user 1. From Fig. 9(b), you can see that user 2 lives and works in area 0, and the data is very concentrated. From Fig. 9(c), we can see that user 3 is mainly concentrated in area 1 during the day and area 0 at night, so the occupation and residence of user 3 are located in area 1 and area 0, respectively.

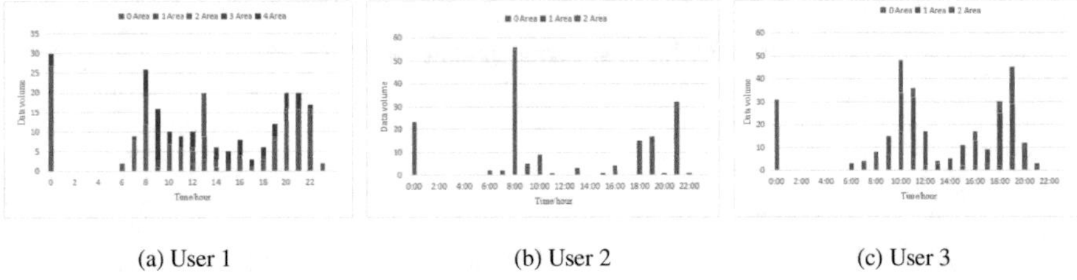

(a) User 1 (b) User 2 (c) User 3

Fig. 9. Statistics of different regions and different time periods

According to the recognition result of the user's occupation and residence, the location of the residence and the work is obtained, and the commuting distance between the two points is calculated. The commute distance of user 1 is 7108.94 m. User 2 has a commuting distance of 720.13 m, and since it is less than 1000 m, user 2 has a commuting distance of 0 m. User 3 has a commute distance of 28364.20 m.

According to the calculation method of 3.4 section, the commuting time of user 1 is 45 min, that of user 2 is 0 min, and that of user 3 is 55 min. Commuting time may be disrupted by other behaviors during the commuting process (eating, shopping, etc.), which is larger than the actual commuting time, but the error range is acceptable.

Table 4. Summary of commuting characteristics

User	Residence	Working place	Distance/(meters)	Time/(minutes)
1	Xi'an suburb	Xi'an city	7108.94	45
2	Xi'an city	Xi'an city	0	0
3	County town	Xi'an suburb	28364.20	55

The commuting characteristics of the users are summarized as shown in Table 4. Combined with the above analysis, the three users have different commuting characteristics. From the results in Table 4, the commute distance of user 1 is moderate, but it can be seen that the user1 commutes to Xi'an city, the traffic problem causes the commute time to be long. In contrast, user 3 has a long commute distance, but the commute time is relatively short because users commute between the town and the outskirts of Xi'an. User 2 can be seen from the commuting distance and commuting hours that the user may be working from home or being a housewife, elderly person, etc.

According to the main commuter characteristics of the user, a simple identification of the user's work, divided into nine-to-five and non nine-to-five types. From the results of clustering, it can be seen that the data of user 1 and user 2 are relatively centralized and user 3 data are relatively scattered. Combined with their respective time analysis, user 1 has a clear time division corresponding to his place of residence and work. So user 1 belongs to nine-to-five type. And user 3 belongs to non nine-to-five type because of the long commuting distance and the corresponding time division of user 3 is 10: 00 am–7: 00 pm. User 2 through the above analysis can be known to belong to the non-nine-to-five type.

6 Conclusion

Spark is used to analyze the user's three commute features, and the results of analysis is used to divide the user into nine-to-five and non-nine-to-five. However, there are some errors in the location of the base station and the mobile signaling data is coarse, so only two kinds of users are divided. The next step is to combine the signaling data of mobile phone with other data, reduce the error, and divide the users belonging to non-nine-to-five type in a more detailed way.

Acknowledgements. This research was supported in part by grants from Shaanxi Provincial Key Research and Development Program (No. 2016KTTSGY01-1).

References

1. Etienne, T., Laurent, M., Sid, L.: Clustering weekly paterns of human mobility through mobile phone data. IEEE Trans. Mob. Comput. **17**(4), 1536–1233 (2018). https://doi.org/10.1109/TMC.2017.2742953
2. Xu, F.L., Lin, Y.Y., Huang, J.X.: Big data driven mobile traffic understanding and forecastig: a time series approach. IEEE Trans. Serv. Comput. **9**(5), 1939–137 (2016). https://doi.org/10.1109/TSC.2016.2599878
3. Jahangiri, A., Rakha, H.A.: Applying machine learning techniques to transportation mode recognition using mobile phone sensor data. IEEE Trans. Intell. Transp. Syst. **16**(5), 2406–2417 (2015). https://doi.org/10.1109/TITS.2015.2405759
4. Wang, L.Y.: Study on the choice and influencing factors of commuting mode of urban residents in China – taking Tianjin as an example. Urban Dev. Res. **23**(7), 108–115 (2016). https://doi.org/10.3969/j.issn.1006-3862.2016.07.016

5. Niu, X.Y., Ding, L., Song, X.D.: Identification of urban spatial structure of Shanghai central city based on mobile phone data. J. Urban Plan. **06**, 61–67 (2014). https://doi.org/10.11819/cpr20150917a
6. Becker, R.A., Caceres, R., Hanson, K., et al.: A tale of one city: using cellular network data for urban planning. IEEE Pervasive Comput. **10**(4), 18–26 (2011). https://doi.org/10.1109/MPRV.2011.44
7. Chen, L., Zhang, W.Z.H., Li, Y.J., et al.: The influence of urban residential space form on commuting mode in Beijing. Geogr. Sci. **36**(5), 697–704 (2016). https://doi.org/10.13249/j.cnki.sgs.2016.05.007
8. Sun, B.D., Dan, B.: Influence of built environment of Shanghai city on residents'choice of commuting mode. J. Geogr. **70**(10), 1664–1674 (2015). https://doi.org/10.11821/dlxb201510010
9. Chen, Y.P., Song, Y., Yi, Z.H., et al.: The influence of urban land use characteristics on resident travel mode – a case study of Shenzhen. Urban Transp. **9**(5), 80–85+27 (2011). https://doi.org/10.13813/j.cn11-5141/u.2011.05.013
10. Han, H.R., Yang, C.H.F., Song, J.P.: Difference of commuting efficiency between public transport and private car travel and its influencing factors – a case study of Beijing metropolitan area. Geogr. Res. **36**(2), 253–266 (2017). https://doi.org/10.11821/dlyj201702005
11. Zhou, J.P., Chen, X.J., Huang, W., et al.: The balance of work and residence and the efficiency of commuting in the big cities of Midwest China – a case study of Xi'an. J. Geogr. **68**(10), 1316–1330 (2013). https://doi.org/10.11821/dlxb201310002
12. Long, Y., Zhang, Y., Cui, C.Y.: Analysis of the relationship between work and residence and commuting in Beijing by using the data of bus credit card. J. Geogr. **67**(10), 1339–1352 (2012). https://doi.org/10.11821/xb201210005
13. Roth, C., Kang, S.M., Batty, M., et al.: Structure of urban movements: polycentric activity and entangled hierarchical flows. PLoS ONE **6**(1), e15923 (2011). https://doi.org/10.1371/journal.pone.0015923
14. Fu, X.: Taxi commuting recognition and spatio-temporal feature analysis based on GPS data. Chin. J. Highw. **30**(7), 134–143 (2017). https://doi.org/10.3969/j.issn.1001-7372.2017.07.017
15. Jiang, B., Yin, J., Zhao, S.: Characterizing human mobility patterns in a large street network. Phys. Rev. **80**(2 Pt 1), 021136 (2009). https://doi.org/10.1103/PhysRevE.80.021136
16. Md, A.U., Joolekha, B.J., Aftab, A., et al.: Human action recognition using adaptive local motion descriptor in Spark. IEEE Access **5**, 21157–21167 (2017). https://doi.org/10.1109/ACCESS.2017.2759225

An Improved Algorithm for Moving Object Tracking Based on EKF

Leichao Hou[2(✉)], Junsuo Qu[1], Ruijun Zhang[2], Ting Wang[2], and KaiMing Ting[3]

[1] School of Automation, Xi'an University of Post and Telecommunications, Xi'an 710121, China

[2] School of Communication and Engineering, Xi'an University of Post and Telecommunications, Xi'an 710121, China
1457633811@qq.com

[3] School of Science, Engineering and Information Technology, Federation University, Ballarat, Australia

Abstract. Kalman filter estimates the desired signal from the amount of measurement related to the extracted signal, which is widely used in engineering due to its simple calculation and easy programming on a computer. However, the basic theory originally proposed by Rudolf E. Kalman is for linear systems only, whereas a realistic physical system is often nonlinear. Extended Kalman Filter (EKF) solves nonlinear filtering problems. In this paper, we focus on issues related with targeted object being occluded We combine EKF and Meanshift to track the moving object. Once the object position is predicted by EKF in the center of the object, then the Meanshift algorithm iterates over the initial value of EKF estimation to track the object. Experiments show that the method reduces the object search time and improves the accuracy of the object tracking.

Keywords: EKF · Nonlinear system · Meanshift · Object tracking

1 Introduction

With the development and application of computer vision, object tracking has become a basic problem in this field. Mean shift algorithm (Meanshift) of the object tracking is a non-parametric feature space analysis technique used to locate the maximum value of the density function, namely the Pattern Search Algorithm (PSA). The algorithm establishes a confidence image in the new image sequence according to the color histogram of the object in the pre-order image, and uses Meanshift to find the peak value of the confidence image near the object position [1]. That is, Meanshift identifies to the real position of the object by continuously iterating the Meanshift vector, and achieves the purpose of tracking, so it is widely used in the field of video objects tracking due to its efficiency in real-time and robustness.

However, Meanshift is easily interfered by external factors when tracking moving object. Based on the global optimal theory [2], when obstacles occlude the target object, the video area is divided into small areas, and the search window scale is adjusted through the area where the center of the global search window is located to

P. Krömer et al. (Eds.): ECC 2018, AISC 891, pp. 483–490, 2019.
https://doi.org/10.1007/978-3-030-03766-6_54

improve the anti-occlusion ability [3]. When there is a large area of the same color interference between the background and the object, the method combination of the optical flow method and the three-frame difference is used to detect the object, then the image is processed for morphological processing [4]. When the system model cannot be linearized, Extended Kalman filter (EKF) is used to linearize the nonlinear system, then it achieves accurate positioning of the mobile robot in the nonlinear system [5]. For multi-objects tracking, Salhi [6] divides the moving object tracking into two parts: detection and tracking; and he compares the advantages and disadvantages of Meanshift, Camshift and Kalman filtering applied to object tracking.

Despite the above advances, accurate object tracking in a complex environment is still a challenge. This paper employs EKF as a motion object position change prediction mechanism. It selects the starting position of the moving object in the initial frame of the video, and then uses Meanshift to track the object when large-area color interference or obstacle occlusion happens. The EKF algorithm is used to predict the position where the moving object may appear in the current frame, and then Meanshift search is used. When the occlusion is severe, the background subtraction is used to detect the position of the moving object, then the position parameter was transmitted to the Meanshift algorithm. Current methods lose track of the targeted object easily when occlusion occurs. We show that the proposed method has no problem tracking the targeted object under the same scenario. In addition, it improves the speed of tracking, and it reduces the number of iterations comparing with the original Meanshift algorithm.

2 Extended Kalman Filter

Kalman filtering is a model-based linear minimum variance estimation for the estimation of stationary or non-stationary multidimensional random signals. Therefore, it has been widely used in random signal processing and motion object trajectory estimation. However, the standard Kalman filter problem assumes that the mathematical model of a physical system is linear, and yet nonlinear systems are often encountered in engineering practice. Extend Kalman Filter performs a Taylor series expansion on system equations and measurement equations of nonlinear systems and preserves linear terms. Then the standard Kalman filter algorithm is used to process the linearized system. This is its way to resolve the problem of system nonlinearity.

The advantage of EKF is that it is simple to calculate and easy to implement. By preserving the first-order linear term of the Taylor expansion of the nonlinear function, ignoring the remaining high-order terms, EKF linearizing the nonlinear problem, and applying the Kalman filter algorithm the linearized system [7]. The extended Kalman filter uses the optimal estimator and real observations from the previous state to predict the current state. Like the Kalman filter algorithm, it can predict the center of the next frame area and update the object area of the current frame in real time. The extended Kalman filter is essentially a set of recursive algorithms. Each recursive cycle contains two processes of time update and measurement update for the estimation. The specific EKF formula is as follows.

Time update equation (predictive equation).

$$\hat{x}_k = f(\hat{x}_{k-1}, u_{k-1}, 0) \tag{1}$$

$$p_k^- = A_k P_{k-1} A_k^T + W_k Q_{k-1} W_K^T \tag{2}$$

State update equation (correction equation).

$$K_k = P_k^T (H_k P_k^- H_k^T + V_k R_k V_k^T)^{-1} \tag{3}$$

$$\hat{x}_k = \hat{x}_{\bar{k}} + K_k(z_k - h(\hat{x}_{\bar{k}}, 0)) \tag{4}$$

$$P_k = (I - K_k H_k) P_k^- \tag{5}$$

3 Improved Meanshift Algorithm

3.1 The Principle of Meanshift Algorithm

The MeanShift algorithm is a gradient-based parameter-free density distribution estimation algorithm, which plays an important role in the object tracking algorithm due to its strong real-time performance. The essence of the algorithm is the process of converge to the probability density maxima through successive iterative offsets from the starting point. The basic principles of Meanshift are as follows:

Given the set of samples $\{x_i\}_{i=1,2\ldots n}$ in the d-dimensional euclidean space R^d, R is the real number field, and $k(x)$ is the kernel function of the space, representing the contribution in the mean value estimation. The expression of $k(x)$ is as follows:

$$k(x) = k(\|x\|^2) \tag{6}$$

And the kernel function k satisfies non-negative, monotonically decreasing, segmented continuous and integrable, i.e. $\int_0^\infty k(x)dx < \infty$. The kernel function, also known as the "window function", plays a smoothing role in kernel estimation.

The point x estimate for the kernel $k(x)$ and the bandwidth matrix H is

$$f(x) = \frac{1}{nh^d} \sum_{i=1}^{n} K(\frac{x - x_i}{h}) \tag{7}$$

Therefore, the kernel function is a weight function, and each sample point in the tracking region is weighted according to the distance from the center point x, the closer the object model center is, the greater the weight. for the pixel points in the edge region, the error increases due to the marginal noise and interfering effects, and the density estimation increases their robustness and improves the anti-interference ability of the tracking.

The probability density of the sample set has been estimated by the kernel function, and the Meanshift algorithm finds the mode of the density distribution of the data set. The Meanshift vector is obtained as:

$$m_{h,G}(x) = \frac{\sum_{i=1}^{n} x_i g\left(\left\|\frac{x-x_i}{h}\right\|^2\right)}{\sum_{i=1}^{n} g\left(\left\|\frac{x-x_i}{h}\right\|^2\right)} - x \tag{8}$$

Substituting the kernel function $g(x) = 1$, the above formula can be rewritten as

$$m_{h,G}(x) = \frac{1}{n}\sum_{i=1}^{n}(x_i - x) \tag{9}$$

3.2 Meanshift Based on Dynamic Kernel Window Width

3.2.1 Traditional Mean Shift Algorithm for Target Tracking

Since Meanshift is a semi-automatic tracking algorithm [8], it is necessary to initialize the tracking object in the first frame when processing the video, that is, manually select the tracking object. This area is also the area where the kernel function acts, and the size of the area is equal to the tracking window. The radius h is the kernel function bandwidth. However, in the traditional Meanshift, the window width h remains unchanged, when the shape size of the object changes. As a result, the real-time performance of the algorithm is poor and often resulting in tracking failure.

3.2.2 Improved Meanshift Algorithm Based on Dynamic Kernel Window Width

In order to improve the real-time performance of the Meanshift algorithm, the tracking window changes as the object scale changes, introducing a dynamic kernel-bandwidth. The initial region of the video frame is called the target model, and the candidate region where the object may exist in each subsequent frame is called the candidate target model. The similarity between the target model and the current candidate target model is measured by the Bhattacharyya coefficient, which is referred to as the BH coefficient. Since the experiment selects the image as a continuous frame sequence with a small time interval, the object size change remains within the controllable range of the adjacent two frames. In this experiment, the color histogram of the candidate model is calculated by using the bandwidth of $h \pm 10\% * h$. Calculate the BH coefficient, select the kernel-bandwidth h with the smallest BH coefficient as the size of the object window, and dynamically update to adapt to the object scale change.

3.3 Tracking Implementation Process of Moving Targets

After the target model and the candidate target model are established, the similarity is compared by the Bhattacharyya coefficient (BH), and the error ε is allowed. the specific algorithm steps are as follows:

(1) Selecting an object to be tracked in the initial frame of the video, and calculating a probability density q_u of the object model, a object initial position x, and a tracking window width h;
(2) Calculating the color probability distribution of the search window;
(3) Performing a meanshift iteration, and updating the kernel window width and performing similarity matching until $\left\| m_{h,G}(x) - x \right\| < \varepsilon$, otherwise return (2).

4 Improved Meanshift and EKF Combined Tracking Algorithm

When the traditional Meanshift is used for video object tracking, the object color histogram is used as the search feature. By continuously iterating the Meanshfit vector, the algorithm converges to the true position of the object, so as to achieve the purpose of object tracking. However, the traditional Meanshift algorithm has the following disadvantages: (1) Due to the fixed tracking window size (fixed bandwidth) during the process of moving object tracking, the tracking effect becomes worse when the tracking object scale changes. In response to this problem, the bandwidth of the kernel function is dynamically processed to adapt to the size change of the tracking object. (2) The histogram feature is slightly scarce in the description of the object color characterization, and the tracking failure occurs when the objects color closed to the background color. Therefore, the Meanshift algorithm only has the tracking function in the object tracking process, but no prediction function. So when the objects color closed to the background color during the process of moving object tracking, the performance of the original object tracking algorithm is greatly reduced, which leads directly to the inaccurate tracking or tracking lost.

In view of its shortcomings, this paper employs the EKF algorithm as the mechanism to predict object position change, and adopts background subtraction to increase the feature used for object matching, improved the performance of traditional algorithms greatly. resolving the tracking lost problem. The specific method is as follows, the object to be tracked is selected in the initial frame of the video, the target model is established, and the back projection view is obtained. Next, (i) the candidate target model in the next frame is processed; (ii) the eigenvalue probabilities of the pixels in the object region and the candidate region are calculated to obtain a description of the target model and the candidate model by the Meanshift algorithm; and (iii) the degree of similarity between the object region and the candidate region is calculated.

Updating the core window width h. If the similarity BH is less than the threshold, this implies that the Meanshift tracking is unreliable. At this time, the prediction function of the EKF is enabled to perform the position update iteration, and the updated position is used as the starting position of the next frame. If the similarity BH is always

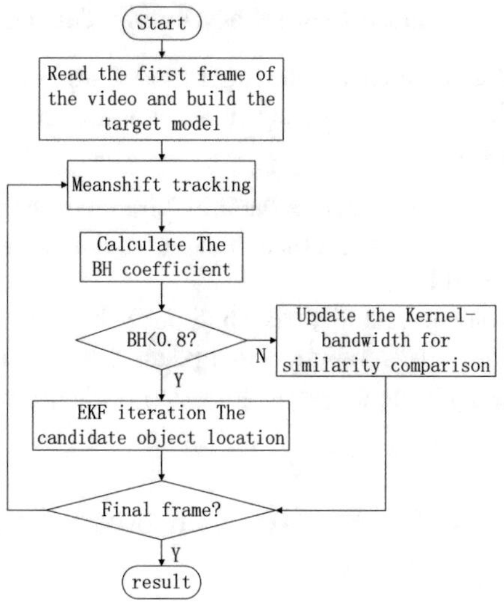

Fig. 1. Improved meanshift algorithm flow chart

above the threshold, the Meanshift iteration is performed all the time. When the video is tracked to the last frame as a cutoff condition, end the operation. Otherwise the EKF outputs the result back to the meanshift tracker after the status update and time update. If the value of the threshold BH is too large, it is easy to cause the tracking failure due to the missing object. If the value is too small, the tracking result will be unreliable. In this paper, the threshold is 0.8. The process is shown in Fig. 1.

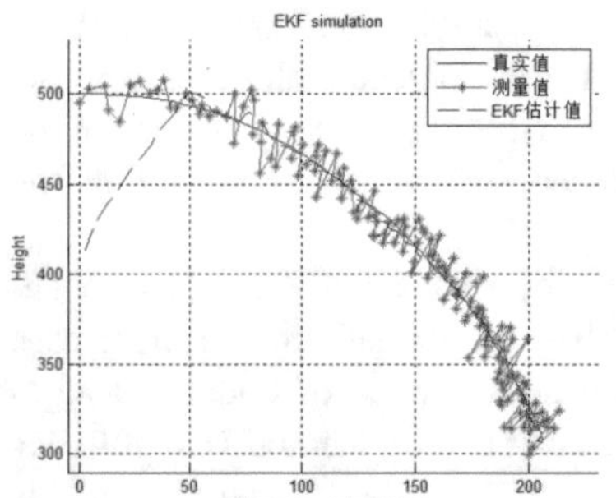

Fig. 2. EKF simulation chart

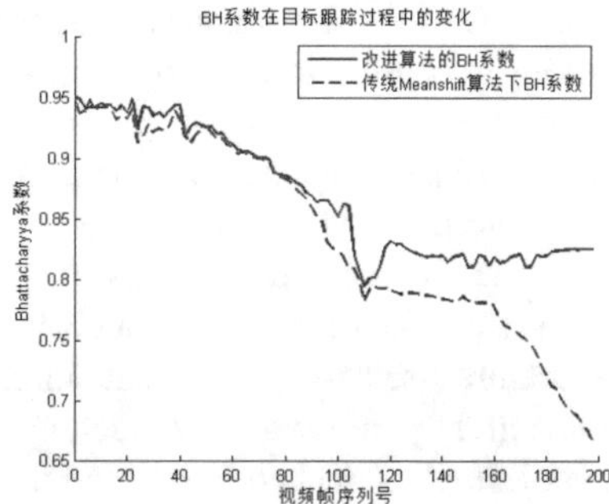

Fig. 3. Comparison of BH coefficient before and after algorithm improvement

5 Experimental Results and Analysis

The estimation results of the extended Kalman filter are simulated as follows. From the simulation results in Fig. 2, it can be seen that the estimated value of EKF at the beginning has a large deviation, since the EKF estimation process is a process of continuously correcting the feedback. After a few iterations, the result tends to the true motion trajectory of the object, and the difference is small, and the measured value varies in a small range around the true value.

The change of the similarity measure BH coefficient in the tracking process, and the improvement is shown in Fig. 3, the red dotted line represents the change of BH coefficient when the object is tracked by the traditional Meanshift algorithm, and the blue solid line represents the BH coefficient change of the improved algorithm. From the above figure, since the front background color is similar, the similarity of the traditional Meanshift is less than 0.8 at 110 frames. The improved algorithm starts the prediction function of the EKF at this time as the initial position of the next frame of the Meanshift iteration. The improved algorithm significantly improves the similarity between the object template and the candidate object model, and increases the tracking accuracy.

In order to verify the effectiveness and accuracy of the extended Kalman filter for the mean shift algorithm, the following scenario was chosen with a video resolution of 320×240 and a frame rate of 25 frames/second. The results of object tracking are as follows.

Figure 4 shows the results of tracking with meanshift alone and using an improved algorithm. Manually select the initial tracking object at frame 50. It can be seen that before the prediction mechanism added, since the tracking object shirt color and the background color are similar, although the tracking is inaccurate when tracking to 110 frames, the real-time performance guarantee. The Meanshift tracking error increases to 162 frames. Figures 4d, e and f show the results of the improved algorithm after

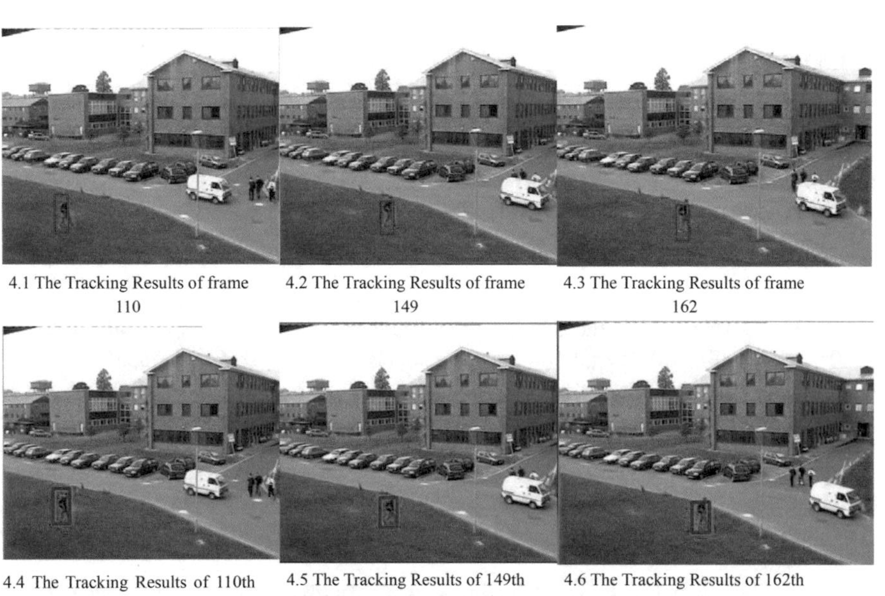

4.1 The Tracking Results of frame 110

4.2 The Tracking Results of frame 149

4.3 The Tracking Results of frame 162

4.4 The Tracking Results of 110th frame by improvement algorithm

4.5 The Tracking Results of 149th frame by improvement algorithm

4.6 The Tracking Results of 162th frame by improvement algorithm

Fig. 4. Comparison of algorithms before and after improvement

Meanshift merges EKF. In this paper, the improvement of the nuclear window width makes the tracking window change with the change of the moving object. At the 149th frame, the similarity is reduced to 0.795, the prediction function of the EKF is started, and the motion trajectory of the object is judged, and 162 frames are completed to complete the tracking of the object. The experimental results show that the EKF-based Meanshift algorithm accurately tracks the object.

6 Conclusion

In this paper, we identify that the object tracking failure is caused by the lack of prediction mechanism in the traditional Meanshift algorithm. EKF is used as the predictor by judging the tracking effect of Meanshift. The EKF prediction update result is used as the initial value of the Meanshift algorithm when the targeted object is slightly occluded. When the occlusion is severe, the tracked object is detected first, and then the detection result is brought into the Meanshift tracker. The experimental results show that the proposed combined application of EKF and Meanshift reduces the object search time and accurately tracks the object.

Acknowledgments. This research was supported in part by grants from the International Cooperation and Exchange Program of Shaanxi Province (2018 K W-026), Natural Science Foundation of Shaanxi Province (2018JM6120), Xi'an Science and Technology Plan Project (201805040YD18C G24(6)), Major Science and Technology Projects of XianYang City (2017k01-25-12), Graduate Innovation Fund of Xi'an University of Posts & Telecommunications (CXJJ2017012, CXJJ2017028, CXJJ 2017056).

References

1. Zhao, H.Y., Zhang, X.L., et al.: Image denoising algorithm based on multi-scale Meanshift. J. Jilin Univ. **44**(5), 1417–1422 (2014). https://doi.org/10.7964/jdxbgxb201405031
2. He, L., Han, B.S., et al.: New definition of filled function applied to global optimization. Commun. Appl. Math. Comput. **30**(1), 128–137 (2016). https://doi.org/10.3969/j.issn.l006-6330.2016.01.011
3. Ou, Y.N., You, J.H., et al.: Tracking multiple objects in occlusions. Appl. Res. Comput. **27**(5), 1984–1986 (2010). https://doi.org/10.3969/j.issn.1001-3695.2010.05.110
4. Li, Z.L.: An visual object tracking algorithm based on improved camshift. Comput. Knowl. Technol. **12**(9X), 150–152 (2016). https://doi.org/10.14004/j.cnki.ckt.2016.3566
5. Jin, G., Zhu, Z.Q.: Improvement and simulation of kalman filter localization algorithm for mobile robot. Ordnance Indus. Autom. (2018). https://doi.org/10.7690/bgzdh.2018.04.017
6. Salhi, A., Moresly, Y., et al.: Modeling from an object and multi-object tracking system. In: Computer and Information Technology. IEEE (2017). https://doi.org/10.1109/gscit.2016.20
7. Rhudy, M., Gu, Y., Napolitano, M.: An analytical approach for comparing linearization methods in EKF and UKF. Int. J. Adv. Rob. Syst. **10**(10), 5870–5877 (2013). https://doi.org/10.5772/56370
8. Wang, B.Y., Fan, B.J.: Adoptive meanshift tracking algorithm based on the combined feature histogram of color and texture. J. Nanjing Univ. Posts Telecommun. **33**(3), 18–25 (2013). https://doi.org/10.14132/j.cnki.1673-5439.2013.03.017

Fatigue Driving Detection and Warning Based on Eye Features

Zhiwei Zhang[1(✉)], Ruijun Zhang[1], Jianguo Hao[1], and Junsuo Qu[2]

[1] School of Communication and Engineering, Xi'an University of Post and Telecommunications, Xi'an 710121, China
13772066729@163.com

[2] School of Automation, Xi'an University of Posts and Telecommunications, Xi'an 710121, China
825459845@qq.com

Abstract. For the aim of reducing the occurrence of traffic accidents caused by fatigue driving, it is of great significance to design a system based on eye features for fatigue driving detection and early warning. The system uses a camera to capture images, using an improved Haar feature cascade classification algorithm to detect the face area, and then uses a Ensemble of Regression Trees (ERT) cascade regression algorithm to detect human eyes and mark 12 points in the area. According to the Eye Aspect Ratio (EAR) algorithm and the blink frequency, the driver's fatigue state can be determined and the alarm can be timely issued,and the image will be uploaded to the cloud platform of the Internet of things.

Keywords: Face detection · Feature extraction · Fatigue driving Eye Aspect Ratio (EAR)

1 Introduction

In recent years, a lot of domestic and foreign research results have been obtained for the detection technology of fatigue driving. Jin LS proposes a method, determining the degree of fatigue of the driver by detecting the state of the steering wheel, but it has no definite criterion, which is prone to misjudgment or missed judgment [1]. Lenskiy has achieved accurate eye location and segmentation based on color and texture features, but this method is slow to detect and it is difficult to guarantee real-time performance [2].

Aiming at the problems existing in the above methods, a fatigue driving detection method based on eye features will be proposed. Firstly, the improved Haar feature cascade classification algorithm can be used to detect the face region; Secondly, the human eye detection is performed in the upper part of the face region, and the aspect ratio of the eye is calculated, by which the eye closure degree is measured. Finally, the fatigue determination and warning are performed according to the given threshold. The method can quickly and accurately detect whether the driver is in a fatigue state, and is suitable for real-time detection of driver fatigue.

P. Krömer et al. (Eds.): ECC 2018, AISC 891, pp. 491–498, 2019.
https://doi.org/10.1007/978-3-030-03766-6_55

2 Haar Feature Detection Plus Tracking Algorithm Principle

Haar feature cascade classification detection algorithm is an iterative algorithm whose core idea is to train different classifiers (weak classifiers) with the same classification ability for the same training set [3]. Then these weak classifiers are superimposed to form a stronger final classification (strong classifier). Take the blink detection as an example. The structure of the blink detector is shown in Fig. 1.

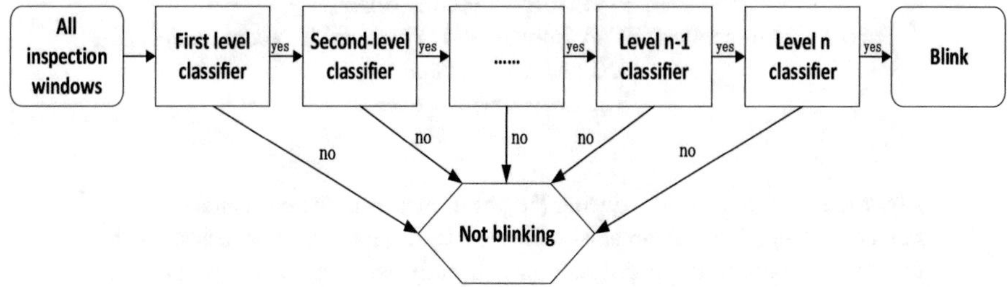

Fig. 1. Broken eye detector structure

The eye positioning flow chart is shown in Fig. 2.

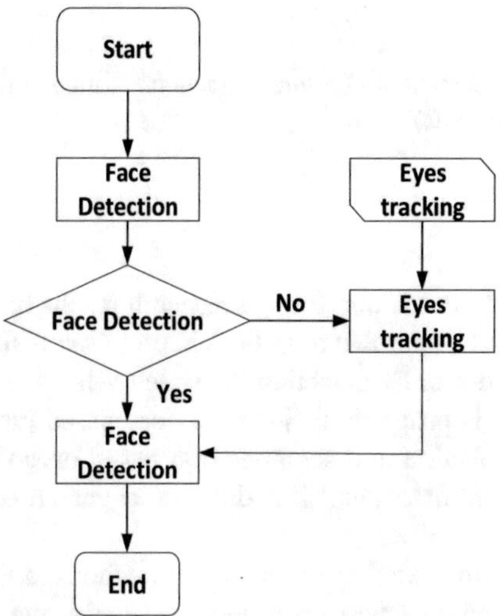

Fig. 2. Human eye positioning flowchart

The results of human eye positioning are shown in Fig. 3.

Fig. 3. Human eye detection after tracking

3 Feature Extraction

3.1 Expression Feature Extraction

In this paper, this kind of 68 face feature point calibration tracking method is used to track the face feature points for a long time. The feature point localization method introduces a method based on deep learning, which can use large data for model training to improve the accuracy of feature point positioning. Experiments show that the adaptive tracking verification method can improve the accuracy of large-scale face feature tracking. Through these 68 markers, the facial state of facial features can be fully extracted to lay the foundation for the recognition of facial expressions. Experimental test as shown below (Fig. 4).

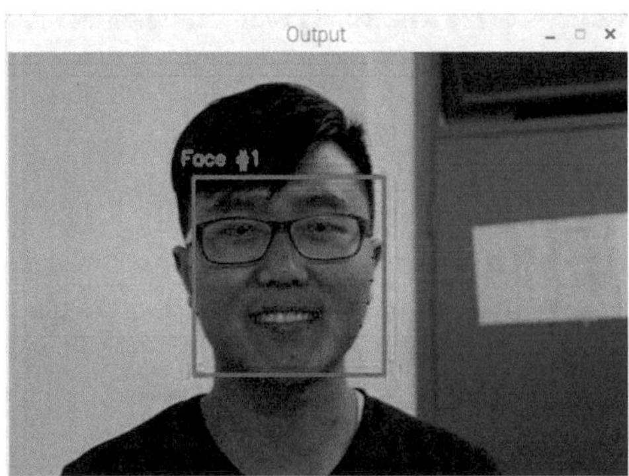

Fig. 4. dlib face 68 points mark

Upon the above, an algorithm based on dlib [4] and EAR [5] combined algorithm is proposed for eye length ratio detection, and this algorithm is different from ordinary eye positioning. The PERCLOS [6] algorithm is used to determine the degree of opening and closing, and finally the EBF algorithm is used to determine whether the eye blinks or not. The eye length ratio algorithm first adds a flag to the person by the dlib landmark, and then the EAR algorithm judges whether the eye blinks or not. This blink detection method is faster, more efficient, and easier to implement.

For eye detection, we need to extract eye signatures. Each eye is represented by six coordinate points, starting from the left corner of the eye, and then rotating clockwise around the rest of the area [7], as Fig. 5 shows.

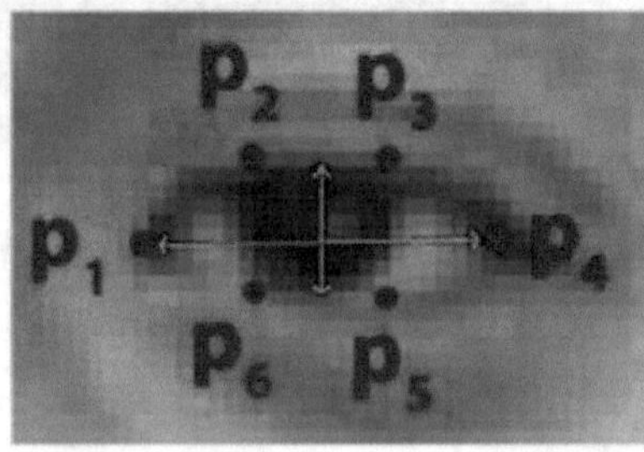

Fig. 5. Open and closed eyes with landmarks pi automatically detected by

When eye features are extracted and marked, the EAR calculation can be performed using the following formula (1),

$$EAR = \frac{\|p2 - p6\| + \|p3 - p5\|}{2\|p1 - p4\|} \tag{1}$$

Where p1, p2, p3, p4, p5 and p6 are the 2D landmark locations, depicted in Fig. 5.

When the eyes open, the EAR value remains essentially unchanged. When the eyes are closed, the EAR value is close to zero. Therefore, when the EAR value turns zero, it is considered that a blink has occurred.

3.2 Eye Opening Degree

The most common method for judging the eye opening degree is the Percentage of Eyelid Closure (PERCLOS) method. This method uses the degree of eye closure to determine the current state of the eye and the driver's expression, as shown in Fig. 6.

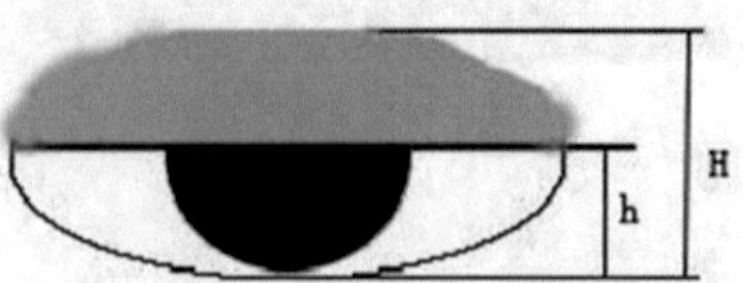

Fig. 6. Eye opening and closing

PERCLOS's formula is shown in formula (2)

$$p = \left(1 - \frac{h}{H}\right) \times 100\% \tag{2}$$

Where p represents the degree of eyelid closure, h represents the height at which the eye is currently open, and H represents the height of the eye when no expression is given. The PERCLOS standard is divided into various types, including P80, P70, P60, etc. Taking P80 as an example, when the calculated P is greater than 80%, the driver is considered to have expressed symptoms, and P70 and P60 are similar, respectively, indicating the calculation when the signs of expression start to appear when P is greater than 70% and 60%.

3.3 Blinking Frequency

For drivers who are driving high-speed moving cars, frequent blinking behaviors caused by factors such as tension, strong light, eye diseases, eye discomfort, contact lenses, foreign matters entering the eyes, and facial expressions, will be a potential danger signal. If necessary, drivers need to be prompted to concentrate for safe driving. Eye Blink Frequency [8, 9] (EBF) has become a very important factor in the detection of expression. This article gives the definition of EBF as shown in Formula (3)

$$EBF = \frac{b}{t} \tag{3}$$

Where b represents the number of blinks, t represents the time required to complete these blinks. Under normal circumstances, people blink about 15 times per minute. The higher the blinking frequency is, the more frequent the blinking will be [10]. In the calculation session, it is necessary to select an appropriate interval Δt to calculate the blinking frequency under normal conditions. Too long or too short time intervals are of little significance for expression detection.

The dlib face marking points with the EAR algorithm are both used in the expression detection to extract the eye features and perform operations on its coordinate points to determine the blink behavior. The EAR detects blinks as shown in Fig. 7.

The experimental system diagram is shown in Fig. 8.

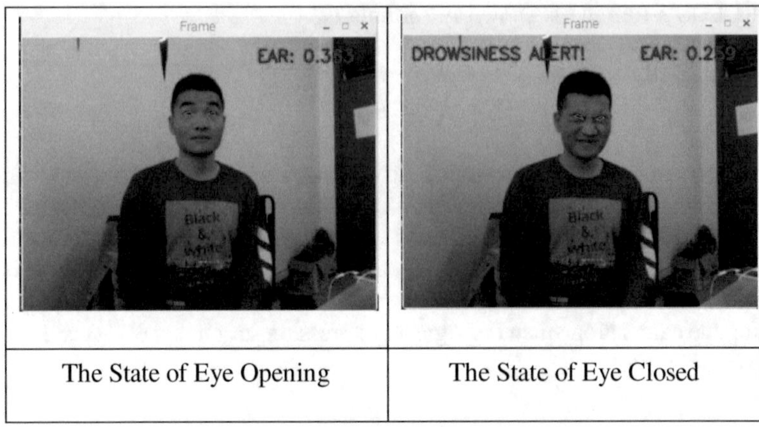

Fig. 7. The EAR detects blinks

Fig. 8. Experimental system diagram

4 Unet Internet of Things Platform Presented

In the existing fatigue detection and warning system, when the fatigue driving behavior is detected, only the local warning (including the local buzzer alarm or the warning light blinks). However, in the market research, it was found that when a driver is driving alone, he often ignores these warnings and chooses to continue his fatigue driving. This poses a great safety risk. For the above issues, we propose a concept of remote warning. In the private car environment, warning information can be sent to the bound guardians through the cloud platform (Fig. 9).

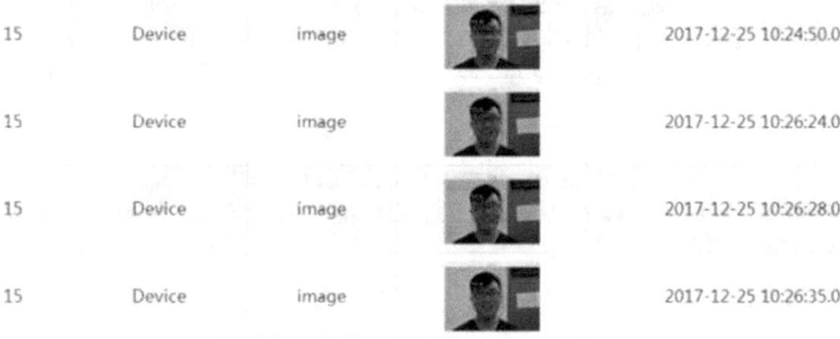

Fig. 9. Fatigue images display

5 Conclusion

In this paper, we discuss the method of fatigue driving detection and early warning based on eye features. It has the advantages of non-contact and strong anti-interference. This method has better face detection effect, and the eye location efficiency is high on this basis. When the system detects fatigue driving, the system will warn and upload the picture through the network to the cloud platform. At this time, the platform can send the dangerous information to the relevant personnel. It can be seen from the experimental results that the accuracy of the fatigue detection method is higher and the effect is better.

Acknowledgments. This research was supported in part by grants from the International Cooperation and Exchange Program of Shaanxi Province (2018KW-026), Natural Science Foundation of Shaanxi Province (2018JM6120), Xi'an Science and Technology Plan Project (201805040YD18CG24(6)), Major Science and Technology Projects of XianYang City (2017k01-25-12), Graduate Innovation Fund of Xi'an University of Posts & Telecommunications (CXJJ2017012, CXJJ2017028, CXJJ2017056).

References

1. Jin, L.S., Niu, Q.N., Hou, H.J., et al.: Driver cognitive distraction detection using driving performance measures. Discrete Dyn. Nat. Soc. **30**(10), 1555–1565 (2012). https://doi.org/10.1155/2012/432634
2. Lenskiy, A.A., Lee, J.S.: Driver's eye blinking detection using novel color and texture segmentation algorithms. Int. J. Control Autom. Syst. **10**, 317–327 (2012). https://doi.org/10.1007/s12555-012-0212-0
3. Lienhart, R., Maydt, J.: An extended set of Haar-like features for rapid object detection. In: International Conference on Image Processing, pp. 900–903. IEEE, Rochester (2002). https://doi.org/10.1109/icip.2002.1038171
4. Xiong, X., De la Torre, F.: Supervised descent methods and its applications to face alignment. In: CVPR, Portland, OR, USA, pp. 532–539 (2013). https://doi.org/10.1109/cvpr.2013.75

5. Uricar, M., Franc, V., Hlavac, V.: Facial landmark tracking by tree-based deformable part model based detector. In: 2015 IEEE International Conference on Computer Vision Workshop (ICCVW), Santiago, Chile, pp. 963–970 (2016). https://doi.org/10.1109/iccvw.2015.127
6. Sommer, D., Golz, M.: Evaluation of PERCLOS based current fatigue monitoring technologies. In: Proceedings of the International Conference on Engineering in Medicine and Biology Society, pp. 4456–4459. IEEE, Buenos Aires (2010). https://doi.org/10.1109/iembs.2010.5625960
7. Asthana, A., Zafeoriou, S., Cheng, S., Pantic, M.: Incremental face alignment in the wild. In: Conference on Computer Vision and Pattern Recognition, Columbus, OH, USA, pp. 1859–1866 (2014). https://doi.org/10.1109/cvpr.2014.240
8. Danisman, T., Bilasco, I.M., Djeraba, C., et al.: Drowsy driver detection system using eye blink patterns. In: International Conference on Machine and Web Intelligence, Algiers, Algeria, pp. 230–233 (2010). https://doi.org/10.1109/icmwi.2010.5648121
9. Zhu, X., Ramanan, D.: Face detection, pose estimation, and landmark localization in the wild. In: Providence, RI, USA, pp. 2879–2886 (2012). https://doi.org/10.1109/cvpr.2012.6248014
10. Lee, W.H., Lee, E.C., Park, K.E.: Blink detection robust to various facial poses. J. Neurosci. Methods, November 2010. https://doi.org/10.1016/j.jneumeth.2010.08.034

Application of Data Mining Technology Based on Apriori Algorithm in Remote Monitoring System

Chenrui Xu[1,2(✉)], Kebin Jia[1,2], and Pengyu Liu[1,2]

[1] Beijing Laboratory of Advanced Information Networks, Beijing 100124, China
xuchenrui_1995@foxmail.com
[2] Department of Information, Beijing University of Technology, Beijing 100124, China

Abstract. At present, the theoretical analysis of gas station oil and gas data is weak, and there is no unified platform for collecting and uploading. In view of these problems, a set of data acquisition and mining scheme is proposed. The Apriori algorithm is used to correlate the current environmental data of oil and gas, focusing on the correlation between oil and gas concentration and liquid resistance pressure, tank temperature, tank pressure, time, and treatment unit emission concentration. In addition, we designed and implemented a remote online monitoring system for oil and gas recovery based on the SSH framework. The results of the application obtained in a gas station in Beijing show that this system can provide the reference basis for the intelligent construction for the gas station to monitor the large oil and gas data. The results of data mining and analysis can provide accurate and objective data support for the monitoring personnel of gas stations, and higher priority monitoring for the heavy point data segment. It has reference value and provides a good technical foundation for the statistics and processing of oil and gas data in the follow-up gas stations.

Keywords: Apriori algorithm · Correlation analysis · Remote detection system Data mining · SSH

1 Introduction

In recent years, with the rapid development of transportation and transportation industry, the number of urban vehicles and urban gas stations has increased rapidly, resulting in the increase of oil and gas emissions from refueling equipment. The heavy emission of oil and gas makes the environment seriously polluted and the content of oil and gas in the air is too high, which greatly increases the probability of safety accidents such as fire at the gas station. Therefore, the problem of oil and gas pollution and the safety of gas stations are becoming more and more serious. How to effectively monitor the oil and gas has been paid much attention by our government and relevant departments. At present, gasoline storage areas at petrol stations and refueling areas of service vehicles are the high incidence areas of oil and gas emissions and leaks. Many gas stations still use traditional manual inspection to monitor the oil and gas in these

P. Krömer et al. (Eds.): ECC 2018, AISC 891, pp. 499–507, 2019.
https://doi.org/10.1007/978-3-030-03766-6_56

areas. This method is not only low in efficiency, high in cost, but also easy to make mistakes. It is a great hidden danger to the safety of personal and property in the gas station. On the other hand, although some gas stations in China have installed sensors to monitor oil and gas, a large number of data reports related to oil and gas produced in the monitoring process are still stored at the station, and there is no unified platform for collecting and displaying the collected oil and gas data, which is not conducive to effective control of such risks by relevant staff and regulatory authorities. In addition, the oil and gas data collected through the sensors at the gas station end generally contain information such as acquisition time, temperature, pressure and oil and gas concentration. There is a certain correlation between these information. If they are used, we can analyze the relationship between various kinds of data and monitor the key data segment with higher priority, so as to improve the efficiency of oil and gas monitoring.

In the field of data mining, association rule mining is an important branch. The strong association rules are deduced from the relationship between different transactions in the data centralization, which helps people intelligently screen the influential factors that have strong correlation with the target data. The most classical algorithm is the first Apriori algorithm, such as Agrawal [1], to analyze the shopping basket. Cui [2] put forward the connection and pruning method to improve the layout of the database. Wang [3] proposed an improved algorithm based on item set bit logic operations: B_Apriori algorithm, and improved the connection and pruning strategy. Zhang [4] and others improved Apriori algorithms are applied to the network audit system, which improves the efficiency of mining and improves the usability of the algorithm. Literature [5] applies the association rule Apriori algorithm to the analysis of students' performance, excavates the relationship between curriculum and curriculum, and seeks the factors that affect students' performance in all aspects. Document [6] proposes a Apriori algorithm based on weight vector matrix reduction, which realizes data dynamic analysis and reduces the scale of the source and candidate sets. The existing association analysis schemes generally exist and do not directly apply to gas station data analysis. Therefore, how to make effective use of the real data collected from gas station sensors to analyze association rules is a problem to be solved.

In view of the above research status and problems, first of all, this paper designs a remote monitoring system for oil and gas information in gas stations. Through the pre designed communication protocol, the system uploads the encrypted packets containing oil and gas information from the gas station to the server side actively. The server is parsed into the database after the server is parsed and interacts with the user by the system that is running on the server. Secondly, this paper uses the Apriori algorithm in the data association rule analysis to carry out association rules mining to find out the most closely related factors when the oil and gas emissions reach the early warning value, which is convenient for the relevant staff or the regulatory authorities to focus on the monitoring.

In the first section of this paper, the architecture and function of remote on-line monitoring system for oil and gas recovery are explained in detail. In the second section, data association rules are analyzed for data related to oil and gas uploaded to server. The third section shows the results of the experiment. The last section summarizes the full text.

2 Data Acquisition and Mining Technology

2.1 Data Acquisition Technology Based on SSH

The remote online monitoring system of oil and gas recovery designed in this paper adopts SSH framework to collect oil and gas data. This framework integrates Spring, Hibernate and SpringMVC to form a combined framework SSH. First, at the Web end, it is implemented through the SpringMVC framework and intersected with the business logic layer through the Spring container management mechanism. Spring is a lightweight container control inversion (IoC) and AOP (face to face) container framework. Hibernate simplifies the operation of the JDBC through the encapsulation of JDBC in the persistent layer. It can automatically implement the operation of the database, simplifies the workload of accessing the database, and is an ideal O/R mapping tool. In the persistence layer, we use Hibernate to realize database interaction. Such a combination can form a clear SSH framework, focusing developers' attention on business logic and reducing the underlying development.

2.2 Association Analysis Data Mining Technology

We can get association rules between data by data mining. Association analysis is a simple and practical analysis technique in data mining technology. It can find the correlation or correlation that exists in a large number of data sets, thus describing the laws and patterns of some attributes in a thing at the same time. Generally, this association does not appear directly in the data, so if there is an association between two things, the association analysis can be used to predict another thing, through one thing [7].

The Apriori algorithm is the most classical and basic algorithm for frequent itemsets of association rules. Because the algorithm has connection step and pruning step, it greatly improves the efficiency of mining. So in this paper, Apriori algorithm is used to select oil and gas data for mining and association analysis. First, we find frequent itemsets in large oil and gas data and generate strong association rules based on frequent itemsets to find correlation analysis among itemsets in massive oil and gas data.

The intensity of association rules can be measured by its support degree and confidence level. Support degree is used to measure the statistical importance of association rules in the whole dataset. The degree of support indicates that the probability of occurrence of item A and B is the ratio of the number of terms contained in A and B at the same time. That is

$$\mathrm{Support}(\mathrm{A} \rightarrow \mathrm{B}) = \mathrm{P}(\mathrm{A} \cup \mathrm{B}) = \frac{Support\_count(A \cup B)}{Total\_count} \tag{1}$$

Confidence measures the credibility of association rules, which is the ratio of the number of items contained in A and B to all items containing A. That is

$$\text{Confidence}(\mathrm{A} \rightarrow \mathrm{B}) = \mathrm{P}(\mathrm{B}|\mathrm{A}) = \frac{Support\_count(A \cup B}{Support\_count(A)} \quad (2)$$

The minimum support degree is the threshold to measure the support degree, which indicates the minimum importance of the item set in the statistical sense; the minimum confidence is the threshold to measure the confidence level, which indicates the minimum reliability of the association rules. At the same time, the rule of minimum support threshold and minimum confidence threshold is called strong association rule.

The implementation of the Apriori algorithm, as shown in Fig. 1, mainly includes 2 steps: (1) finding all the frequent itemsets, that is, all the items that satisfy the minimum support threshold. In this step, the connection step and the pruning step are fused to get the largest frequent itemsets. (2) from the frequent item set in the last step, all the rules of high confidence are extracted, that is, the strong association rules are produced by the frequent itemsets (the minimum support and the minimum confidence level).

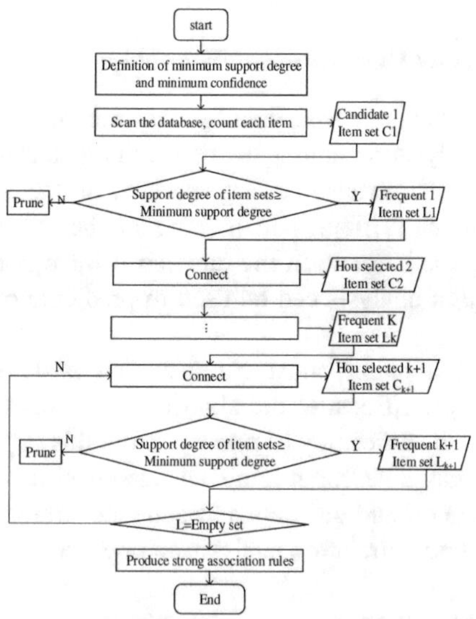

Fig. 1. Apriori algorithm flow

Through the organic combination of the above technology, the remote on-line detection system for oil and gas recovery is designed and built in this paper, and the oil and gas data are deeply excavated and analyzed.

3 Application and Verification of Data Mining Technology in Association Analysis

In view of the key technologies and methods mentioned above, the system is built to collect and obtain the large data of oil and gas, and based on the correlation analysis of the oil and gas data collected in the on-line monitoring system, the final data mining is analyzed as follows.

3.1 Association Analysis Data Mining Technology

The analysis data are derived from the oil and gas data collected from a gas station in Beijing. Through this system, the data are uploaded to the server database, and there are 20160 data collected on average every 30 s within the range of 0 h and 0 s, 0 min and 0 s in September 1st, 2016, and 30 s in the range of 23:59 in September 7th, 2016. After data cleaning and data transformation pretreatment, effective data satisfying conditions are selected. Then, based on the correlation analysis of the environmental data of oil storage area, the more important factors affecting oil and gas concentration are obtained.

(1) Data Preprocessing
The original data contains some incomplete or not strong correlation with the target factors, which can not be directly used for data mining, or the results of the mining are poor. In order to improve the quality of data mining, we need to use data preprocessing technology. He is a very important link in the process of data mining. Experience shows that if the data preparation work is done very carefully, a lot of energy will be saved in the modeling phase of the model [8].

The main task of data cleaning is to fill in the missing data value [9] and delete those data values which are weakly correlated with the research targets. In this article, we use the method of ignoring the tuple to delete oil and gas data with low oil and gas concentration or high oil and gas concentration. Specifically, when the oil and gas concentration is less than 1 or more than 2, we delete it, because these too small data are weakly associated with the target factors on the one hand, and on the other hand, some of these data are produced. The reason is a systematic error, which will have a certain impact on data mining. For the convenience of display, we select 1100 of the data to draw the area map. As shown in Fig. 2, we can see that most of the data are distributed between 1 and 2, and a small number of data are beyond this range. After data cleaning, the total number of records is 12677.

Data transformation is mainly to normalize data and transform data into a unified format for data mining. Logical data are needed in the association analysis mining of oil and gas data, so the data of liquid resistance pressure, tank pressure, tank temperature, oil and gas concentration in the unloading area and the discharge concentration of the treatment device should be converted to Boolean representation. To achieve this goal, we first visualize the data we have selected. we categorize data and classify them through different data segments. Among them, T indicates time, L indicates liquid resistance pressure, S indicates oil tank pressure, A indicates temperature, and D indicates emission concentration. Further processing of the data after the

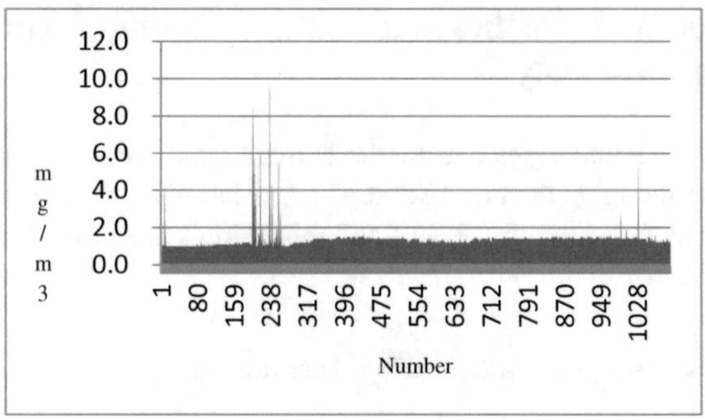

Fig. 2. The original data of oil and gas statistics

visualization, "1" represents the appearance of such factors, and "0" represents that this data is not affected by such factors.

(2) Correlation Analysis of Oil and Gas Environmental Data in Oil Storage Area
After several experiments in the earlier period, 5 types of oil and gas data are selected from the above data set to select the hydraulic pressure, tank pressure, storage tank temperature, oil and gas concentration in the unloading area, and the discharge concentration of the treatment device. Apriori algorithm is used to study the correlation between oil and gas concentration and other factors. The minimum support is 0.2 and the minimum confidence is 0.5, and 322 rules of association rule are obtained. The first 5 association rules from high to low are taken respectively. The three factor analysis results of oil and gas concentration and other factors, such as Table 1, are expressed, and the first 3 association rules of the support degree from high to low are taken respectively, and the five factor analysis results of oil and gas concentration and other factors are obtained, such as Table 5 [10].

In Table 1, the following conclusions can be obtained:

(1) in the time period of 11:30 ~ 15:45, when the hydraulic pressure is between 500 ~ 1600 N and the oil tank pressure is between 0 ~ 600 N, the possibility of the oil and gas concentration to reach the early warning value is the most, the support is up to 46.60%, and the reliability of this rule is 100%.
(2) in the time period of 15:50 ~ 19:55, when the hydraulic pressure is between 1600 ~ 2700 N and the oil tank pressure is between 1800 ~ 2400 N, the possibility of the oil and gas concentration to reach the early warning value is 26.21%, and the reliability of this rule is 100%.
(3) in the time period of 19:55 ~ 23:55, the possibility of the oil and gas concentration to reach the early warning value is about 22.33% when the pressure of liquid resistance is between 1600 ~ 2700 N and the pressure of the oil tank is between 1800 ~ 2400 N, and the reliability of this rule is 92%.
(4) in the time period of 19:55 ~ 23:55, the possibility of the oil and gas concentration to reach the early warning value is about 22.33% when the pressure of liquid resistance is between 1600 ~ 2700 N and the pressure of the oil tank is between 1200 ~ 1800 N, and the reliability of this rule is 92%.

Table 1. Correlation analysis of three factors of oil and gas concentration and other factor

Number	Factor A	Factor B	Factor C	Oil and gas concentration	
				Support degree/%	Confidence degree/%
1	Time 11:30~15:45	Liquid resistance pressure/N 500~1600	Storage tank temperature/N 1200~1800	21.359	100
2	Time 19:55~23:55	Liquid resistance pressure/N 1600~2700	Storage tank temperature/N 1800~2400	22.33	92
3	Time 19:55~23:55	Liquid resistance pressure/N 1600~2700	Storage tank temperature/N 1200~1800	22.33	92
4	Liquid resistance pressure/N 500~1600	Storage tank temperature/N 0~600	Time 11:30~15:45	46.602	100
5	Time 15:50~19:55	Liquid resistance pressure/N 1600~2700	Storage tank temperature/N 1800~2400	26.21	100

(5) in the time period of 11:30 ~ 15:45, the possibility of the oil and gas concentration to reach the early warning value is about 21.359% when the pressure of liquid resistance is between 500 ~ 1600 N and the pressure of the oil tank is between 1200 ~ 1800 N, and the reliability of this rule is 100%.

In the association rules obtained above, 5 factors of time, liquid resistance pressure, tank pressure, storage tank temperature, and treatment device emission concentration have a great relationship with the concentration of oil and gas. This research can find the information correlation between various data, and provide the decision basis for the monitoring personnel from the scientific point of view. It is of positive significance to the on-line monitoring of oil and gas.

4 Conclusions

(1) For many gas stations, the traditional manual inspection method is still used to monitor the oil and gas situation, which has the problems of low efficiency and high cost. In this paper, a system of remote monitoring of oil and gas information for gas stations is designed and implemented. The encrypted data packets containing oil and gas information are uploaded to the server from the end of the gas station to the server. The user can monitor the oil and gas data directly on the computer to achieve accurate and efficient data acquisition through the web server set.

(2) The main use of the remote online monitoring system for oil and gas recovery is the SSH framework. The results show that the framework can be more flexible with the database, and it is convenient to design the front-end page. It can display the oil and gas related data intuitively, and help the system data collection and the follow-up data mining analysis.
(3) In this paper, data collection and data mining of large oil and gas data are carried out, and Apriori algorithm is used to analyze the correlation of oil and gas concentration. The results show that the 5 factors are strongly related to the concentration of oil and gas and time, pressure of liquid resistance, pressure of tank, temperature of tank and the concentration of discharge of treatment device. For example, during the 11:30 ~ 15:45 period at noon, the fluid resistance pressure reaches 1600 N and the oil tank pressure reaches 600 N, the regulators need to focus on monitoring the oil and gas data to prevent accidents such as oil and gas leakage. At 19:55 in the evening, the hydraulic pressure is between 1600 ~ 2700 N, the pressure of the tank is between 1800 ~ 2400 N, the concentration of the discharge is at 10 ~ 30 mg/m3, the concentration of oil and gas in the storage area is relatively low, and the gas station can reasonably distribute the staff for monitoring.

To sum up, this paper designs and implements a remote online monitoring system for oil and gas recovery, which has been applied and verified in a gas station in Beijing. The application results show that the system can provide reference for intelligent monitoring of oil and gas big data in gas stations, and improve the efficiency of oil and gas monitoring. In addition, the results and analysis of data mining in this paper can provide accurate and objective data support for the monitoring personnel of gas stations. It has a reference value for higher priority monitoring of the key data segments, and provides a good technical basis for the statistics and processing of oil and gas data in the subsequent gas stations.

Acknowledgment. This paper is supported by the Project for the National Natural Science Foundation of China under Grants No. 61672064, 81370038, the Beijing Natural Science Foundation under Grant No. 4172001.

References

1. Agrawal, R., Srikant, R., et al.: Fast algorithms for mining association rules. In: International Conference on Very Large Data Bases, pp. 487–499. Morgan Kaufmann Publishers Inc. (1994)
2. Cui, G., Li, L., Wang, K., et al.: Research and improvement of Apriori algorithm in association rule mining. Comput. Appl. **11**, 2952–2955 (2010)
3. Wang, W.: Research and improvement of Apriori algorithm in association rules. In: Ocean University of China (2012)
4. Zhang, J., Li, T.: Application and research of Apriori algorithm in network audit system. Sci. Technol. Field Vis. **11**, 42–46 (2015). https://doi.org/10.3969/j.issn.2095-2457.2015.29.028
5. Wang, C.: Student achievement analysis based on Apriori algorithm of association rules. Value Eng. **5**, 171–173 (2018)

6. Yang, Q., Sun, H.: Apriori algorithm based on weight vector matrix reduction. Comput. Eng. Des. **3**, 25–32 (2018). 7
7. Bai, J., Tian, R., Zhang, X.: Application of Apriori algorithm in user characteristic association analysis. Comput. Netw. **12**, 70–72 (2016). https://doi.org/10.3969/j.issn.1008-1739.2016.12.065
8. Tan, Q.: Application of association rule Apriori algorithm in the analysis of test results. J. Xinyang Normal Univ. Nat. Sci. Edn. **2**, 300–303 (2009). https://doi.org/10.3969/j.issn.1003-0972.2009.02.038
9. Li, S., Jiao, B., Qu, S., et al.: Data mining research based on campus smart card system. China Educ. Inf. **3**, 227–302 (2018). https://doi.org/10.3969/j.issn.1673-8454.2018.02.020
10. Jia, K., Li, H., Yuan, Y.: Application of data mining in mobile medical system based on Apriori algorithm. J. Beijing Univ. Technol. **3**, 394–401 (2017). https://doi.org/10.11936/bjutxb2016120059

A Fast and Efficient Grid-Based K-means++ Clustering Algorithm for Large-Scale Datasets

Yang Yang[1(✉)] and Zhixiang Zhu[2]

[1] School of Computer Science and Technology,
Xi'an University of Posts and Telecommunications, Xi'an 710121, China
mr.yangy@foxmail.com

[2] Institute of IOT & IT-Based Industrialization,
Xi'an University of Posts and Telecommunications, Xi'an 710121, China

Abstract. In the k-means clustering algorithm, the selection of the initial clustering center affects the clustering efficiency. Currently widely used k-means++ can effectively improve the speed and accuracy of k-means. But k-means cluster algorithm does not scale well to massive datasets, as it needs to traverse the data set multiple times. In this paper, based on k-means++ clustering algorithm and grid clustering algorithm, a fast and efficient grid-based k-means++ clustering algorithm was proposed, which can efficiently process large-scale data. First, the N-dimensional space is granulated into disjoint rectangular grid cells. Then, the dense grid cell is marked by statistical gird cell information. Finally, the modified k-means++ clustering algorithm is applied to the meshed datasets. The experimental results on the simulation dataset show that compared with the original k-means++ clustering algorithm, the proposed algorithm can quickly obtain the clustering center and can effectively deal with the clustering problem of large-scale datasets.

Keywords: K-means · K-means++ · Grid-based clustering algorithm
Large-scale datasets

1 Introduction

Clustering is an unsupervised pattern recognition method widely used in data mining and artificial intelligence. It discovers potential similar patterns from data sets and groups data sets without any a priori information. Classification results require that the similarities in the same class are as large as possible, and the differences between the classes are as large as possible. In recent years, data mining has been widely used in many fields, such as images [1], medicine [2], aviation [3], etc. This makes the amount of data grow rapidly [4]. Due to the large amount of data and complex data types, improving the efficiency of data mining has become an important challenge for data mining. With the rapid growth of datasets and diversity of data source, the traditional clustering algorithms cannot solve the requirements of practical applications. How to quickly find the cluster center in the clustering process and finally obtain effective and

P. Krömer et al. (Eds.): ECC 2018, AISC 891, pp. 508–515, 2019.
https://doi.org/10.1007/978-3-030-03766-6_57

accurate clustering results is the main problem that the current clustering algorithm needs to solve when applied to large-scale dataset.

The K-means algorithm is widely used in data mining [5], but the algorithm relies on the selection of initial center points, which ultimately leads to intensive computational process and low time efficiency. K-means++ is improved on the basis of k-means [6]. The D2-sampling adaptive sampling algorithm is used to select the initial random clustering points, which makes the clustering efficiency significantly improved. In addition, there are related algorithms such as k-means based on genetic algorithm [7] and k-means by introducing penalty factors [8]. Grid-based clustering algorithm has an important role in spatial information processing and has been widely used in many fields. Grid clustering quantizes the space into a finite number of grids, and implements a clustering algorithm according to the spatial grid distribution [9]. Compared with other clustering methods, the grid-based clustering method has a faster processing speed [10], and the time complexity of the algorithm is determined by the number of grid cells rather than the size of the data set [11]. The grid clustering algorithm can effectively manage large-scale spatial data and has good scalability.

These grid-based clustering methods have greatly improved clustering accuracy and algorithm complexity, but they are not very effective when applied to data sets with complex topologies and noisy datasets [12]. In this paper, a grid-based k-means++ clustering algorithm is proposed. The algorithm granulates the data through the grid cells and denotes the number of data points within the grid cell as the grid density. Then the modified k-means++ clustering algorithm is applied to the meshed datasets. The algorithm can solve the problem of central point selection failure caused by local density non-uniformity and effectively improve the clustering efficiency. So the proposed algorithm is suitable for processing large-scale data.

2 Related Definitions

In this section, we formally define k-means, k-means++ and Grid-Based K-means++ Clustering algorithms

2.1 The K-means Algorithm

In the k-means clustering algorithm, we are given an integer k and a set of all data points X. For any finite set $c \in C$, we define

$$d(x, C)^2 = min_{c \in C} \left\| x - c^2 \right\| \tag{1}$$

The purpose of the k-means clustering algorithm is to find the set C of k cluster center points so as to minimize the function $\varnothing_C(X)$,

$$\varnothing_C(X) = \sum\nolimits_{x \in X} d(x, C)^2 \tag{2}$$

K-means algorithm flow is as follows:

1. Arbitrarily select k samples from the dataset as initial clustering centers.
2. For each x_i in the dataset, calculate its distance to the k cluster centers and assign it to the class corresponding to the cluster center with the smallest distance.
3. For each $i \in \{1, \ldots, k\}$, recalculate the clustering center c_i:

$$c_i = \frac{1}{|c_i|} \sum\nolimits_{x \in c_i} x \tag{3}$$

4. Repeat Steps 2 and 3 until C no longer changes.

It is standard practice of k-means clustering algorithm to randomly choose k clustering centers from X. As long as steps 2 and 3 are continued, the loop can be ended. Because they make local improvement to the cluster and reduce $\varnothing_C(X)$ until it no longer changes.

2.2 The K-means++ Algorithm

The k-means++ algorithm proposed a specific method for selecting clustering centers based on the k-means algorithm. Let *D(x)* denote the shortest distance from the data point x to the nearest clustering center we have selected. The algorithm flow is as follows:

(1) Select a center c_1 randomly from X.
(2) Select $x \in X$ as the new center c_i with probability $p(x|X)$.

$$p(x|X) = \frac{D(x)^2}{\sum_{x \in X} D(x)^2} \tag{4}$$

(3) Repeat Step 1 until k centers were taken.
(4) Execute the k-means algorithm's Steps 2–4.

The improved idea of k-means++: a point farther away from the existing cluster center has a greater probability of being selected as the next cluster center. This improvement can be intuitively understood as the fact that the k initial cluster centers should be separated from each other as possible.

2.3 The Grid-Based K-means++ Clustering Algorithm

To facilitate the description of the algorithm, the following definitions are introduced

Definition 1: Grid Cell
After meshing the datasets space, the most basic partitioning cell is called grid cell g.

Definition 2: Grid Cell Density
Grid cell density is the number of data points included in the grid cell g, which is denoted by *Den(g)*.

Definition 3: Dense Grid Cell

If the density *Den(g)* is greater than the set density threshold *Minpts*, the grid cell g is a dense grid cell.

Definition 4: Dense Grid Cell Center

If the dense grid cell g contains n data points $\{x_1, x_2, \ldots, x_n\}$, then let b denote the dense grid cell center of g, we define

$$b = \frac{\sum_{i=1}^{i=n} x_i}{n} \tag{5}$$

Let B denote a set of all dense grid cell centers.

Definition 5: Free Data

If the grid cell g is not a dense grid cell, the data points in the grid cell g are free data.

3 Algorithm Model and Analysis

3.1 Algorithm Model

Based on the dataset grid structure, the grid-based k-means++ algorithm first analyzes the dataset to determine the density of each grid cell. Grid cells whose density exceeds the density threshold *Minpts* are marked as dense grid cells. And the data point within the grid cell whose density is below the density threshold *Minpts* is marked as free data. If *Minpts* is too large, it can't effectively reduce the clustering computing cost. On the other hand, if the value is too small, it is possible that the two clusters that were originally separated are merged, which easily leads to the loss of clusters. Based on the characteristics of mean calculation, the grid-based k-means++ algorithm uses adaptive strategy to distinguish grid cells.

From the analysis and research, we know that the grid clustering performance is mainly based on the selection of the density threshold *Minpts*. *Minpts* are set too large or too small, which will affect the accuracy and efficiency of the algorithm. So we first sort $Den(g_i)$ in descending order to get the result $SD(g_i)$. If the difference between $SD(g_M)$ and $SD(g_{M+1})$ changes significantly, the density threshold is determined as follows:

$$Minpts = \left[\frac{\sum_{i=1}^{M} RD(g_i)}{M}\right]^{\frac{1}{2}} \tag{6}$$

The detailed steps of the algorithm are given below:

(1) Mesh the dataset space.
(2) Classify data objects of the dataset into corresponding grid cells.
(3) Calculate the density of each grid cell, mark the grid cell with its density greater than *Minpts* as a dense grid cell, and mark data points in grid cells with their density less than *Minpts* as free data.

(4) Randomly select one dense grid cell center as the initial cluster center point c_1.
(5) Let d(b, C) denote the shortest distance from a dense grid cell center b to the closest clustering center we have already chose. Selecting a dense grid cell center $b \in B$ as new center c_i with probability $p(b|B)$.

$$p(b|B) = \frac{d(b,B)^2}{\sum_{b' \in B} d(b',B)^2} \tag{7}$$

(6) Repeat Step 5 until we have taken k centers altogether.
(7) For each free data and dense grid cell center, calculate its distance to k cluster centers and assign it to the cluster with the smallest distance.
(8) For each i $\in \{1,\ldots,k\}$, let n denote the number of free data points contained in c_i, let m denote the number of dense grid cell centers contained in c_i. Recalculate the clustering center c_i:

$$c_i = \frac{\left(\sum_{j=1}^{m} x_{ij}\right) + \left(\sum_{j=1}^{n} \left(Den(g_{ij}) \cdot b_{ij}\right)\right)}{m + \sum_{j=1}^{n} Den(g_{ij})} \tag{8}$$

(9) Repeat Steps 7 and 8 until C no longer changes.

3.2 Time Complexity Analysis

The time complexity of the proposed algorithm mainly depends on dividing the grid, calculating the grid cell information and the distance from the free data and the dense grid cell center to the cluster center point. Dividing grid cells is to divide all the data points into the corresponding grid cell, so we need to traverse the entire dataset, so the time complexity is $O(n)$. Calculating the grid cell information requires scanning the data once and updating the statistics of each grid cell. The time complexity is also $O(n)$.

The number of dense grid cell centers and free data points is denoted by S, and the number of iterations is denoted by T. Assigning dense grid cell centers and free data points to k clusters requires traversing the entire dataset T times. We need to calculate their distance to each cluster center and then assign them to the nearest cluster center. The time complexity of this step is $O(S \cdot T \cdot K)$. Therefore, the total time complexity of the proposed algorithm is:

$$T_{all} = O(n) + O(n) + O(S \cdot T \cdot K) \tag{9}$$

Due to the proposed algorithm meshes the original data set, the number of iterations T and the number of points S that need to be clustered will be significantly reduced, especially when applied for large-scale data mining. So the algorithm can effectively reduce the time complexity compared with the traditional k-means algorithm.

4 Experimental Results

In this section, all our experiments were executed in Python 3.5.2 environment running on Windows 7 with Intel Core i3-3240 3.4 GHz CPU and 4 GB RAM

Experimental data were selected from the UCI dataset, including Iris, Wine, New-thyroid, Diabetes, and segment. We used these datasets to verify the correctness and validity of clustering algorithms. The properties of each dataset are shown in Table 1.

Table 1. Properties of each dataset.

Datasets	Number of classes	Number of attributes	Number of instances
Iris	3	4	150
Wine	3	13	178
New-thyroid	3	5	213
Diabetes	2	8	768
Segment	7	19	2310

The experiment compares the k-means algorithm, the k-means++ algorithm and the proposed algorithm in the paper. The comparison of the running time and clustering accuracy of algorithms is shown in Table 2.

Table 2. Clustering performance of each algorithm.

Datasets	Accuracy/%			Time/s		
	K-means	K-means++	Proposed algorithm	K-means	K-means++	Proposed algorithm
Iris	80.67	84.67	79.33	0.328	0.136	0.113
Wine	66.98	68.54	66.30	0.367	0.175	0.126
New-thyroid	78.40	78.40	76.53	0.582	0.436	0.352
Diabetes	62.76	64.06	62.92	2.536	2.023	0.521
Segment	59.87	63.03	60.17	3.153	2.361	1.182

It can be seen from Table 2 that compared with the traditional clustering algorithm, the clustering speed of the proposed algorithm is improved, and the local optimization problem caused by randomly selecting the initial clustering points can be avoided. However, compared with the traditional K-means++ algorithm, the accuracy of the proposed algorithm is reduced. The root cause is that the algorithm reduces the data size by dividing the grid, and at the same time reduces the grid resolution, that is, sacrificing precision in exchange for time reduction.

In order to test the performance of the proposed algorithm in processing large-scale data, we have artificially extended the number of data sets Iris and Wine to 1000 times, respectively Iris* and Wine*. We retested the K-means algorithm, the K-means++ algorithm and the proposed algorithm. The comparison of the running time and clustering accuracy of algorithms is shown in Table 3.

Table 3. Performance of algorithms applied on large-scale datasets

Datasets	Accuracy/%			Time/s		
	K-means	K-means++	Proposed algorithm	K-means	K-means++	Proposed algorithm
Iris*	80.23	82.45	79.69	52.32	36.97	5.356
Wine*	66.57	67.34	66.20	68.35	51.86	7.578

In Fig. 1, we compare the time performance of three algorithms on different scale data sets

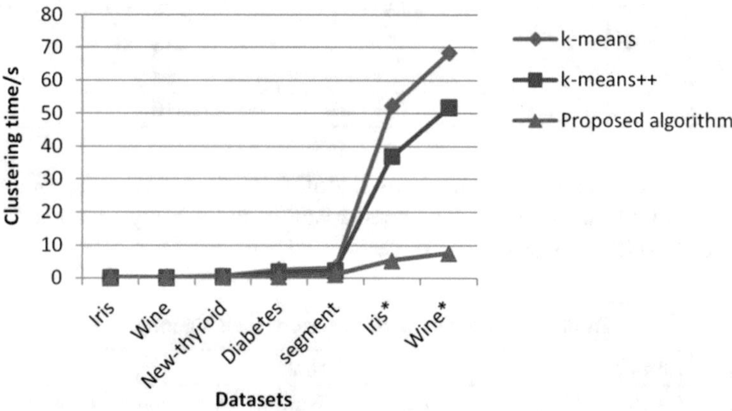

Fig. 1. Time performance of the three algorithms on different scale datasets.

As shown in Table 3 and Fig. 1, the algorithm has significant advantages in dealing with large-scale data. Considering comprehensively, although the proposed algorithm is slightly less accurate than traditional k-means clustering algorithm, it can handle coarse-grained large-scale datasets and has faster processing speed. In contrast, k-means cannot scale well to large datasets, and slow processing speed is its main disadvantage. Therefore, the algorithm proposed in this paper has practical application value when it is necessary to process large-scale datasets quickly.

5 Conclusion

In this paper, we propose a fast and efficient grid-based k-means++ clustering algorithm for large-scale datasets. The proposed algorithm improves the traditional k-means algorithm, adjusts the method of selecting initial clustering centers and reallocating data objects in the iterative process, which can reduce the number of iterations and quickly find clustering centers. The proposed algorithm eliminates the dependence of k-means algorithm on the initial random clustering centers, solves the problem that k-means algorithm falls into the local optimal solution due to improper selection of initial clustering centers, and accelerates the convergence of k-means++ algorithm. So the proposed algorithm can be used to process large-scale datasets. Under some degree of time limit constraint, the proposed algorithm achieves satisfactory results and implements a clustering algorithm model for quickly processing large-scale dataset. The next step is to study the impact of different grid granularity on clustering results and how to improve the clustering accuracy of the proposed algorithm while maintaining faster computational speed.

References

1. Chen, Y.S., Chen, B.T.: Efficient fuzzy c-means clustering for image data. J. Electron. Imaging **14**(1), 013017 (2005). https://doi.org/10.1117/1.1879012
2. Lavrač, N.: Selected techniques for data mining in medicine. Artif. Intell. Med. **16**(1), 3–23 (1999). https://doi.org/10.1016/S0933-3657(98)00062-1
3. Nazeri, Z., Bloedorn, E., Ostwald, P.: Experiences in mining aviation safety data. In: ACM SIGMOD Record, vol. 30, No. 2, pp. 562–566. ACM (2001). https://doi.org/10.1145/376284.375743
4. Lynch, C.: Big data: How do your data grow? Nature **455**(7209), 28 (2008). https://doi.org/10.1038/455028a
5. Hartigan, J.A., Wong, M.A.: Algorithm AS 136: a k-means clustering algorithm. J. Roy. Stat. Soc. Ser. C (Appl. Stat.) **28**(1), 100–108 (1979). https://doi.org/10.2307/2346830
6. Arthur, D., Vassilvitskii, S.: k-means ++: the advantages of careful seeding. In: Proceedings of the Eighteenth Annual ACM-SIAM Symposium on Discrete Algorithms, pp. 1027–1035. Society for Industrial and Applied Mathematics, Philadelphia (2007). https://doi.org/10.1145/1283383.1283494
7. Anusha, M., Sathiaseelan, J.G.R.: Feature selection using k-means genetic algorithm for multi-objective optimization. Procedia Comput. Sci. **57**, 1074–1080 (2015). https://doi.org/10.1016/j.procs.2015.07.387
8. Li, M.J., Ng, M.K., Cheung, Y.M., Huang, J.Z.: Agglomerative fuzzy k-means clustering algorithm with selection of number of clusters. IEEE Trans. Knowl. Data Eng. **20**(11), 1519–1534 (2008). https://doi.org/10.1109/TKDE.2008.88
9. Berger, M., Rigoutsos, I.: An algorithm for point clustering and grid generation. IEEE Trans. Syst. Man Cybern. **21**(5), 1278–1286 (1991). https://doi.org/10.1109/21.120081
10. Bhatnagar, V., Kaur, S., Chakravarthy, S.: Clustering data streams using grid-based synopsis. Knowl. Inf. Syst. **41**(1), 127–152 (2014). https://doi.org/10.1007/s10115-013-0659-1
11. Park, N.H., Lee, W.S.: Statistical grid-based clustering over data streams. ACM Sigmod Record **33**(1), 32–37 (2004). https://doi.org/10.1145/974121.974127
12. Yue, S., Wei, M., Wang, J.S., Wang, H.: A general grid-clustering approach. Pattern Recogn. Lett. **29**(9), 1372–1384 (2008). https://doi.org/10.1016/j.patrec.2008.02.019

Research on Technology Innovation Efficiency of Regional Equipment Manufacturing Industry Based on Stochastic Frontier Analysis Method

Taking the New Silk Road Economic Belt as an Example

Yang Zhang, Lin Song(✉), and Minyi Dong

School of Economics and Finance,
Xi'an Jiaotong University, Xi'an 710061, China
15249248421@163.com

Abstract. Based on SFA model measurement, the paper examines the overall and phased efficiency of the technological innovation of equipment manufacturing industry in 'the New Silk Road Economic Belt' from 2007 to 2015, and further utilizes the space panel model to test the spatial spillover effect. The results show that the technological innovation efficiency of equipment manufacturing industry in this region presents a development trend of "double core and periphery" during the observation period; there is room for improvement of total technical innovation efficiency and scale efficiency; the economic efficiency of this economic zone has accelerated after 2013. Finally, the paper draws conclusions based on empirical results and proposes corresponding policy suggestion.

Keywords: Equipment manufacturing industry · New silk road economic belt Efficiency of technological innovation

1 Introduction

The development strategy of the 'New Silk Road Economic Belt' has opened up new opportunities for international production capacity investment cooperation in the international arena, and such strategy also provided an opportunity for the strategic transformation and upgrading of the domestic equipment manufacturing industry. The area along the new Silk Road in China has jumped from the hinterland to the forefront of opening up, and thus allows local equipment manufacturers to enjoy broader market prospect. In today's increasingly internationalized equipment market, innovation drive has become the only way to break the development difficulties of regional equipment manufacturing industry, especially in the historical reviv al of the 'Silk Road'. The investigation of regional technological innovation efficiency of equipment manufacturing industry for the realization of synergetic development of equipment manufacturing industry and innovative resources distribution in the 'New Silk Road Economic Belt' area.

P. Krömer et al. (Eds.): ECC 2018, AISC 891, pp. 516–523, 2019.
https://doi.org/10.1007/978-3-030-03766-6_58

At present, researches on the technological innovation efficiency of equipment manufacturing industry mainly focus on method selection of efficiency measurement and the efficiency measurement of full-calibre equipment manufacturing industry. While the researches on the innovation efficiency of regional equipment manufacturing industry are relatively few. The research on the innovation efficiency and spillover effect of regional equipment manufacturing mainly includes the following two aspects.

First, there are some researches adopt different methods to measure the technical innovation efficiency of sample objects. Some scholars used the SFA method to measure the technical efficiency of the innovative manufacturing activities of equipment manufacturing industry and obtained the conclusion that the improvement of total technical efficiency is at relatively low efficiency level [1]. Some other scholars utilize the DEA model to measure the innovation efficiency of the provincial equipment manufacturing industry [2]. In addition, some scholars use both the DEA model and the Malmquist index method to measure the innovation efficiency of regional specific industries and explore the degree of spatial pattern change [3, 4]. Second, there exist a strand of literature conducting in-depth analysis of the differences in technological innovation efficiency of sample objects in the time dimension. Previous researches generally attribute the deep-seated reasons for innovation of equipment manufacturing industry to the division of market and resources caused by institutional policies [5, 6]. Based on foregoing statement, Tang Xiaohua [7] studied the growth motivation of China's equipment manufacturing industry from the perspective of supply. With in-depth study, some scholars propose that the deepening of international trade can be considered as one of the most important reasons of the lack of innovation motivation [6]. Combing the existing literature, we find that the regional background of the development of the 'Belt and Road' is rarely considered when measuring the innovation efficiency of the equipment manufacturing industry, and the investigation of integration between equipment manufacturing innovation and the 'New Silk Road Economic Belt' still needs to be further deepened.

2 Research Method

2.1 Model Setting

Stochastic Frontier Analysis Method Based on Output Distance Function. The model utilized in this paper is based on the single-stage stochastic frontier model proposed by Battese [8] and the output distance function created by Coelli [9]. The stochastic frontier model is a logarithmic random frontier model obtained after adding a random term:

$$\begin{aligned} \mathrm{Iny}_{\mathrm{Mit}} &= -\,\mathrm{TL}(\mathrm{x}_{\mathrm{it}}, \mathrm{y}^{*}_{\mathrm{it}}; \theta) + \mathrm{v}_{\mathrm{it}} - \mu_{\mathrm{it}} \\ &= -\,\alpha - \sum_{k=1}^{\mathrm{K}} \alpha_{\mathrm{k}} \mathrm{Inx}_{\mathrm{kit}} - \frac{1}{2} \sum_{k=1}^{\mathrm{K}} \sum_{l=1}^{\mathrm{K}} \alpha_{\mathrm{kl}} \mathrm{Inx}_{\mathrm{kit}} \mathrm{Inx}_{\mathrm{lit}} - \sum_{k=1}^{\mathrm{K}} \sum_{m=1}^{\mathrm{M-1}} \rho_{\mathrm{km}} \mathrm{Inx}_{\mathrm{kit}} \mathrm{Iny}^{*}_{\mathrm{mit}} \\ &\quad - \sum_{m=1}^{\mathrm{M-1}} \beta_{\mathrm{k}} \mathrm{Iny}^{*}_{\mathrm{mit}} - \frac{1}{2} \sum_{m=1}^{\mathrm{M-1}} \sum_{n=1}^{\mathrm{M-1}} \beta_{\mathrm{mn}} \mathrm{Iny}^{*}_{\mathrm{mit}} \mathrm{Iny}^{*}_{\mathrm{nit}} - \mathrm{v}_{\mathrm{it}} - \mu_{\mathrm{it}} \end{aligned} \tag{1}$$

Where t represents the period and i denotes the industry. Y is the normalized output vector, and x represents the element input vector, $\theta = [\alpha\beta\rho]$ is a series of parameters that need to be evaluated. The random disturbance term consists of two parts: u_{it} is a technical inefficiency term, representing the technical efficiency level of the leading edge; v is a random error term, including random factors such as measurement error, obeying $\mathrm{N}\left(0,\, \sigma_{\mathrm{v}}^{2}\right)$ distribution and independent of u_{it}.

According to Banker [9], the total technical efficiency (TE^{CRS}) is decomposed into pure technical efficiency (TE^{VRS}) and scale efficiency (SE) methods to deal with SE, TE^{CRS}, namely:

$$\mathrm{TE}^{\mathrm{CRS}} = \mathrm{TE}^{\mathrm{VRS}} \times \mathrm{SE} \tag{2}$$

The formula for calculating the scale efficiency based on the transcendental logarithmic output distance function is:

$$\mathrm{SE}^{\mathrm{t}}(\mathrm{x}_{\mathrm{it}}, \mathrm{y}_{\mathrm{it}}) = \exp\left[-\frac{\left(\epsilon_{0}^{\mathrm{t}}(\mathrm{x}_{\mathrm{it}}, \mathrm{y}_{\mathrm{it}}) - 1\right)^{2}}{2\alpha}\right] \tag{3}$$

Among them, $\epsilon_{0}^{\mathrm{t}}(\mathrm{x}_{\mathrm{it}}, \mathrm{y}_{\mathrm{it}})$ is the local scale elasticity beyond the logarithmic function, item $\epsilon_{0}^{\mathrm{t}}(\beta^{*}\mathrm{x}_{\mathrm{it}}, \beta^{*}\mathrm{y}_{\mathrm{it}})$ denotes the value of β^{*} need to satisfy condition that the scale elasticity at $\epsilon_{0}^{t}(\beta^{*}x_{it}, \beta^{*}y_{it})$ is equal to 1, namely:

$$(x_{it}, y_{it}) =_{0}^{t} (\beta^{*}x_{it}, \beta^{*}y_{it}) - (\alpha In\beta^{*}) \tag{4}$$

Variable Selection and Data Processing.

Innovation Input. Considering the availability of data, this study selects the total internal expenditure (RD) of scientific and technological activities to measure the capital investment in the innovation process. In terms of personnel input in the research and development process, this paper selects the number of scientific and technological personnel (RDP) as a measurement. In addition, the New Product Development Funding (NPRD) can assess the investment capacity of industrial innovation achievements, the paper thus incorporates it into the indicator system of innovation input.

Since innovative production activities are uninterrupted, capital investment should be translated into corresponding stock indicators. This paper uses the perpetual inventory method to calculate the capital stock of research and development, that is, the formula of the research and development capital stock RD_{i0} of the base period (i.e. 2006) is presented as follow:

$$RD_{i0} = I_{i0}/(g_i + h) \tag{5}$$

Where g_i is the average annual growth rate of actual scientific and technological activities in the i-th industry during the sample period, I_{i0} represents the actual economic activity expenditure of the i-th industry in the base period, accompanied with a depreciation rate of h = 15%. The same method can be used to calculate the new product development capital stock.

Innovation Output. In terms of innovation output, patents are generally deemed as direct output of R&D activities, and the output value of new products provides a good measure for the transformation ability of industrial innovation activities. Therefore, this study employs new product output (NPV) and patent applications (PAT) as output indicators for innovative production activities.

This paper selects PAT as dependent variable and normalizes NPV. The standardized variables are denoted by the superscript "*", and the stochastic frontier model of Eq. (1) can be decomposed into the formula (6), namely:

$$\begin{aligned} InPAT_{it} = & -\alpha_0 - \alpha_1 InRD_{it-1} - \alpha_2 InRDP_{it-1} - \alpha_3 InNPRD_{it-1} - \frac{1}{2}\alpha_{11}(InRD_{it-1})^2 \\ & - \frac{1}{2}a_{22}(InRDP_{it-1})^2 - \frac{1}{2}a_{33}(InNPRD_{it-1})^2 - \alpha_{12} InRD_{it-1} InRDP_{it-1} \\ & - \alpha_{13} InRD_{it-1} InNPRD_{it-1} - \alpha_{23} InRDP_{it-1} InNPRD_{it-1} - \beta_1 InNPV^*_{it} \\ & - \frac{1}{2}\beta_{11}(InNPV^*_{it})^2 - \rho_{11} InRD_{it-1} InNPV^*_{it} - \rho_{21} InRDP_{it-1} InNPV^*_{it} \\ & - \rho_{31} InNPRD_{it-1} InNPV^*_{it} + v_{it} - \mu_{it} \end{aligned} \tag{6}$$

3 Empirical Results and Analysis

3.1 Technical Innovation Efficiency Estimation Results

Model Estimation Result. In this paper, the maximum frontier model is conducted to estimate the stochastic frontier model of Eq. (9), which is obtained by Frontier4.1 software. The output distance function and efficiency estimation results are shown in Tables 1 and 2.

As presented in Table 1, the estimated values of other coefficients of the output distance function have good statistical significance, and the overall effect of model regression is better. The LR statistic and γ are both statistically significant at the 1% level, indicating that the inefficiency effect exists in Eq. (1), and the variance term possesses a distinct composite structure, implying the random frontier model can be employed to analyze the innovation efficiency of equipment manufacturing industry

Table 1. Stochastic frontier model estimation results

Output distance function estimation					
Variable	Coefficient	Value	Variable	Coefficient	Value
Intercept	a_0	−20.57 (−1.53)	InNPV*	β_1	−8.73** (−2.17)
$InRD_{t-1}$	a_1	1.723* (1.68)	σ^2		0.16* (1.91)
$InRDP_{t-1}$	a_2	−1.77*** −2.96)	γ		0.70*** (3.11)
$InNPRD_{t-1}$	a_3	6.23*** (4.14)	Log likelihood function value		70.52

Table 2. Equipment manufacturing innovation technology efficiency

	2007	2008	2009	2010	2011	2012	2013	2014	2015	Mean
Shaanxi	0.57	0.63	0.68	0.73	0.78	0.82	0.85	0.87	0.89	0.76
Gansu	0.48	0.55	0.61	0.66	0.72	0.76	0.80	0.84	0.86	0.70
Ningxia	0.20	0.27	0.34	0.41	0.48	0.55	0.61	0.67	0.77	0.48
Qinghai	0.15	0.21	0.28	0.35	0.42	0.50	0.57	0.63	0.68	0.42
Chongqing	0.76	0.80	0.83	0.85	0.88	0.90	0.92	0.93	0.94	0.87
Yunnan	0.28	0.36	0.43	0.50	0.56	0.63	0.69	0.73	0.77	0.55
Xinjiang	0.68	0.11	0.16	0.23	0.30	0.37	0.44	0.51	0.58	0.38
Guangxi	0.38	0.46	0.53	0.59	0.65	0.70	0.75	0.79	0.83	0.63
Sichuan	0.46	0.51	0.56	0.61	0.68	0.62	0.69	0.71	0.85	0.63
Economic Belt	0.45	0.48	0.44	0.5	0.56	0.62	0.67	0.72	0.77	0.58
Nationwide	0.34	0.42	0.49	0.56	0.62	0.67	0.72	0.77	0.80	0.60

along the 'New Silk Road Economic Belt'. At the same time, γ is statistically significant with a value of 0.7, indicating that the stochastic frontier production model is effectively estimated overall.

Based on the output distance function and the estimation of the inefficiency equation, the technical innovation efficiency of equipment manufacturing industry along the 'New Silk Road Economic Belt' from 2007 to 2015 can be calculated following Eq. (2). The software Frontier4.1 directly gives the calculation results, which are shown in Table 2.

As a general comment, the whole level of technical innovation efficiency of equipment manufacturing industry of the 'New Silk Road Economic Belt' from 2007 to 2015 increased rapidly, but the internal differences were large and the regional technical efficiency was extremely uneven. As can be observed from Table 2, the equipment manufacturing industry in the selected regions has increased steadily from 0.45 in 2007 to 0.58 in 2015, with an enhancement of 0.13. Specifically speaking, three features can be captured from the estimated results: First, from the perspective of changing trend, the technical innovation efficiency of the selected provinces has not increased much, but it shows a rising trend in general. The provinces with high technical efficiency grow slower than the provinces with low technical efficiency. Second, analyzing from regional results, technological innovation efficiency of each province varies a lot. For instance, Xinjiang's equipment manufacturing industry has the lowest innovation efficiency, which is only 0.38, remaining plenty room for improvement; while Chongqing possesses the highest technological innovation efficiency, which is 0.87, indicating that the province's innovation activities are close to the level of technology at the forefront. Third, from the perspective of time-segmented technical efficiency, the efficiency value presents a rising trend over time. Since 2013, the provinces have continuously expanded cooperation areas and built new support platform according to the 'Belt and Road' strategy. The research thus sets year 2013 as time watershed and constructs two sub-sample periods, which correspondingly are 2007–2012 and 2013–2015, to conduct sample test separately. The significance of F-statistic is 0.0012, which is less than 0.005, indicating significant difference of sample variance between two sub-samples; the significance of T-statistic is 0.002, which is also less than 0.05, implying the difference between sub-samples are statistically significant. Hence, the results provide robust evidence that there exists certain difference between the innovative efficiency of equipment manufacturing industry of the 'New Silk Road Economic Belt' before and after year 2013.

Decomposition of Technological Innovation Efficiency. Based on the estimation results of the output distance function in Table 2, this paper calculates the total technological innovation efficiency and scale efficiency of the 'New Silk Road Economic Belt' according to the abovementioned equations. The total technical efficiency is decomposed into two parts: scale efficiency and pure technical efficiency. Table 3 gives the decomposition results of the innovation efficiency of the equipment manufacturing industry along the 'New Silk Road Economic Belt'.

It can be seen from Table 3 that the total technical efficiency and scale efficiency of the innovative production in selected provinces are all at low efficiency levels, and the average range of the two changes is 0.1–0.3, 0.1–0.3. Consistent with the results of technological innovation efficiency, Chongqing owns the highest scale efficiency and the lowest is in Inner Mongolia; the total innovation efficiency remains the same. Through the analysis of the total technical efficiency of equipment manufacturing industry along the economic belt area, it can be concluded that the reason of low conversion efficiency between innovation input and output in the developing process of R&D and innovation of the equipment manufacturing industry in the western region is that the scale efficiency is not the most important indicator in the frontier. In addition, the empirical results show that the scale elasticity of the equipment manufacturing industry in selected provinces is almost greater than 1, indicating that the scale of innovation in the western region is small. Therefore, in order to improve the technological innovation of the equipment manufacturing industry in the western region, it is necessary to expand the scale structure of innovation.

Table 3. Equipment manufacturing innovation efficiency decomposition results

Province	Technological innovation efficiency	Scale efficiency	Total technical efficiency	Province	Technological innovation efficiency	Scale efficiency	Total technical efficiency
Shaanxi	0.76	0.34	0.26	Yunnan	0.55	0.26	0.14
Gansu	0.70	0.31	0.22	Xinjiang	0.38	0.16	0.06
Ningxia	0.48	0.21	0.10	Guangx	0.63	0.32	0.20
Qinghai	0.42	0.18	0.08	Sichuan	0.83	0.37	0.30
Chongqing	0.87	0.38	0.33	Economic belt	0.62	0.28	0.18

4 Conclusion

This paper adopts the stochastic frontier analysis method and utilizes the spatial panel model to measure the technical efficiency of R&D and production activities of equipment manufacturing industry along the 'New Silk Road Economic Belt' based on the panel data ranged from 2007 to 2015. Moreover, the paper further analyzes the spillover effect of innovation efficiency of equipment manufacturing industry of the 'New Silk Road Economic Belt'. The main conclusions of this paper are as follows:

First, the technological innovation efficiency of equipment manufacturing industry of the 'New Silk Road Economic Belt' presents a 'double-cores-and-double-peripheries' development trend during the observation period. The innovation efficiency of equipment manufacturing industry in Chongqing and Shaanxi province has been continuously improved. Although the innovation efficiency of equipment manufacturing industry in Sichuan, Ningxia, Gansu and Xinjiang has also improved, the growth rate of overall efficiency is rather slow. Guangxi and Yunnan have the potential to develop modern equipment manufacturing industry, and the development of Qinghai's equipment manufacturing industry has a relatively late start, it is also necessary to continuously enhance the ability of its industrial innovation.

Second, there is still room for improvement in the total technological innovation efficiency, as well as technical efficiency of equipment manufacturing industry of the 'New Silk Road Economic Belt'. In terms of provincial efficiency, the total technical efficiency and scale efficiency of innovation production in selected provinces are generally at low level; in terms of stage efficiency, the improvement degree of both is greater than the improvement degree of technological innovation efficiency. It can be seen that the reason why the total technological innovation efficiency of equipment manufacturing industry of the 'New Silk Road Economic Belt' is not high is that the scale structure is not perfect.

Third, the technological innovation efficiency in equipment manufacturing industry of the 'New Silk Road Economic Belt' has accelerated since 2013. It can be seen from the estimated SFA results that the technological innovation efficiency of the provinces in the economic zone has accelerated after 2013, and the annual growth efficiency of each province is better than the previous growth efficiency and has been greatly improved. This is primarily benefited from the gradual maturity of the national strategy of the 'New Silk Road Economic Belt' and the increasing efficiency of technological innovation.

References

1. Niu, Z.D., Zhang, Q.X.: Technological innovation efficiency of china's equipment manufacturing industry. Quant. Econ. Econ. Res. **65**(11), 102–116 (2012). https://doi.org/10.13653/j.cnki.jqte.2012.11.009
2. Zou, L., Zeng, G., Cao, X.Z.: ESDA-based R&D investment spatial differentiation characteristics and time-space evolution of the yangtze river delta urban agglomeration. Econ. Geogr. **43**(3), 67–79 (2015). https://doi.org/10.15957/j.cnki.jjdl.2015.03.011
3. Wang, Z.B., Sun, C.: An empirical study on the efficiency of technology R&D in China's equipment manufacturing industry. China Sci. Technol. Forum **78**(8), 24–38 (2007). https://doi.org/10.13580/j.cnki.fstc.2007.08.01
4. Gui, H.B.: Spatial econometric analysis of china's high-tech industry innovation efficiency and its influencing factors. Econ. Geogr. **39**(6), 75–91 (2014). https://doi.org/10.15957/j.cnki.jjdl.2014.06.007
5. Chen, A.Z., Liu, Z.B., Zhang, S.J.: The binary division of labor network restriction in china's equipment manufacturing industry innovation. J. Xiamen Univ. (Philos. Soc. Sci. Ed.) **78**(6), 65–79 (2016). https://doi.org/10.13510/j.cnki.jit.2011.04.010
6. Chen, A.Z., Zhong, G.Q.: Does China's equipment manufacturing international trade promote its technological development? Economist **76**(5), 67–84 (2014). https://doi.org/10.16158/j.cnki.51-1312/f.2014.05.018
7. Tang, X.H., Li, S.D.: Empirical study on china's equipment manufacturing industry and economic growth. Chin. Ind. Econ. **57**(12), 63–77 (2010). https://doi.org/10.19581/j.cnki.ciejournal.2010.12.003
8. Gong, B.H., Sickles, R.C.: Finite sample evidence on the performance of frontiers and data envelopment analysis using panel data. J. Econ. **51**(1–2), 259–284 (1992). https://doi.org/10.1016/0304-4076(92)90038-S
9. Battese, G.E., Coelli, T.J.A.: Prediction of firm-level technical efficiencies with a generalized frontier production function and panel data. J. Econ. **3**(38), 387–399 (1995). https://doi.org/10.1016/0304-4076(88)90053-X

SAR Image Enhancement Method Based on Tetrolet Transform and Rough Sets

Wang Lingzhi(✉)

Automation School, Xian University of Posts and Telecommunications, Xi'an 710121, China
wlzmary@126.com

Abstract. SAR image enhancement is one of the key issues on SAR image processing. In this paper, a new SAR image enhancement method is presented. Firstly, SAR image is abstracted into a knowledge system by rough sets, and obtained the approximate subsets of the edge and texture respectively. And then the introduction of tetrolet transformation, edge subset and texture subset is so represented sparsely that the signal energy is more concentrated. In Tetrolet transform domain, edge subset is refined by margin adjustment and texture subset is enhanced by threshold method. Finally, the edge and the texture subset processed are inversed by tetrolet transform, and weighted them to obtain the enhanced results. Experimental results show the proposed method that has better performance to retain detail information and suppress Speckle noise, superior to the traditional wavelet transform and contourlet transform method.

Keywords: Rough sets · Tetrolet transform · Synthetic aperture radar
Image enhancement

1 Introduction

Synthetic aperture radar (synthetic aperture radar, SAR) is an all-day, all-weather imaging radar, which makes SAR images have very important application value in military and commercial application due to its characteristics of multi-polarization, multi-angle, multi-angle data capture ability, high resolution and strong penetration. In SAR images, coherent spot noise interference is inevitable because of the imaging processes of SAR systems which seriously affects the visual quality of SAR images and bring difficulties to the subsequent application of SAR images processing. Therefore, SAR image enhancement technology becomes the key of SAR images processing [1–5]. SAR image enhancement needs to preserve texture and orientation information of SAR images better when suppressing coherent spot noise. However, the rich and complex content of SAR images makes the SAR image enhancement technology very challenging, it has become a new research topic.

SAR image enhancement methods can be roughly divided into two categories: one is the multi-view smoothing pretreatment before imaging, and the other is coherent spot noise filtering technology after imaging. Multi-look processing technology is the average of *L*-lookl SAR images after superposing, which will reduce the resolution of the system. Therefore, in recent years, people have turned to the research of filtering

P. Krömer et al. (Eds.): ECC 2018, AISC 891, pp. 524–532, 2019.
https://doi.org/10.1007/978-3-030-03766-6_59

denoising technology, and this method can be roughly divided into two categories: spatial filtering and transform domain filtering. Spatial filtering, based on the local statistical features of the image, select the appropriate window size adjustment and filtering functions, but this method is on the edge of isotropic filtering, blurring image structure and detail information, resulting in enhanced effect is not satisfactory. In recent years, wavelet transform has been widely used in SAR image enhancement [6–10], which obtains better results. However, the wavelet transform can only represent the point feature of the image optimally, and cannot depict the texture and the edge of the image very well, while the texture and edge features of the SAR images are very rich. So, the wavelet transform cannot capture the texture and edge features of SAR images more accurately. To solve this problem, a multi-scale geometric analysis (Mutilscale analysis, MGA) transformation is appeared, represented by Contourlets, Rigelets, Brushlets, Curvelets and Bandelets [11, 12]. One important feature of this kind of transformation method is using the anisotropic base function in its construction process. Relative to the wavelet transform, they can more sparsely approximate a particular singular curve or surface, and detect the line singularity and surface singularity in the image while detecting the singular point of the image, that is the image edge and texture. Texture and edge contain the main structural information of the image, so MGA can take full advantage of the specific geometric characteristics of the image data to sparsely characterize the image, which is widely used in the SAR image enhancement technology [13]. However, most of the SAR image enhancement methods based on threshold estimation set the speckle noise coefficient as zero in the transform domain, and the edge coefficients are enlarged. Although they have a better effect on spot suppression, they don't approach the edges of the image very well owing to finiteness of directional filter group direction description in MGA transform. So they have some limitations on the sparse representation of the image and require a high accuracy of the threshold estimation. As a result, they will have insufficient reservation of the detail texture information after SAR image enhancement and produce some false contours. For example, SAR image enhancement method using wavelet transform will cause more serious glitches and box effects on the edge details. In 2009, Krommweh proposed a new adaptive multi-scale geometric transform based on Haar-type wavelet transform, named as Tetrolet transform [14]. The transform uses local adaptive technique to obtain good local texture approximation, so as to simulate the texture and edge more accurately, making the transformed coefficients more sparser and energy more concentrated.

Based on the adaptive Tetrolet transform, this paper proposes a new SAR image enhancement method. Firstly, the rough sets of SAR images are divided to obtain the edge subsets and texture subsets. And then the edge enhancement and the texture filtering are respectively performed by utilizing Tetrolet coefficients. Finally, we obtain the enhanced results by weighting. Experimental results show that the proposed method has better results in detail information retention and speckle reduction.

2 SAR Image Enhancement Based on Tetrolet Transform

SAR image enhancement requires not only effective suppression of speckle noise but more importantly, need to better preserve and enhance the orientation and texture of SAR image and other important information. In this paper, the SAR image is abstracted into a knowledge system, and the rough set is used to approximate the edges and texture subsets of the image, and then the two subsets are respectively subjected to Tetrolet transformation for optimal sparse representation and local texture approximation. The edge subsets are subjected to refinement of direction adjustment and enhance the edge subgraph by inverse Tetrolet transformation. The texture subset is thresholded in the Tetrolet transform domain and the enhanced texture subgraph is obtained by inverse Tetrolet transform. Finally, the enhanced edge subgraph and texture subgraph are weighted combined to achieve the purpose of SAR image enhancement.

2.1 Rough Set Classification

In suppressing process the noise of the SAR image, it is difficult to distinguish the noise and the edge texture from the high-frequency information. In order to avoid losing the edge texture information while suppressing the noise, we need to objectively and accurately distinguish the noise from the edge texture. Moreover, the human visual characteristics show that as the gradient value increases, the visibility of the noise gradually decreases. Therefore, where the gradient is large, the influence of noise is relatively small, whereas in the small gradient, the influence of noise is relatively large. For better enhancement, we compute edge subsets and texture subsets in the image content, and process them separately.

Define SAR image U as approximate space $K = (U, R)$, where $R = \{R_1, R_2\}$ is the equivalent relationship on U. In this equivalent relationship, the attribute of A is defined as: among the pixels in any area, there may be only noise pixels and edge pixels, but some pixels must belong to noise. So for the rough edge of the concept, its upper approximation is $R_1(X)$ and the next approximation $\underline{R}_1(X)$ is the noise subset. So, we can get the edge subset:

$$A_1 = R_1(X) - \underline{R}_1(X) \tag{1}$$

Similarly, the attribute of A is defined as: Among the pixel points in any area, noise pixels and texture pixels may be included but a few pixels must belong to noise. So for the rough concept of texture, the upper approximation is $R_2(X)$, and the lower approximation $\underline{R}_2(X)$ is the noise subset. So, we can get a subset of the texture:

$$A_2 = R_2(X) - \underline{R}_2(X) \tag{2}$$

For noise, its characteristics in the image are mainly reflected in the contrast with its neighborhood contrast, so we can define:

$$\begin{aligned} R_{noise}(X) &= \underline{R}_1(X) = \underline{R}_2(X) \\ &= \bigcup_i \bigcup_j \{S_{ij} \, |\mathrm{int}| \, \overline{m}(S_{ij}) - \overline{m}(S_{i\pm 1, J\pm 1})| > Q\} \end{aligned} \quad (3)$$

where S_{ij} represents a 3 × 3 image block, $S_{i\pm 1, j\pm 1}$ represents a neighborhood image block of $S_{ij} \cdot \overline{m}(S_{ij})$ represents the mean of image block S_{ij}. Since this step only initially suppresses the noise, and in order to completely preserve the edge and texture information, we choose a larger value of Q, in this paper, $Q = 2.25\, std(S'_{i,j})$, $S'_{i,j}$ are 9 × 9 image blocks consisting of S_{ij} and $S_{i\pm 1, j\pm 1}$, respectively. To obtain subsets $R_1(X)$ and $R_2(X)$, we calculate gradient map $I(u, v)$ of the SAR image. $P = k \times \max(I)$, P is the image gradient threshold, the size of the coefficient can choose a value between 0 and 1 according to the image texture, where $k = 0.0286$.

$$R_1(X) = \{X|: I(u, v) > P\} \quad (4)$$

$$R_2(X) = \{X|: I(u, v) > P\} \quad (5)$$

2.2 Tetrolet Transformation

In the process of obtaining the edge subsets and texture subsets, although the rough set is used to approximate the edge subsets and texture subsets of the SAR image and the initial suppression of the noise, since in the previous step, in order to completely preserve edge and texture information, we choose a larger threshold so that there are residual noise in edge subset and texture sub region. To further enhance the SAR image, we use the Tetrolet transform to decompose the edge and texture subsets by three levels respectively. Using the excellent features of texture approximation, sparse representation and multi-directional capturing of Tetrolet transform, the edge and texture subsets are sparsely represented so as to refine their edge distribution and enhance the detail texture locally.

Figure 1 shows a three-layer Tetrolet decomposition of an 2048 × 2048-size SAR image. As we can see from the figure, the coefficients after the transformation of Tetrolet are very sparse. The energy mainly concentrates on a few coefficients, which are represented as a few scattered white spots in the Tetrolet transform coefficient graph. Moreover, the adaptive coverage obtained in the decomposition process can well approximate the texture information of 4 × 4 the image block, as shown in the enlarged view in the figure.

2.3 Edge Enhancement

For people perception of the image firstly comes from the edge information in the image, so the edge information is often dominant. However, speckle noises in SAR images will inevitably affect the edge structure information, which may change the edge contour of an object when the speckle appears exactly on the edge of the object. In the process of obtaining the edge subset, the edge of the object may be discontinuous or not smooth due to the existence of noise. Therefore, we also need to adjust the direction

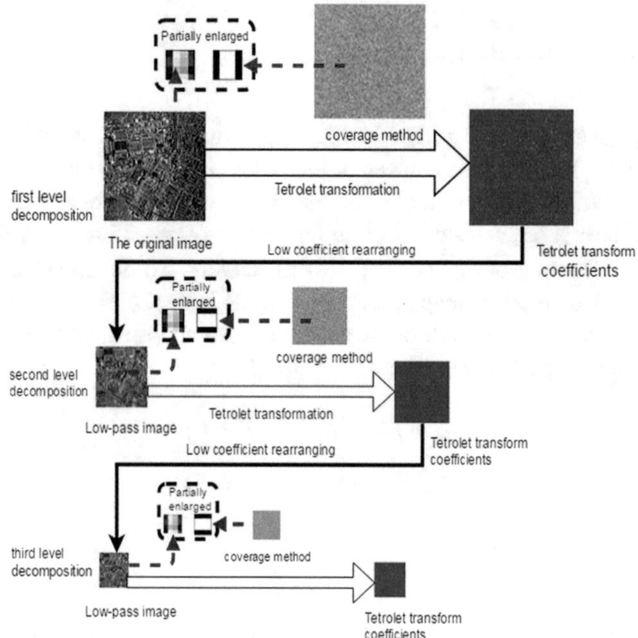

Fig. 1. SAR image through the three-tier Tetrolet decomposition diagram

of the edge while suppressing the noise and preserving the edge texture, to correct those edges direction affected by the noise. And the adjustment is based on the fact that the edge of object magnified to a certain size is continuous and smooth. The Tetrolet transformation provides us with a powerful tool to deal with this problem. In Fig. 2, we consider that the geometric edges shown in (a) are contaminated by speckle noises, and then compute the edge subsets to obtain (b), where (a) and (b) are magnified partial images of size 4×12. We simulate (a) and (b) separately using the matrices of (c) and (d). After locally adaptive Tetrolet transform, the coverage mode (e) of (a) and the coverage mode (f) of (b) are obtained respectively, and the texture directions respectively depicted by (g) and (h). We can see that due to the pollution of speckle noise, the edges are not smooth, and the texture direction of the Tetrolet transform is not continuous, as shown in (h).

Taking into account the continuity of the edge, B1, B2 and B3 must be able to smooth the transition, we adjust the direction of B2 and B1 and B2 in the same direction, and then reverse Tetrolet transform—data (d) using coverage method (f) Tetrolet transform and then using coverage method (e) Tetrolet transform, as shown in Fig. 3.

As we can see from Fig. 3, we can change the coverage method of the inverse Tetrolet transform in order to achieve the purpose of adjusting the edge smoothness, so that the edge becomes continuous smooth. And its essence is to adjust the direction of the edge by adjusting the distribution of energy, to improve the discontinuous and unsmooth situation of object edge. For such a principle, we modify the direction of the

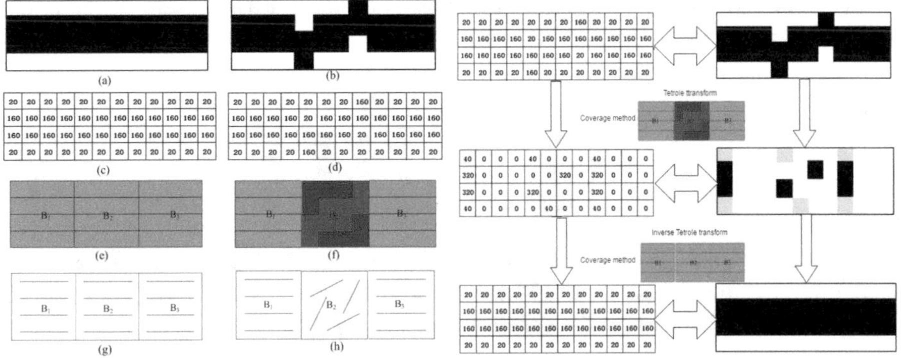

Fig. 2. Simulation of the edge of the Tetrolet transform schematic

Fig. 3. Edge adjustment diagram

edge Tetrolet transform coefficients and adjust the energy distribution so that the edges are smooth and continuous, reducing the influence of speckle noise.

Record $B_{i,j}$ as an image block of 4×4 with subscript (i, j), whose direction is denoted as $\theta_{i,j}$ after Tetrolet transformation, and $\widehat{\theta}_{i,j}$ as the direction after adjustment for $B_{i,j}$, then $\widehat{\theta}_{i,j}$ should satisfy the minimum of (6):

$$\min\left[\left|\widehat{\theta}_{i,j} - \theta_{i-1,j}\right| + \left|\widehat{\theta}_{i,j} - \theta_{i-1,j}\right|,\ \left|\widehat{\theta}_{i,j} - \theta_{i-1,j}\right| + \left|\widehat{\theta}_{i,j} - \theta_{i-1,j}\right|\right] \tag{6}$$

After the edge is adjusted in the direction of the Tetrolet transform domain, an inverse Tetrolet transform is performed to obtain an enhanced edge subgraph.

2.4 Texture Filtering

For the texture subset, mainly include the texture area, flat area and residual noise, in order to expand the contrast dynamic range of the SAR image, we will enhance the texture region and the flat region to different degrees and suppress the residual noise. Firstly, the texture area, flat area and residual noise are divided by using the Tetrolet coefficient of the texture subset. The criteria are as follows.

$$\begin{cases} \text{texture}, & \text{if } \mathrm{X}_c > k_1\sigma_x \\ \text{flat size}, & \text{if } \mathrm{X}_c > k_2\sigma_x \; and \, \mathrm{X}_c > k_1\sigma_x \\ \text{residual size}, & \text{if } \mathrm{X}_c \le k_2\sigma_x \end{cases} \tag{7}$$

X_c is the coefficient of the texture subset after the Tetrolet transform, σ_x is the standard deviation of different scales, $k_1 = 3$, $k_2 = 1$. The coefficient corresponding to each pixel after the judgment is as follows.

$$Y_c = \begin{cases} \max((\frac{k\sigma_x}{|X_c|})^p, 1)X_c, & \text{texture} \\ X_c, & \text{flat size} \\ 0, & \text{residual size} \end{cases} \tag{8}$$

where Y_c is the enhanced coefficient, $p = 1$, after inverse Tetrolet transformation of Y_c, an enhanced texture subgraph is obtained. Finally, the enhanced SAR image is obtained by weighted combination of the enhanced edge subgraphs.

3 Experimental Results

In this section, two SAR images with more severe speckle are enhanced. It is compared with the wavelet transform method [6] and the Contourlet transform method [7]. Wavelet transform select 'DB4' for three decomposition. Contourlet Transform Select 'maxflat' Non-subsample tower decomposition. The non-directional sampling filter bank chooses 'dmaxflat7' and decomposes three layers with 2, 4 and 8 directions respectively. Tetrolet transform adaptive selection filter bank carries on three layer decomposition. The following is a comparison of experimental results.

From the experimental results of Fig. 4, it can be seen that the method proposed in this paper has some improvements over the methods of wavelet transform and Contourlet transform in the direction information enhancement and background noise suppression. It is obvious from Fig. 4(b) that the edge detail has the glitch phenomenon and the square effect. This is exactly the defect of the wavelet transform in the singularity information of the directionality. The wavelet transform uses square basis function to approximate the linear singularity feature. On the one hand, it increases the number of basic functions, on the other hand, it will seriously affect the visual effects. However, due to the limitation of Contourlet basis function in the Contourlet transform method, directional scratches are generated in the image, as shown in Fig. 4(c). The method in this paper has achieved good results in speckle suppression and direction preserving, and has fine visual effects. It is obvious that the edge is relatively smooth, the detail texture is relatively clear, and there are no obvious scratches.

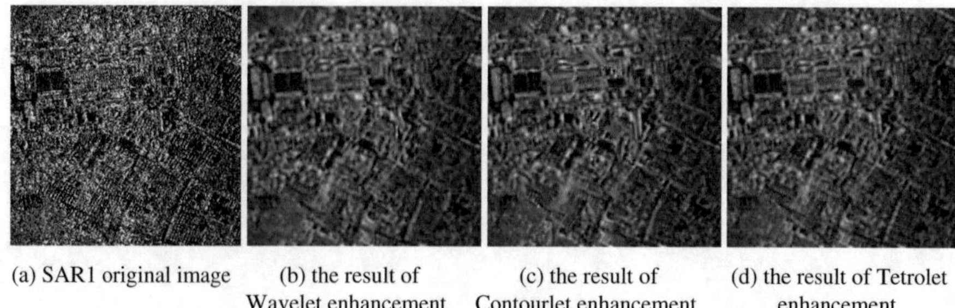

(a) SAR1 original image (b) the result of Wavelet enhancement (c) the result of Contourlet enhancement (d) the result of Tetrolet enhancement

Fig. 4. SAR1 image contrast experimental results

Evaluating indicator commonly used for image enhancement method are Background Variance (BV) and Detailed Variance (DV). The effective enhancement algorithm is to increase the detail variance while keeping the variance of background variance small. Since enhancement and restoration of SAR images are equivalent to a blind process, PSNR cannot be used to evaluate the smoothness. Normally equivalent visual number (ENL) is used to measure the smoothness of SAR images. Defined as μ^2/σ^2, where μ, σ^2 corresponds to the mean and variance of the SAR image respectively. ENL is theoretically equal to the number of views of SAR images. By comparing the ENL, it can be measured approximately the smoothness effect of SAR images in an objective way. The ideal objective evaluation result is to get better edge retention when the ENL is controlled within a certain range.

It can be seen from Table 1 that the method in this paper has been significantly improved in the control of BV and the improvement of DV, and the wavelet transform method with equivalent visual number and Contourlet transform method have also been improved.

Table 1. Comparison of different methods of experimental results

		SAR1
Original image	ENL	2.678 4
	BV	0.012 7
	DV	0.024 5
Wavelet	ENL	7.515 8
	BV	0.014 8
	DV	0.032 1
Contourlet	ENL	7.541 9
	BV	0.015 1
	DV	0.041 0
Method in this article	ENL	7.905 1
	BV	0.015 7
	DV	0.047 8

4 Conclusion

In this paper, a SAR image enhancement method based on Tetrolet transform is proposed. The rough set theory is used to approximate the edge subsets and texture subsets, and the edge subsets and texture subsets are respectively transformed by Tetrolet to complete the sparse representation of the images. The edge and texture subsets are adjusted respectively in the Tetrolet transform domain and the threshold is enhanced, and then the inverse Tetrolet transform is performed to obtain enhanced edge and texture subgraphs. At last, the obtained enhancer subgraphs are weighted and combined to achieve the purpose of SAR image enhancement. The experimental results show that the proposed method in this paper outperforms the traditional wavelet transform and Contourlet transform in preserving details and suppressing speckles, and has better visual effects.

Acknowledgement. This work was supported by Scientific Research Plan Projects of Shannxi Education Department (Grant No. 16JK1690).

References

1. Fengkai, L., Jie, Y., Deren, L.: Polarimetric SAR image adaptive enhancement lee filtering algorithm. Acta Geod. Cartogr. Sin. **43**(7), 690–697 (2014). https://doi.org/10.13485/j.cnki-2089.2014.0112
2. Li, Y., Hu, J., Jia, Y.: Automatic SAR image enhancement based on nonsubsampled contourlet transform and memetic algorithm. Neurocomputing **134**, 70–78 (2014). https://doi.org/10.1016/j.neucom.2013.03.068
3. Zhang, B., Wang, C., Zhang, H., Wu, F.: An adaptive two-scale enhancement method to visualize man-made objects in very high resolution SAR images. Remote Sens. Lett. **6**(9), 725–734 (2015). https://doi.org/10.1080/2150704x.2015.1070313
4. Tan, G., Pan, G., L, W.: SAR image enhancement based on fractional fourier transform. Open Autom. Control Syst. J. **6**(01), 503–508 (2014). https://doi.org/10.2174/1874444301406010503
5. Jin, G., Zhang, J., Huang, G.: Enhancement of airborne SAR images without antenna pattern. Acta Geodaetica Cartogr. Sin. **42**(4), 554–558+567 (2013)
6. Sveinsson, J.R., Benediktsson, J.A.: Almost translation invariant wavelet transformations for speckle reduction of SAR images. IEEE Trans. Geosci. Remote Sens. **41**(10), 2404–2408 (2003). https://doi.org/10.1109/TGRS.2003.817844
7. Sha, Y., Liu, F., Jiao, L.: SAR image enhancement based on nonsubsampled contourlet transform. J. Electron. Inf. Technol. **31**(07), 1716–1721 (2009)
8. Chen Jiayu, X., Xin, S.H., Bao, G.: SAR image point target detection based on multiresolution statistic level. Syst. Eng. Electron. **27**(02), 205–209 (2005)
9. Sudan, L., Guangxia, L., Cui, Z., Zhengzhi, W.: Multiscale edge detection of SAR images. Syst. Eng. Electron. **26**(03), 307–320 (2004)
10. Chengling, M., Shouhong, W., Lihua, Y., Yu, X.: A SAR image edge detection algorithm based on double tree complex wavelet transform. J. Univ. Chin. Acad. Sci. **31**(02), 238–242 +248 (2014)
11. Romberg, J.K.: Multiscale Geometric Image Processing. Ph. D. thesis. Rice University (2003). https://doi.org/10.1117/12.509903
12. Do, M.N., Martin, V.: The contourlet transform: an efficient directional multiresolution image representation. IEEE Trans. Image Processing **14**(12), 2091–2106 (2005). https://doi.org/10.1007/978-3-319-73564-1_44
13. Haiyan, J., Licheng, J., Fang, L.: SAR Image denoising based on curvelet domain hidden markov tree model. Chin. J. Comput. **30**(3), 491–497 (2007)
14. Krommweh, J.: Tetrolet Transform: a new adaptive haar wavelet algorithm for sparse image representation. Annal. Appl. Stat. **21**(4–21), 364–374 (2010). https://doi.org/10.1016/j.jvcir.2010.02.011

A Fast Mode Decision Algorithm for HEVC Intra Prediction

Hao-yang Tang(✉), Yi-wei Duan, and Lin-li Sun

School of Automation, Xi'an University of Posts and Telecommunications, Xi'an 710121, China
tanghaoyang@xupt.edu.cn, 446542386@qq.com

Abstract. Compared to H.264/AVC, High Efficiency Video Coding (HEVC) can achieve bits rate savings of up to 50% while maintaining the same subjective quality. However, this great advance is obtained at the expense of significantly increased encoder complexity. In this paper, a fast mode decision algorithm is proposed for HEVC intra prediction. Firstly, the depth information of the CU block is used to skip some of impossible selected modes, assuming that there is no need to perform a strong search for large CUs. Secondly, the number of rough mode decision (RMD) is adjusted to further reduce the number of modes that need to be evaluated. Experimental results show that the proposed algorithm can achieve about 28.8% encoder time saving on average under the all intra configuration compared with HM 10.0.

Keywords: High Efficiency Video Coding (HEVC) · Intra prediction
Fast algorithm · RMD

1 Introduction

With the rapid development of video applications, the high-definition (HD) video (720P and 1080P) is popular and the ultra-high-definition (UHD) video (4K and 8K) has emerged. The video coding standard H.264 published in 2003 has been unable to meet the requirement of the storage and the transmission of HD and UHD video, and ITU-T and ISO/IEC have jointly finalized the High Efficiency Video Coding (HEVC) [1] standard in 2013 to obtain a better video coding efficiency. Compared to the H.264 [2], HEVC can achieve up to 50% bits rate saving while providing the same subjective video quality.

In HEVC coding standard, a leaf node of a CTU is defined as a coding unit, and the encoder recursively divides the CU into a plurality of coding units in a quadtree manner according to the complexity of the video content inside the CU. The size of the CU can be recursively divided from the largest 64 × 64 into a minimum of 8 × 8 and its size is expressed in recursive depth, which details as follows depth = 0, CU size is 64 × 64: Depth = 3, CU size is 8 × 8 [3]. These new features affect the complexity extremely, making it difficult for real-time implementation.

There are angle prediction modes in HEVC, up to 35 types. Compared with the 9 modes in H.264/AVC. As the amount of mode increases, the accuracy is higher and the residual is smaller, which significantly increases the compression efficiency. However,

P. Krömer et al. (Eds.): ECC 2018, AISC 891, pp. 533–540, 2019.
https://doi.org/10.1007/978-3-030-03766-6_60

the extra prediction mode and recursive coding structure take a lot of time to choose the best mode. As a result, encoder complexity becomes intolerable in power limited and real-time applications. This is critical for mobile applications. For example, if the video is captured on a smartphone, the additional benefits of HEVC may be unpopular due to its high power [4]. Therefore, it is important to find a better way to reduce computational complexity and improve coding performance.

In order to solve the above problems, possible solutions are available. Zhou et al. [5] adopt the depth of the spatiotemporal neighbor CTUs to predict the depth of the current CTU and use the depth of the adjacent CU to reduce the depth search of sub-CTUs. Wang et al. [6] proposed a three-step fast intra prediction algorithm where they used CU split prediction and a low precision RMD to speed up the intra coding. Ozcan et al. [7] proposed a computation reduction hardware implementation based on Sum of Absolute Transformed Difference (SATD). Some other earlier fast algorithms also show complexity reductions for HEVC [8, 9].

This paper proposes a fast mode decision algorithm based on intra prediction for HEVC, the depth information is used to skip certain modes to reduce the number of modes calculated in the RMD, after fewer modes are selected as candidates to mitigate the computational burden in the RDO process.

2 Intra Prediction in HEVC

HEVC standard adopts a new concept for image representation. For coded block structures, HEVC uses a tree structure, which can be divided into smaller blocks by quadtree partitioning. Each frame of the image is divided into a number of coding tree units, and each CTU can be divided into CUs of different depths in a recursive manner. Among them, CU is defined as a square unit, which has 8×8, 16×16, 32×32, and 64×64 for 4 sizes [3], but also increases the set of prediction modes to 35.

As shown in Fig. 1, intra coding needs to select the best mode for prediction from among 35 PU prediction modes. In the 35 modes, the number 0 mode (Planner mode) is applied to the area where the pixel value changes slowly, the number 1 mode (DC mode) is applied to the large area flat area, and the number of 2– 34 mode is the angle prediction mode, respectively. A texture direction, where the angle prediction mode numbered 2–18 is applied to the area where the texture is generally biased to the horizontal direction, and the angle prediction mode of 19–34 is applied to the area where the texture is generally biased to the vertical direction. In the reference model HM of HEVC, the prediction residual Hadamard cost of all 35 modes is first calculated, and then the first few prediction modes of the Hadamard cost from small to large are composed of the rate distortion cost candidate mode list, and then selected from the list. The prediction mode with the lowest rate distortion cost is used as the intra prediction mode. Since each CTU recursively traverses all depth CUs in the encoding process, and each recursion needs to calculate the Hadamard cost of all PU prediction modes and the rate distortion cost of the candidate prediction modes, the encoding calculation time is greatly increased.

The prediction mode in 35 covers only half of the plane, and the bottom and right sides are always unavailable in a fixed coding order. HEVC is very time consuming

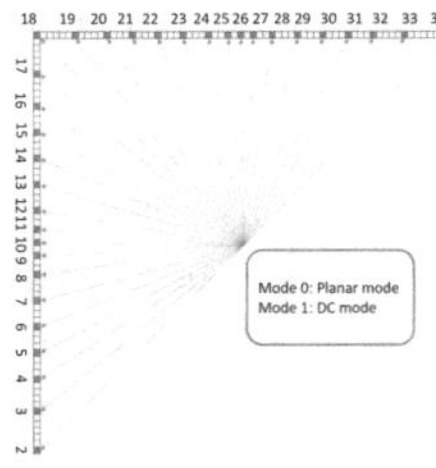

Fig. 1. Intra prediction modes in HEVC.

due to the many evaluation possibilities. This fact allows us to reduce the time consuming by deep prediction and then reducing the number of patterns traversed in the angle prediction mode.

The rest of this paper is organized as follows. Section 3 presents an introduction of the proposed algorithm. Section 4 gives the experimental results and conclusion is given in Sect. 5.

3 Proposed Method

3.1 CU Depth Range Decision

When using spatial prediction, the neighboring CU has a strong correlation with the current CU. It is known that the intra frame coding makes use of the encoded above, left, left-above and right-above CUs as shown in Fig. 2 to reduce computational burden in the current CU.

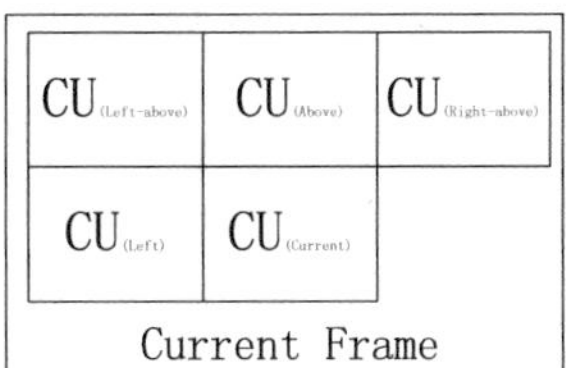

Fig. 2. Current CU and its neighbor depth decision.

Before the CU is divided, the traversal range can be determined in advance, which can save a lot of coding time. There are temporal and spatial correlations between successive video content, and we can take advantage of their spatial correlation, and the predicted depth formula (1) is as follows.

$$\text{Depth}_{pre} = \sum_{i=0}^{N} \text{Depth}_i \times \alpha_i \tag{1}$$

$Depth_{pre}$ is the weighted prediction depth value of $CU_{current}$, i represents the number of candidate CUs, and the number of candidate CUs is N. Under the algorithm, N takes 4. $Depth_i$ is the depth value of the i-th candidate CU, and α_i indicates the first candidate CU depth. By encoding a large number of test sequences, the correlation between $CU_{current}$ and its neighboring CUs in the time-space domain is calculated, and the correlation coefficient between CUs as shown in the Table 1, and the weights associated with the space-time domain allocation are given according to the data in the Table 1.

Table 1. Candidate CU depth value and α_i value.

$Depth_i$	0	1	2	3
α_i	0.35	0.35	0.15	0.15

In this paper, the area where the space-time domain neighboring candidate CU depth values are the same is called a smooth area. For the smooth region, the depth values of the temporally spatially adjacent coding blocks CU_{left} and CU_{above} are selected from the prediction candidate sets to predict the $CU_{current}$ depth range. The specific implementation process is as follows:

1. If the depth values of both candidate CUs are 0, it indicates that the areas near the CU of the two frames before and after are relatively stationary. $CU_{current}$ terminates partitioning early, and the $CU_{current}$ depth value is 0;
2. If the depth values of both candidate CUs are 1, the $CU_{current}$ depth prediction range DR is [0, 2];
3. If the depth values of the two candidate CUs are both 2, the $CU_{current}$ depth prediction range DR is [1, 3];
4. If the depth values of the two candidate CUs are both 3, the $CU_{current}$ depth prediction range DR is [2, 3].

3.2 A Fast Mode Decision Algorithm

The frame of the video is divided into a series of coding tree units, and the partitioning of the unit is largely dependent on the texture features. For those that do not contain much information, the texture and uniformly flat area, we tend to choose a small depth, DC, Planar mode, and other simple directions (horizontal mode and vertical mode) may be selected as the best mode. At the same time, it is preferable to further divide the complex and uneven regions, and it is easy to select a large depth. Therefore, there is no need to search through large CUs in detail. Formally, we define the prediction mode set A, B represents the planner and DC modes, and the remaining 33 angle prediction modes constitute the C set, while set A is used for different CU depths as follows in Fig. 3.

Figure 4 shows the example of corresponding prediction modes set selection introduced above. It is noted that the number of directions for intra prediction for smaller depth CU can be significantly reduced. After RMD process, the neighbor

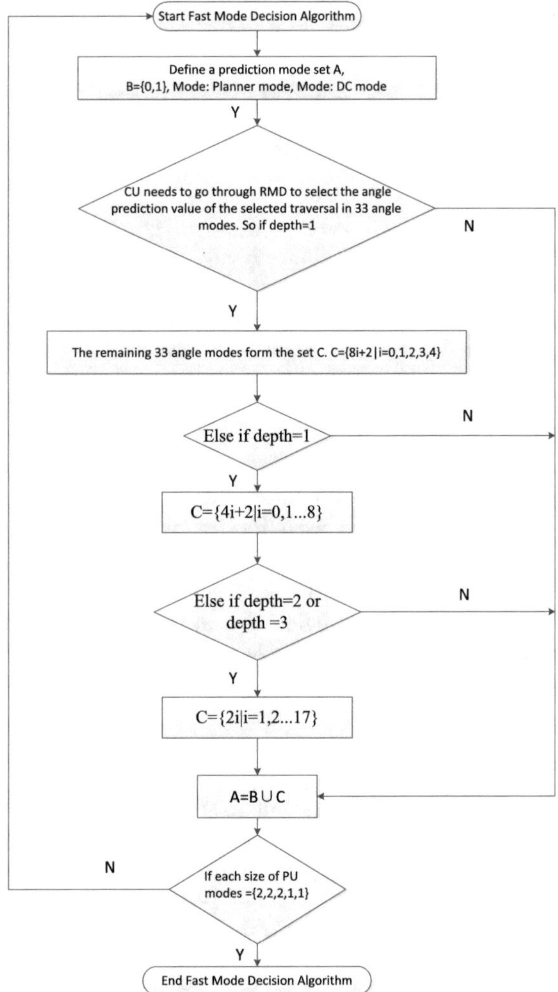

Fig. 3. A fast mode decision algorithm.

modes of the selected mode are checked, and the candidates list is updated by comparing the HAD cost of neighboring modes.

At the same time, the default HEVC reference selects the $\{8, 8, 3, 3, 3\}$ PU modes, which is 4×4, 8×8, 16×16, 32×32, 64×64. These selected modes are further evaluated in the RDO process to obtain the final end result, which will result in a rather expensive time consumption. Since the HAD cost is an approximate cost of the RD, it can be assumed that the best mode RMD process is selected to be highly relevant to the final decision and the complete RDO process selection mode. Moreover, a mode with low HAD costs may also have low RD costs, so it is the best mode that may be selected as the RDO process. Therefore, the proposed method selects only PUs of each size in the $\{2, 2, 2, 1, 1\}$ modes to further alleviate the burden processing of the RDO.

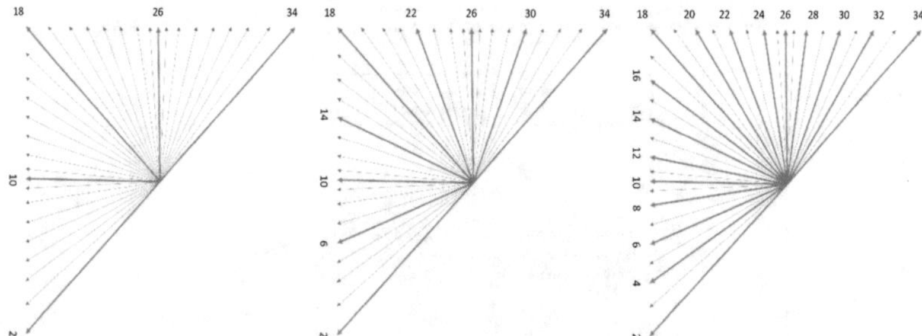

Fig. 4. (a) Left represent when depth = 0 the number of traversal modes under 33 angle prediction (b) Middle, when depth = 1, (c) Right, when depth = 2 or 3.

4 Experimental Results

An experiment has been conducted to demonstrate the effectiveness of the integrated algorithm. The sequence BQMall is used for testing and analyzing the amount of patterns evaluated during the RMD and RDO processes. The results are shown in Table 2. The number of RMD_{HM} modes represents the default HEVC calculation, RMD_{pro} represents number of modes calculated in proposed algorithm, RMD_{red} represents the percentage of reduction. RDO is the same as the RMD representation method in Table 2.

Table 2. RMD and RDO comparison of the number of modes calculated.

QP	Depth	RMD_{HM}	RMD_{pro}	RMD_{red}	RDO_{HM}	RDO_{pro}	RDO_{red}
22	0	35	8.0	77.2%	4.8	2.7	43.8%
	1	35	13	63.9%	4.8	2.7	43.8%
	2	35	21	40.0%	4.8	3.7	22.9%
	3	35	22	37.2%	9.6	3.8	60.5%
27	0	35	7.7	88.0%	4.8	2.7	43.8%
	1	35	12.4	64.6%	4.8	2.7	43.8%
	2	35	20.2	32.3%	4.8	3.7	22.9%
	3	35	20.7	40.9%	9.7	3.8	60.5%
32	0	35	7.5	78.6%	4.8	2.7	43.8%
	1	35	12.3	64.9%	4.8	2.7	43.8%
	2	35	20.4	41.7%	4.8	3.7	22.9%
	3	35	20.5	41.4%	9.7	3.8	60.5%
37	0	35	7.2	79.5%	4.7	2.7	43.8%
	1	35	12.3	64.9%	4.8	2.7	43.8%
	2	35	20.4	41.8%	4.8	3.7	22.9%
	3	35	20.5	41.5%	9.6	3.8	60.5%
Ave.		35	15.4	56.2%	6.0	3.2	42.3%

The proposed fast algorithm is implemented on the HEVC reference software HM10.0 reference software [10] and was run on a platform with Intel® Core TM i7-7700 CPU 16.0 GB RAM size. The test objects are Category 5 HEVC Standard Test Sequence Class A (2560 × 1 080), Class B (1920 × 720), Class C (832 × 480), Class D (416 × 240), and Class E (1280 × 720) [11]. Experiments were implemented for QP values of 22,27,32,37 with all intra encoding configuration. BD-rate [12] and encoder time saving ΔTime are used for evaluation. The results are shown in Table 3. The ΔTime is defined as follows (2):

$$\Delta\text{Time} = \frac{\text{Time}_{HM} - \text{Time}_{pro}}{\text{Time}_{pro}} \times 100\% \tag{2}$$

where Time_{HM} denotes the time consuming of the default HM 10.0 and Time_{pro} represents the time consumed by the proposed method. The RMD and RDO in the original HEVC program in Table 2 are significantly reduced in number compared to our proposed algorithm, with an average reduction ratio of 42.3%. This proves that our algorithm can achieve high efficiency coding with less coding time. As shown in Table 3, on average, the proposed algorithm can achieve about 28.8% encoder time saving for all intra encoding.

Table 3. BD-rate loss and time reduction of proposed algorithm.

Class	Sequence	Y	U	V	ΔTime%
A	Traffic	0.2%	0.1%	0.1%	27.8%
(2560 × 1600)	PeopleOnStreet	0.4%	−0.2%	0.1%	29.1%
B	BasketballDrive	0.6%	−0.3%	−0.4%	28.8%
(1920 × 1080)	BQTerrace	0.7%	0.1%	0.1%	29.5%
C	BasketballDrill	0.5%	−0.3%	−0.2%	27.9%
(832 × 480)	RaceHorces	0.3%	0.1%	0.2%	29.3%
D	BQSquare	0.8%	−0.2%	−0.3%	28.4%
(416 × 240)	BlowingBubble	0.7%	−0.1%	−0.2%	28.2%
E	Vidyol	1.0%	-0.1%	−0.2%	29.6%
(1280 × 720)	KristenAndSara	0.8%	−0.1%	−0.2%	29.4%
Ave.		0.6%	−0.1%	−0.1%	28.8%

5 Conclusion

This paper proposes a fast mode decision algorithm for HEVC intra prediction. Depth prediction and adjusted angle prediction are used to skip some unnecessary pattern searches to reduce the computational burden. The experimental results show that under the frame configuration of HM10.0, the method can save about 28.8% of the encoder time.

Acknowledgement. This work was supported by Xi'an Science and Technology Bureau Project, under Grant 201805040YD18CG24(1) and the Department of Education Shaanxi Province, China, under Grant 2013JK1023.

References

1. Bross, B., Han, W.-J.: High efficiency video coding (HEVC) text specification draft 6. (JCT-VC) ITU-T VCEG and ISO/IEC MPEG, San Jose, CA, USA, April 2012. Doc
2. Wiegand, T., Sullivan, G.J., Bjontegaard, G., Luthra, A.: Overview of the H.264/AVC video coding standard. IEEE Trans. Circuits Syst. Video Technol. **13**(7), 560–576 (2003). https://doi.org/10.1109/TCSVT.2003.815165
3. Lainema, J., Bossen, F., Han, W.J., Min, J., Ugur, K.: Intra coding of the HEVC standard. IEEE Trans. Circuits Syst. Video Technol. **22**(12), 1792–1801 (2012). https://doi.org/10.1109/TCSVT.2012.2221525
4. Garcia, R., Kalva, H.: Subjective evaluation of HEVC and AVC/H.264 in mobile environments. IEEE Trans. Consum. Electron. **60**(1), 116–123 (2014). https://doi.org/10.1109/TCE.2014.6780933
5. Zhou, C., Zhou, F., Chen, Y.: Spatio-temporal correlation based fast coding unit depth decision for high efficiency video coding. J. Electron. Imaging **22**(4) (2013). https://doi.org/10.1117/1.JEI.22.4.043001
6. Wang, Y., Fan, X., Zhao, L., Ma, S., Zhao, D., Gao, W.: A fast intra coding algorithm for HEVC. In: 2014 IEEE International Conference on Image Processing (ICIP), pp. 4117–4121. IEEE (2014). https://doi.org/10.1109/ICIP.2014.7025836
7. Ozcan, E., Kalali, E., Adibelli, Y., Hamzaoglu, I.: A computation and energy reduction technique for HEVC intra mode decision. IEEE Trans. Consum. Electron. **60**(4), 745–753 (2014). https://doi.org/10.1109/TCE.2014.7027351
8. Shen, X., Yu, L., Chen, J.: Fast coding unit size selection for HEVC based on Bayesian decision rule. In: Picture Coding Symposium (PCS), 2012, pp. 453–456. IEEE (2012)
9. Teng, S.-W., Hang, H.-M., Chen, Y.-F.: Fast mode decision algorithm for residual quadtree coding in HEVC. In: 2011 IEEE Visual Communications and Image Processing (VCIP), pp. 1–4. IEEE (2011). https://doi.org/10.1109/VCIP.2011.6116062
10. Joint Collaborative Team on Video Coding Reference Software, ver. HM 10.0. https://hevc.hhi.fraunhofer.de/svn/svnHEVCSoftware/
11. Wang, Y.L., Shen, J.X., Liao, W.H., et al.: Automatic fundus images mosaic based on sift feature. In: International Congress on Image and Signal Processing, Yantai, China, 16–18 October 2010 pp. 2747–2751 (2010)
12. Bjontegaard, G.: Calculation of average PSNR differences between RD Curves. Doc. VCEG-M33 ITU-T Q6/16, Austin, TX, USA, 2–4 April 2001 (2001)

An Analysis and Research of Network Log Based on Hadoop

Wenqing Wang(✉), Xiaolong Niu, Chunjie Yang, Hongbo Kang, Zhentong Chen, and Yuchen Wang

School of Automation, Xi'an University of Posts and Telecommunications, Xi'an 710121, Shaanxi, China
WWQ@xupt.edu.cn, 1175296815@qq.com

Abstract. With the rapid development of the internet technology, we have entered the era of big data, people product massive amount of data on the internet. Through the analysis and data mining of Web logs, we can dig out valuable information such as user's behavior preferences. But handling massive amounts of data, the traditional single machine can no longer meet the requirements. With the continuous development of big data technology, massive Hadoop log data can be analyzed through the framework of big data. In this paper, the Hadoop large data platform is built, the MapReduce programming model is used to preprocess the network log, and the Hive data warehouse is used to analyze the processed data in multi dimension. The analysis results have good guiding significance for mastering the user browsing behavior, promoting the promotion effect, optimizing the structure and experience of the website.

Keywords: Big data · Data mining · Hadoop · MapReduce
Hive data warehouse

1 Introduction

The rapid development of Internet technology has caused the amount of information carried by the web to show an explosive growth trend. Therefore, the amount of data in web logs bigger than before and these massive amounts of network log data contain a lot of valuable information, how to store and process large-scale data becomes a new challenge. With the development of cloud computing big data technology, the transformation from the original single-machine processing data to the multi-node processing of data in the network has become a new solution. Hadoop is an open source computing framework that enables massive amounts of data to be distributed across multiple computers.

Based on this, this paper is based on the Hadoop big data platform, according to the specific needs of data analysis and mining, First, data cleaning is performed on the network log to filter data that does not conform to rules and is meaningless, redundant, or abnormal, and the data that meets the requirements is processed. Multi-dimensional analysis of data through Hive data warehouse, which analyses of user's key time indicators such as website time, click volume, user volume, hotspot page, and link to the link, and use Sqoop to export the analyzed results to MySQL for later data display.

P. Krömer et al. (Eds.): ECC 2018, AISC 891, pp. 541–548, 2019.
https://doi.org/10.1007/978-3-030-03766-6_61

By analyzing the results of the display, it provides a reference for mastering the user's browsing behavior, improving the promotion effect, and optimizing the structure and experience of the website.

2 Main Related Technology Introduction

2.1 Introduction to Hadoop

Hadoop [1] is an open source software framework for the Java language implementation of the Apache Software Foundation [2], Hadoop is both a software and a big data ecosystem. Hadoop provides a programming framework that allows massive amounts of data to be distributed across multiple computers. Their cores components are the HDFS [3] distributed file system (for distributed storage of massive amounts of data), YARN (job scheduling and cluster task resource management), MapReduce [4] (distributed computing programming framework). Hadoop has the following advantages.

1. Capacity expansion: Hadoop is the storage and calculation of data between computer clusters, the number of calculations in the cluster can be increased to thousands.
2. Low cost: The Hadoop cluster can be built using a general server, so that the cost is very low.
3. High efficiency: Distributed data is distributed to different machines for distributed calculation, which makes the calculation speed greatly improved.
4. Reliability: Hadoop clusters have a copy mechanism that copies the data that users need to process to different machines to ensure that data is not lost, and can be automatically redeployed and calculated if the task fails during the calculation.

As the lowest-level storage service, the HDFS distributed file system mainly solves the storage problem of massive data. The file can be located in a unified namespace. The files in HDFS are physically stored in blocks. The files in the HDFS are physically stored in blocks and adopt the master/slave master-slave architecture [5]. Usually, an HDFS cluster has multiple Name Node master nodes and multiple data node slave nodes. The name node is responsible for maintaining the directory number structure in the file and the information of each file block. The Data Node is responsible for file block storage [6], and each one the blocks have copies, and the two nodes perform their own functions to jointly complete the distributed storage of data. The Hadoop distributed file system is shown in Fig. 1.

MapReduce is a distributed programming framework for computing programs. The core idea is divided and conquers [7]. The core function is to integrate the user-written business logic code and its own default components into a complete distributed computing program, which runs concurrently on the Hadoop cluster.

The input data type of the MapReduce parallel computing framework is in the form of <key, value> key-value pairs, the key is the offset of reading a row of data and the value is a line of content of the read file. Each time a key-value pair is read, the user-defined Map method is called once, and the logical operation is performed in the Map method and output to the cache as a key-value pair. Partition and sort all key-value

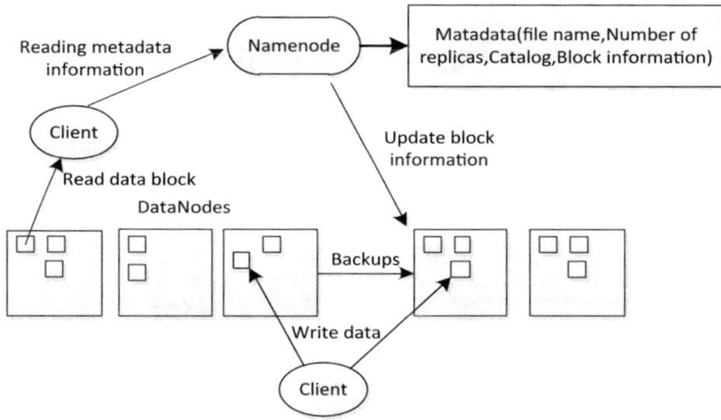

Fig. 1. Hadoop distributed file system architecture.

pairs in the cache, then generate temporary files and finally merge the temporary files to form multiple temporary files.

The MapReduce framework consists of the control node Job Tracker and the slave node Task Tracker [8]. The task submitted by the user is called Job. At the same time, The Job Tracker receives the task submitted from the client and initializes it and allocates the required resources for the task, and schedules the execution of the task. The Job Tracker monitoring the entire execution process and managing all task nodes. The Task Tracker is usually deployed on the unified machine with the data node to receive tasks assigned by the master node and to be responsible for the specific execution of the task. If a Task Tracker is down, Job Tracker will transfer the task of the machine to another idle Task Track to re-run, so Hadoop can build a highly reliable distributed computing framework [9].

2.2 Data Warehouse Hive

Hive is a data warehouse tool based on the Hadoop file system, which can map structured data files into a database table and provide SQL query functions, the essence of Hive is to convert SQL statements into MapReduce programs, the main purpose is to use offline analysis. Hive uses HDFS to store data, uses MapReduce to query and to analyze data is more efficient than uses MapReduce directly.

3 Log Analysis Overall Architecture Design

In this paper, three PCs are used to build a Hadoop distributed platform. One PC is called Name Node and Job Tracker, the main node of Hive cluster also runs on this machine. The other two machines are used as Data Node and Task Tracker, and the slave nodes of Hive also run on these two machines. The whole system consists of a log collection module, a data preprocessing module, an indicator analysis module, data storage and a display module. The log collection module uploads the collected user log data to the HDFS distributed file system. The MapReduce distributed computing framework reads data from the HDFS and performs cleaning and format conversion. It saves the preprocessed data to the Hive data warehouse for metric analysis. and the metric data analyzed by Hive is exported to the MySQL database through Sqoop for display on the web side. The system architecture diagram is shown in Fig. 2.

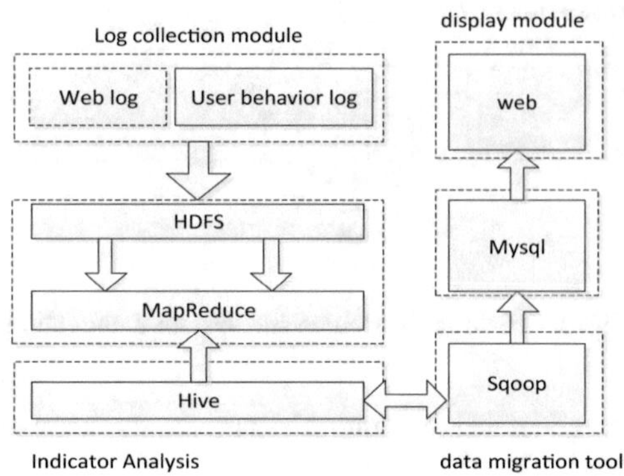

Fig. 2. System architecture diagram.

3.1 Experimental Data and Content

The data used in this article is derived from log data on a blog server. Through the processing of blog log data, then performing metric analysis and user behavior mining, and providing the results to the operator for decision-making operation. The data selection blog one day log data is analyzed, and the sample data is shown in Table 1.

Table 1. Data format description

Field	Example
Addr	222.44.41.33
TimeStr	18/Sep/2017:07:00:31
Request	/wp-content/themes/Silesia/images/bullets/5.gif HTTP/1.1
Referer	http://blog.fens.me/wp-content/themes/silesia/style.css
Agent	Mozilla/5.0 (X11;Ubuntu;Linux x86_64;rv:20.0)Gecko/20100101 Firefox/20.0

3.2 Data Preprocessing

The Hadoop programming framework MapReduce is used to clean the original access logs uploaded to HDFS; it filters out non-compliant and meaningless data and format the data. The main process is divided into two phases: Map phase and Reduce phase.

Map stage: In the map phase, the default read data component LineRecordReader is used to read a row of log file data, and the data of each row is parsed into <key, value>, the key is the offset position of the read text content, and the value is the text content of the line. In the map function, the text is divided according to the pre-set separator, the real PV (Pageview) request is filtered out, the conversion time format is filled, the default value is filled in the missing field, and the normal data is added to the field valid abnormal data to add the field invalid A useful URL classifier for loading a website from an external configuration file, it used to filter the log data and splicing multiple fields into a complete record and outputting it to a file. The Map stage process is shown in Fig. 3(a).

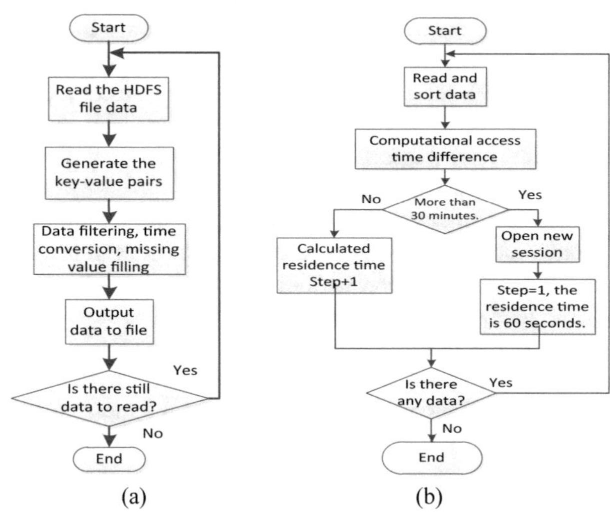

Fig. 3. (a) Map stage process, (b) Reduce stage process.

Reduce phase: In the Reduce phase, the session needs to be identified and an IP address is used to identify a user. First, the time of all the access records of a user is sorted, so that the time difference between each access URL and the time spent on the current web page can be calculated. Then sort each user's access records by time difference and add a new field step to mark the current operation as the first step. If the time difference between the two access records exceeds 30 min, the current user is considered to have opened a new session and resets step to 1, which represents the first operation of the current session. In order to distinguish different sessions, when a new session is generated, a field SessionId is usually added and a randomly generated string is used as the value of the field, so that the page click flow model Pageview is sorted out. The page click flow model can be further processed to obtain the click flow model. The click flow model is shown in Table 2. The Reduce stage process is shown in Fig. 3(b).

Table 2. Page click flow model

Session	Address	Interview time	Access page	Resident time	Steps
SessionId	Addr	TimeStr	Request	StayTime	Step

3.3 Indicator Analysis

The metric analysis uses the Hive data warehouse tool, Hive provides HQL similar to the SQL query language, which essentially converts the HQL language into a MapReduce program.

Pageview: Generally speaking, the PV value is recorded by the user every time one website page is opened. The user repeatedly opens the same page PV multiple times. The popular explanation is the total number of times the page has been loaded.

Unique Pageview: The number of unique users visiting the website within 1 day (based on browser cookies), the same visitor visited the website multiple times a day only once.

Hotspots: The number of pages visited during the day is the most, and you can analyze which content of the site is preferred by users.

Visiting path: The last link to visit a website or page, through which the path can be analyzed to find out which link or website comes in, so that you can conduct business promotion such as website promotion on these websites.

New users: How many of the users in the site visited the site for the first time and analyzes how attractive the site is to users.

Independent visitors: You can count the number of users in these three dimensions based on the time, day, and month dimensions. Count the most popular pages of the day Top10.

4 Results and Analysis

Through six sets of different size log file data in the stand-alone model and the cluster model execution time comparison, when the size of the data to be processed is 1 MB, the processing time is basically the same, even the cluster processing time is more longer than the single processing time, this is because when the cluster processes data, it first starts the task and reads the data, In this process, the overhead time of the task is often greater than the actual data processing time, thus causing this phenomenon. When the processed log reaches 15 MB or even reaches 30 MB, the advantage of your parallel computing is revealed. The time for processing data by the cluster is significantly lower than the time for processing data by a single machine. The execution time comparison is shown in Fig. 4.

Through the analysis of Hive's indicators, the number of page views from 2017-09-18 to 2017-09-24 is shown in Fig. 5.

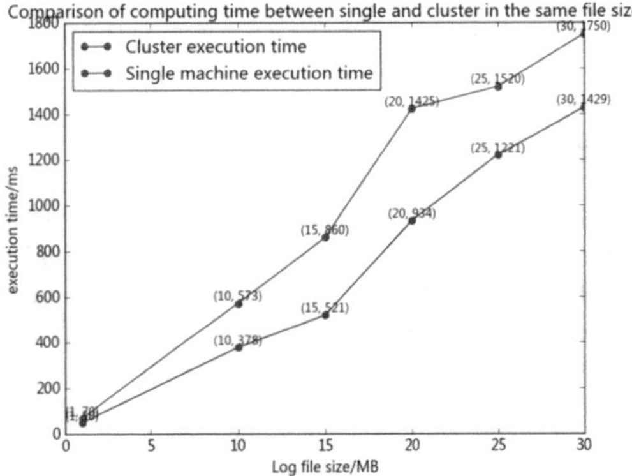

Fig. 4. Comparison of computing time between single and cluster in the same size.

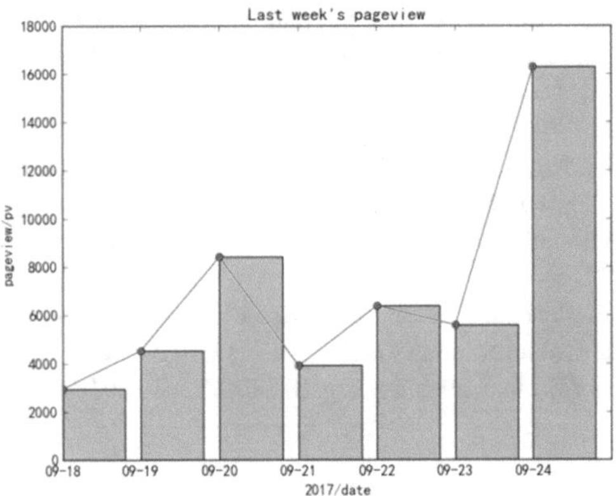

Fig. 5. Last week's pageview.

5 Conclusion

In view of the low efficiency of data storage and data processing in the traditional single machine when dealing with massive network log data, this paper designs a massive log data processing model based on Hadoop, which makes full use of the advantages of big data computing framework to process massive data in parallel and to solve the problem of the single-mode mode. The Hadoop programming framework is used to clean and preprocess the web log data, and the key indicators required by the administrator are distributed. At the same time, the corresponding indicators are visually displayed on the web side, and the analysis results are used to understand the user browsing behavior, improve the promotion effect, and optimize. Website structure and experience have good guiding significance.

References

1. Edwards, M.F., Rambani, A.S., Zhu, Y.T., et al.: Design of hadoop-based framework for analytics of large synchrophasor datasets. Procedia Comput. Sci. **12**(4), 254–258 (2012)
2. Chansler, R., Kuang, H., Radia, S., Shvachko, K.: The hadoop distributed file system. In: 2010 IEEE 26th Symposium on Mass Storage Systems and Technologies (MSST 2010) (MSST), Incline Village, NV, pp. 1–10 (2010). https://doi.org/10.1109/msst.2010.5496972
3. Dean, J.F., Ghemawat, S.S.: MapReduce: simplified data processing on large clusters. ACM **51**(1), 107–113 (2008). https://doi.org/10.1145/1327452.1327492
4. Kotiyal, B.F., Kumar, A.S., Pant, B.T., et al.: Big data: mining of log file through hadoop. In: International Conference on Human Computer Interactions, pp. 1–7. IEEE (2014). https://doi.org/10.1109/ich-ci-ieee.2013.6887797
5. Wang, C.H., Tsai, C.T., Fan, C.C., et al.: A hadoop based weblog analysis system (2014). https://doi.org/10.1109/u-media.2014.9
6. Suguna, S.F., Vithya, M.S., Eunaicy, J.I.C.: Big data analysis in e-commerce system using Hadoop MapReduce. In: International Conference on Inventive Computation Technologies, pp. 1–6 (2017). https://doi.org/10.1109/inventive.2016.7824798
7. Du, J.F., Zhang, Z.S., Zhao, C.T.: Analysis on the digging of social network based on user search behavior. Int. J. Smart Home **10**(5), 297–304 (2016). https://doi.org/10.14257/ij-sh.2016.10.5.27
8. Dewangan, S K, Pandey, S., Verma, T.: A distributed framework for event log analysis using MapReduce. In: International Conference on Advanced Communication Control and Computing Technologies, pp. 503–506. IEEE (2017). https://doi.org/10.1109/icaccct.2016.7831690
9. He, G.F., Ren, S.S., Yu, D.T., et al.: Analysis of enterprise user behavior on hadoop. In: Sixth International Conference on Intelligent Human-Machine Systems and Cybernetics, pp. 230–233. IEEE (2014). https://doi.org/10.1109/ihmsc.2014.158

Deep Learnings and Its Applications in All Area

One-Day Building Cooling Load Prediction Based on Bidirectional Recurrent Neural Network

Ye Xia(✉)

Fujian Province University Key Laboratory of New Energy and Energy-Saving in Building, Fujian University of Technology, Fuzhou 350108, China
2751808409@qq.com

Abstract. Short-term building cooling load prediction is very important for building energy management tasks. Traditional way relies on physical principles. Due to the nonlinearity of the features of the data, it is a challenge for prediction. This work applies the Bidirectional Recurrent Neural Network (BRNNs) in prediction of 24-h ahead building cooling load profiles. The results show that BRNNs have good performance in prediction on building cooling load prediction. The mode can predict the building cooling load profiles effectively.

Keywords: Building cooling load · Short-term prediction · BRNNs

1 Introduction

At present, energy conservation and emission reduction is a trend. The energy consumption of buildings accounts for the majority of the total energy consumption of the whole society. The greater the cooling load of air conditioning, the greater the energy consumption of air conditioning system. Building energy management first focuses on the prediction of air-conditioning cooling load.

AE Ben-Nakhi, MA Mahmoud designed General regression neural networks (GRNN) and trained it to investigate the feasibility of using this technology to optimize HVAC thermal energy storage in public buildings as well as office buildings [1]. Zhijian Hou, Zhiwei Lian et al. presented a novel method integrating rough sets (RS) theory and an artificial neural network (ANN) based on data-fusion technique to forecast an air-conditioning load [2]. Simon S.K.Kwok, Richard K.K.Yuen et al. discussed the use of the multi-layer perceptron (MLP) model, one of the artificial neural network (ANN) models widely adopted in engineering applications, to estimate the cooling load of a building [3]. Cheng Fan, Fu Xiao et al. investigated the potential of one of the most promising techniques in advanced data analytics, i.e., deep learning, in predicting 24-h ahead building cooling load profiles [4]. Yongjun Sun, Shengwei Wang et al. developed a simplified online cooling load prediction method for a super high-rise building in Hong Kong [5].

This paper attempts to apply the intelligent prediction method BRNNs [6] to the cooling load prediction of buildings.

P. Krömer et al. (Eds.): ECC 2018, AISC 891, pp. 551–557, 2019.
https://doi.org/10.1007/978-3-030-03766-6_62

2 Bidirectional Recurrent Neural Network (BRNNs)

Given an input sequence x = (x_1, …, x_T), a standard recurrent neural network(RNN) computes the hidden layer sequence h = (h_1, …, h_T) and output layer sequence y = (y_1, …, y_T) by iterating the following equations from t = 1 to T:

$$h_t = \Phi(W_{xh}x_t + W_{hh}h_{t-1} + b_h) \tag{1}$$

$$y_t = W_{hy}h_t + b_y \tag{2}$$

Where the W is weight matrices, the b is bias vectors and Φ is the hidden layer function. Φ is usually a sigmoid function with elementwise. Long Short-Term Memory (LSTM) architecture uses purpose-built memory cells to store information and is better at finding and exploiting long range context. Figure 1 [7] illustrates a single LSTM memory cell. In this work, Φ is implemented by the following composite equations:

$$i_t = \sigma(W_{xi}x_t + W_{hi}h_{t-1} + W_{ci}c_{t-1} + b_i) \tag{3}$$

$$f_t = \sigma\left(W_{xf}x_t + W_{hf}h_{t-1} + W_{cf}c_{t-1} + b_f\right) \tag{4}$$

$$c_t = fc_{t-1} + i_t \tan h(W_{xc}x_t + W_{hc}h_{t-1} + b_c) \tag{5}$$

$$o_t = \sigma(W_{xo}x_t + W_{ho}h_{t-1} + W_{co}c_t + b_c) \tag{6}$$

$$h_t = o_t \tan h(c_t) \tag{7}$$

where σ denotes the sigmoid function, and i, f, o and c are the same size as the hidden vector h and respectively the input gate, forget gate, output gate and cell activation vectors.

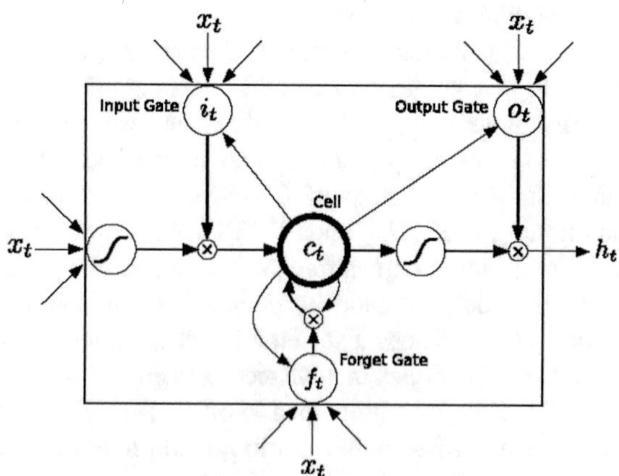

Fig. 1. Long short-term memory cell

One shortcoming of conventional RNNs is that they only make use of previous context. Bidirectional RNNs (BRNNs) [8] are proposed by applying the data in both direction with two separate hidden layers, which are then fed forwards to the same output layer. As illustrated in Fig. 2 [7], a BRNN have the forward hidden sequence $\overrightarrow{h}$, the backward hidden sequence $\overleftarrow{h}$ and the output sequence y by iterating the backward layer from t = T to 1, the forward layer from t = 1 to T and then updating the output layer:

$$\overrightarrow{h}_t = \Phi\left(W_{x\overrightarrow{h}} x_t + W_{\overrightarrow{h}\overrightarrow{h}} \overrightarrow{h}_{t-1} + b_{\overrightarrow{h}}\right) \tag{8}$$

$$\overleftarrow{h}_t = \Phi\left(W_{x\overleftarrow{h}} x_t + W_{\overleftarrow{h}\overleftarrow{h}} \overleftarrow{h}_{t-1} + b_{\overleftarrow{h}}\right) \tag{9}$$

$$y_t = W_{\overrightarrow{h}y} \overrightarrow{h}_t + W_{\overleftarrow{h}y} \overleftarrow{h}_t + b_y \tag{10}$$

The hybrid BRNNS which combine BRNNS with LSTM can access long-range context in both directions.

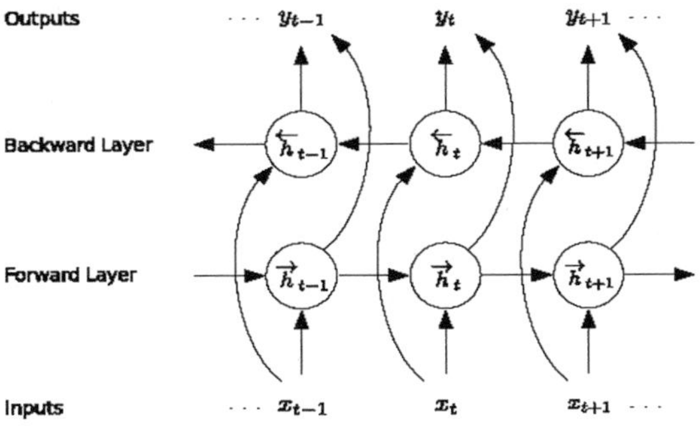

Fig. 2. Bidirectional RNN

3 Evaluation

Among the factors that influence the cold load of buildings, there are two major factors, one is the building occupancy, and the other is the outdoor conditions. The building occupancy is usually fixed and correlated with time. Outdoor conditions can be well-described using outdoor dry-bulb temperature, outdoor relative humidity, wind direction and speed, outdoor luminance, etc. Basic features set contains five time variables (i.e., Month, Day, Hour, Minute and Day type), add the outdoor temperature and the outdoor relative humidity at time T. These seven features [4] are taken as model inputs to predict building cooling load at time T. Additional information of building cooling load, outdoor temperature and RH during the past 24-h are added. After feature extraction, there are 121 variables as the input training data.

The entire dataset is divided into training, and test data with proportions of 70%, and 30% respectively. The model hyper parameters are determined through experiments.

Hidden size is 200, training step is set to 4000, and batch size set to 200. Minimizing cross-entropy error is used as the optimization objective during the BRRNNs model training procedure. The cross-entropy indicates the distance between the probability distributions of network outputs and target labels. The cross-entropy error is defined in the following:

$$Cross - entropy\ Err(CEE) = -\sum_{i} \widehat{y_i} \log(y_i) \tag{11}$$

Where $\widehat{y_i}$ is the predicted probability of value of class i, and y_i is the true probability for that class. The activation function uses Tanh and sigmoid).

The prediction performance is evaluated using three metrics which is defined in Eqs. (12)–(14). While y_k and $\widehat{y_k}$ are the actual and perdition at time k respectively). MAE and RMSE are scale-dependent metrics is a quantification of the prediction error. Previous studies use CV-RMSE for model evaluation, it is specified that if CV-RMSE is below 30% when using hourly data, the model is sufficiently close to physical reality for engineering purpose [9].

$$RMSE = \sqrt{\frac{\sum_{k=1}^{n} (y_k - \widehat{y_k})^2}{n}} \tag{12}$$

$$MAE = \sqrt{\frac{\sum_{K=1}^{N} |y_k - \widehat{y}_k|}{n}} \tag{13}$$

$$CV - RMSE - \frac{RMSE}{MEAN(y_k)} \tag{14}$$

The accuracy and cross-entropy is shown in Fig. 3:

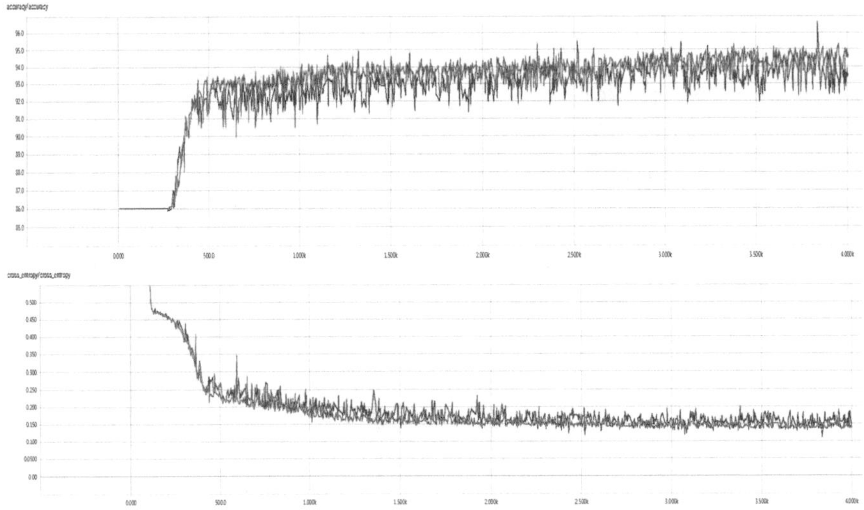

Fig. 3. The accuracy and cross-entropy of the model

From Fig. 3, it illustration that the cross-entropy of train and test are very close and imply that the model has good generalization. The test accuracy can reach 94%. It can predict unseen data effectively.

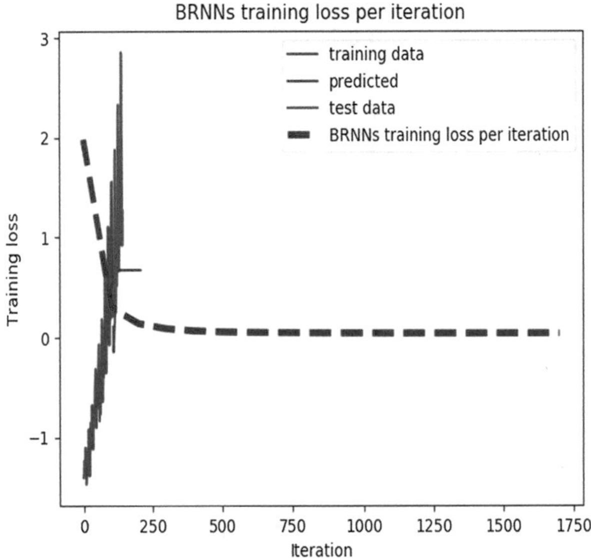

Fig. 4. BRNNS training loss per iteration

The training loss per iteration of the model is shown as Fig. 4. From Fig. 4, It shows that the loss is close 0 after the 250th iteration step. The test loss per iteration is also shown in Fig. 6, which shows the test loss is below 0.25.

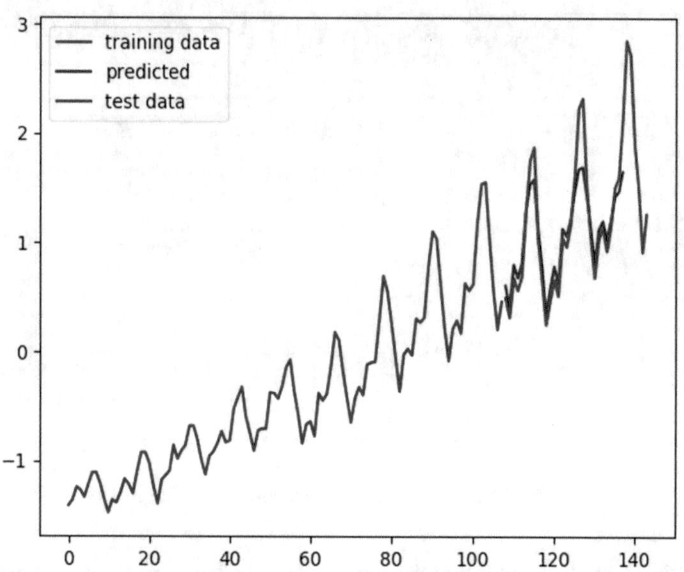

Fig. 5. The cooling load prediction using BRNNs

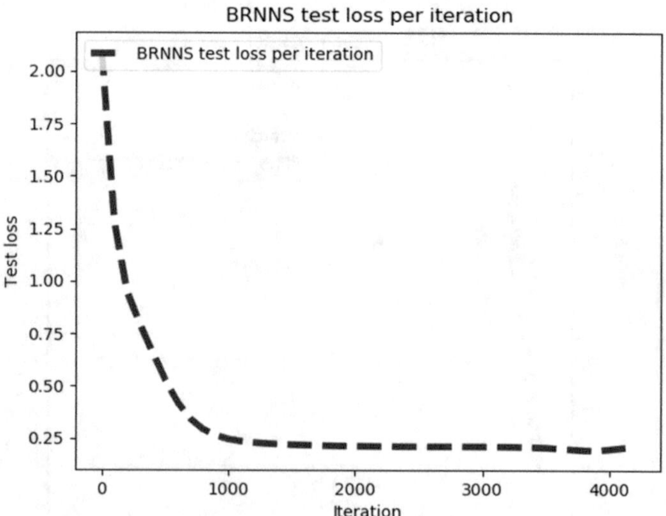

Fig. 6. The test loss per iteration of the model BRNNS

The cooling load prediction result is shown in Fig. 5; the red line is the predicted line which is fit the test data perfectly. It shows the model can predict the building cooling load effectively.

4 Summary

This paper investigates the BRNNs to predict the building cooling load. The research results indicate that BRNN can enhance prediction performance due to the BRNNs can make use both of the previous data and future data context. The BRNNs model proposed in this work can achieve accurate and reliable prediction of one-day ahead building cooling load profiles. It is the foundation for many building operation management tasks. It can be used to optimal control strategies, as well as fault detection and diagnosis methods.

References

1. Ben-Nakhi, A.E., Mahmoud, M.A.: Cooling load prediction for buildings using general regression neural networks. Energy Convers. Manag. **45**, 2127–2141 (2004)
2. Hou, Z., Lian, Z., et al.: Cooling-load prediction by the combination of rough set theory and an artificial neural-network based on data-fusion technique. Appl. Energy **83**, 1033–1046 (2006)
3. Kwok, S.S.K., Yuen, R.K.K., et al.: An intelligent approach to assessing the effect of building occupancy on building cooling load prediction. Build. Environ. **46**, 1681–1690 (2011)
4. Fan, C., Xiao, F., et al.: A short-term building cooling load prediction method using deep learning algorithms. Appl. Energy **195**, 222–233 (2017)
5. Sun, Y., Wang, S., et al.: Development and validation of a simplified online cooling load prediction strategy for a super high-rise building in Hong Kong. Energy Convers. Manag. **68**, 20–27 (2013)
6. Schuster, M., Paliwal, K.K.: Bidirectional recurrent neural networks. Signal Process. **45**, 2673–2681 (1997)
7. Graves, A., Mohamed, A., et al.: Speech Recognition with Deep Recurrent Neural Networks. https://arxiv.org/abs/1303.5778
8. Su, Y., Huang, Y., et al.: On Extended Long Short-term Memory and Dependent Bidirectional Recurrent Neural Network. https://arxiv.org/abs/1803.01686
9. Reddy, T.A., Maor, I., et al.: Calibrating detailed building energy simulation programs with measured data-Part II: application to three case study office buildings (RP-1051). HVAC&R **13**, 221–241 (2007)

Prediction of Electrical Energy Output for Combined Cycle Power Plant with Different Regression Models

Zhihui Chen[1,2], Fumin Zou[2(✉)], Lyuchao Liao[1,2], Siqi Gao[1,2], Meirun Zhang[1,2], and Jie Chun[1,2]

[1] Beidou Navigation and Smart Traffic Innovation Center of Fujian Province, Fujian University of Technology, Fuzhou 350118, Fujian, China

[2] Fujian Key Laboratory for Automotive Electronics and Electric Drive, Fujian University of Technology, Fuzhou 350118, Fujian, China

achao@fjut.edu.cn

Abstract. Prediction of electrical output is beneficial to energy saving and financial interests. The electrical energy output of Combined Cycle Power Plant (CCPP) is predicted with four different models in this paper. The analysis reveals that some attributes in CCPP have a high linear relation with the energy output but the other has multi-collinearity. Therefore, the output is decided by attributes in a specific combination which means the output can be precisely predicted by a suitable models. We input four attributes to train models with 5×2 cross-validation for tuning hyper-parameters, and four machine learning methods are compared with Multi-linear Regression, Support Vector Regression, Backward propagation neural network and CART based algorithm XGBoost. The result shows that XGBoost has the best fitting in output with the lowest variance and bias, which is based on boosting algorithm and ensemble learning with a root mean square error of 2.752, mean absolute error 1.938 and a R^2 of 0.9748.

Keywords: Combined cycle power plant (CCPP) · Electrical output prediction
Machine learning · XGBoost

1 Introduction

With the rapid development of economy and industry, improving efficiency of generating and utilization, decreasing pollution is becoming a main mission in nowadays society. A kind of new ecosystem is emerging in future days, Combined cycle power plant (CCPP) is a structure of thermal power plant which largely improve the heat efficiency and effectively solve the pollution problem [1]. For the CCPP contains a vast energy related to safety of staff and its economy benefit, the energy output of CCPP should be accurately and precisely predicted, the algorithms need to be designed in advanced to increase the safety and economy degree.

In the past time, conventional power prediction models are based on environment detection data and white-box pattern to predict the behavior and system but in complicated systems which is hard to complete. Recently, as the theory and machine

P. Krömer et al. (Eds.): ECC 2018, AISC 891, pp. 558–566, 2019.
https://doi.org/10.1007/978-3-030-03766-6_63

learning technology raised, the predicting problem can be easily solved by well trained model. The machine learning algorithm can analyze power behavior of alliances or buildings, which including building consumption, transformer output, individual household consumption, PV system, distributed generators output, vehicles trade, climate trend, etc. Pinar et al. used 17 different regression models to predict full load electrical power output, the experiment use 5×2 folds cross validation to compare performance of different models and use 1 to 4 attributes with different combination to modeling, but the study did not give the portion of train set data and the hyper-parameters of each regression models. [2] Aydin et al. evaluated seven models in which k-NN, Linear regression and RANSAC regressions achieved the best performance with a error rate 0.59%, but the experiment simply divided the data into training set and test set which unable to fully evaluate the model performance [3]. In paper [4], Guo proposed a three classes decision tree to predict the behavior of CCPP which has a 97% accuracy, due to structure of the model, the model works worse in high precision.

2 Data Description

The dataset used in this paper has 5 attributes which including 9568 instances collected from a combined cycle power plant over 72 months (2006–2011), when the power plant was in a full load working condition .8568 instances were used to training models and the rest are testing data. Features consist of hourly average ambient variables Temperature (T), Ambient pressure (AP), Relative humidity (RH), and Exhaust Vacuum (V). In the following Table 1 shows all variables description of analysis.

Table 1. All variables in this paper

Variables	Description	Unit
AT	Temperature range 1.81to 37.11	°C
AP	Ambient pressure range 992.89 to 1033.30	Milibar
RH	Relative humidity range 25.56 to 100.16	%
V	Exhaust vacuum in teh range 25.36 to 81.56	cm Hg
EP	Net hourly electrical energy output range 420.26 to 495.76	MW

3 Data Analysis

Dataset was created by Namik Kemal University and by Heysem Kaya, Department of Computer Engineering, Boğaziçi University, Turkey from UCI machine learning repository https://archive.ics.uci.edu/ml/datasets/Combined+Cycle+Power+Plant [5].

Here we calculate the Pearson coefficient (also known as Pearson product-moment correlation coefficient) which is a kind of linear coefficient of association. Pearson correlation coefficient is a type of statistical magnitude for reflect linear correlation degree of two variables, as shown in the following formula:

$$r = \frac{1}{n-1}\sum\nolimits_{n-1}^{n}\left(\frac{X_i - \bar{X}}{s_X}\right)\left(\frac{Y_i - \bar{Y}}{s_Y}\right) \tag{1}$$

We calculate Pearon correlation coefficients between each attributes the relation revealed in the following figure.

As shown in Figs. 1 and 2, we found that the attributes of influencing PE has a correlation at least 0.39, feature AT which represents ambient temperature has most negative correlation while the AP has the most positive correlation. The effect of Pearson correlation will be reflected in the building of models.

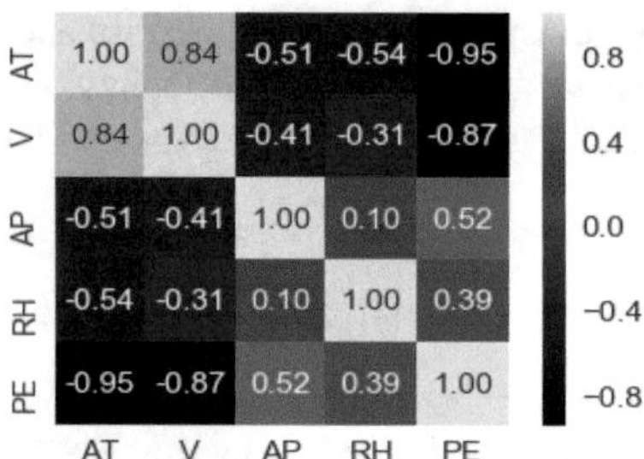

Fig. 1. Pearson correlation

4 Machine Learning Models and Experiment

4.1 Multiple Linear Regression Model

Multiple Linear Regression (MLR) is used for building up a model which has multiple input based on attributes and one output. On the other hand, MLR has a high readability. The significance of the attributes can be easily interpreted in the trained models. The multiple linear regression uses multi-dimension variables X_b to predict output Y_o as shown in Eq. 2 [6]:

$$Y_0 = w_0 + w_1X_1 + w_2X_2 + \ldots + w_nX_n \tag{2}$$

In the equation above, given the w_i as a column, the best fitting vector β can be express as (least square method):

$$\beta = \left(X^TX\right)^{-1}X^TY \tag{3}$$

Elements in vector β interpret the significance of each variables to the output (positive coefficient represent a positive effect to the result vice versa).

After training models with different loss functions (all models used grid searching to find out best hyper-parameters), the result is given in the follow Table 2. We use RMSE and R^2 to evaluate the models performance. As can be seen from the table,

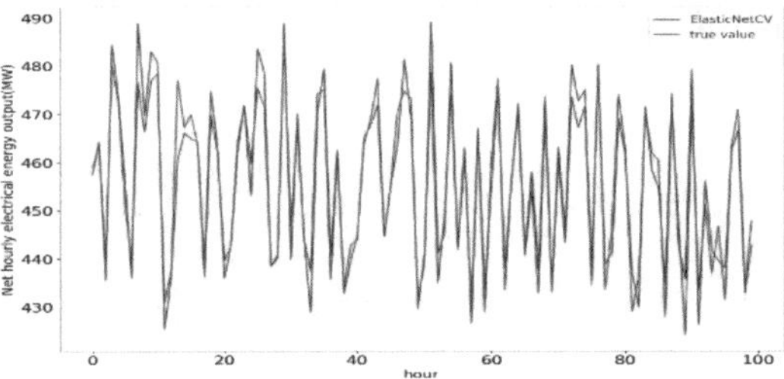

Fig. 2. Prediction of multiple regression with ElasticNet loss function

different loss functions make tiny differences to the predicting result, ElasticNet performing best in multiple linear regression in this dataset.

Table 2. RMSE and R^2 of five loss function

Models	RMSE	R^2
Least square method	4.559	0.9284359
ElasticNet	4.559	0.9284307

The net hourly electrical energy output (EP) mainly influenced by ambient condition factors, the relationship of predicting value to attributes with ElasticNet loss function can be represented as the following equation:

$$Y_o = 457.24 - 1.96X_1 - 0.23X_2 + 0.06X_3 - 0.16X_4 \tag{4}$$

Input variables $X_1X_2X_3X_4$ are attributes AT V AP RH and the output Y_o regards as variable EP. From the conclusion of Pearson coefficient, the coefficient of the model input variables related to the importance of itself to the output variable Y_o. As we can see in the equation above, the conclusion is similar to the Pearson analysis, but X_3X_4 contrary to the previews conclusion, so it is likely to exist multicollinearity between output and attribute.

4.2 Support Vector Regression

Support vector regression (SVR) is used for fit a continuous series which is a popular algorithm in machine learning. It is algorithm based on support vector machines (SVM) which use multiple data X to classify output Y. In SVR, the training data is features with high dimensions, using the training data to build up models for prediction. With the spacing by kernel functions, the features in the high dimensions can be easily computed and understanded. With the compared to the expression of linear regression, the SVR allows machines to fitting a curve instead a line.

In $\varepsilon - SVR$ (Margin soft vector regression), the goal in SVR is to find out a line or curve to best fit the output in a minimize error condition, given the goal function:

$f(x) = W^T x + b$, the constraint condition with epsilon insensitive factor can be express as following condition [7]:

$$\min_{w,b,\xi,\xi*} \frac{1}{2} w^T w + C \sum\nolimits_{i=1}^{l} \xi_i + C \sum\nolimits_{i=1}^{l} \xi_i^* \quad (5)$$

$$\begin{aligned} \text{s.t}: \quad & w^T \emptyset(x_i) + b - z_i \leq \in + \xi_i, \\ & z_i - w^T \emptyset(x_i) - b \leq \in + \xi_i^*, \\ & \xi_i, \xi_i^* \geq 0, i - 1, \ldots, l. \end{aligned}$$

The objective is decreasing the fluctuation between the predicted value and the true value. These constraints make sure the model best fitting with the true values which means the best condition of points fall within the range of accuracy accepted. However there are still a number of deviation of points are huge, in order to decrease the effect, we can use a way of soft margin which brings in different relaxation factor ξ_i, ξ_i^*. High dimension, small sample, and nonlinear problem can be easily solved by SVR, and there are still unknown approach to decreasing effect of overfitting and curse of dimensionality to research.

We use two different kernel functions to predict CCPP output, Table 3 summarizes two kernel functions perform in SVR. Figure 2 shows the detail prediction results for each model. We can easily observe that RBF performs better than the other.

Table 3. MSE and R^2 of 3 different kernel function

Kernel	RMSE	R^2
Linear	4.571	0.9305
Rbf	4.118	0.9436

4.3 BP Neural Network Regression

The general used BP neural network structure is triple layer structure which the propagation process divided into three layers namely the input layer, hidden layer and output layer. The determination of neural number is still a researching hotspot, the best neural number in hidden layers mainly figuring out by empirical formula in resent studies. In the hidden layers, neurons should be determined reasonable which means the prediction models will be under-fitting if neurons too few besides too large will lead to over–fitting and the fitting process needs longer [8]. To find out a specific number of neurons in hidden layer, we use usually use empirical formula to calculate neurons. According to the formula and the input attributes, the hidden layer compose of 3 to 13 neurons. In this paper, we design a 4-X-1 structure BP neural network to predict CCPP electrical output with 4 attributes.

The experiment use grid searching to find out the best neurons number. Figure 3 shows the loss descent. After try 28 different number of neurons in hidden layer, we found out that when the amount of neurons in hidden layer is 4,the R^2 has a value of 0.9345 which is the highest of all the try (Table 4).

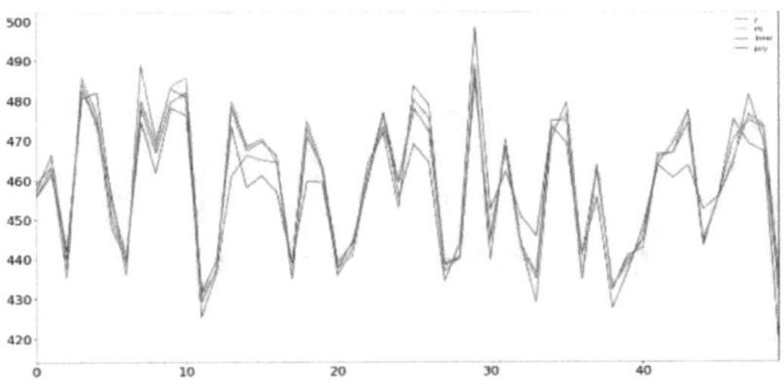

Fig. 3. Prediction of CCPP output using two different kernel function

From Fig. 4, we can find we can find the orange prediction curve and the test curve are basically coincident, which indicates the BP neural network performance great in regression problem.

4.4 Extreme Gradient Boosting Regression

Extreme gradient boosting (XGBoost) is an algorithm evolved from boosting tree. Gradient boosting is a machine learning technique in regression and classification problem [10]. It based on iteration algorithm which using forward distributing. In the iteration of gradient boosting decision tree, we assume the strong learner getting from previous round is $f_{t-1}(x)$ and the loss function is $L(\mathrm{y} - f_{t-1}(x))$. Our goal in the current round is to find a weak learner $h_t(x)$ of CART regression tree model to optimize the loss of current round. In other words, we need to find a decision tree which can bring largest descent to the samples [9]. In XGBoost algorithm, complexity of a weak learner can be represented as a number vector in order to optimizing the loss function in mathematical way. The objective has the form:

$$\mathrm{obj}^{(t)} = \sum_{j=1}^{T} \left[G_j w_j + \frac{1}{2}(H_j + \lambda) w_j^2 \right] + \gamma T$$

Where G_j H_j are the accumulation of gradient descent of leaves $\partial_{\hat{y}_i^{(t-1)}} l(y_i, \hat{y}_i^{(t-1)})$ and $\partial^2_{\hat{y}_i^{(t-1)}} l(y_i, \hat{y}_i^{(t-1)})$. G_j and H_j can be computed in a parallel way which accelerate the training procedure.

The experiment using cross validation to tuning model hyperparameters. The best combination of parameters and the result list in the following chart (Table 5).

The result shows gradient boosting has a very effective performance in regression tasks. XGBoost has a better performance than GBRT algorithm with a RMSE 2.752 and R^2 0.9748, the scatter plot of the XGBoost which predicts CCPP output, is denoted in Fig. 5.

Table 4. RMSE and R^2 of 3 best neurons amount in hidden layer

Amount	RMSE	R^2
4	4.301	0.93458

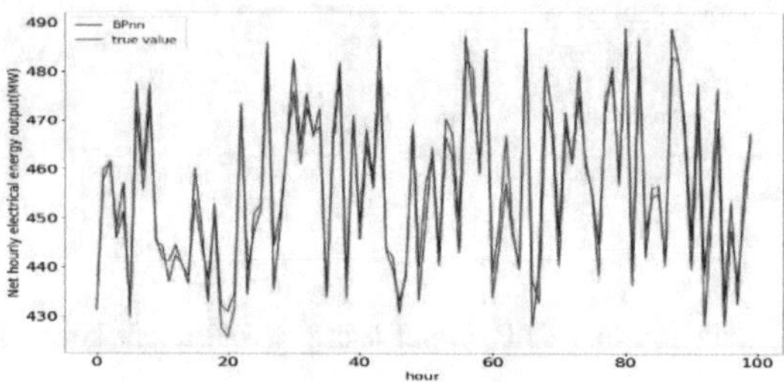

Fig. 4. Prediction of BP neural network

Table 5. Hyperparameters in XGBoost

l_rate	n_estimators	m_depth	m_c_weight	Seed
0.05	1580	8	1	0
Subsample	c_bytree	gamma	r_alpha	r_lambda
0.9	0.8	0.2	0.1	2

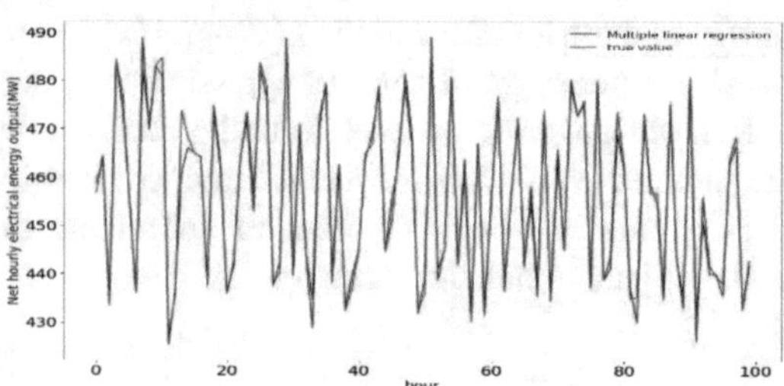

Fig. 5. Prediction of tuned XGBoost

Table 6. RMSE and R^2 of GBRT and XGBoost

Model	RMSE	R^2
GBRT	2.920	0.9716
XGBoost	2.752	0.9748

Table 7. RMSE and R^2 of XGBoost

Models	RMSE	R^2	MAE
MLR	4.559	0.9284	1.959
SVR	4.118	0.9436	3.321
BPNN	4.301	0.9345	3.613
XGBoost	**2.752**	**0.9748**	**1.938**
BREP [2]	3.787	/	2.977

5 Conclusions

This paper use four different algorithm to predict output, with training multiple linear regression model, support vector regression model, BP neural network model and boosting algorithm XGBoost. Table 6 is the summary performance of different algorithm in predicting CCPP output,the result shows that the best algorithm in predicting CCPP output is XGBoost with a RMSE 2.752 and R^2 0.9748. In addition XGBoost has a great performance in predicting with multicollinearity attributes (Table 7).

Acknowledgments. Our deepest gratitude goes to financial support from CERNET innovation Project (NGII20170625) for energy project research, and supporter of dataset for modeling Pinar Tüfekci (email: ptufekci '@' nku.edu.tr), ÇorluFaculty of Engineering, Namik Kemal University is also acknowledged for supporting data. Greatly appreciate your reviewing which contribute to a better paper.

References

1. Kesgin, U., Heperkan, H.: Simulation of thermodynamic systems using soft computing techniques. Int. J. Energy Res. **29**(7), 581–611 (2005)
2. Tüfekci, P.: Prediction of full load electrical power output of a base load operated combined cycle power plant using machine learning methods. Int. J. Electr. Power Energy Syst. **60**, 126–140 (2014)
3. Islikaye, A.A., Cetin, A.: Performance of ML methods in estimating net energy produced in a combined cycle power plant. In: 2018 6th International Istanbul Smart Grids and Cities Congress and Fair (ICSG), Istanbul, Turkey, pp. 217–220. IEEE (2018)
4. Zhandos, A., Guo, J.: An approach based on decision tree for analysis of behavior with combined cycle power plant. In: 2017 International Conference on Progress in Informatics and Computing (PIC), Nanjing, pp. 415–419. IEEE (2017)
5. UCI machine learning repository https://archive.ics.uci.edu/ml/datasets/Combined+Cycle+Power+Plant
6. Izzah, A., Sari, Y.A., Widyastuti, R., Cinderatama, T.A.: Mobile app for stock prediction using improved multiple linear regression. In: 2017 International Conference on Sustainable Information Engineering and Technology (SIET), Malang, pp. 150–154. IEEE (2017)
7. Shengwei, W., Yanni, L., Jiayu, Z., Jiajia, L.: Agricultural price fluctuation model based on SVR. In: 2017 9th International Conference on Modelling, Identification and Control (ICMIC), Kunming, pp. 545–550. IEEE (2017)

8. Zhang, X., Fang, C., Wang, Z., Ma, H.: Prediction of urban built-up area based on RBF Neural network—comparative analysis with BP neural network and linear regression. Res. Environ. Yangtze Basin **22**(6), 691–697 (2013)
9. Friedman, J.H.: Greedy function approximation: a gradient boosting machine. Ann. Stat. 1189–1232 (2001)
10. Chen, T., Guestrin, C.: Xgboost: a scalable tree boosting system. In: Proceedings of the 22nd ACM SIGKDD International Conference on Knowledge Discovery and Data Mining, pp. 785–794. ACM (2016)

Application Research of General Aircraft Fault Prognostic and Health Management Technology

Liu Changsheng[1,2(✉)], Li Changyun[2], Liu Min[1], Cheng Ying[1], and Huang Jie[1]

[1] Changsha Aeronautical Vocational and Technical College, Changsha 410014, Hunan, China
30623297@qq.com

[2] Hunan Key Laboratory of Intelligent Information Perception and Processing Technology, Zhuzhou 412007, Hunan, China

Abstract. General aircrafts are very complicated to analyze from the perspective of structure and system, which means the maintenance task is very heavy. The traditional "broken-then- repair" and "planned repair" methods have serious shortcomings in dealing with the ever-changing new situation. "Maintenance depending on the situation" and "predictive maintenance" will nip the fault in the bud and become the direction of future system maintenance strategy development. This research studies three key technologies: general aircraft intelligent monitoring technology, general aircraft health assessment and prediction method based on multi-source big data fusion, and general aircraft operation and maintenance process visualization evolution simulation technology. Built on these technologies, a General Aviation Health Supervision platform is developed. This supervision platform is of great significance to improve the safety and reliability of general aviation aircraft, reduce operation and maintenance costs, and promote the development of the local navigation industry. The research outcome is tested on the Ararat SA60L light sport aircraft manufactured by Hunan Shanhe Technology Co., Ltd. The test confirms that the general aviation health supervision platform, successfully provides real-time, systematic and intelligent solution for the monitoring and health supervision of general aviation aircrafts. It is expected that the new platform will create a revenue of more than 50 million yuan in the first three years of commercialization, with an annual growth rate of over 20%.

Keywords: General aircraft · Fault prognostic · Health management

1 Introduction

With the increase in the structural complexity, level of integration and level of Informatization, the cost of development, production and especially maintenance of general-purpose aircrafts is increasing. At the same time, as a result of the complex relationship between the different components and the strong coupling, the probability of failure and malfunction also increases. In the event of a failure, it often leads to

P. Krömer et al. (Eds.): ECC 2018, AISC 891, pp. 567–575, 2019.
https://doi.org/10.1007/978-3-030-03766-6_64

unpredictable casualties and losses, even devastating consequences. Prognostics and health management (PHM) technology has become a key technology to support equipment to achieve cost-effective maintenance and proactive health management [1, 2]. China's "National Medium- and Long-Term Science and Technology Development Plan (2006 ~ 2020)" clearly states technologies that can predict the useful life of major and facilities are the key to improve operational reliability, safety and maintainability. Using the PHM technology, once a failure is predicted, the problem that will lead to the failure can be repaired immediately. This can ensure that no catastrophic failure occurs and also avoid over-maintenance, truly realizing the "maintenance depend on the situation" and "on-demand maintenance" practices. The technology can also provide guidelines to develop maintenance plans, allocate maintenance resources, improve the system reliability, maintenance and repair efficiency, and save cost throughout the life cycle. Successful applications will enable an effective management and operation of navigation companies, bringing higher economic benefits and significant industry advantages. As a result, the PHM technology, which is based on predictive technologies, has received increasing attention and application. The PHM technology is important for flight safety and flight operations, and there is an urgent need to increase the research and development of this technology.

2 Review of Relevant Researches in China and Overseas

2.1 Research Progress in China

Zhang Baozhen of China Aviation Information Center proposed the concept of intelligent monitoring and health supervision technology [3]. Research institutes such as the Institute of Reliability Engineering of Beijing University of Aeronautics and Astronautics and Aviation Institute No. 634 have conducted more follow-up studies on equipment health decline paternts, fault prediction models, and health supervision techniques [4]. Beijing University of Aeronautics and Astronautics cooperated with the CALCE Center of the University of Maryland to conduct PHM research on electronic equipment. They completed the structural block diagram of the PHM software and hardware systems, outlined the key technical elements required, and the implementation plan of development, and promoted the concept of systematic health management complex systems [5]. Lu Feng studied information fusion technology for fault diagnosis [6]; Huang Weibin, Yan Yunfeng et al. studied the airborne adaptive real-time estimation model for health supervision [7, 8]. Zuo Hongfu carried out the analysis of oil abrasive grain analysis research on engine fault diagnosis [9]. Bo [10] of the Air Force Engineering University summarized the implementation strategies of four types of electronic system fault prediction methods, including the characteristic parameter method, the early warning circuit, the cumulative damage model method and the comprehensive method.

2.2 Research Progress Out of China

Research on health supervision technology started much earlier Western countries. Since the 1980s, the helicopter health and usage monitoring system (HUMS) has been applied to helicopter health supervision to improve the safety and reliability of helicopters. The Smiths aerospace industry in the UK has been working on HUMS technology research and has developed HUMS from its initial state of monitoring only to comprehensive functions such as condition assessment and data management [11].

NASA launched the flight safety plan AVSP in 1999 in response to the US government's goal of reducing the flight accident rate by 80% in 10 years and 90% in 25 years. As the main technical means to improve safety, the engine health supervision system is the key to achieving AVSP. The program has studied aviation system modeling and monitoring techniques, model-based engine control and fault diagnosis, engine vibration fault analysis, component stability monitoring, and advanced sensor technology [12].

NASA is currently working with other research institutes on an Integrated System Health Management (ISHM) program. The research program has imposed a series of requirements for the expected technologies for real-time performance assessment, fault prediction and diagnosis of the overall system and the subsystems (such as the power and the control system). Meanwhile, it requires the performance and the fault prediction and diagnosis modules to have features like "plug and play" and fault source identification [13].

Sponsored by the US Navy, a comprehensive aircraft health monitoring program IAHM was led by Boeing and involved, the University of Hawaii and Referential Systems Incorporated. The project proposed the establishment of a multi-platform military and commercial engine health regulatory data processing and analysis system, which was designed to improve the reliability, security, maintainability and affordability of aircraft systems and to improve combat performance and operational performance. IAHM collects and stores key data during flight, monitors and analyzes them and makes plans for the maintenance of aircraft and major components [14, 15].

The Air China B747-400 fleet has been using the Boeing Aircraft Health Supervision System since December 2009. The number of abnormal events such as repetitive faults and delays of the fleet have been significantly reduced. The current average daily flight hours have reached 14-15FH [16].

2.3 Technology Development Trends

The aviation engine health supervision technology has a good foundation in Western countries after decades of development; however, the development in China, falls behind the world, and it has the following limitations: (1) Most existing health supervision systems are designed for specific models, and the framework is not flexible enough to be put into operation. It is difficult to add new monitoring objects after installation; (2) Most existing health monitoring methods use limited number of modeling parameters, only focus on operational status information, have insufficient attention to production data and maintenance records. This results in low reliability in health assessment and prediction; (3) Most of the existing mature health supervision technologies use "broken-then-repair" or "planned maintenance" strategies, which results in high cost, long cycle and poor reliability in equipment maintenance.

In order to improve the reliability of general-purpose aircrafts and reduce maintenance cost, it is necessary to monitor the operating conditions and implement health supervision for key components. The General Aircraft Health Supervision concept is proposed, which refers to the use of integrated techniques or means to detect, diagnose and predict general aircraft systems, components and accessories. It builds on the identification, acquisition, processing and integration of collected information, proactively analyzes the health of the aircraft, predicts performance trends, component failures and remaining useful life of the complete machine or components, and takes necessary measures to improve availability and safety. It integrates equipment management rules and procedures, business processes, and closely integrates information such as condition monitoring, maintenance, use, and environment to comprehensively diagnose factors related to equipment health; and optimize maintenance activities.

3 Research Framework

The project team has carried out research work on system state intelligent monitoring health assessment and prediction, and visualization of the operation and maintenance of general aviation aircrafts. The framework of this research project is shown in Fig. 1:

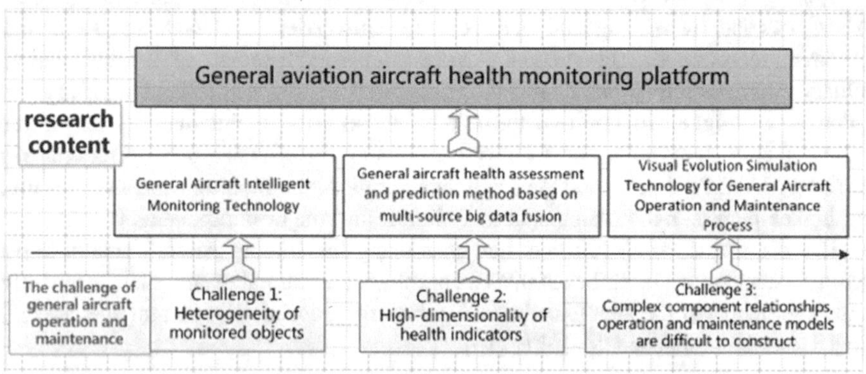

Fig. 1. Project research framework

4 Key Technologies

4.1 General Aircraft Intelligent Monitoring Technology

General aircraft intelligent monitoring technology framework. Because the monitoring parameters are heterogeneous, the performance of it is different, the scale is different, the monitoring frequency is different, and the presentation requirements are different. Therefore, it is necessary to construct an open and flexible intelligent monitoring framework to facilitate hierarchical structure expression, personalized attribute construction, and diversified implementation of performance calculation, display of different model status.

The structure of the intelligent monitoring framework is shown in Fig. 2. Firstly, the collection center is responsible for the collection and data analysis of the underlying sensor data. Different types of sensor data analysis rules can be maintained by the monitoring center. The collection center can automatically update the local data analysis rules from the server. After the data is parsed, it will be transmitted to the data center through the network for unified storage. In order to realize the real-time display of the status, the user can maintain different data display templates for different sensor types and components through the monitoring center. The monitoring center calls the corresponding data to fill the corresponding template from the data center based on the component type, to realize the real-time monitoring of the status. This solution makes the entire framework compatible with different sensors and general-purpose aircraft, enabling an open intelligent monitoring framework.

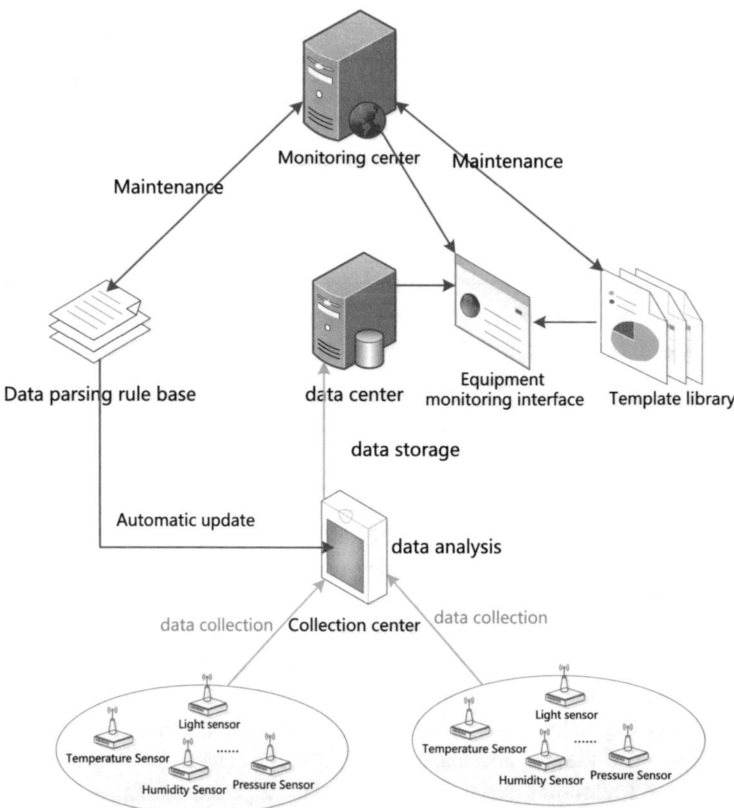

Fig. 2. General aircraft intelligent monitoring framework

In the research of general aircraft structure, attributes and performance expression and storage methods, firstly, in order to support different types of general aviation aircraft express clearly its hierarchical structure, the project intends to use a tree structure to model its system; secondly, in order to achieve scalability and availability attributes, the attributes are divided into two categories: general attributes and unique attributes, where the general attributes mainly store the basic information of the parts such as number, name, category, uptime, MTTR, MTBF, current status, standard energy consumption, Information on service life, etc.; unique attributes are closely related to equipment type, such as engine speed, maximum thrust, etc.; due to different general aircraft have different performance models, the implementation of the model needs to have strong experience in related fields. Thus the project is intended to be implemented by means of an external interface call, and the user writes the corresponding performance model himself when building the system model. All data is stored using structured data storage.

The project is intended to provide two different ways: for the expression and storage implementation of parameter associations, the regular expressions and user-defined interface functions. The regular expressions can be directly stored as structured data, the user-defined interface functions can be through the dll library, Jar package, groovy script or web service to achieve, directly storage with file. In order to reduce the false alarm rate and invalid alarm, the alarm threshold of the process parameters is optimized. Firstly, based on historical data, the kernel density method in non-parametric statistics is used to estimate the alarm state of the process parameters, then establish the optimization model of the process parameter alarm threshold, and use the alarm threshold of the process parameters as the manipulated variable to minimize false alarms and leak alarms. Constructs the objective function from the angle of minimize false negatives and probability of false alarms, and to solve the optimal alarm threshold by the method of numerical optimization is used.

4.2 General Aircraft Health Assessment and Prediction Method Based on Multi-source Big Data Fusion

An efficient fusion method for multi-source heterogeneous big data. Due to the diverse sources of data resources, complex structure, large capacity, and information islands and faults, it is difficult to conduct knowledge reasoning, sharing and interoperability. Therefore, an effective multi-source heterogeneous big data fusion method is needed to reduce reasoning the degree of ambiguity, realize automatically analysis and synthesis, offer the knowledge timely and accurately, and improved decision-making ability.

In the implementation of multi-source big data fusion technology, it is proposed to adopt multi-source big data fusion health assessment method based on grey relational analysis and evidence theory, and select three feature vectors: product quality, operating condition and historical state. The above data is fused to achieve quantitative detection of components, providing technical support for health assessment and prediction.

In the implementation of health assessment technology, it is proposed to establish a health status assessment index system based on information such as product quality, operating conditions and historical status, and establish a health assessment model by combining the improved manifold learning algorithm with the hidden semi-Markov model. Come up with a method based on the multi-distance morphological similarity assessment (M-DSSE) method, from the extraction of state feature information, the health status is evaluated by M-DSSE method, and the health index is calculated to achieve multi-level assessment of health status.

In terms of life expectancy and health prediction technology, a life prediction method based on reliability and failure analysis is proposed. According to the distribution characteristics of life, a Weibull model for predicting the life loss of equipment is constructed. From the characteristics of Weibull distribution, the parameter estimation is carried out, and the shape parameters of the Weibull model of the equipment life loss are calculated. The rest life of the equipment is obtained and the remaining usage time is determined. Aiming at the health state prediction problem, the prediction method of correlation vector machine (RVM) is proposed. The RVM regression model is used to predict the engine health index to predict the engine health trend and provide important technical support for the final predictive maintenance.

4.3 Visual Evolution Simulation Technology for General Aircraft Operation and Maintenance Process

Multi-agent operation and maintenance evolution simulation of general aviation aircraft. Because of the performance calculation methods and maintenance methods of aircraft and components are different, it is necessary to design a flexible and universal mechanism to support user-defined maintenance methods and performance calculation methods of different components, so as to realize multi-agent evolutionary simulation operation for different aircraft operation and maintenance processes.

Under the different constraints, the evolution of operation and maintenance process simulation is implemented based on the multi-agent simulation engine of Repast open source project. Firstly, the Agent class is designed according to the attributes and functions of the general aircraft and its components, so that each type of Agent has corresponding evolution rules and behavior characteristics. Then, build the simulation environment, including the physical structure and logical structure. Finally, simulate the operation and maintenance activities, traverse the physical structure tree in each simulation step, and evolve and deduct the status, funding requirements and performance of each agent.

In terms of the general operation and maintenance performance calculation method, the working principle and structure of various general-purpose aircraft are different. This project is to study the implementation method of component performance calculation. Component-specific properties are defined by dynamic properties. Component-specific performance calculation methods are defined in scripts. Each Agent can load and execute a custom performance calculation script. The script can access the dynamic properties of the corresponding component.

In the network environment, the visualization and simulation of the operation and maintenance process is implemented by Unity3D. Firstly, build a scenario by modeling generic aircraft and components and importing the 3D model into the Unity3D editor. Then, design a logic manager (a global script) to control the interaction between the public transaction and the model within the entire module. The logic manager is responsible for parsing the simulation data files and driving each component to update its state. For example, healthy parts are displayed in green; components with high risk of failure are displayed in red; and the health status of the components is displayed. The visualization module interface is intended to be implemented using UGUI.

5 Conclusion

A General Aviation Health Supervision system has been developed, which can significantly improve the safety and reliability of general aviation aircrafts, reduce the operation and maintenance cost, and promote the development of the local navigation industry. The system has been deployed and tested on the Ararat SA60L light sport aircraft manufactured by Hunan Shanhe Technology Co., Ltd., providing a real-time, systematic and intelligent solution for the intelligent monitoring and health supervision of general aviation aircrafts. It is expected that the research outcome will create 50 million yuan revenue in the first three years in the market with an annual growth rate of over 20%.

Acknowledgement. About the Author: Liu Changsheng, professor/doctor, main research areas: computer application technology, intelligent manufacturing technology, higher vocational education.

Project Funding: Hunan Natural Science Fund–Science and Education Joint Project (2017JJ5054), Research and Application of Key Technologies for General Aircraft Fault Prediction and Health Management.

References

1. Tsui, K.L., Chen, N., Zhou, Q., et al.: Prognostics and health management: a review on data driven approaches. Math. Probl. Eng. (2015). https://doi.org/10.1155/2015/793161
2. Esteves, M.A.M., Nunes, E.P.: Prognostics health management: perspectives in engineering systems reliability prognostics. Saf. Reliab. Complex Eng. Syst., 2423–2431 (2015)
3. Baozhen, Z.: Development and application of forecasting and health supervision technology. Measur. Control Technol. **27**(2), 5–7 (2008). https://doi.org/10.19708/j.ckjs.2008.02.002
4. Zeng, S., Pechi, M., Wu, J., et al.: Current status and development of fault prediction and health supervision (PHM) technology. J. Aviation **26**(5), 626–632 (2005)
5. Bo, S., Zhao, Y., Wei, H., et al.: Case study of electronic product health monitoring and fault prediction methods. Syst. Eng. Electron. **29**(6), 1012–1016 (2007)
6. Lu, F.: Research on Fusion Technology of Aeroengine Fault Diagnosis. Nanjing University of Aeronautics and Astronautics (2009)
7. Huang, W.: Adaptive airborne real-time model for engine health supervision. Nanjing University of Aeronautics and Astronautics (2007)
8. She, Y., Huang, J., Lu, F.: Performance estimation of gas path components of turboshaft engine based on Kalman filter. J. Changchun Univ. Sci. Technol. (Nat. Sci. Edn.) **3**, 33–36 (2010)
9. Zuo, H.: Engine Wear State Monitoring and Fault Diagnosis Technology. Aviation Industry Press (1996)

10. Bo, J., Yifeng, H., Jianye, Z.: Current status and development of avionics system fault prediction and health supervision technology. J. Air Force Eng. Univ. (Nat. Sci. Ed.) **11**(6), 1–6 (2010)
11. Larder, B., Azzam, H., Trammel, C., et al.: Smith industries HUMS: changing the M from monitoring to management. In: Aerospace Conference Proceedings, pp. 449–455. IEEE, Montana (2000)
12. Zuniga, F.A., Maclise, D.C., Romano, D.J.: NASA Aviation safety program aircraft engine health management data mining tools roadmap. In: Data Mining and Knowledge Discovery: Theory, Tools and Technology II, Orlando Florida, USA, pp. 292–298 (2000)
13. Safety, W.S.: The Military Aircraft Joint Strike Fighter Prognostics & Health Management. AIAA 98-3710, pp. 1–7 (1998)
14. Zhang, B., Wang, P.: Application of prediction and health supervision (PHM) technology in new generation fighter engines abroad. In: 2008 Aviation Test and Testing Technology Summit, Nanchang, Jiangxi, China, pp. 220–225 (2008)
15. Clark, G.J., Vian, J.L., West, M.E., et al.: Multi-platform airplane health management. In: IEEE Aerospace Conference Proceedings, MT, United states, pp. 1–13 (2007). https://doi.org/10.1109/aero.2007.352944
16. Gong, J.: Apply Boeing AHM system to ensure the safe operation of Air China B747-400 fleet. Chin. Civil Aviation **6**, 65–67 (2010)

Support Vector Regression with Multi-Strategy Artificial Bee Colony Algorithm for Annual Electric Load Forecasting

Siyang Zhang[1], Fangjun Kuang[1(✉)], and Rong Hu[2]

[1] School of Information Engineering, Wenzhou Business College, Wenzhou 325035, China
kfjztb@126.com

[2] Fujian University of Technology, Fuzhou 350118, China

Abstract. A novel support vector regression (SVR) model with multi-strategy artificial bee colony algorithm (MSABC) is proposed for annual electric load forecasting. In the proposed model, MSABC is employed to optimize the punishment factor, kernel parameter and the tube size of SVR. However, in the MSABC algorithm, Tent chaotic opposition-based learning initialization strategy is employed to diversify the initial individuals, and enhanced local neighborhood search strategy is applied to help the artificial bee colony (ABC) algorithm to escape from a local optimum effectively. By comparison with other forecasting algorithms, the experimental results show that the proposed model performs higher predictive accuracy, faster convergence speed and better generalization.

Keywords: Support vector regression · Annual load forecasting Multi-strategy · Artificial bee colony algorithm · Parameter optimization

1 Introduction

With the rapid development of China's electric power industry, Electric power industry plays a vital role for the national economic and social stability. To a certain extent, the annual electric load forecasting can affect the development trends of the electric power industry, and provide reliable guidance for power grid operation and power construction planning [1]. However, annual electric loads have complex and non-linear relationships with some factors such as the political environment, human activities, and economic policy [2], making it is quite difficult to accurately forecast annual electric loads. In recent years, many artificial intelligence forecasting techniques are presented for load forecasting, such as fuzzy-neural [3], artificial neural networks (ANNs) [4], grey model GM(1,1) [5] and support vector regression (SVR) [6, 7].

As influenced by various factors, an annual load curve shows a non-linear characteristic, which demonstrates that the annual load forecasting is a non-linear problem. Support vector regression (SVR) is proven to be useful in dealing with non-linear forecasting problems. However, the forecasting performance of SVR model largely depends on its parameters, so some evolutionary algorithms have been applied to select

P. Krömer et al. (Eds.): ECC 2018, AISC 891, pp. 576–585, 2019.
https://doi.org/10.1007/978-3-030-03766-6_65

the appropriate parameters of SVR, including genetic algorithm (GA) [6], particle swarm optimization (PSO) [8], differential evolution algorithm (DE) [2] and artificial bee colony algorithm [7].

Numerical comparisons demonstrated that the performance of the ABC algorithm is competitive to other evolutionary algorithms with the advantage of employing fewer control parameters [9]. Due to its simplicity and ease of implementation, the ABC algorithm has captured much attention and has been applied to solve many practical optimization problems [7, 10]. However, it is very regretfully finds that few researchers use ABC algorithm for SVR parameters optimization problem in load forecasting. In this paper, a novel multi-strategy ABC algorithm (MSABC) is proposed to improve the optimization ability of standard ABC and optimize the appropriate parameters of SVR for improving the model's forecasting accuracy.

The rest of the paper is organized as follows: Section 2 introduces SVR. A novel multi-strategy ABC algorithm (MSABC) is proposed in Sect. 3. In Sect. 4, a hybrid forecasting model combined MSABC and SVR is discussed in detail. In Sect. 5, annual electric load forecasting experiment is presented and further comparison and discussion are also presented. Section 6 presents the conclusions.

2 Support Vector Regession Model

The basic concept of support vector regression (SVR) model is to nonlinearly map the original data x into a higher dimensional feature space. Hence, given a set of data $T = \{(x_i, y_i)\}_{i=1}^{N}$, where $x_i \in R^n$ is n dimension input vector, $y_i \in R^1$ is the actual value, and N is the total number of data patterns, the SVR function.

$$f(t) = w\varphi(x_i) + b \tag{1}$$

where $\varphi(x_i)$ is the feature of inputs, w is the weight vector and b is the threshold value, which are estimated by minimizing the following regularized risk function.

$$R(C) = (C\frac{1}{N})\sum_{i=1}^{N} e(f(x), y) + \frac{1}{2}\|w\|^2 \tag{2}$$

$$e(f(x), y) = \begin{cases} 0, & |f(x) - y| \le \varepsilon \\ |f(x) - y| - \varepsilon, & otherwise \end{cases} \tag{3}$$

where $\frac{1}{2}\|w\|^2$ measures the flatness of the function, $e(f(x), y)$ is the ε-insensitive loss function, C is a positive constant which determines the trade-off between the empirical loss and the inner product, Both C and ε are user-determined parameters.

In order to obtain the estimated values of w and b, the optimization formulation can be transformed into a dual problem by introducing the Lagrange multiplier coefficients α_i and α_i^*, which are bounded by a user-specified constant C.

$$\text{minimize} \quad -\frac{1}{2}\sum_{i=1}^{N}\sum_{j=1}^{N}(\alpha_i - \alpha_i^*)(\alpha_j - \alpha_j^*)k(x_i \cdot x_j) + \sum_{i=1}^{N} y_i(\alpha_i - \alpha_i^*) - \varepsilon\sum_{i=1}^{N}(\alpha_i + \alpha_i^*)$$
$$\textit{subject to} \sum_{i=1}^{N}(\alpha_i - \alpha_i^*) = 0,\ 0 \le \alpha_i,\ \alpha_i^* \le C \tag{4}$$

At the optimal solution from (4), the regression function can be expressed in the following form:

$$f(t) = \sum_{i=1}^{m}(\alpha_i - \alpha_i^*)K(x_i, x_j) + b \tag{5}$$

where $K(x_i, x_j) = \exp(-\gamma\|x_i - x_j\|^2)$ is RBF kernel function in this study.

It is well known that the forecasting accuracy of SVR model depends on a suitable setting of the punishment factor C, kernel parameter γ and the tube size ε. Therefore, Multi-strategy artificial bee colony (MSABC) algorithm is used to optimize the three major parameters C, γ and ε of SVR model.

3 Multi-Strategy Artificial Bee Colony Algorithm

3.1 Artificial Bee Colony Algorithm

ABC algorithm was applied to multidimensional and multimodal function optimization. The swarm is divided into employed bees, scouts and onlookers. The food sources are produced randomly within the range of the boundaries of the variables.

$$X_i^j(0) = X_{\min}^j + R(X_{\max}^j - X_{\min}^j) \tag{6}$$

where $i = 1, 2, \cdots, SN, j = 1, 2, \cdots, D$, SN is the number of food sources and equals to half of the colony size. D is the dimension of the problem, representing the number of parameters to be optimized, $X_{\max}^j, X_{\min}^j$ is upper and lower bounds of the jth parameter, respectively. The fitness of food sources will be evaluated.

In the employed bees' phase, a number of employed bees, set as the number of the food sources and half the colony size, are used to find new food sources using (7).

$$V_i^j(t) = X_i^j(t) + \Phi(X_i^j(t) - X_k^j(t)) \tag{7}$$

where $i = 1, 2, \cdots, SN, j$ is a randomly selected number in $[1, D]$ and D is the number of dimensions, Φ_{ij} is a random number uniformly distributed in the range $[-1, 1]$, and k is the index of a randomly chosen solution, and $k \neq j$.

Onlooker bees next choose a random food source according to the selection probability. If a food source cannot be improved for a predetermined number of cycles, referred to as *Limit*, this food source is abandoned. The employed bee that was exploiting this food source becomes a scout that looks for a new food source.

3.2 Tent Chaotic Opposition-Based Learning Initialization Strategy

Population initialization is a crucial task in evolutionary algorithms because it can affect the convergence speed and the quality of the final solution. If no information about the solution is available, then random initialization is then most commonly used method to generate initial population. Owing to the randomness and sensitivity dependence on the initial conditions of chaotic maps, the chaotic maps have been used to initialize the population. Therefore, Tent chaotic opposition-based learning strategy [10] is used to initialize the population, so that the initial population can increase diversity and preserve individual randomness.

3.3 Tournament Selection Strategy

Onlooker bees in the improved algorithm select food source by using the tournament selection strategy [10]. It is a process based on local competition which only refers to the relative value of individuals. Tournament selection probability is as follow:

$$P_i(t) = c_i(t)/(\sum\nolimits_{i=1}^{N} c_i(t)) \tag{8}$$

where c_i is the score of an individual.

3.4 Enhanced Local Neighborhood Search Strategy

In solving complex optimization problems, how to achieve the balance between local exploitation and global exploration is still the key to improve the performance of artificial bee colony algorithm. As the ABC algorithm realizes local search by employed bees and onlooker bees, the global search is realized by onlooker bees and scout bees to balance global exploration and local exploitation ability of the algorithm. Therefore, take into account the leading role of individual X_i and local best solution X_{best}, a novel enhanced local neighborhood search strategy is proposed, which introduces adaptive step to enhance the local search ability in the later period, and enhance the local neighborhood search. Enhanced local neighborhood search is as follow:

$$V_i^j(t) = \lambda X_i^j(t) + (1-\lambda)X_{best}^j(t) + \Phi((1-\lambda)(X_{best}^j(t) - X_k^j(t)) \tag{9}$$

where, $\lambda = 1 - \lambda_{\max}/(1 + (\frac{\lambda_{\max}}{\lambda_{\min}} - 1)e^{gen}), gen,$ is the local iteration number, $\lambda_{\max} = 1, \lambda_{\max} = 0.001, \alpha = 0.1$. It is known by (9) that λ is gradually reduced from $1 - \lambda_{\min}$ to 0, which makes the weight of the current optimal solution X_{best} gradually increase and the weight of individual X_i gradually decreases, in order to realize the balance between global exploration and local exploitation ability of the algorithm.

4 MSABC for Parameters Selection of SVR Model

By means of the MSABC algorithm, the three major parameters C, γ and ε of SVR model, can be optimized, which a potential solution is comprised of a vector $(\mathrm{C}, \gamma, \varepsilon)$, $D = 3$. The parameter optimality is measured by means of fitness functions that are defined in relation to the considered optimization problem. Therefore, the fitness function is employed the mean absolute percentage error (MAPE). The MAPE is shown as (10), which serves as the forecasting accuracy index for identifying suitable parameters in the SVR model.

$$MAPE = \frac{1}{n}\sum_{i=1}^{n}\left|\frac{f_i - \hat{f}_i}{f_i}\right| \times 100\% \tag{10}$$

where n is the number of forecasting periods; f_i and $\hat{f}_i$ represent the actual value and the forecast value at period i, respectively. The MSABC algorithm is used to seek a better combination of the three parameters in the SVR so that a smaller MAPE is obtained during forecasting iteration. The detail procedure of the MSABC algorithm for the parameters selection of SVR model (MSABC-SVR) is introduced as follows:

Step 1: Initial the food sources and computation conditions include population of bee colony N, number of employed bees $SN = (N/2)$, upper and lower boundaries of every decision variable, the maximum iteration $G_{\max}$, *Limit* and chaotic local search iteration number $C_{\max}$, the number of parameters D is set as 3 in this study.
Step 2: Set iteration $iter = 0$, generate the SN vectors X_i with D dimensions as food sources according to chaotic opposition-based learning initialization method.
Step 3: Sent SN employed bees to food sources. Initialize the flag vector $trial(i) = 0$, which is recorded the cycle number of a food source.
Step 4: Produce new solutions V_i using employed bees by (7), and calculate the fitness value using (10).
Step 5: If the fitness value of V_i is better than that of X_i, then $X_i = V_i$, $trial(i) = 0$; Else X_i is maintained, $trial(i) = trial(i) + 1$.
Step 6: Calculate the probability values P_i of food sources by (8) applying tournament selection.
Step 7: Onlooker bees choose the food sources by probabilities P_i until all of them have a corresponding food source, and produce new solutions V_i. Calculate the fitness value using (10).
Step 8: If the fitness value of V_i is better than that of X_i, then $X_i = V_i$, $trial(i) = 0$; Else X_i is maintained, $trial(i) = trial(i) + 1$.
Step 9: If $trial(i) > Limit$, there is an abandoned solution for the scout then replace it with a new food source V_i, which will be reinitialized by carrying out enhanced local neighborhood search strategy.
Step 10: Memorize the best solution found so far.
Step 11: Update $iter = iter + 1$. If the maximum iteration cycle is not reached yet, then go to step 4. Otherwise, return best solution.

5 Example Computation and Discussion

5.1 Data Set and Preprocessing

The selected data set was the annual total electricity consumption of China between 1978 and 2017, shown in Table 1, where 1978–2007 load data from Literature [11], and 2008–2017 load data from National Energy Administration of China. The data are divided into the training data and testing data. According to a series of experiment, when the last six load data put into the SVR model with the default parameters to forecast the current load, the satisfied performance is achieved. Therefore, the last six load data $L_{n-6}, L_{n-5}, L_{n-4}, L_{n-3}, L_{n-2}, L_{n-1}$ as the input variables of the MSABC-SVR model and the output variable is L_n. Due to using last six load data to forecast, the training set is started in 1984 and ended in 2010, the testing set is from 2011 to 2017.

Table 1. Annual electric load of China between 1978 and 2017 (unit: 109 kWh)

Year	Electric load	Year	Electric load	Year	Electric load	Year	Electric load	Year	Electric load
1978	246.53	**1986**	451.03	**1994**	926.04	**2002**	1633.15	**2010**	4192.30
1979	282.02	**1987**	498.84	**1995**	1002.34	**2003**	1903.16	**2011**	4692.80
1980	300.63	**1988**	547.23	**1996**	1076.43	**2004**	2197.14	**2012**	4959.10
1981	309.65	**1989**	587.18	**1997**	1128.44	**2005**	2494.03	**2013**	5322.30
1982	327.92	**1990**	623.59	**1998**	1159.84	**2006**	2858.80	**2014**	5523.30
1983	351.86	**1991**	680.96	**1999**	1230.52	**2007**	3271.18	**2015**	5550.00
1984	377.89	**1992**	759.27	**2000**	1347.24	**2008**	3451.00	**2016**	5919.80
1985	411.90	**1993**	842.65	**2001**	1463.35	**2009**	3643.20	**2017**	6307.70

In the training stage, a roll-based data processing procedure is used. Firstly, the top six load data (from 1978 to 1983) of the data series are fed into the MSABC-SVR model, and then the first electric load forecasting value of 1984 is obtained. Secondly, the actual electric load value of 1984 in the series is employed for the next processing process, in other words, the next roll-top six load data (from 1979 to 1984) are substituted into the MSABC-SVR model, and the forecasting value of 1985 is gotten. Similarly, the processes are cycling until all the load forecasting values (from 1984 to 2010) are produced. Finally, the three parameters are evolved generation by generation, and until the MSABC-SVR gets the stopping criterion, the three parameters are finally determined from the best solution in the terminated population, and are applied to forecast annual electric load. Because of the roll-based data processing procedure, the value of n in (10) equals to 33 for the training dataset, while n is 7 for the testing dataset.

5.2 Forecasting Results and Discussions

In order to confirm the annual electric load forecasting result of the MSABC-SVR model, several other electric load forecasting models were selected. The single SVR model with default parameters ($C = 1000$, $\gamma = 1$, $\varepsilon = 0.001$), SVR model combined with ABC algorithm (ABC-SVR), SVR model combined with Logistic chaotic ABC algorithm (CABC-SVR), MFO-GM(1,1) [5], SVR model combined with PSO algorithm (PSO-SVR), SVR model combined with GA algorithm (GA-SVR), and GRNN [12] model are also employed for comparison. For SVR model with ABCs, the population size is set to 20, the number of food sources, employed bees and onlooker bees is half of the population size. Limit time of food source cannot be improved is 30; maximum iteration cycle number G_{max} is 100; chaotic search iteration number C_{max} is 30. The input and output data are also the same as those of the MSABC-SVR model. The accuracy sets 10^{-4} and the max generation is 100. The experimental environment includes Matlab 2016b, libsvm 3.22 toolbox, the PC with the Intel(R) Core (TM) i7-6700K 4.0 GHz CPU, 16 GB RAM and the Windows 10 operating system.

The forecasting results and the suitable parameters (C, γ, ε) for the MSABC-SVR, CABC-SVR and ABC-SVR models are (23.8764, 4.2689, 0.0009), (63.9806, 4.6917, 0.0014), (99.2461, 2.1431, 0.0022), respectively. In the GRNN model, the spread parameter value is set as 0.2. Figure 1 shows the test set forecasting results of these nine models. Table 2 gives the annual electric load testing set forecasting results and the relative errors with these comparison models.

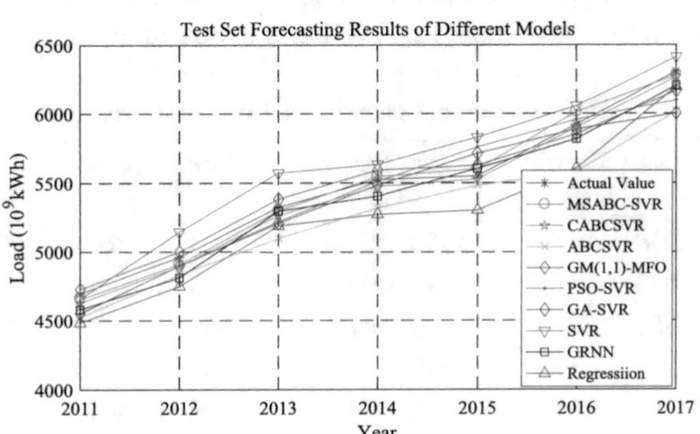

Fig. 1. Test forecasting results of different models

From Table 2 and Fig. 1, the deviations between the forecasting results of these nine forecasting models and the actual values can be captured. The relative error ranges [−3%, +3%] and [−1%, +1%] are always considered as a standard to assess forecasting results, the range is also used to measure the performance of the nine forecasting models. Firstly, the relative errors of the proposed MSABC-SVR model are all in the range [−3%, +3%], and the maximum and minimum relative errors are 1.07802% in 2012 and −0.86856% in 2011, respectively. In addition, five out of seven points means that 71% of the forecasting points are in the scope of [−1%, +1%]. Secondly, in the CABC-SVR model, three forecasting points are in the scope of [−1%, +1%]. Thirdly, the ABC-SVR model has three forecasting point that exceeds the relative error range [−3%, +3%]. However, there is one forecasting point in the scope of [−1%, +1%]. For MFO-GM(1,1) are all in the range [−3%, +3%]. However, there is two forecasting point in the scope of [−1%, +1%]. For the single SVR model, there are three forecasting point exceed the scope of [−3%, +3%]. However, all the forecasting points exceed the scope of [−1%, +1%]. The maximum relative error of regression model is −4.54131%, which is the largest error among these nine forecasting models. In additional, the proposed MSABC-SVR model has the best performance in MAPE, while Regression model has the maximum MAPE value in the testing set. The results proved that the parameters determined by MSABC algorithm can efficiently improve the forecasting accuracy of the SVR.

Table 2. Forecasting results of different models (Unit (Results: 109kWh, Error: %))

Year		Model									
		Actual value	MSABC-SVR	CABC-SVR	ABC-SVR	MFO-GM(1,1)	PSO-SVR	GA-SVR	SVR	GRNN	Regression
2011	Result	4692.8	4652.04	4625.64	4522.99	4725.64	4548.84	4552.37	4594.25	4580.12	4483.71
	Error	-	−0.86856	−1.43113	−3.61852	0.699795	−3.06768	−2.99246	−2.10003	−2.40113	−4.45555
2012	Result	4959.1	4905.64	4889.56	4830.94	4999.56	4896.77	4895.1	5145.4	4809.75	4746.63
	Error	-	−1.07802	−1.40227	−2.58434	0.815874	−1.25688	−1.29056	3.75673	−3.01164	−4.28445
2013	Result	5322.3	5299.12	5263.76	5269.23	5383.76	5220.26	5208.85	5571.95	5296.58	5189.55
	Error	-	−0.43553	−1.0999	−0.99713	1.154764	−1.91722	−2.1316	4.690641	−0.48325	−2.49422
2014	Result	5523.3	5544.3	5472.8	5318.41	5592.8	5510	5484.26	5633.19	5403.74	5272.47
	Error	-	0.380207	−0.91431	−3.70956	1.258306	−0.2408	−0.70682	1.989571	−2.16465	−4.54131
2015	Result	5550	5606.05	5622.78	5480.82	5626.78	5757.53	5712.69	5728.03	5604.12	5352.67
	Error	-	1.00991	1.311351	−1.24649	1.383423	3.739279	2.931351	3.207748	0.975135	−3.5555
2016	Result	5919.8	5964.96	5898.3	5821.66	5852.3	5955.58	5887.18	6054.98	5815.7	5704.89
	Error	-	0.762864	−0.36319	−1.65783	−1.14024	0.604412	−0.55103	2.283523	−1.75851	−3.63036
2017	Result	6307.7	6281.18	6267.16	6016.45	6168.16	6098.39	6062.55	6412.15	6206.78	6197.56
	Error	-	−0.42044	−0.64271	−4.61737	−2.21222	−3.31833	−3.88652	1.655913	−1.59995	−1.74612
MAPE (%)			0.6012	1.8577	2.3644	1.0856	1.2796	1.6684	2.6521	2.6926	3.8997

In conclusion, the proposed MSABC-SVR model outperforms other eight models in annual load forecasting. Compared with the SVR model, the MSABCSVR model which uses MSABC algorithm to optimize the parameters of SVR can improve the forecasting accuracy effectually.

6 Conclusions

Electricity load forecasting plays an important role to operate the power system reliably and economically. In this paper, a hybrid forecasting model using multi-strategy artificial bee colony algorithm (MSABC) to select the parameters of SVR model is proposed for annual electric load forecasting. In the proposed MSABC algorithm, Tent chaotic opposition-based learning initialization strategy is employed to diversify the initial individuals, and enhanced local neighborhood search strategy is applied to help the artificial bee colony (ABC) algorithm to escape from local optimum effectively. With the proposed MSABC applied to optimize the parameters of SVR model, a novel forecasting model, MSABC-SVR, is presented to forecast the annual electric load forecasting. The experiment results show that the MSABC can select the appropriate parameters of SVR model, which can effectively improve the forecasting accuracy of SVR. The intelligence load forecasting model has better performance than the regression model, the reason is the intelligence forecasting models has good non-linear fitting capacity. However, the SVR forecasting model has stability performance in the small sample forecasting.

In the future work, extensive experiment will be studied in other forecasting problems, compare more extensively with other models, and develop more efficient forecasting methods.

References

1. Li, L.H., Mu, C.Y., Ding, S.H., et al.: A robust weighted combination forecasting method based on forecast model filtering and adaptive variable weight determination. Energies **9**(1), 20–42 (2016). https://doi.org/10.3390/en9010020
2. Wang, J.J., Li, L., Niu, D.X., et al.: An annual load forecasting model based on support vector regression with differential evolution algorithm. Appl. Energy **94**(6), 65–70 (2012)
3. Chen, T.: A collaborative fuzzy-neural approach for long-term load forecasting in Taiwan. Comput. Ind. Eng. **63**(3), 663–670 (2012). https://doi.org/10.1016/j.cie.2011.06.003
4. Bozkurt, Ö.Ö., Biricik, G., Tayşi, Z.C.: Artificial neural network and SARIMA based models for power load forecasting in Turkish electricity market. PLoS ONE **12**(4), e0175915 (2017). https://doi.org/10.1371/journal.pone.0175915
5. Zhao, H.R., Zhao, H.R., Guo, S.: Using GM(1,1) optimized by MFO with rolling mechanism to forecast the electricity consumption of inner Mongolia. Appl. Sci. **6**(1), 20–38 (2016). https://doi.org/10.3390/app6010020
6. Wu, Q.: Hybrid model based on wavelet support vector machine and modified genetic algorithm penalizing Gaussian noises for power load forecasts. Expert Syst. Appl. **38**(1), 379–385 (2011). https://doi.org/10.1016/j.eswa.2010.06.075

7. Hong, W.C.: Electric load forecasting by seasonal recurrent SVR (support vector regression) with chaotic artificial bee colony algorithm. Energy **36**(9), 5568–5578 (2011). https://doi.org/10.1016/j.energy.2011.07.015
8. Kuang, F.J., Zhang, S.Y., Jin, Z.: A novel SVM by combining kernel principal component analysis and chaotic particle swarm optimization for intrusion detection. Soft. Comput. **9**(5), 1187–1199 (2015). https://doi.org/10.1007/s00500-014-1332-7
9. Karaboga, D., Basturk, B.: A comparative study of artificial bee colony algorithm. Appl. Math. & Comput. **214**(1), 108–132 (2009). https://doi.org/10.1016/j.amc.2009.03.090
10. Kuang, F.J., Zhang, S.Y.: A novel network intrusion detection based on support vector machine and tent chaos artificial bee colony algorithm. J. Netw. Intell. **2**(2), 195–204 (2017)
11. China National Bureau of Statistics: China Energy Statistical Yearbook 2011. China Statistics Press, Beijing (2011)
12. Amiri, M., Davande, H., Sadeghian, A., et al.: Feedback associative memory based on a new hybrid model of generalized regression and self-feedback neural networks. Neural Netw. **23**(9), 892–904 (2010). https://doi.org/10.1016/j.neunet.2010.05.005

Congestion Prediction on Rapid Transit System Based on Weighted Resample Deep Neural Network

Rong Hu[1,2](✉)

[1] Fujian Province Key Laboratory of Automotive Electronics and Electric Drive, Fujian University of Technology, Fuzhou 350108, China
ronghu0910@gmail.com
[2] Department of Civil Engineering and Engineering Mechanics, University of Arizona, Tucson 85718, USA

Abstract. Investigating congestion in train rapid transit system (RTS) in today's urban is demanded by both the operators and the public. Increase traffic data availability can be obtained from travel smart card and allowed to investigate the congestion of RTS. Artificial neural network are employed to do prediction on traffic. However the imbalance of data is a challenge to make an efficient prediction on congestion of RTS. This work proposes a Weighted Resample Deep Neural Network (WRDNN) model to predict the congestion level of RTS. The case study of RTS of one city of US indicate that the model introduced in this work can effectively predicting the congestion level of RTS with the 90% accuracy..

Keywords: Congestion prediction · Deep neural networks · Data imbalance Rapid transit system

1 Introduction

With the population density rising in urban cities, transportation planners often construct transit systems (RTS) as a first step. Yet with population growth and the increased complexity of train lines, planners are confronted with the difficulty of predicting commuter ridership, route choices, and also the various outcomes of the RTS during disruptions [1]. Increased station and train crowdedness in RTS lead to congestion, commuter discomfort, and trip delays. For this reason, it is very important to distribute the congestion information to the commuters timely. So they can change their trip plan or change the departure time to avoid congestion. On the other hand, the planners can explore effective approaches to remission congestion.

Large-scale data analytics into commuter travel behavior are gained through smart card ticketing in RTS. Some regression models have been proposed to identification of boarded trains [2], estimate commuter's patio-temporal density [3], travel patterns [4, 14], and transit use variability [5]. Most works on predicting traffic have focused on predicting crowd flows [6, 7].

P. Krömer et al. (Eds.): ECC 2018, AISC 891, pp. 586–593, 2019.
https://doi.org/10.1007/978-3-030-03766-6_66

Some works focus on predicting traffic congestion. Such as, Wanli Min et al. [8] propose an adaptive data-drive real-time congestion prediction method to identify different traffic patterns. Ma et al. [9] extend deep learning theory into large-scale transportation. They combine a deep restricted Boltzmann Machine and Recurrent Neural Network architecture to model and predict traffic congestion evolution. However, most of the approaches have some limitation. Especially for those data-driven methods, the accuracy of prediction is very low due to the data imbalance. In classification or prediction tasks, data imbalance problem is frequently observed when most of instances belong to one majority class [10]. This work we use Weight Resample Deep Neural Network to predict the congestion of RTS. Commonly, the history data is severe imbalance. Because the congestion or crowdedness only occurred on peak time, other time there is no congestion on RTS. If we use the imbalance data as the training data, it is very common for model to be overfitting by those majorities no congestion data and fail to gain good generalization on the unseen severe congestion data. Thus this kind of model cannot predict the true congestion and crowdedness of RTS timely. To tackle this problem, a Weighted Resample Deep Neural Network (WRDNN) is proposed to predict the congestion level timely. The Rapid Transit System on San Francisco of US is studied as a case. The experiments show the model outperforms model without weighted resample. The overall accuracy can be reach 90%.

2 Model WRDNN

2.1 DNN Architecture

The Artificial Neural Network consists of a number neurons arranged in a series of consecutive layers. Typically, it consists of an input layer, a hidden layer and an output layer [11]. Among data-drive models, the artificial neural network (ANN) models received a great attention during decades. Each neuron receives an array of inputs and produces a single output. The first layer is input layer using the training data as the input data. The second layer is the hidden layer which uses the output of input layer as its input data and output data to the output layer. Each neuron in all layers is activated by a function. A DNN, like a multi-layer perception (MLP), consists of an input layer, several hidden layers and an output layer [12]. Each layer has a fixed number of nodes and each sequential pair of layers is fully connected with a weight matrix. The nodes on a given layers are computed by transforming the output of the previous layer with the corresponding weight matrix: $a^{(i)} = M^{(i)} X^{(i-1)}$. The output of a given layer is calculated by applying an activation function: $X^{(i)} = h^{(i)} a^{(i)}$. An example of DNN architecture is shown as Fig. 1. Commonly the activation function uses the sigmoid, the hyperbolic tangent, rectified linear units and even a simple linear transformation.

The types of activation function applied for output layer depends on what the DNN is used for. If DNN is trained as a regression, then a linear function is applied and the mean squared error is applied as optimize function. If the DNN is trained as a classifier then the soft-max and the cross entropy is used as optimize function. For a classifier, each class is represented by an output node of the DNN classifier which is estimated by the posterior probability of the class given the input data.

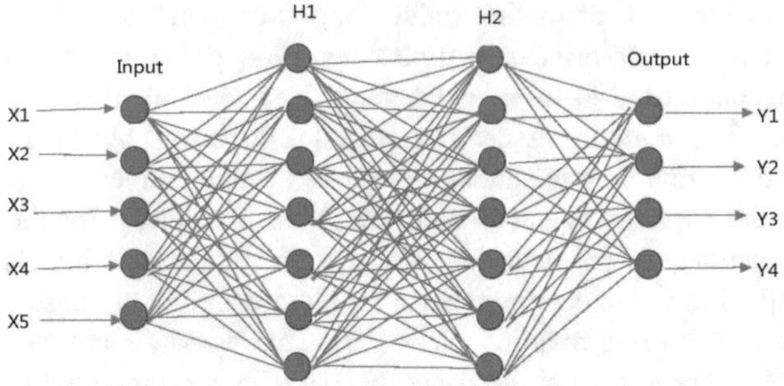

Fig. 1. Example DNN architecture

2.2 WRDNN for Congestion Detection

Commonly, the large scale data used as training data is imbalance. This work studies the RPT system of San Francisco. The dataset contains days ranging from January 1, 2017 to November 30, 2017. The data consists of three categories of variables: demand, supply and day attributes. Through test numerous variable, 45 variables are included in the final model. The total number of training data items is 625140. According understanding, the congestion is defined based on the average amount of passengers per train car to 4 levels: congestion, moderate, light and normal. Generally, data used as the training data is imbalance due to the congestion only occurred in peak time or special events day. In this case study, there 86% of the data indicate that it is normal without crowdedness. If this kind of data used as training data be directly input to the model, it will lead to overfitting on the majority class like normal and can't catch the good presentation of minor congestion data. Thus the model is useless. In order to solve this problem, a Weighted Resample Deep Neural Network is proposed. The procedure of Weight Resample is shown in Fig. 2.

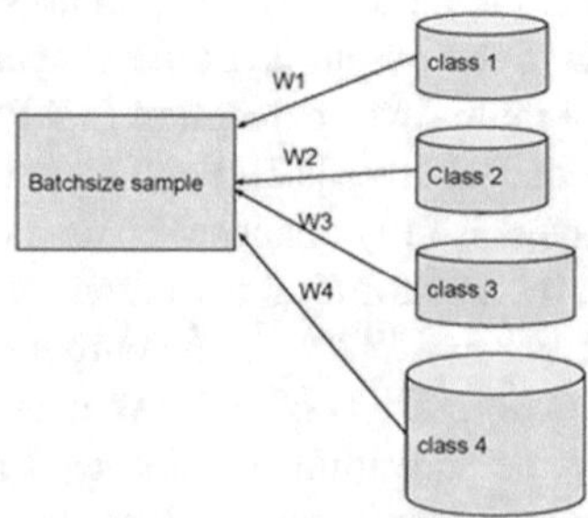

Fig. 2. Weighted Resample from different sub-dataset every batchsize sample

As shown in Fig. 2, for every training step, the input data weighted resample from different sub-dataset. W_i is the resample weight. Suppose the mini-batch size is B calculated by the following formula:

$$
\begin{aligned}
& B = \sum_{i=1}^{n} w_i \cdot B \qquad (i = 1, 2, 3 \ldots) \\
& s.t. \sum_{i=1}^{n} w_i = 1
\end{aligned} \tag{1}
$$

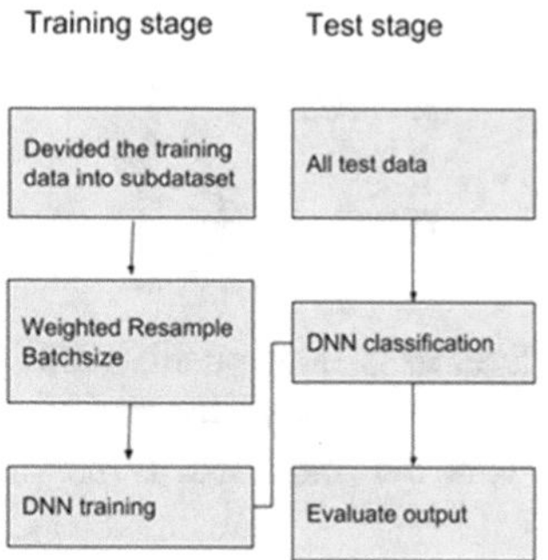

Fig. 3. Flowchart of the proposed WRDNN

Every training step, resample w_iB data from the ith class sub-dataset respectively as the input date. The flowchart of the proposed WRDNN is shown as Fig. 3.

3 Experimental Result and Evaluate

We use RTS of one city of US as a case study. The dataset contains days ranging from Jan. 1, 2017 to Nov. 30, 2017. The data consists of three categories of variables: demand, supply and day attributes which include 45 variables in the final model. Each period is defined to be 20 min in the modeling process. The operating hours is from 4:00 to Midnight on weekdays, from 6:00 am to Midnight on Saturday, from 8:00 am to Midnight on Sunday. Each period is assumed as and data, so there are 48 data one day on average. In total, the dataset consists of 8333520 data.

To validate the effectiveness and efficiency of the proposed model WRDNN, the minimizing cross-entropy error is used as the optimization objective during the model train procedure. The cross-entropy indicates the distance between the probability distributions of network outputs and target labels. The cross-entropy error is defined as the formula (2):

$$
CEE = -\sum_{i} \widehat{y}_i \log(y_i) \tag{2}
$$

While CEE is the Cross-entropy, $\widehat{y}_i$ is the predicted probability of value of class i. and y_i is the true probability for class i. A confusion matrix is a specific table layout that visualizes the performance of the model. Each row of the matrix indicates that the instances in a predicted class while each column signifies the instance in a true class [13].

In classification, precision, recall is indicators of the performance of model. Precision (positive predictive value) is the fraction of the relevant instances among the retrieved instances and recall (sensitivity) is the fraction of the relevant instances retrieved over the total amount of relevant instances. Precision and recall is calculated as the formula (3)

$$\begin{aligned} precision &= \frac{tp}{tp + fp} \\ recall &= \frac{tp}{tp + fn} \end{aligned} \tag{3}$$

While tp is the true positive, fp is false positive and fn is false negative.

F1 score is a measure that combines precision and recall. It is the harmonic mean of precision and recall. In this work the traditional F-measure is used:

$$F = 2 \cdot \frac{precision \cdot recall}{precision + recall} \tag{4}$$

This measure is known as the F1 measure, because it evenly weights recall and precision. In order to identify the structure of model which includes how many layers, and some other hyper parameters, some experiments are tried. Through experiments three hidden layers are deployed. The first hidden layer has 150 nodes, the second hidden layer has 100 nodes, and the third hidden layer has 50 nodes. The learning rate is set to 0.001, iterate step is set to 2000, batch size is set to 1000. The run environment of the model is Intel core i7 @ 3.40 GHz with 16 GB installed memory and 64-bit Operating system. To validate the performance of the proposed model, we run the model set with different weights of resample. The experiments of results are shown as Fig. 4.

From the Fig. 4(1), although the total test accuracy is 92%, but the recall of congestion is only 34%, also the recall of moderate is 51%, and the recall of the light is 61%. This shows the model failed to gain a good presentation of the minority data and it is useless when the mode is built without resample. From Fig. 4(3), we can see that when the resample is weighted by average the congestion level and normal level prediction obtain a good performance. From Fig. 4(2). When the weighted rates resample of congestion and moderate are increased a little more, the prediction accuracy on that class is increase lightly respectively. The Fig. 4(4) shows that the same weighted rate of resample, the test data is from total test data. For same reason, due to the imbalance of the test data, the results on precision are not so good. Only because if a little fraction of majority data (here is 89% of data is normal) are misclassified into congestion level, the result is overwhelmed by it (here only 2% of data is congestion). It is not reasonable to evaluate the model only considering the precision indicator. This

```
[[   1279     624     222    1593]
 [    130    2500    2071     197]
 [     29     919   13191    6207]
 [    805      29    2958 175626]]
Test Accuracy: 0.92425376
             precision    recall  f1-score   support

 congestion       0.57      0.34      0.43      3718
   moderate       0.61      0.51      0.56      4898
      light       0.72      0.65      0.68     20346
     normal       0.96      0.98      0.97    179418

avg / total       0.92      0.92      0.92    208380
```

Figure 4(1) The confusion matrix and report without resample

```
[[262  34   0   4]
 [ 29 241  29   1]
 [  3  46 144   7]
 [  6   0  15 179]]
             precision    recall  f1-score   support

 congestion       0.87      0.87      0.87       300
   moderate       0.75      0.80      0.78       300
      light       0.77      0.72      0.74       200
     normal       0.94      0.90      0.92       200

avg / total       0.83      0.83      0.83      1000
```

Figure 4(2): Weighted resample (Congestion is 0.3; Moderate is 0.3; Lighte is 0.2; Normal is 0.2)with the same weighted test data

```
[[214  32   1   3]
 [ 19 202  26   3]
 [  1  55 183  11]
 [  5   1  26 218]]
             precision    recall  f1-score   support

 congestion       0.90      0.86      0.88       250
   moderate       0.70      0.81      0.75       250
      light       0.78      0.73      0.75       250
     normal       0.93      0.87      0.90       250

avg / total       0.82      0.82      0.82      1000
```

Figure 4 (3): Weighted resample (Congestion is 0.25; Moderate is 0.25; Lighte is 0.25; Normal is 0.25) with the same weighted test data

```
[[  3153     533      12      20]
 [   380    4007     498      13]
 [   144    4477   14886     839]
 [ 12702     573   17638  148505]]
             precision    recall  f1-score   support

 congestion       0.19      0.85      0.31      3718
   moderate       0.42      0.82      0.55      4898
      light       0.45      0.73      0.56     20346
     normal       0.99      0.83      0.90    179418

avg / total       0.91      0.82      0.85    208380
```

Figure 4(4) : Weighted resample(Congestion is 0.25; Moderate is0.25;Lighte is 0.25; Normal is 0.25) with all test data

Fig. 4. The confusion matrix and the performance reports

can be concluded from the confusion matrix of Fig. 4(4). It shows that 12702 (which are only 7% of normal data) are misclassified into congestion level (which is only 2% of total data).

Table 1. Report of performance compared with different parameter model

model	Weighted Rate(C,M,L,N)	Precision(C,M,L,N)	Recall(C,M,L,N)	F1 (C,M,L,N)	Support(C,M,L,N)
Hidden 1 , size 200	0.2,0.2,0.3,0.3	0.86,0.69,0.76,0.89	0.83,0.69,0.79,0.89	0.84,0.69,0.78,0.8	200,200,300,300
Hidden 1 , size 200	0.25,0.25,0.25,0.25	0.86,0.73,0.74,0.90	0.84,0.72,0.77,0.89	0.85,0.72,0.75,0.89	250,250,250,250
Hidden 2 , size (150,150)	0.25,0.25,0.25,0.25	0.88,0.72,0.76,0.92	0.87,0.79,0.75,0.86	0.87,0.75,0.76,0.89	250,250,250,250
Hidden 3, size(250,200,150	0.25,0.25,0.25,0.25	0.87,0.68,0.77,0.92	0.80,0.78,0.75,0.90	0.84,0.73,0.76,0.91	250,250,250,250
Hidden 3, size(250,200,150	0.3,0.3,0.2,0.2	0.87,0.75,0.77,0.94	0.87,0.80,0.72,0.90	0.87,0.78,0.74,0.92	300,300,200,200
Hidden 3, size(100,50,50)	0.2,0.2,0.3,0.3	0.23,0.42,0.50,0.99	0.82,0.83,0.70,0.87	0.36,0.56,0.59,0.92	3718,4898,20346,179418
Hidden 3, size(100,50,50)	0.25,0.25,0.25,0.25	0.89,0.75,0.78,0.79	0.82,0.80,0.80,0.87	0.85,0.77,0.99,0.88	255,255,255,255
Hidden 3,size(100,50,50)	0.3,0.3,0.2,0.2	0.86,0.73,0.74,0.94	0.84,0.82,0.72,0.81	0.85,0.77,0.73,0.87	300,300,200,200
Hidden 3,size(100,50,50)	0.2,0.3,0.3,0.2	0.90,0.76,0.78,0.90	0.81,0.85,0.8,0.8	0.86,0.8,0.79,0.85	200,300,300,200
Hidden 3,without resample	random sampling	0.43,0.61,0.72,0.96	0.34,0.51,0.65,0.98	0.43,0.56,0.68,0.97	3718,4898,20346,179418

In order to validate the performance, some other experiments are explored and the comparison is shown in Table 1.

From Table 1, it shows the model performance with different hidden layers and different weighted rate of resample. It indicates that when the hidden layer is 3, the performance is better than 1 layer hidden layer and 2 layer hidden layer. When the resample are evenly weighted by 0.25, and hidden layer is 3 with hidden size is 100, 50, and 50 with respectively, the model gain a good result. The recall of every class is higher than 80% which indicate that the model can predict congestion level effectively. We also notice that when the weighted rate of resample is increased on some class, the mode can get a little more accuracy of prediction on that corresponding class. At the same time, the model with weighted resample outperformance that without resample, which can be shown from the last row of Table 1. Although the training data contain only 2% of congestion level data, the model proposed can still effectively predict the congestion level on 90% of accuracy, which is shown from the second line from the bottom of the Table 1.

4 Summary

Congestion in Rapid Transit Systems has presented a major problem in many cities. If the traffic information including congestion prediction of RTS can be distributed to public timely, it benefits both of passengers and planners. The passengers can change their trip plan to avoid the congestion. Also the city planners or the operator of RTS can plan their lines or some other control ways to address congestion. However data-drive mode always cannot effectively predict the unseen date due to the imbalance of the training data. This work proposes a Weighted resample Deep Neural Network (WRDNN) to predict the congestion level of RTS. The study case shows that carefully weighted resampling can increase the performance of the prediction compared with random sampling. The model can predict the congestion level with 90% accuracy based on only 2% training date of that.

Acknowledgments. This research is funded by Fujian Provincial Department of Science and Technology (Granted No. 2017J01729) and the China Scholarship Council.

References

1. Othman, N.B., Legara, E.F., Selvam, V., Monterola, C.: A data-driven agent-based model of congestion and scaling dynamics of rapid Transit systems. J. Comput. Sci. **10**, 338–350 (2015)
2. Kusakabe, T., Iryo, T., Asakura, Y.: Estimation method for railway passengers' train choice behavior with smart card transaction data. Transportation **37**(5), 731–749 (2010)
3. Sun, L., Lee, D.-H., Erath, A., Huang, X.: Using smart card data to extract passenger's spatio-temporal density and train's trajectory of MRT system. In: Proceedings of the ACM SIGKDD International Workshop on Urban Computing, pp. 142–148. ACM (2012)
4. Ma, X., Liu, C., Wen, H., Wang, Y., Wu, Y.-J.: Understanding commuting patterns using transit smart card data. J. Transp. Geogr. **58**, 135–145 (2017)
5. Kusakabe, T., Asakura, Y.: Behavioural data mining of transit smart card data: a data fusion approach. Transp. Res. Part C: Emerg. Technol. **46**, 179–191 (2014)
6. Zhang, J., Zheng, Yu., Qi, D., Li, R., Yi, X., Li, T.: Predicting citywide crowd flows using deep spatio-temporal residual networks. Artif. Intell. **259**, 147–166 (2018)

7. Polson, N.G., Sokolov, V.O.: Deep learning for short-term traffic flow prediction. Transp. Res. Part C: Emerg. Technol. **79**, 1–17 (2017)
8. Min, W., Wynter, L.: Real-time road traffic prediction with spatio-temporal correlations. Transp. Res. Part C: Emerg. Technol. **19**(4), 606–616 (2011)
9. Ma, X., Yu, H., Wang, Y., Wang, Y.: Large-scale transportation network congestion evolution prediction using deep learning theory. PLoS ONE **10**(3), e0119044 (2015)
10. Kim, M.-J., Kang, D.-K., Kim, H.B.: Geometric mean based boosting algorithm with over-sampling to resolve data imbalance problem for bankruptcy prediction. Expert Syst. Appl. **42**(3), 1074–1082 (2015)
11. Shoaib, M., Shamseldin, A.Y., Melville, B.W., Khan, M.M.: A comparison between wavelet based static and dynamic neural network approaches for runoff prediction. J. Hydrol. **535**, 211–225 (2016)
12. Richardson, F., Reynolds, D., Dehak, N.: Deep neural network approaches to speaker and language recognition. IEEE Signal Process. Lett. **22**(10), 1671–1675 (2015)
13. Powers, D.M.: Evaluation: from precision, recall and F-measure to ROC, informedness, markedness and correlation (2011)
14. Hu, R., Xia, Y.: Traffic condition recognition based on vehicle trajectory big data. J. Internet Technol. **18**(7), 1587–1596 (2017)

Visual Question Answer System Based on Bidirectional Recurrent Networks

Haoyang Tang(✉), Meng Qian, Ziwei Sun, and Cong Song

Xi'an University of Posts and Telecommunications, Xi'an 710121, China
tanghaoyang@xupt.edu.cn

Abstract. Visual Question Answer (VQA) system is the task of automatically answering natural language questions based on the content of reference image. A commonly approach for VQA is to extract image feature and question feature by convolution neural network (CNN) and long short-term memory network (LSTM) respectively, and then combine them to infer the answer through attention mechanism such as the stacked attention networks (SAN). However, the CNN ignores the information between adjacent image regions and the LSTM just memorizes the past contextual information of the question. In this paper, we propose a model based on two bidirectional recurrent networks (BiSRU and BiLSTM) to improve the accuracy of feature extraction. The BiSRU is used to allow the adjacent local region vectors of the image to maintain information each other. The BiLSTM is used to encode the question feature, which obtains past and future contextual information meanwhile when the question is very complex. The feature of image and question obtained by bidirectional recurrent networks is used to predict the answer precisely. Experiment result shows that our model get better performance on four datasets.

Keywords: Visual question answer system · BiSRU network
BiLSTM network

1 Introduction

As the science and technology have been developed rapidly, machine vision has been emerged as the active research area that includes natural language processing (NLP) [1], artificial intelligence (AI), visual question answering (VQA) [2], and so on. The goal of VQA system is to automatically answer natural language question according to the content of a reference image [3], as shown in Fig. 1.

Recently, most of the VQA models are based on neural networks [4]. Convolution neural network (CNN) is used to extract local regional vectors for local regions. Long short-term memory network (LSTM) [5] is used to encode feature vectors for the corresponding question. The attention mechanism such as the stacked attention networks (SAN) is used to locate the regions that are highly relevant to the answer by forming a refined query vector to query the image feature.

However, the answer obtained by these attention mechanisms such as SAN are not completely correct when the answer is consist of two adjacent local regional in the

P. Krömer et al. (Eds.): ECC 2018, AISC 891, pp. 594–602, 2019.
https://doi.org/10.1007/978-3-030-03766-6_67

Q: Are people skiing in the picture? Q: How many people are there in the photo?

Fig. 1. Sample images and question in VQA data sets. Note that commonsense knowledge is needed along with a visual understanding of the scene to answer many questions. Each question can be answered in simple vocabulary.

image and the question is a complex sentence. Thus we proposed a model to solve these problems. The main contributions of our work are as follows:

First, we applied BiSRU to maintain the information from adjacent local regions, which can obtain a more refined query vector to predict the potential answer.

Second, we applied BiLSTM to the question model, which can maintain contextual information from past and future and extract a semantic vector.

Third, we performed comprehensive evaluations on four image QA benchmarks, demonstrating that the experimental results of our model are superior to the previous model.

2 Related Work

Up to now there are many methods to address VQA [2, 6]. Most of the methods based on deep learning and extract image feature by CNN and encode question feature by LSTM.

However, there are two problems in extracting the image features by using CNN. First, the local regional vectors extracted from CNN cannot have global information. Without global information, their representational power is quite limited, which leads to the inaccuracy of the attention mechanism to locate the image regions. Second, the information of adjacent regions is interconnected. However, the CNN is not considered the information between adjacent local regional vectors. To solve these, we add the BiSRU behind the CNN. We input a series of local region vectors from CNN into BiSRU. Its principle is to obtain the global information and maintain the sequence information by statistical moving average.

LSTM can just memorize the past contextual information from the question and not use the following information of the question. These create errors when extracting the question feature. In NLP, BiLSTM network is used to capture information of sequential dataset and maintain contextual features from past and future. So we applied the BiLSTM to question feature extraction. The BiLSTM is used to obtain more representative question feature by accumulating context information adaptively through memory unit.

The SAN use semantic representation of a question as query to search for the regions in an image that are related to the answer. However, the query vector will be inaccurate when the input image and question feature are not highly representative.

3 Approach

The overall architecture of our model is shown in Fig. 2. First, we extract the image feature by using CNN and BiSRU. The CNN is used to obtain one vector for each image region. The BiSRU is used to make the connection in adjacent local regional vectors and output the image feature. Second, we extract a semantic vector of the question by using BiLSTM. Finally, given the image feature and question feature, SAN is used to combine them to infer the answer.

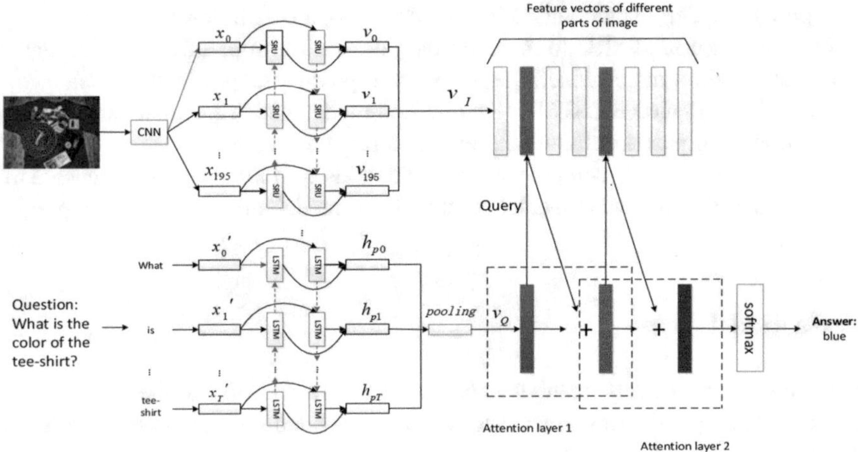

Fig. 2. The structure of our model for Visual Question Answering.

3.1 Image Feature Extraction

CNN is used to extract the image feature, which is based on pre-trained VGG-19 [7]. We first rescale the image to be 448 × 448 pixels and then take the feature from the last pooling layer, which therefore have a dimensions of 512 × 14 × 14, as shown in Fig. 3. Here 14 × 14 is the number of regions in the image and 512 is the dimension of the feature vector for each region. We use x_i, $i \in [0, 195]$ to represent the feature vector of each region, and $X = [x_0, x_1, \ldots, x_{195}]$ to represent the feature matrix of whole image.

However, the adjacent local regions are connected to each other. The image feature obtained only by using CNN is unable to express the relationship between adjacent local regions. To make the connection show in adjacent local regional vectors, the BiSRU is adopted to process image feature from CNN. A SRU chain is used to maintain the vector sequences information of local regions by moving averages of

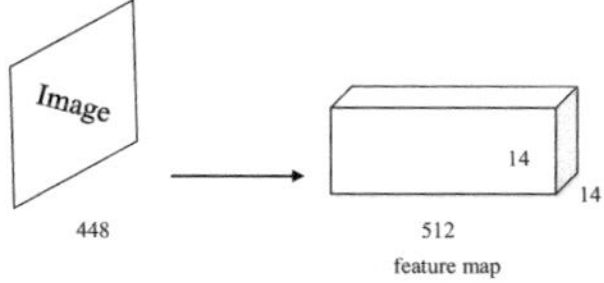

Fig. 3. CNN based image model.

statistics in multiple scales. SRU can receive moving averages and recurrent statistics by analyzing vector sequences [8]. We detail the equation for the SRU:

$$r_t = ReLU(W^r \mu_{t-1} + b^r) \tag{1}$$

$$\varphi_t = ReLU(W^{\varphi} r_t + W^x x_t + b^{\varphi}) \tag{2}$$

$$\forall \alpha \in A, \mu_t^{\alpha} = \alpha \mu_{t-1}^{\alpha} + (1 - \alpha)\varphi_t \tag{3}$$

$$o_t = ReLU(W^o \mu_t + b^o) \tag{4}$$

Where W and b are weighted parameters. Here $x_t, r_t, \varphi_t, \mu_t$ and o_t are input of SRU, previous date, recurrent statistics, moving averages and output of the network. So we input each image feature vector x_i to each SRU, and then the whole network output $v_p \in R^{512 \times 196}$, $v_p = (o_0, o_1, \ldots, o_{195})$.

Since SRUs process sequences in temporal order, it is not considered future information. So we adopt the BiSRU which consists of a forward SRU and a backward SRU. These two parallel layers can obtain information from past and future vectors to propagate each other. Finally, the output of BiSRU, image feature, equals to the addition of outputs both of forward SRU and backward SRU, which can be expressed as

$$v_I = \overrightarrow{v_p} + \overleftarrow{v_p} \tag{5}$$

Where $\vec{v_p}$ is the output of the forward SRU and $\overleftarrow{v_p}$ is the output of the backward SRU. And v_I is output fact of the BiSRU, where $v_I \in R^{512 \times 196}$ and its i-th column v_i is the visual feature vector for the region indexed i.

3.2 Question Feature Extraction

BiLSTM is used to extract question feature, which is intended to obtain information of sequential dataset and maintain contextual information from past and future. BiLSTM neural network is similar to LSTM network because both of them are constructed with LSTM units.

The basic structure of LSTM unit is composed of three gates and a cell state: input gate i_t, forget gate f_t, output gate o_t and memory cell c_t. The essential structure of a LSTM unit is a memory cell c_t which reserves the state of a sequence. At each step, the LSTM unit takes one input vector $x_t^{\prime}$ and updates the memory cell c_t, then output a hidden state h_t. The detailed update process is as follows:

$$i_t = \sigma\left(W_{xi}x_t' + W_{hi}h_{t-1} + b_i\right) \tag{6}$$

$$f_t = \sigma\left(W_{xf}x_t' + W_{hf}h_{t-1} + b_f\right) \tag{7}$$

$$o_t = \sigma\left(W_{xa}x_t' + W_{hf}h_{t-1} + b_o\right) \tag{8}$$

$$c_t = f_t c_{t-1} + i_t \tanh\left(W_{xc}x_t' + W_{hc}h_{t-1} + b_c\right) \tag{9}$$

$$h_t = o_t \tanh(c_t) \tag{10}$$

Where i, f, o, c are input gate, forget gate, output gate and memory cell. And W is weight matrix for input part and recurrent part of different gates. Here σ is non-linear function sigmoid.

Given the question $q = [q_0, \ldots q_T]$, where q_t is the hot vector representation of word at position t. We first embed the words to a vector space through an embedding matrix $x_t' = W_e q_t$. So the question translates into a matrix $X' = (x_0', x_1', \ldots, x_T')$. Finally, we feed the question matrix into LSTM, and then output $H = [h_0, h_1, \ldots, h_T]$.

Different from LSTM network, BiLSTM network has two parallel layers propagating in two directions. The internal structure of the forward and backward layers is the same. The output of BiLSTM, question feature, equals to the addition of outputs both of forward LSTM and backward LSTM, which can be expressed as

$$H_P = \overrightarrow{H_p} + \overleftarrow{H_p} \tag{11}$$

Where $\overrightarrow{H_p}$ is the output of the forward LSTM and $\overleftarrow{H_p}$ is the output of the backward LSTM. And H_P is output fact of the BiLSTM.

To produce a salient information from H_P, we process H_P through using max-pooling, average-pooling and min-pooling. In other words, we can obtain the maximum, average, and minimum values of $H(r, \cdot)$.

$$h\_\max(r) = \max\left[H_p(r,1), H_p(r,2), \ldots H_p(r,T)\right] \tag{12}$$

$$h\_\text{avg}(\text{r}) = \frac{1}{n}\sum_{j=0}^{n-1} H_p(r,j) \tag{13}$$

$$h\_\min(r) = \min[H_p(r,1), H_p(r,2), \ldots H_p(r,T) \tag{14}$$

Where $1 \le r \le T$. Finally, the representation vector for the question v_Q is composed with $h\_\max$, h_avg and $h\_\min$. Selecting tanh as the activation function:

$$h_p^* = [h\_\max^T, h\_\text{avg}^T, h\_\min^T]^T \tag{15}$$

$$v_Q = \tanh\left(h_p^*\right) \tag{16}$$

3.3 Stacked Attention Model

Given the Image feature matrix v_I and the question feature vector v_Q, SAN predicts the answer via multi-step reasoning [6]. The SANs is used to iterate the query-attention process using multiple attention layers, each extracting more fine-grained visual attention information for answer prediction. Formally, the SANs take the following formula: for the k-th attention layer, we compute:

$$h_A^k = \tanh\left(W_{I,A}^k v_I \oplus \left(W_{Q,A}^k u^{k-1} + b_A^k\right)\right) \tag{17}$$

$$p_I^k = softmax\left(W_P^k h_A^k + b_P^k\right) \tag{18}$$

$$\tilde{v}_I^k = \sum_i p_i^k v_i \tag{19}$$

$$u^k = \tilde{v}_I^k + u^{k-1} \tag{20}$$

Where u^k is exact query vector, which is computed by $\tilde{v}_I^k$ and u^{k-1}. Here u^0 is initialized to be v_Q. That is, we compute a new query vector u^k by combing question and image vector u^{k-1}, We repeat this K times and then use the final u^K to infer the answer:

$$p_{ans} = softmax\left(W_u u^K + b_u\right) \tag{21}$$

4 Experiment

4.1 Datasets

We evaluate the following common public Image QA networks for benchmark datasets such as DAQUAR-ALL, DAQUAR-REDUCED, COCO-QA and VQA. They collected question-answer pairs from existing image datasets and the answers are basically words or phrases.

COCO-QA dataset contains 78,736 training questions and 38,948 testing questions in the dataset. These questions are based on 8,000 and 4,000 images respectively.

DAQUAR-ALL dataset contains 6,795 training questions and 5,673 testing questions. In different contexts, 795 training images and 654 testing images were generated.

DAQUAR-REDUCED is a reduced version of DAQUAR-ALL. There are 3,876 training samples and 297 testing samples.

VQA dataset includes 248,349 training questions, 121,512 validation questions, 244,302 testing questions, and a total of 6,141,630 question answers pairs.

4.2 Evaluation Metrics

DAQUAR and COCO-QA employ both classification accuracy and its relaxed version based on word similarity, WUPS [9]. It uses threshold Wu-Palmer similarity based on WordNet taxonomy to compute the similarity between words. We measure all the models in terms of accuracy (Acc), WUPS 0.9(0.9), and WUPS 0.0(0.0).

VQA dataset provides open-ended task and multiple choice task for evaluation. For open-ended task, the answer can be any word or phrase while an answer should be chosen out of 18 candidate answers in the multiple-choice task. In both cases, answers are evaluated by accuracy reflecting human consensus.

4.3 Results and Analysis

The dataset includes DAQUAR-ALL, DAQUAR-REDUCED, COCO-QA and VQA are used to test the performance of our model. Our improved model can be called EnSANs. The model has been tested better than the original experimental model (SANs). Since we use the BiSRU to maintain the information between adjacent local regional of the image, and the BiLSTM maintain the contextual information from past and future, these are the specific advantage of our model. The experiment results in Tables 1, 2 and 3 show that the EnSANs gives the best results across all datasets.

Table 1. DAQUAR–ALL, DAQUAR-REDUCED and COCO-QA results.

Methods	COCO-QA			DAQUAR-ALL			DAQUAR-REDUCED		
	Acc	0.9	0.0	Acc	0.9	0.0	Acc	0.9	0.0
SANs [2]	61.6	71.6	90.9	29.3	35.1	68.6	46.2	51.2	85.1
EnSANs	63.8	73.5	91.6	31.1	36.2	69.9	47.9	53.1	85.7

Table 1 shows results on the COCO-QA, DAQUAR-ALL and DAQUAR-REDUCED datasets. On COCO-QA, our proposed EnSANs model outperforms the SANs in terms of accuracy, WUPS 0.9 and WUPS 0.0 by 2.2%, 1.9% and 0.7%. We also observe significant improvements on DAQUAR-ALL and DAQUAR-REDUCED.

Table 2 shows the results of our model and SANs on COCO-QA dataset. Compared to SAN, the biggest improvement is in the question type of *Object*, which reach 1.7%, followed by 0.9% in *Number*, 1.1% in *Color* and 2.7% in *Location*. The possible reason is BiSRU is used to handle the image feature is helpful for focusing image regions more relevant to answer to some extent.

Table 2. COCO-QA accuracy per category.

Methods	Objects	Number	Color	Location
SANs [2]	64.5	49.8	57.9	54.0
EnSANS	66.2	50.7	59.0	56.7

Table 3. VQA results on the official server.

Methods	Test-dev				Test-std All
	All	Yes/No	Number	Other	
SANs: [2]	58.7	79.3	36.6	46.1	58.9
EnSANS	59.3	80.7	38.4	47.7	59.6

Table 3 shows the result of our model and SANs on the VQA dataset. We also observe significant improvements on VQA dataset. Our model outperforms the SANs by 1.4% in the type of *Yes/No*, followed by 1.8% and 1.6% in *Number* and *Other*. The superior performance of our model across four datasets the effectiveness of using bidirectional recurrent network for input modules.

5 Conclusion

In this paper, we presented an improved model that BiSRU and BiLSTM were respectively used to extract more accurate image features and problem features. Experimental results shows that our model express best performance in COCO-QA, which reach 2.2% in term of accuracy. Our model is very effective for extracting question feature of complex problems, because the BiLSTM can process historical information and future information simultaneously, which well on sequential modeling problems. The BiSRU can maintain the information from neighboring image patches and capture long term information in the sequence. Since we propose several improvements to input modules, we improve the accuracy of the predicted answer.

Acknowledgement. This work was supported by Xi'an Bureau of Science and Technology Program (No. 201805040 YD18CG24 (1)).

References

1. Karpathy, A.: Deep visual-semantic alignments for generating image descriptions. IEEE Trans. Pattern Anal. Mach. Intell. **39**(4), 664–676 (2015). https://doi.org/10.1109/TPAMI.2016.2598339
2. Kulkarni, G.: Baby talk: understanding and generating simple image descriptions. IEEE Trans. Pattern Anal. Mach. Intell. **35**(12), 2891–2903 (2013). https://doi.org/10.1109/TPAMI.2012.162
3. Yao, J.: Describing the scene as a whole: joint object detection, scene classification and semantic segmentation. In: 6th IEEE Conference on Computer Vision and Pattern Recognition, pp. 702–709. IEEE, Providence, RI, USA (2012). https://doi.org/10.1109/cvpr.2012.6247739
4. Zhang, H.: Static correlative filter based convolutional neural network for visual question answering. In: 1th IEEE International Conference on Big Data and Smart Computing, pp. 526–529. IEEE, Shanghai (2018). https://doi.org/10.1109/bigcomp.2018.00087

5. Chowdhury, I.,: A cascaded long short-term memory (LSTM) driven generic visual question answering (VQA). In: 1th IEEE International Conference on Image Processing (ICIP), pp. 1842–1846. IEEE, Beijing (2017). https://doi.org/10.1109/icip.2017.8296600
6. Yang, Z.: Stacked attention networks for image question answering. In: 6th IEEE Conference on Computer Vision and Pattern Recognition, pp. 21–29. IEEE, Las Vegas, NV, USA (2016). https://doi.org/10.1109/cvpr.2016.10
7. LeCun, Y.: Gradient-based learning applied to document recognition. Proc. IEEE **86**(11), 2278–2324 (1995). https://doi.org/10.1109/5.726791
8. Rohrbach, M.: Translating video content to natural language descriptions. In: 3th IEEE International Conference on Computer Vision, pp. 433–440. IEEE, Syd-ney, NSW, Australia (2013). doi: https://doi.org/10.1109/iccv.2013.61
9. Wu, Z.: Verbs semantics and lexical selection. In: 7th Meeting on Association for Computational Linguistics, pp. 133–138. Acl Proceedings of Annual Meeting on Association for Computational Linguistics, Sydney, Australia (1994). https://doi.org/10.1162/ling.1994.00012

Multi-target Tracking Algorithm Based on Convolutional Neural Network and Guided Sample

You Zhou, Yujuan Ma, Guijin Han(✉), and Linli Sun

Xi'an University of Post and Telecommunications, Xi'an 710121, China
hgjin123@126.com

Abstract. In order to reduce the number of samples when using the convolutional neural network to train the moving target template online and improve the validity of samples, a sample selection method based on guided samples is proposed and applied to the fast multi-domain convolutional neural network tracking algorithm. The basic idea of the sample selection method is as follows, the initial samples are determined firstly by the sample filtering method of frame level detection and nonlinear regression model, and then the similarity between the initial samples and the target template are calculated, the samples with the similarity greater than a certain threshold are finally used as the guidance sample. The experimental results show that the tracking time of the proposed tracking algorithm is greatly reduced compared with the fast multi-domain convolutional neural network, the proposed tracking algorithm can speed up the tracking speed, improve the accuracy and robustness in complex environments.

Keywords: Target tracking · Convolutional neural network
Similarity measure

1 Introduction

The general idea of target tracking is to detect the position of the target by analyzing the video frame and the target initial bounding box, but there are still have a series of problems in achieving fast and robust target tracking under complex conditions. Object tracking algorithm with multiple instance learning [1] uses a strong classifier, which is constructed by Haar-like features, to determine the target position. Scale adaptive object tracking based on multiple features integration [2] combines three feature together to construct training samples, and use them to target modeling, the algorithm have high precision and can cope with complex scene changes, but the real-time performance is not very good. The TLD algorithm [3] uses the combination of tracking and detection strategies to identify targets in real time, but it can not to track the targets that disappear briefly. The KCF algorithm [4] uses the cyclic matrix to greatly reduce the amount of computation and thus improve the tracking speed, but the tracking target frame will drift due to the change of the target size. The tracking algorithm based on deep learning has been very successful in learning target feature representation. The full convolutional network [5] uses the deconvolution layer to upsample the feature

P. Krömer et al. (Eds.): ECC 2018, AISC 891, pp. 603–613, 2019.
https://doi.org/10.1007/978-3-030-03766-6_68

map of the convolutional layer, it can accept input images of any size and achieve end-to-end target tracking, but it is insensitive to image details, and the tracking accuracy is not good. GOTURN [6] uses a deep regression network for off-target tracking, the tracking speed is fast, but the accuracy is more accurate. The detection-based MDNet tracking algorithm [7] uses a new CNN structure to learn the common feature representations of different sequences, and updates the network weights by combining long-term updates and short-term updates, but online tracking takes a long time. Fast MDNet [8] uses a multi-domain network to train, the computing speed of network is quickened by the pooling layer, but the random sampling increases the number of network activations, and then results in slower tracking.

Almost all of the traditional tracking algorithms based on deep learning have a common defect, that is the training samples are sampled randomly by using the Gaussian distribution, in which the target position of the previous frame is used the centre of Gaussian distribution. This leads to the number of samples is too many, so that the network computation speed is too slow. For overcoming the defect, a Fast multi-domain convolutional neural networks based on guided sample (Guided Fast MDNet) is proposed in this paper, it uses the detection module of the TLD tracker and the ridge regression model of the KCF tracker to collect samples, and then calculate their similarity with the initial target, a guided sample can be obtained by comparing the similarity with a threshold. This method can solve time-consuming problem of fast multi-domain convolutional neural networks (Fast MDNet) tracking algorithm, and can improve the accuracy and robustness of tracking.

2 Fast Multi-domain Convolutional Neural Network Tracking Algorithm

2.1 Network Structure

The structure of the Fast MDNet network is shown in Fig. 1. The network including 5 convolution layers (conv1-conv5), one RoI pooling layer and two fully connected layers (Fc6-7), the input of network is 600 $\times$ 1000 RGB image, the 8 hidden layers complete feature extraction and classification, The K branches of the fully connected layer (Fc8) are correspond to K domains respectively, and each field corresponds to the video sequence training data of the target location. The image features are extracted by using five convolutional layers, the feature map information are extracted by the RoI pooling layer generates fixed-size feature vectors, and each feature vector is an input of two fully connected layers. Fc6 and Fc7 have 4096 neural units, which is combined with the ReLUs activation function and the Dropouts method to avoid linearization and overfitting. The two output neurons of the K branch layers classify the target and background in each branch, the output $[\phi(x), 1 - \phi(x)]^T$ is the probability of the target and the background in the input frame.

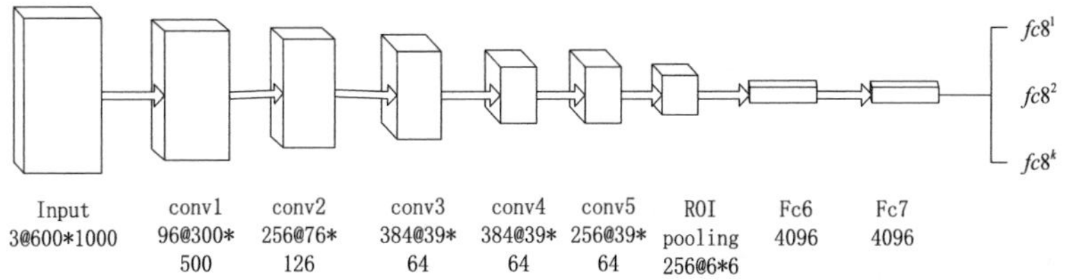

Fig. 1. Fast MDNet network structure

2.2 Fast Multi-domain Convolutional Neural Network Tracking Algorithm

The fast multi-domain convolutional neural network tracking algorithm includes two processes, offline training and online tracking. Offline training is used to extract feature information of the target, online tracking uses the previously tracked target position to predict the current frame target position.

Offline Training. The common information can be extracted from different domains by the training of fast multi-domain convolutional neural networks. The convolutional layer extracts only common information and ignores the personality information in each domain, the difference information will concentrates to the multi-domain layer by multi-domain training.

The samples are collected from the video sequence and trained in a small batch of Stochastic Gradient Descent (SGD), there are only one multi-domain branch is activated to participate in the calculation at each iteration, the branch is trained with the corresponding video data. The subsequent iterations will activate next multi-domain branch and the remaining branches do not participate in the calculation. In the Kth iteration, there is only one branch fc8 is used to update the weight. The training will be ended after the above process is iterated 100 times or the training error converges to a certain threshold.

Online Tracking. After offline training. The current frame target position can be predicted from the samples acquired in the previous frame. The tracking process is as shown in Fig. 2.

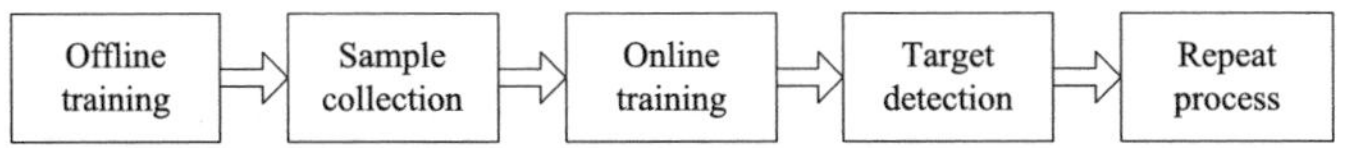

Fig. 2. Tracking process

Step 1. Detecting the contour of the person by using the network model constructed after the offline training.
Step 2. Collecting samples by Gaussian random sampling, which centre is the target position in Last frame.
Step 3. Calculating the sample confidence by entering the resulting sample into the network.

Step 4. Detecting the current frame target position by Using the network confidence.
Step 5. Repeating the above process

3 Multi-target Tracking Algorithm Based on Convolutional Neural Network and Guided Sample

Since the samples for predicting the current frame target in Fast MDNet tracker is randomly sampled, the difference between these samples and target is too large to guarantee their validity, and the number of samples is too large, that results in slow network calculation. In response to this problem, a sample selection method based on guidance sample is proposed in this paper. The sample is collected by frame level detection and nonlinear regression model, and then the similarity between these samples and the initial target is calculated by the similarity measure, those samples with the similarity greater than the specific threshold are used as the guidance sample. Compared with the random sampling, this method selects only part of them with higher similarity, so this method not only can reduce the number of samples, but also can increase sample validity. The sample collection method based on the guidance sample is used to Fast MDNet tracking algorithm, and a Guided Fast MDNet tracking algorithm is proposed.

3.1 Sample Selection Method Based on Guidance Samples

The process of selecting of guided sample is as shown in Fig. 3. The initial samples are gotten firstly by frame level detection and nonlinear regression model, in which the frame level detection is to use a sliding window to acquire candidate regions for each frame of image, and then use a variance filter and random pixel comparison to filter out more non-target regions, and the nonlinear regression model uses a ridge regression classifier to get regions similar to the target. Secondly, the similarity between all of the initial samples and the target from the last frame are calculated. Finally, the sample with higher similarity is retained as the guidance sample.

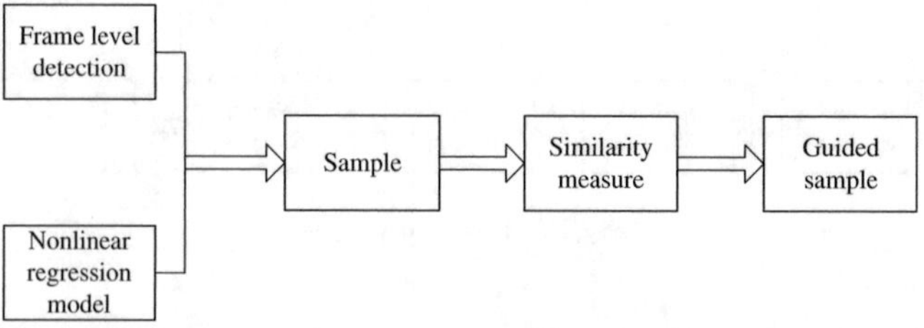

Fig. 3. Guided sample estimation

Sample Collection Based on Frame Level Detection. The target data based on the first frame mark includes a positive sample and a negative sample. Creating sliding windows of different sizes by marking the data and scanning the images separately to detect any possible position of the target in the frame, using a variance filter to screen out a large number of background areas. That is, the variance of the gray value of the candidate region and the target region is compared. the candidate region smaller than half of the variance of the target region is filtered out. The mean value of the gray value of the image area is $E(x)$. The variance of the gray value of the image area is

$$D(\mathrm{x}) = E(x^2) - E^2(x) \tag{1}$$

The obtained candidate region is input into the random forest set classifier, and the basic classifier on the scan window detects the pixel points of the candidate region according to the pixel determined by the first frame, comparing the difference between the pixel point and the gray level and generating a binary code x, probability of detection is

$$P_i(y|\mathrm{x}) = \frac{p}{p+n} \tag{2}$$

p and n are the number of positive and negative samples, $y \in (0, 1)$, and when the average of the posterior probability of the basic classifier is greater than 50% and the set classifier accepts the sliding window.

Sample Collection Based on Nonlinear Regression Model. Taking the first frame target position as the center using the cyclic matrix dense sampling, shifting the sample size to 2.5 times the target size, extracting the positive and negative sample direction gradient histogram features, and weighting the feature picture cosine window to alleviate the image caused by the boundary shift not smooth. The input histogram feature vector x is mapped to the feature space $\varphi(x)$, the parameter w is the linear combination of the input, and the coefficient of the ridge regression model is α.

$$w = \sum_i \alpha_i \varphi(x_i) \tag{3}$$

$$\alpha = (K + \lambda I)^{-1} y \tag{4}$$

I is the unit matrix and λ is the regular term. Calculating the mapped kernel matrix

$$K_{ij} = k(x_i, x_j) \tag{5}$$

Converting a circulant matrix to a Fourier transform

$$\hat{\alpha} = \frac{\hat{y}}{\hat{k}_{ij} + \lambda} \tag{6}$$

Where K_{ij} is the first row element of kernel matrix K and $\hat{K}_{ij}$ is the discrete Fourier transform of K_{ij}. Calculating the probability that the candidate region is the target location

$$\mathrm{y} = F^{-1}\left(\hat{k}_{xz} \cdot \hat{\alpha}\right) \tag{7}$$

In the previous frame, the target position is sampled, and the obtained sample feature vector is input into the regression model to obtain the ridge regression coefficient α to calculate the regression target y. A candidate region with a higher obtained value is selected as a sample.

Sample Similarity Measure. The sample similarity measure is as follows.

The resulting samples together form a data structure M, where s_i^+ is a positive sample, s_i^- is a negative sample, m is the number of positive samples, and n is the number of negative samples.

$$M = \{s_1^+, s_2^+ \ldots, s_m^+, s_1^-, s_2^- \ldots, s_n^-\} \tag{8}$$

The similarity measure defines the similarity between two samples s_i and s_j, where NCC is the normalized correlation coefficient, and the similarity between s_i and s_j is

$$S(s_{i,} s_j) = 0.5(NCC(s_{i,} s_j) + 1) \tag{9}$$

Negative nearest neighbor similarity can be expressed as

$$S^-(s, M) = \max_{(s_i^- \in M)} S\left(\mathrm{s}, \mathrm{s}_i^-\right) \tag{10}$$

50% positive nearest neighbor similarity is

$$S^+_{50\%}(s, M) = \max_{\left(s_i^+ \in M \wedge i \leq \frac{m}{2}\right)} S\left(\mathrm{s}, \mathrm{s}_i^+\right) \tag{11}$$

Conservative similarity S^c indicates that the sample is the first 50% probability of the nearest neighbor positive sample, and the conservative similarity is

$$S^c = \frac{S^+_{50\%}}{S^+_{50\%} + S^-} \tag{12}$$

When the conservative similarity of the sample is higher than the threshold T_{n}, it is defined as the guidance sample, and when the conservative similarity of the sample is less than T_{n}, it is randomly sampled.

3.2 Multi-target Tracking Algorithm Flow

The algorithm flow consists of five steps to complete the detection and tracking of the target.

Step 1. Collecting 128 samples every 4 frames from the video sequence, the samples, which overlap rate with the target are greater than 0.7, are selected as positive samples, and those samples that overlap ratio less than 0.5 are selected as negative samples. It uses SGD iterative training, in which each domain does not interfere with each other. Only one multi-domain branch is activated to participate in calculating in each iteration, and the corresponding video data is used to train the branch. The subsequent iteration activates the next multi-domain branch and the other branches do not participate in the calculation. In the Kth iteration, only one branch is used to update the weight. the training will be ended after 100 iterations or the training error converges to a certain threshold.

Step 2. Replacing the multi-branch output generated by the pre-training phase with a new branch output to accommodate the current tracking task.

Step 3. Randomly generating 1000 samples according to the initial bounding box given in the first frame, dividing each batch into 256 samples, iterating 5 times, training the Bounding-Box regression model, and fine-tuning the boundary regression weight parameters.

Step 4. For predicting the current frame target position, the samples are sampled by the frame level detection and the nonlinear regression model. Those samples with similarity higher than the threshold T_{n} is defined as the guidance sample, but if the overall similarity of samples is less than T_{n}, random sampling is performed. Then, he target confidence level $f^{+}(x^{i})$ and the background confidence level $f^{-}(x^{i})$ of the guidance samples are calculated by the forward propagation, the average value of the corresponding positions of the higher five samples of the confidence ranking is taken as the initial position of the current frame prediction target.

$$x^{*} = \arg\max_{x^{i}} f^{+}(x^{i}) \tag{13}$$

The candidate samples are sampled by the predicted target position, the sample confidence is calculated through the network, and the sample feature with the highest confidence is used as the input of the Bounding-Box regression model, and the Bounding-Box regression model is used to adjust the target position to complete the tracking.

Step 5. Saving the history tracked target, collecting samples based on the target location, and inputing the network to update the full connectivity layer parameters every 10 frames to complete the network update.

4 Experiment and Result Analysis

The ALOV300++ [9] and VOT2016 [10] benchmark data sets are used to the experiment in this paper. The tracking algorithm is run on a computer with a 3.20 GHz Intel Xeon E3-1225 CPU and 16 GB RAM, the program is implemented in python 2.7.

4.1 AlOV300++ Data Set Evaluation

The ALOV300++ data set includes 300 videos, and 60 videos are selected as experimental samples in this paper. Every 5 videos represent a kind of properties, including lighting, surface occlusion, specular reflection, transparency, shape changes, etc. The ALOV300++ data set uses F-Score as the evaluation indicator.

The evaluation method using the classification model represents the accuracy and survival curve of the tracker. F-Score is defined as the harmonic mean of the accuracy and recall rate, its range of value is from 0 to 1, where 0 is the worst and 1 is the best. Calculating the F-Score of the test video and plotting the corresponding survival curve. The curve data represents the performance of the tracker on the data set.

It can be seen from Fig. 4 that the survival curves of Guided Fast MDNet and Fast MDNet are basically identical, but the Guided Fast MDNet has better accuracy. Figure 5 compares the average *F-Score* score for each attribute. The experimental results show that Guided Fast MDNet has good adaptability under the conditions of illumination change, transparency, smooth movement, coherent movement, background chaos, occlusion, zoom and so on. From Table 1, it can be seen that Fast MDNet takes about 26 h in the total time of 60 video tracking of ALOV300++ dataset, while Guided Fast MDNet only takes 16.5 h. The experimental results show that Guided Fast MDNet has obvious advantages in reducing time consumption.

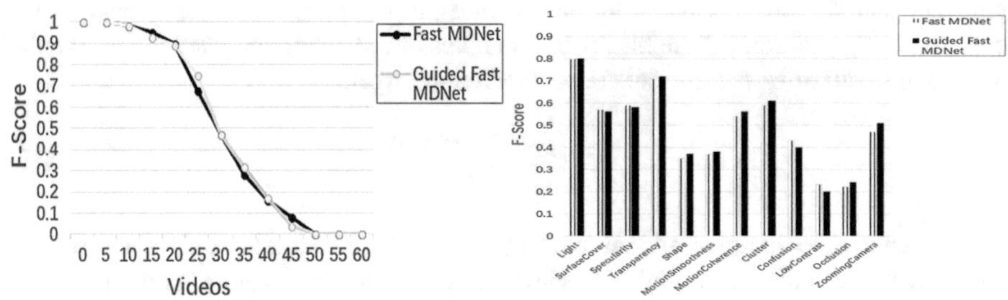

Fig. 4. Survival curve

Fig. 5. Average F-Score value

Table 1. Online tracking rate and time.

Iterm	Guided Fast MDNet	Fast MDNet
Tracking rate (ms/frame)	35.32	55.66
Total tracking time (hours)	16.50	26.00

4.2 VOT2016 Data Set Evaluation

The VOT2016 data set consists of 60 videos and examining the accuracy and robustness of the 15 videos of the VOT2016 data set. Accuracy is defined as the overlap rate between the predicted bounding box and the target location, and robustness represents the number of failures of the tracker. In order to test the tracking effect of Guided Fast MDNet, some typical trackers were selected for comparison tests, such as MDNet, DeepSRDCF [11], MUSTer [12], MEEM [13], SAMF [14], DSST [15] and KCF. The accuracy level and robustness level of the tracker are compared. Figure 6 shows the tracker accuracy and robustness of the A-R ranking obtained by the VOT2016 evaluation experiment. It can be seen from Fig. 6 that the robustness of Guided Fast MDNet ranks first. The robustness of Guided Fast MDNet is greatly improved compared to the MDNet tracker, and its accuracy is also improved. The ranking list of average expected overlap is as shown in Fig. 7, the average expected overlap can evaluate the overall performance of the tracker. Figure 7 shows that Guided Fast MDNet has better tracking performance. Table 2 shows that the total time of the 15 video tracking of the data set, Fast MDNet takes about 7 h, and Guided Fast MDNet only 4.5 h, you can see that Guided Fast MDNet tracking speed is faster.

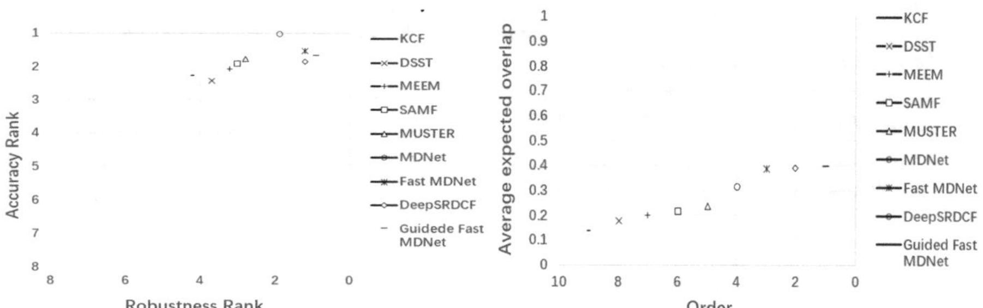

Fig. 6. A-R diagram of the VOT2016 data set

Fig. 7. Expected overlap analysis

Table 2. Online tracking rate and time.

Iterm	Guided Fast MDNet	Fast MDNet
Tracking rate (ms/frame)	37.41	59.43
Total tracking time (hours)	4.50	7.00

5 Conclusion

Aimed at the problem of the long calculating time of Fast MDNet tracking algorithm, the sample selection method is optimized from Gaussian random sampling to the guidance sample. The scanning grid filter and the ridge regression model are used to filter out a large number of background regions, the samples with higher similarity are selected to forecast the target location. The experimental results show that the Fast MDNet tracking time is greatly reduced, the tracking rate is faster, and the accuracy and robustness are higher in complex environments.

Acknowledgments. This work was supported by the Department of Education Shaanxi Province, China, under Grant 2013JK1023.

References

1. Li, N., Li, D.X., Liu, W.H., Liu, Y.: Object tracking algorithm with multiple instance learning. J. Xi'an Univ. Posts Telecommun. **19**, 43–47 (2014). https://doi.org/10.13682/j.issn.2095-6533.2014.02.007
2. Li, K., Liu, Y., Li, N., Wang, W.J.: Scale adaptive object tracking based on multiple features integration. J. Xi'an Univ. Posts Telecommun. **21**, 44–50 (2016). https://doi.org/10.13682/j.issn.2095-6533.2016.06.009
3. Kalal, Z., Mikolajczyk, K., Matas, J.: Tracking-learning-detection. IEEE Trans. Pattern Anal. Mach. Intell. **34**, 1409–1422 (2012). https://doi.org/10.1109/TPAMI.2011.239
4. Henriques, J.F., Rui, C., Martins, P., Batista, J.: High-speed tracking with kernelized correlation filters. IEEE Trans. Pattern Anal. Mach. Intell. 583–596 (2014). https://doi.org/10.1109/tpami.2014.2345390
5. Wang, L., Ouyang, W., Wang, X., Lu, H.: Visual tracking with fully convolutional networks. In: IEEE International Conference on Computer Vision, pp. 3119–3127. IEEE Press, Santiago (2016). https://doi.org/10.1109/ICCV.2015.357
6. Held, D., Thrun, S., Savarese, S.: Learning to track at 100 fps with deep regression networks, pp. 749–765 (2016). https://doi.org/10.1007/978-3-319-46448-0_45
7. Nam, H., Han, B.: Learning multi-domain convolutional neural networks for visual tracking. In: Computer Vision and Pattern Recognition, pp. 4293–4302. IEEE Press, Las Vegas (2016). https://doi.org/10.1109/cvpr.2016.465
8. Qin, Y., He, S., Zhao, Y., Gong, Y.: RoI pooling based fast multi-domain convolutional neural networks for visual tracking. In: International Conference on Artificial Intelligence and Industrial Engineering, (2016). https://doi.org/10.2991/aiie-16.2016.46
9. Smeulders, A.W.M., Chu, D.M., Cucchiara, R., Calderara, S., Dehghan, A., Shah, M.: Visual tracking: an experimental survey. IEEE Trans. Pattern Anal. Mach. Intell. 1442–68 (2013). https://doi.org/10.1109/tpami.2013.230
10. Kristan, M., Leonardis, A., Matas, J.: The visual object tracking VOT 2016 challenge results. In: IEEE International Conference on Computer Vision Workshops, pp. 98–111. IEEE Press (2013). https://doi.org/10.1007/978-3-319-48881-3_54
11. Danelljan, M., Hager, G., Khan, F.S., Felsberg, M.: Convolutional features for correlation filter based visual tracking. In: IEEE International Conference on Computer Vision Workshop, pp. 621–629. IEEE Press, Santiago (2016). https://doi.org/10.1109/iccvw.2015.84

12. Hong, Z., Chen, Z., Wang, C., Mei, X., Prokhorov, D., Tao, D.: MUlti-store tracker (muster): a cognitive psychology inspired approach to object tracking. In: Computer Vision and Pattern Recognition, pp. 749–758. IEEE Press, Boston (2015). https://doi.org/10.1109/cvpr.2015.7298675
13. Zhang, J., Ma, S., Sclaroff, S.: MEEM: robust tracking via multiple experts using entropy minimization, pp. 188–203. Springer (2014). https://doi.org/10.1007/978-3-319-10599-4_13
14. Li, Y., Zhu, J.: A scale adaptive kernel correlation filter tracker with feature integration. In: Lecture Notes in Computer Science, pp. 254–265 (2014). https://doi.org/10.1007/978-3-319-16181-5_18
15. Danelljan, M., Hager, G., Khan, F.S., Felsberg, M.: Learning spatially regularized correlation filters for visual tracking. In: 2015 IEEE International Conference on Computer Vision (ICCV), pp. 4310–4318. IEEE Press, Santiago (2016). https://doi.org/10.1109/iccv.2015.490

Face Recognition Based on Improved FaceNet Model

Qiuyue Wei, Tongjie Mu, Guijin Han(✉), and Linli Sun

Xi'an University of Posts and Telecommunications, Xi'an 710121, China
hgjin123@126.com

Abstract. The convolutional neural networks (CNN) is one of the most successful deep learning model in the field of face recognition, the different image regions are always treated equally when extracting image features, but in fact different parts of the face play different roles in face recognition. For overcoming this defect, a weighted average pooling algorithm is proposed in this paper, the different weights are assigned to the abstract features from different local image regions in the pooling operation, so as to reflect its different roles in face recognition. The weighted average pooling algorithm is applied to the FaceNet network, and a face recognition algorithm based on the improved FaceNet model is proposed. The simulation experiments show that the proposed face recognition algorithm has higher recognition accuracy than the existing face recognition methods based on deep learning.

Keywords: Face recognition · Deep learning · FaceNet
Convolutional neural networks · Weighting coefficient

1 Introduction

As one of the research hotspots in the field of computer vision, face recognition is an intelligent identity verification technology that uses a computer to complete facial feature extraction and identification. Face recognition is widely used in many fields, such as criminal detection, human-computer interaction and so on, because of its advantages of convenience, reliability and non-contact [1]. According to the difference of characterization methods, the existing face recognition technologies can be divided into face recognition method based on traditional features and face recognition methods based on deep learning.

The face recognition method based on traditional features is mainly based on artificially constructing facial features for classification and recognition. In the early days, the two approaches based on prior knowledge and geometric structure were the main approaches [2]. Subsequently, classical approaches, such as the method based on subspace analysis [3], elastic graph matching [4] and model-based [5] were emerging. Although the above-mentioned method can achieve relatively good results in a most ideal experimental environment, it is very sensitive to the internal and external factors, such as expression, posture, illumination, and pixels.

Nevertheless, the face recognition method based on deep learning uses the Convolutional Neural Networks (CNN) [6] to train the original image directly to obtain

P. Krömer et al. (Eds.): ECC 2018, AISC 891, pp. 614–624, 2019.
https://doi.org/10.1007/978-3-030-03766-6_69

high-level abstract features or low-dimensional representations of the face, so which can be further divided into two methods based on the intermediate layer classification and spatial distance discrimination. Among them, the method based on the intermediate layer classification uses face images to complete the training of the networks firstly, then takes the intermediate layer output as the high-level abstract features of the face, and finally sends the features into the classifier to classify. This method is represented by the DeepFace [7] and DeepID series [8–11] approaches. The disadvantage of this method is that its indirectness and the bottleneck representation cannot generalize well to new faces. The method based on spatial distance discrimination refers to the direct training network mapping the face image to the vector space, to get the low dimensional representation of the face, and to classify end to end using the space distance, which is represented by the FaceNet method [12]. The above described face recognition methods based on deep learning obtain good recognition performance in an uncontrolled environment.

The present face recognition based on FaceNet has achieved good recognition results, but it has a shortcoming, that is it treat equally different local image areas, which is not conformity with human habits, the different local image areas of face have different contributions when people identifying human face. For overcoming the above defects, a weighted average pooling algorithm is proposed and applied it to the FaceNet network, and then a face recognition algorithm based on improved FaceNet model is designed.

2 FaceNet Model

Figure 1 is a block diagram of training the FaceNet model, which includes two modules: preprocessing and low-dimensional representation extraction. First, the preprocessing module uses the Multi-task Cascaded Convolutional Networks (MTCNN) [13] to detect and align the sample set. Secondly, the low-dimensional representation extraction module consists of a batch input layer and a deep CNN followed by L2 normalization, which results in the face embedding. This is followed by the triplet loss during training.

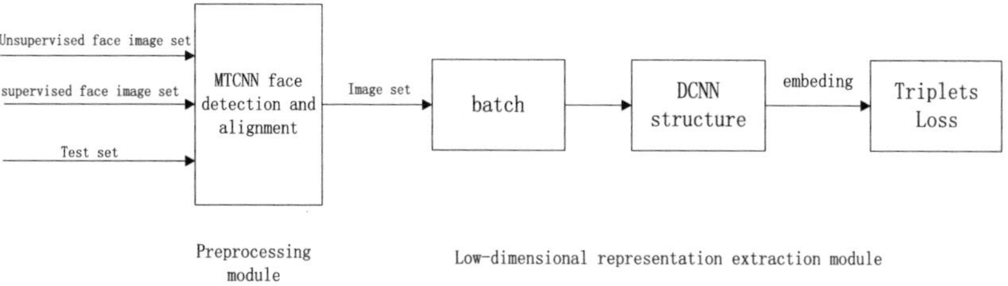

Fig. 1. The training block diagram of FaceNet model.

2.1 Model Structure

FaceNet [12] mainly discusses two different core architectures based on convolutional neural networks. The first category adds 1 × 1 × d convolutional layers between the standard convolutional layers of the Zeiler & Fergus architecture, and then get a model 22 layers NN1 model. The second category is Inception models based on GoogLeNet, which also is focused in this paper. Figure 2 is the network structure of an Inception module. It has 4 branches from left to right. The first branch is a 1 × 1 convolution, the second branch is a 3 × 3 convolution, the third branch is a 5 × 5 convolution, the fourth branch is a 3 × 3 max pooling, and each branch uses a 1 × 1 convolution to reduce time complexity.

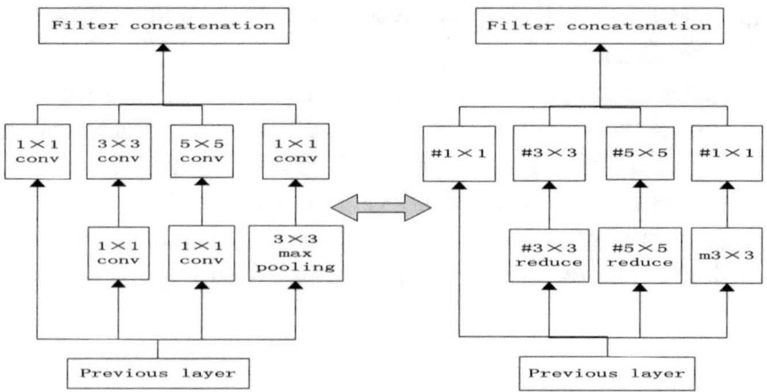

Fig. 2. Inception module.

2.2 Triplet Loss

For the input sample image $x \in \mathbb{R}^{H \times W \times D}$ and its corresponding sample embedding $f(x) \in \mathbb{R}^d$, there are:

$$f : \mathbb{R}^{H \times W \times D} \rightarrow \mathbb{R}^d \tag{1}$$

It indicates that an image x is embedded into a d-dimensional Euclidean space. Additionally, we constrain this embedding to live on the d-dimensional hypersphere by L2 normalization, i.e. $||f(x)||_2^2 = 1$. When training the network model by using the triple loss, all the triples generated by the training set should satisfy the requirements of Eq. (2).

$$||x_i^a - x_i^p||_2^2 + \alpha < ||x_i^a - x_i^n||_2^2, \forall (x_i^a, x_i^p, x_i^n) \in \mathrm{T} \tag{2}$$

where x_i^a(anchor) represents an image of a specific person, x_i^p(positive) represents an image of the same person, x_i^n(negative) represents an image of any other person. Besides, α is a margin between positive and negative pairs, which belongs to the empirical value and is set to $\alpha = 0.2$. T is the set of all possible triplets in the training set and has cardinality N.

However, the most triplets generating from training dataset are easily satisfied with Eq. (2), which results in a slower convergence. Here, given x_i^a, we want to select an x_i^p (hard positive) such that $argmax_{x_i^p} ||f(x_i^a) - f(x_i^p)||_2^2$ in a mini-batch, which is trained to satisfy that

$$||x_i^a - x_i^p||_2^2 < ||x_i^a - x_i^n||_2^2 \tag{3}$$

The goal of network training is to make the loss fall as small as possible in the training iteration. To ensure the closer the anchor sample is to the positive sample, the better the distance from the negative sample. Thus, the loss function of the training network by transforming Eq. (2), which is being minimized, is given by

$$L = \sum_i^N \max\left\{0, ||f(x_i^a) - f(x_i^p)||_2^2 - ||f(x_i^a) - f(x_i^n)||_2^2 + \alpha\right\} \tag{4}$$

3 Face Recognition Based on Improved FaceNet Model

The existing face recognition based on FaceNet model ignores the inequality of local features extracted from different regions of the sample image, and directly performs the unified pooling operation on the local features of the sample. As shown in Fig. 3, even if the sample image has been preprocessed, there is still an edge region partially containing no face information, and the different facial areas containing face information also plays different roles in face recognition. For overcoming these deficiencies, this paper introduces the learnable pooling weights, proposes a weighted average pooling algorithm, applies it to FaceNet network, and designs a face recognition algorithm based on improved FaceNet model.

Fig. 3. The sample images after preprocessing.

3.1 Weighted Average Pooling

The max pooling and average pooling methods are the two most commonly used pooling methods. Among them, the max pooling is suitable for extracting image local texture information, which is often used in the initial pooling layer of the model; the average pooling is suitable for extracting the global information of the image and is commonly used in the last pooling layer of the model. The two methods can be represented by Eqs. (5) and (6), respectively. Where Z_j^l is the *j*-th neuron of the *l*-th layer in the convolutional neural network, and y_i^{l-1} is the set of neurons in a kernel size region of the *l*-1th layer in the neural network corresponding to the neuron Z_j^l.

$$Z_j^l = \max\{y_i^{l-1}\} \quad (5)$$

$$Z_j^l = mean\{y_i^{l-1}\} \quad (6)$$

Unfortunately, the above pooling operation ignores the different contribution intensity difference of the local feature information of the feature image obtained by convolution. Motivated by this observation, a pooling method for weighted averaging of local features is proposed and the local and global information of the image is extracted according to the learned pooling weights. Figure 4 is a schematic diagram of a simple convolutional neural networks structure, replacing the maximum pooling of the initial pooling layer in the traditional convolution structure with a weighted average pooling, comparing the traditional convolution structure with the improved convolution structure.

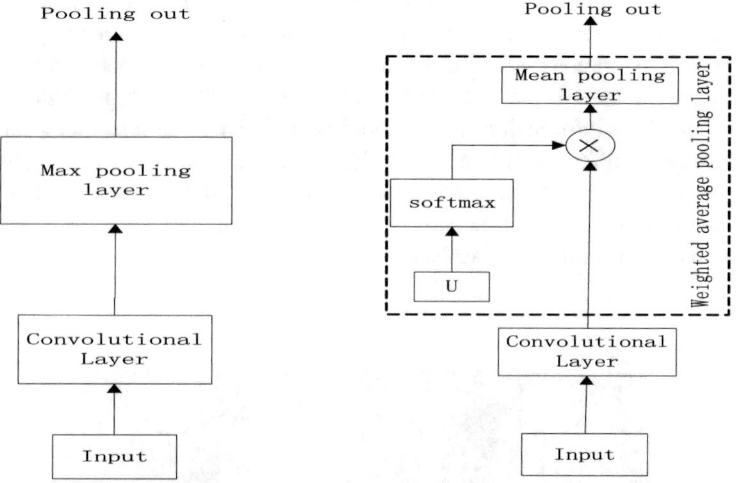

Fig. 4. A simple convolutional neural networks structure. **Left**: Traditional convolutional networks. **Right**: Improved convolutional networks.

Combined with Fig. 4, we present the specific construction process of the *l*-th layer weighted average pooling method as follows:

Step 1. Construct a Local Feature Set of the Sample. The local features of the sample are obtained from the previous *l*-1th layer convolution output to construct the local feature set to be pooled:

$$\mathbf{Y} = \left\{y_1^{l-1}, y_2^{l-1}, \cdots, y_i^{l-1}\right\}, i = \{1, 2, \cdots, n\} \tag{7}$$

Step 2. Determine the local feature initial weight coefficient. The truncated normal distribution function is used to generate random initial weights in the range [0, 1], and the corresponding stochastic initial weight vector is:

$$\mathbf{U} = \left\{u_1^l, u_2^l, \cdots, u_i^l\right\}, 0 \leq u_i \leq 1 \tag{8}$$

where, the mean value $u = 0.5$, the standard deviation $\sigma = 0.25$, u_i^l represents the initial weight coefficient of the i-th local feature of the sample to be pooled in the l-th layer.

Step 3. Construct a weighted pooling kernel vector. To ensure that the pooling weight is always within the value range of [0, 1], the softmax function is added to the pooled kernel, and the final pooling weight of each local feature in the pooled kernel is obtained.

$$\beta_i^l = softmax\left(u_i^l\right) = \frac{e^{u_i^l}}{\sum_{i=1}^{k} e^{u_i^l}}, \qquad 0 < \beta_i < 1, \ \& \sum_{j=1}^{k} \beta_i^l = 1 \tag{9}$$

where, β_i^l is computed in a softmax fashion following Eq. (9), u_i^l is a parameter need to be learned and k represents the size of the pooling window, i.e. the size of the pooling kernel. And the corresponding weighted pooling kernel vector is denoted as:

$$\mathbf{B} = \left\{\beta_1^l, \beta_2^l, \cdots, \beta_i^l\right\}, i = \{1, 2, \cdots, n\} \tag{10}$$

Step 4. Establish a weighted average pooling representation. The weighted pooled kernel vector is multiplied by the local feature set to calculate the final pooled abstract feature representation, as shown in Eq. (11). Where Z_j^l represents the j-th abstract feature representation after weighted average pooling of the l-th layer.

$$Z_j^l = k \times mean\left\{\beta_i^l y_i^{l-1}\right\} \tag{11}$$

3.2 Face Recognition Based on Improved FaceNet Model

In order not to increase the model complexity excessively, only the initial max pooling layer is replaced with the weighted average pooling layer in the improved model. Table 1 is a network structure for improving the NN3 model.

The Convolutional layers in Table 1 have similar meanings, such as conv1 (7 × 7 × 3, 2) indicates that the size of the first convolution kernel is 7 × 7, the channels is 3 and the stride is 2. According to Fig. 2, it can be seen that #1 × 1 represents the first branch in the corresponding Inception module; #3 × 3 and #3 × 3 reduce represent the second branch in the corresponding Inception module; #5 × 5 and #5 × 5 reduce represent corresponding to the third branch in the Inception module;

m in branch 4 indicates that the pooling type is the max pooling, w indicates that the pooling type is weighted average pooling, L2 indicates that the pooling type is L2 pooling, 2 indicates that the step size is 2,and 128P indicates that the channel is reduced to 128. In addition, the pooling is always 3 × 3 (aside from the final average pooling) and all convolution layer activation functions are modified linear units (ReLU), as expressed by

Table 1. Improved NN3 model structure and specific parameters

Type	Output size	#1 × 1	#3 × 3 reduce	#3 × 3	#5 × 5 reduce	#5 × 5	Pool proj(p)
conv1 (7 × 7×3,2)	80 × 80 × 64	–	–	–	–	–	–
weight pool + norm	40 × 40 × 64	–	–	–	–	–	w 3 × 3,2
Inception (2)	40 × 40 × 192	–	64	192	–	–	–
norm +max pool	20 × 20 × 192	–	–	–	–	–	m 3 × 3,2
inception(3a)	20 × 20 × 256	64	96	128	16	32	m,32p
inception(3b)	20 × 20 × 320	64	96	128	32	64	L2,64p
inception(3c)	10 × 10 × 640	0	128	256,2	32	64,2	m 3 × 3,2
inception(4a)	10 × 10 × 640	256	96	192	32	64	L2,128p
inception(4b)	10 × 10 × 640	224	112	224	32	64	L2,128p
inception(4c)	10 × 10 × 640	192	128	256	32	64	L2,128p
inception(4d)	10 × 10 × 640	160	144	288	32	64	L2,128p
inception(4e)	5 × 5 × 1024	0	160	256,2	64	128,2	m 3 × 3,2
inception(5a)	5 × 5 × 1024	384	192	384	48	128	L2,128p
inception(5b)	5 × 5 × 1024	384	192	384	48	128	m,128p
avg pool	1 × 1 × 1024	–	–	–	–	–	–
fully conn	1 × 1 × 128	–	–	–	–	–	–
L2normalization	1 × 1×128	–	–	–	–	–	–

$$f(x) = max(0, x) = \begin{cases} x, \ x > 0 \\ 0, \ x \leq 0 \end{cases} \tag{12}$$

Because the weighted average pool is introduced into initial pooling layer of the improved FaceNet model, the initial pooling layer parameters are also included in the process of training by BP back propagation. Similarly, the following ternary loss function is used as an objective function for improving FaceNet model training.

$$\begin{cases} L = \sum_{i}^{N} \max\left\{0, \|x_i^a - x_i^p\|_2^2 - \|x_i^a - x_i^n\|_2^2 + \alpha\right\} \\ s.t. \|x_i^a - x_i^p\|_2^2 + \alpha < \|x_i^a - x_i^n\|_2^2, \forall (x_i^a, x_i^p, x_i^n) \in T \end{cases} \quad (13)$$

4 Experimental Results

In order to verify the effectiveness of the proposed method, two face databases of CASIA-WebFace [14] and LFW (Labeled Faces in the Wild) [15] were used for experiments.

4.1 Sample Preprocessing and Parameter Setting

This paper carries out the same preprocessing for all training and test samples. First, the MTCNN [13] algorithm is used to perform face detection and locate five key points for each sample image. Then, a similar transformation is performed according to the position of the key points that are located, and finally all faces are cropped into pictures of a certain size (refer to Table 4).

In order to verify the feasibility and effectiveness of the improved FaceNet model, the CASIA Web Face database is selected as the training samples, the LFW dataset is selected as the test samples. And the training of the face model is completed by using the triplet loss function of Eq. (13). The main parameters are set as follows: the initial learning rate is 0.05, the weight attenuation is 0.0005, the training batch size is 100, and the max number of iterations is 90,000.

4.2 Performance on LFW

For verifying the feasibility of improved FaceNet model, the NN3 model is selected firstly to test. After replacing the max pooling of the initial pooling layer in the NN3 model with the weighted average pooling, the improved NN3 model can be gotten. The recognition accuracy of the NN3 model and the improved NN3 model are shown in Fig. 5(a). In addition, because the softmax function is added to the NN3 model to deal with the pooling weights, it is necessary to prove the importance of the softmax function. The experimental results of the improved NN3 model with or without the softmax function are shown in Fig. 5(b).

It can be seen from Fig. 5(a) that the recognition accuracy of the improved NN3 model has been significantly improved in the case of the same training set and parameter setting. Especially in the beginning (20,000 times of iterations), the accuracy is improved by 2.17%, which shows that the improved model has a faster convergence speed. And as can be seen from Fig. 5(b), the softmax function has a great impact on the accuracy of the model.

Combined with the data in Table 2, we can determine that the improved NN3 model can effectively improve the recognition accuracy of the original model without increasing the computational complexity overly. In the same way, the initial pooling

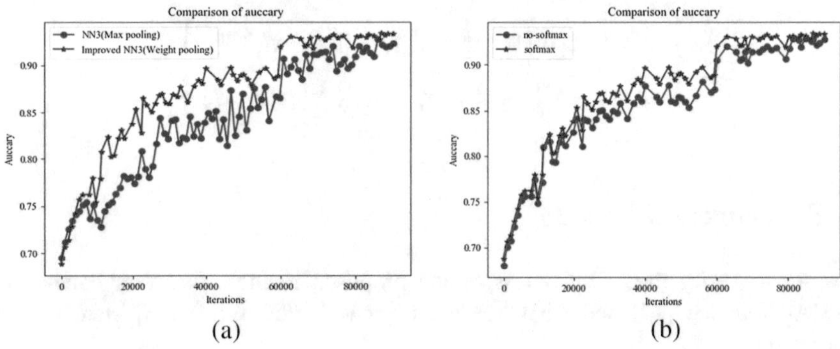

Fig. 5. (a) Comparison between the NN3 model and the improved NN3 model in accuracy. (b) Comparison of the improved NN3 model with and without softmax function in accuracy.

layer of the models such as NN2, NN4 and NNS2 is replaced, and the corresponding improved model structure is obtained. Comparison of the recognition accuracy before and after improvement of each model on the LFW test set is listed in Table 3. The recognition accuracy of each model after the improvement has increased by nearly 1%, which demonstrated that the strategy of the improved model proposed in this paper is effective.

Table 2. The recognition accuracy comparison among the NN3 model, the improved NN3 model and the improved NN3 model with and without softmax function.

Initial pooling layer	Accuracy (%)
Max pooling	92.30 ± 1.71
Weight pooling (softmax)	93.35 ± 2.40
Weight pooling(no-softmax)	92.76 ± 1.32

Finally, the pre-training model is introduced to complete the face recognition method based on the improved FaceNet model, and the experiment is compared with other face recognition methods based on deep learning, as shown in Table 4. The FaceNet model and the improved FaceNet model use the NNS2 model and the improved NNS2 model, respectively, trained on the CASIA-WebFace face database as the pre-training model, are fine-tuned on the LFW standard training set, and then are tested on the LFW standard test set. And it can be seen from Table 4 that the accuracy of the improved FaceNet model proposed in this paper has been slightly increased on the LFW database. Compared with the DeepID2 method, although our method has a slight gap in recognition accuracy, the number of networks used is far less than that of DeepID2.

Table 3. Comparison of recognition accuracy before and after improvement of each model.

Network architecture	Original model (%)	The improved model (%)
NN2(Inception 224 × 224)	95.05 ± 1.62	96.13 ± 1.37
NN3(Inception 160 × 160)	92.30 ± 1.71	93.35 ± 2.40
NN4(Inception 96 × 96)	91.27 ± 1.52	92.31 ± 1.70
NNS2(tiny Inception 224 × 224)	95.57 ± 1.43	96.68 ± 1.15

Table 4. Comparison of the face recognition performance on LFW.

Method	Training set	Number of network	Accuracy
DeepFace	SFC	1	97.35%
DeepID1	CelebFaces+	60	97.20%
DeepID2	CelebFaces+	25	99.15%
Face ++ v2014	–	–	97.30%
VGGFace	2.6 M	1	98.95%
FaceNet	CASIA-WebFace	1	98.47%
Improved FaceNet	CASIA-WebFace	1	99.07%

5 Conclusion

This paper proposes a weighted average pooling algorithm, applies it to the FaceNet network and designs a face recognition algorithm based on improved FaceNet model. The local features are extracted differential in the initial pool layer by introducing the contribution intensity coefficient, which can make up for the defect that the texture and details cannot be considered effectively in traditional model. The experimental results show that the method can increase the occupancy rate of effective information, reduce the loss of useless information, and obtain more distinguishable effective feature information without excessively increasing the parameter amount. At the same time, a higher recognition rate can be obtained.

Acknowledgements. This work was supported by the Shaanxi Natural Science Foundation (2016JQ5051) and the Department of Education Shaanxi Province (2013JK1023).

References

1. Mao, Y.: Research on Face Recognition Algorithm Based on Deep Neural Network. Master, Zhejiang University (2017)
2. Jing, C., Song, T., Zhuang, L., Liu, G., Wang, L., Liu, K.: A survey of face recognition technology based on deep convolutional neural networks. Comput. Appl. Softw. **35**(1), 223–231 (2018). https://doi.org/10.3969/j.issn.1000-386x.2018.01.039
3. Belhumeur, P.N., Hespanha, J.P., Kriegman, D.J.: Eigenfaces vs. Fisherfaces: recognition using class specific linear projection. IEEE Trans. Pattern Anal. Mach. Intell. **19**(7), 711–720 (1997). https://doi.org/10.1109/34.598228

4. Lades, M., Vorbruggen, J.C., Buhmann, J., et al.: Distortion invariant object recognition in the dynamic link architecture. IEEE Trans. Comput. **42**(3), 300–311 (1993). https://doi.org/10.1109/12.210173
5. Qin, H., Yan, J., Li, X., Hu, X.: Joint training of cascaded CNN for face detection. In: 29th IEEE Conference on Computer Vision and Pattern Recognition, pp. 3456–3465. IEEE, Las Vegas (2016). https://doi.org/10.1109/cvpr.2016.376
6. LeCun, Y., et al.: Backpropagation applied to handwritten zip code recognition. Neural Comput. **1**(4), 541–551 (1989). https://doi.org/10.1162/neco.1989.1.4.541
7. Taigman, Y., Yang, M., Ranzato, M., Wolf, L.: DeepFace: Closing the gap to human-level performance in face verification. In: 27th IEEE Conference on Computer Vision and Pattern Recognition, pp. 1701–1708. IEEE, Columbus (2014). https://doi.org/10.1109/CVPR.2014.220
8. Sun, Y., Wang, X., Tang, X.: Deep learning face representation from predicting 10,000 classes. In: 27th IEEE Conference on Computer Vision and Pattern Recognition, pp. 1891–1898. IEEE, Columbus (2014). https://doi.org/10.1109/cvpr.2014.244
9. Sun, Y., Chen, Y., Wang, X., Tang, X.: Deep learning face representation by joint identification-verification. In: 28th Annual Conference on Neural Information Processing Systems 2014, pp. 1988–1996. Neural information processing systems foundation, Montreal (2014)
10. Sun, Y., Wang, X., Tang, X.: Deeply learned face representations are sparse, selective, and robust. In: 28th IEEE Conference on Computer Vision and Pattern Recognition, pp. 2892–2900. IEEE, Boston (2015). https://doi.org/10.1109/cvpr.2015.7298907
11. Sun, Y., Ding, L., Wang, X., Tang, X.: DeepID3: Face recognition with very deep neural networks. arXiv:1502.00873 (2015)
12. Schroff, F., Kalenichenko, D., Philbin, J.: FaceNet: a unified embedding for face recognition and clustering. In: 28th IEEE Conference on Computer Vision and Pattern Recognition, pp. 815–823. IEEE, Boston (2015). https://doi.org/10.1109/cvpr.2015.7298682
13. Zhang, K., Zhang, Z., Li, Z., Qiao, Y.: Joint face detection and alignment using multitask cascaded convolutional networks. IEEE Signal Process. Lett. **23**(10), 1499–1503 (2016). https://doi.org/10.1109/LSP.2016.2603342
14. Yi, D., Lei, Z., Liao, S., Li, S. Z.: Learning face representation from scratch. arXiv:1411.7923 (2014)
15. Huang, G.B., Ramesh, M., Berg, T., Learned-Miller, E.: Labeled faces in the wild: a database for studying face recognition in unconstrained environments. Technical report, University of Massachusetts, Amherst (2007)

Robot and Intelligent Control

Sliding Window Type Three Sub-sample Paddle Error Compensation Algorithm

Chuan Du, Wei Sun(✉), and Lei Bian

School of Aerospace Science and Technology,
Xidian University, Xi'an 710118, China
wsun@xidian.edu.cn

Abstract. In the speed update process of the Strap-down Inertial Navigation System (SINS), an improved paddle error compensation algorithm is proposed to solve the problem that increasing the number of subsamples will reduce the system speed update frequency, but increasing the sampling frequency will increase the hardware burden. The gyroscope and accelerometer sample values of the first two cycles and the current cycle gyroscope and accelerometer sample values form a window, and the sliding window is used for paddle error compensation. In this paper, the principle of the three-sample paddle error compensation algorithm based on sliding window is discussed in detail. The performance of the proposed algorithm and the traditional three-subsample algorithm are compared and tested. The experimental results show that the proposed algorithm can increase the speed update frequency by 2 times without increasing the sampling frequency, which has good application value.

Keywords: Strap-down Inertial Navigation System · Paddle error Compensation algorithm · Speed update

1 Introduction

Speed update is an important part of the Strap-down Inertial Navigation System (SINS) algorithm. In the environment of high dynamic operation or severe vibration, the non-exchangeability of rigid body finite rotation will bring great negative effects. Cone effect will occur in attitude update and the paddle effect is generated in the speed update. The corresponding error compensation algorithm is called the cone error compensation algorithm and the paddle error compensation algorithm [1]. In the process of speed calculation, due to the linear motion and angular motion of the carrier, when the speed increment is used for calculation, the system's paddle error and its compensation must be considered.

In the papers [3–7], these existing paddle error compensation algorithm has greatly improved the accuracy of the algorithm, but there are problems: increasing the number of subsamples at the same sampling frequency will reduce the SINS speed update frequency, but to maintain the speed update frequency, it is necessary to increase the sampling frequency, which will increase the navigation processor hardware burden. Due to hardware limitations, the system's maximum sampling frequency is usually fixed. Therefore, based on the literature [5, 6], this paper proposes a sliding window

P. Krömer et al. (Eds.): ECC 2018, AISC 891, pp. 627–634, 2019.
https://doi.org/10.1007/978-3-030-03766-6_70

type three sub-sample paddle error compensation algorithm. Compared with the traditional paddle error compensation algorithm, the speed update frequency is improved without changing the sampling frequency of the system.

2 The Principle of Paddle Error Compensation Algorithm

According to the literature [1], the specific force of the previous moment can be written as:

$$\begin{aligned}\Delta v_{sfk}^{b_{k-1}} &= v_k + \frac{1}{2}(\sigma_k \times v_k) + \frac{1}{2}\int_{t_{k-1}}^{t_k}[\sigma(\tau) \times f^b(\tau) - \omega^b(\tau) \times v(\tau)]d\tau \\ &= v_k + \Delta_{rot} + \Delta_{vsul}\end{aligned} \quad (1)$$

$$\Delta_{vsul} = \frac{1}{2}\int_{t_{k-1}}^{t_k}[\sigma(\tau) \times f^b(\tau) - \omega^b(\tau) \times v(\tau)]d\tau \quad (2)$$

Δ_{vsul} is called the paddle error compensation term for speed. That is, when there are angular vibrations and line vibrations of the same frequency in the same phase, a paddle error occurs.

The paddle error is discussed below. From the literature [8], the three-sample algorithm for paddle error under linear fit is:

$$\Delta\widehat{V}_{scul} = k\frac{BC}{\Omega}[\sin\lambda(4k_2 - k_1) + \sin 2\lambda(2k_1 - k_2) - k_1\sin 3\lambda] \quad (3)$$

In the formula, k_1 and k_2 are the optimal coefficients.

The exact calculation of the paddle error is:

$$\Delta V_{scul} = k\frac{BC}{\Omega}[T - \frac{1}{\Omega}\sin\Omega\tau] \quad (4)$$

For the three-subsample algorithm, $T = 3\Delta T$, so:

$$\Delta V_{scul} = k\frac{BC}{2\Omega}[3\lambda - \sin 3\lambda] \quad (5)$$

Therefore, the three sub-like paddle errors is

$$\begin{aligned}\Delta\widehat{V}_{scul} &= \Delta V_{scul} - \Delta\widehat{V}_{scul} \\ &= k\frac{BC}{\Omega}[-\frac{\lambda^3}{3!}(12k_1 + 12k_2 - \frac{27}{2}) + \frac{\lambda^5}{5!}(-180k_1 - 60k_2 + \frac{243}{2}) + \cdots]\end{aligned} \quad (6)$$

Let the coefficients of the third power and the fifth power be zero, and the optimization coefficients $k_1 = \frac{27}{20}$ and $k_2 = \frac{9}{40}$ in the three-subsample algorithm can be solved, which can minimize the paddle error of the three subsample algorithm. which is:

$$\Delta\widehat{V}_{scul} = \tfrac{9}{20}[\Delta\theta_m(1)\times\Delta V_m(3)+\Delta V_m(1)\times\Delta\theta_m(3)]+\tfrac{27}{40}[\Delta\theta_m(1)\times\Delta V_m(2) + \Delta\theta_m(2)\times\Delta V_m(3)+\Delta V_m(1)\times\Delta\theta_m(2)+\Delta V_m(2)\times\Delta\theta_m(3)] \tag{7}$$

3 The Algorithm of Error Compensation of Paddle with Sliding Windows

In terms of three-sample algorithm, it can be known from Eqs. (1), (7) that in the traditional algorithm of error compensation of paddle with sliding windows, it needs three sampled-numbers to calculate a term of the error compensation of paddle $\Delta\widehat{V}_{scul}$. Assuming that the period of sampling is θ_k, and the compensation period of paddle error is h_k, then $h_k = 3\theta_k$. That is why the period of speed updating becomes slower. Figure 1 shows the traditional algorithm of error compensation of paddle.

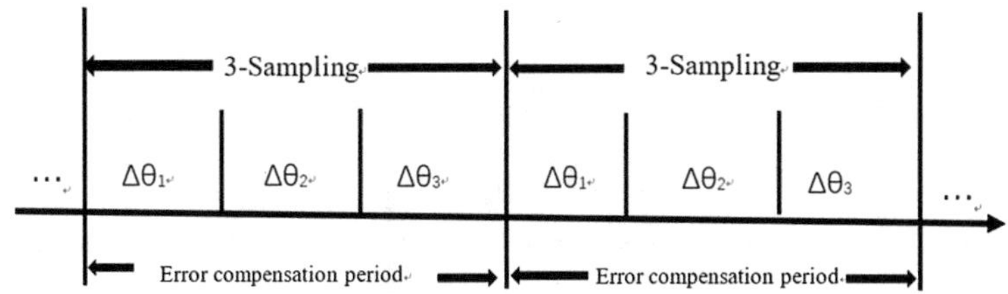

Fig. 1. The traditional algorithm of error compensation of paddle with sliding windows

Based on the traditional three-sample algorithm, this paper proposes an improved way. The sampling period θ_{n+1} with θ_{n-1} and θ_n which are previous forms a new three-subsample, forms a new window and ends the period of the compensation. The window slides back a time-interval when a sample is taken. It needs three sampling periods because the calculation of paddle error must depend on three subsamples. Therefore, the calculation of the paddle error is separated from the period of compensation. One third of the compensation of the paddle error in the every calculation cycle is used as the compensation of paddle error of the first sampling period in the window. Since the calculation of the paddle error must rely on three subsamples, it needs three sampling periods. In this way, the sampling period is consistent with the compensation period of the paddling error as shown in Fig. 2.

It shows in the formula (1) of the paddling error which is rewritten by the formula (8).

$$\Delta v_{sfk}^{*b_{k-1}} = v_k + \Delta_{rot} + \frac{1}{3}\Delta_{vsul} \tag{8}$$

At the same time, the period of speed updating (t_{k-1}, t_k) in the entire system becomes the sampling period of the gyroscope and acceleration, as well as $h_k = 3\theta_k$. The period of the compensation of the acceleration $\Delta v_{g/cork}$ and the compensation of the rotation effect Δ_{rot} become one third of the original ones. Figures 3 and 4 show the contrast between the improved process of speed updating and the previous process of speed updating.

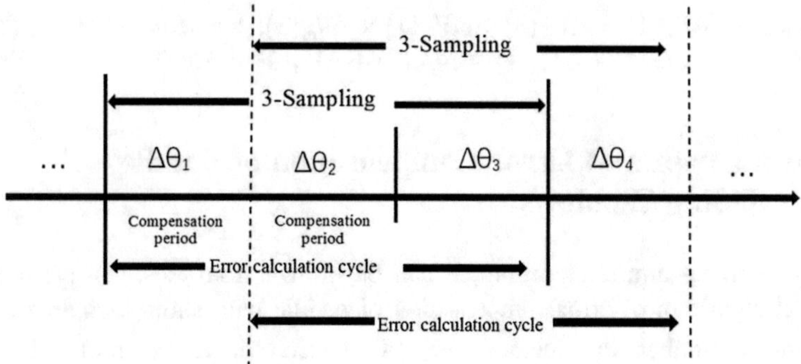

Fig. 2. The improved algorithm of error compensation of paddle

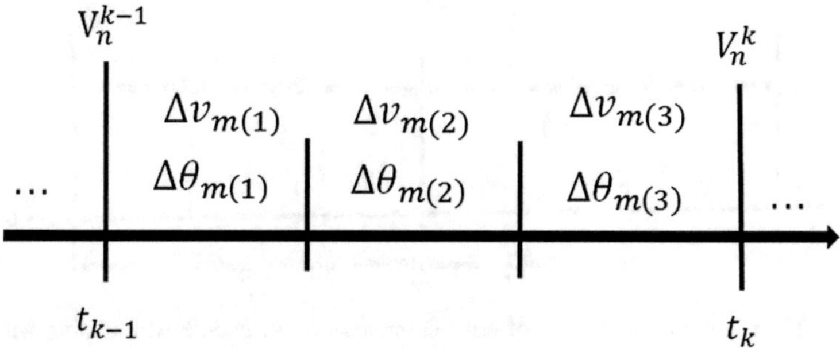

Fig. 3. The traditional algorithm

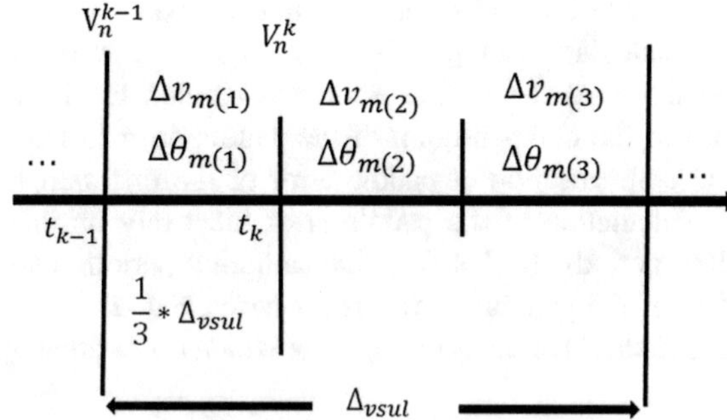

Fig. 4. The improved algorithm

It can be seen from Figs. 1, 2, 3 and 4 that the frequency of the system's speed updating is tripled by using the algorithm of error compensation of paddle with sliding windows without changing the system's sampling frequency.

4 The Test of the Algorithm Performance

4.1 The Comparison About the Speed Details Measurement and Response Speed

The sensor is attached to the car which is running at low speed, and the X axis is facing the running direction of the car. Table 1 shows the model and main parameters of the sensor.

Table 1. Sensor parameters

Parameter	Value
Acc Range(g)	$\pm$ 18
Acc Accuracy(g)	0.03
Gyro Range(dps)	$\pm$ 450
Gyro Bias Repeatability(1yr)(dps)	0.2
Attitude Update Frequency(Hz)	100

The car starts running with the speed of 0 m/s on the horizontal road, accelerates rapidly to the speed of 3.5 m/s, and then stabilizes at the speed of 3.5 m/s until the time reaches 160 s. Then, the car stops immediately. The response speed test of the algorithm is applied with this way of fast start and stop. Figure 4 shows the images of the carrier speed calculated by two algorithms (The blue one is calculated by the improved algorithm, and the red one is calculated by the traditional algorithm).

Figure 5 shows that both algorithms detect the motion curve of the carrier very well. Figure 6 shows the curve of the improved algorithm is more delicate and smooth which is due to the improved frequency of the new algorithm. In the initial acceleration process, the speed calculated by the improved algorithm is firstly increased. In the final braking process, the speed calculated by the improved algorithm is firstly reduced to zero. This shows the improves in the response speed of the algorithm.

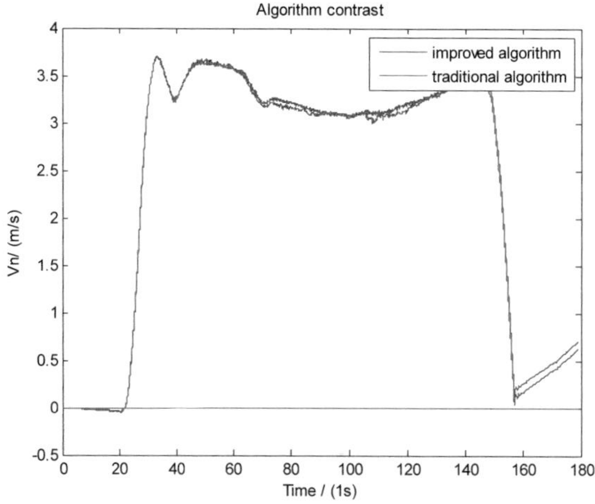

Fig. 5. The comparison about the algorithms' response speed

4.2 Algorithm Error Test

The sensor is fixed to the car, the Y-axis is facing the direction of the car, the x-axis is toward the right of the car, the Z-axis is vertical, the initial speed of the car is 10 m/s, and the car is accelerated to 18 m/s in 80 s and then decelerated to 10 m/s and keep at a constant speed, the system calculates the carrier speed image as shown in Fig. 6.

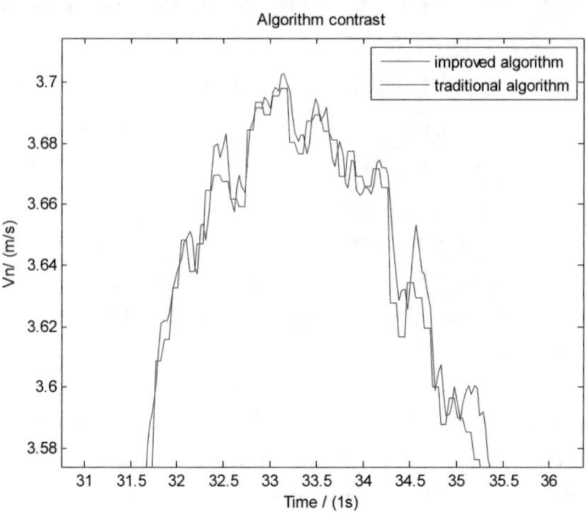

Fig. 6. The comparison about the algorithms' details

It can be seen from Fig. 7 that the system still detects the change of the speed of the car very well. The Z and X axis offsets do not exceed 1.8 m/s, the Y axis average speed is 9.32 m/s, and the error is 0.68 m/s. In fact, due to the deviation of the car's forward direction from the starting direction, the speed is decomposed into a part of the X-axis. Therefore, the actual speed measurement error is less than 0.68 m/s.

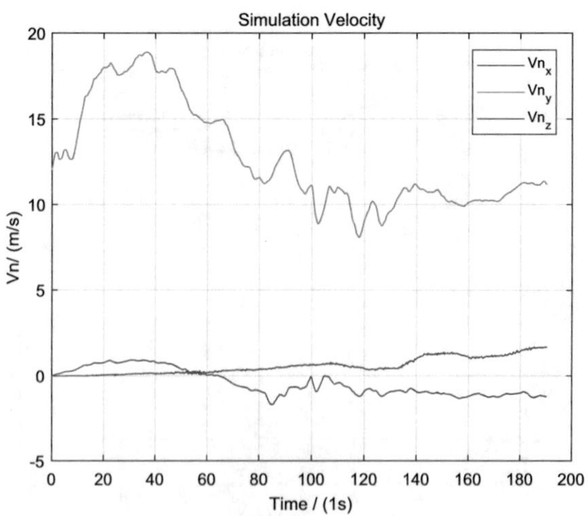

Fig. 7. Acceleration and deceleration test

The car advances at a high speed in the north direction at a speed of about 20 m/s. The test results are shown in Fig. 8.

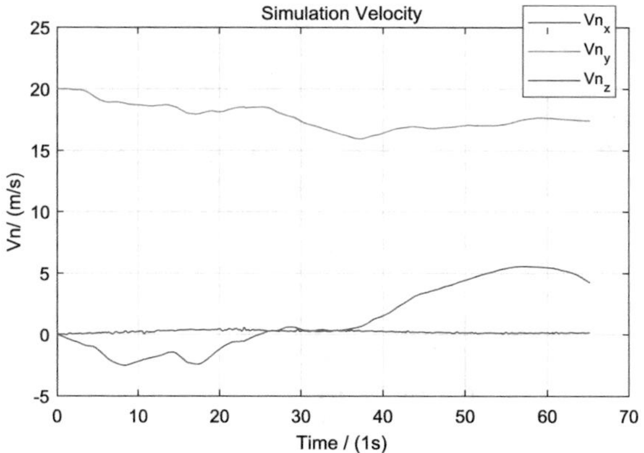

Fig. 8. High-speed status test

It can be seen from Fig. 8 that in the high speed state, the average speed of the carrier calculated by the whole phase system is 19.67 m/s, and the error is 0.33 m/s. The speed is also decomposed on the X-axis due to the deviation of the car from the initial position. Therefore, the system speed measurement error should be less than 0.33 m/s.

5 Conclusion

In the speed update process of the Strapdown Inertial Navigation System, the speed of the paddle error compensation algorithm determines the speed of the speed update algorithm. In this paper, a three-sample paddle error compensation algorithm for sliding window is proposed, and the response speed and error of the algorithm are tested. Experiments show that The algorithm's speed update frequency is tripled, and the response speed is improved without increasing the sampling frequency and without increasing the hardware burden. It has certain reference significance for improving the speed of the Strapdown Inertial Navigation System.

Acknowledgements. This work was supported by National Nature Science Foundation of China (NSFC) under Grants 61671356, 61201290.

References

1. Xiaoping, H.: Autonomous Navigation Technology. National Defence Industry Press, Beijing (2016)
2. Dongliang, S., Yongyuan, Q., Li, S.: Research on the paddle effect compensation algorithm of strapdown inertial navigation system. J. Proj. Archery Guid. **26**(2), 727–730 (2006)
3. Chaofei, Z., Mengxing, Y., Mingqiang, W.: An improved strapdown inertial navigation system optimization algorithm. Mod. Def. Technol. **4**(5), 80–85 (2012)
4. Wei, Z., Lixin, W., Yujing, Z., et al.: SINS cone error compensation based on spacer rotation vector. Piezoelectric Sound Light. **37**(1), 158–161 (2015)
5. Miller, R.B.: A new strap-down attitude algorithm. J. Guid. Control Dyn. **6**(4), 287–291 (1983)
6. Lee, J.G., Yoon, Y.J., Mark, J.G.: Estension of strapdown attitude algorithm for high-frequency base motion. J. Guid. Control Dyn. **13**(4), 738–743 (1990)
7. Xue-yuan, L., Jian-ye, L., Wei, Z.: Improved rotation vector attitude algorithm. Journal of Southeast Univ. **33**(2), 182–185 (2003). (Natural Science Edition)
8. Chuanye, T.: Research on Strapdown Algorithm and Combined Filtering Technology in SINS/GPS Combined Measurement. Southeast University (2016)
9. Yongyuan, Q.: Inertial Navigaion. Science Press, Beijing (2006)
10. Yang, Y., Hong-yue, Z.: Two interval sculling compensation algorithm based on duality principle. J. Beijing Univ. Aeronaut. Astronaut. **35**(3), 326–329 (2009)

Design and Control of Robot Car for Target Tracking and Recognition Based on Arduino

Yunsheng Li[1(✉)] and Chen Dongyue[2]

[1] Department of Intelligent Science and Technology, Xi'an University of Posts and Telecommunications, Xi'an, Shaanxi, China
ylee@xupt.edu.cn

[2] Department of Measurement Technology and Instrument, Xi'an University of Posts and Telecommunications, Xi'an, Shaanxi, China
1254207308@qq.com

Abstract. This paper focuses on the study of robot car with multi-sensors, using Arduino UNO R3 development board to achieve doing information acquire, and marching strategy apply under various environmental conditions. The robot car can track and complete the obstacle avoidance. To successfully use the inexpensive microcontrollers, machine vision functions are achieved, and the application space of low-end MCUs is expanded. Image data been uploaded to the workstation real time using Wi-Fi communication module to processing. Image processing techniques are used to generate command signals from the real-time video stream, then guide the robot to act in a specified direction. The interaction with the robot car can be achieved by using round objects or specific colors (red, yellow, green). The difficulty of controlling the robot car has been reduced by this system. Experiments show that the automatic obstacle avoidance and target tracking system of the robot car designed in this paper can correctly realize the requirements of the subject.

Keywords: OpenCV · Arduino · Color recognition · Hough circle transform Robot car

1 Introduction

Robot car can be considered as a type of smart self-propelled robot and be often referred to as smart cars [1]. Robots can adapt many different environments, because of their construction differences from organism. These characteristics make it possible to perform tasks in place of humans in environments where humans have no way to enter or threaten to survive [2]. It is precisely because of its strong environmental adaptability that robotic vehicles have great application potential in many fields, such as military and civilian, therefore have a good prospect for development.

The main contents in this paper include the design and manufacture of the hardware structure of the robot car, the trajectory tracking and obstacle avoidance software and the core intelligent control algorithm design. The core hardware of this thesis uses a microcontroller with high reliability, strong anti-interference ability and good real-time performance as the control platform. In the hardware, there are power supply, sensor,

P. Krömer et al. (Eds.): ECC 2018, AISC 891, pp. 635–643, 2019.
https://doi.org/10.1007/978-3-030-03766-6_71

drive chip board of motor and servo, MCU system control module, keyboard and wireless debugging, etc. They can realize different functions such as input signal and output execution. The software part contains the system initialization program, sensor acquisition program, display program, etc. It is mainly used to implement the basic input and output and configuration of the system, as well as trajectory tracking and obstacle avoidance.

2 Design of System

2.1 Composition of System

In order to realize the target recognition based on the USB camera on the premise of strictly controlling the cost, a communication module is added to transmit image data and command data.

Use a CMOS camera to capture the image. Recognize specific target and send the moving instruction back to the microcontroller to control car's movement.

The implementation of the project is divided into three major parts: infrared tracking, ultrasonic obstacle avoidance, target identification and tracking. Figure 1 briefly shows the structure of the entire system. The first thing to solve is infra-red tracking. After solving this problem, the car's movement problem will be Solved. In addition to the infrared tracking, the ultrasonic obstacle avoidance function is added to enable the robot carto have the ability to avoid obstacles in addition to tracing [3]. After the above two functions are realized, the camera in front of the robot is used to capture the target that needs to be recognized, and the video stream obtained is uploaded to the workstation through the communication module. Workstation running image recognition algorithm to identify targets; According to the type of target, sent selected instruction to the robot car [4]. The physical structure of the robot is shown in Fig. 2, part A is the structure diagram, and parts B and C are the physical maps.

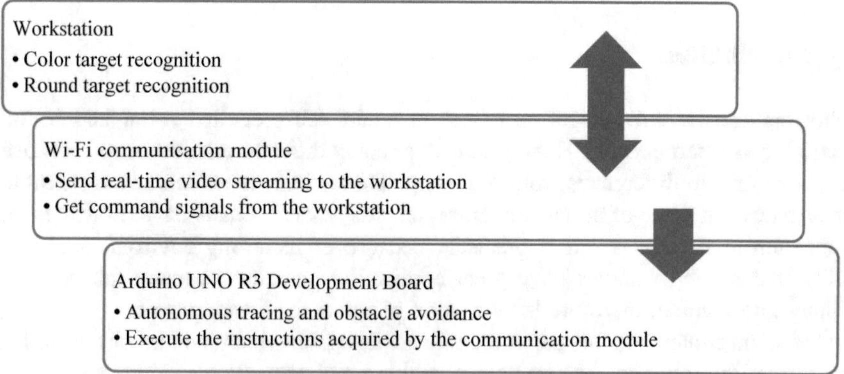

Fig. 1. System architecture

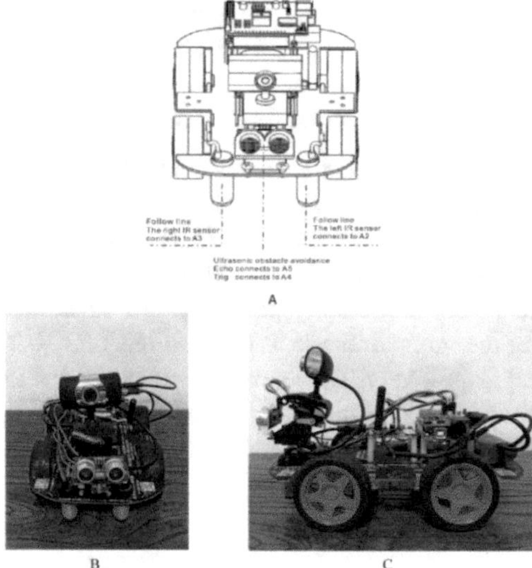

Fig. 2. Physical structure of the robot

2.2 Design of Control Methods

Autonomous tracking and obstacle avoidance. Because the Arduino UNO R3 development board does not have image processing capabilities, it just need to complete marching strategy execution. A separate development board can only perform autonomous tracing and obstacle avoidance functions.

As shown in Fig. 3, the initialization of the program mainly defines the corresponding pins and controls the definition of the motor output. After the program is started, the ultrasonic obstacle avoidance section is preferentially executed so that the robot car does not touch the obstacle [5]. This is important because although there are many requirements for the design of robot cars, safety is always at the top of the list. After a fixed obstacle is detected, the vehicle stops when the distance reaches 20 cm. After detecting an intrusion obstacle, turn the steering gear to detect if there is any obstacle on the left or right side without rotating the vehicle. If no obstacle is detected on one side, turn the body and move to the side without obstacles. If an obstacle is detected on both sides, the motor is driven to retreat and rotate the vehicle. Get rid of obstacles in this way.

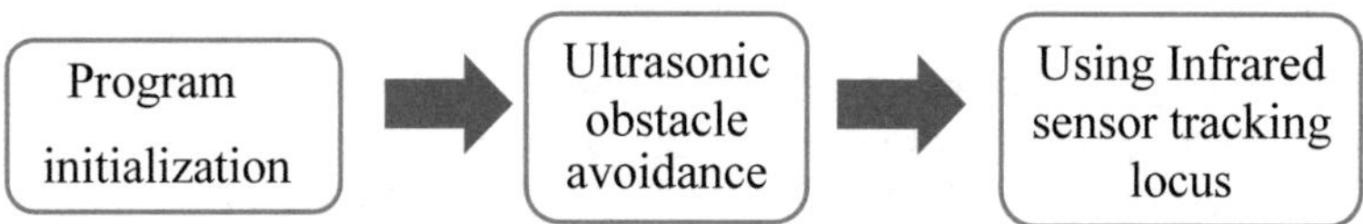

Fig. 3. Autonomous tracking obstacle avoidance schematic diagram

After the ultrasonic obstacle avoidance procedure is performed, the infrared tracking process is entered. Robot car straight ahead until the two infrared sensors installed in the front of the robot detect the trajectory identified by the black line. When both the infrared sensors do not detect the black line, continue to move straight ahead.

Round Target Tracking. The original image is converted to a grayscale image, which removes unnecessary color information. Gaussian blur is used to reduce image noise and reduce the level of detail, further reducing image complexity. Finally, the Hough circle transformation is performed and the edge detection is performed using the Canny operator algorithm to obtain a boundary binary image. Then the Sobel operator calculates the gradient of the original image. It has two sets of 3 × 3 matrices, which are horizontal and vertical respectively. They are plane-convolved with the binary image to obtain the horizontal and vertical brightness difference approximations. Let A be the original image, and $\boldsymbol{G_x}$ and G_y respectively represent the images detected by the horizontal and vertical edges.

$$G_x = \begin{bmatrix} -1 & 0 & 1 \\ -2 & 0 & 2 \\ -1 & 0 & 1 \end{bmatrix} * \mathrm{A} \quad G_y = \begin{bmatrix} -1 & -2 & -1 \\ 0 & 0 & 0 \\ 1 & 2 & 1 \end{bmatrix} * \mathrm{A} \tag{1}$$

The approximate horizontal and vertical gradients of each pixel of the image can be converted to a variable G by the following formula to calculate the gradient size.

$$G = \sqrt{G_x^2 + G_y^2} \tag{2}$$

The gradient direction is given by Eq. (3).

$$\theta = arctan\left(\frac{G_y}{G_x}\right) \tag{3}$$

Along the gradient direction and the opposite direction, all non-zero points in the edge binary graph are drawn (the direction of the gradient is the normal direction of the arc), the starting point and length of the line segment are determined by the allowable radius range. The point where the line segment passes is counted in the accumulator. Record these points and sort them from big to small. Calculate the distance of non- zero points from the center of the circle in all the borders and sort them from small to large. Starting from the minimum radius r, the points whose length are within the set range are regarded as the same circle. Record the number of non-zero points n under the radius; then enlarge the radius, repeat the same step repeatedly, until the radius exceeds The allowable range of the parameter. In this way obtaining the optimal radius (The method for judging the credibility is to find the linear density of the points (linear density = n/r). The higher the linear density, the more likely the Sobel operator is to use this radius.).

The image processing part is completed in the upper computer [6]. In order to make the robot car have the ability to identify the target. A webcam is used to capture the real-time video stream of the car's perspective and generate commands for the robot from the Real-time image. Image frames are treated as input and using image processing technique on the PC. When the target is circular object, the extracted information is the coordinates of the center of the circle and the radius of the circle.

Color Target Recognition. As shown in Fig. 4, Histogram equalization is carried out in HSV color space in order to improve the contrast and gray changes and make the image clearer. Then when doing image thresholds, ensure that each channel's value is within the specified threshold range. In this way, a binary image of the wanted color can be obtained. Then remove the noise to ensure that it does not mistakenly identify the non-existent color. Connect some connected domains, and finally determine the size of the connected domain (ensuring that it will not be erroneously identified). When targets are red, yellow, and green, the information extracted from the image is the HSV value and the area occupied by the color in the video.

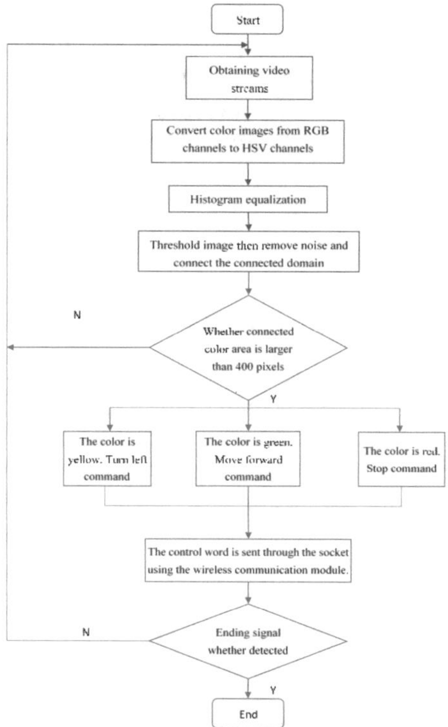

Fig. 4. Color target recognition

According to the processed image detection, the relevant command transmission control signal is selected, and the control signal is transmitted to the robot car through the wireless network connection [7]. Robot cars move on the ground within five basic commands (straight, stop, rewind, left turn, right turn).

Before implementing communication, a Socket connection should be established, so a pair of sockets need create. One of them runs on the client, called Client Socket; the other runs on the server, called Server Socket. Since this topic uses a Wi-Fi communication module with transparent serial transmission, it only needs to consider the actual data to be transmitted, and the instruction transmission can be realized without setting the Wi-Fi communication module.

The control words corresponding to these commands are encapsulated by the socket and transmitted to the Arduino development board through the wireless network. Arduino development board executes the corresponding travel strategy after acquiring the control word through the wireless communication module with the serial port transparent transmission set in advance [8].

The command parsing of the Arduino development board is similar to the reverse process of the workstation sending commands. In this paper, the control signal is divided into four parts, the header, the type bit, the command bit, and the tail. The header and the end of the packet are 0xff. These two parts are used to tell the lower computer when to start parsing the data and when to stop parsing. The type bit (0x13 or 0x00) tells the analytic function which type of command to enter the different parsing part (is the target tracking or mode switching); the command bit identifies what command is sent, is forward, left Turn, turn right, stop, back (0x00, 0x01, 0x02, 0x03, 0x04); autonomous tracing, or idle waiting (0x00, 0x01), which of these groups of control words.

3 Experimental Results

3.1 Autonomous Tracking and Obstacle Avoidance

There is a knob on the infrared sensor that can adjust the sensitivity, which can be adjusted according to the different materials of the track to maximize the sensitivity of the sensor. As shown in Fig. 5, the direction of the trajectory is judged based on the signal sent back by the sensors, when both infrared sensors do not detect the black line, the robot continue to moving straight ahead. The infrared tracking function depends entirely on the infrared sensor. Because of the size of the robot cars itself and the size of the sensor, sensors are only installed on both sides of the vehicle head. This makes it sometimes impossible for a sensor to give a corresponding signal in time under all conditions. When the radius of the turn is less than 65 cm it is possible to run out trajectories.

As shown in Fig. 6, a part shows the car stops when the fixed obstacle's distance to car reaches 20 cm (because the distance of the fixed obstacle cannot be less than the present value, the distance between the trolley and the obstacle reaches the preset value). B to D part shows after detecting an intrusion obstruction, turn the steering gear and check whether there are any obstacles on the left or right without rotating the car. If no obstacle is detected, turn the body and move it to the side without obstacles. If obstructions are detected on both sides, the motor is driven so that the cart retreats and turns. Get rid of obstacles in this way.

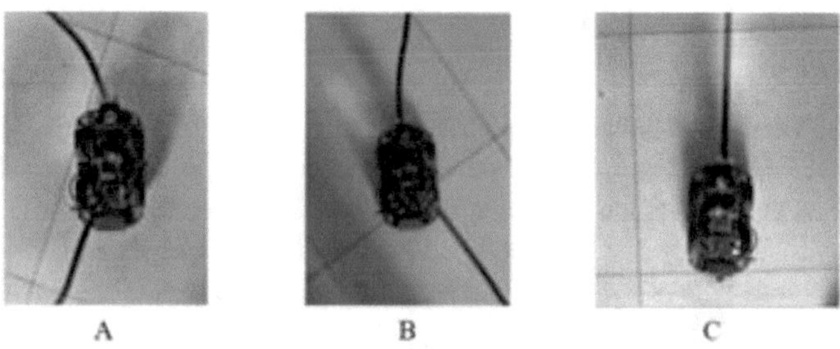

Fig. 5. Trajectory tracking turn left

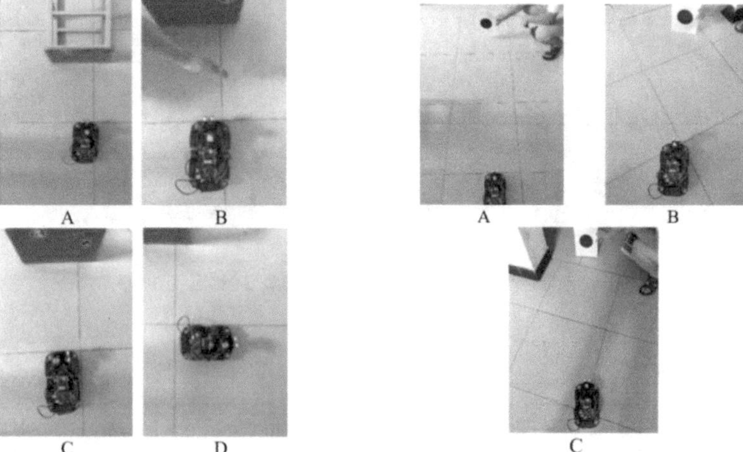

Fig. 6. Ultrasonic obstacle avoidance

Fig. 7. Round target tracking

3.2 Round Target Tracking

As shown in Fig. 7, the robot car tracks a black circle printed on A4 paper. The car follows the target moving straight, turn or stop until the radius of the circle caught in the camera is 85 mm larger and less than 100 mm smaller. When the radius of the circle captured in the camera is greater than 100 mm, the robot car will move back until the radius of the circle captured in the camera is between 85 mm and 100 mm then stop. Because the camera capture target of sight is limited, if the circular object and the location of the car camera round is beyond the scope of view camera, which means when the angle between the target and the camera is greater than 35 degrees target will be lost.

3.3 Color Target Recognition

In the experiment, it was found that the actual color read by the computer is not necessarily the same as the actual color, it will vary with light and angle. Therefore, an experiment of color recognition was performed in a room with sufficient illumination.

In Fig. 8, a part shows the robot car recognizes the color and judges that the color is green, goes straight. B part shows recognizes the color and judges that the color is red, so stopped; C part shows the car in parking state recognizes the color and judges the color; D part shows, turn left immediately after judge the color as yellow. Also the angle of object with the camera should be within 35 degrees to ensure that camera can obtain the target. When the color target is not acquired, the Robot cars will be stopped.

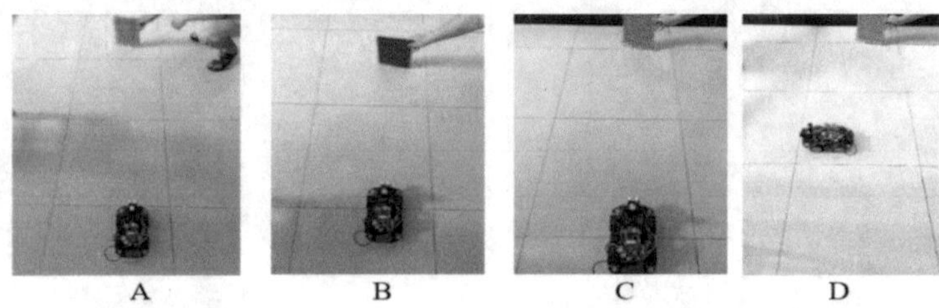

Fig. 8. Color target recognition

4 Conclusions

In this paper, a design method for economical and practical intelligent robot car was realized with autonomous tracking obstacle avoidance, round target identification and color target identification system. The hardware design and software development of the robot car were completed, and the system design was also successfully implemented. Below is a detailed of what's completed:

(a) Completed the design and assembly of the hardware part.
(b) Completed the software development of autonomous tracking and obstacle avoidance function.
(c) Completed the development of the image processing program on the workstation.
(d) Completed software development for communication between workstations and robot car.
(e) Realized the collaborative debugging of the hardware and software parts, the robot car can autonomous tracking locus and obstacle avoidance, tracking round targets and identifying the color target experiments were performed in the actual environment, achieving the desired goal.
(f) Successfully used inexpensive microcontrollers to achieve machine vision functions.

However, there are still some deficiencies. This paper uses the wireless communication module which built in AR9331 chip, can consider using the higher speed wireless network communication module in the future to reduce the delay in the video and order transmission process.

Acknowledgements. Research supported by the Foundation of the Shaanxi Provincial Department of Education (project number: 16JK1703).

References

1. Liu, C.H., Dong, Z.: Design of reversing radar system based on Arduino. Mod. Electron. Tech. **17**(37), 150–153 (2014). https://doi.org/10.16652/j.issn.1004-373x.2014.17.010
2. Wang, Z.R., Pang, J.T.: Design and realization of control system for data acquisition intelligent car based on Arduino control board. Comput. Technol. Autom. **36**(1), 66–73 (2017). https://doi.org/10.3969/j.issn.1003-6199.2017.01.014
3. Zhu, S.C.: Design of autonomous tracking for intelligent robot car preventing rear-end collision. Guizhou Sci. **32**(2), 31–34 (2014). 1003-6563(2014)02-0031-34
4. Harish, K.K., Vipul, H.: Gesture controlled robot using image processing. IJARAI **2**(5), 69–77 (2013)
5. Zhou, Y., Wang, X.: Design of obstacle avoidance car system based on Arduino platform. Comput. Knowl. Technol. **13**(18), 180–181 (2017). https://doi.org/10.14004/j.cnki.ckt.2017.1790
6. Lakshay, G., Tanmay, M.: Arduino based shape, color and laser follower using computer vision. IJSRET **2**(9), 542–545 (2013)
7. Wang, F., Yang, J.J.: Design of serial communication based on visual basic and Arduino's intelligent car control system. Sci. Technol. Innov. **1**(1), 73+75 (2016). https://doi.org/10.15913/j.cnki.kjycx.2016.01.073
8. Xu, Y.W., Zhang, J.J.: Design of wireless environment detection car based on Arduino. Jisuanji Yu Xiandaihua **6**, 126–199 (2015). https://doi.org/10.3969/j.issn.1006-2475.2015.06.026

Reliability Analysis of the Deployment of Astro-Mesh Antenna

Min-juan Wang(✉), Qi Yue, and Jing Guo

Xi'an University of Posts and Telecommunications, Xi'an 710121, China
wangminjuan@xupt.edu.cn

Abstract. The deployment reliability of the Astro-Mesh of a large satellite antenna is investigated. By contrasting the difference of movement principium, a mechanical analysis model is presented. Considering the effects of the randomness of dimensional errors and the space environment factors, the failure models of two performance functions are established by means of the random functional movement method and the second moment method. The corresponding computational expressions for reliability analysis are derived by using the first order second moment method. Furthermore, the theoretical basis and reference are provided for the deployment reliability of a large satellite antenna, and some useful conclusions are obtained.

Keywords: Satellite antenna · Astro-Mesh antenna · Random variable
Reliability analysis

1 Introduction

Due to the limitation of carrier loading equipment and dimension, larger Satellite Antenna commonly is folded in the launch stage, and then the antenna is deployed by remote control after satellite being in the orbit. The smooth deployment of Antenna is related with its normal work in space. So the reliability prediction of antenna deployment has become an important bottleneck in several factors of affecting the performance index of Satellite electronic equipment.

With the continuing exploration of space, various types of satellite antenna were developed [1]. Compared to the structural reliability analysis, it is more difficult to analyze mechanism system reliability for its complex structure, alterable shape and multiform failure mechanism. In recent years, the most studies of the large satellite antenna are concentrated on the structural design [2], the thermal analysis [3], deployment dynamics and the control of satellite into orbit position [4], etc. [5]. However, the reliability analysis of satellite antenna is still at the initial stage. At present, the studies on the type of satellite antenna are mostly focused on Hoop Truss Deployable Antenna, radially-rib, etc. Compared to the hoop truss deployment antenna, the triangle rod for increasing stiffness is simplified out, the intertwist of frame and cables can be restrain, and the structure simple, the store-volume smaller, lightweight, the large-diameter antenna is achieved easily.

The reliability function of the deployment mechanism and the calculating formulae for the internal forces of Antenna are constructed. Under the condition of considering

P. Krömer et al. (Eds.): ECC 2018, AISC 891, pp. 644–650, 2019.
https://doi.org/10.1007/978-3-030-03766-6_72

the uncertainty of friction coefficient of shaft sleeve, radius of joint axis, radius of gears axis and the traction force as random variables, simultaneously, the reliability index is derived by using the method of second-order moment. The result illustrates that the reliability analysis has important reference value and significance in enhancing the reliability prediction of a large deployment mechanism.

2 Astro-Mesh Antenna Deployment Mechanism

Figure 1 shows that it is different from the hoop-trus in the structure forms [6], the truss elements is composed by rectangular frame and diagonal bars. The referred satellite antenna with motor cables composing paraboloid contour, have a caliber of 17 m, number of truss elements is n [7]. The deployment principium of Astro-Mesh antenna is that the continuous deployable cables can be obtained through the diagonal bars (sleeve structure) of every truss elements [8].

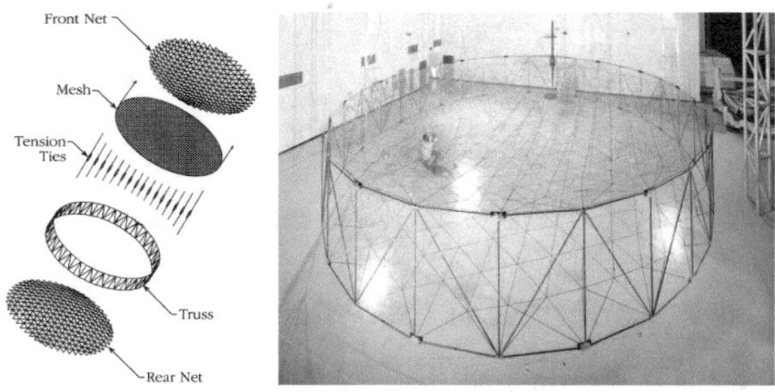

Fig. 1. Astro-Mesh deployment antenna.

The deployment process of Astro-Mesh antenna is divided into three stages: unlock stage, deployment stage and self-locking stage. In the unlock stage, the truss structure can open at an angle from self-locking position under the action of torsion spring when the antenna cable bundle is fell off. In the deployment stage, the cables get through scalable diagonal bars (sleeve structure) of truss elements, and achieve shrinkage under motor drive to the winding wheel rotation. With the cable taut, the length of scalable diagonal bars is shortened by motor, and then the truss is deployed slowly. In the self-locking stage [9], the motor is stopped under the spring lock in sleeve of diagonal bars and the joint limit, and antenna mechanism get to the appointed position and self-locking

3 The Mechanical Analysis of Astro-Mesh Antenna in Process of Deployment

3.1 Determining the Space Geometry Relationship of Bars

The deployment movement of antenna is very slow, we can be approximated that each of instantaneous state of the deployment system are in static balance. Figure 2 shows the force structure diagram of two frame elements. When the angle between L_1 and L_2 is θ, L_1 and L_3 is α, we can obtain the expression as follows:

$$L_3 = \sqrt{L_1^2 + L_2^2 - 2L_1L_2\cos(\theta)} \quad (1)$$

$$\alpha = \arcsin(L_2\sin(\theta)/L_3) \quad (2)$$

$$\beta = (n-2)\pi/n \quad (3)$$

$$L_2 = d\cos(\beta/2) \quad (4)$$

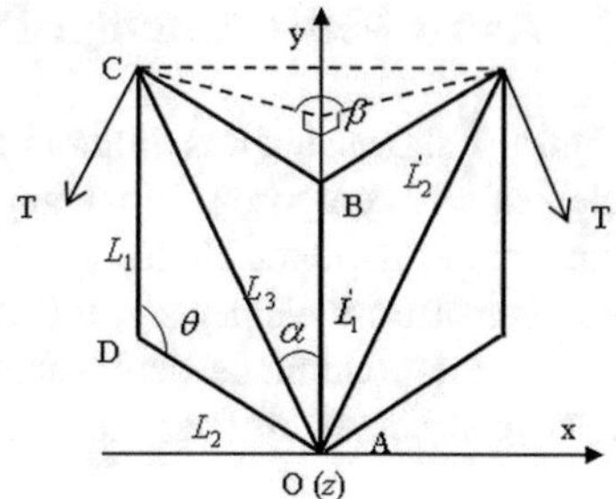

Fig. 2. Diagram of two frame elements.

where β is the angle of XOZ coordinate plane, n is the number of frame elements.

3.2 Determining Internal Force of Every Bar

For the section method, the internal force of L_1 and L_2 are analyzed, respectively. According to the principle of counterbalance, we have:

$$2T\cos(\alpha) + N_1 + 2N_2\cos(\pi - \theta) = 0 \quad (5)$$

$$2T\sin(\alpha)\cos(\beta/2) + 2N_2\sin(\pi - \theta)\cos(\beta/2) = 0 \quad (6)$$

$$N_1 = 2T\sin(\alpha)\cos(\pi - \theta)/\sin(\pi - \theta) - 2T\cos(\alpha) \quad (7)$$

$$N_2 = -T\sin(\alpha)/\sin(\pi - \theta) \quad (8)$$

where T is the traction of motor, N_1 and N_2 are the internal force of L_1 and L_2.

4 Deployment Reliability Analysis of Astro-Mesh Antenna

The friction resistance moment is determined by the positive pressure on joint axis and friction coefficient. Other resistance moment is estimated to be k_m times of friction resistance moment, with k_m being comprehensive influence coefficient. Considering the component's machining, the assembling errors and the environmental uncertainty, we regard the traction of motor T, radius of joint axis or gears axis r and friction coefficient of shaft sleeve f as random variables. To ensure antenna deploy successfully, it must satisfy one of the following two movement functions.

(1) In the process of deployment, active moment is always larger than resistance moment. The limit state equation:

$$Z_M = M_a(\theta) - M_r(\theta) = 0 \tag{9}$$

where Z_M, $M_a(\theta)$ and $M_r(\theta)$ is the performance function of moment form. θ is the variational angles in the process of deployment.

The total resistance moment and the total active moment can be expressed as

$$M_r = (1 + k_m) \times n \times (2|f_1 N_2 r_1| + 2|f_1 N_2 r_2| + |2T f_2 r_3 \cos(\alpha)|) \tag{10}$$

$$M_a = n \times |2TL_3 \cos(\alpha) \sin(\alpha)| \tag{11}$$

where f_1 and f_2 are the static friction coefficients of pulley axis and joint axis, respectively. r_1 and r_2 are the radius of pulley axis and joint axis respectively and r_3 is the radius of gears axis. Then we have $r_1 = r_2 = R_1$, $r_3 = R_2$, $f_1 = f_2 = \mu$, we can get the performance function as follows:

$$Z_M = n \times [TA(\theta) - T\mu R_1 B(\theta) - T\mu R_2 C(\theta)] \tag{12}$$

where $A(\theta)$, $B(\theta)$ and $C(\theta)$ are the certain part of deployment angle θ. μ, R_1, R_2 and T are random variables. The mean value μ_{Z_M} and variance $\sigma^2_{Z_M}$ of Z_M in principal coordinates can be can be expressed as:

$$\mu_{Z_M} = n \times [\mu_T A(\theta) - \mu_T \mu_\mu \mu_{R_1} 2B(\theta) - \mu_T \mu_\mu \mu_{R_2} C(\theta)] \tag{13}$$

$$\begin{aligned} \sigma^2_{Z_M} = & n^2 \times ((A(\theta) - \mu_\mu \mu_{R_1} B(\theta) + \mu_\mu \mu_{R_2} C(\theta))^2 \cdot \\ & \sigma^2_T + \mu^2_T \mu^2_\mu B(\theta)^2 \sigma^2_{R_1} + (\mu_T \mu_\mu C(\theta))^2 \sigma^2_{R_2} + \\ & (\mu_T \mu_{R_1} B(\theta) + \mu_T \mu_{R_2} C(\theta))^2 \sigma^2_\mu \end{aligned} \tag{14}$$

The reliability index β_M of Z_M can be obtained from the following equations.

$$\beta_M = \mu_{Z_M} / \sigma_{Z_M} = (\mu_{M_a} - \mu_{M_r}) / (\sigma^2_{M_a} + \sigma^2_{M_r})^{1/2} \tag{15}$$

Then the reliability P_M can be written as:

$$P_M = P(M_a - M_r > 0) = \Phi(\beta_M) \tag{16}$$

(2) In process of deployment, the work done by active moment is always larger than the work done by resistance moment. The limit state equation:

$$Z_w = W_a(\theta) - W_r(\theta) = 0 \tag{17}$$

where Z_w, $W_a(\theta)$ and $W_r(\theta)$ are the performance function of moment acting work, accumulated active work and resistance work respectively. Similarly, θ is the variational angles in the process of deployment.

When the Rotary Joint structure is deployed from θ_0 to θ, the accumulated resistance work and the accumulated active work can be expressed as:

$$W_r(\theta) = \int_{\theta_0}^{\theta} (1 + k_m) \times n \times \bar{M}_r(\theta) d\theta \tag{18}$$

$$W_a(\theta) = \int_{\theta_0}^{\theta} n \times \bar{M}_a(\theta) d\theta \tag{19}$$

From the Eq. (9), (18) and (19), the performance function of work can be expressed as:

$$\begin{aligned} Z_W &= W_a(\theta) - W_r(\theta) \\ &= TG(\theta) - T\mu R_1 H(\theta) - T\mu R_2 K(\theta)/R_3 \end{aligned} \tag{20}$$

Therefore, the mean value μ_{Z_W} and variance $\sigma^2_{Z_W}$ of Z_W can be obtained. The reliability index β_W of Z_W can be obtained from the Eqs. (16) and (17).

5 Reliability Calculation of Astro-Mesh Antenna Deployment

The design parameters of Astro-Mesh satellite antenna are listed as follows: The antenna aperture is $d = 17$ m. The number of frame element is n = 24. The height of Astro-Mesh antenna is $L_1 = 3395.673$ mm. Radius means value of joint axis is $\mu_{R_1} = 2.5$ mm. Radius means value of gear axis is $\mu_{R_2} = 3$ mm. Radius means value of pulley is $\mu_{R_3} = 10.5$ mm. The variation coefficient of geometric dimensions is $v_{R_1} = v_{R_2} = v_{R_3} = 0.1$. The comprehensive influence coefficient is $k_m = 1.5$. Mean value of static friction coefficients is $\mu_\mu = 0.02$.

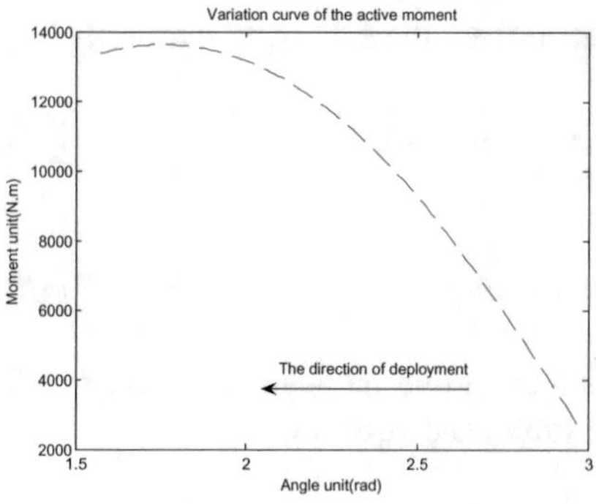

Fig. 3. Variation curve of active moment.

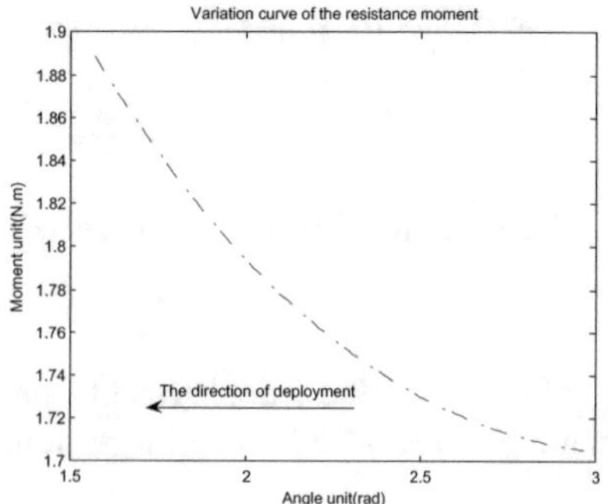

Fig. 4. Variation curve of resistance moment.

The variation coefficient of static friction coefficients is $v_{\mu_1} = 0.1$. Mean value of traction tension is $\mu_T = 150N$, and the corresponding coefficient of variation is $v_T = 0.1$. Figures 3, 4 and 5 show variation curve of the active moment, the resistance moment and the reliability index of moment in process deployment, respectively. Figure 6 shows variation curve the active work and the resistance work in process of antenna deployment. Figure 7 shows the reliability index of work.

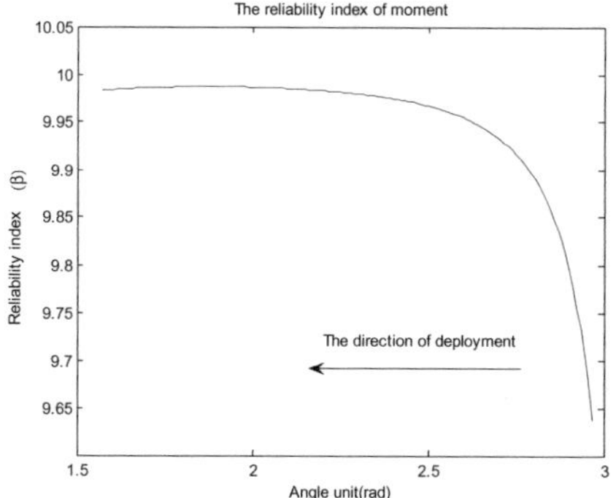

Fig. 5. Variation curve of reliability index

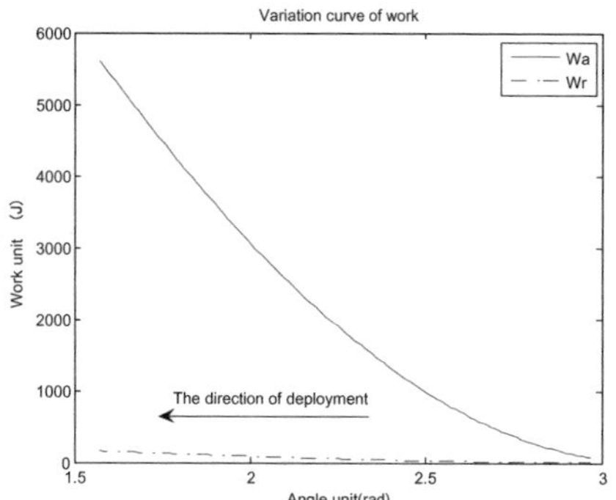

Fig. 6. Variation curve of work.

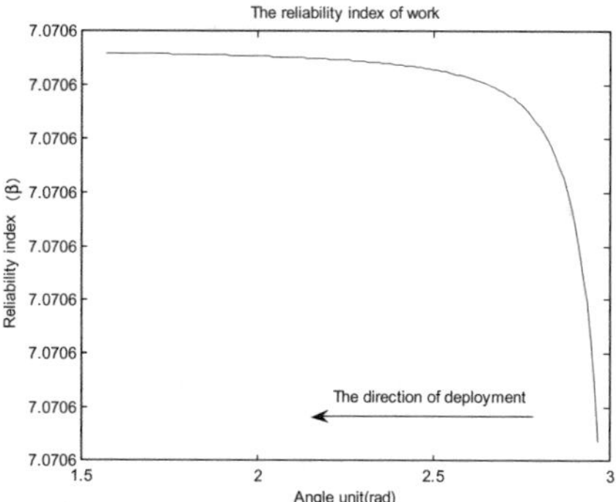

Fig. 7. Variation curve of reliability index.

It can be seen from the simulation results of Figs. 3 and 4, mean value of the active moment is $M_a \in [2767.6, 1.36 \times 10^4]$, mean value of the resistance moment is $M_r \in [1.7054, 1.8887]$. The corresponding reliability index of moment is $\beta_M \in [9.6375, 9.9872]$. In the initial stage of antenna deployment, it is difficult to deployment for the self-locking effect of antenna structure, so β_M is very small. With the resistance moment increasing, the angle θ at 105° and the reliability index β_M starts to decline slightly.

From Fig. 6 we can seen that the active work and the resistance work are $W_a \in [73.5305, 5609.7]$ and $W_r \in [1.9117, 159.77]$ respectively. The reliability of work changed a little, it is approximately equal to 7.0706. At the initial angle of deployment mechanism is 10°, it ascends to around 7 quickly, and then increases smoothly.

6 Conclusion

The prediction results indicate that reliability index β_M and β_W become larger and larger with the changing of deployment angle, which shows that the method of this paper is reasonable and effective. A possible choice for reliability prediction of large mechanism is provided based on the probabilistic method in the paper; it is useful for the reliability analysis and design of a large mechanism.

Acknowledgement. This work was supported by the Department of Education Shaanxi Province, China, under Grant 2013JK1073.

References

1. Tibert, G.: Deployable Tensegrity Structures for Space Applications. Royal Institute of Technology, Stockholm (2002)
2. Tanaka, H.: Design optimization studies for large-scale contoured beam deployable satellite antennas. Acta Astronaut. **58**(9), 443–451 (2006). https://doi.org/10.1016/j.actaastro.2005.12.015
3. Vu, K.K., Liew, J.Y.R., Anandasivam, K.: Deployable tension-strut structures: from concept to implementation. J. Constr. Steel Res. **62**(3), 195–209 (2006). https://doi.org/10.1016/j.jcsr.2005.07.007
4. Werner, U.: Influence of electromagnetic field damping on forced vibrations of induction rotors caused by dynamic rotor eccentricity. J. Appl. Math. Mech. **97**(1), 38–39 (2017). https://doi.org/10.1002/zamm.201500285
5. Wettergren, J., Bonnedal, M., Ingvarson, P., Wästberg, B.: Antenna for precise orbit determination. Acta Astronaut. **65**(11–12), 1765–1771 (2009). https://doi.org/10.1016/j.actaastro.2009.05.004
6. Zhu, Z.-q., Chen, J.-j., Liu, G.-l., et al.: Reliability analysis for the deployment mechanism of a large satellite antenna based on unascertained information. J. Xidian Univ. **36**(5), 909–915 (2009)
7. Patel, J., Ananthasuresh, G.K.: A kinematic theory for radially foldable planar linkages. Int. J. Solids Struct. **44**(18), 6279–6298 (2007). https://doi.org/10.1016/j.ijsolstr.2007.02.023
8. Lin, L., Chen, J., Liu, G., et al.: Fault analysis of deployment mechanism systems of satellite antennas based on the grey relation method. Chin. High Technol. Lett. **20**(9), 905–910 (2010)
9. Omer Soykasap.: Analysis of tape spring hinges. Int. J. Mech. Sci. **49**(2), 853–860 (2007). https://doi.org/10.1016/j.ijmecsci.2006.11.013

Design of Malignant Load Identification and Control System

Wei Li, Tian Zhou(✉), Xiang Ma, Bo Qin, and Chenle Zhang

Xi'an University of Posts and Telecommunications, Xi'an 710121, China
wli@xupt.edu.cn, 1830927268@163.com

Abstract. The potential security risks existing in high-power illegal electric appliances is a problem that university power management faces. The traditional electrical identification system has disadvantages of low accuracy, high complexity and high hardware cost. In order to solve this problem, a malignant load intelligent identification and control system based on SoC (RN8302) chip combined with STM32F105 processor is designed, moreover, the principles of algorithm and hardware circuit are given in detail. The real-life test proves that the system has accurate measurement and high load identification rate. The RN8302 chip-based intelligent load identification and control system has certain practical application value for its simple design and low cost.

Keywords: RN8302 · STM32F105 · Intelligent identification

1 Introduction

With the development of social economy, the housing conditions of college students' apartment have been gradually improved. These requirements also lead to the corresponding change of power supply mode of students' apartment. The past electricity management of student apartment mainly adopts power management mode of limited current and intermittent power supply, which has been unable to meet the learning requirements of college students. However, due to the extensive use of electric appliances, the electricity safety problems frequently arise in dormitory [1, 2]. According to the relevant data, the high-power resistivity loads, such as fast heater and induction cooker, are the main causes of the electricity problems in students' apartment. Therefore, high power resistivity loads are also called malignant loads.

Against above problems, this paper designed a kind of malignant load intelligent identification and control system based on SoC chip RN8302 [3], this system can monitor electrical appliances in real-time and cut off the power supply in time through the relay, and power is delivered again after a certain time delay. When the malignant loads still exist after being detected for many times, it will be reported to the platform and manual closing is required after inspection by relevant personnel. This system not only realizes the remote control and detection and greatly reduces the hardware cost through the use of high integration SoC chip, meanwhile, reducing equipment volume and shortening the development cycle, therefore, this system has very high application value.

P. Krömer et al. (Eds.): ECC 2018, AISC 891, pp. 651–657, 2019.
https://doi.org/10.1007/978-3-030-03766-6_73

2 Principle and Algorithm of Load Identification

The algorithm of load identification plays a decisive role in the malignant load identification system. The mainstream malignant load identification methods mainly include the time domain and frequency domain method [4, 5]. Time domain method mainly includes methods of the total power limit, instantaneous power increase and the waveform comparison; Frequency domain method is mainly neural network identification algorithm. There are some problems existing in these two kinds of algorithm, the time domain and frequency domain method have problem of low load identification rate and the difficulty to implement the algorithm respectively. At present, these algorithms have been applied to identify whether the category of a single appliance is resistive or inductive. However, when the resistive load is mixed with the malignant load, it is difficult to judge by the above algorithm alone. Therefore, this paper proposes an area-based algorithm to judge whether there is any malignant load in the hybrid electric equipment and thus implements it with the hardware combined.

First of all, conducting an analysis to a single appliance, the current waveform of linear load is sine wave and its amplitude is assumed to be 1. For the integral of the current waveform in $[0, \pi]$, its area is:

$$S_{LinerLoad} = \int_0^{\pi} \text{sinxdx} = 2 \tag{1}$$

The non-linear load current waveform with rectifying equipment is pulsed and is analyzed by Fourier. According to the fundamental and harmonic wave proportion, the pulse waveform can be expressed by the following formula:

$$y = 0.36 \sin x - 0.25 \sin 3x + 0.1 \tag{2}$$

The formula is also integrated in $[0, \pi]$ and the area of the nonlinear load is:

$$S_{NonlinearLoad} = \int_0^{\pi} ydx = 0.36 \times 2 - 0.25 \times \frac{2}{3} + 0.15 \times \frac{2}{5} - 0.1 \times \frac{2}{7} = 0.6 \tag{3}$$

Differentiate on the basis of the difference of area between the linear load and nonlinear load at the same time. The current waveform of Fig. 1 contains linear and nonlinear load, for such a current waveform, we set the amplitude of the pulsed current waveform to Z_1 and the linear load current waveform amplitude to Z_2, the sum of the two waveforms is the active line in Fig. 1. The current waveform is integrated in $[0, \pi]$:

$$S_{\sum} = Z_1 S_3 + Z_2 S_3 = 0.6 Z_1 + 2 Z_2 \tag{4}$$

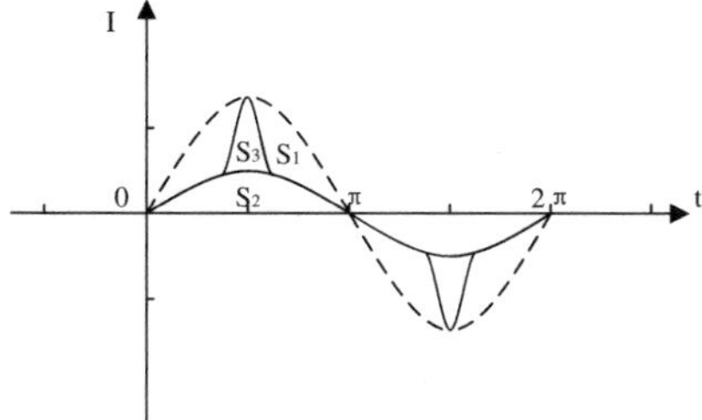

Fig. 1. Mixed current waveform of linear load and nonlinear load

Assuming that there is no nonlinear load in the electrical system, for the sine wave with the amplitude of (Z_1+Z_2) (as shown in the dotted line of Fig. 1), the integral is performed on the alignment of $[0, \pi]$:

$$S = 2(Z_1 + Z_2) \tag{5}$$

Set the difference between the two areas to ΔS:

$$\Delta S = S - S_{\sum} = 1.4Z_1 \tag{6}$$

The amplitude of both can be obtained by the above formula:

$$Z_1 = \frac{\Delta S}{1.4}, Z_2 = \frac{S}{2} - \frac{\Delta S}{1.4} \tag{7}$$

Through the above formula, the amplitude Z_1 and Z_2 of the nonlinear load and linear load can be obtained respectively. By which the power of the linear load and nonlinear load can be easily obtained so that we can easily determine whether the electricity system contains high power malignant load, and the malignant load identification system can control it in real time.

3 Hardware Circuit Design

3.1 Overall Functional Design

Control system consists of three-phase multifunctional anti-theft electric measurement chip RN8302 and STM32F105 single chip microcomputer [6], current sampling module, alarm module, the NB-IOT data transmission module and relay control circuit module [7, 8], the specific block diagram is shown in Fig. 2.

The whole system realized the separation between strong and weak electricity through the relay module, voltage transformer and current transformer with optical coupling isolation, which ensures the safety and stability of the system. Voltage and current sampling module sampled and extracted the voltage signal and input the resulting differential signal into RN8302 containing 7 channels of 24-bit ADC, reading

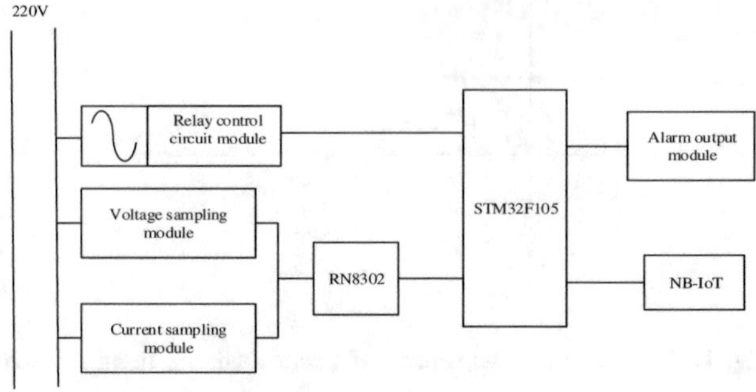

Fig. 2. System structure block diagram

the related electrical parameters through STM32F105 microcontroller and determining whether a circuit contains malignant load through the algorithm. When the system detects a malignant load in circuit, the circuit will be cut off in time and automatically closed after 10 min. If the malignant load is detected in the circuit for many times, the system will not be closed, it will be reported to the platform through NB-IOT and inform relevant personnel to check and manually close the switch.

3.2 Design of Sampling Circuit

Current sampling circuit part [9] uses the open type current transformer ZDKCT38 M with the working temperature being in 85 °C to 40 °C, which confirms to the daily work environment and avoids the influence of temperature on the accuracy of measurement. In addition, when the current signal is collected by the current transformer, the standard voltage is generated when it is sent to the high-precision resistors R3 and R4 (Fig. 3).

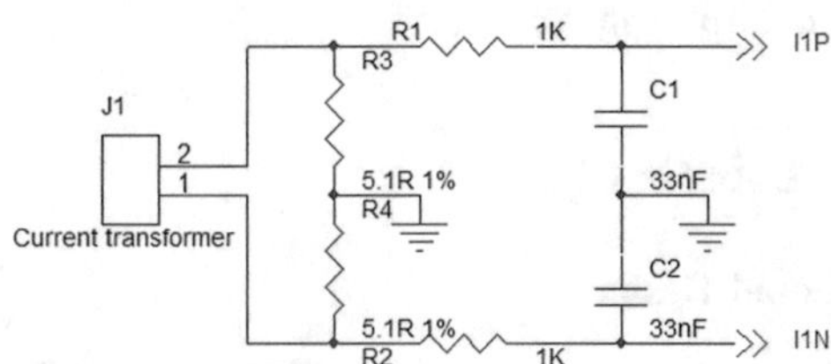

Fig. 3. Principle of current sampling

Because the current transformer has a low temperature coefficient and the resistance sampling accuracy is 1%, the system's requirements for the high precision of the current transformer are reduced and thus the cost is saved. The converted voltage is transferred to 24-bit ADC of RN8302. The current conversion coefficient is as follows, among which I_B is the rated current of the inductor coil and U_i is the voltage at both ends of the sampling resistance.

$$K_i = \frac{I_B \times 0.8^{27}}{U_i} \tag{8}$$

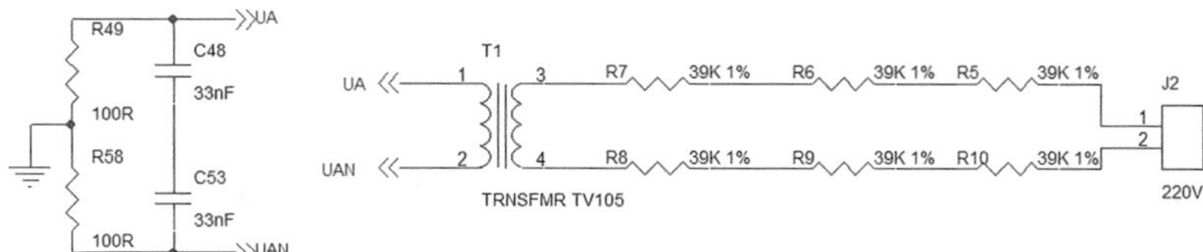

Fig. 4. Voltage sampling resistance

Voltage sampling circuit part [10], by using the current voltage transformer, realized the isolation between the high and low voltage, and thus guaranteed the safety and stability of operation of the system. Besides, adopting 6 patch resistors with 1812 encapsulation precision of 1% to divide 220 V voltage and using the voltage sampling circuit in Fig. 4 to prevent the problems of strong electric surge and guarantee the system's pressure resistance, which is suitable for the sampling resistor R7 to R10. RC low-pass filter consisting of R49 and C48 so as to prevent the interference of high frequency signal to the system. Because RN8302 voltage channel input voltage is 100–200 mV, using the following formula, including 100Ω transformer impedance measured.

$$U_{\text{The input voltage}} = \frac{220}{R_{\text{The sampling resistor}} + R_{\text{Internal resistance of the transform}}} \times R_{\text{Internal resistance of the transform}} \tag{9}$$

The input voltage is 180 mV based on the calculation, this circuit design conforms to the requirement of the input voltage range of RN8302.

4 Test Results

This system first conducted a judgment test to the individual electric appliance to judge whether this system has the wrong operation to the normal electric appliance. In this experiment, different brands of laptops were judged, and then high-power malicious loads such as electric stoves and quick heater were judged. The test data are as follows (Table 1):

Table 1. Test records of an individual electric appliance

Load	Laptop 1	Laptop 2	Electric Stove	Quick Heater
The Power Factor	0.7324	0.6376	0.9999	0.9999
Effective Power/W	22.5162	24.1653	1431.3249	1220.6279
Relay Operation	closed	closed	Switch off	Switch off

The second set tested whether the system can judge the malignant load in the mixed load and cut off power. In this test, laptop is the first load and connected to the electric stove, hot and electric heaters in turn for parallel test (Table 2).

Table 2. Test records for mixed loads

Load	Laptop + Quick Heater	Laptop + Electric Stove	Laptop + Electric Heater
The Power Factor	0.9998	0.9996	0.9998
Effective Power/W	1267.2784	1479.6734	1687.7841
Relay Operation	Switch off	Switch off	Switch off

The above experimental results show that in the case of using safe electric appliance, the system fails to cut off the power; While in a mixed load, the system can identify the malignant load equipment with high power and timely cut off the power.

5 Conclusion

The malignant load identification system based on RN8302 can independently measure the three circuits of electricity, and by using the area based malignant load identification method, it can accurately judge the high-power malignant load in the mixed load, the power circuit with potential risks can be cut off in time through the relay. High integration SoC chip and modular design reduce system volume, shorten development cycle and ensure the stability of system operation. It can also be used for the electricity detection of residents by cutting out the system's functions. Through mobile APP, users can know their own electricity-use situation and potential safety risks in real time. At present, the prototype development and test of the system show that the system owns good reliability and accuracy and has broad market prospect.

Acknowledgements. This work was supported by Shanxi Province Technical Innovation Guide Special project (2018SJRG-G-03). This work also supported by Shanxi education department industrialization project (16JF024).

References

1. Ying, C.: Design of intelligent detection system for illegal electric appliances with malicious load in campus power grid. Sci. Technol. Bull. **29**(4), 61–63 (2013). https://doi.org/10.13774/j.cnki.kjtb.2013.04.010
2. Cui, J., Li, P.: Design of high precision intelligent electric meter based on ATT7022B. Electron Technol. **23**(2), 46–48 (2010). https://doi.org/10.16180/j.cnki.issn1007-7820.2010.02.003
3. Liu, M., Bo, H.: Design and implementation of a new comprehensive monitoring linkage function model. Autom. Instrum. **27**(11), 31–34 (2012). 10.19-557/j.cnki.1001-9944.2012.11.009
4. Chen, W., Deng, X., Lu, T.: Design and implementation of a new power grid voltage monitor based on STC12C5A32AD. Instrumentation **20**(9), 41–43 (2009). https://doi.org/10.19432/j.cnki.issn10062394.2009.0-9.015
5. Du, J., Wan, S., Zhu, Z.: Research on the auxiliary decision-making function of integrated monitoring system based on case reasoning. J. Qingdao Univ. **26**(4), 39–42 (2011). 10.1330-6/j.10069798.2011.04.008
6. Linna, W., Meng, X.: A new sinusoidal signal distortion evaluation method. J. Electron. Meas. Instrum. **19**(3), 67–71 (2005). https://doi.org/10.13382/j.je-mi.2005.03.007
7. Chen, D., Han, J.: MATLAB based design method for fixed-point DSP wavelet transform program. Data Acquis. Process. **21**(5), 86–89 (2006). 10.163-37/j.1004-9037.2006.s1.040
8. Zhao, C., He, M.: Harmonic detection algorithm based on complex wavelet transform phase information. China J. Electr. Eng. **1**(25), 38–40 (2005). https://doi.org/10.13334/j.02588013.pcse-e.2005.01.008
9. Su, Y., Jade, W.: Modeling of non-contact power supply phase-shifting control systems. J. Electron. Technol. **23**(7), 92–97 (2008). 0.19595/j.cnki.1000-6753.tces.2008.07.016
10. Zhao, W., Chen, S., Lu, W.: Research on intelligent identification methods of malignant load. Sci. Technol. Sq. **3**(3), 48–50 (2014). https://doi.org/10.13838/j.cnki.kj-gc.2014.03.013

Fractional-Order in RC, RL and RLC Circuits

Yang Chen(✉) and Guang-yuan Zhao

School of Automation, Xi'an University of Posts and Telecommunications,
Xi'an, China
cyxiaowanzi@163.com

Abstract. In mathematics, differential equations with fractional-order derivatives have a long history, for example, the "one in third" derivative, but haven't gotten tremendous use in applied science and engineering. While applications do exist in several modeling specific phenomena, such as semi-infinite lossy transmission, which are difficult to model, and there exist some extensions of control in fractional-order PID, everyday use of fractional order modeling is more and more common. In this paper, the basic principles of the conventional RC and RL circuits in fractional-order way and a fractional differential equation are studied in the electrical RLC circuit. We consider the order of the derivative ($0 < \gamma \leq 1$). In order to keep the dimensionality of the physical quantities R, L and C, an auxiliary parameter σ is introduced.

Keywords: Fractional calculus · Caputo derivative · Fractional-Order circuit Simulation of Fractional-Order response

1 Introduction

Fractional calculus is an old topic that has a long history. Although the number of applications which fractional calculus has been used grows rapidly. The "0.5" order of the derivative has been described by Leibniz in a letter to L'Hospital in 1695. Fractional calculus is the natural generalization of the classical integer calculus [1–5]. Several physical phenomena have "intrinsic" fractional-order description, so Fractional calculus is necessary to explain them. In many applications, Fractional calculus provides more accurate models of the physical systems than ordinary calculus do. It has become a vital tool in several areas such as physics, chemistry, mechanics, engineering, finances and bioengineering [6–12]. Basic principle of physical considerations based on derivatives of fractional order are given in [13–15]. The Hamilton and Lagrangian formulation of electromagnetic and dynamics field in the way of fractional calculus has been proposed in [16–21]. Modeling in fractional order proves to be useful for systems particulary, which memory or the properties of hereditary play a vital role. This is due to the fact that an integer order derivative is a local operator which considers the nature of the function of its neighborhood and at that instant, while a fractional derivative takes the past history of the function from some earlier point in time into account.

Tremendous efforts have been made to generalize the conventional, basic principles into fractional-order ways. In circuit designs, the general principles of fractional order oscillators and filters are introduced by numerical analyses, analytical conditions,

P. Krömer et al. (Eds.): ECC 2018, AISC 891, pp. 658–666, 2019.
https://doi.org/10.1007/978-3-030-03766-6_74

circuit simulations, and experimental results [12, 15]. The fundamentals of the conventional LC tank circuit showing new responses are proposed in [20], which exist in fractional-order way only. In addition, the stability analysis of the fractional order RLC circuit is presented for independent fractional-orders in [21].

This paper is organized as follows. basic definitions of fractional calculus are proposed first in Sect. 2. Then, fractional order RC, RL circuit are introduced in Sect. 3. Additionally, the simulation and comparison of fractional order and conventional circuits in Sect. 4. The fractional order RLC circuit are proposed in Sect. 5. Finally, Sect. 6 presents conclusions and future work.

2 Fractional Calculus

In order to analyze a fractional dynamical system, it is necessary to use an appropriate definition of fractional calculus. In fact, there are several definitions of the fractional order derivative, including: Grünwald-Letnikov, Riemann-Liouville, Weyl, Riesz and the Caputo representation.

$$ {}_aD_t^r = \begin{cases} d^r/dt^r & \Re(r) > 0 \\ 1 & \Re(r) = 0 \\ \int_a^t (d\tau)^{-r} & \Re(r) < 0 \end{cases} $$

${}_aD_t^r$–fractional order fundamental operator, where r is the order of the operation, generally, r could be a complex number and a and t are the lower and upper bounds of integration, respectively.

2.1 Definition of Fractional Differentiation and Integration

Two definitions used for the general fractional calculus are the Riemann-Liouville (RL) definition and the Grunwald-Letnikov (GL) definition. The RL definition is given as

$$ {}_aD_t^r f(t) = \frac{1}{\Gamma(n-r)} \frac{d^n}{dt^n} \int_a^t \frac{f(\tau)}{(t-\tau)^{r-n+1}} d\tau, \tag{1} $$

for $(n-1 < r < n)$ and $\Gamma(.)$ is the Gamma function [4]. The GL definition is given as

$$ {}_aD_t^r f(t) = \lim_{h \to 0} h^{-r} \sum_{j=0}^{\left[\frac{t-a}{h}\right]} (-1)^j \binom{r}{j} f(t - jh). \tag{2} $$

where [.] is the integer part.

The Laplace Transform Method Is Aimed to Solve Engineering Problems. the Formula for the Laplace Transform of the RL Fractional Derivative (1) Has the Following Form

$$\int_0^{\infty} e^{-st}{}_aD_t^r f(t)dt = s^r F(s) - \sum_{k=0}^{n-1} s^k{}_aD_t^{r-k-1} f(t)|_{t=0}, \tag{3}$$

For $(n - 1 < r < n)$, which $s \equiv jw$ is the Laplace transform variable.

In Caputo case, the derivative of a constant is zero and thus we can define the initial conditions for the fractional differential equations that can be handled by using an analogy with the ordinary derivative. Caputo derivative emphasis a memory effect by means of a convolution between a power of time and the integer order derivative. Caputo fractional derivative [4]:

$$\frac{d^r}{dt^r} f(t) = {}_0^cD_t^r f(t) = \frac{1}{\Gamma(n-r)} \int_0^t \frac{f^{(n)}(\tau)}{(t-\tau)^{r-n+1}} d\tau, \tag{4}$$

Where $r \in R$ is the order of the fractional derivative, $n - 1 < r \leq n \in N = \{1, 2, \ldots\}$, and

$$f^{(n)}(\tau) = \frac{d^n f(\tau)}{d\tau^n}, \Gamma(x) = \int_0^{\infty} e^{-t} t^{x-1} dt, \tag{5}$$

are the common derivative and the Gamma function, respectively.

The Laplace transform of the caputo fractional derivative is given as (4)

$$L\left\{\frac{d^r f(x)}{dx^r}\right\} = s^r F(s) - \sum_{m=0}^{n-1} s^{r-m-1} f^{(m)}(0), n - 1 < r < n. \tag{6}$$

The Caputo derivative operator satisfies the following relations

$${}_0^cD_t^r[f(t) + g(t)] = {}_0^cD_t^r f(t) + {}_0^cD_t^r g(t). {}_0^cD_t^r C = 0, \tag{7}$$

In recent years, the Mittag-Leffler function has caused enormous interest among scientist due to its role played in describing realistic dynamic systems with memory and delay. The Mittag-Leffler function is defined by the series expansion as [8]

$$E_a(t) = \sum_{m=0}^{\infty} \frac{t^m}{\Gamma(am+1)}, (a > 0) \tag{8}$$

where $\Gamma(.)$ is the Gamma function, when a = 1, from (8) we have

$$E_1(t) = \sum_{m=0}^{\infty} \frac{t^m}{\Gamma(m+1)} = \sum_{m=0}^{\infty} \frac{t^m}{m!} = e^t. \tag{9}$$

Thus, the M-L function generalizes the exponential function.

3 Fractional-Order RC, RL, RLC Circuit

3.1 Fractional-Order Series RC Circuit

The fractional order α affects real and imaginary parts of the fractional impedance with nonlinear functions, which changes the frequency response heavily for all circuit. Table 1 discusses how the fractional order α affects the impedance. If $\alpha > 1$, the real part of the impedance spans the range from $-\infty$ at $\omega \rightarrow 0$ up to R at $\omega \rightarrow \infty$, which the zero value is included. So, the impedance can be pure imaginary at specific frequency.

Table 1. Impedance properties of the fractional RC circuit for some non-integer order

		$\alpha \rightarrow 0$	$\alpha = 0.5$	$\alpha = 1$
Impedance	Real part	$R + \frac{1}{C}$	$R + \frac{1}{\sqrt{2\omega C}}$	R
	Imaginary part	0	$-\frac{1}{\sqrt{2\omega C}}$	$-\frac{1}{\omega C}$
Magnitude	Range	constant	(R, ∞)	(R, ∞)
Phase	Range	0	$(-\pi/4, 0)$	$(-\pi/2, 0)$

Correspondingly, the phase response of the fractional RC circuit has many interesting properties and is different totally. The effect of frequencies is very small relative to the huge dependence for large values of $\alpha \gg 1$. Also, it is clear that a critical minimum exists at certain α, and with frequency changing, its location changes.

3.2 Fractional-Order Series RL Circuit

The fractional-order RL circuit can be generalized which the fractional order is negative, where the magnitude increases as the frequency increases. The frequency-dependent fractional impedance $Z(j\omega, \beta)$, the magnitude $\|Z.(j\omega, \beta)\|$.and the phase $\phi(Z(j\omega, \beta))$ are given in Table 1.

Similarly, the fractional impedance of RL is pure imaginary (acting as an integer inductor), which is given from

$$\omega_{pi} = \left(\frac{-R/L}{\cos\left(\frac{\beta\pi}{2}\right)}\right)^{\frac{1}{\beta}} \tag{10}$$

$$Z\left(j\omega_{pi}\right) = -jR\tan\left(\frac{\beta\pi}{2}\right) = j\omega_{pi}L_{eq} \rightarrow \frac{L_{eq}}{L_{cir}} = \omega_{pi}^{\beta-1}\sin\left(\frac{\beta\pi}{2}\right) \tag{11}$$

3.3 Fractional-Order Parallel RC and RL Circuits

Similarly, the analysis of the fractional-order parallel RC and RL circuits can be acquired which the admittances are shown in Table 2.

Table 2. Fractional order parallel RC and RL circuits

Circuit	Fractional order admittance
Parallel fractional order RC circuit	$Y(j\omega) = \frac{1}{R} + (j\omega)^{\alpha} C$ $Y(j\omega, \alpha) = \frac{1}{R} + \cos\left(\frac{\alpha\pi}{2}\right)\omega^{\alpha} C + j\sin\left(\frac{\alpha\pi}{2}\right)\omega^{\alpha} C$
Parallel fractional order RL circuit	$Y(j\omega) = \frac{1}{R} + \frac{1}{(j\omega)^{\beta} L}$ $Y(j\omega) = \frac{1}{R} + \frac{\cos\left(\frac{\beta\pi}{2}\right)}{\omega^{\beta} L} - j\frac{\sin\left(\frac{\beta\pi}{2}\right)}{\omega^{\beta} L}$

4 The Simulation and Comparison Between Fractional-Order and Conventional Circuits

All theories of integer circuit were studied under the assumption of resistor, inductor, and capacitor. The main features of interest in fractional circuits are the ability to control the impedance in the frequency domains and the time.

In Table 3, the generalized equations of the frequency-dependent impedance, phase response, magnitude response, and for both circuits are listed in the third column in Table 3. The real and imaginary parts are functions of the operating angular frequency as $\omega^{-\alpha}$ and ω^{β} for the fractional RC and RL, respectively. It is so clear that, with the fractional order approaching two, the circuit behavior will approach the known second-order system (Figs. 1, 2, 3, 4, 5, 6, 7, 8, 9 and 10).

Table 3. Comparison between the conventional and fractional order RC and RL circuits

		Conventional	Fractional order
RC	Impedance	$Z(j\omega) = R - j\frac{1}{\omega C}$	$Z(j\omega, \alpha) = R + \frac{\cos\left(\frac{\alpha\pi}{2}\right)}{\omega^{\alpha} C} - j\frac{\sin\left(\frac{\alpha\pi}{2}\right)}{\omega^{\alpha} C}$
	Magnitude	$\frac{\lVert Z(j\omega)\rVert^2}{R^2} = 1 + \frac{1}{(\omega RC)^2}$	$\frac{\lVert Z(j\omega,\alpha)\rVert^2}{R^2} = 1 + \frac{1}{(\omega^{\alpha} RC)^2} + \frac{2}{\omega^{\alpha} RC}\cos\left(\frac{\alpha\pi}{2}\right)$
	Phase	$\phi(Z(j\omega)) = -\tan^{-1}\left(\frac{1}{\omega RC}\right)$	$\phi(Z(j\omega, \alpha)) = -\tan^{-1}\left(\frac{\sin\left(\frac{\alpha\pi}{2}\right)}{\cos\left(\frac{\alpha\pi}{2}\right) + \omega^{\alpha} RC}\right)$
RL	Impedance	$Z(j\omega) = R + j\omega L$	$Z(j\omega, \beta) = R + \omega^{\beta} L\cos\left(\frac{\beta\pi}{2}\right) + j\omega^{\beta} L\sin\left(\frac{\beta\pi}{2}\right)$
	Magnitude	$\frac{\lVert Z(j\omega)\rVert^2}{R^2} = 1 + \left(\frac{\omega L}{R}\right)^2$	$\frac{\lVert Z(j\omega,\beta)\rVert^2}{R^2} = 1 + \left(\frac{\omega^{\beta} L}{R}\right)^2 + 2\frac{\omega^{\beta} L}{R}\cos\left(\frac{\beta\pi}{2}\right)$
	Phase	$\phi(Z(j\omega)) = \tan^{-1}\left(\frac{\omega L}{R}\right)$	$\phi(Z(j\omega, \beta)) = \tan^{-1}\left(\frac{\sin\left(\frac{\beta\pi}{2}\right)}{\cos\left(\frac{\beta\pi}{2}\right) + \frac{R}{\omega^{\beta} L}}\right)$

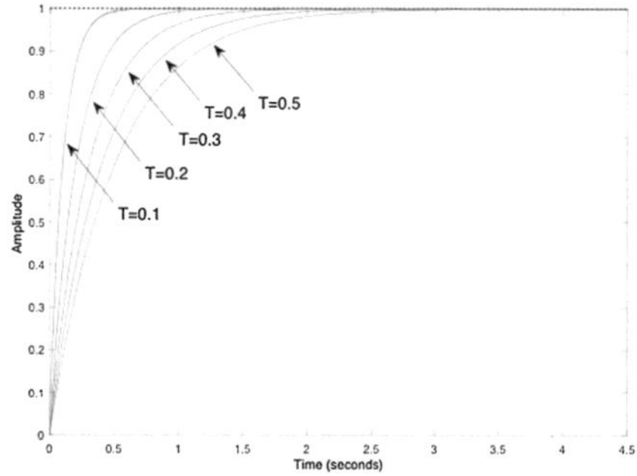

Fig. 1. Step response of 1/(Ts + 1) RC circuit (integer)

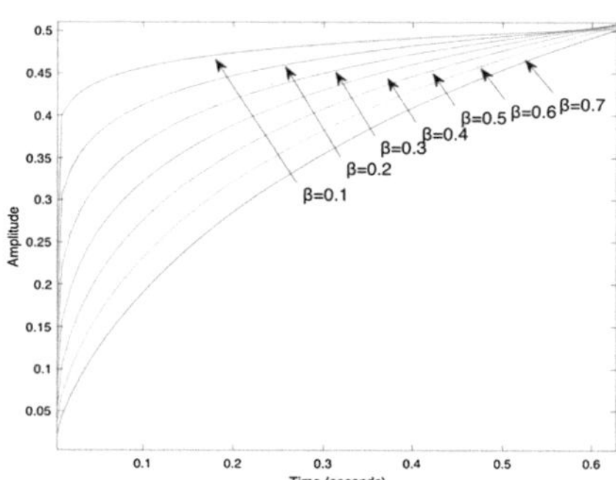

Fig. 2. Step response of 1/(Ts^ β +1) RC circuit(fractional-order)

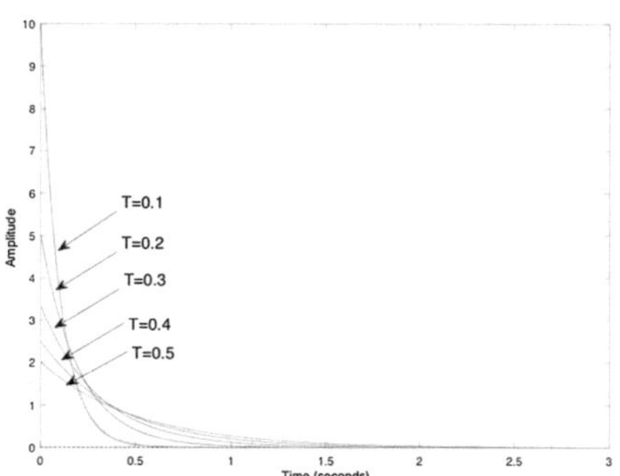

Fig. 3. Impulse response of 1/(Ts + 1) RC circuit(integer)

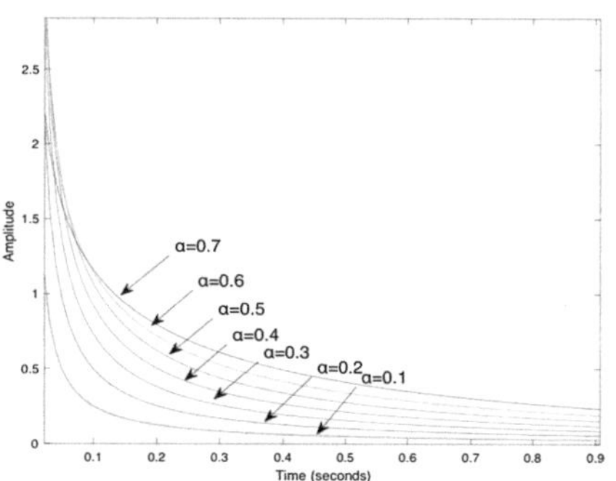

Fig. 4. Impulse response of 1/(Ts^ α +1)RC circuit(fractional-order)

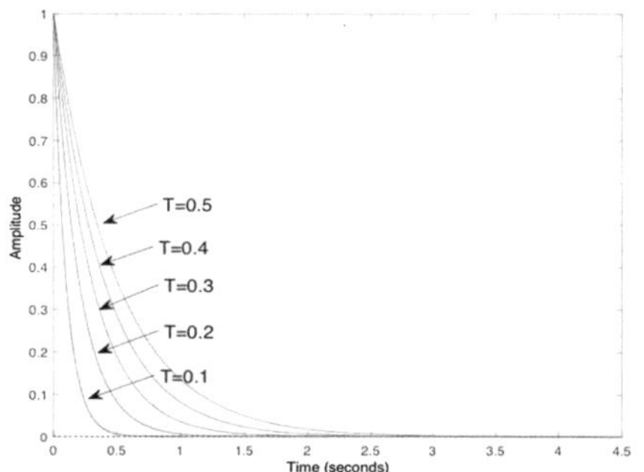

Fig. 5. Step response of Ts/(Ts + 1) RL circuit (integral)

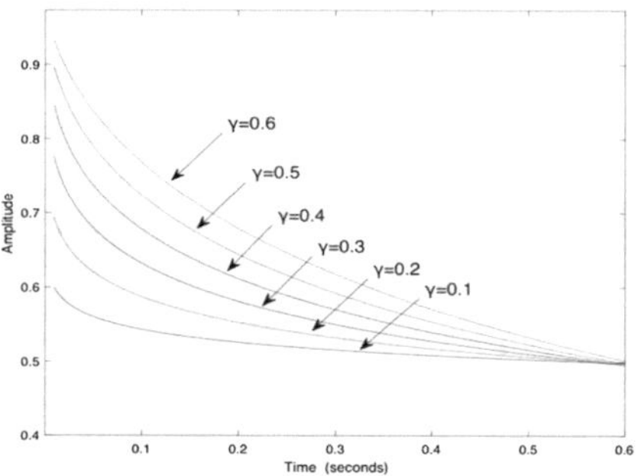

Fig. 6. Step response of Ts^ γ/(Ts^ γ +1) RL

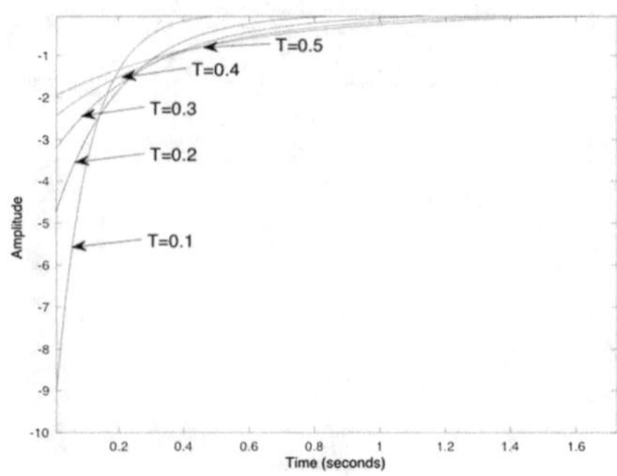

Fig. 7. Impulse response of Ts/(Ts + 1) RL circuit(integer)

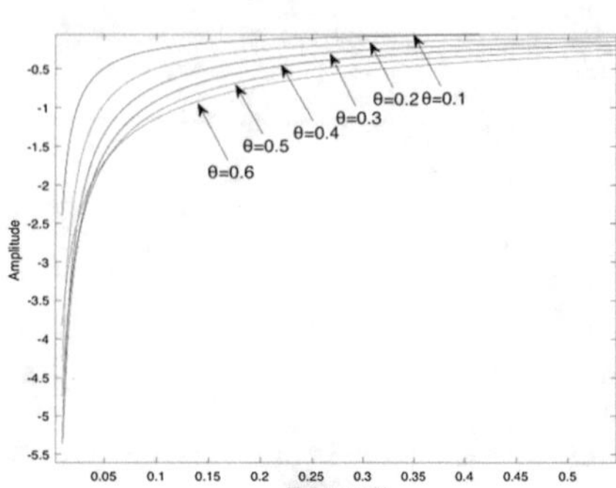

Fig. 8. Impulse response of Ts^ θ/(Ts^ θ +1) RL circuit(fractional-order)

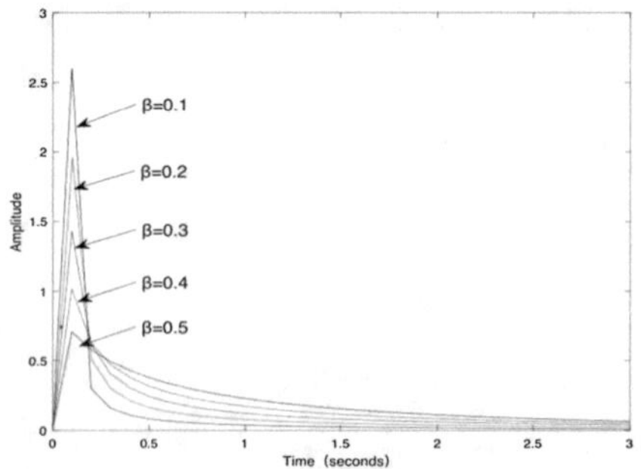

Fig. 9. Step response of RLC circuit (fractional-order)

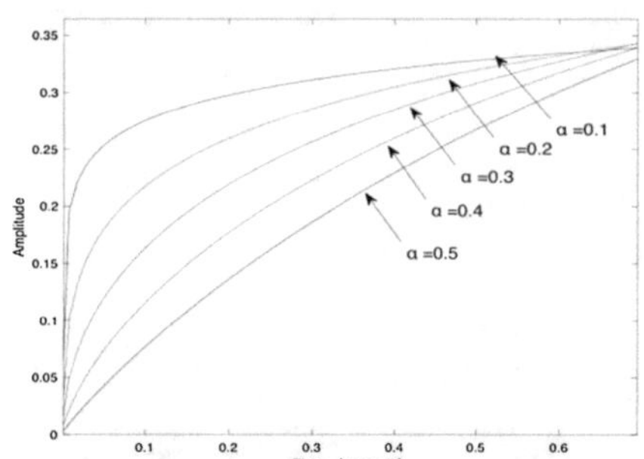

Fig. 10. Impulse response of RLC circuit (fractional-order)

5 Fractional RLC Circuit

In general, an oscillating circuit in series is an electrical circuit making up of three kinds of circuit elements: a resistor with a resistance R, an inductor with an inductance L, and a capacitor with capacitance C. The change with respect to time of the electric charge $q(t)$ in the shell of the capacitor is described by the homogeneous differential equation

$$L\frac{d^2q(t)}{dt^2}+R\frac{dq(t)}{dt}+\frac{q(t)}{C}=0 \tag{12}$$

The main goal of this work is to study the differential Eq. (15) from the view of the fractional calculus. It was presented the transition of the ordinary derivative operator to the fractional operator as follows:

$$\frac{d}{dt} \to \frac{1}{\delta^{1-\gamma}} \frac{d^{\gamma}}{dt^{\gamma}} \tag{13}$$

Where the auxiliary parameter δ has dimension of seconds and γ is an arbitrary parameter representing the order of the fractional derivative, and when $\gamma = 1$, it becomes the integer derivative. using (13) corresponding to (12),the fractional differential equation is given by

$$\frac{L}{\delta^{2(1-r)}} \frac{d^{2r}q}{dt^{2r}} + \frac{R}{\delta^{1-r}} \frac{d^{r}q}{dt^{r}} + \frac{q(t)}{C} = 0, 0 < r \leq 1. \tag{14}$$

Where the fractional derivative is Caputo derivative. The solution of the Eq. (14) may be acquired by applying direct and inverse Laplace transform. The solution is given by

$$q_{rRLC}(t) = q_0 E_r\left\{ -\frac{R\delta^{1-r}}{2L} t^r \right\} \times E_{2r}\left\{ -\left[\frac{1}{LC} - \frac{R^2}{4L^2} \right] \delta^{2(1-r)} t^{2r} \right\}, \tag{15}$$

In the case $r = 1$, we have the well-known result from (15)

$$q_{RLC}(t) = q_0 e^{-\frac{R}{2L}t} \cos\left(\sqrt{\frac{1}{LC} - \frac{R^2}{4L^2}} t \right) \tag{16}$$

6 Conclusions and Future Work

We have proposed analysis of the RC, RL and RLC circuit from the point of the fractional calculus. We also simulated them in step response and impulse response to see whether there have some differences between integer order.

We hope to study other aspects of the fractional modified circuits in future. We also hope that it can give some new insights from promising topics for future research such as fractional filters.

Acknowledgments. Yang Chen acknowledges fruitful discussion with Prof. Zhao and Hao-yu Li and has been supported by Graduate Innovation Fundation (Key No. 114-602080146).

References

1. Oldham, K.B., Spanier, J.: The Fractional Calculus. Academic Press, New York (1974)
2. Miller, K.S., Ross, B.: An Introduction to the Fractional Calculus and Fractional Differential Equations. Wiley, New York (1993)
3. Samko, S.G., Kilbas, A.A., Marichev, O.I.: Fractional Integrals and Derivatives Theory and Applications. Gordon and Breach Science Publishers, Langhorne (1993)
4. Podlubny, I.: Fractional Differential Equations. Academic Press, New York (1999)
5. Baleanu, D., Diethelm, K., Scalas, E., Trujillo, J.J.: Fractional Calculus Models and Numerical Methods Nonlinearity and Chaos. Series on Complexity. World Scientific, Singapore (2012)
6. Agrawal, O.P., Tenreiro-Machado, J.A., Sabatier, I.: Fractional Derivatives and Their Applications: Non-linear Dynamics. Springer, Berlin (2004)
7. Hilfer, R.: Applications of Fractional Calculus in Physics. World Scientific, Singapore (2000)
8. West, B.J., Bologna, M., Grigolini, P.: Physics of Fractional Operators. Springer, Berlin (2003)
9. Gómez-Aguilar, F., Rosales-García, J., Guía-Calderón, M., Bernal-Alvarado, J.: UNAM XIII, vol. 3 (2012)
10. Rosales, J.J., Guía, M., Gómez, J.F., Tkach, V.I.: Discontinuity, Nonlinearity Complexity **4**, 325 (2012)
11. Gómez, F., Bernal, J., Rosales, J., Córdova, T.: J. Electr. Bioimpedance **3**, 2 (2012)
12. Magin, R.L.: Fractional Calculus in Bioengineering. Begell House Publisher, Roddin (2006)
13. Caputo, M., Mainardi, F.: Pure. appl. Geophys. **91**, 134 (1971)
14. Westerlund, S.: Causality. University of Kalmar, Rep. No. 940426 (1994)
15. Baleanu, D., Günvenc, Z.B., Tenreiro Machado, J.A.: New Trends in Nanotechnology and Fractional Calculus Applications. Springer, London (2010)
16. Riewe, F.: Phys. Rev. E **53**, 4098 (1996)
17. Herzallah Mohamed, A.E., Muslih Sami, I., Baleanu, D., Rabei Eqab, M.: Nonlinear Dynam. **66**, 4 (2011)
18. Golmankhaneh Alireza, K., Yengejeh Ali, M., Baleanu, D.: Int. J. Theor. Phys. **51**, 9 (2012)
19. Alireza, K.G., Lambert, L.: Investigations in dynamics: with focus on fractional dynamics. Academic Publishing, Germany (2012)
20. Sami, I.M., Saddallah, M., Baleanu, D., Rabei, E.: Rom. J. Phys. **55**, 7 (2010)
21. Baleanu, D., Muslih Sami, I., Rabei Eqab, M.: Nonlinear Dynam. **53**, 1 (2008)

State Representation Learning for Multi-agent Deep Deterministic Policy Gradient

Zhipeng Li and Xuesong Jiang(✉)

College of Information, Qilu University of Technology (Shandong Academy of Sciences), Jinan 250353, China
jxs@qlu.edu.cn

Abstract. Multi-Agent Deterministic Policy Gradient (MADDPG) is a very useful algorithm in multi-agent domains. We analyze the algorithm and find that MADDPG uses deep neural networks (DNNs) as a Q function model. One advantage of DNNs is that they can build very complex processing functions to handle high-dimensional input. However, the disadvantage of this end-to-end learning is that it usually requires a lot of data, which is not always available for real-world control applications. In this paper, a new algorithm, State Representation Learning Multi-Agent Deep Deterministic Policy Gradient (SRL-MADPPG), is proposed that combines MADDPG with state representation learning which uses DNNs as a function fitting. i.e., model learning network is used to pre-train the first layer of the actor and critic networks, then the actor and critic learn from the state representation instead of the raw observations. Simulation result shows that the SRL-MADDPG algorithm improves the final performance in comparison with the end-to-end learning.

Keywords: MADDPG · DNNs · State representation learning SRL-MADDPG

1 Introduction

In recent years, with the rapid development of artificial intelligence, deep neural networks (DNNs) has been widely used [1]. What's more, they can use statistical learning methods to extract high-level features from raw sensory data and obtain effective representation of input space in a large amount of data, so deep neural networks (DNNs) have also been applied to the field of reinforcement learning. DNNs was usually used as a function approximation [2, 3] in Q-learning. This technique, by improving the Deep Deterministic Policy Gradient (DDPG) [4] algorithm, adds some additional information of other agents in the Critic, and introduces a policy that uses a centralized training decentralized operation to form the Multi-Agent Deep Deterministic Policy Gradient (MADDPG) [5] algorithm which can complete multi-agent cooperation and competition.

In the face of complex and changeable environments, how multi-agents make rational decisions is also a hot topic that people have been researching for many years [6]. The agents interact with the environment so as to collect various information related to the task, and use a state estimator to establish a state vector from a set of

P. Krömer et al. (Eds.): ECC 2018, AISC 891, pp. 667–675, 2019.
https://doi.org/10.1007/978-3-030-03766-6_75

observation data for state control. However, these observations need to be processed so that the mapping of the observation data to the state information is not dependent on the state estimator designed by the engineer. In fact, this work can also be given to the machine to complete for us. Before solving the RL task, we called this mapping from the observed state as state representation learning [7]. which has been applied in robotic priors [7], Slow Feature Analysis (SFA) [8], auto-encoder network [9] etc. However, these applications are based on visual learning. They do not combine these methods with reinforcement learning and DNNs. In [10], the state representation learning is combined with Deep Deterministic Policy Gradient Control (DDPG) [4] and DNNs to achieve continuous control. However, this is only combined with a single agent and DNNs. Learning from others' experience, this paper introduces a state a priori learning to the multi-agent field and combines it with MADDPG to form a new algorithm called state representation learning multi-agent deep deterministic policy gradient (SRL-MADDPG), in which the MADDPG algorithm using DNNs as function approximator is combined with state representation learning. The state representation learning can learn from the observations of high-dimensional multi-agent continuous control, whose observations seldom apply to other algorithms.

The SRL-MADDPG algorithm that multi-agent learns the model network based on the concept of prediction prior. It assumes that the next state and reward should take the current state and the actions taken in this state which is predictable into account. The way of constructing the model network is to learn the mapping of the observed state through the prediction error of the back propagation, so as to get an inherent and predictable state representation. Both the actor and the critic learn from the state of state, not from the original observation.

The end to end learning [11] reduces manual preprocessing and subsequent processing, making the model as much as possible from the original input to the final output, providing a larger space for the model to automatically adjust the data and increase the overall adaptability of the model. However, the end-to-end learning's input is the original data, and the output is the final result. The original input is not the direct raw data, but is the feature extracted from the original data. This is particularly prominent in the image problem because there are too many image pixels [3]. The high dimensionality of the data creates a dimensional disaster, so the original idea was to manually extract some key features of the image. This is actually a process of dimension reduction, then the question comes from that how to mention the characteristics. The criticality of feature extraction is even more critical than learning algorithms. This means that features need sufficient experience to design, which is more and more difficult in the case of increasingly large amounts of data. This article introduces the state representation learning which uses neural networks, so that the network itself learn how to capture features well. This method is more flexible to the data fitting, so the new algorithm ML-MADDPG was born.

In the remainder of the paper, we fist introduce MADDPG algorithm and state representation learning, then introduce the new algorithm SRL-MADDPG and the result of the experiment before summarizing some current and future lines of research.

2 Related Work

The work of this article is based on previous work, including MADDPG algorithm and state representation learning. This section will detail the previous work.

2.1 Multi-agent Deterministic Policy Gradient

Last July, researchers from OpenAI company invented a new algorithm which is named as MADDPG, which is suitable for centralized learning and decentralized execution in multi-agent environments. The MADDPG algorithm allows agents to learn to cooperate and compete. It is a simple extension of the actor-critic policy gradient approach, and criticism adds additional information about other agent policies. The model of this algorithm is based on the centralized training with decentralized execution framework. Critic's network structure is: input observations and actions, output the Q value of each agent. Each actor has a crisp that collects the observations and actions of all agents and outputs the Q of a single actor to update the actor. For each agent, sampling a random mini batch of S samples (x^j, a^j, r^j, x'^j) from the reply buffer D into the Critic network. The Q value of the output is fed back to the actor, whose objective function is:

$$y^j = r_i^j + \gamma Q_i^{\mu'}(x'^j, a'_1, \ldots, a'_N)|_{a'_k = \mu'_k(o_k^j)} \tag{1}$$

The Q in the formula is output by the above critic network. r is a single agent reward and is used to update the loss of the critic network. The loss formula is

$$\mathcal{L}(\theta_i) = \frac{1}{s}\sum\nolimits_j (y^j - Q_i^{\mu}(x'^j, a'_1, \ldots, a'_N))^2 \tag{2}$$

After the actor receives the Q value, it uses its own network to calculate the policy gradient for optimizing the parameters. The optimization formula is

$$\nabla_{\theta_i} J \approx \frac{1}{s}\sum\nolimits_j \nabla_{\theta_i \mu_i}(o_i^j)\nabla_{a_i} Q_i^{\mu}(x^j, a_1^j, \ldots, a_N^j)|_{a_i = \mu_i(o_i^j)} \tag{3}$$

After actor and critic have finished updating, in order to prevent θ from updating too quickly, so we do some tweaking:

$$o'_i \leftarrow \tau\theta_i + (1 - \tau)\theta'_i \tag{4}$$

τ represents the trade-off between learning speed and stability. For an overview of the algorithm for MADDPG, see [5].

2.2 State Representation Learning

In the reinforcement learning literature, the terms we use define an environment ε, where one agent performs an action $a_t \in A$ at a time step t, where A is the action space. Each action will cause the agent to switch from the real state $\tilde{s}_t$. The agent obtains an observation $o_t \in \vartheta$ from the sensor at time step t , which makes learning possible by interacting with the behavior. Typically, agents can receive rewards r_t by performing an action in a given state. The reward is given by the reward function that the agent tries to maximize. The reward function is designed to guide the agent in the specific behavior of solving the task [12] (see Fig. 1).

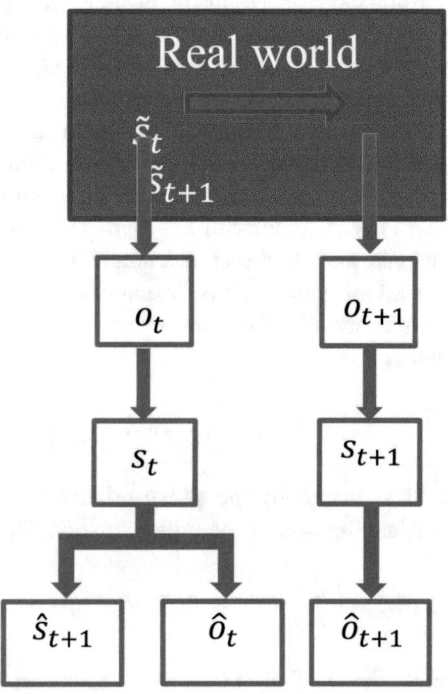

Fig. 1. The notation organization ($\tilde{s}_t$ represents the true state. s_t is the learned state that computed from the observation o_t, and $\hat{s}_{t+1}$ the estimation or prediction of the learned state, e.g.)

In RL, state is not always directly accessible, but it needs to be constructed from a set of observations. Such an observed state diagram can be a result feature engineering where engineers choose to observe and design the map, but this can also be learned from the data. The SRL task is learning to find the mapping of observation history to the current state $s_t = \phi(o_1{:}t)$. Please note that action a_{t1} and reward r_{t1} can also be added to the parameters [13].

State prior learning is a form of unsupervised. Therefore, our mapping of learning observed states involves making some assumptions about the structure of states. In [14, 15], an observer is compressed using an automatic encoder into a low dimensional state vector and trained to construct a map of observed states.

Another unsupervised method is Slow Feature Analysis (SFA) [16], which uses information from the time signal to learn linear factor models of invariant features. The basic idea is that the important features of a scene are basically unchanged. In [17], this assumption is used to learn the mapping between visual observations and state representations that gradually change over time.

In [7, 10], these methods and assumptions are merged into the robot prior. Expressing the state learning representation as an input to the network rather than as an input from the original result. This article shows an increase in performance.

The state prior method in this article consists of two parts. The observed state obtains observations from the real world. Given the initial state of each agent i and the actions in that state, it can predict the next state and reward of the agent by training. The reward help predict transformation before converging to a meaningful representation of a given task.

2.3 State Representation Learning Multi-agent Deep Deterministic Policy Gradient

In this paper, we construct the SRL-MADDPG algorithm consisting of three DNNs (a model network, a critic network, and an actor network). First use the SRL model to train and then combine some weights with the other two DNNs. In the experiment, the three model networks were performed separately, which allowed us to train the MADDPG and SRL-MADDPG algorithms in the same data set and the same multi-agent environment.

In [12], the author summarized some of the SRL methods (such as Auto-Encoder [16], Forward Model [17], Inverse Model [18], Model with prior [13] (The architecture is shown in Fig. 2) and hybrid model [19]) used in recent years. The first layer of DNNs in this paper uses the method of model with prior. Using prior knowledge of functions, dynamics or world physics, such as SRL processing, the continuity or causality principle of time, generally reflects the object or environment. Priors are defined as an objective function or loss function $\mathcal{L}$, applied on a set of states $(s_1, \ldots, s_n)$ that is minimized (or maximized) under certain conditions C. The loss is

$$\text{loss} = \mathcal{L}_{prior}(s_1, \ldots, s_n; \theta_\phi | C) \tag{5}$$

The first DNNs of SRL-MADDPG uses state representation learning to obtain the predicted value of the next state and reward $(\hat{s}_{i,t+1}, \hat{r}_{i,t+1})$ from the observations-action tuple $(o_{i,t}, a_{i,t})$. Train by the following minimization objective function

$$\mathcal{L}_{SRL} = \left\| s_{i,t+1} - \hat{s}_{i,t+1} \right\|_2^2 + \lambda_{SRL} \left\| r_{i,t+1} - \hat{r}_{i,t+1} \right\|_2^2 \tag{6}$$

where λ_{SRL} represents a compromise between the predicted reward and the next state. Its entire network architecture is shown in Fig. 3, in which the circle represents the network layer containing n neurons. These lines are tensors which represent n-dimensional signals input. It is observed that the mapping of the state is encoded on the first level of the network, whose input is observations and output is states. This state and action consist of second levels of input. Finally, two parallel linear output layers are generated to predict the next state and reward respectively.

Before the actor and the Critic network, the state prior learning network is first trained. The weights of the state representation learning model network are fixed. By

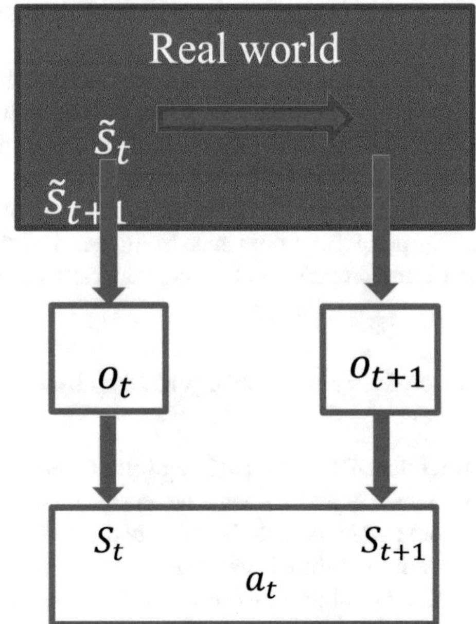

Fig. 2. The structure of model with prior

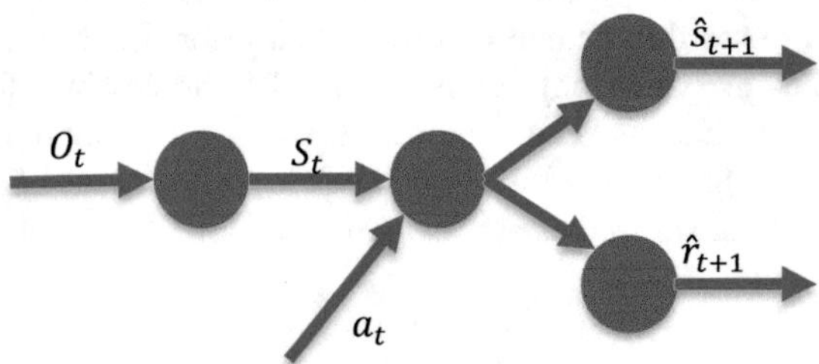

Fig. 3. The neural network model structure of state representation learning

replicating the first layer network weights to the actor and the Critic network, the actor and the Critic network weights are updated (Fig. 4).

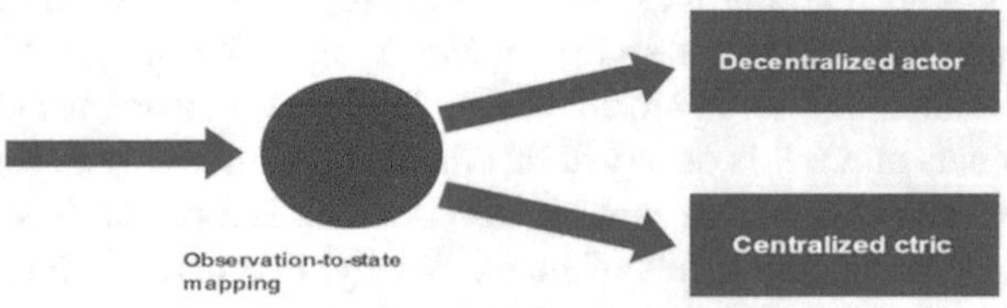

Fig. 4. The model of SRL-MADDPG

3 Experiments

In order to implement our experiments, we used the grounded communication environment proposed in [5]. It consists of N agents and L landmarks. It exists in a two dimensional world with continuous space and discrete time. In order to compare the performance of the algorithm, we took the pursuit of escape as an example, selected three predator agents, and completed the encircling of prey by cooperation.

In order to assess the quality of the policy that is being learned in the competitive environment, this article compares the SRL-MADDPG policy agent with the MADDDPG policy agent and compares the success of the agent and the opponent in the environment. For SRL-MADDPG and MADDPG algorithms, the algorithm is performed on the same database. The SRL-MADDPG algorithm uses the learning rates $\alpha_{i,SRL} = 0.01, \alpha_{i,a} = 0.01$ and $\alpha_{i,c} = 0.01$ to update the weights respectively. The hidden layer of the first layer of the model has 400 neurons. Meanwhile, the hidden layer of the actor and the Critic network has 300 neurons, and we record the mean episodes every 500 episodes. Iteratively evaluate various metrics, train our models until convergence, and we draw the learning curve of 60000 episodes of the various methods in Fig. 5. As can be seen from the Fig. 5, the average rewards for the two rounds are similar within the first 10,000 episodes. As the episode increases, after more than 27,000 rounds, the mean episode reward of SRL-MAPPG algorithm is higher than MADDPG algorithm.

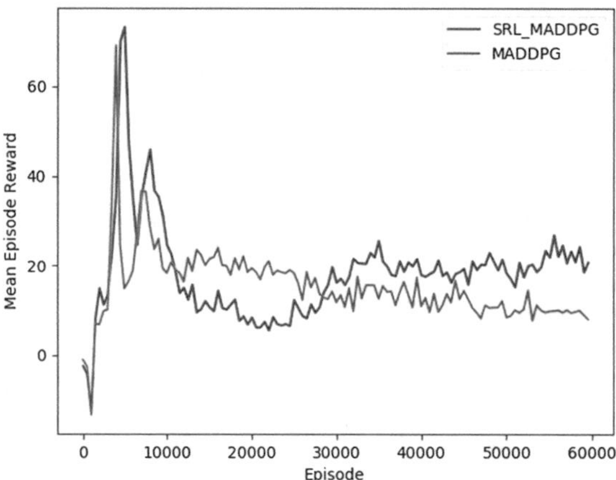

Fig. 5. Comparison of the mean episode reward of the two algorithms

4 Conclusion

This paper proposes a new algorithm RSL-MADDPG, which he uses prior knowledge to complete the training before learning the RL task. One of the characteristics of using prior knowledge is that the agent does not need to follow a specific policy. Therefore, the agent can be created from any Learn in ERD (Experience Replay Database) [31].

This state indicates that learning methods and multi-agent collaboration deserve further study. In particular, how to improve the efficiency of cooperation between agents, and increase the number of agents that are worth researching. Other work will be further studied in the future.

Acknowledgments. This work was supported by Key Research and Development Plan Project of Shandong Province, China (No. 2017CXGC0614)

References

1. Sze, V., Chen, Y.H., Yang, T.J.: Efficient processing of deep neural networks: a tutorial and survey. Proc. IEEE **5**(12), 2295–2329 (2017)
2. Riedmiller, M.: Neural fitted Q iteration – first experiences with a data efficient neural reinforcement learning method. In: European Conference on Machine Learning, pp. 317–328. Springer (2015). https://doi.org/10.1007/1156409632
3. Mnih, V., Kavukcuoglu, K., Silver, D.: Human-level control through deep reinforcement learning. Nature. **518**, 529–533 (2015). https://doi.org/10.1038/nature14236
4. Lillicrap, T.P.: Continuous control with deep reinforcement learning. Comput. Sci. **8**(6), A187 (2015)
5. Lowe, R.: Multi-agent actor-critic for mixed cooperative-competitive environments. In: Neural Information Processing Systems (NIPS) (2017)
6. Foerster, J.N.: Learning to communicate with deep multi-agent reinforcement learning. In: NIPS 2016: Proceedings of the Thirtieth Annual Conference on Neural Information Processing Systems (2016)
7. Jonschkowski, R., Brock, O.: State representation learning in robotics: using prior knowledge about physical interaction. In: Robotics: Science and Systems (2014)
8. Wiskott, L., Sejnowski, T.J.: Slow feature analysis: unsupervised learning of invariances. Neural Comput. **14**(4), 715–770 (2002)
9. Jetchev, N., Lang, T., Toussaint, M.: Learning grounded relational symbols from continuous data for abstract reasoning. In: Workshop on Autonomous Learning, International Conference on Robotics and Automation (ICRA) (2013)
10. Munk, J., Kober, J., Babuška, R.: Learning state representation for deep actor-critic control. In: Decision and Control IEEE (2016). https://doi.org/10.1109/cdc.2016.7798980
11. Miao, Y., Gowayyed, M., Metze, F.: End-to-end speech recognition using deep RNN models and WFST-based decoding. In: Proceedings of ASRU (2015). https://doi.org/10.1109/asru.2015.7404790
12. Lesort, T.: State Representation Learning for Control: An Overview. arXiv:1802.04181 (2018)
13. Jonschkowski, R., Brock, O.: Learning state representations with robotic priors. Auton. Robots **39**(3), 407–428 (2015). https://doi.org/10.1007/s10
14. Finn, C.: Learning Visual Feature Spaces for Robotic Manipulation with Deep Spatial Autoencoders (2015)
15. Lange, S., Riedmiller, M., Voigtlander, A.: Autonomous reinforcement learning on raw visual input data in a real world application. In: International Joint Conference on Neural Networks, pp. 1–8 (2012). https://doi.org/10.1109/ijcnn.2012.6252823
16. Vincent, P., Larochelle, H., Bengio, Y.: Extracting and composing robust features with denoising autoencoders. In: International Conference on Machine Learning, pp. 1096–1103. ACM (2008)

17. Klyubin, A.S., Polani, D., Nehaniv, C.L.: Empowerment: a universal agent-centric measure of control. In: Congress on Evolutionary Computation, pp. 128–135 (2005). https://doi.org/10.1109/cec.2005.1554676
18. Pathak, D.: Curiosity-driven exploration by self-supervised prediction, pp. 488–489 (2017). https://doi.org/10.1109/cvprw.2017.70
19. Watter, M.: Embed to control: a locally linear latent dynamics model for control from raw images. In: International Conference on Neural Information Processing Systems, pp. 2746–2754. MIT Press (2015)

Small Moving Object Detection Based on Sequence Confidence Method in UAV Video

Dashuai Yan and Wei Sun(✉)

School of Aerospace Science and Technology, Xidian University, No. 2 South Taibai Road, Xi'an 710071, Shaanxi, China
wsun@xidian.edu.cn

Abstract. In the small object detection on the UAV (Unmanned Aerial Vehicle) platform, the confidence description of the moving object is proposed to improve the accuracy, robustness and reliable tracking method of the object detection. Due to the low resolution and slow motion of small moving object in aerial video, and the image is easily subject to illumination and camera jitter noise, and the correlation between video sequences is neglected, it is prone to false detection of moving object and low detection accuracy, the characteristics of poor robustness. For the UAV video with small moving object, the algorithm uses the ORB operator to extract reliable global feature points for each frame of the video, and then performs global motion compensation on the motion background through the affine transformation model and calculates the difference image. The energy accurately detects the small object, and then describes the confidence of the moving object. The n-step back-off method is used to increase the correlation information between the video sequences. The proposed method is to evaluate the video captured on the airborne aircraft, and has done a lot of experiments and tests. For the object as small as 25 pixels, the method still has better performance, and our method can be realized by parallel computing. Real-time, processing 1280 × 720 frames at around 45 fps.

Keywords: UAV · Aerial video · Sequence confidence
Small object detection

1 Introduction

Moving object detection in complex background is a key and difficult point in the field of video application. The diversity of background and the complex motion of the camera increase the difficulty of object detection. The conventional human-monitoring surveillance systems have been progressively replaced by intelligent video analyses such as motion detection, object tracking and event analysis in Ref. [1, 2], and the stationary cameras show their limitation on field of view and immobility. In the conventional video surveillance system with stationary cameras, approaches such as background subtraction or frame difference in Ref. [3, 4] have been widely utilized for detecting moving foreground since the background is static. Generally, the approaches for detecting moving objects in aerial videos can be classified into two categories, optical-flow based and frame-difference based. The frame-difference based approaches

P. Krömer et al. (Eds.): ECC 2018, AISC 891, pp. 676–683, 2019.
https://doi.org/10.1007/978-3-030-03766-6_76

are faster in Ref. [5, 6], but usually cause a moving object comes to pieces especially when the object's color distribution is homogenous.

Recently, advances in apparently unrelated areas have given tracking a fresh impulse: Specially, progress in the definition of features invariant to various imaging transformations in Ref. [9, 10], object detection in Ref. [11, 12], and online learning in Ref. [13, 14] have spawned the approach of tracking by detection in Ref. [15], in which a object identified by the user in the first frame is described by a set of features, motion constraints restrict the space of boxes to be searched for the object. In order to improve the accuracy of object detection by using the correlation between aerial video sequences, and to find a more optimal decision-making method under dynamic systems with high uncertainty and randomness, it can be considered as Markov decision problem of sequence correlation.

As discussed above, within the innovation framework, the proposed method incorporates at least three novel elements should be highlighted: the improved ORB, efficient solution of object scale calculation and low computational complexity. The paper is organized as follows. In Sect. 2, the procedure of moving object detection is discussed in detail. Section 3 will discuss the proposed Confidence analysis; In Sect. 4, some results on different sequences are presented. Through the aerial video in different application scenarios, a large number of comparative experiments and arguments have been made on the accuracy and frame rate as evaluation points.

2 Small Moving Object Detection

Since the resolution of moving small object in aerial images is low and the images are easily interfered by illumination and noise, the accuracy of background motion vector estimation during background motion compensation will directly affect the subsequent positioning. The corresponding points obtained by improved ORB are utilized in this stage.

2.1 Camera Model

To solve the interference problem of camera motion noise, we must start from the camera self-motion model. Given a feature point P_i and two successive image frames a and b whose viewpoint are at O_a and O_b, respectively, we denote their projection on the images are $P_{a,i}$ and $P_{b,i}$, respectively. Considering the homography geometry, in homogeneous system, we can obtain,

$$P_a = \begin{bmatrix} x_a \\ y_a \\ 1 \end{bmatrix} P_b = \begin{bmatrix} x_b \\ y_b \\ 1 \end{bmatrix} H_{ab} = \begin{bmatrix} h_{11} & h_{12} & h_{13} \\ h_{21} & h_{22} & h_{23} \\ h_{31} & h_{32} & h_{33} \end{bmatrix} \tag{1}$$

Through $P_b = H_{ab}P_a$, Given a set of matched feature points $\{(P_i, Q_i)\}$, where P_i or Q_i is the homogeneous coordinate of the matched points in the frame I_P and I_Q, and a planar parametric transformation of the form

$$Q_i = HP_i \quad (2)$$

Where H is the 3 × 3 homography, the central aspect of our approach to affine image registration is the use of least mean square technique to efficiently solve Eq. (2) for a robust affine estimate, we then use the parameters to warp successive images.

2.2 Background Compensation

As discussed in Sect. 2.1, and the best value for Eq. (2) can be obtained by minimize the sum of squared residuals

$$E = \sum \|HP_i - Q_i\|^2 \quad (3)$$

Due to the existence of inaccurate matched points on the moving object, the precision of H is influenced and reduced. The RANSAC algorithm in Ref. [14] can be introduced to overcome this problem. It can improve the precision by kick out the outliers. We make use of the improved RANSAC algorithm in Ref. [1] to speed up this process.

In Fig. 1(a) and (b) show the original image. Image Fig. 1(c) shows the registration result, Fig. 1(d) shows the difference of pixel intensities between the reference image and the aligned image.

(a) (b) (c) (d)

Fig. 1. Motion compensation. (a) Reference image; (b) Adjacent image; (c) Aligned image; (d) Difference image.

2.3 Moving Object Detection

Consider a aerial video image sequence at times $t1, t2, t3, \ldots, tn$ and represented as $f(x, y, t1), f(x, y, t2), \ldots, f(x, y, tn)$. Let the first image frame $f(x, y, t1)$ be the reference image. An accumulative difference image is obtained by comparing the reference image in the sequence. A count for each pixel location in the accumulative image is incremented every time. Therefore, we use the interval sequence multi-frame sampling grouping of image sequences, and accumulate multiple sets of differential image methods to achieve accurate detection of small object.

In Fig. 2(a), the differencing image is figured out by adding four the difference images to obtain a sum image, the opening operation on (a) is given in Fig. 2(b) which the noise is eliminated, after canny edge detecting, the moving object is figure out in Fig. 2(c) and labeled in Fig. 2(d).

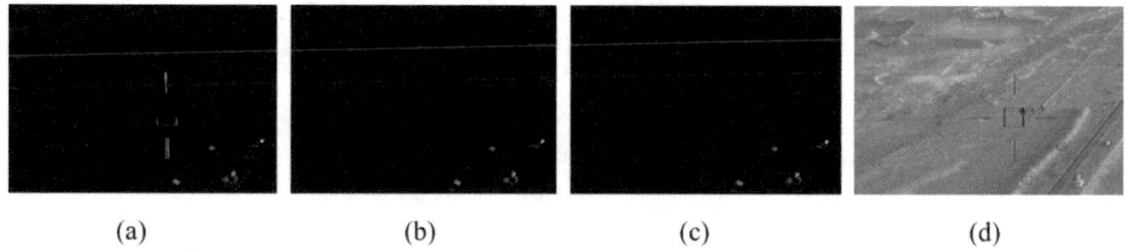

(a) (b) (c) (d)

Fig. 2. Moving object detection. (a) Sum of difference image; (b) Opening operation on (a); (c) Canny edge detection; (d) Labeled image.

3 Confidence Analysis and Detection

The object motion model in the aerial video can be modeled as a uniform motion. Some abnormal points such as bright spots or obvious edges appearing in the image, a large probability is detected as a moving object, and the correlation between video sequences is neglected, and the false detection of the moving object is easy to occur, the detection accuracy is low, and the robustness is low. The characteristics of poor robustness. The main advantages of tracking by detection come from the flexibility and resilience of its underlying representation of appearance. In order to solve the sequence image, the interval step is n multi-frame sampling group, and the m group differential image method is added to realize the accurate detection of the small moving object.

3.1 Confidence Analysis

A frame difference accumulation video sequence of sequence length n can be expressed as $I = \{I_{t-n}, \ldots, I_{t-2}, I_{t-1}, I_t\}$, I_t is Current frame difference, Adding the corresponding frame difference to the moving object set of the video sequence can be expressed as $O = \{O_{t-n}, \ldots, O_{t-2}, O_{t-1}, O_t\}$, The t frame accumulating video sequence includes a object set of n moving object and pseudo-moving object represented as

$$O_t = \{o_{t,1}, o_{t,2}, o_{t,3}, \ldots, o_{t,k}, \ldots, o_{t,n}\} \quad (4)$$

The accuracy of the object frame is described by adding confidence, and the influence of the historical sequence on the current sequence is increased by the long-term value expectation between sequences, and coordinate information and size information. The k th object is expressed as $\{x_{t,k}, y_{t,k}, w_{t,k}, h_{t,k}, c_{t,k}, v_{t,k}\}$. The object's constraint is defined as

$$D(o_{t,k}) = \frac{1}{w_{t,k}h_{t,k}} \sum\nolimits_{i=1}^{w_{t,k}} \sum\nolimits_{j=1}^{h_{t,k}} I_{t,i,j}/\sqrt{d} > \frac{1}{\beta^2 w_{t,k}h_{t,k}} \sum\nolimits_{i=1}^{\beta w_{t,k}} \sum\nolimits_{j=1}^{\beta h_{t,k}} I_{t,i,j}/\sqrt{d} \quad (5)$$

Where d is the distance from the center of the left point of the object point set, and $\beta = 1.3$ is the extended area with the object as the boundary.

Under the condition that the t-th frame accumulates the k-th object in the object set corresponding to the video sequence, the long-term value expectation of the n-step fallback is expressed as:

$$v_{t,k} = E\left[\sum_{i=1}^{n} \gamma^i v_{t-i,k} | I = I_t, O_t = o_{t,k}\right] \tag{6}$$

Where γ is the discount factor, the γ larger, the representation history accumulated video sequence has greater impact on the current frame object determination. In order to ensure the absolute advantage of the value expectation of the moving object in the calculation process, in the actual test, because the number of pseudo-moving object in the object set is the majority, in order to suppress the pseudo-object from being amplified in the long-term value expectation, a nonlinear suppression function *tanh* is introduced.

$$c_{t,k} = \frac{D(o_{t,k})}{D(\beta o_{t,k})} tanh(v_{t,k}) \tag{7}$$

Finally, in the set of moving object, the accurate object of the object is selected according to the object confidence.

3.2 Procedures of the Proposed Algorithm

The Kalman filter uses the dynamic information of the object to remove the influence of noise and obtain the position estimate and future position estimate of the object at this moment. As discussed above, the procedures of the proposed algorithm are given as follows:

1. Extract features P_0 from the object image, Kalman filter is initialized by the initial state of object.
2. Capture the next frame image I_k, using the predicted state of object, we get the searching region of the object defined as I_1.
3. Extract features P_1 from the searching region, using I-ORB descriptors P_0 to find similar regions in I_k', and M corresponding point is matched.
4. Calculate the object set O_t and Confidence $c_{t,k}$.
5. Get the centroid of the matched points $(x_{k,i}, y_{k,i}), i = 1, 2 \ldots M$ in frame I_k.
6. When the object is detected, the Kalman filter first predicts its state at the current video frame, and then uses the newly detected object location (x_k, y_k) to correct its state. This produces a filtered location.
7. According to $Scale_{i \in [1,M]}$ of the matched points.
8. Repeat the step 3–7, we can get the tracking object's trajectory.

4 Experiments

To demonstrate the effectiveness of the proposed algorithm, we do a lot of experiments in the desert, suburbs, towns and other aerial videos, In Fig. 3, the resolution of the video is 1280 × 720 and 1920 × 1080, and in the suburbs and towns aerial video of the mobile vehicles on the road as small as 5 × 5 pixels. Experiments show that the accuracy and robustness of detection are greatly improved by introducing the confidence description of moving object.

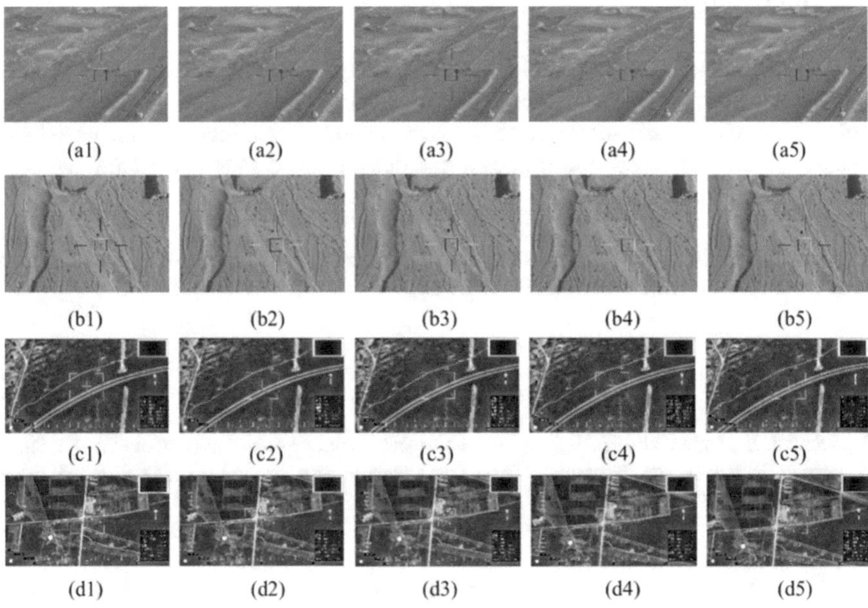

Fig. 3. Detection result. Frame of Desert. 1 video clip, (a1) Frame 10; (a2) Frame 40; (a3) Frame 80; (a4) Frame 120; (a5) Frame 190. Frame of Desert. 2 video clip, (b1) Frame 10; (b2) Frame 60; (b3) Frame 90; (b4) Frame 200; (b5) Frame 300. Frame of Suburbs video clip, (c1) Frame 10; (c2) Frame 50; (c3) Frame 300; (c4) Frame 600; (c5) Frame 900. Frame of Town video clip, (d1) Frame 10; (d2) Frame 80; (d3) Frame 300; (d4) Frame 500; (d5) Frame 800.

By contrasting the selection of SURF and ORB global feature points in background compensation, blob and connected domain detection method, and discount factor γ which affect sequence correlation. The algorithm is evaluated by the accuracy P and the frame rate F to obtain the experimental results (Table 1).

Table 1. Performance of Detection algorithms evaluated by precision *P* and Frame rate *F*.

Sequence	Frame	Size	SURF-Blob	ORB-Blob	ORB-ConD	YanD $\gamma = 0.8$	YanD $\gamma = 0.9$
Desert. 1	205	720p	0.63/15.0	0.55/30.2	0.49/51.1	0.94/45.1	0.89/45.2
Desert. 2	329	1080p	0.67/8.6	0.63/16.1	0.59/34.0	0.93/25.4	0.88/25.6
Suburbs	962	1080p	0.52/7.8	0.47/16.0	0.39/35.8	0.69/25.9	0.83/26.3
Town	863	1080p	0.56/8.0	0.45/15.8	0.35/34.6	0.71/25.5	0.83/25.8

5 Conclusions

In this paper, the task was defined as detection, we have presented a technique to detect and segment moving objects in complex dynamic scenes shot by possibly moving cameras. Introducing the sequence confidence method, which increases the correlation of aerial video sequences, and improves the detection accuracy and frame rate compared to the traditional method.

Acknowledgements. This work was supported by National Nature Science Foundation of China (NSFC) under Grants 61671356, 61201290.

References

1. Annaka, K., Munakata, H., Kanamura, K.: Electrochemical evaluation of Li4Ti5O12 single particle at various temperatures. IEEE Trans. Image Process. Publ. IEEE Signal Process. Soc. **6**(9), 1281–1295 (2012)
2. Andrew, A.M.: Multiple view geometry in computer vision. Kybernetes **30**(9/10), 1865–1872 (2004)
3. Yilmaz, A., Li, X., Shah, M.: Contour-based object tracking with occlusion handling in video acquired using mobile cameras. IEEE Trans. Pattern Anal. Mach. Intell. **26**(11), 1531–1536 (2004)
4. Saleemi, I., Shafique, K., Shah, M.: Probabilistic modeling of scene dynamics for applications in visual surveillance. IEEE Trans. Pattern Anal. Mach. Intell. **31**(8), 1472–1485 (2009)
5. Ludington, B., Reimann, J., Vachtsevanos, G.: Target tracking and adversarial reasoning for unmanned aerial vehicles. In: IEEE Aerospace Conference, pp. 1–17 (2007)
6. Mejías, L., Saripalli, S., Campoy, P., et al.: Visual servoing of an autonomous helicopter in urban areas using feature tracking. J. Field Robot. **23**(3–4), 185–199 (2006)
7. Bell, W., Felzenszwalb, P., Huttenlocher, D.: Detection and long term tracking of moving objects in aerial video (1999)
8. Kanade, T., Amidi, O., Ke, Q.: Real-time and 3D vision for autonomous small and micro air vehicles. In: IEEE Conference on Decision and Control (CDC), vol. 2, pp. 1655–1662. IEEE (2004)
9. Mondragon, I.F., Campoy, P., Correa, J.F., et al.: Visual model feature tracking for UAV control. In: IEEE International Symposium on Intelligent Signal Processing, pp. 1–6. IEEE (2008)

10. Ali, S., Shah, M.: COCOA: tracking in aerial imagery. Society of Photo-optical Instrumentation Engineers. International Society for Optics and Photonics, pp. 105–114
11. Ali, S., Reilly, V., Shah, M.: Motion and appearance contexts for tracking and re-acquiring targets in aerial videos. In: IEEE Conference on Computer Vision and Pattern Recognition, CVPR 2007, pp. 1–6. IEEE (2007)
12. Chung, Y.C., He, Z.: Low-complexity and reliable moving objects detection and tracking for aerial video surveillance with small UAVS. In: IEEE International Symposium on Circuits and Systems, pp. 2670–2673. IEEE (2007)
13. Ross, D.A., Lim, J., Lin, R.S., et al.: Incremental learning for robust visual tracking. Int. J. Comput. Vision **77**(1–3), 125–141 (2008)
14. Babenko, B., Yang, M.H., Belongie, S.: Robust object tracking with online multiple instance learning. IEEE Trans. Pattern Anal. Mach. Intell. **33**(8), 1619–1632 (2011)
15. Rublee, E., Rabaud, V., Konolige, K., et al.: ORB: an efficient alternative to SIFT or SURF. In: International Conference on Computer Vision, pp. 2564–2571. IEEE (2012)

Research on Multi-robot Local Path Planning Based on Improved Artificial Potential Field Method

Bo Wang[1(✉)], Kai Zhou[1], and Junsuo Qu[2]

[1] School of Communication and Engineering,
Xi'an University of Posts and Telecommunications, Xi'an 710121, China
bowangmatcha@163.com

[2] School of Automation, Xi'an University of Post and Telecommunications,
Xi'an 710121, China

Abstract. Aiming at solving the dynamic real-time obstacle avoidance problem of multi-robots, the artificial potential field method is adopted. The traditional artificial potential field method has local minimum and target unreachable ability problems. Redefining the repulsion function makes the robot not mistakenly use a additional point, so that reaching the target point ,it can solve the problem that the multi-robot performing collision avoidance in real time in the case of dynamic obstacles, connected realizing the optimal path planning from the source coordinate point to the target coordinate point, and through the simulation verification method accuracy and feasibility.

Keywords: Dynamic obstacle avoidance · Artificial potential field
Target unreachable · Local minimum

1 Introduction

The purpose of this paper is to solve the collision avoidance problem of the multi-robot, so that the multi-robot system can complete the task safely and collision-free. The core idea of the algorithm is to adopt the concept of physics field [1], the target point generates some kind of attraction which attracts the robot to move to it [2], objecting in the unreachable area generates some repulsive force to each single robot in the multi-robot system, leading it to reject the proximity of the robot [3]. The gravitational force interacts with the repulsive force and it creates a resultant force along the target point. It is improved for the traditional artificial potential field method, and the safe distance between the robot and the obstacle is increase by improving the repulsion field function [4], at the same time, the distance function relationship is introduced from the target point.

P. Krömer et al. (Eds.): ECC 2018, AISC 891, pp. 684–690, 2019.
https://doi.org/10.1007/978-3-030-03766-6_77

2 Improved Artificial Potential Field Method

The root cause of the robot's inability to reach the target position is that the point at the target position is not 0 [5]. Due to the obstacle is present around the target position and the size of repulsion is constantly changing, its value is also greater than gravity, so we need to improve the repulsion potential field function. It multiply the cube of the robot from the target point position in the traditional force field function [6]. When the robot approaches the target position, the repulsion and gravity will become smaller and the distance of the robot from the destination point is 0 at the target position. Leading to the values both of the gravitational and repulsive forces are 0, so the value of the resultant force is also zero.

$$U_{chi}(q) = \frac{1}{2}\eta W_2^2(q, q_o)||q - q_o||^3 = \frac{1}{2}\eta(\frac{1}{||q - q_o||} - \frac{1}{q_*})^2||q - q_o||^3 \tag{1}$$

the repulsive force F_{chi} is:

$$F_{chi} = \frac{3}{2}\eta(\frac{1}{||q - q_o||} - \frac{1}{q_*})^2||q - q_o||^2 - \eta(\frac{1}{||q - q_o||} - \frac{1}{q_*})\frac{1}{||q - q_o||^2}||q - q_o||^3 \tag{2}$$

As shown in Fig. 1, the repulsion is broken down into two parts. When the robot keeps approaching the target position, the two parts of the repulsion function are decreasing and approaching 0 [7]. At this time, as a result of the gravitational force, the robot can approach the target position and effectively solve the problem which the robot can reach the target point.

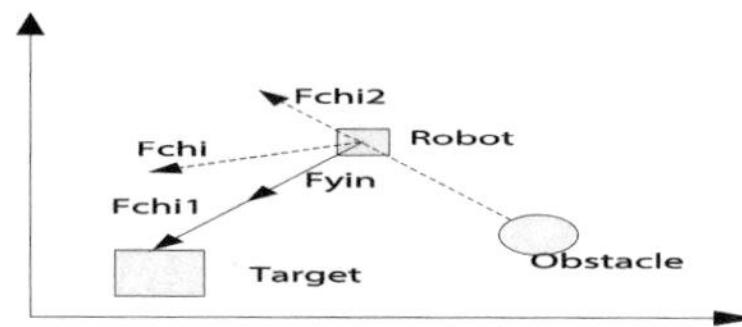

Fig. 1. Improved repulsion function repulsion force analysis

2.1 Adding Additional Force

The generation of falsely identified point is due to the position of the non-target point is 0, the resultant force is 0, and the robot is trapped in the local extreme point, so that the position is mistaken as the target point [8], that is

$$\frac{F_{yin} - \Sigma_{i=1}^{m} F_{chi}}{F_{yin}} < \varepsilon \tag{3}$$

And $|\varepsilon| \to 0$

$$\cos(\angle F_{yin} - \angle F_{chi}) \in [-1, 0] \tag{4}$$

When the robot satisfies the above formula (3) and formula (4), it will stop moving or oscillate back and forth at a certain position. For the aim of preventing this phenomenon, a local minimum point occurred, a certain force F_{add} is added [9], and the direction is the target point that the robot pointed. Therefore, when the local minimum value occurred, the direction of the resultant force is consistent with the direction of the gravitation due to the increased F_{add}, so that the robot move in the direction of the destination point. That is

$$F_{add} = \frac{\beta}{q_o^{\lambda_1} + q_{obs}^{\lambda_2}} \tag{5}$$

Where $\beta, \lambda_1, \lambda_2$ are the gain coefficient and q_{obs} is the distance from the robot to the nearest surrounding obstacle [10]. The magnitude of the resultant force is the same as the direction and the additional force.

2.2 Local Minimum Detection

The fundamental principle of local minimum detection is whether the distance of the detected robot is still within a certain distance after a period of time. For each single robot, a certain local influence distance *R* is given. At a given time point *t*, the position coordinated point of the robot is detected every time point *t*, and the distance of the point at the moment from the last calculated point is calculated, and the calculation result and the influence distance are performed and compared, as shown in Fig. 2:

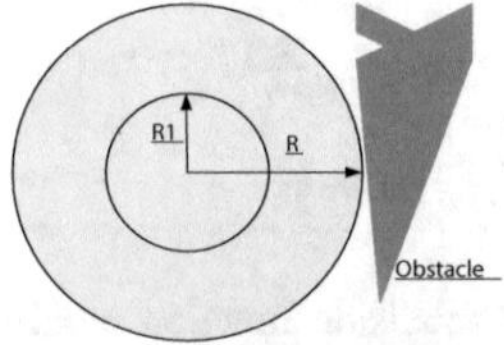

Fig. 2. Detection of local influence range

Relative to each robot, to record the location of the current point is (x_0, y_0), Record the position of the robot every time point *t*, the position the next time interval is (x_0, y_0), therefore, $R_1 = \sqrt{(x_0 - y_0)^2 + (x_1 - y_1)^2}$, compared the size of R_1 with *R*, when R_1 less than *R*, the robot is in a local minimum value and cannot continue to go forward, otherwise, the robot is considered that it is not in oscillating state, then continue to repeat the above steps into the next time interval of detection. For the detection of local minimum, additional force is added to get rid of the state.

The entire algorithm flow chart is as follows (Fig. 3):

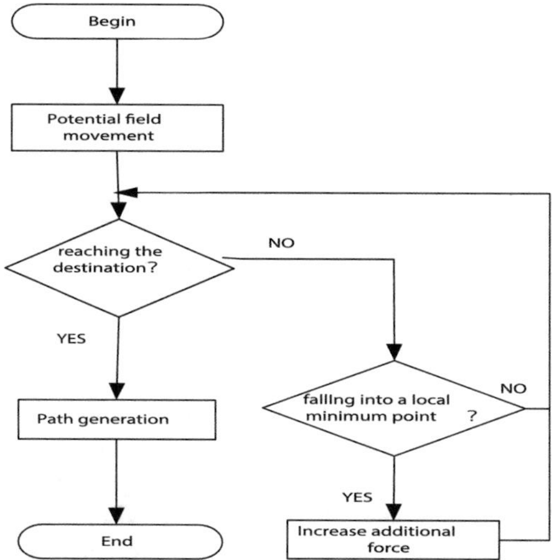

Fig. 3. Improved flow chart of artificial potential field method

3 Simulation Results and Analysis

The simulation of the position is randomly generated, all obstacles effects: (1) whether to reach the target accurately (2) whether the curve is smooth or appear point superposition phenomenon. The simulation experiment was carried out by Matlab. Parameter setting of artificial potential field method, set the repulsion gain to $m = 5$, and set the gain of gravity value to $k = 2$, set the index to $n = 7$, blank black circles represent obstacles, the relative position of robot expressed as (0, 0) and the movement track of robots is expressed as black curve (the setting of the initial and final position here is a fixed constant value to demonstrate the main effect of obstacle avoidance). The obstacle avoidance path simulation diagram of the unimproved artificial potential field method is shown in Fig. 4.

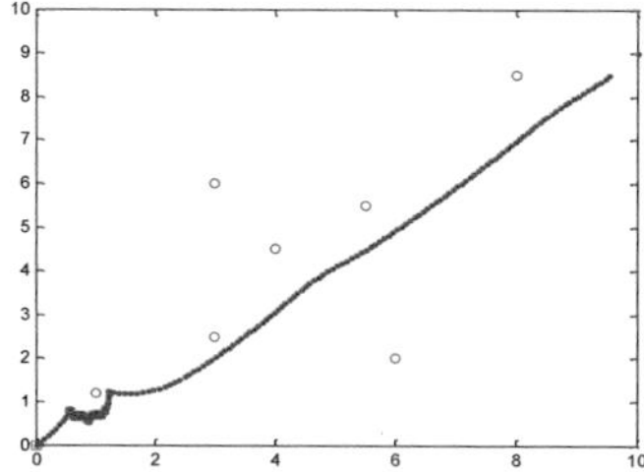

Fig. 4. Behavior diagram of unimproved artificial potential field robot

It can intuitively find that the robot has not reached the target point. To verify the improved theoretical accuracy, only additional forces are added without redefining the repulsive function. The same parameter settings are adopted.

It can be seen from the figure that the problem of unreachable target point occurs when only additional force is added, and the phenomenon of back-and-forth oscillation occurs near the first obstacle. Using the redefined repulsive function without adding additional force, the path graph is obtained through simulation, as shown in Fig. 6.

The figure shows that the robot can reach the target in the case of redefining the repulsive force function without the additional force, but it appears the oscillating phenomenon back and forth. The black lines in the picture generated superimposed phenomenon, showing that the robot oscillates back and forth at this position. It explains that the problem of target point" unreachable can be solved by redefining the repulsion function, but the problem of "local minimum value "will not be.

Compared the simulation of Figs. 5 and 6, it can be known that adding additional force can solve the problem of local minimum value and the robot will not "missing the target position", and it will not consider a certain point as the point which will reach. it can be solved by redefine the repulsive force function for the problem of the target location point unreachable, Through the improved algorithm, the robot can effectively avoid obstacles in real time and realize the collision-free path planning, so that it can obtain a smooth path graph. The improved artificial potential fields can be obtained by adding additional force and redefining repulsion function, and the simulation results are shown in Fig. 7.

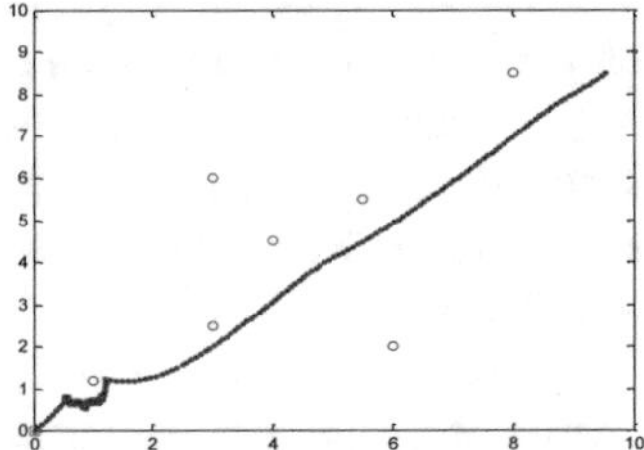

Fig. 5. Path diagram without redefined repulsive function

The figure shows that the improved artificial potential field method can efficiently solve the problem the target cannot be reached, and it is able to enter to the target coordinate points for the robots with a smooth path way. it does not appear the phenomenon the black lines superimposed, and it was able to reach the target accurately.

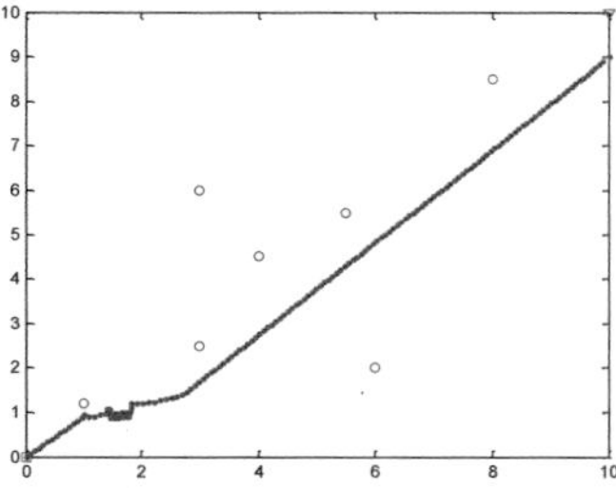

Fig. 6. Path diagram without additional force added

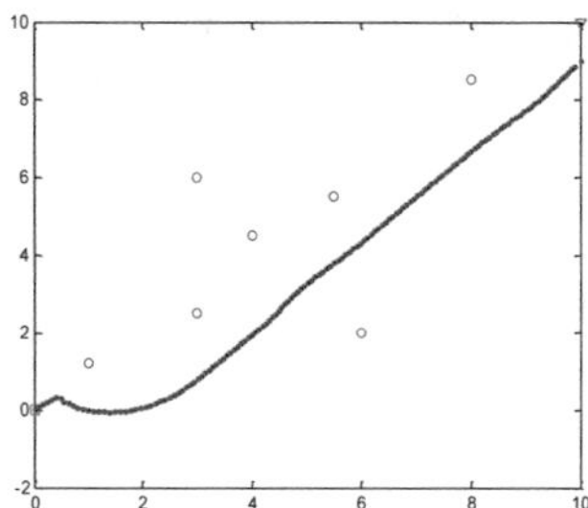

Fig. 7. Improved obstacle avoidance path of artificial potential field robot

4 Conclusion

The improved artificial potential field method is proved that it can effectively solve the problems of "local minimum value" and "target" unreachable, and verify the accuracy of the theory. Therefore, the improved artificial potential field can effectively solve the problem which is the real-time dynamic obstacle avoidance of multi-robot system and realize local path planning.

Acknowledgments. This research was supported in part by grants from the International Cooperation and Exchange Program of Shaanxi Province (2018KW-026), Natural Science Foundation of Shaanxi Province (2018JM6120), Xi'an Science and Technology Plan Project (201805040YD18CG24(6)), Major Science and Technology Projects of XianYang City (2017k01-25-12), Graduate Innovation Fund of Xi'an University of Posts & Telecommunications (CXJJ2017012, CXJJ2017028, CXJJ2017056).

References

1. Tao, Z., Liu, Z., Sun, Y.: Research on design and path planning of handling robot based on embedded system. J. Comput. Measur. Control **24**(8), 215–217 (2016). https://doi.org/10.16526/j.cnki.11-4762/tp.2016.08.059
2. Montiel, O., Orozco-Rosas, U., Sepúlveda, R.: Path planning for mobile robots using Bacterial Potential Field for avoiding static and dynamic obstacles. Expert Syst. Appl. **42** (12), 5177–5191 (2015). https://doi.org/10.1016/j.eswa.2015.02.033
3. Li, C.-M., Gong, J., Niu, W.C., et al.: Combinatorial optimization of spray painting robot tool trajectory based on improved membership cloud models ant colony algorithm. J. Shanghai Jiaotong Univ. **49**(3), 387–391 (2015). https://doi.org/10.16183/j.cnki.jsj-tu.2015.03.024
4. Liu, J., Yang, J., Liu, H., et al.: Robot global path planning based on ant colony optimization with artificial potential field. Trans. Chin. Soc. Agric. Mach. **46**(9), 18–27 (2015). https://doi.org/10.6041/j.issn.1000-1298.2015.09.003
5. Ding, J., Du, C., Zhao, Y., et al.: UAV path planning algorithm based on improved artificial potential field method. J. Comput. Appl. **36**(1), 287–290 (2016). https://doi.org/10.11772/j.issn.1001-9081.2016.01.0287
6. Shi, W.R., Huang, X.H., Zhou, W.: Path planning of mobile robot based on improved artificial potential field. J. Comput. Appl. **30**(8), 2021–2023 (2010). https://doi.org/10.3724/SP.J.1087.2010.02021
7. Wu, Z., Tang, N., Chen, Y., et al.: AUV path planning based on improved artificial potential field method. Chem. Ind. Autom. Instrum. **12**, 1421–1423 (2014). https://doi.org/10.3969/j.issn.1000-3932.2014.12.021
8. Gu, J., Meng, Huizhen, Xia, Hongmei, et al.: Research on multi-robot path planning based on improved ant colony algorithm. J. Hebei Univ. Technol. **45**(5), 28–34 (2016). https://doi.org/10.14081/j.cnki.hgdxb.2016.05.005
9. Xu, F.: Research on robot obstacle avoidance and path planning based on improved artificial potential field method. Comput. Sci. **43**(12), 293–296 (2016). https://doi.org/10.11896/j.issn.1002-137x.2016.12.054
10. Best, G., Faigl, J., Fitch, R.: Multi-robot path planning for budgeted active perception with self-organising maps. In: IEEE/RSJ International Conference on Intelligent Robots and Systems. IEEE (2016). https://doi.org/10.1109/iros.2016.7759489

Study on Intelligence Recognition Technology of Pedestrians Based on Vehicle Images

Yunsheng Li(✉) and Jiaqing Liu

Department of Intelligent Science and Technology,
Xi'an University of Posts and Telecommunications, Xi'an, Shaanxi, China
ylee@xupt.edu.cn, liujiaqing140421@163.com

Abstract. This paper is about pedestrian automatic recognition research on vehicle assisted driving in the field of intelligent transportation. Pedestrian detection provides important technical support for multiple areas of intelligent transportation such as video surveillance. Pedestrian detection technology is mainly from the perspective of machine learning, and the problem is transformed into a classification that determine the appropriate feature points with a large number of training samples, through a machine learning method to obtain a suitable classifier for pedestrian detection. In the VS+OpenCV environment, the HOG feature extraction method and the Support Vector Machine (SVM) are used to identify dynamic target pedestrians. The recognition and effective detection of pedestrians are realized in vehicle video images. The system has friendly interface and high real-time detection. It also studies the pedestrian recognition in different poses. In this paper, the experimental verification shows that the designed pedestrian identification system can correctly recognize and represent the pedestrians in the vehicle video and realize the requirements of the subject.

Keywords: Intelligent transportation · Machine learning · OpenCV
HOG feature · SVM · Vehicle image · Pedestrian detection

1 Introduction

Pedestrian identification and detection provide important technical support for many areas of intelligent transportation such as video surveillance, vehicle-assisted driving, and intelligent robots. The technical realization of pedestrian detection is mainly from the perspective of machine learning. The pedestrian detection problem is transformed into a classification problem, the appropriate feature points are determined, and a large number of training samples are used to obtain a classification suitable for identifying and detecting pedestrians through machine learning [1].

In this research, first of all, due to the diversity of clothing in the real life, the diversity of morphological changes, the diversity of transportation backgrounds, the diversity of illumination, and the occlusion of pedestrians, there are many uncertain problems in pedestrian detection. Second, due to the large amount of video information, slow processing speed, and at the same time to meet real-time, it is necessary to

P. Krömer et al. (Eds.): ECC 2018, AISC 891, pp. 691–699, 2019.
https://doi.org/10.1007/978-3-030-03766-6_78

consider efficient vehicle image processing algorithms with using HOG Feature and SVM in pedestrian detection.

2 Design of the System

2.1 Structure Diagram of System

The system is mainly divided into the following modules as shown in Fig. 1.

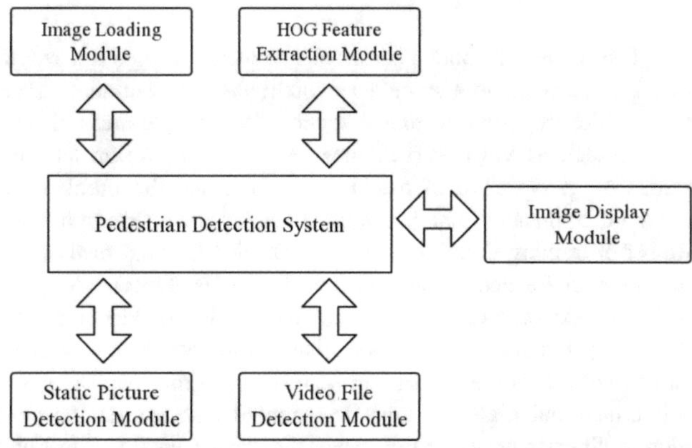

Fig. 1. Structure diagram of each module in system

(1) Image Loading Module: The system can open the picture or video to be processed by opening the local file or by adjusting the camera via USB;
(2) HOG Feature Extraction Module: The system realizes the detection of pedestrians by means of HOG feature extraction, and module visualizes HOG feature points;
(3) Static Picture Detection Module: The module only analyzes and detects the loaded static picture, and displays the identified pedestrians through a red rectangular frame;
(4) Video File Detection Module: The module only analyzes and detects the loaded video file, and displays the pedestrians detected in the video through a red rectangular frame;
(5) Image Display Module: This module mainly displays the final result of the recognition.

2.2 Structure of Car Video Capture System

The information collection unit is mainly composed of a camera, and the camera can capture high-definition video information. The schematic diagram of the information collection structure is shown in Fig. 2.

Fig. 2. Acquisition structure diagram

The camera is installed in the middle of the top of the windshield in the car, about 170 cm from the ground, and the left and right field of view is about 120°. The camera information is shown in Table 1.

Table 1. Camera parameters list

Type	G30
Master	LingTong6624
Camera lens	GC0308
Display	2.4 inchTFT
Angle	120°
Resolution	720P
Compression format	Motion.jpeg
Light sensing element	MOS

3 HOG+SVM Pedestrian Recognition

3.1 HOG Feature Extraction Algorithm Implementation

First, the image is divided into a number of relatively small connected intervals according to a certain proportion of pixel size. We call it a cell; then calculate the gradient vector corresponding to each pixel in each cell unit along the 360° direction and obtained gradient vectors are weighted and summed to form an edge direction histogram; finally, the vector summation of the gradient values calculated by each cell is the HOG feature. Under the guidance of statistical theory, each small cell unit represents the direction histogram of the point only when the target area is infinitely subdivided, even when the target area of the image to be detected is large, even two the images that are not identical, the extracted HOG features are not very different. On the contrary, this possibility is very small [2–4].

HOG feature extraction is a summation feature of the gradient histogram. The specific implementation method is to first divide an image into 9 intervals according to different gradient directions (0–2π), and then divide the image to be extracted into 20 blocks of 16 * 16 as shown in Fig. 3. Each slider is then subdivided into 4 cell units with a size of 8 * 8. The calculation for each cell unit is to calculate the gradient mode and gradient value of the point in each direction, and finally to construct a high-

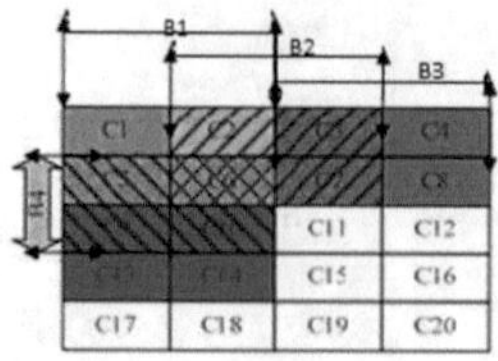

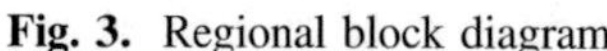

Fig. 3. Regional block diagram

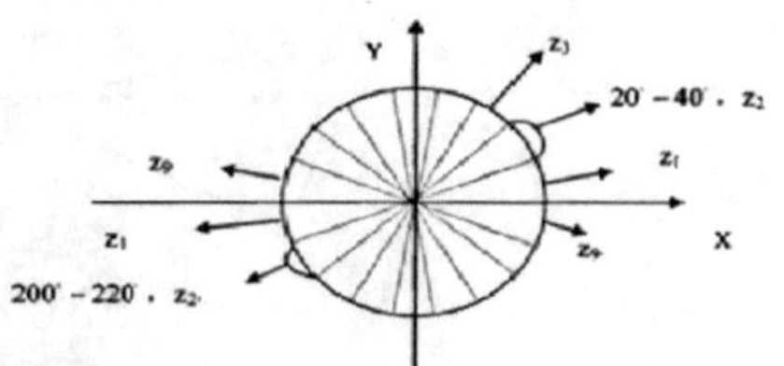

Fig. 4. A division map of each cell in the 360°

dimensional vector by summing all the obtained sliding blocks and the gradient histogram of the subdivided cell unit. Characteristics, this vector feature is the HOG feature [5–8].

The specific process of HOG feature extraction is as follows

(1) Standardized gamma space and color space
The input image is first normalized to reduce the impact of illumination intensity on subsequent processing. For the processing of image textures, the pixels of the local surface are the primary features. Since the HOG feature extraction has little effect on the color requirements of the image, it can be converted into a grayscale image for subsequent processing;
(2) Calculating the graphical gradient
Secondly, calculate the gradient size and direction of the abscissa and ordinate directions of the pixel where each cell element is located in the normalized image; the gradient information in the image can ignore some noise effects caused by illumination on the extraction of pedestrian features;
(3) Gradient histogram for each cell unit [9]
First divide the input image into a number of "cells" according to the normalized standard, and then divide each cell into 360° directions into 9 blocks as shown in Fig. 4, and then perform all the pixels in each cell. Weighted projection to obtain a gradient histogram and a feature vector;
(4) Combine each cell into a large block, and the cells in each block are normalized gradient histograms
Each cell connected to the edge region is constructed as a large interval block that is connected to each other. Thus, the HOG feature of the block is obtained by summing all the element feature vectors in the interval. The set of intervals that overlap each other is referred to as a HOG descriptor;
(5) Collecting HOG features
In order to obtain the HOG feature vector, all overlapping blocks in the sliding detection window are weighted and summed.

3.2 Linear SVM Classifier Design

HOG+SVM is a pedestrian detection method with good real-time performance and stability in pedestrian detection. As mentioned in the SVM classifier in the previous section, our goal is to achieve classification of pedestrians. Therefore, the classification criteria of classifiers are to separate pedestrians from the background [10–14], and training SVM classifiers is needed. The training sample is the HOG feature. Table 2 shows the HOG parameter description.

Table 2. HOG parameters description

Parameter	Effect
Sample	3-channel RGB, size 96 * 160
Detection Window	64 * 128, Consistent with the training sample
Block Size	16 * 16, one sample contains 105 blocks, each 36D HOG
Step	8 units in the horizontal and in the vertical direction
Cell	8 * 8, each 4 cells form a block, each cell has 9D characteristics
Gradient direction	9 will divide 0−π into 9 directions and perform statistics in each cell unit

Here is the flow of the SVM classifier pedestrian training test:

(1) Prepare sufficient positive and negative sample sets. The positive samples here are pedestrians with a size of 96 * 160 pixels, and the negative samples are random backgrounds;
(2) Crop the collected training samples, adjust all positive samples to the same size; set the parameters of the HOG feature extraction;
(3) Extracting and marking the HOG features of all positive samples;
(4) Extracting and marking the HOG features of all negative samples;
(5) Input all HOG features and markers of positive and negative samples to the SVM trainer for training;
(6) Save the trained results;
(7) Read in a still picture or video file for pedestrian recognition.

4 Analysis of Pedestrian Recognition Results in Vehicle Images

4.1 Color Target Recognition

We changed the different HOG characteristic parameters according to Table 2 and tested it through the data set on INRIA, and obtained the results shown in Figs. 5 and 6.

Figures 5 and 6 show the pedestrian false detection rate and the detection time required by the system for different slider sizes. The 1 * 1 size detection window is the best and the highest accuracy, but the detection time required is also the longest.

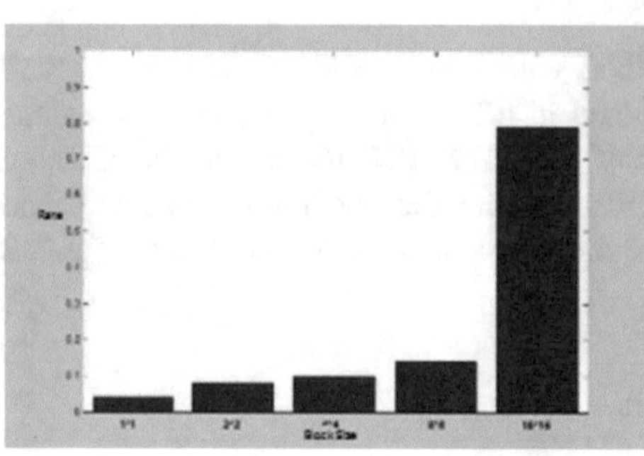

Fig. 5. False detection rate of different sizes

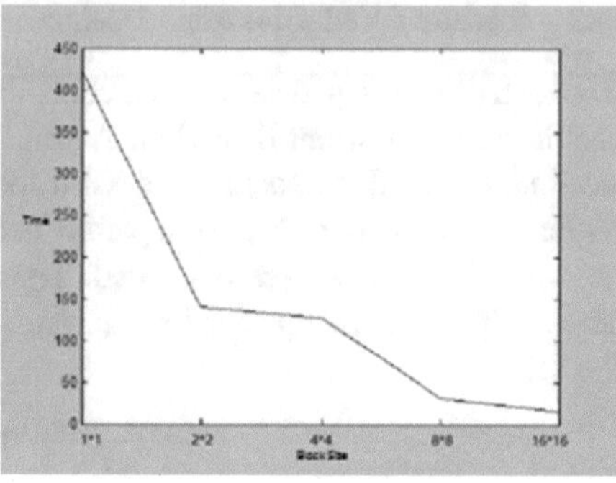

Fig. 6. Detection time

Therefore, for the detection of video files, we must ensure the accuracy of the detection and ensure the real-time detection of the system, so we select the 8 * 8 sliding block as the detection window, which not only has better detection rate. The guarantee is also an improvement in the detection of real-time.

4.2 The Effect of Different Detection Distances on Pedestrian Recognition Results

In the experiment, our test drive recorder was 4 min and 10 s long, with a total of 7500 frames and a resolution of 720 * 576 video.

Table 3 shows the results of pedestrian identification of the system under different detection distances in a real test environment. It can be clearly seen that when the system recognizes pedestrians at medium and close distances, the recognition result is above 88.9%, due to the system. The relationship between hardware conditions and detection algorithms is only 42.9% for long-distance pedestrian recognition.

Table 3. Test results in video at different distances

Distance (S)	5 m–20 m (Close)	20 m–40 m (Middle)	>40 m (Long)
Correctly Identifies	38	24	6
Number of Pedestrians	40	27	14
Recognition Rate (%)	95	88.9	42.9

4.3 The Influence of Pedestrians with Different Postures on the Recognition Results

In this experiment, we tested the classification ability of pedestrians' posture and posture behavior. We used a video from the test video and a picture in the pedestrian sample data set to perform a pedestrian recognition test.

Through experiments, it is found that when detecting pedestrians in the "walking" posture, the recognition rate of the system is about 95%; when detecting pedestrians in the "riding" posture, the recognition rate of the system is about 80%; in detecting the "running" posture When the pedestrian is pedestrian, the recognition rate of the system is about 46%; when detecting the pedestrian in the "bent-shouldered" posture, the recognition rate of the system is about 15%.

4.4 Pedestrian Recognition in the Video

In our test video, the vehicle travels at a speed of 5 to 50 km/h during the journey, and according to the video, the car passes through a red light, enters a congested section, and is surrounded by an empty section. The test samples in the experiment are all from the pedestrian video images of the real traffic roads collected by them. They contain a variety of pedestrians with different postures and costumes. At the same time, the test sample set also contains occlusions. The pedestrian images are missing or incomplete.

We test in the actual real environment as shown in Fig. 7(a). Figure 7(b) and (c) show that the pixels of pedestrians are between 220–350 pixels in the range of 10 m in close range, between 50–120 pixels in 20 m–50 m, and the pedestrian pixels in 50 m are low. In the test environment where the weather conditions are consistent and the road condition information is consistent, the distance from the camera is far away, the pixels of the pedestrians in the resulting image will be less, and the feature points are not fully extracted, and the classifier will regard it as the background separation. However, pedestrians who are closer are more likely to extract feature points because they have more pixels.

(a) (b) (c)

Fig. 7. Recognition results of pedestrians

From the experimental results combined with the detection results of the video, there are still some cases of missed detection and misjudgment, mainly in the trees in the surrounding roads and the high buildings in the city. The experimental results show that the classifier designed in the close-range and complete information of the pedestrian achieves the ideal result, and the pedestrian identification of the complete distance information is within the tolerance range that can be accepted. In addition, the sample library can be re-selected according to the detection requirements of the special environment, and the classifier will greatly improve the recognition performance for a specific scene.

5 Conclusions

In this paper, the pedestrians in the vehicle video can been correctly recognized and represented by the designed pedestrian identification system. as results, we observe:

(1) Further research on the extraction of HOG features had been made to visualize the information extracted by HOG;
(2) The accuracy and real-time performance of pedestrians recognition was combined by the HOG parameter interface in OpenCV;
(3) In the aspect of vehicle video, the real-time performance of pedestrian detection was optimized, and the average detection time was reduced, so that it has greater application value in embedding hardware;
(4) Preliminary studies were conducted on pedestrians with different characteristics and different postures. In the data sample set, a certain number of samples were collected for pedestrians with indefinite posture.

From the current development of information technology, pedestrian detection technology is a hot topic in the field of computer vision academics, and it is a crucial technology in the field of intelligent transportation and driverless technology. Although the research on pedestrian recognition has made some progress, there are still some problems that need to be solved, mainly:

(1) Since the background of the image information in the in-vehicle recorder is not fixed, the existing background extraction algorithm cannot be used to abstract the foreground and perform pedestrian detection;
(2) There is no classifier algorithm that can be used for commercial maturity due to the change of human body, the change of clothing appearance and the change of weather.

Acknowledgements. Research supported by the Foundation of the Shaanxi Provincial Department of Education (project number: 16JK1703).

References

1. Li, Z.: Pedestrian detection method based on SVM-AdaBoost algorithm. Ind. Instrum. Autom. **04**, 117–120 (2016)
2. Chen, N., Chen, W.-N., Zhang, J.: Fast detection of human using differential evolution. In: Signal Processing, p. 110 (2015)
3. Chenglong, W., Guangcai, C., Zhiming, C.: Pedestrian detection method based on Opencv. J. Huizhou Univ. **36**(03), 55–57 (2016)
4. Gavriilidis, A., Velten, J., Tilgner, S., Kummert, A.: Machine learning for people detection in guidance functionality of enabling health applications by means of cascaded SVM classifiers. J. Frankl. Inst. (2017)
5. Cong, X., Wenwu, W.: Research on pedestrian detection technology based on DPM model. Electron. Des. Eng. **22**(23), 172–173 (2014)
6. Xianxian, T., Wei, B., Cheng, X.: A pedestrian detection algorithm for improving HOG characteristics. Comput. Sci. **41**(09), 320–324 (2014)

7. Li, Y., Zhang, M., Du, F.: The study on automatic recognition system of road traffic marking. In: Proceedings of the 2017 IEEE International Conference on Real-time Computing and Robotics, Okinawa, Japan, p. 465–470 (2017)
8. Lanfang, D., Yuzhong, S., Jin, W., Jianfu, W.: Research on GPU-based HD video image pedestrian detection method. Electron. Technol. **43**(03) 2014
9. Rui, S., Nengqian, H., Jun, C.: Fast pedestrian detection based on feature fusion and cross-core SVM. Photoelectr. Eng. **41**(02), 53–62 (2014)
10. Ailian, J., Xingyu, Y.: Pedestrian detection based on AdaBoost-SVM cascade classifier. Comput. Eng. Des. **34**(07), 47–65 (2013)
11. Conde, C., Moctezuma, D., De Diego, I.M., Cabello, E.: HoGG: Gabor and HoG-based human detection for surveillance in non-controlled environments. Neurocomputing 100 (2013)
12. Guangli, W., Mianxu, W.: Pedestrian detection based on HOG features and SVM classifier. China Cable TV **12**, 13–15 (2017)
13. Yuan, X., Xiaoliang, X., Cainian, L., Mei, J., Jianguo, Z.: Pedestrian detection based on SVM classifier and HOG feature extraction. Comput. Eng. **42**(01), pp. 56–60+65 (2016)
14. Li, Y., Takezawa, Y., Suzuki, H., et al: Prediction of fish motion by neural network. In: Proceedings of the 3rd International Symposium on Autonomous Minirobots for Research and Edutainment (AMiRE 2005). Springer Berlin Heidelberg, pp. 217–222 (2006)

Design of Self-balancing Vehicle System

Wenqing Wang(✉), Yuan Yan, Ruyue Zhang, Li Zhang, Hongbo Kang, and Chunjie Yang

School of Automation, Xi'an University of Posts and Telecommunications, Xi'an 710121, Shaanxi, China
wwq@xupt.edu.cn, 980260826@qq.com

Abstract. Intelligent balancing two-wheeled vehicle system is designed to maintain balance and move automatic in the specified path. The balancing vehicle system adopts modular design idea, including MCU minimum system, sensor module, drive unit, power supply unit and human-computer interaction module. The accelerometer and gyroscope output data as well as obtaining the accurate vehicle model posture are fused by complementary correction algorithm. Through image recognition algorithm, the image collected by photoelectric sensor is processed and the road information is obtained. The PID controller is designed to achieve dynamic balancing and directional control of two-wheeled balancing vehicles. In this paper, the detailed description of hardware design and algorithm description are given. After the actual track test, the model runs reliably and steadily, it can adapt to various track elements to complete independent operation, which provides a practical and good solution for the operation of the driverless balancing vehicle.

Keywords: Intelligent vehicle · Microcontrollers · Self-tracking

1 Introduction

With the rapid development, change and upgrade of transportation modes in China, means of transportations are also showing diversified development. Except the high-income groups who take the vehicle consumption as oriented in large and medium-sized cities, the larger consumer groups have an urgent need to upgrade from bicycles [1]. As a short distance walking tool, the self-balance scooter has the advantages of portability, green and easy operation [2]. Meanwhile, as the development direction of future vehicles, intelligent vehicles will play an important role in reducing traffic accidents, developing automation technology, improving comfort and many other aspects.

The design of the system is based on the given model, build the mechanical structure of the vehicle model according to the demand, and design the hardware circuit system of each unit. Use a suitable control algorithm to realize the function of the vehicle model in the track that contains various elements, such as avoiding obstacles, running upright and autonomously and finally stopping at the starting line.

P. Krömer et al. (Eds.): ECC 2018, AISC 891, pp. 700–707, 2019.
https://doi.org/10.1007/978-3-030-03766-6_79

2 Working Principle

Two-wheeled balancing vehicle is a natural and unstable system [3]. The inverted pendulum system is used to explain the principle of two-wheeled dynamic vertical balancing. It is simplified as an inverted pendulum system model, which can move wheels with a height of L and a mass of M. Figure 1 is a simplified model for vehicle models.

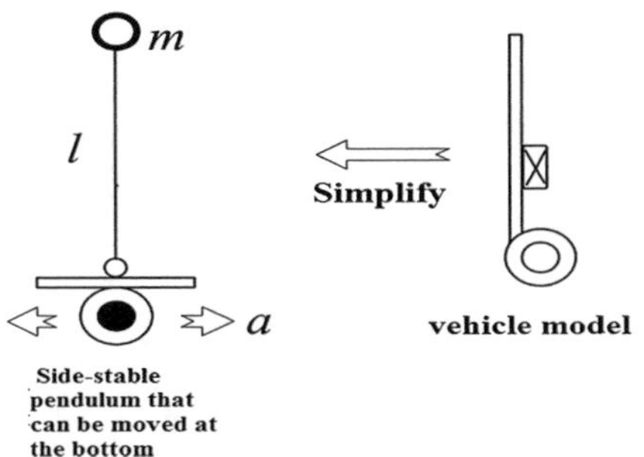

Fig. 1. Simplified model for inverted pendulum system

Conduct the force analysis of the model as is shown in Fig. 2. Because the direction of force is the same as the direction of displacement, it can't be stabilized in vertical position under natural conditions. The inverted pendulum system is subjected to the force opposite to the displacement direction by the external force. At this time, the inverted pendulum system can be stabilized in the vertical position, which is called the restoring force. Standing on the vehicle (non-inertial system, with the wheel as the origin of coordinates) for force analysis, make accelerated motion by controlling the wheel, so that the vehicle model receives additional inertial force. The force is opposite to the acceleration of the wheel and it is proportional to its size [3]. Meanwhile, the inverted pendulum system is subjected to restoring force, as shown in formula (1), and its value is:

$$F = mg\sin\theta - ma\cos\theta \approx mg\theta - mp_1\theta \tag{1}$$

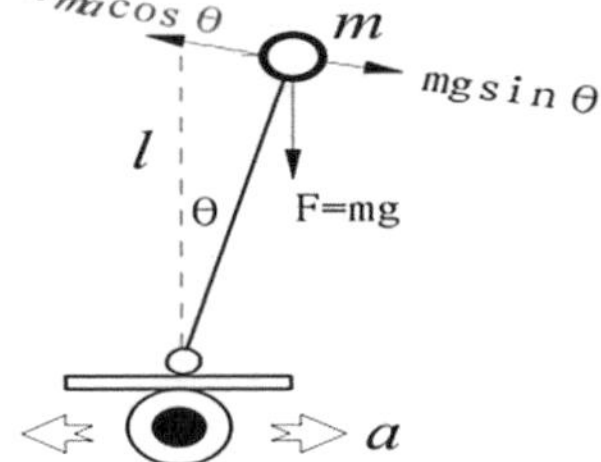

Fig. 2. Force analysis of vehicle model

When θ is small, the formula can be simplified to the upper right side. It is assumed that the wheel acceleration a is proportional to the pitch angle θ of the inverted pendulum system, and the ratio is p_1. When $p_1 > g$, the vehicle model is subjected to restoring force, g is the acceleration of gravity.

What's more, in order to stabilize the inverted pendulum system vertically, an additional force is added to the formula (1). The force is proportional to the velocity θ' of the inverted pendulum system, the ratio is p_2, the direction is opposite, the effect is similar to the damping effect. Therefore, formula (1) can be improved to:

$$F = mg\theta - mp_1\theta - mp_2\theta' \tag{2}$$

Once the inverted pendulum system moves away from the vertical position, it will be restored and eventually stabilized. Thus, formula (3) is a condition for the acceleration of wheels.

$$a = p_1\theta + p_2\theta' \tag{3}$$

In the above formula, θ is the pitch angle of the vehicle model; θ' is the angular velocity of the vehicle's pitch: p_1, p_2 are all proportionality coefficients. When pitch angle θ is small and $p_1 > g$, $p_2 > 0$ is guaranteed, the model can be maintained in a vertical position similar to a pendulum to achieve upright.

3 Overall System Design

The flow chart design of the vertical balancing vehicle system which is divided into intelligent balancing vehicle on-board system and remote monitoring platform is shown in Fig. 3. The vehicle mounted system includes sensor module, driving unit and human-machine interaction module. The sensor module includes image acquisition unit, motor speed measuring unit and posture detection unit [5]. The image acquisition unit collects road pictures through the OV7725 camera. The motor speed measuring unit uses the 512 wire Mini encoder to measure the speed of the motor. The posture detection unit uses the MPU-6050 nine-axis motion processor. Calculate the pitch angle of the vehicle model by the output of the gyroscope and accelerometer [7]. The human-computer interaction module includes display unit, dial switch and wireless communication unit, which can directly observe the running state of the vehicle model and change the algorithm in real time. The remote monitoring platform includes wireless communication unit which is used to communicate with vehicle mounted system and PC host computer which can display the speed or the pitch angle of the motor and the image in real time.

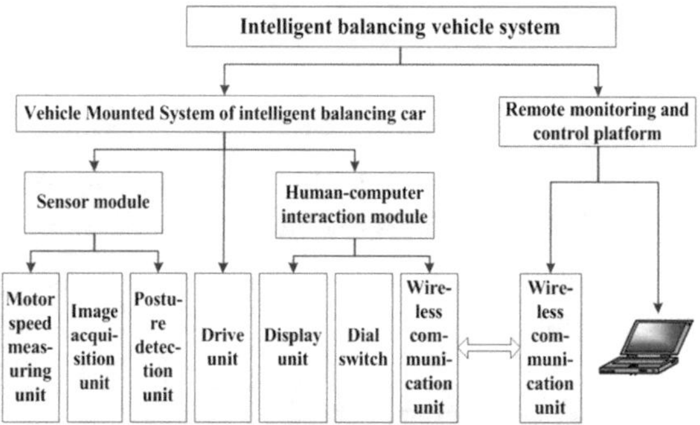

Fig. 3. Overall system design

4 System Hardware Design and Implementation

The whole hardware circuit includes three modules: sensor unit, driving unit and human-computer interaction unit [4]. The sensor unit is divided into gyroscope and accelerometer. The sensor unit is for calculating angle and angular velocity, the encoder speed measuring unit is for collecting motor speed and the photoelectric sensor is for obtaining track information. The human-computer interaction unit is divided into Bluetooth unit to communicate with host computer, OLED display screen to display some parameters and dialing switch to change control strategy. The overall hardware circuit design is shown in Fig. 4.

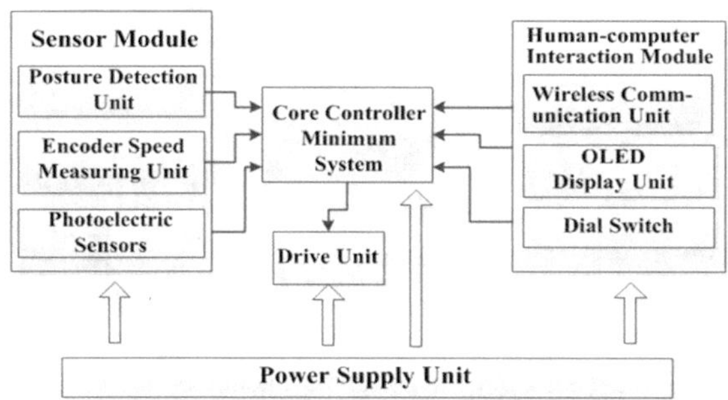

Fig. 4. Overall hardware circuit design

5 System Software Design and Implementation

5.1 Software Function and Overall Framework

The intelligent balancing vehicle software system mainly realizes the following functions [6, 8]:

1. Data acquisition and processing of each sensor in the sensor unit.

2. Vehicle mode running control: Angle control, speed control, direction control and motor PID control.
3. Vehicle mode process control: Peripheral initialization, vehicle mode state detection, etc.
4. Human-computer interaction control: OLED parameter display, wireless information transmission, dial switch control and so on.

Among them, 1, 2, and 3 require precise timing, fixed control cycle, and timing interrupt to control execution.

5.2 Main Control Functions Description

This section will introduce the design idea and implementation of the main control functions combined with the control block diagram.

Angle Calculation Program Design. The program calculates the real-time posture of the vehicle model by reading the data of Z-axis accelerometer and Y-axis gyroscope in MPU-6050 motion processor. Because of the zero bias error of sensor data, the zero bias error (the two-channel data of vehicle stationary and stable position) must be subtracted from the data of gyroscope and accelerometer. Zero bias error is often inaccurate, and it will make the model to a certain direction to accelerate, this can be eliminated through the speed error behind.

The accelerometer data need to be normalized to −90°–90°, the gyroscope data are normalized according to the data acquisition scale GYRO_ANGLE_RATIO.

The calling cycle of this function is 15 ms, and the algorithm block diagram is shown in Fig. 5. The flow chart of the program is shown in Fig. 8(a).

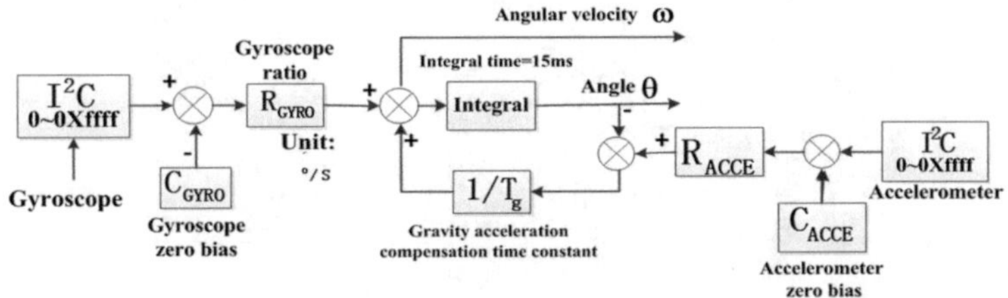

Fig. 5. Block diagram of angle calculation algorithm

The gyroscope scale factor R_{GYRO} can be obtained from the data manual, it needs to be adjusted by itself. The compensation time constant of gravity accelerometer determines the effect of angle calculation. Both parameters need to be adjusted and debugged. The flow chart of angle calculation program is shown in Fig. 8(a).

Angle Control Program Design. After obtaining the accurate angular θ and angular speed ω, the vehicle model is balanced in a stable position and upright through the PD controller. The block diagram of the angle control algorithm is shown in Fig. 6, and its program flow chart is shown in Fig. 8(b).

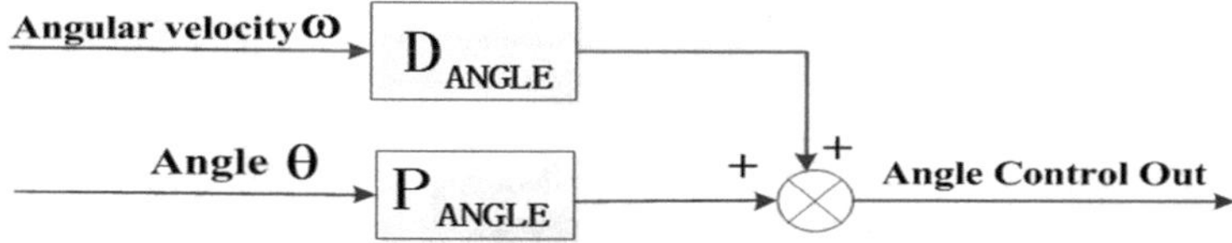

Fig. 6. Block diagram of angle control algorithm

In this program, angle control proportional coefficient D_{ANGLE} and angle control differential coefficient P_{ANGLE} need to be adjusted in debugging. With the increase of the proportional coefficient, the model will swing back and forth near the stable position. The increase of the differential coefficient can make the model stable in the balanced position as soon as possible.

Speed Control Program Design. According to the motor speed and expected speed obtained by encoder, the program realizes the closed-loop control of vehicle model operation by PI controller. The polarity of the speed control motor is opposite to that of the angle control output motor. Both speed control and direction control will affect the angle control to maintain the model upright. Therefore, the expected speed calculated by the speed control function is evenly distributed with 20 angle control cycles that is 300 ms output. The block diagram of speed control algorithm is shown in Fig. 7. Figure 8(c) is the flow chart of speed control program.

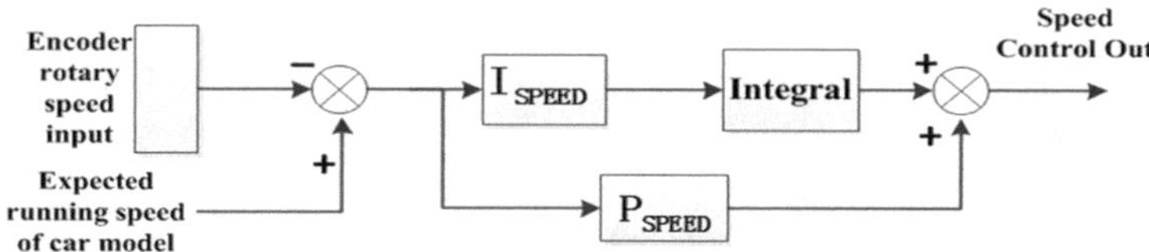

Fig. 7. Block diagram of speed control algorithm

In this program, 20 angle control cycles are obtained, and the average speed of the left and right motors is used as deviation control. The speed control proportional coefficient P_{SPEED} and the speed control integral coefficient I_{SPEED} are set by the debugging link.

6 Debugging and Analysis

In Fig. 9, the straight line has a desired pitch angle of 6° (for the convenience of display, the angle is multiplied by 100), and the curve is the actual pitch angle of the vehicle model. The vehicle model is stable at a position and the angle control effect is good.

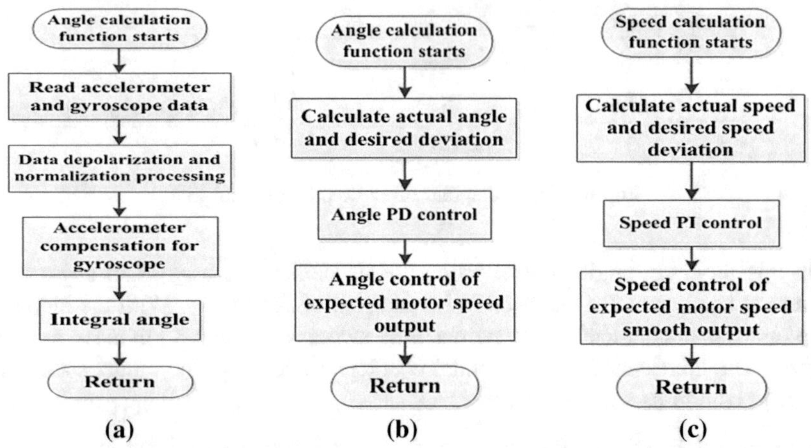

Fig. 8. Flow chart of angle and speed calculation program

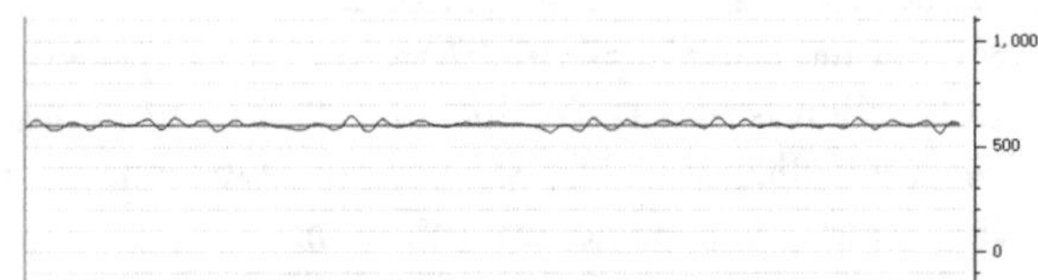

Fig. 9. Comparison of angle control effect

7 Conclusion

This paper introduces an intelligent balancing vehicle system for track recognition and autonomous operation in detail. It is from the three aspects of hardware structure, algorithm implementation and software design to improve the balance of intelligent vehicle. Meanwhile, due to the limited knowledge of the author, there are still some problems in the design, such as sensor selection, and the angle calculation needs to be further improved.

Acknowledgments. This work is supported by Design and Development of Intelligent Collection Platform of Energy Consumption based on Internet of Things, Industrialization project of Shaanxi Provincial Education Department (16JF024) and the Shaanxi Education Committee project (14JK1669) Shaanxi Technology Committee Project (2018SJRG-G-03).

References

1. Babu, S.S.: Design and implementation of two-wheeled self-balancing vehicle using accelerometer and fuzzy logic. In: 2nd International Conference on Computer and Communication Technologies, pp. 45–53. Springer, Xiamen (2016). https://doi.org/10.1007/978-81-322-2526-3_6
2. Sun, H.: Design of two-wheel self-balanced electric vehicle based on MEMS. In: 4th IEEE International Conference on Nano/Micro Engineered and Molecular Systems, NEMS 2009, pp. 143–146. IEEE Computer Society, Shenzhen (2009). https://doi.org/10.1109/NEMS.2009.5068545
3. Gao, M.X.: The research of self-balancing vehicle based on posture sensor system. In: 4th International Conference on Frontiers of Manufacturing Science and Measuring Technology, ICFMM 2014, pp. 735–738. Trans Tech Publications Ltd, Gui Lin (2014). https://doi.org/10.4028/www.scientific.net/AMM.599-601.735
4. Almeshal, A.M.: Robust PD-PID control of a new configuration of two-wheeled machines under various operating conditions. In: 15th International Conference on Climbing and Walking Robots and the Support Technologies for Mobile Machines, CLAWAR 2012, pp. 673–680. World Scientific Publishing Co. Pvt Ltd. (2012)
5. Pu, B.J.: Based on the research of self-balancing vehicle posture sensor system. In: 4th International Conference on Intelligent Systems Design and Engineering Applications, ISDEA 2013, pp. 193–196. IEEE Computer Society, Hunan (2013). https://doi.org/10.1109/isdea.2013.448
6. Feng, J.: Research on a new approach of auto self-balancing for rescue robot on uneven terrain. In: 2017 IEEE International Conference on Robotics and Biomimetics, ROBIO 2017, pp. 74–79. Institute of Electrical and Electronics Engineers Inc, Beijing (2017). https://doi.org/10.1109/ROBIO.2017.8324397
7. Ping, J.: Market situation and development prospect of electric balance scooter in China. Shanxi Agric. Econ. **2016**(03), 106–109 (2016). https://doi.org/10.16675/j.cnki.cn14-1065/f.2016.03.69
8. He, C.: Design of two wheel self balancing vehicle. In: 3rd International Conference on Advances in Energy Resources and Environment Engineering, ICAESEE 2017. Institute of Physics Publishing, Harbin (2017). https://doi.org/10.1088/1755-1315/113/1/012074

Intelligent Water Environment Monitoring System

Wenqing Wang(✉), Ruyue Zhang, Chunjie Yang, Hongbo Kang, Li Zhang, and Yuan Yan

School of Automation, Xi'an University of Posts and Telecommunications, Xi'an 710121, Shaanxi, China
WWQ@xupt.edu.cn, 772346730@qq.com

Abstract. In recent years, water pollution in China has become a serious problem to be solved. The intelligent water environment monitoring device is designed to solve the problem of water quality information monitoring in our country. Intelligent water environment monitoring device consists of monitoring device unit and remote monitoring unit. The monitoring unit includes micro-controller module, wireless communication module, information acquisition module, GPS positioning module, driving control module, power management and so on. The remote monitoring unit takes the computer as the upper computer terminal and realizes the remote monitoring of the ship hull main control unit through the Internet. The intelligent water environment monitoring system designed by the subject is stable, running well, saving energy and reducing consumption. It provides a good solution for the water quality monitoring system of our country, and has strong innovative and practical value.

Keywords: Autonomous mobile · Environmental monitoring
Internet of Things

1 Introduction

In recent years, the monitoring and treatment of water pollution problems have received extensive attention. Due to the wide distribution of waters, environmentalists often use periodic inspections to monitor water environment information. This method does not provide immediate access to water pollution information, which is extremely harmful to the local ecological environment. At the same time, there are 24 basic standards for water quality quantitative standards in China, and the analysis methods for each project are not identical [1]. A large number of test projects and complex processes have brought a lot of difficulties to the staff in the analysis of water quality.

This study designed an automatic intelligent water environment monitoring system that integrates single-chip microcomputer, embedded technology, Internet of Things technology and sensor technology. It not only monitors various sources of water pollution in real time, but also analyzes and filters data. The data is transmitted to the server through the wireless network, which makes the water quality information more accurate, and the location of the pollution source is more convenient, providing a faster and more accurate way for water quality monitoring.

P. Krömer et al. (Eds.): ECC 2018, AISC 891, pp. 708–714, 2019.
https://doi.org/10.1007/978-3-030-03766-6_80

2 The Composition and Principle of System

The intelligent water environment monitoring system is mainly composed of intelligent monitoring terminals and remote monitoring terminals (see Fig. 1). The monitoring terminal is mainly composed of an ultra-low power MSP430F149 chip [2], a wireless communication module (SIM module), a GPS positioning module, and a low-power information acquisition sensor. The remote monitoring terminal is mainly composed of a host computer, a database, and a web page. The data acquired by the collection terminal can communicate with the server in two-way data through the GPRS function of the SIM module. The server performs data mining and big data processing to send data to the mobile terminal. Monitors can view data on the management platform and monitor water information in real time. At the same time, record the location of the equipment, avoid obstacles, plan the route of travel, and realize the environmental monitoring function of the waters.

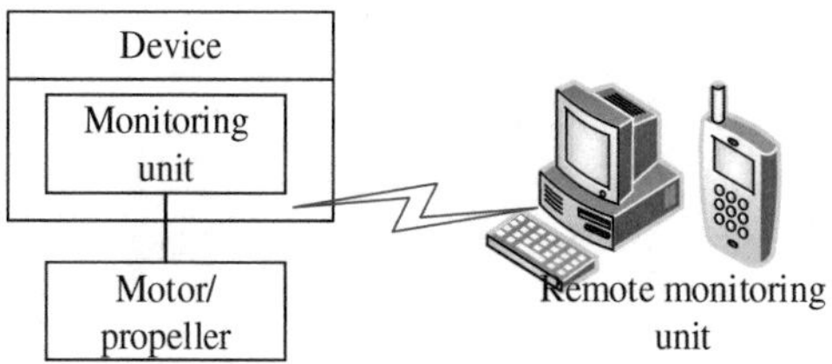

Fig. 1. System structure

3 Hardware Design

The device adopts MSP430F149 microprocessor as the main control module, and integrates information acquisition module, drive control module, wireless communication module, GPS positioning module and power management module. The system hardware block diagram is shown in Fig. 2.

3.1 Main Control Board Circuit Design

The minimum system of the MSP430F149 main controller guarantees smooth operation of the master and stability of each module. The design is based on the principle of reliability and stability. The main control board circuit is shown in Fig. 3.

3.2 Information Acquisition Circuit Design

Using ultrasonic ranging module [3], DHT11 [4], PH value acquisition module [5], turbidity acquisition module and oxygen content acquisition module (see Fig. 4) are used to obtain equipment obstacle avoidance information, temperature, humidity, PH value, turbidity and oxygen content.

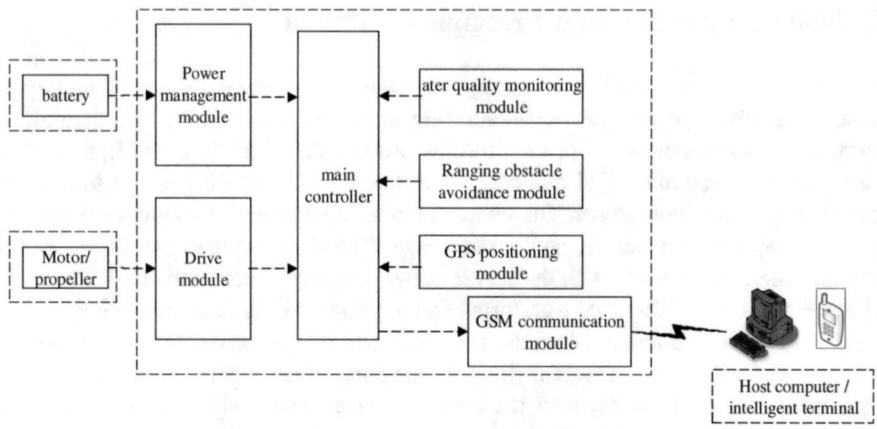

Fig. 2. The system hardware block diagram

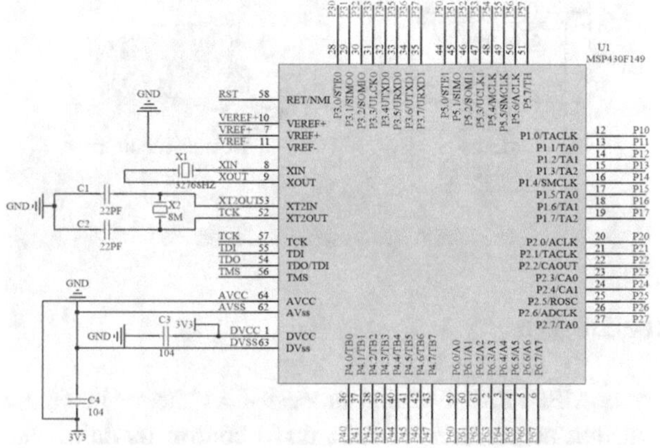

Fig. 3. The main control board circuit

3.3 Drive Module Design

The device needs to be moved automatically in the water and the design of the drive module is particularly important. The device uses two L298N-type driver chips [6] to drive two motors separately, allowing the device to run smoothly.

3.4 Wireless Communication Circuit Design

The system needs to make a data connection between the hardware and the software host, and finally use the SIM module by comparison and selection.

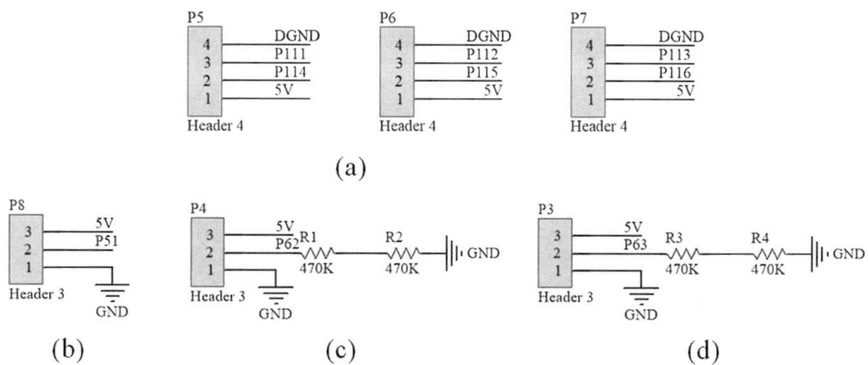

Fig. 4. (a) Ranging circuit, (b) Temperature and humidity acquisition circuit, (c) PH value acquisition circuit, (d) Turbidity acquisition circuit

3.5 GPS Circuit Design

The GPS module model is UBLOX_NEO-M8N, which has excellent performance and stability, and is often used in mobile phone positioning, child positioning, and intercom positioning.

3.6 Power Management Circuit Design

The system is powered by a 7.2 V lithium battery. Due to the different voltage requirements of the various modules of the system, as well as the complex power supply structure in the system. To meet system power requirements, the LM2596S and LM1117 chips are used respectively to convert to 5 V and 3.3 V. The voltage regulator circuit design is shown in Fig. 5.

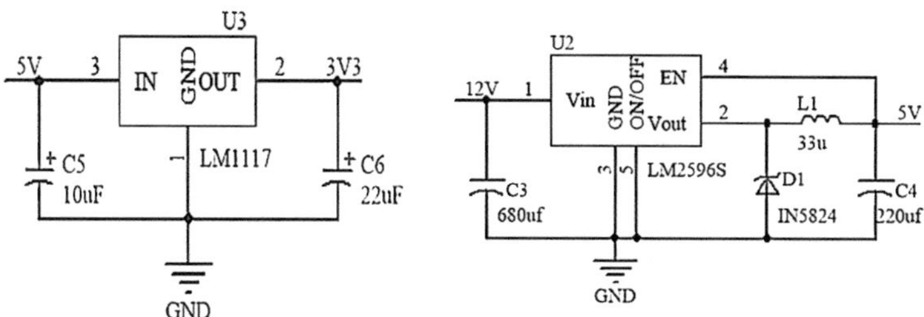

Fig. 5. The voltage regulator circuit

3.7 Mechanical Design

The mechanical structure of the intelligent water environment monitoring device adopts 3D printing technology. The main body of the device is shown in Fig. 6. Two hollow tubes at the rear of the unit are used to mount the propeller, and three projections on the upper level are used to mount the sensor in the water quality monitoring section, with the battery and system board placed on top.

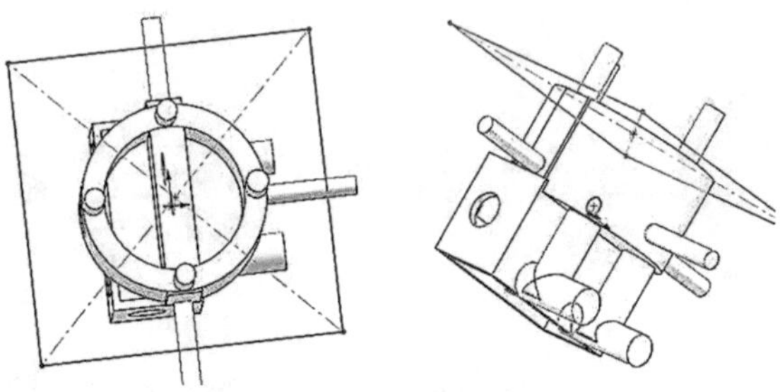

Fig. 6. Mechanical design

4 Software Design

The main program workflow is as follows: system self-test and initialization of each part (serial port, watchdog, timer, interrupt, A/D conversion initialization) [7]. The main control module detects whether a timing command is passed in and judges the instruction. Then jump to each subroutine to run. The ultrasonic sensor and water quality monitoring module started working. Each module collects information and returns the information to the main control chip for data processing. After the data is processed, the information is sent to the database. The main program flow chart is shown in Fig. 7.

The software design of the device is done in the IAR compilation environment. In the compilation process, we first understand the operation flow of each hardware module, edit each part of the module program, and finally integrate with the main program. During the debugging process, the resource utilization of the master chip is improved by optimizing the register configuration and the overall system program architecture. At the same time, the response speed of each part of the program is improved, so that the working efficiency of the system is improved.

5 System Testing and Analysis

This section is about the test plan for the intelligent water monitoring device. The main task is to debug the operation of each module, check its stability and data calibration of the sensor. Perform a single commissioning of each module and then coordinate the system with the goal of implementing the intended function of the intelligent water environment monitoring device. In order to verify the system, water quality monitoring experiments were carried out in an artificial lake with uniform water quality. Make the device move autonomously in the water and monitor the data in real time. The test results are shown in Table 1.

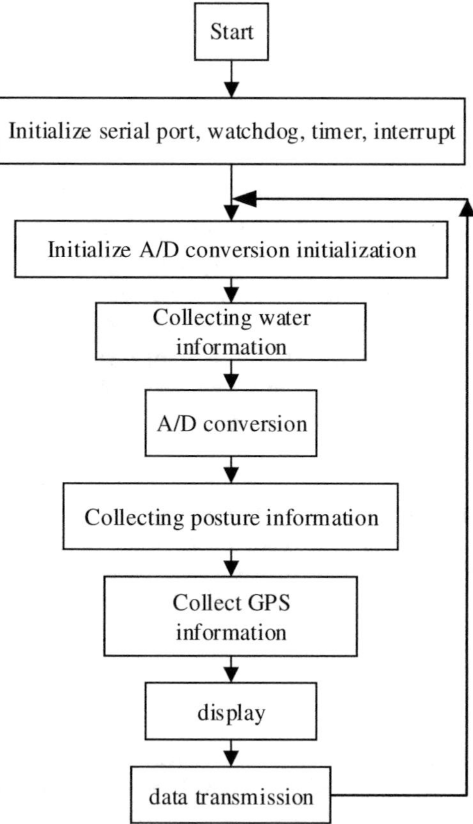

Fig. 7. The main program flow chart

Table 1. Test results

Data number	Temperature/(°C)	PH value	Oxygen content (mg/L)	Turbidity (v)
1	28	6.7	8.24	1.96
2	27	6.7	8.25	2.43
3	27	6.6	8.25	2.00
4	27	6.8	8.26	2.02
5	28	6.5	8.29	2.06
6	26	6.8	8.25	2.30
7	27	6.7	8.26	1.84
8	27	6.6	8.25	1.83
Mean	27.125	6.675	8.256	2.055

It can be derived from Table 1. When the test environment is certain, the water quality in the same water area is not exactly the same, so the data will fluctuate a little, but the monitoring data of the 8 groups are all near the average value, which meets the requirements of water quality monitoring.

6 Conclusion

This study overcomes the shortcomings of traditional water quality monitoring. The design of intelligent water environment monitoring devices was completed using multi-sensors, Internet of Things and wireless communication technologies. The system enables real-time monitoring of water quality (temperature and humidity, pH, turbidity, oxygen content). The device can move and avoid obstacles in a certain water area, perform multi-point testing, and send test data to the server for remote monitoring. The experimental data shows that the system is relatively stable and meets the requirements of water quality monitoring, which can provide good technical support for the water quality monitoring department.

Acknowledgments. This work is supported by Design and Development of Intelligent Collection Platform of Energy Consumption based on Internet of Things, Industrialization project of Shaanxi Provincial Education Department (16JF024) and the Shaanxi Education Committee project (14JK1669) Shaanxi Technology Committee Project (2018SJRG-G-03).

References

1. Said, A., Stevens, D.K., Sehlke, G.: An innovative index for evaluating water quality in streams. Environ. Manag. **34**(3), 406–414 (2004). https://doi.org/10.1007/s00267-004-0210-y
2. Nagy, C.: A Sample Application. Elsevier Inc., 15 June 2003. https://doi.org/10.1016/b978-075067623-6/50014-3
3. Rahman, M.F., Sasongko, R.A.: Obstacle avoidance for quadcopter using ultrasonic sensor. J. Phys. Conf. Ser. **1005**(1) (2018). https://doi.org/10.1088/1742-6596/1005/1/012037
4. Gay, W.W.: DHT11 Sensor. Apress, 15 June 2014. https://doi.org/10.1007/978-1-4842-0181-7_22
5. Jiang, P., Xia, H., He, Z., Wang, Z.: Design of a water environment monitoring system based on wireless sensor networks. Sensors **9**(8), 6411 (2009). https://doi.org/10.3390/s90806411
6. Yin, L., Han, S., Wang, F., Li, Y., Sun, W., Li, C., Wang, Q.: Design and application of DC motor speed control system based on L298N. Inf. Technol. **06**, 104–106, 111 (2017). https://doi.org/10.13306/j.1006-9798.2000.01.003
7. Peng, Y., Yu, M.: The research and application of embedded database. Energy Procedia **13**, 6540–6544 (2011). https://doi.org/10.1016/j.egypro.2011.12.342

Pattern Recognition Technologies and Applications

Face Alignment Based on K-Means

Yunhong Li(✉) and Qiaoning Yuan

Xi'an Polytechnic University, Xi'an 710048, China
hitliyunhong@163.com, 13572863104@163.com

Abstract. In order to address the generalization problem when Active Appearance Model (AAM) is applied to unseen subjects and images. In this paper, an accurate face alignment algorithm based on K-means is proposed to tackle the generalization problem of AAM. Firstly, the original AAM is reformulated as a sparsity-regularized problem. Then, for an input facial image, we learn a strong localized shape and appearance prior through exploiting its K-similar patterns to further approximate sparse representation problem. Finally, learning many localized linear face model instead of a global non-linear face model. Through numerical experiments, this approach is shown to outperform some common existing methods on the task of generic face fitting.

Keywords: Face alignment · K-means · K-nearest neighbors

1 Introduction

Face alignment is critical for practical applications such as face tracking in video [1], facial expression recognition [2–4], 3D facial reconstruction [5–7] and medical image interpretation. Among sizable literatures on face alignment, there are three main kinds of approaches: Parameterized Appearance Models, Regression based Models and Constrained Local Models.

The first model is one of the earliest approaches. It includes Active Shape Model (ASM) [8] and Active Appearance Model (AAM) [9]. The power of AAM stems from statistically modeling the shape and appearance variations simultaneously through principal component analysis (PCA) on a set of labeled data. It didn't perform well when it is trained on a large data. We assert that this is mostly due to the fact that a single PCA model cannot well capture the non-linear appearance variation of a large training set.

Explicit Shape Regression (ESR) and Supervised Descend Method (SDM) [10] are the representative approaches of Regression based Models. ESR directly learns a vector regression function to infer the whole facial shape from the image [11]. The shape constraint is realized by a linear combination of all training shapes. Compared with ESR, SDM aims to solve the nonlinear optimization problem by learning gradient descend directions.

The last type of face alignment is Part-based Deformable Models. Constrained Local Models (CLMs)) is one of it [12]. CLMs are considered to be more robust to partial occlusion and global lighting than the holistic approaches, due to their part-based

P. Krömer et al. (Eds.): ECC 2018, AISC 891, pp. 717–725, 2019.
https://doi.org/10.1007/978-3-030-03766-6_81

modeling. However, the local detectors of CLMs are imperfect and have been shown to result in detection ambiguities in testing.

In recent years, deep learning has also been applied to face alignment. The most commonly used is the Convolutional Neural Network (CNN) [13]. However, CNN has several drawbacks, e.g., lack of end-to-end training, handcrafted features and slow training speed [14]. The deep convolutional network work well in characterizing the nonlinear relationship between the facial appearance and the face shape. However, the several reasons make it work not well on face alignment. Frist, face alignment is a task for forecasting structural. Nevertheless, off-the-shelf large networks are used to train image classification. Second, it is easy to over-fitting. For face alignment with large poses up to 90°, Zhu et al. proposed the 3D Dense Face Alignment (3DDFA) algorithm. A dense 3D face model is fitted to the image via CNN [15].

In this paper, inspired by the empirical observations of Zhao et al. [16], we first cluster the training face set by k-means according to the Histogram of Oriented Gradients (HOG) feature. Then find its K-nearest neighbors (K-NN) [17] as the shape and appearance prior. Finally, the global non-linear face is approximated by many localized linear models, which are specific to the input images and result in a faster convergence.

2 Review of Active Appearance Model

Typically, the fitting of AAM is to minimize the difference between the image $I(W(u;\alpha)$ and the synthesized image $A_0(u) + \sum_{i=1}^{L} \beta_i A_i(u)$. It can be expressed as follows:

$$\min_{\alpha,\beta} \sum_{u \in s_0} [A_0(u) + \sum_{i=1}^{L} \beta_i A_i(u) - I(W(u;\alpha))]^2 \tag{1}$$

where l is the test image, $A_0(u)$ is the mean appearance, $A_i(u)$ is the ith appearance basis. $W(u;\alpha)$ is usually defined as a piecewise-affine warp from the mean shape s_0 to the shape of facial image. The coefficients α is the shape parameter to control the variation of shape. The coefficient is the appearance parameter to control the variation of appearance.

In [16], it is assumed that the shape and appearance of faces can be sparsely represented by faces of a large training set. Due to the multiple iterations and high dimension l_l optimization problem, the process of sparsity-regularized AAM will take a lot of time. In order to address above problem, locality constraint is introduced by us for a fast approximation. The locality constraint is expressed as follows:

$$\min_{\alpha,\beta}\{\sum_{u \in s_0} [\sum_{i=1}^{N} \beta_i A_i(u) - I(W(u;\alpha))]^2 + \lambda_1 \|d \odot \alpha\|^2 + \lambda_2 \parallel d \odot \beta \parallel^2\} \tag{2}$$

where N is the number of faces in the training set, λ_1, λ_2 are the regularization coefficients. Specifically, d are the distances between the input image I and the appearance bases $A(u)$, which add larger weights to the appearance bases nearer to I. Symbol $\odot$ denotes the element-wise multiplication.

In practice, Eq. (2) is fast implemented by directly setting the coefficients of the distant faces to be zero, i.e., selecting the K-nearest neighbors of I as the shape and appearance bases, which is formulated as:

$$\min_{\alpha,\beta}\{\sum_{u\in s_0}[\sum_{i=1}^{K}\beta_i A_i(u) - I(W(u;\alpha))]^2\} \tag{3}$$

where K is the number of selected nearest neighbors and means selecting the K-nearest neighbors of I as the shape and appearance bases. It is worth noting that $\overline{s_0}$ is the mean shape of the selected K-nearest neighbors. Then can be solved by gradient decent algorithm or the IC-AAM algorithm.

3 The Proposed Method

In our algorithm, there are two improvements for face alignment. The first improvement is the use of K-means. Before running a K-means learning algorithm on our input data set, it is useful to consider the factors that we have covered, which:

(1) It is necessary to normalize inputs data.
(2) A major factor in the success of K-means is the dimensionality of the input data and should be as low as possible.
(3) It is important that the choice of cluster center.
(4) Take certain measures to avoid empty clusters and improve the stability of the algorithm.

We first extract HOG features of training set and testing set in the same way as LC-AAM does. Then we apply PCA to reduce the feature vector from 3600 to 112. In [16], the Euclidean distance was adopted to characterize the similarity of two faces. However, Euclidean distance only calculates the distance between two points, and does not learn directional edge features [18]. The k-means algorithm is applied to cluster the HOG feature of the training set, due to its simplicity and good performance. The feature for those faces at a similar pose and expression should also be similar. We use k-means clustering to learn the feature, furthermore, classify training set according to similarity. The classic K-means clustering algorithm finds cluster centroids that minimize the distance between data points and the nearest centroid [18].

The second improvement is about the K-NN algorithm. For a test image, we select the most similar images from training set as the prior of shape and appearance. We didn't select specific number of neighbors because that an inappropriate number of neighbors can lead to greater error. So we select the number of each test image neighbors according to the distance of the current image to training sets. If this distance is within our tolerance error criterion, we select the training image as the shape and appearance bases of this test image. By contrast, if the distance greater than we expected, we discard it.

Having described the major improvements for our method, we will present our method in detail as illustrated in Fig. 1.

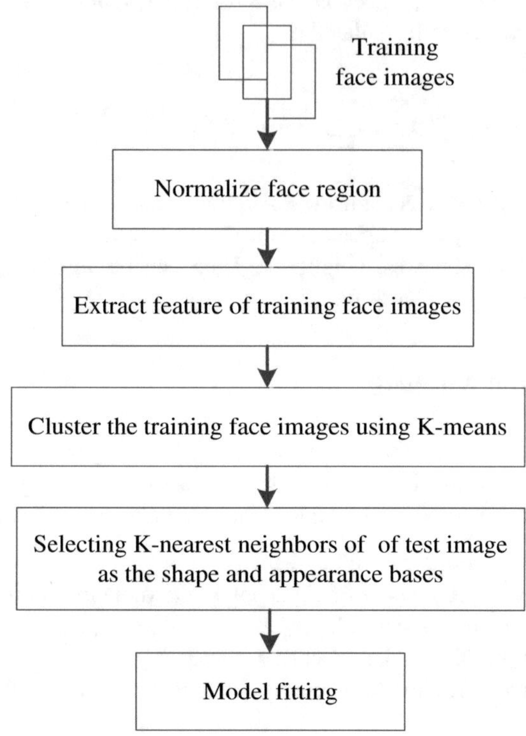

Fig. 1. The procedure of our method.

4 Experiment

In this section, we evaluate the effectiveness of our algorithm through comparing it with the LC-AAM method and other state-of-the-art face alignment methods.

4.1 Databases and Evaluation Metric

To establish the practical efficiency of our algorithm, we implemented it and tested its performance on two publicly available face data set. We randomly collected 7000 images from CMU-PIE [19], FRGCv1 [20], FERET [21], CAS-PEAL [22], and PubFig [23] face databases as training images named T1.

Set 2 is selected from 300 W face data base [24], noted as T2. We use the training part of LFPW, HELEN and the whole AFW for training, and perform testing on three parts: the test samples from LFPW and HELEN as the common subset, the 135-image IBUG as the challenging subset, and the union of them as the full set (689 images in total) [15]. The algorithm was implemented with Matlab 2016Ra.

Following the standard [25, 26], we use the RMSE (Root Mean Square Error) as the evaluation metric. The performance can be expressed in terms of the normalized root-mean-squared error (NRMSE) that calculated by dividing the root mean square error by the distance between the centers of the two eyes. The NRMSE is given as a percentage. The cumulative distribution function (CDF) of NRMSE is used to evaluate the performance of face alignment algorithm. Given the ground-truth, the normalized error is computed as follows:

$$e_i^k = \frac{d[(x_i^k, y_i^k), (\tilde{x}_i^k, \tilde{y}_i^k)]}{IOD} \tag{4}$$

where $d[(x_i^k, y_i^k), (\tilde{x}_i^k, \tilde{y}_i^k)]$ indicates the Euclidean distance between the *kth* ground-truth and the estimated landmark of the *ith* person. *IOD* here represents the Inter-Ocular Distance, which is defined as the distance between two eye pupils. When $e_i^k < 0.1$ can be taken as an acceptable error criterion;

4.2 Experimental Results

Comparisons with respect to different dimensionality of input feature. To show how our method is affected by the number of dimensionality of input feature. We conduct our experiments on data set using dimensionality = 1866, 144, 121,100 respectively. From the experimental results, as shown in Fig. 2, it is observed that dimensionality = 121 is better than the other parameters. In addition, when a larger dimensionality is selected, the alignment results become worse. So, in the following experiments, the number of dimensionality is set to 121.

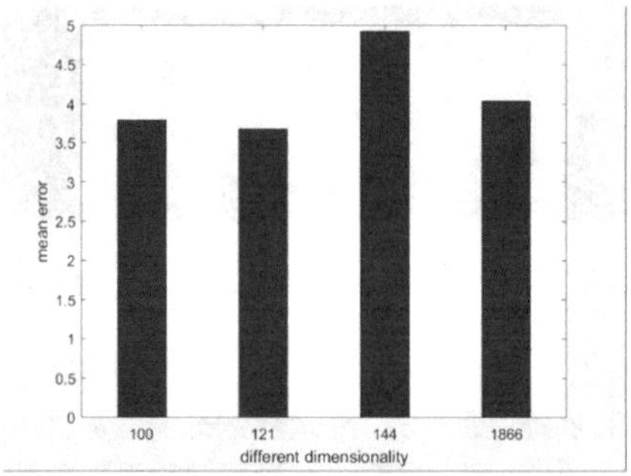

Fig. 2. Comparison of the accuracy of different dimensionality.

Comparisons with respect to different number of clusters. We conduct our experiments on T1 using K = 4, 10, 50 respectively. Experimental results, shown in Fig. 3, demonstrate that K = 4 is best than the other parameters.

Comparisons with LC-AAM In this subsection, we compare our algorithm with LC-AAM algorithm on the T1. Specifically, for LC-AAM, the number of nearest neighbors is set to 20 and the HOG feature is exploited to calculate the similarity of image pairs. Moreover, the iteration number in the fitting procedure of all these methods is set to 25. For our method, the dimensionality of input feature is set to 121 and the clusters is set to 4. The comparison results is shown in Fig. 4, which demonstrate that our method greatly outperforms the LC-AAM method. Visualized example images of our method are shown in Fig. 5.

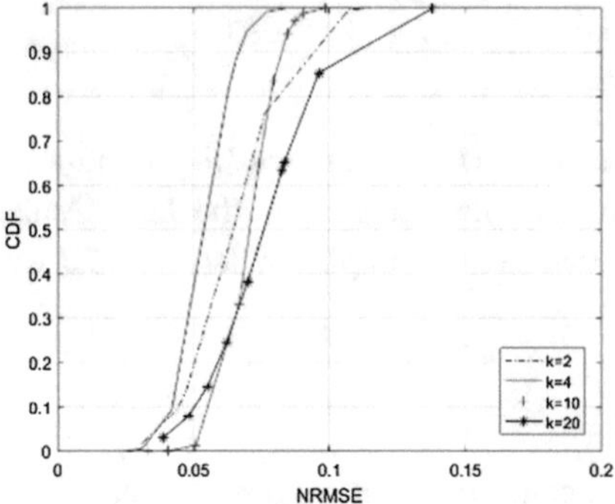

Fig. 3. Effect of clusters number.

Fig. 4. Comparison results on T1 data set.

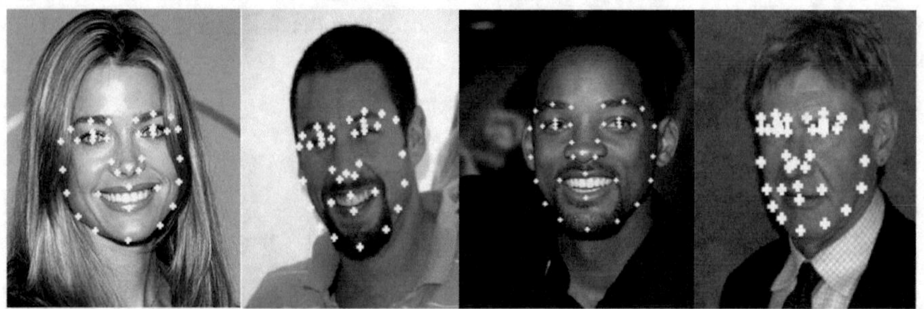

Fig. 5. Visualized example images of our method.

Table 1. The NME(%) of different methods on 300 W dataset.

Method	Common	Challenging	Full
ESR [11]	5.28	17.00	7.58
SDM [10]	5.57	15.40	7.5
RCPR [27]	6.18	17.26	8.35
LBF [28]	4.95	11.98	6.62
TSPM [29]	8.22	18.33	10.20
Proposed method	5.63	10.87	6.98

Comparisons with state-of-the-art Methods. We compare our algorithm with other algorithm on the T2. The comparison results is shown in Table 1, which demonstrate that our method greatly outperforms the other method.

5 Conclusion and Future Work

In this paper, a more efficient and accurate algorithm has been proposed, which can improve facial alignment performance in several aspects. These include the introduction of K-means and adaptive selecting the number of neighbors according the distance. Among these improvements, the major contribution is that instead of Euclidean distance with a K-means. Therefore, our method can align facial features more robustly and accurately when a face is under the variations of pose, expression, and shape. Our future work will focus on further improving the robustness of our model against illumination variations and occlusions by facial hair and glasses.

References

1. Chrysos, G.G., Antonakos, E., Snape, P., Asthana, A., Zafeiriou, S.: A comprehensive performance evaluation of deformable face tracking "In-the-Wild". Int. J. Comput. Vision **126**(2–4), 198–232 (2018). https://doi.org/10.1007/s11263-017-0999-5
2. Chen, J., Chen, Z., Chi, Z., Fu, H.: Facial expression recognition in video with multiple feature fusion. IEEE Trans. Affect. Comput. **9**(1), 38–50 (2018). https://doi.org/10.1109/taffc.2016.2593719
3. Li, H., Ding, H., Huang, D., Wang, Y., Zhao, X., Morvan, J.M., Chen, L.: An efficient multimodal 2D + 3D feature-based approach to automatic facial expression recognition. Comput. Vis. Image Underst. **140**, 83–92 (2015). https://doi.org/10.1016/j.cviu.2015.07.005
4. Jung, H., Lee, S., Yim, J., Park, S., Kim,J.: Joint fine-tuning in deep neural networks for facial expression recognition. In: IEEE ICCV, pp. 2983–2991. Santiago, Chile (2015). https://doi.org/10.1109/ICCV.2015.341
5. Liu, F., Zeng, D., Zhao, Q., Liu, X.: Joint face alignment and 3D face reconstruction. In: Leibe, B., Matas, J., Sebe, N., Welling, M. (eds.) Computer Vision – ECCV 2016. Lecture Notes in Computer Science, vol 9909, pp. 545–560. Springer, Cham (2016). https://doi.org/10.1007/978-3-319-46454-1_33
6. Dou, P., Shah, S.K., Kakadiaris, I.A.: End-to-end 3D face reconstruction with deep neural networks. In: 2017 IEEE Conference on Computer Vision and Pattern Recognition (CVPR), Honolulu, Hawaii, USA, pp. 1503–1512 (2017). https://doi.org/10.1109/cvpr.2017.164
7. Roth, J., Tong, Y., Liu, X.: Adaptive 3D face reconstruction from unconstrained photo collections. In: IEEE CVPR, 2017. Las Vegas, NV, USA, pp. 2127–2141 (2016).https://doi.org/10.1109/cvpr.2016.455
8. Cootes, T.F., Taylor, C.J., Cooper, D.H., Graham, J.: Active shape models-their training and application. Comput. Vis. Underst. **61**(1), 38–59 (1995). https://doi.org/10.1006/cviu.1995.1004&%23xB7
9. Cootes, T.F., Edwards, G.J., Taylor, C.J.: Active appearance models. IEEE Trans. Pattern Anal. Mach. Intell. **23**(6), 681–685 (2001).https://doi.org/10.1109/34.927467

10. Xiong, X., Torre, F.D.L.: Supervised descent method and its applications to face alignment. In: 2013 IEEE Conference on Computer Vision and Pattern Recognition, pp. 532–539. IEEE, Portland (2013). https://doi.org/10.1109/cvpr.2013.75
11. Cao, X., Wei, Y., Wen, F., Sun, J.: Face alignment by Explicit Shape Regression. In: 2012 IEEE Conference on Computer Vision and Pattern Recognition, pp. 177–190. IEEE, Providence (2012). https://doi.org/10.1109/cvpr.2012.6248015
12. Jin, X., Tan, X.: Face alignment in-the-wild: a survey. Comput. Vis. Image Underst. **162**, 1–22 (2017). https://doi.org/10.1016/j.cviu.2017.08.008
13. Bulat, A., Tzimiropoulos, G.: Two-stage convolutional part heatmap regression for the 1st 3D Face Alignment in the Wild (3DFAW) challenge. In: Hua G., Jégou H. (eds.) Computer Vision – ECCV 2016 Workshops, ECCV 2016. Lecture Notes in Computer Science, vol 9914, pp. 616–624. Springer, Cham (2016). https://doi.org/10.1007/978-3-319-48881-3_43
14. Jourabloo, A., Ye, M., Liu, X., Ren, L.: Pose-invariant face alignment with a single CNN. In: ICCV, pp. 3219–3228. IEEE, Venice (2017). https://doi.org/10.1109/iccv.2017.347
15. Zhu, X., Zhu, L., Liu, X., Shi, H., Li, S.: Face alignment across large poses: a 3D solution. In: 2016 IEEE Conference on Computer Vision and Pattern Recognition (CVPR), pp. 146–155. IEEE, Las Vegas (2016). https://doi.org/10.1109/cvpr.2016.23
16. Zhao, X., Shan, S., Chai, X., Chen, X.: Locality-constrained active appearance model. In: Lee, K.M., Matsushita, Y., Rehg, J.M., Hu, Z. (eds) Computer Vision – ACCV 2012. Lecture Notes in Computer Science, vol 7724, pp. 636–647. Springer, Heidelberg (2012). https://doi.org/10.1007/978-3-642-37331-2_48
17. Kanungo, T., Mount, D.M., Netanyahu, N.S., Piatko, C.D., Silverman, R., Wu, A.Y.: An efficient k-means clustering algorithm: analysis and implementation. IEEE Trans. Pattern Anal. Mach. Intell. **24**(7), 881–892 (2002). https://doi.org/10.1109/tpami.2002.1017616
18. Coates, A., Ng, A.Y. Learning feature representations with K-Means. In: Montavon, G., Orr, G.B., Müller, KR. (eds) Neural Networks: Tricks of the Trade. Lecture Notes in Computer Science, vol 7700, pp. 561–580. Springer, Heidelberg (2012).https://doi.org/10.1007/978-3-642-35289-8_30
19. Sim, T., Baker, S., Bsat, M.: The CMU Pose, Illumination, and Expression (PIE) database. In: Proceedings of Fifth IEEE International Conference on Automatic Face Gesture Recognition, p. 0053. IEEE, Washington, DC (2002). https://doi.org/10.1109/AFGR.2002.1004130
20. Phillips, P.J., Flynn, P.J., Scruggs, T., Bowyer, K.W., Chang, J., Hoffman, K., Marques, J. Min, J., Worek, W.: Overview of the face recognition grand challenge. In: 2005 IEEE Computer Society Conference on Computer Vision and Pattern Recognition (CVPR 2005), (CVPR), pp. 947–954. IEEE, San Diego (2005). https://doi.org/10.1109/cvpr.2005.268
21. Phillips, P.J., Wechsler, H., Huang, J., Raussa, P.J.: The FERET database and evaluation procedure for face-recognition algorithms. Image Vis. Comput. **16**(5), 295–306 (1998). https://doi.org/10.1016/s0262-8856(97)00070-x
22. Gao, W., Cao, B., Shan, S.: The CAS-PEAL large-scale chinese face database and baseline evaluations. IEEE Trans. Syst., Man, Cybern. Part A Syst. Hum. **38**(1), 149–161 (2008). https://doi.org/10.1109/tsmca.2007.909557
23. Kumar, N., Berg, A., Belhumeur, P., Nayar, S.K.: Attribute and simile classifiers for face verification. In: Proceedings/ IEEE International Conference on Computer Vision, pp. 365–372. IEEE, Kyoto (2009).https://doi.org/10.1109/iccv.2009.5459250
24. Sagonas, C., Antonakos, E., Tzimiropoulos, G., Zafeiriou, S., Panticac, M.: 300 Faces In-The-Wild Challenge: database and results * **. Image Vis. Comput. **47**, 3–18 (2016). https://doi.org/10.1016/j.imavis.2016.01.002
25. Cao, C., Weng, Y., Lin, S., Zhou, K.: 3D shape regression for real-time facial animation. ACM Trans. Graph. **32**(4), 1–10 (2013). https://doi.org/10.1145/2461912.2462012

26. Belhumeur, P.N., Jacobs, D.W., Kriegman, D.J., Kumar, N.: Localizing parts of faces using a consensus of exemplars. IEEE Trans. Pattern Anal. Mach. Intell. **35**(12), 545–552 (2011). https://doi.org/10.1109/cvpr.2011.599560
27. Burgos-Artizzu, X.P., Perona, P., Dollár, P.: Robust face landmark estimation under occlusion. In: 2013 IEEE International Conference on Computer Vision, pp. 1513–1520. IEEE, Sydney (2013). https://doi.org/10.1109/iccv.2013.191
28. Ren, S., Cao, X., Wei, Y., Sun, J.: Face alignment at 3000 FPS via regressing local binary features. In: 2014 IEEE Conference on Computer Vision and Pattern Recognition, pp. 1685–1692. IEEE, Columbus (2014). https://doi.org/10.1109/cvpr.2014.218
29. Zhu, X., Ramanan, D.: Face detection, pose estimation, and landmark localization in the wild. In: 2012 IEEE Conference on Computer Vision and Pattern Recognition, pp. 2879–2886. IEEE, Providence (2012). https://doi.org/10.1109/cvpr.2012.6248014

Image Denoising Method Based on Weighted Total Variational Model with Edge Operator

Hong Zhang[1], Xiaoli Zhou[1(✉)], Weixiao Zhan[2], and Fuhua Yu[1]

[1] Xi'an University of Post and Telecommunications, Xi'an 710121, China
zxlsunnyd@163.com

[2] China Academy of Information and Communications Technology, Beijing, China

Abstract. In order to eliminate image noise effectively, the weighted total variation algorithm based on edge detection is proposed. By calculating the amplitude of the edge operator of the image, accurate estimates of edge weights are achieved, and then the weight of the canny operator is used to weigh the Lagrangian multiplier, which is no longer a global variable, so that the filter has a better edge protection feature. Theoretical analysis and experimental results show that the method can remove noise while preserving the edge details of the image more completely. The step effects of the total variation model is effectively suppressed, and has a better performance in terms of structural similarity and the visual effect of image.

Keywords: Image denoising · Weighted total variation model
Canny · Structural similarity

1 Introduction

During the image acquisition and transmission process, the image is inevitably contaminated by noise, which has an unfavorable influence on image recognition and the application of subsequent process [1–3]. So removing noise is very important. There are many methods for image denoising. In the frequency domain processing, there is wavelet denoising [4]. In the spatial domain processing, there are linear filtering represented by Gaussian filtering [5], and nonlinear filtering represented by median filtering [6]. Gaussian filtering is a widely used denoising method because of its simplicity and low computational complexity. However, it performs the same degree of processing on noise and signal, which makes it denoise and weakens the contrast of the image, and also causes images edge drift. Therefore, many nonlinear denoising methods are proposed for this problem. Image denoising based on partial differential method is a representative method, which solves the problem of traditional denoising method in noise reduction and detail protection.

There are two classical methods for image denoising based on partial differential equation (PDE): one is the nonlinear diffusion filter model (PM) proposed by Perona et al. [7]; the other is the total variation (TV) model proposed by rudin et al., which uses the TV regularization method [8]. The TV model can not only remove noise but also protect the edge texture and other details of the image. Various improvements to

P. Krömer et al. (Eds.): ECC 2018, AISC 891, pp. 726–735, 2019.
https://doi.org/10.1007/978-3-030-03766-6_82

the TV model have been proposed. Huber equation is used to reconstruct the high-order total variational optimization equation [9], and the random absolute difference ranking value between neighboring pixels is used to estimate the position of the noise [10], and a dual-formulation-based adaptive TV (ATV) regularization method is applied to solving the TV regularization. The parameter adaptation of the TV regularization is performed based on the noise level estimated via wavelets [11], constructing weighted functions with different gradient and variance characteristics in different regions of the image, in the regular term of the TV model to control the diffusion intensity [12]. However, these algorithms always use the same Lagrangian multiplier in the denoising process, which will lead to the loss of some texture details and step effects.

To solve this problem, a new improved algorithm is proposed in this paper. Based on the TV model, we introduce a weighted function to the Lagrangian multiplier by calculating the edge operator magnitude of the image, and obtain the adaptive Lagrangian multiplier of the edge region and the smooth region, respectively. The proposed algorithm can effectively suppress the step effects and better maintain the details of the edge texture of the image.

2 Total Varation Model

The TV model was proposed in 1992. For the first time introduced into the field of image denoising, image denoising based on TV is to represent the image energy as a function. Under the constraint condition, TV denoising algorithm minimizing the TV norm, it removes the noise effectively in flat regions, simultaneously preserving the boundaries. In this way, the problem of image denoising is transformed into an energy minimization problem.

f_0 represents the observed image, Ω denotes the domain of definition of the image, $f_0(x,y) = f(x,y) + n(x,y)$, where $f(x,y)$ is the original clear image, $n(x,y)$ represents white noise with a variance of σ^2. Then the TV of the f is defined as:

$$TV(f) = \int_{\Omega} \sqrt{|\nabla f|^2} dxdy \tag{1}$$

Image contamination by noise is a forward problem, and the original image data is obtained based on the observed image data, the constraints imposed, or assumptions made. The denoising process is an inverse problem, which may lead to the non-uniqueness or instability of the solution. In mathematics, regularization is usually adopted to solve such problems to constrain the solution space. Therefore, image denoising based on TV can be formulated as the following minimization problem:

$$\min E(f) = \frac{\lambda}{2} \int_{\Omega} (f - f_0)^2 dxdy + \int_{\Omega} \sqrt{|\nabla f|^2} dxdy \tag{2}$$

where $\frac{\lambda}{2} \int_{\Omega} (f - f_0)^2 dxdy$ is the data fidelity term, which represents the degree of approximation between the restored image and the noisy image, and mainly plays the role of preserving the characteristics of the original image and reducing the distortion of the image. $\int_{\Omega} \sqrt{|\nabla f|^2} dxdy$ is a regularization term, which can reflect the edge information of the image very well. The necessary condition for the minimum value of the energy functional is that it satisfies the Euler-Lagrangian equation, which is:

$$-\nabla \cdot \left(\frac{\nabla f}{|\nabla f|} \right) + \lambda (f - f_0) = 0 \tag{3}$$

where the diffusion coefficient is $\frac{1}{|\nabla f|}$. At the edge of the image, $|\nabla f|$ is large and the diffusion coefficient is small, hence, the diffusion along the edge region is weak, thus the edge is retained; in the smooth region, $|\nabla f|$ is small and the diffusion coefficient is large, hence, the diffusion ability in the image smooth region is strong, thus the noise is removed.

3 Proposed Method

From formula (3), we can see that λ in the TV model is a global variable, but in fact, the signal-to-noise ratio of different parts of the image is different. In denoising, using the same λ as the fidelity term coefficient is not good enough to highlight the texture details of the image. Therefore, by designing the weight function for the Lagrangian multiplier, the method proposed in this paper has a better boundary retention performance, eliminates the step effect, and avoids the phenomenon of boundary oscillation.

The edge weights proposed in [12] are based on the variance within the window. However, the region with a large variance does not correspond to the edge region. The edge value of the image detected by the edge operator can be used instead of the local variance to achieve an accurate estimation of the normalization factor. The most commonly used methods for edge detection include Log operator [13], Sobel operator [14] and Canny operator [15]. The Log operator still has a large influence on the image because the scale factor cannot be adaptively adjusted. Sobel operator has a low computational complexity. However, it is easy to detect the false edge. The Canny operator is one of the best operators for detecting the step edge effect in the first-order traditional differential, and has a better edge detection performance. In this paper, Canny operator is used to detect the edge of the image, and the edge weighted [16] $\varphi(p)$ is defined as follows:

$$\varphi(p) = \frac{1}{N} \sum_{p'}^{N} \frac{C(p) + \gamma}{C(p') + \gamma} \tag{4}$$

where the Canny edge value corresponding to the point P is $C(P)$, N is the total number of pixels, the γ value is 0.065536, there are significant differences in the values of Canny edge values for different images. In this paper $\varphi(p)$ is normalized to make a linear map to the range of [0, 1]. The final result is $\overline{\varphi(p)}$, and the energy functional of the improved denoising model is as follows:

$$TV(f) = \int_{\Omega} \sqrt{|\nabla f|^2} dxdy + \frac{\lambda}{2}\overline{\varphi(p)} \int_{\Omega} (f - f_0)^2 dxdy \tag{5}$$

Using Euler-lagrange principle and gradient descent method leads to

$$\begin{cases} \frac{\partial f(x,y,t)}{\partial t} = -\nabla \cdot \left(\frac{\nabla f}{|\nabla f|}\right) + \lambda \overline{\varphi(p)}(f - f_0) \\ f(x,y,t)|_{t=0} = f_0(x,y) \end{cases} \tag{6}$$

which is discretized by finite difference scheme, where $f_{i,j}$ denotes the gray value of the image f at the pixel point $x_i = ih$, $y_i = jh$, $i,j = 0, 1...N$, $Nh = L$, L for the length of the image, Δt as the time step, h for the space step, n as the number of iterations, $f_{i,j}^n$ represents the $n-th$ iteration value, boundary conditions satisfied: $f_{N,j}^n = f_{N-1,j}^n$ $f_{i,0}^n = f_{i,N}^n = f_{i,N-1}^n$, so we have:

$$\begin{cases} (f_x)_{i,j}^n = \frac{f_{i+1,j}^n - f_{i-1,j}^n}{2h} \\ (f_y)_{i,j}^n = \frac{f_{i,j+1}^n - f_{i,j-1}^n}{2h} \\ (f_{xx})_{i,j}^n = \frac{f_{i+1,j}^n - 2f_{i,j}^n + f_{i-1,j}^n}{h^2} \\ (f_{yy})_{i,j}^n = \frac{f_{i,j+1}^n - 2f_{i,j}^n + f_{i,j-1}^n}{h^2} \\ (f_{xy})_{i,j}^n = \frac{f_{i+1,j+1}^n - f_{i-1,j+1}^n - f_{i+1,j-1}^n + f_{i-1,j-1}^n}{4h^2} \\ div\left(\frac{\nabla f}{|\nabla f|_\varepsilon}\right) = \frac{f_{xx}f_y^2 - 2f_xf_yf_{xy} + f_x^2f_{yy}}{\left(f_x^2 + f_y^2 + \varepsilon^2\right)^{3/2}} \end{cases} \tag{7}$$

The discrete format is

$$\frac{f_{i,j}^{n+1} - f_{i,j}^n}{\Delta t} - div\left(\frac{\nabla f_{i,j}^n}{\left|\nabla f_{i,j}^n\right|_\varepsilon}\right) - \lambda^n \overline{\varphi(x,y)}\left(f_{i,j}^n - f_{i,j}^0\right) \tag{8}$$

In the smooth region, the edge weight $\overline{\varphi(p)}$ of the pixel point $p(x,y)$ tends to be zero, then the regularization term has its main function. Therefore, diffusion occurs at a faster speed, which can effectively remove the noise. In the vicinity of the boundary, the edge weight $\overline{\varphi(p)}$ of the pixel point $p(x,y)$ tends to be one, then the fidelity term has its main function, the diffusion speed is reduced, the diffusion velocity can be reduced and the edge feature can be preserved.

4 Experimental Results and Analysis

We test our proposed denoising algorithm on three images: Barbara, Cameraman and Lake. The three images have different grayscale distribution, edge and texture features. With noise standard deviation ranging from 15 to 30, performance measures which have been considered in evaluation are peak signal-to-noise ratio (PSNR), mean square error (MSE) and structural similarity (SSIM).

The results of denoising on Barbara, Cameraman and Lake images with noise standard deviation 15. The proposed method is compared with the improved TV model of reference [12] and the NLM filtering, and the results are shown in Figs. 1, 2 and 3.

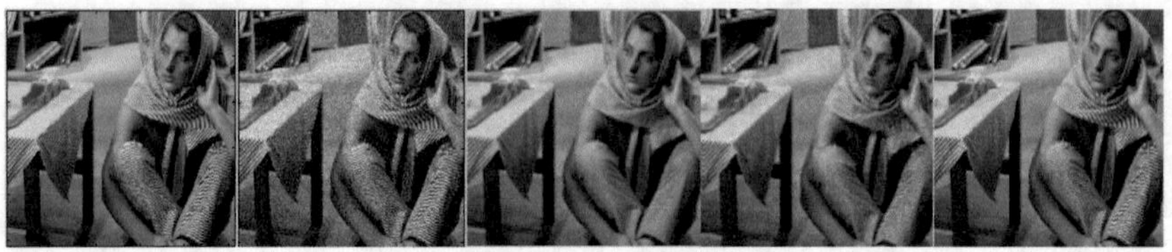

Fig. 1. Visual comparisons of denoising results for Barbara. (a) original image. (b) noisy image. (c) improved TV model of reference [12] denoised. (d) NLM denoised. (e) the proposed denoised.

From Figs. 1, 2 and 3, it is found that when the noise level is small, the difference of denoising effect of each algorithm is not very obvious, because a small amount of noise does not affect the structure of the image greatly, and the structure information of the orginal image is still reserved. Hence, in order to better verify the denoising effect of the proposed algorithm, we choose the three images of Couple, Man and Boat with size, the Gaussian noise with a mean variance of 30 is added, and the experimental results are shown in Figs. 4, 5 and 6.

Figures 4, 5 and 6 shows that the advantages of the proposed algorithm are reflected when the noise level is increased. It can be observed that although TV model of reference [12] can suppress certain noise, and it excessively smoothes the details in the image, so that Couple's clothes texture and curtains are excessively smooth. And from the Boat image, the background brightness of the image becomes dark after denoising, and the denoising effect is not ideal. The NLM method is used to denoise incompletely, and the edges of the image are blurred, and there is distortion, such as the face of the Couple, the flowers on the table, and the background wall, which paintings are blurred. After denoised by the proposed method, the image is smooth and natural, and its edge and detail information are more complete. Such as Couple's hair and clothes, Man's headwear and tablecloth clothes texture are relatively clear. Boat's mast and hull are more visible. Compared with the TV model of reference [12] and NLM denoising algorithms, the proposed method has a better denoising effect and better protects the edge texture and detail information of the image.

To further objectively evaluate the method of this paper, we choose PSNR, MSE and SSIM as the evaluation metrics, which are defined as follows:

$$P_{PSNR} = 10\lg \frac{255^2 MN}{\sum_{i=1}^{M}\sum_{j=1}^{N}[f(i,j) - g(i,j)]^2} \tag{9}$$

Fig. 2. Visual comparisons of denoising results for Cameraman. (a) original image. (b) noisy image. (c) improved TV model of reference [12] denoised. (d) NLM denoised. (e) the proposed denoised.

Fig. 3. Visual comparisons of denoising results for Lake. (a) original image. (b) noisy image. (c) improved TV model of reference [12] denoised. (d) NLM denoised. (e) the proposed denoised.

Fig. 4. Visual comparisons of denoising results for Couple. (a) original image. (b) noisy image. (c) improved TV model of reference [12] denoised. (d) NLM denoised. (e) the proposed denoised.

Fig. 5. Visual comparisons of denoising results for Man. (a) original image. (b) noisy image. (c) improved TV model of reference [12] denoised. (d) NLM denoised. (e) the proposed denoised.

$$M_{MSE} = \frac{1}{MN} \sum_{i=1}^{M} \sum_{j=1}^{N} [f(i,j) - g(i,j)]^2 \tag{10}$$

Fig. 6. Visual comparisons of denoising results for Boat. (a) original image. (b) noisy image. (c) improved TV model of reference [12] denoised. (d) NLM denoised. (e) the proposed denoised.

$$S_{SSIM} = \frac{(2u_f u_g + c_1)(2\sigma_{fg} + c_2)}{(u_f^2 + u_g^2 + c_1)(\sigma_f^2 + \sigma_g^2 + c_2)} \tag{11}$$

where $f(i,j)$ represents a noisy image, M and N are the numbers of rows and columns of the image, respectively, u_f, u_g, σ_f^2, σ_g^2, and $2\sigma_{fg}$ respectively represent the mean value, variance and covariance of images f and g. c_1 and c_2 are the stable constants. PSNR represents the difference in noise level between the original image and the denoised image, and can reflect the degree to which the image approaches the original clear image after processing. The larger the PSNR value, the closer the denoised image is to the original image in the average sense. MSE is the expectation of the difference between the original image and the denoised image. The smaller the MSE value, the better the image denoising effect. SSIM directly uses the structural similarity of the two images as the evaluation standard. The larger the SSIM value, the more similar the denoised image is to the original image, and the better the visual effect of the image after denoising. In this paper, the standard deviations are set to 15, 20, 25 and 30, respectively, and the effects of noise degree on denoising methods are observed by using of reference [12] TV denoising method, NLM filtering method and the improved denoising method in this paper, such as PSNR, MSE and SSIM. The experimental results of test images are listed in Tables 1, 2 and 3:

From Table 1, with the increase of σ the PSNR of the three methods are gradually decreased. Among the three methods, whether it is images of Couple and Man with rich texture information, or images of Boat with rich edge information, our method can achieve better results in terms of PSNR and MSE than the TV model of reference [12] and NLM algorithm.

From Table 3, when the standard deviation $\sigma = 15$, the SSIM values of the three methods are not different significantly, and when $\sigma = 30$, the SSIM value obtained by the method in this paper is reduced by an average of 0.184 compared to the SSIM value obtained at $\sigma = 15$, while the SSIM values obtained by the other three methods are reduced an average of 0.288, and with the increase of the value σ, the SSIM value obtained by this method is the largest. It is shown that the denoising image obtained by proposed method in this paper has the best visual effect.

Table 1. PSNR comparisons of image denosing.

Image	σ	Reference [12]	NLM	Proposed method
Couple	15	32.21	31.90	34.70
	20	28.99	28.49	30.62
	25	26.20	27.82	29.71
	30	25.55	25.48	27.51
Man	15	33.44	32.05	33.88
	20	29.81	29.96	30.66
	25	27.12	27.74	28.03
	30	26.91	26.82	27.99
Boat	15	32.50	32.11	32.89
	20	30.97	29.96	30.97
	25	28.99	29.63	29.37
	30	27.57	27.26	27.96

Table 2. MSE comparisons of image denoising results.

Image	σ	Reference [12]	NLM	Proposed method
Couple	15	2.41	2.11	0.83
	20	4.22	3.51	1.43
	25	6.25	4.1	1.62
	30	7.23	4.59	2.21
Man	15	1.11	0.49	0.15
	20	2.54	1.18	0.38
	25	4.98	1.45	1.11
	30	6.89	1.81	1.24
Boat	15	1.21	0.92	0.67
	20	2.54	1.14	0.93
	25	4.81	1.57	1.12
	30	6.39	1.86	1.28

Table 3. SSIM comparisons of image denoising results.

Image	σ	Reference [12]	NLM	Proposed method
Couple	15	0.803	0.812	0.837
	20	0.721	0.734	0.736
	25	0.665	0.673	0.713
	30	0.510	0.524	0.653
Man	15	0.821	0.813	0.862
	20	0.765	0.767	0.786
	25	0.634	0.653	0.666
	30	0.507	0.511	0.612
Boat	15	0.819	0.824	0.832
	20	0.723	0.713	0.738
	25	0.657	0.642	0.685
	30	0.527	0.571	0.614

5 Conclusion

In order to solve the problem of partial texture edge loss caused by the same fidelity coefficient in TV denoising, a new method based on edge detection weighted TV is proposed without changing the linear complexity of TV. By modifying the approximation term with the normalized edge operator amplitude, the adaptive fidelity coefficients of the texture region and the smoothing region of the denoised image are obtained, and the self-adaptability of the algorithm is enhanced and the step effect produced by traditional TV model is suppressed. Numerical examples show that our method can obtain better results in terms of PSNR and SSIM. The visual comparisons show that our proposed model has a stronger ability to process the details and texture of image, preserving more useful image information.

Acknowledgements. This work was supported by the Shaanxi Natural Science Foundation (2016JM8034) and Scientific research plan projects of Henan Education Department (12JK0791).

References

1. Wang, Z., Hou, G., Pan, Z., Wang, G.: Single image dehazing and denoising combining dark channel prior and variational models. IET Comput. Vis. **12**(11), 393–402 (2018). https://doi.org/10.1049/iet-cvi.2017.0318
2. Vazquez-Corral, J., Bertalmío, M.: Angular-based preprocessing for image denoising. IEEE Signal Process. Lett. **25**(11), 219–223 (2018). https://doi.org/10.1109/LSP.2017.2777147
3. Dou, Z., Song, M., Gao, K., Jiang, Z.: Image smoothing via truncated total variation. IEEE Access **5**(11), 27337–27344 (2017). https://doi.org/10.1109/access.2017.2773503
4. Habib, W., Sarwar, T., Siddiqui, A.M., Touqir, I.: Wavelet denoising of multiframe optical coherence tomography data using similarity measures. IET Image Proc. **11**(11), 64–79 (2017). https://doi.org/10.1049/iet-ipr.2016.0160

5. Kwon, S., Lee, H., Lee, S.: Image enhancement with Gaussian filtering in time-domain microwave imaging system for breast cancer detection. Electron. Lett. **52**(5), 342–344 (2016). https://doi.org/10.1049/el.2015.3613
6. Zhang, S., Li, X., Zhang, C.: Modified adaptive median filtering. In: 2018 International Conference on Intelligent Transportation. Big Data & Smart City (ICITBS), Xiamen (2018). https://doi.org/10.1109/icitbs.2018.00074
7. Perona, P., Malik, J.: Scale-space and edge detection using anisotropic diffusion. IEEE Trans. Pattern Anal. Mach. Intell. **12**(7), 629–639 (1990). https://doi.org/10.1109/34.56205
8. Rudin, L., Osher, S., Fatemi, E.: Nonlinear total variation based noise removal algorithms. Physica Section D. **60**, 259–268 (1992). https://doi.org/10.1016/0167-2789(92)90242-F
9. Hu, Y., Zhong, C.X., Cao, M.Y., Zhao, G.S.: A Fast High order Total variational Image denoising method based on augmented Lagrangian multiplier. Syst. Eng. Electron. Technol. **39** (12): 2831–2839 (2017). https://doi.org/10.3969/j.issn.1001-506x.2017.12.29
10. Said, B.A., Foufou, S.: Modified total variation regularization using fuzzy complement for image denoising. In: 2015 International Conference on Image and Vision Computing New Zealand (IVCNZ), Auckland, pp. 1–6 (2015) https://doi.org/10.1109/ivcnz.2015.7761561
11. Zhao, Y., Liu, G.J., Zhang, B., Hong, W., Wu, Y.R.: Adaptive total variation regularization based SAR image despeckling and despeckling evaluation index. IEEE Trans. Geosci. Remote Sens. **53**(5), 2765–2774 (2015). https://doi.org/10.1109/tgrs.2014.2364525
12. Yan, N.L., Jin, C.: Total variation image denoising model based on weighting function. Electron. Measur. Technol. **41**(07), 58–63 (2018). https://doi.org/10.19651/j.cnki.emt.1701305
13. Mallick, A., Chaudhuri, S.S., Roy, S.: Optimization of Laplace of Gaussian (LoG) filter for enhanced edge detection: new approach. In: Proceedings of the 2014 International Conference on Control. Instrumentation, Energy and Communication (CIEC), pp. 658–661 (2014). https://doi.org/10.1109/ciec.2014.6959172
14. Amara, B.A., Pissaloux, E., Atri, M.: Sobel edge detection system design and integration on an FPGA based HD video streaming architecture. In: 2016 11th International Design & Test Symposium (IDT), Hammamet, pp. 160–164 (2016). https://doi.org/10.1109/idt.2016.7843033
15. Baştan, M., Bukhari, S.S., Breuel, T.: Active Canny: edge detection and recovery with open active contour models. IET Image Proc. **11**(12), 1325–1332 (2017). https://doi.org/10.1049/iet-ipr.2017.0336
16. Ao, J.S., Zong, K., Ma, C.B.: Underwater image enhancement algorithm based on weighted guided filtering. J. Guilin Univ. Electron. Sci. Technol. **36**(02), pp. 113–117 (2016). https://doi.org/10.16725/j.cnki.cn45-1351/tn.2016.02.006

A Recognition Method of Hand Gesture Based on Stacked Denoising Autoencoder

Miao Ma[1,2(✉)], Ziang Gao[2], Jie Wu[2], Yuli Chen[2], and Qingqing Zhu[2]

[1] Key Laboratory of Modern Teaching Technology Ministry of Education, Xi'an, China
[2] School of Computer Science, Shaanxi Normal University, Xi'an, China
{mmmthp, chenyuli}@snnu.edu.cn

Abstract. In order to avoid the complex preprocessing, this paper proposes a recognition method based on stacked denoising autoencoder (SDAE), in which the structure and the strategies including the number of hidden units, the number of hidden layers, the level of noise and the regularization are carefully considered and analyzed for American Sign Language Dataset (ASL). Specifically, with the increasing number of hidden units and hidden layers, the optimal structure of SDAE is gradually determined, whose performance is simply measured by the recognition accuracy on testing samples. And then, the influences of the noisy strength and the regularization methods on the performance of the designed SDAE are analyzed and compared. Finally, an effective SDAE network is suggested for ASL Dataset. Experiment results show that, compared with stacked autoencoder (AE), deep belief network (DBN) and convolutional neural network (CNN) etc., the designed SDAE shows a better performance, the accuracy in ASL Dataset is up to 98.07% while the training time is reduced to 1 h.

Keywords: Gesture recognition · Stacked denoising autoencoder
Network structure · Regularization

1 Introduction

As one of the most widely used communication methods in daily life, hand gesture has become one of the most important ways for human-computer interaction. Classically, the gesture recognition procedure could be divided into two stages: preprocessing (such as segmentation, location etc.) and recognition.

In 2010, Vincent et al. proposed a stacked denoising autoencoder in which a deep neural network was formed with the stacking of multiple Denoising Autoencoders (DAEs) to extract useful features [1]. Owing to the effect of the autoencoder on the representation of information, it has been widely used in many aspects. However, there are few analyses on the construction of the network structure, especially for a specific task.

In this paper, we suggest a recognition method based on SDAE for ASL Dataset, in which the number of first hidden units was gradually increased until the optimal number of units was selected, while the number of the hidden layer was sequentially increased. By repeating the above stages, an efficient structure of SDAE for ASL Dataset was determined. Moreover, the effects of two strategies including the noisy

P. Krömer et al. (Eds.): ECC 2018, AISC 891, pp. 736–744, 2019.
https://doi.org/10.1007/978-3-030-03766-6_83

strength and the regularization on the performance of designed SDAE were analyzed and compared. And then, our designed SDAE was compared with some traditional neural networks.

2 Stacked Denoising Autoencoder

To learn some complicated functions that can represent high-level features, a deeper architecture is required, which is composed of multiple levels of non-linear operations, such as in neural network with many hidden layers. Actually, to extract some useful information from a natural image, the raw pixel should be transformed gradually to be a more abstract representation, e.g., starting from the presence of edges, the detection of more complex but local shapes, up to the parts of the image of recognized objects, and putting all these together to capture enough understanding of the scene to answer questions about it. Therefore, it often stacks multiple DAEs to extract more complex and abstract features in the form of deep network structure, which is Stacked DAE (SDAE).

Generally, the output of last layer can be considered as a representation of the input signal. So, a classifier is usually added at the end of the last layer for the application of the classification. Moreover, to make full use of the label, the back-propagation process is always adopted to fine-tune the network to make a better performance on the classification task.

Depending on the label information, the performance of the constructed SDAE can be adaptive fine-tuned via a supervised learning. Generally speaking, supervised learning can be viewed as the minimization of the object function which contains two terms, i.e. the data fidelity and the regularization. Usually, regularization helps to get a more reasonable result, so an available object function can be defined as follows:

$$\arg\min_{w} \sum_{i} \|y_i - f(x_i, w)\|^2 + \lambda\Omega(w) \tag{1}$$

where the first item indicates the sum of differences between the ground-truth *yi* and the predicted value *f*(*xi, w*), *i* represents the *i*-th sample, the second item is the regularization with which some prior information can be easily introduced to solve the problem and λ is the coefficient of the regularization. Practically, the meaning of the regularization can be considered as the constraint added on the data model or some useful prior knowledge about the data's distribution, such as the sparse [2], low rank [3] and some other properties about the problems. In the literature, there are many different regularization which are added on the different relative variations to constraint the data model, such as L1-norm, L2-norm and so on.

3 A Recognition Method of Gesture Based on Stacked Denoising Autoencoder

In this paper, given the effectiveness of SDAE, a specialized SDAE is designed for the problem of the hand gesture recognition. Note that, a set of soft-max classifiers are added to the end of the SDAE to complete the task. The structure of our method is shown in Fig. 1.

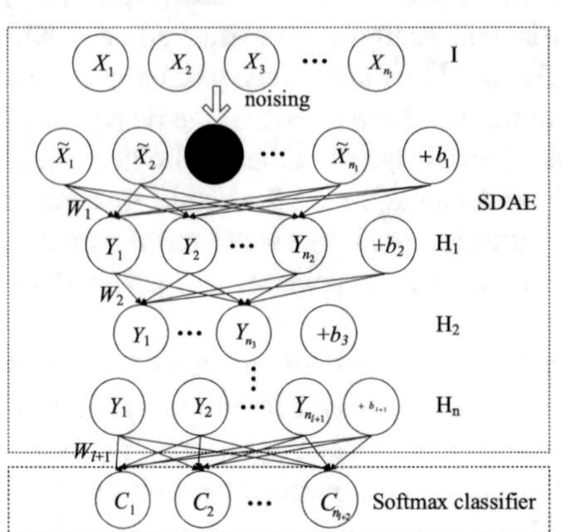

Fig. 1. Network structure diagram.

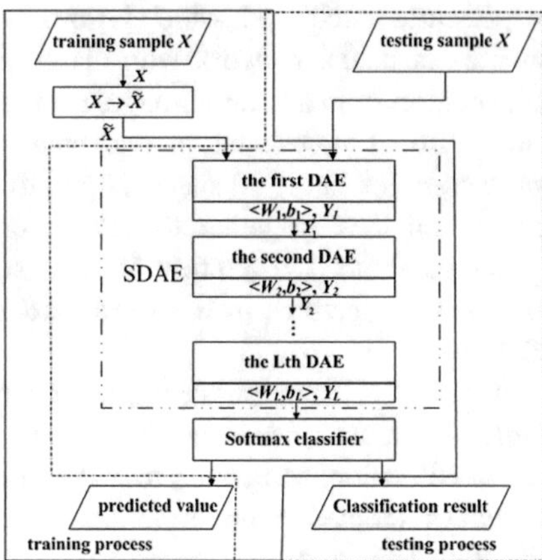

Fig. 2. Diagram of our method.

Specifically, the input is sample X with n_1 dimensions, and then the impulse noise is added on X to obtain the noised signal $\tilde{X}$ which will be inputted into a SDAE. Hereafter, to make a better explanation, the number of hidden layer is named as l while the number of units used in each layer is n_i. Moreover, $<W_i, b_i>$ are the parameters used between the layer i and layer $i + 1$. Obviously, W1 is a matrix whose size is $n_1 \times n_2$ and b_1 is a bias vector whose length is n_1. For $<W_{l+1}, b_{l+1}>$, their sizes are respectively $n_1 \times n_2$ and n_{l+1}. The corresponding flow chart is shown in Fig. 2.

In the training process (shown in Fig. 2), each input sample X is firstly corrupted with the impulse noise. Then, with the help of SAE, some features are formed while the discrimination of the feature is improved via the back-propagation with the corresponding label. When the maximum iteration number or a pre-defined error is achieved, the obtained SDAE model is adopted for testing (shown as the part surround with dot-dash line in Fig. 2). Note that there is no noise adding process for the input of testing samples.

4 Experiment on American Sign Language Dataset

American Sign Language (ASL) Dataset contains about 60000 signs indicating twenty-four English letters (except the letters ‘j’ and ‘z’), and each sign is recorded via the Kinect from 5 different persons under the different light conditions and backgrounds.

Most gestures represented by a single hand are located in the middle of the image. Due to the similar color between the hand and the other parts of the body, the similarity between the gestures of different letters (e.g. 'a', 'e', 'm', 'n', 's' and 't'), and the inter-variation of a letter expressed by different people, the gesture recognition is still a complex problem. To make a full analysis on the performance of SDAE neural network on ASL Dataset, the structure and some strategies used in SDAE neural network are studied in this section.

Note that, in the following experiments, 56400 images containing 470 samples of each sign and recorded from 5 different persons are used, among which 6000 images including 150 images of each sign are randomly selected as testing sample while the rest consists of the training sample. Each result is obtained as the average of ten-repeated experiments. Moreover, all images are transformed into gray images whose sizes are adjusted as 32×32.

4.1 The Structure of SDAE

Intuitively, the structure of SDAE is determined by the number of layers and the number of units used in each layer. Usually, using more units or a deeper architecture, the efficiency of SDAE on the representation of different objects will be enhanced. However, the bias and variance dilemma should be carefully considered. It means the structure of SDAE should be carefully designed for a special task. In this subsection, for hand gesture recognition in ASL Dataset, the structure of SDAE is analyzed, where the accuracy obtained on testing sample is used as the criterion and the number of iteration is fixed as 200. Since all input images were resized to 32×32, the number of the units in the first layer is fixed as 1024. For the rest layers, they are sequentially added while the number of units in each layer is tested when the corresponding layer has been added.

The results are given in Fig. 3, from which we find that the performance obtained by SDAE with two hidden layers (shown in Fig. 3(b)) is very similar as that of three hidden layers (shown in Fig. 3(c)). Obviously, the difference between the best performances of the two cases is less than 1%, while about one more hour is needed for the latter case to achieve the best performance. Thus, the SDAE with two hidden layers is adopted in our method. Besides, the number of the units used in each layer is also being tested in Fig. 3. For the first hidden layer (shown in Fig. 3(a)), we can see that improvement of the performance is decreased with the increasing of the number of units. By considering the efficient and effectiveness, 600 is adopted as the number of units used in the first hidden layer. For the second hidden layer (shown in Fig. 3(c)), we find that the highest performance is achieved with 200 units. Thus, in our method, 600 units are used in first hidden layer while 200 units are selected in the second hidden layer.

To make a further comparison, the number of the units in the first layer of SDAE ($n_2 = 200$) is set to 500, 600 and 700 respectively, their corresponding error rates are 8.72%, 6.32% and 7.45%. Thus, the SDAE with $n_1 = 600$ and $n_2 = 200$ is appropriate for ASL Dataset.

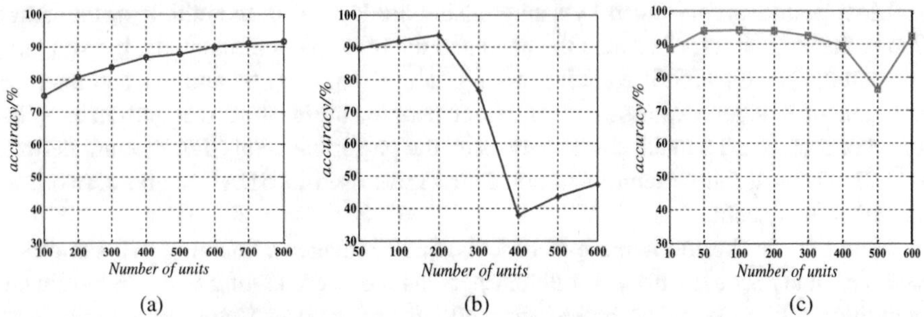

Fig. 3. Curves of the accuracy obtained with SDAE by sequentially adding (a) the first hidden layer, (b) the second hidden layer and (c) the third hidden layer. Note that, the curve is only related with the number of units of the added hidden layer while the number of units of the other layers is fixed.

4.2 Testing on the Effect of Noisy Strength

As described in [4], by adding the some noise, the robustness of the features extracted by SDAE is largely improved. However, for a special task, the strength of the noise added on the input signal is not properly analyzed. Thus, in this subsection, the influence of the noisy strength used in SDAE is analyzed, whose structure is fixed as in Sect. 4.1. The results are summarized in Fig. 4.

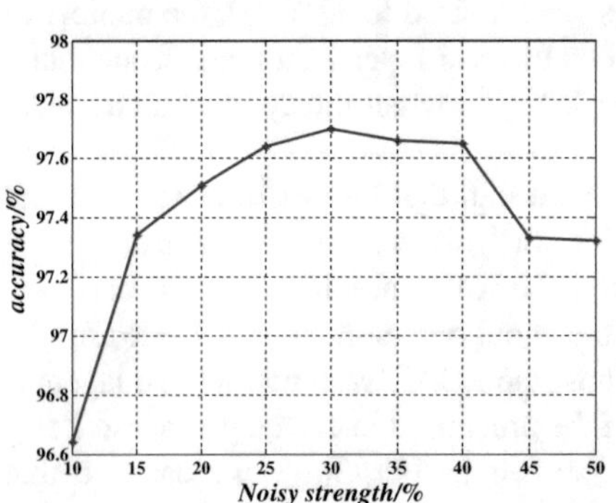

Fig. 4. Effect of noise level on accuracy.

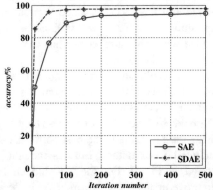

Fig. 5. Experimental results of SAE and SDAE.

In Fig. 4, the curve about the performance of our method obtained with different noisy strength (e.g. 10%, 15%, 20%, 25%, 30%, 35%, 40%, 45%, 50%) is given. Visually, the accuracy is improved with the increasing of the noise strength. After 30%, the accuracy begins to decrease. This means the noisy strength should be set properly to get a better performance. In later experiments, the noisy strength is fixed as 30%. Moreover, to prove the effectiveness of the noisy addition, the performances of SAE and SDAE are also given in Fig. 5 where the curves about the accuracy on the different iterations are given. Visually, the curve is stable after 100 interaction for SDAE, while the curve of SAE is stable after 200-interaction. Moreover, a higher accuracy is achieved by SDAE. Thus, with the addition of noisy, the performance of SAE is largely improved.

4.3 Testing on the Effect of Regularization Term

As shown in Eq. (1), to introduce the regularization, an important factor λ is always adopted to balance the function of the data fidelity and that of the regularization. So, the value of λ is firstly selected by trial and error. In the experiments, the values of λ are fixed as le-6 for L1-norm and le-5 for L2-norm. The results are shown in Fig. 6.

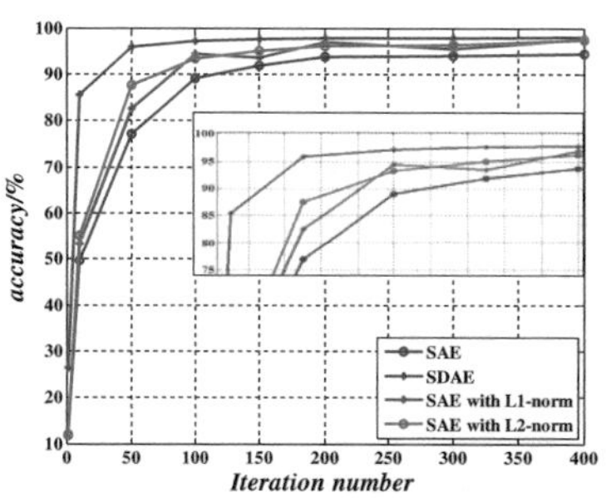

(a) Results of SAE, SDAE and Regularized SAE

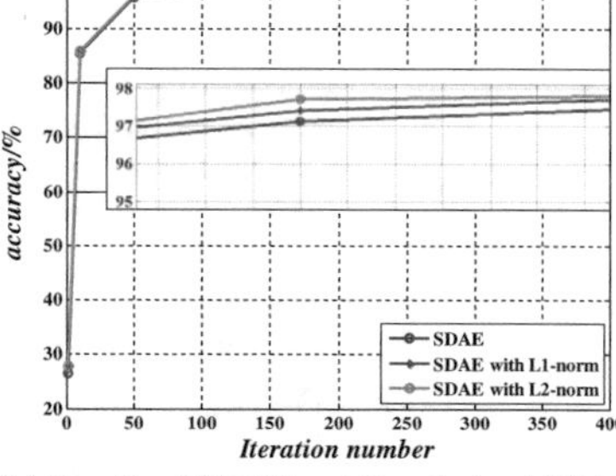

(b) Results of SDAE and Regularized SDAE

Fig. 6. Testing on the effect of regularization term.

In Fig. 6(a), the performances of SAE, SDAE, SAE with L1-norm and SAE with L2-norm are plotted with the number of iteration respectively. Obviously, by using the noisy adding process and the regularization term, the performance of SAE is clearly improved. And, the best performance is achieved by SDAE. It means the benefit of adding noise is more than that of the regularization. Moreover, the curve obtained with L2-norm looks smoother than that obtained with L1-norm. Besides, their performances is very closer after 400-iteration. This means, with more iteration, the performances obtained by the noisy addition and the regularization are the same. In Fig. 6(b), to distinguish the function between the noisy addition and the regularization, the performances of SDAE, SDAE with L1-norm and SDAE with L2-norm are plotted. With a careful comparison, we find that there is some differences among the comparison methods and the performance of SDAE with L2-norm achieves the best. This means the function of the noisy addition and that of the regularization is similar but not the same. With all above analysis, a two hide layers based SDAE is designed with L2-norm for ASL Dataset.

4.4 The Stability Test of the Networks

In this subsection, using Eq. (2), the variation of the reconstruction error (measured by mean square error, MSE) corresponding to different iterations is plotted for the designed SDAE in Fig. 7.

$$g(x+1) = 0.99 \times g(x) + 0.01 \times e \tag{2}$$

where $g(x)$ is the MSE summarized at x-th iteration, e is the MSE obtained at $(x + 1)$-th iteration. Due to the usage of weighted summarization, the variation of $g(x)$ is smoother. This is good for analyzing the performance of the designed method. Note that, the reconstruction errors obtained at the two hidden layers are give respectively in Fig. 7(a). Also, the labels' difference on the training samples is shown in Fig. 7(b).

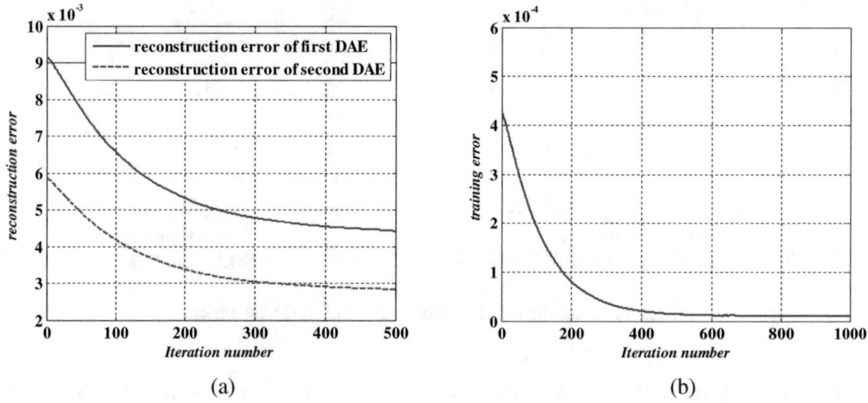

Fig. 7. With different iteration numbers, the curves about (a) the reconstruction errors at two hidden layers and (b) the labeling error of the training samples.

From Fig. 7(a), it can be seen that the reconstruction error is decreased in each hidden layer. For the first hidden layer, the reconstruction error is lower than 0.5% after 500 iterations, while the reconstruction error becomes lower than 0.3% after 500 iterations for the second hidden layer. Moreover, the labeling error (shown in Fig. 7(b)) is smoothly decreasing and becomes stable after 400-iteration.

4.5 Performance Comparison Between Various Recognition Methods

In this subsection, to make a further analysis, the accuracy obtained for each category is also given, where the number of test samples of each category is fixed as 250. Note that the accuracy is obtained as the average of ten-repeated experiments. The result is shown in Fig. 8.

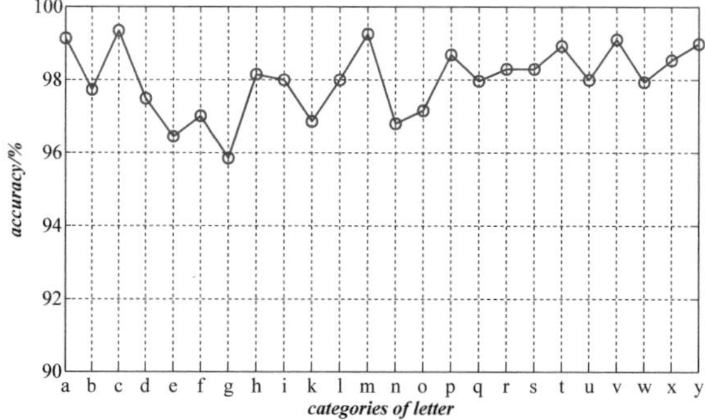

Fig. 8. The recognition rate of the corresponding letters of the 24 kinds of gestures.

From Fig. 8, it can be found that the accuracy obtained on the signs of 'e' and 'g' is lower. With a careful comparison, we find that the letter 'e' is prone to be considered as the letter 'd', while the letter 'g' is easily to be classified as the letter 'h'. The main reason is that the hand gestures about the letters 'e' and 'd' ('g' and 'h', shown in Fig. 9) are very similar to each other.

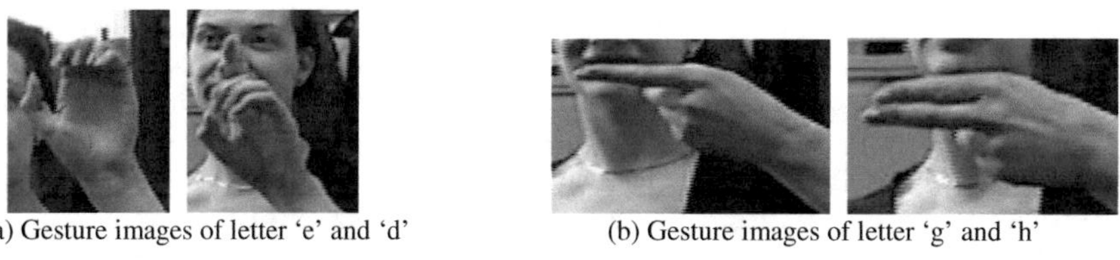

(a) Gesture images of letter 'e' and 'd' (b) Gesture images of letter 'g' and 'h'

Fig. 9. Gesture images of the letters 'e', 'd', 'g' and 'h'.

Besides, using HSF-RDF method [5], SIFT-PLS method [6], MPC method [7], DBN and CNN as the comparison methods, the performance of our method on the hand gesture recognition task in ASL image database is also analyzed. For HSF-RDF method, RGB images and depth images were captured with Kinect and OpenNI (a third-party library developed by Kinect) was used to detect and track gesture, while random forest classifier was used as classifier. In SIFT-PLS method, SIFT feature was directly extracted and partial least squares (PLS) based classifier was used for gesture recognition. For MPC method, Blob and Crop operators were used to extract the region of interest, while Sobel operator was adopted subsequently to extract the gesture region whose centroid and area are used as feature for gesture recognition. The DBN network contains 2 hidden layers whose hidden units are fixed as 600 and 200 respectively. The structure of CNN is consisted of 2 convolution layers, 2 down-sampling layers, 1 full connection layer and convolution kernel size is 5×5. The accuracy results provided by HSF + RDF, SIFT-PLS, MPC, DBN, CNN, and SDAE are respectively 75%, 71.51%, 90.19, 84.65%, 88.22%, and 98.07%. On the other hand, the time consuming of SDAE is shorten to 1.5 h, while DBN and CNN needs 1.8 h and 7 h respectively.

5 Conclusion

In this paper, for the hand gesture recognition task on ASL Dataset, the strategies used in SDAE, including the number of layers, the number of hidden units, the level of corruption and the regularization, are carefully analyzed and a specially designed SDAE is proposed. Using the SDAE method, the accuracy on the testing samples achieves 98.07%. Note that, the gray image is only used in the experiments of this paper. Thus, the color information and the depth information of the image will be studied in our future work, which will be a multi-channel based SDAE.

Acknowledgment. This work is supported by National Natural Science Foundation of China under grants 61501286, 61501287, 61601274 and 61877038, the Natural Science Basic Research Plan in Shaanxi Province of China (2018JM6068), the Fundamental Research Funds for the Central Universities of Shaanxi Normal University (GK201703054 and GK201703058) and The Key Science and Technology Innovation Team in Shaanxi Province of China (2014KTC-18).

References

1. Vincent, P., Larochelle, H., Bengio, Y., Manzagol, P.: Extracting and composing robust features with denoising autoencoders. In: Proceedings of International Conference on Machine Learning, pp. 1096–1103 (2008). https://doi.org/10.1145/1390156.1390294
2. Zhou, X., Zhu, M., Leonardos, S., Daniilidis, K.: Sparse representation for 3D shape estimation: a convex relaxation approach. IEEE Trans. Pattern Anal. Mach. Intell. **39**, 1648–1661 (2017). https://doi.org/10.1109/tpami.2016.2605097
3. Zhang, Z., Mei, X., Xiao, B.: Abnormal event detection via compact low-rank sparse learning. IEEE Intell. Syst. **31**, 29–36 (2016). https://doi.org/10.1109/MIS.2015.95
4. Vincent, P., Larochelle, H., Lajoie, I., Bengio, Y., Manzagol, P.: Stacked denoising autoencoders: learning useful representations in a deep network with a local denoising criterion. J. Mach. Learn. Res. **11**, 3371–3408 (2010)
5. Pugeault, N., Bowden, R.: Spelling it out: real-time ASL fingerspelling recognition. In: IEEE International Conference on Computer Vision Workshops, pp. 1114–1119 (2011). https://doi.org/10.1109/iccvw.2011.6130290
6. Estrela, B., Cámara-Chávez, G., Campos, M., Schwartz, W., Nascimento, E.: Sign language recognition using partial least squares and RGB-D information. In: Proceedings of the IX Workshop de Visao Computacional (2013)
7. Pansare, J., Gawande, S., Ingle, M.: Real-time static hand gesture recognition for American Sign Language (ASL) in complex background. J. Signal Inform. Process. **3**, 364–367 (2015). https://doi.org/10.4236/jsip.2012.33047

Long-Term Tracking Algorithm Based on Kernelized Correlation Filter

Na Li[1,2,3(✉)], Lingfeng Wu[1,2,3(✉)], and Daxiang Li[1,2,3]

[1] Center for Image and Information Processing, Xi'an University of Posts and Telecommunications, Xi'an 710121, China
linall4@xupt.edu.cn, lfwu1822@163.com

[2] Key Laboratory of Electronic Information Application Technology for Scene Investigation, Ministry of Public Security, Xi'an 710121, China

[3] International Joint Research Center for Wireless Communication and Information Processing, Xi'an 710121, China

Abstract. KCF (Kernelized Correlation Filter) is a classical tracking algorithm based on correlation filter, which has good performance in short-term tracking. But when the object is partially or fully occluded, or disappeared in the view, KCF doesn't work well. In this paper, a long-term tracking algorithm based on KCF is proposed. HOG (Histogram of Oriented Gradient) and LAB three-channel color information are employed to represent the object, and a re-detection module is added into the KCF tracking procedure. The peak ratio is introduced to control the start of the re-detection module, and a correlation filter model based on SURF feature points is re-learned to continuously track the occluded object. Experimental results on OTB dataset show that our algorithm has higher tracking accuracy than other five trackers, and is suitable for long-term tracking.

Keywords: Long-term object tracking · Correlation filter · Re-detection

1 Introduction

Object tracking technique is widely used in intelligent surveillance, traffic monitor, vehicle navigation and unmanned aerial vehicle, and it is one of the hot topics in computer vision research. The tracking of moving targets is to acquire the trajectory of the tracked objects in a continuous video sequence, so that it can be further analyzed and processed in high level. The target object's actual coordinate position of the first frame of the image is initialized for the candidate position of the target calculation in the next frame. In the process of motion, the target cannot always be in an ideal state, and some appearance changes may occur in posture, scale and illumination, as well as occlusion or deformation, often causing certain difficulties for the object tracking. Recently, the research focus of the tracking algorithm still lies in how to better solve the problem of object tracking mismatch and make the algorithm more robust.

Based on the particle filter framework, Wang et al. measured the reliability of candidate targets via the typical correlation between image sub-regions, and solved the global information appearance model sensitive to external object occlusion [1]. They

P. Krömer et al. (Eds.): ECC 2018, AISC 891, pp. 745–755, 2019.
https://doi.org/10.1007/978-3-030-03766-6_84

then proposed a self-refactoring algorithm that splits and merges the particle filter tracker [2], reducing the probability of losing targets, but unable to recover the target objects and continue tracking after losing. Jiang et al. proposed a soft-feature-based target predecessor tracking method to achieve long-term tracking, but the target motion trajectory and its frequency domain transform need to be continuous, and thus it is not suitable for object tracking applications under high-speed motion scenarios [3]. Therefore, how to effectively solve occlusion and achieve long-term object tracking is still the keys of current object tracking technique.

We provide a re-detection mechanism to solve the problem of occlusion, and the experimental results show that the algorithm can deal with the problem effectively. In this paper, Sects. 2 and 3 provided the related work and principles of proposed long-term tracking algorithm, respectively; Sect. 4 quantitatively compared them experimentally; and Sect. 5 summed up this work.

2 Related Work

2.1 Correlation Filter

The classifier training process is a ridge regression problem. The ridge regression function behaves as well as the SVM for searching a regression function $f(z) = \omega^T z$ that minimizes the error function as Eq. (1):

$$\min_{\omega} \sum_{i} (f(x_i) - y_i)^2 + \lambda \|\omega\|^2 \tag{1}$$

In Eq. (1), x_i is the training sample, y_i is the regression label, ω is the column vector weight coefficient, and λ is a regularization parameter ($\lambda \geq 0$). Equation (1) has a closed-form solution as illustrated in Eq. (2).

$$\omega = \left(X^T X + \lambda I\right)^{-1} X^T y \tag{2}$$

The introduction of the kernel function transforms the solution of the weight vector ω into the coefficient α, and the optimal solution is illustrated in Eq. (3)

$$\alpha = (K + \lambda I)^{-1} y \tag{3}$$

Where X is the sample matrix, I is the identity matrix, y is the regression value corresponding to each sample, and K is the kernel function matrix

With the nature of the circulant matrix, the above solution process is transformed to the Fourier domain, and the final result is illustrated in Eq. (4)

$$\hat{\alpha} = \frac{\hat{y}}{\hat{k}^{xx} + \lambda} \tag{4}$$

In which, the symbol ^ represents the frequency domain transform; $\hat{k}^{xx}$ is the first row element vector of the autocorrelation Gaussian kernel matrix of the training sample x.

The Minimum Output Sum of Squarel Error (MOSSE) is a new type of correlation filter originally proposed by [4] for object tracking. Afterwards, the papers [5, 6] proposed CSK tracker and KCF tracker, respectively. CSK introduced a circulant matrix to generate samples, which aided to train a better classifier. KCF made further improvements on the basis of CSK, which introduced multi-channel features and HOG features.

KCF solves the problem of too few samples by using circulant matrix displacement. At the same time, it introduces kernel regression to perform training and detection in the frequency domain, which greatly reduces the amount of computation with speed up to several hundred FPS. However, there are two obvious disadvantages: First, there is no scale estimation for the size of the fixed target template. If the target size changes, the target template drifts easily and the tracking error fails to be accumulated. Second, the object cannot be well covered in the tracking process. For the first deficiency of KCF, both SAMF [7] and DSST [8] tracking algorithms proposed solutions. SAMF fuses HOG with CN features, using scale pool technology to estimate the optimal target size. The tracking accuracy is significantly better than KCF tracker, but the speed is slower, about one-tenth that of KCF. DSST tracker decomposes the object tracking task into two parts: translation and scale estimation. After finding the best target translation, a one-dimensional scale pyramid classifier is trained to independently estimate the optimum scale. The algorithm has the advantages of excellent performance and good portability, and its disadvantage is that the speed is slow and unsatisfied with the real-time requirements.

2.2 Long-Term Object Tracking Algorithm

Reference [9] proposed a long-term Tracking-Learning-Detection (TLD) framework, which mainly contains tracking, learning and detection. The improved tracking mechanism continuously updates the tracking module and the detection module, effectively solving the deformation and partial occlusion of the target objects during the tracking process. The references [10–12] made full use of spatio-temporal context information and utilized the Bayesian framework to model the temporal and spatial relationship between the object and its local context. The confidence map of the target position in the next frame based on the statistical correlation obtained depends largely on the spatio-temporal relationship between the foreground and the background. When the target is completely occluded, this spatio-temporal relationship is not well established. The reference [13] introduced a random fern classifier, which could re-detect the object in the case of failure to meet long-term tracking, but the calculation is time-consuming and the speed is too slow to meet the real-time application requirements.

3 Our Approach

3.1 Feature Extraction and Scale Estimation

The color feature owns rotation and deformation robustness [14]. Our algorithm uses a 34-dimensional fusion feature and stitches the three-channel information of the LAB color space based on the 31-dimensional HOG feature. Compared with the RGB color model, the LAB has a wider color gamut and is more colorful. It can fully utilize the color information of the video frame and effectively handle the tracking of the object in scenes such as plane rotation and deformation.

The reference [7] proposed an adaptive scale estimation method for defining a scaling pool. In [8], a more accurate scale-space is proposed. The scale estimation is performed independently using a one-dimensional correlation filter with 33-layer pyramid features. In this paper, the method of adding a scale pool $S = \{1, 0.985, 0.99, 0.995, 1.005, 1.01, 1.015\}$ with seven scales is also used. The fixed target template is multiplied by the factors in the scale pool in order to obtain seven candidate areas with different scales and seven maximum response values $y_{\hat{s}}$. Selecting a maximum value $\max y_{\hat{s}}$ in it to find out the position of the corresponding target center point and updating the coordinate position.

Using the Eq. (5) for weighted update when updating normally.

$$\begin{cases} \alpha^t = (1-\eta)\alpha^{t-1} + \eta\hat{\alpha} \\ x^t = (1-\eta)x^{t-1} + \eta\hat{x} \end{cases} \tag{5}$$

Where η is the model update rate ($0 < \eta < 0.1$). The larger the weight, the faster the model update; x^{t-1} and α^{t-1} are the templates and coefficients of the target appearance model of the previous frame, respectively; $\hat{x}$ and $\hat{\alpha}$ are the target matching templates and coefficients of the current frame, respectively; x^t and α^t are updated filter templates and coefficients, respectively.

3.2 Re-detection Module

Once the object undergoes occlusion, resulting in deviations in the tracking results, the re-detection module should be considered starting to effectively track the object over a long period of time. At this point, the characteristics of the object can no longer be well described. In this paper, SURF feature points are used to describe the target appearance, and a new correlation filter is generated as a candidate model to continue tracking the object. After the object is restored to the field of vision, replacing it with the original relevant model to continue tracking. SURF is a local feature point, which has scale invariance and robustness, and is much faster than SIFT feature. When the object begins to be occluded, the key points of most of the object and its surrounding background area do not change significantly, which can describe the current target appearance very well and make up for the deficiencies of the previous model. If it is known that the size of the object search region is R, a SURF appearance model with a fixed 64-dimensional vector is obtained. When training the classifier, a regression function $h(z) = \omega^T z$ is also required to find the smallest error function.

The maximum response value max $y_{\hat{s}}$ reflects the reliability of the tracking result. According to the change of the peak ratio K_0 of the two frames before and after the confidence map, a threshold T_r is set to determine whether the credibility of the current object tracking result is lower, so as to determine whether it is necessary to start the re-detection module. When the object is tracked normally, the confidence maps of the two frames before and after change little, and the peak ratio should be within the range of (0, 1). If $K_0 < T_r$, indicating that the fluctuation degree of the confidence map is large and the credibility is low. In this case, the re-detection module should be considered starting.

Figures 1 and 2 give the maximum response curve and peak ratio curve of the tracking result both using the video sequence Girl2 as an example without starting the re-detection mechanism. We can see that the change trend of the maximum response value and the change of the peak ratio are similar. In Fig. 1(a), there is a dashed line around the 110th frame. At this time, the maximum response value suddenly drops to a certain threshold, and the object is likely to be at the moment of occlusion. The actual situation is shown in Fig. 1(b), in which the object is a little girl. The girl begins to experience occlusion at the 105th frame, and it is already in the full occlusion at the 111th frame. By the way, the six color tracking boxes represent the tracking results of six different algorithms in the frame, respectively (see Fig. 5 for the specific correspondence). Besides, in the range of the same frame number in Fig. 2, the peak ratio also has a significant downward trend, indicating the change of the peak ratio is related to the object occlusion.

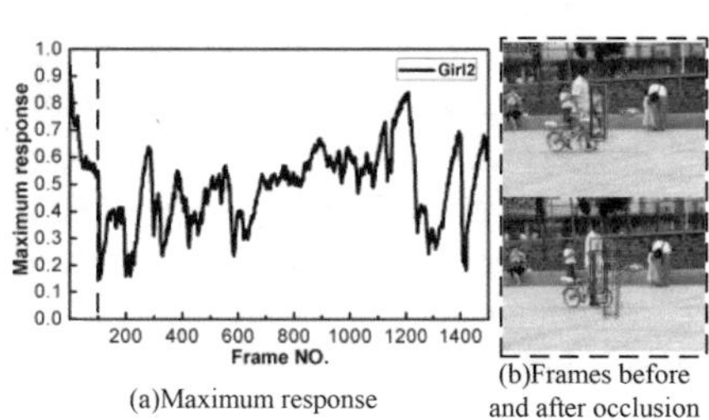

(a)Maximum response

(b)Frames before and after occlusion

Fig. 1. Maximum response curve

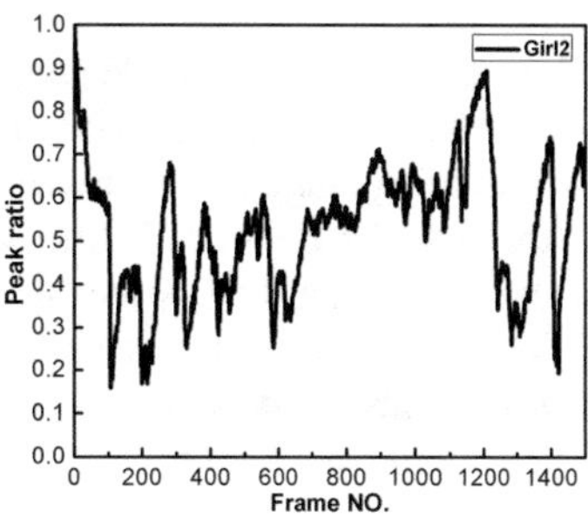

Fig. 2. Peak ratio curve

3.3 Tracking Procedure

We use the fusion features of HOG and LAB color histograms, increasing the scale estimation and re-detection module, which can effectively ensure the long-term tracking of the object. The specific algorithm steps are as follows:

Algorithm 1: Proposed tracking algorithm.

Input: The first frame of a video sequence.

Output: The target center point position pos_t for each frame in the video sequences.

Translation Estimation:

1: Initializing the target size and center point position for the first frame of the image and generating a correlation filter.

2: Extracting the features of the tth ($t \geq 2$) frame image and training the relevant models.

Scale Estimation:

3: With the seven scale factors of the scale pool, calculating seven maximum response values $y_{\hat{s}}$ and finding out the $\max y_{\hat{s}}$ and the corresponding target center position pos.

Target Re-detection:

4: Determining whether $K_0 < T_r$. If true, then start re-detection module; otherwise, go to step (6).

5: The SURF feature of the previous frame search region is extracted to perform related operations, a new model is established, the relevant frame is trained using the current frame, and the maximum response value $\max y_{\hat{s}}$ and its corresponding target center point position pos are obtained.

Model Update:

6: Updating the model using (5).

4 Experiments

All experiments were conducted on computers with Intel Xeon (CPU E3-1220 v3 @ 3.10 GHz, 8 GB RAM, 64-bit operating system) using the C and MATLAB mixed programming language. From the OTB [15] dataset, eight challenging sequences (Bolt, Coke, Girl, Girl2, Human9, Liquor, Singer1 and Soccer) were selected for comparison. At the same time, five state-of-the-art trackers (KCF [6], SAMF [7], CN [16], DSST [8], and CSK [5]) were selected for comparison experiments. The code is downloaded from the author's personal home page.

Assuming the target size is $w \times h$, the search region size is $2.5 \times w \times h$, the regularization parameter is set to $\lambda = 10^{-4}$, the learning rate in (5) is set to $\eta = 0.02$, and the scale pool is set to $S = \{1, 0.985, 0.99, 0.995, 1.005, 1.01, 1.015\}$. All test sequences parameters remained unchanged during the experiment.

4.1 Threshold Selection

We explored the setting of the threshold T_r and presented the histograms of the eight sequences whose precision is taken as the average as shown in Fig. 3. It can be seen that when $T_r = 0.231 (0.225 \leq T_r \leq 0.237)$, the tracking performance is the best.

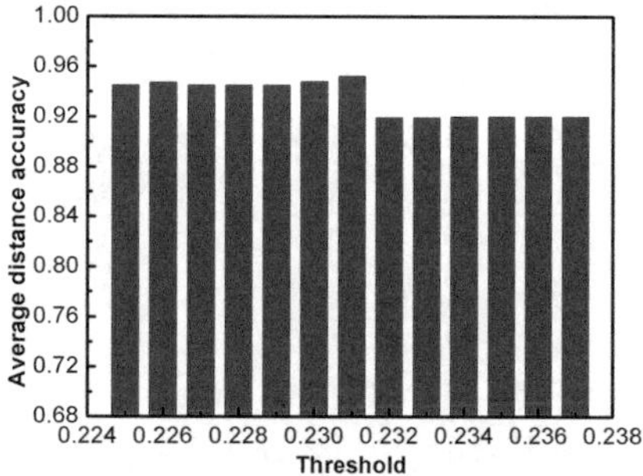

Fig. 3. Average distance accuracy for different thresholds

4.2 Quantitative Evaluation

Figure 4 shows the distance precision (DP) [15] of the six trackers on the eight video sequences. The x-axis indicates the center position error threshold, and the y-axis indicates the tracking accuracy. When the threshold is 20 pixels, the tracking accuracy of each tracker is given in the legend. It can be seen that proposed MKCF method can achieve the optimal or suboptimal results, and its average DP is shown in Table 1. Among them, the optimal and suboptimal results are shown in bold and italic, respectively.

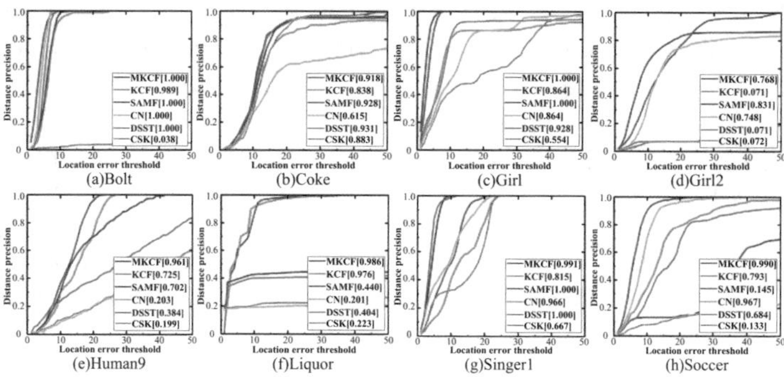

Fig. 4. Tracking accuracy curve

Table 1 shows that the average DP of the tracking results of our algorithm on the eight test sequences is 0.952, which is 19.3% higher than that of the KCF tracker. Compared with SAMF, CN, DSST and CSK, the MKCF method improves by 19.6%, 25.6%, 27.7%, and 60.6%, respectively.

The video processing time for the six trackers is shown in Table 2. Among them, the optimal and suboptimal results are shown in bold and italic, respectively.

Table 2 shows that the CSK tracker uses the original pixels to describe the appearance of the target, and the processing speed is the fastest, reaching 224.15 FPS.

Table 1. Average distance accuracy

Sequences	Algorithms					
	KCF	SAMF	CN	DSST	CSK	MKCF
Bolt	*0.989*	**1.000**	**1.000**	**1.000**	0.038	**1.000**
Coke	0.838	*0.928*	0.615	**0.931**	0.883	0.918
Girl	0.864	**1.000**	0.864	*0.928*	0.554	**1.000**
Girl2	0.071	**0.831**	0.748	0.071	0.072	*0.768*
Human9	*0.725*	0.702	0.203	0.384	0.199	**0.961**
Liquor	*0.976*	0.440	0.201	0.404	0.223	**0.986**
Singer1	0.815	**1.000**	0.966	**1.000**	0.677	*0.991*
Soccer	0.793	0.145	*0.967*	0.684	0.133	**0.990**
Average	*0.759*	0.756	0.696	0.675	0.346	**0.952**

Table 2. The average frame rate (unit: FPS)

Algorithms	KCF	SAMF	CN	DSST	CSK	MKCF
Average	*164.86*	13.65	104.46	25.22	**224.15**	19.16

Both the KCF tracker and the CN tracker use a single feature to describe the appearance of the target with the speed of 164.86 and 104.46 FPS, respectively, and it is second only to the CSK tracker. The DSST tracker has increased the 33-dimensional scale estimation, obtaining the speed of 25.22 FPS with a large amount of calculation. Based on the increased scale estimation, the SAMF tracker uses HOG to describe the appearance of the target and feature fusion, which is the slowest, only 13.65 FPS. The proposed MKCF tracker considers the LAB color histogram, scale estimation and re-detection module whose speed reaches 19.16 FPS, meeting the real-time requirement. This is because the scale estimation of the algorithm in this paper is less time-consuming than the SAMF tracker due to the reduction of computational complexity. Our MKCF algorithm achieves better tracking results without sacrificing time and can better balance the tracking accuracy and time loss.

4.3 Qualitative Evaluation

The qualitative analysis was performed on video sequences containing various complex conditions such as illumination variation, scale variation, occlusion, deformation, rotation, and motion blur. The results of the tracking of each algorithm are shown in Fig. 5.

The operation results of six trackers on the test sequences are shown in Fig. 5. The improved algorithm can handle challenges in a variety of complex scenarios and achieve better tracking results.

Take Girl, Girl2, Human9, Liquor, and Soccer in Fig. 5 as examples for detailed analysis. The object (Girl) in the video has undergone many rotations during the whole

Fig. 5. Comparison of experimental results of each algorithm

tracking process (e.g. the 80th–127th frames, the 162th–295th frames are 360° rotation, and the 296th–339th frames are in-plane rotation), only our MKCF tracker and the SAMF tracker effectively dealing with this kind of problem. The remaining 4 algorithms lead to complete failure as long as there are large deviations that result in excessive accumulation of errors. The object (Girl2) mainly experienced two full occlusions (e.g. the 105th–134th frames, the 1380th–1403rd frames, etc.). After the first cover completely left the target, except our algorithm continuing to track the object normally, the remaining five trackers misidentified the cover as the target. The object (Human9) undergoes large-scale changes such as deformation, motion blur, and fast motion (e.g. the 140th frames, the 217th frames, the 254th frames, etc.), and the general algorithm is difficult to handle multiple challenges simultaneously. From Fig. 4, we can see that the average test results of all trackers in the video Human9 are 0.961, 0.725, 0.702, 0.203, 0.384 and 0.199, respectively. The CSK, CN and DSST tracker successively failed to track the object. The tracking results of the KCF and the SAMF tracker were better, but our proposed algorithm was the best, owning the highest tracking accuracy. The object (Liquor) experienced insufficient visibility and was occluded several times by other bottles (e.g. the 350th–369th frames, the 383th–409th frames, the 501st–511th frames, the 722th–737th frames, the 767th–780th frames, etc.). The CN tracker is unable to cope when the object is occluded for the first time. The CSK tracker cannot deal with the problem that the object begins to have insufficient vision to reappears in the field of view, while the remaining trackers can continue to track the target normally. When the object experiences a third partial occlusion by a bottle with similar background, the DSST tracker mistakenly uses the cover as a object for tracking. When the same cover occludes the object again, the SAMF tracker also suffers from the dirfting problem. Only our MKCF tracker and the KCF tracker keep track of the normal object. The object (Soccer) experienced various complex scenes such as occlusion, fast motion, and similar background interference (e.g. the 54th–87th frames, the 90th–100th frames, the 101st–213th frames, etc.). During the rapid movement of the object and occlusion, only the proposed algorithm and the CN tracker can track the object normally. When the object has similar background interference, the SAMF tracker and the CSK tracker completely lose the ability to track the object normally, while our algorithm and the other three trackers can still better cope with the current situation. Combining with Table 1 and Fig. 2, it can be seen that the re-detection module of the MKCF algorithm can effectively recover after occlusion, and has good tracking results for deformation, rotation, fast motion, motion blur, background clutter and other scenes.

According to the quantitative and qualitative analysis, the proposed MKCF tracker has greatly improved performance, and the tracking performance is the best, compared with the other five trackers. At the same time, the processing speed is 19.16 FPS, which is closer to real-time tracking. The proposed MKCF algorithm greatly improves the overall performance of the tracking results, instead sacrificing the time to achieve the purpose for improving the tracking accuracy. It effectively balances the tracking accuracy and robustness of this group of contradictions, as well as takes into account the real-time nature of the algorithm.

5 Conclusions

In this paper, a long-term tracking algorithm based on KCF is proposed. When occlusion occurs, a re-detection module based on SURF is started, which takes over the tracking task. Moreover, the re-detection algorithm can be added into the existing trackers, which is helpful to improve the tracking performance. However, our algorithm is insufficient to deal with background clutter and low resolution. Our future work is to make full use of spatio-temporal context to further improve tracking performance.

Acknowledgments. The work is sponsored by the Shaanxi International Cooperation Exchange Funded Projects (2017KW-013, 2017KW-016), Graduate Creative Funds of Xi'an University of Posts and Telecommunications (CXJJ2017004).

References

1. Wang, Y.X., Zhao, Q.J., Zhao, L.J.: Robust object tracking based on FREAK and P3CA. Chin. J. Comput. **38**, 1188–1201 (2016). https://doi.org/10.11897/SP.J.1016.2015.01188
2. Wang, Y.X., Zhao, Q.J., Cai, Y.M., et al.: Tracking by auto-reconstruction particle filter trackers. Chin. J. Comput. **39**, 1294–1306 (2016). https://doi.org/10.11897/SP.J.1016.2016.01294
3. Jiang, W.T., Liu, W.J., Yuan, H., et al.: Research of object tracking based on soft feature theory. Chin. J. Comput. **39**, 1334–1355 (2016). https://doi.org/10.11897/SP.J.1016.2016.01334
4. Bolme, D.S., Beveridge, J.R., Draper, B.A.: Visual object tracking using adaptive correlation filters. In: Computer Vision and Pattern Recognition, pp. 2544–2550. IEEE (2010). https://doi.org/10.1109/cvpr.2010.5539960
5. Henriques, J.F., Rui, C., Martins, P., et al.: Exploiting the Circulant Structure of Tracking-by-Detection with Kernels. Lecture Notes in Computer Science, vol. 7575, 702–715 (2012). https://doi.org/10.1007/978-3-642-33765-9_50

6. Henriques, J.F., Caseiro, R., Martins, P., et al.: High-speed tracking with kernelized correlation filters. IEEE Trans. Pattern Anal. Mach. Intell. **37**, 583–596 (2015). https://doi.org/10.1109/TPAMI.2014.2345390
7. Li, Y., Zhu, J.A.: A scale adaptive kernel correlation filter tracker with feature integration. In: European Conference on Computer Vision, pp. 254–265. Springer, Cham (2014). https://doi.org/10.1007/978-3-319-16181-5_18
8. Danelljan, M., Häger, G., Fahad, S.K., et al.: Accurate scale estimation for robust visual tracking. In: British Machine Vision Conference, pp. 1–5. BMVA Press (2014). https://doi.org/10.5244/c.28.65
9. Kalal, Z., Mikolajczyk, K., Matas, J.: Tracking-learning-detection. IEEE Trans. Pattern Anal. Mach. Intell. **34**, 1409–1422 (2012). https://doi.org/10.1109/TPAMI.2011.239
10. Zhang, K., Zhang, L., Yang, M.H., et al.: Fast visual tracking via dense spatio-temporal context learning. In: European Conference on Computer Vision, pp. 127–141. Springer, Cham (2014). https://doi.org/10.1007/978-3-319-10602-1_9
11. Xu, J.Q., Lu, Y.: Robust visual tracking via weighted spatio-temporal learning. Acta Automatica Sinica **41**, 1901–1912 (2015). https://doi.org/10.16383/j.aas.2015.c150073
12. Liu, W., Zhao, W.J., Li, C.: Long-term visual tracking based on spatio-temporal context. Acta Optica Sinica **36**, 179–186 (2016). https://doi.org/10.3788/AOS201636.0115001
13. Ma, C., Yang, X., Zhang, C., et al.: Long-term correlation tracking. In: Computer Vision and Pattern Recognition, pp. 5388–5396. IEEE (2015). https://doi.org/10.1109/cvpr.2015.7299177
14. Zhao, G.P., Shen, Y.P., Wang. J.Y.: Adaptive feature fusion object tracking based on circulant structure with kernel. Acta Optica Sinica **37**, 0815001. https://doi.org/10.3788/aos201737
15. Wu, Y., Lim, J., Yang, M.H.: Online object tracking: a benchmark. In: IEEE Conference on Computer Vision and Pattern Recognition, pp. 2411–2418. IEEE (2013). https://doi.org/10.1109/cvpr.2013.312
16. Danelljan, M., Khan, F.S., Felsberg, M., et al.: Adaptive color attributes for real-time visual tracking. In: Computer Vision and Pattern Recognition, pp. 1090–1097. IEEE (2014). https://doi.org/10.1109/cvpr.2014.143

Food Recognition Based on Image Retrieval with Global Feature Descriptor

Wei Sun[1(✉)] and Xiaofeng Ji[2]

[1] School of Aerospace Science and Technology, Xidian University, No. 2 South Taibai Road, Xi'an 710071, China
wsun@xidian.edu.cn

[2] School of Aerospace Science and Technology, Xidian University, Xi'an 710100, China

Abstract. This paper proposes a simple and effective non-parametric approach to solve the problem of food images parsing and label images with their categories. Firstly, the proposed approach works by six types of global image features: CEDD, FCTH, BTDH, EHD, CLD and SCD to matching with global image descriptors, labeling image with their categories, and the distance for each descriptor are fused to get the likelihood probability of each class, then efficient Markov random field (MRF) optimization is proposed for incorporating neighborhood context, besides optimization minimization are used Iterated Conditional Modes (ICM) algorithms. And this paper also introduces a non-parametric, data-driven approaches framework. This approach requires no training, just prior distribution and joint distribution are taken into account, and it can easily scale to data sets with tens of thousands of images and hundreds of labels. At last, the experiments show that the proposed method is significantly more accurate and faster at identifying food than existing methods.

Keywords: Automatic food recognition · Global image descriptors
Markov random field

1 Introduction

Diet is one of the most important and regular activities, healthy dietary habits have been highly emphasized in recent years. If nutrition information can be automatically extracted from images, it will relieve the user from the burden of finding and recording such information manually. In recent years, interest on capturing and processing one's daily lives is growing [1–3] and such research field is called "life-log". People are logging their lives in various ways. But it is not easy to find the images with food inside, because it is huge and too redundant.

As an alternative to manual logging, we will investigate methods for automatically recognizing foods based on their appearance. Unfortunately, there has been relatively little work on the problem of food recognition. These include approaches based on local features such as the SIFT [11] descriptor or global features such as color histograms or GIST [12] features; Yang et al. [5] introduced the PFID dataset for food recognition and two baseline algorithms: color histogram and bag of SIFT [11] features.

P. Krömer et al. (Eds.): ECC 2018, AISC 891, pp. 756–764, 2019.
https://doi.org/10.1007/978-3-030-03766-6_85

Felzenszwalb proposes techniques [7] to represent a deformable shape using triangulated polygons. Leordeanu et al. [10] proposed an approach that recognizes category using pairwise interaction of simple features. A recent work [8] learns a mean shape of the object class based on the thin plate spline parameterization. These approaches all require detecting meaningful feature points. But precise features like these are typically not available in images of food.

Recently, several researchers have begun advocating nonparametric, data-driven approaches. Liu et al. [15] try to retrieve the most similar training images and transfer the desired information from the training images to the query. However, inference via SIFT Flow is currently very complex and computationally expensive. In this paper, we used the global descriptors CCD and fused to the likelihood probability of each class. The main contribution of the food recognition can be given as following:

1. We implement a nonparametric solution without training, makes use of a retrieval set of food images. It can easily scale to ever larger image collections and sets of labels.
2. With some statistics and analysis on food dataset, we use forms of context that can be cast in an MRF framework and optimization by ICM algorithms [17, 18].

The remainder of this paper is organized as follows. The overview and some details of this system are explained in Sect. 2. Section 3 discusses the details of feature selection, match and the algorithm of food recognition; Sect. 4 describes the experimental methodology and presents food recognition results on the dataset. Finally, Sect. 5 presents our conclusions and proposes directions for future work in this area.

2 System Overview

2.1 Wearable Device

We have developed a wearable computing system to recognize food for calorie monitoring (called "eButton" [4–6]), as shown in Fig. 1, which is embedded with gyroscope, accelerometer, GPS and camera.

(a) (b)

Fig. 1. (a) eButton Prototype; (b) A person wearing an eButton during eating

2.2 Framework of the Proposed System

We implement our food recognition based on image retrieval technical, executes the search using the similarity matching technique recommended for each descriptor. Then arranges the images contained in the index file according to their proximity to the query image, and presents the results. It's also used for improving retrieval results from large probably distributed inhomogeneous databases; Fig. 2 shows this system.

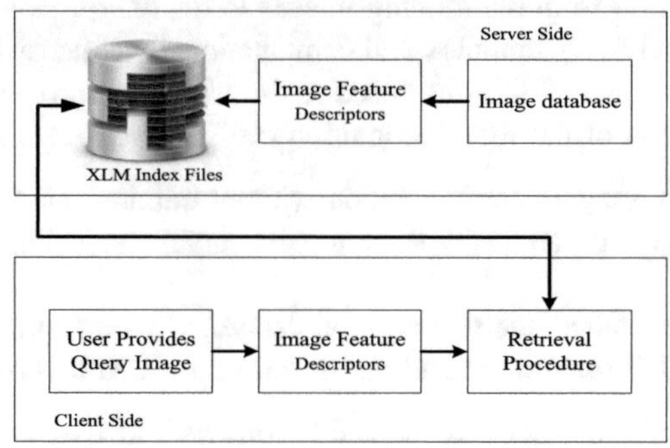

Fig. 2. The proposed framework for food recognition

It presents content-based image description and retrieval, covering image preprocessing, features extraction, similarity matching and evaluation. an image search from XML-based index files, extracting the comparison features in real-time.

3 Proposed Food Recognition Algorithm

We implement a sum of low level features (color, texture and shape) in visual similarity image retrieval. For the food recognition we used Bayes' theorem and introduce a nonparametric, data-driven approaches framework in place of the training approach.

3.1 Descriptors and Feature Distance

Compact Composite Descriptors (CCDs) contains 3 different types: CEDD, FCTH and BTDH. To extract texture information, CEDD uses MPEG-7 to form 6 texture areas. FCTH form 8 texture areas. When an image block interacts with the system it simultaneously goes across 2 units: color unit and texture unit see Fig. 3.

Classification employ a supervised, distribution-free approach known as the minimum (mean) distance classifier. In this study, we input an image and based on certain global features, then the system brings up similar images. So, we use the following method to calculate the distance between two descriptors

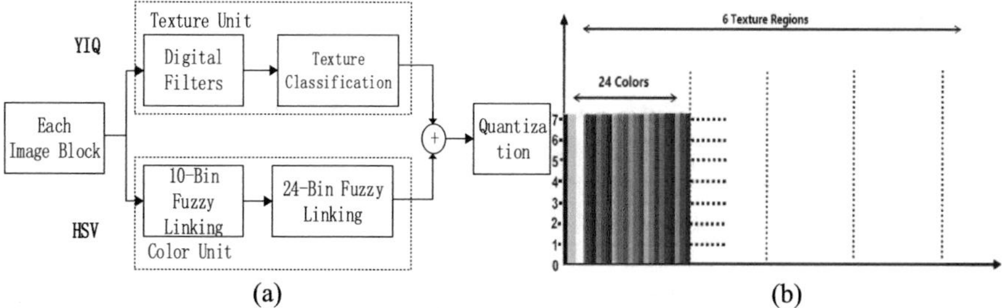

Fig. 3. The proposed framework for food recognition. (a) shows the process of image block interacts with the system; (b) the content of units.

$$T_{i,j} = t(x_i, x_j) = \frac{x_i^T x_j}{x_i^T x_i + x_j^T x_j - x_i^T x_j} \tag{1}$$

3.2 Likelihood

We will give some definitions such as retrieval set and training set in this paper. For each feature type, we rank all database images in increasing order of distance given in Eq. (1) and take the minimum of per-feature ranks amounts to taking the top matches. The likelihood score for food class c and segmented image s_i is obtained by all the independent features with Naive Bayes assumption

$$\mathrm{L}(s_i|\mathrm{c}) = P(s_i|\mathrm{c}) = \frac{1}{N}\sum\nolimits_{k=1}^{N} \omega_k P(f_i^k|c) \tag{2}$$

Where f_i^k is the feature vector of the kth type for s_i, ω_k is the weight coefficient of kth type feature vector. $\mathrm{i} = 1 : \mathrm{Q}, \mathrm{Q}$ is defined as the total number of input query images. Specifically, let D denote the set of all food images in the retrieval dataset, and N_i^k denote kth distance from f_i^k is below a fixed threshold h_k in D, then we have

$$P(f_i^k|c) = n(c, N_i^k)/n(c, D) \tag{3}$$

Where $n(c, D)$ is the number of food in set D with class label c.

We calculate the standard deviation of the likelihood probability of each descriptor

$$\omega_k = \frac{1}{N}\sum\nolimits_{i=1}^{N} \left(P(f_i^k|c) - \overline{P(f_i^k|c)}\right) \tag{4}$$

Where N is the kind of foods in breakfast, and we make $N = 6$. k is the kind of descriptors used in image retrieval, $k = 1, 2, \ldots, 6$. $\overline{P(f_i^k|c)}$ is the mean of $P(f_i^k|c)$.

3.3 Contextual Inference

As discussed in Sect. 3.2, we can obtain an initial labeling of the food image by maximizes Eq. (2). But as a meal, we would like to enforce contextual constraints used by Markov Random Fields (MRF) on the food class.

Let G denote the set of c_j food images in the menu, and M_k^i denote the set of c_j food images eat with c_i in the menu, then we have

$$P(c_i - c_j) = n(c_i, M_k^i)/n(c_j, G) \tag{5}$$

In keeping with our nonparametric philosophy and emphasis on scalability, we restrict ourselves to contextual models that require minimal training and that can be solved efficiently. Therefore, we minimization of a standard MRF energy function defined over the field of food image labels c = $\{c_i\}$.

$$J(c) = \sum\nolimits_{s_i \epsilon SP} E_{data}(s_i, c_i) + \lambda \sum\nolimits_{(s_i, s_j) \epsilon A} E_{smooth}(c_i, c_j) \tag{6}$$

where SP is the set of food images, A is the set of food pairs and λ is the smoothing constant. We define the data term as

$$E_{data}(s_i, c_i) = 1 - \omega_i L(s_i, c_i) \tag{7}$$

$$E_{smooth}(c_i, c_j) = 1 - \left[\left(P(c_i|c_j) + P(c_j|c_i)\right)/2\right] \tag{8}$$

3.4 Optimization

Optimization in an MRF problem given in Eq. (6) and we employ Iterated Conditional Modes (ICM) to minimization. It iterates over each node and calculates the value that minimizes the energy given the current values for all the variables. Then update the value and begins the next iteration until convergence. An example of this technique in action can be seen below and we will show the algorithm steps in the next chapter.

4 Experiment and Results

Similarly to several other data-driven methods [7, 12, 14, 15], we evaluate feature selection on the Dataset and compare the search results on several new features; The Dataset is a collection of Asia food images and we focus on the set of 6 categories (Baozi, Egg, Steam Bun, Milk, Noodle, Youtiao), as given in Fig. 4. Each food category contains 31 different instances of the food.

This system extracts food images by automatic detection. In this paper, we bring into effect a number of new as well as state of the art descriptors, and execute an image search from XML-based index files or directly from a folder containing image files, extracting and comparison features in real time.

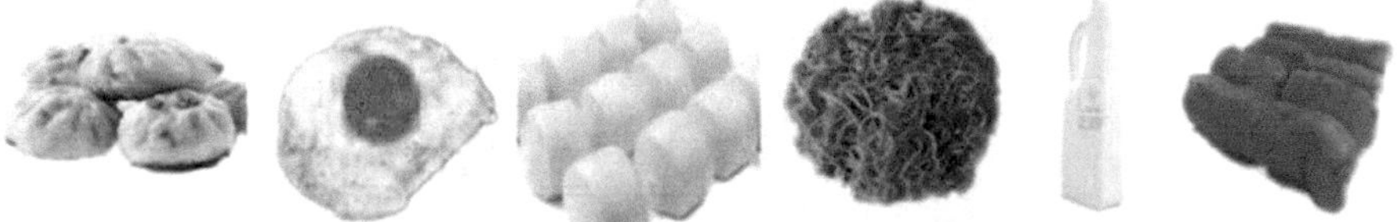

Fig. 4. Breakfast foods of Asia.

The menu of Xiaoming's breakfast is given in Table 1 (Baozi = B, Egg = E, Steam Bun = S, Milk = M, Noodle = N, Youtiao = Y).

Table 1. Xiaoming's Breakfast list.

Day	Mon	Tue	Wed	Thu	Fri	Sat	Sun
Food1	**B**	**Y**	**S**	**N**	**B**	**Y**	**S**
Food2	**E**	**E**	**E**		**E**	**E**	**E**
Food3	**M**	**M**	**M**		**M**	**M**	**M**

To attempt to capture this kind of similarity, we use six types of global image features: CEDD, FCTH, BTDH, EHD, CLD and SCD. For each feature type, we take the minimum and get it's amounts to taking the top 9 matches according to descriptor. Intuitively, taking just one best scene matches from the global descriptors leads to poor recognition results. So we will fuse all results in Table 2 to get more accurate results.

Table 2. Initial results for food recognition by image retrieval.

Descriptor	B	E	S	M	N	Y	Std
CEDD	0.3333	0.5556	0.1111	0	0	0	0.2304
FCTH	0.3333	0.5556	0.1111	0	0	0	0.2304
BTDH	0	0.3333	0.2222	0.4444	0	0	0.1956
EHD	0.1111	0.1111	0.3333	0.2222	0.2222	0	0.1165
CLD	0.2222	0.4444	0.1111	0.1111	0	0.1111	0.1532
SCD	0.2222	0.3333	0.2222	0.1111	0	0.1111	0.1165

Observing these descriptors results, it is easy to ascertain that in some of the queries, better retrieval results are achieved by using CEDD, while in others by using FCTH. Given that the CED, FCTH descriptor as well as other descriptor for each image is available in the index file that the retrieval system uses, the descriptors should be combined as Eq. (2) to achieve better retrieval results.

The final likelihood probability of the input food images (see Fig. 5) is given as following, from left column to right are $P(s_1|c)$, $P(s_2|c)$, $P(s_3|c)$:

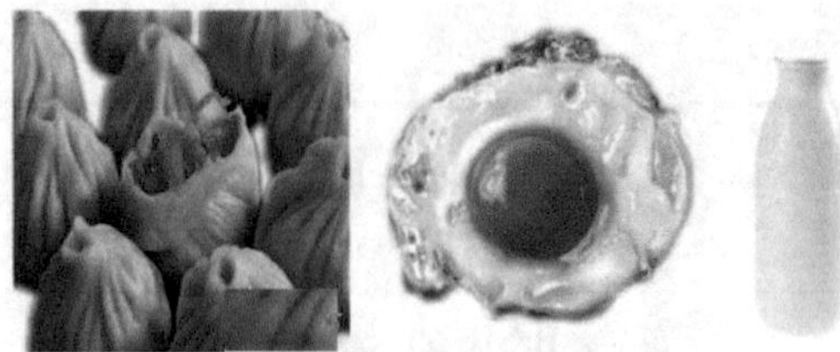

Fig. 5. Query image get from Xiaoming's breakfast.

$$P(s_1|c) = \{0.2185, 0.0700, 0.1700, \mathbf{0.2684}, 0.0757, 0.1974\}$$
$$P(s_2|c) = \{0.1131, \mathbf{0.4716}, 0.0725, 0.1497, 0.1374, 0.0556\}$$
$$P(s_3|c) = \{0.0362, 0.1757, 0.0598, 0.1415, \mathbf{0.5332}, 0.0536\}$$

Table 3. The joint probability of different kinds of foods.

$P(c_i \mid c_j)$	B	E	S	M	N	Y
B	0	1/3	0	0	1/3	0
E	1	0	1	0	1	1
S	0	1/3	0	0	1/3	0
N	0	0	0	0	0	0
M	1	1	1/3	0	0	1
Y	0	1	0	0	1/3	0

From Eq. (5), we can get $P(c_i|c_j)$, and the probability are given as follows in Table 3:

So, we can calculate Eq. (6) by the data given in Tables 2 and 3.

In this study, one of the most successful MRF called Iterated Conditional Modes (ICM) algorithms is used specifically for MRF minimization optimization. The detail of ICM Algorithm is given as follows:

1. Start at an initial configuration $P(s_i|c)P(c_i|c_j)$ and set k = 0.
2. For each configuration which differs at most in one element from the current configuration, compute the energy according to Eq. (6).
3. From the configurations in $J(\mathrm{c}, s_i)$, select the one which has a minimal energy.
4. Go to Step 2 with k = k +1 until convergence is obtained.

After three iterations, the energy of each food class is given as

$$J(\mathrm{c}, s_1) = \{\mathbf{2.2902}, 5.9449, 2.3388, 5.7490, 5.9392, 2.3113\}.$$
$$J(\mathrm{c}, s_2) = \{6.3207, \mathbf{1.5185}, 4.1906, 6.2006, 6.8626, 4.2074\}.$$
$$J(\mathrm{c}, s_3) = \{6.3977, 6.8243, 4.2033, 6.2088, \mathbf{1.4569}, 4.2095\}.$$

and we can get the results of input images as: B, E, M.

The method supports the XML files containing information from 133 images selected by the user. Descriptors and the index file size for the food database are given in Table 4. Time calculations were made on an Intel core 2 Quad 2.8 GHz, 2 GB ram.

Table 4. Descriptors and time calculations.

Descriptor	XML size (KB)	Extraction time (sec)	Retrieval time (sec)	Extraction time (sec)
CEDD FCTH BTDH SCD CLD EHD	384	18.52	0.65	76.8
SIFT [7]	5032	321.49	4.43	62.5
Bag of Words [9]	8125	1126.83	14.21	65.9

5 Conclusions

Food recognition is a new but growing area of exploration. The end goal of this work is to extract information that facilitates providing people beneficial information about their dietary habits.

In this paper, we have presented a simple and effective nonparametric approach to the problem of food image parsing by labeling image with their categories and demonstrated that personalized menu can contribute to improving the food balance estimation. This approach requires no training, it can easily scale to data sets with tens of thousands of images and hundreds of labels. Our experiments show that the proposed representation is significantly more accurate at identifying food than existing methods.

Acknowledgements. This work was supported by National Nature Science Foundation of China (NSFC) under Grants 61671356, 61201290.

References

1. Chen, M., Dhingra, K., Wu, W., Yang, L., Sukthankar, R., Yang, J.: PFID: pittsburgh fast-food image dataset. In: ICIP, pp. 289–292 (2009)
2. Chatzichristofis, S.A., Zagoris, K., Boutalis, Y.S., Papamarkos, N.: Accurate image retrieval based on compact composite descriptors and relevance feedback information. Int. J. Pattern Recognit Artif Intell. **24**(02), 207–244 (2010)
3. Jiang, T., Jurie, F., Schmid, C.: Learning shape prior models for object matching. In: CV PR, pp. 845–855 (2009)
4. Lazebnik, S., Cordelia, S., Ponce, J.: Beyond bags of features: Spatial pyramid matching for recognizing natural scene categories. In: CVPR, pp. 2169–2178 (2006)
5. Lin, M.T., Haksar, A., Peron, F.G.: Beyond local appearance: category recognition from pairwise interactions of simple features. In: CVPR (2007)

6. Matsuda, Y., Hoashi, H., Yanai, K.: Recognition of multiple-food images by detecting candidate regions. In: Multimedia and Expo (ICME), pp. 25–30 (2012)
7. Duan, P., Wang, W., Zhang, W., Gong, F., Zhang, P., Rao, Y.: Food Image recognition using pervasive cloud computing. In: Green Computing and Communications (GreenCom), pp. 1631–1637 (2013)
8. Kusumoto, R., Han, X. H., Chen, Y. W.: Sparse model in hierarchic spatial structure for food image recognition. In: BMEI, pp. 851–855 (2013)
9. Kawano, Y., Yanai, K.: Real-time mobile food recognition system. In: Computer Vision and Pattern Recognition Workshops (CVPRW), pp. 1–7 (2013)
10. Oliveira, L., Costa, V., Neves, G., Oliveira, T., Jorge, E., Lizarraga, M.: A mobile, lightweight, poll-based food identification system. Pattern Recognit. **47**(5), 1941–1952 (2014)
11. Shroff, G., Smailagic, A., Siewiorek, D. P.: Wearable context-aware food recognition for calorie monitoring. In: Wearable Computers (ISWC), pp. 119–120 (2008)
12. Fischer, W.J., Fischer, W.J.: Food intake recognition conception for wearable devices. In: ACM MOBIHOC Workshop on Pervasive Wireless Healthcare, pp. 7. ACM (2011)
13. Yüksel, B.: Automatic food recognition and automatic cooking termination by texture analysis method in camera mounted oven. In: Signal Processing and Communications Applications Conference (SIU), pp. 1987–1990 (2014)
14. Wu, W., Yang, J.: Fast food recognition from videos of eating for calorie estimation. In: Multimedia and Expo (ICME), pp. 1210–1213 (2009)
15. Datta, R., Joshi, D., Li, J., Wang, J.Z.: Image retrieval: ideas, influences, and trends of the new age. ACM Comput. Surv. **40**(2), 1–60 (2008)
16. Lux, M., Chatzichristofis, S. A.: Lire: lucene image retrieval - an extensible java CBIR library. In: ACM International Conference on Multimedia, pp. 1085–1087 (2008)
17. Chatzichristofis, S.A., Boutalis, Y.S.: Cedd: Color and edge directivity descriptor a compact descriptor for image indexing and retrieval. In: 6th International Conference on Computer Vision Systems ICVS, pp. 312–322(2008)
18. Chatzichristofis, S. A., Boutalis, Y. S.: Fcth: fuzzy color and texture histogram - a low level feature for accurate image retrieval. In: Ninth International Workshop on Image Analysis for Multimedia Interactive Services, pp. 191–196 (2008)

Wavelet Kernel Twin Support Vector Machine

Qing Wu(✉), Boyan Zang, Zongxian Qi, and Yue Gao

School of Automation, Xi'an University of Posts and Telecommunications, Xi'an 710121, China
xiyouwuq@126.com, boyanzang@126.com

Abstract. To enhance model's ability of reflecting data distribution details and further improve speed of data training, a new wavelet kernel is introduced. It is orthonormal approximately and can save more data distribution details. Based on this kernel, a wavelet twin support vector machine (WTWSVM) and a wavelet least square twin support vector machine (WLSTSVM) are presented respectively. The theoretical analyses and experiment results show WTWSVM and WLSTSVM have better performance and faster speed than those in the existing works.

Keywords: Twin support vector machine · Wavelet kernel · Least square Nonlinear

1 Introduction

The support vector machine (SVM) based on statistical learning theory is an universal machine learning technique proposed by Vapnik [1], which is applied to both regression and pattern classification. The standard SVM is aimed to find the optimal hyperplane between the positive and negative data. SVM is successfully applied to a wide variety of fields, such as biological recognition [2], reliability analysis [3] and handwriting recognition [4] etc. For nonlinear situation, SVM maps training data from low-dimensional input space to high-dimensional feature space using kernel mapping, in which the problem becomes linearly separable.

Multiple kernel learning has recently become much popular since they can get better performance in high dimensional and heterogeneous data. Combined the wavelet technique with SVMs, wavelet support vector machines (WSVMs) was proposed [5], where a basic wavelet, namely the Modulated Gaussian, was used in the simulations. The use of frames has been proposed in [6]. Computational results show WSVMs are the feasibility and validity.

Jayadeva et al. [7] proposed a twin support vector machine(TWSVM) for binary classification, motivated by GEPSVM [8]. TWSVM generates two nonparallel hyperplanes such that each hyperplane is closer to one of the two classes and farther from the other. Difference between TWSVM and SVM is that TWSVM solves two smaller sized quadratic programming problems (QPPs) instead of a large one in the conventional SVM, which means the leaning speed of TWSVM is about four times faster than that of standard SVM. In TWSVM, the inequality constraints are transformed into equality constraints, then the least squares support vector machine

P. Krömer et al. (Eds.): ECC 2018, AISC 891, pp. 765–773, 2019.
https://doi.org/10.1007/978-3-030-03766-6_86

(LSTSVM) is put forward [9]. LSTSVM has higher classification efficiency than TWSVM.

Motivated by ideas and principles from multi-resolution and wavelet theory, we introduce the wavelet technique to TWSVM and LSTSVM and present a wavelet TWSVM (WTWSVM) and a wavelet LSTSVM (WLSTSVM) respectively in this paper. The wavelet kernel is a multidimensional wavelet function which can approximate arbitrarily nonlinear functions. And the goal of the WTWSVM and WLSTSVM is to find the optimal hyperplane in the space spanned by multidimensional wavelet kernels. The theoretical analyses and experimental results show the feasibility and validity of WTWSVM and WLSTSVM in classification.

2 Twin Support Vector Machine

To improve training speed of SVM, Javadeva et al. proposed Twin Support Vector Machine (TWSVM) inspired by SVM and GEPSVM [8]. TWSVM considers a binary classification problem of m_1 positive class data and m_2 negative class data. Suppose that data points belonging to positive class are denoted by $A \in R^{m_1 \times n}$, where each row $A_i \in R^n$ represents a data point, similarly, $B \in R^{m_2 \times n}$ represents all of negative points.

2.1 Linear Twin Support Vector Machine

For the linear case, TWSVM determines two nonparallel hyperplanes:

$$\begin{cases} f_+(x) = w_1^T x + b_1 = 0 \\ f_-(x) = w_2^T x + b_2 = 0 \end{cases} \tag{1}$$

where $w_1, w_2 \in \mathrm{R}^n$, $b_1, b_2 \in \mathrm{R}$, here each hyperplane is close to one of two classes and is at least one distance from the other class. A data point belongs to positive class or negative class depending on the distance of the point from two hyperplanes. Formally, for finding the positive and negative hyperplane, the TWSVM solves following QPPs:

$$\begin{array}{ll} \min & \frac{1}{2}(Aw_1 + b_1)^T(Aw_1 + b_1) + c_1 e_2^T \xi_1 \\ s.t. & -(Bw_1 + b_1) + \xi_1 \geq e_2, \xi_1 \geq 0 \end{array} \tag{2}$$

$$\begin{array}{ll} \min & \frac{1}{2}(Bw_2 + b_2)^T(Bw_2 + b_2) + c_2 e_1^T \xi_2 \\ s.t. & -(Aw_2 + b_2) + \xi_2 \geq e_1, \xi_2 \geq 0 \end{array} \tag{3}$$

where $c_1, c_2 > 0$ are the pre-specified penalty factor, and e_1, e_2 are vectors of ones of appropriate dimensions. By introducing Lagrangian multipliers, the wolfe dual QPPs can be represents respective as follows:

$$\begin{array}{ll} \max & e_2^T \alpha - \frac{1}{2}\alpha^T G(H^T H)^{-1} G^T \alpha \\ s.t. & 0 \leq \alpha \leq c_1 e_2 \end{array} \tag{4}$$

$$\begin{array}{ll} \max & e_1^T\beta - \frac{1}{2}\beta^T H(G^T G)H^T\beta \\ s.t. & 0 \le \beta \le c_2 e_1 \end{array} \tag{5}$$

where $G = [B, e_2]$, $H = [A, e_1]$, and $\alpha = R^{m_2}$, $\beta = R^{m_1}$ are Lagrangian multipliers.

By defining $v_1 = [w_1, b_1]$ and $v_2 = [w_2, b_2]$, we can obtain them from the solutions α and β of (4) and (5)

$$v_1 = -\left(H^T H\right)^{-1} G^T\alpha \tag{6}$$

$$v_2 = \left(G^T G\right)^{-1} H^T\beta \tag{7}$$

The new simple point is assigned to negative class or positive class depending on the function as follows:

$$i = \arg\min_{k=1,2} \frac{\left|w_k^T x + b_k\right|}{\|w_k\|} \tag{8}$$

2.2 Nonlinear Twin Support Vector Machine

For the nonlinear case, TWSVM uses kernel function to map training data from low dimensional input space to high dimensional space like SVM. Two nonparallel hyperplanes are as follows:

$$\begin{cases} f_+ = K(x^T, C^T)u_1 + b_1 = 0 \\ f_- = K(x^T, C^T)u_2 + b_2 = 0 \end{cases} \tag{9}$$

where $C = [A, B]^T$, and $K(\cdot)$ is kernel function. One of two nonparallel hyperplanes are can be obtained by solving the following QPP:

$$\begin{array}{ll} \min & \frac{1}{2}(K(A, C^T)u_1 + e_1 b_1)^T (K(A, C^T)u_1 + e_1 b_1) + c_1 e_2 \xi_1 \\ s.t. & -(K(B, C^T)u_1 + e_2 b_1) + \xi_1 \ge e_2, \xi_1 \ge 0 \end{array} \tag{10}$$

The dual problem of (10) can be represented as follows:

$$\begin{array}{ll} \max & e_2^T\alpha - \frac{1}{2}\alpha^T R(S^T S)^{-1} R^T\alpha \\ s.t. & 0 \le \alpha \le c_1 e_2 \end{array} \tag{11}$$

where $S = [K(A, C^{\mathrm{T}}), e_1]$, $R = [K(B, C^{\mathrm{T}}), e_2]$.

Define $z_1 = [u_1, b_1]$ and $z_2 = [u_2, b_2]$. We can solve dual problems (11) and get the following result:

$$z_1 = -(S^T S)^{-1} R\alpha \tag{12}$$

Similarity, the other QPP is solved and the result is as follows:

$$z_2 = (R^T R)^{-1} S\beta \tag{13}$$

3 Least Squares Twin Support Vector Machine

Suykens proposed least square vector machine (LSSVM) in 1999 [10]. Now LSSVM has been got much attention because it has faster training speed than SVM.

In this paper, for nonlinear case, LSTWSVM need to find two nonparallel hyperplanes based on kernel function as follows:

$$\begin{cases} K(x^T, C^T)u_1 + b_1 = 0 \\ K(x^T, C^T)u_2 + b_2 = 0 \end{cases} \tag{14}$$

where $C^T = [A, B]^T$, $K(\cdot)$ is kernel function. And the QPP_S are:

$$\begin{array}{ll} \min & \frac{1}{2}\|K(A, C^T)u_1 + e_1 b_1\|^2 + \frac{1}{2}\xi_1^2 \\ s.t. & -(K(B, C^T)u_1 + e_2 b_1) + \xi_1 = e_2 \end{array} \tag{15}$$

$$\begin{array}{ll} \min & \frac{1}{2}\|K(B, C^T)u_2 + e_2 b_2\|^2 + \frac{1}{2}\xi_2^2 \\ s.t. & (K(A, C^T)u_2 + e_1 b_2) + \xi_2 = e_1 \end{array} \tag{16}$$

Construct the Lagrangian functions as follows:

$$\begin{cases} \frac{1}{2}\|K(A, C^T)u_1 + e_1 b_1\|^2 + \frac{c_1}{2}\|e_2 + (K(B, C^T)u_1 + e_2 b_1)\|^2 \\ \frac{1}{2}\|K(B, C^T)u_2 + e_2 b_2\|^2 + \frac{c_2}{2}\|e_1 - (K(A, C^T)u_2 + e_1 b_2)\|^2 \end{cases} \tag{17}$$

Then the optimal solution can be determined according to KKT conditions as follows:

$$\begin{cases} (u_1, b_1)^T = -\left(H^T H + \frac{1}{c_1} G^T G\right)^{-1} H^T e_2 \\ (u_2, b_2)^T = \left(G^T G + \frac{1}{c_2} H^T H\right)^{-1} G^T e_1 \end{cases} \tag{18}$$

where $G = [K(A, C^T), e_1]$, $H = [K(B, C^T), e_2]$.

4 Wavelet Kernel Twin Support Vector Machine

Now multi-kernel learning has become a research hotspot for its superior performance in multi-view learning since many kinds of information from multiple views can easily be combined. Wavelet kernel is a kind of multidimensional wavelet function that can almost approach arbitrary function. All wavelet kernels must obey the following theorems [11].

Theorem 1 [6]: let $h(x)$ be a mother wavelet, and let a and c denote the dilation and translation, respectively. $x, a, c \in R$, If $X, X' \in R^N$, then dot-product wavelet kernel are:

$$K(X, X') = \prod_{i=1}^{N} h\left(\frac{x_i - c_i}{a}\right) h\left(\frac{x_i' - c_i'}{a}\right) \tag{19}$$

and translation-invariant wavelet kernels that satisfy the translation invariant kernel theorem are:

$$K(X, X') = \prod_{i=1}^{N} h\left(\frac{x_i - x_i'}{a}\right) \tag{20}$$

Without loss of generality, we construct a translation-invariant wavelet kernel by a wavelet function adopted in [11]:

$$h(x) = \cos(1.75x)\exp\left(-\frac{x^2}{2}\right) \tag{21}$$

Theorem 2 [6]: Given the mother wavelet (19) and the dilation $a, c, x \in R$, If $X, X' \in R^N$, then the wavelet kernel of the mother wavelet is:

$$\begin{aligned} K(X, X') &= \prod_{i=1}^{N} h\left(\frac{x_i - c_i}{a}\right) \\ &= \prod_{i=1}^{N} \left(cos\left(1.75 \times \frac{(x_i - x_i')}{a}\right) exp\left(\frac{\|x_i - x_i'\|^2}{2a^2}\right)\right) \end{aligned} \tag{22}$$

That is a kind of multidimensional wavelet kernel. TWSVM with wavelet kernel determines two nonparallel hyperplanes as follows:

$$\begin{cases} K(X, C^T)u_1 + b_1 = \sum_{i=1}^{l} u_1(i) \prod_{j=1}^{N} h\left(\frac{x^j - x_i^j}{a_i}\right) + b_1 \\ K(X, C^T)u_2 + b_2 = \sum_{i=1}^{l} u_2(i) \prod_{j=1}^{N} h\left(\frac{x^j - x_i^j}{a_i}\right) + b_2 \end{cases} \tag{23}$$

Now, the decision function of WTWSVM for classification is given:

$$i = \arg\min_{k=1,2} \frac{\left| \sum_{i=1}^{l} u_i \prod_{j=1}^{n} h\left(\frac{x^i - x_j^i}{a}\right) + b \right|}{\|w_k\|} \tag{24}$$

Similarly, we can obtain the decision function of WLSTSVM for classification, which is similar to Eq. (24) to WTSVM.

5 Experiments

To test the performance of our proposed approaches, we compare numerically WTWSVM and WLSTSVM with Gaussian kernel TWSVM (GTWSVM) and Gaussian kernel LSTSVM (GLSTSVM) respectively on a synthetic dataset and 12 datasets from UCI Repository [12]. All experiments were implemented by using MATLAB 8.4 on a personal computer with 1.6 GHz and 4 GB RAM.

In first experiment, a synthetic dataset "Cross-planes" is been classified. Classification accuracy of each algorithm is measured by tenfold cross-validation method. From a and b in Fig. 1, we can see WTWSVM has better classification performance than GTWSVM. From c and d in Fig. 1, it is easily seen that the classification performance of WLSTSVM is superior to that of GLSTSVM. Figure 1 shows WTWSVM has best performance in four approaches.

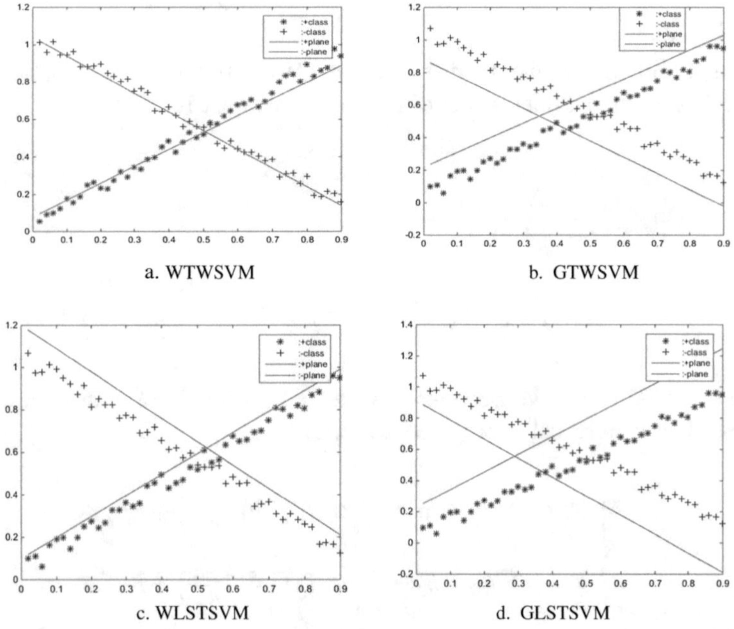

a. WTWSVM

b. GTWSVM

c. WLSTSVM

d. GLSTSVM

Fig. 1. Learning results of the four algorithms on the Cross-planes data set (a) WTWSVM (b) GTWSVM (c) WLSTSVM (d) GLSTSVM

The second experiment is designed to demonstrate the effectiveness of WTWSVM and WLSTSVM. All the datasets are available from UCI Repository [12]. The selected datasets are listed in the Table 1.

Table 1. Attribute characteristics of the UCI datasets

Dataset	Dimension	Number	Dataset	Dimension	Number
australian	14	690	ionosphere	34	351
breast	9	277	pima	8	768
bupa	6	345	sonar	60	268
diabetes	8	768	vote	15	435
german	24	1000	wdbc	31	569
heart	13	270	wpbc	33	198

Table 2 compares the performance of the WTWSVM classifier with that of GTWSVM. Table 2 indicates that WTWSVM has not higher classification precision than GTWSVM, but WTWSVM is faster obviously than GTWSVM on the vast majority of datasets. The results in Table 3 demonstrate that WLSTSVM has higher training speed than GLSTSVM.

Table 2. Experimental result for WTWSVM and GTWSVM on the UCI datasets

Dataset	WTWSVM		GTWSVM	
	Accuracy(%)	Time(s)	Accuracy(%)	Time(s)
australian	80.75	67.258	85.61	557.28
Breast	73.25	41.78	84.21	111.47
Bupa	70.15	53.836	69.56	148.7
Diabetes	82.63	859	78.57	711.69
german	78.61	72.74	76	1248
heart	81.81	87.362	87.27	104.7
ionosphere	94.37	99.73	97.18	168.571
pima	77.27	195.44	85.32	713.83
sonar	88.10	50.23	95.34	78.604
vote	97.73	107.66	97.72	236.63
wdbc	98.12	722.64	97.39	447.04
wpbc	90.02	58.81	85.36	70.91

In the two experiments, WTWSVM has better classification results than GTWSVM on most datasets, especially on high-dimensional datasets, which verifies that wavelet kernel can save more data distribution details and have better classification results than Gaussian kernel.

Table 3. Experimental result for WLSTWSVM and GLSTSVM on the UCI datasets

Dataset	WLSTSVM		GLSTSVM	
	Accuracy(%)	Time(s)	Accuracy(%)	Time(s)
australian	62.59	8.601	84.17	368.55
breast	78.57	1.846	71.92	56.07
bupa	64.29	1.382	75.36	79.98
diabetes	65.34	3.711	74.63	395.1
german	76.62	31.132	74.5	730
heart	56.63	3.056 6	85.45	51
ionosphere	80.28	21.936	95.77	91.14
pima	69.88	3.867 1	79.87	395.19
sonar	60.23	39.463	86.04	35.62
vote	69.32	6.150	96.59	167
wdbc	61.22	30.723	96.52	300.31
wpbc	70.73	11.764	85.36	65.89

6 Conclusion

In this paper, a new wavelet kernel is proposed. Compared with Gaussian kernel, it is orthonormal or orthonormal approximately. Based on this construction, a WTWSVM and a WLSTSVM are introduced respectively. The theoretical analyses and experiment results show the feasibility and validity of the WLSTSVM and WTWSVM.

Acknowledgments. This work was supported in part by the National Natural Science Foundation of China under Grants (61472307, 51405387), the Key Research Project of Shanxi Province (2018GY-018) and the Foundation of Education Department of Shaanxi Province (17JK0713).

References

1. Vapnik, V.N.: The nature of statistical learning theory. Technometrics **38**(4), 409 (1995). https://doi.org/10.1007/978-1-4757-2440-0
2. Schölkopf, B., Tsuda, K., Vert, J.P.: Support Vector Machine Applications in Computational Biology, pp. 71–92. Publisher, Cambridge (2004)
3. Zhao, H., Ru, Z., Chang, X., et al.: Reliability analysis of tunnel using least square support vector machine. Tunn. Undergr. Space Technol. Incorporating Trenchless Technol. Res. **41**, 14–23 (2014). https://doi.org/10.1016/j.tust.2013.11.004
4. Adankon, M.M., Cheriet, M.: Model selection for the LS-SVM. Application to handwriting recognition. Pattern Recognit. **42**(12), 3264–3270 (2009). https://doi.org/10.1016/j.patcog.2008.10.023
5. Zhang, L., Zhou, W., Jiao, L.: Wavelet support vector machine. IEEE Trans. Syst. Man Cybern. Part B Cybern. **34**(1), 34–39 (2004). https://doi.org/10.1109/TSMCB.2003.811113

6. Gao, J., Harris, C., Gunn, S.: On a class of support vector kernels based on frames in function hilbert spaces. Neural Comput. **13**(9), 1975–1994 (2001). https://doi.org/10.1109/TSMCB.2003.811113
7. Jayadeva, Khemchandani, R., Chandra, S.: Twin support vector machines for pattern classification. IEEE Trans. Pattern Anal. Mach. Intell. **29**(5), 901–905 (2007). https://doi.org/10.1109/tpami.2007.1068
8. Mangasarian, O.L., Wild, E.W.: Multisurface proximal support vector machine classification via generalized eigenvalues. IEEE Trans. Pattern Anal. Mach. Intell. **28**(1), 69–74 (2006). https://doi.org/10.1109/TPAMI.2006.17
9. Xie, X.: Regularized multi-view least squares twin support vector machines. Appl. Intell. **17**, 1–8 (2018). https://doi.org/10.1007/s10489-017-1129-3
10. Suykens, J.A.K., Vandewalle, J.: Least squares support vector machine classifiers. Neural Process. Lett. **9**(3), 293–300 (1999). https://doi.org/10.1023/a:1018628609742
11. Szu, H.H., Telfer, B.A., Kadambe, S.L.: Neural network adaptive wavelets for signal representation and classification. Opt. Eng. **31**(9), 1907–1916 (1992). https://doi.org/10.1117/12.59918
12. UCI Machine Learning Repository. http://archive.ics.uci.edu/ml/index.php. Accessed 10 July 2018

A Non-singular Twin Support Vector Machine

Wu Qing(✉), Qi Shaowei, Zhang Haoyi, Jing Rongrong, and Miao Jianchen

School of Automation, Xi'an University of Posts and Telecommunications, Xi'an 710121, China
xiyouwuq@126.com

Abstract. Due to high efficiency, twin support vector machine (TWSVM) is suitable for large-scale classification problems. However, there is a singularity in solving the quadratic programming problems (QPPs). In order to overcome it, a new method to solve the QPPs is proposed in this paper, named non-singular twin support vector machine (NSTWSVM). We introduce a nonzero term to the result of the problem. Compared to the TWSVM, it does not need extra parameters. In addition, the successive overrelaxation technique is adopted to solve the QPPs in the NSTWSVM algorithm to speed up the training procedure. Experimental results show the effectiveness of the proposed method in both computation time and accuracy.

Keywords: Twin support vector machine · Singularity · Classification

1 Introduction

Support vector machine (SVM) is now a well-known machine learning tool, which has been introduced by Vapnik in 1990s [1–3]. It has already been applied in a wide variety of fields [4–8]. Recently, much research has been done to improve the effectiveness and accuracy. The most popular SVM is "Maximum margin" one that attempts to reduce generalization error by maximizing the margin between two parallel hyperplane [9, 10]. The regularization term is introduced to implement the structural risk minimization principle.

Mangasarian and Wild proposed a nonparallel plane classifier for binary data classification, which they termed the generalized eigenvalue proximal support vector machine (GEPSVM) [11]. And then, Jayadeva and his co-workers introduced the twin support vector machine (TWSVM) [12]. The TWSVM aims to generate two nonparallel planes such that each plane is closer to one of the two classes and is as far as possible from the other. The SVM aims at generating a single convex quadratic problem and has all data points included in the constraints. However, the TWSVM aims to solve a pair of small quadratic programming problems, and all data points are distributed into two classes. The computing complexity of the TWSVM in the training phase is 1/4 of the standard SVM and TWSVM is 4 times faster than the standard SVM. Hence TWSVM is suitable for large-scale classification problems.

It is well known that one significant advantage of the SVM is the implementation of the structural risk minimization principle [13]. The empirical risk is only considered in

P. Krömer et al. (Eds.): ECC 2018, AISC 891, pp. 774–783, 2019.
https://doi.org/10.1007/978-3-030-03766-6_87

the primal problems of the TWSVM, so that the inverse matrices $(H^T H)^{-1}$ and $(G^T G)^{-1}$ appear in the dual problems. The TWSVM often assumes that the inverse matrices exist. However, in fact, they may not satisfy the extra prerequisite. Yuan-Hai Shao, et al. proposed a twin bounded support vector machines (TBSVM) in [14]. The TBSVM has overcome the singularity of the TWSVM, however, it introduced several extra parameters into formulation compared to the TWSVM. Therefore, the selection of the parameters becomes a new problem for TBSVM.

In order to overcome singularity of the TWSVM and avoid the selection of extra parameters, in this paper, there are some improvements on the TWSVM. We propose a non-singular TWSVM to overcome singularity. Similar to the TWSVM, the NSTWSVM constructs two nonparallel hyperplanes by solving two smaller QPPs. However, the advantages of NSTWSVM are obvious: (1) in the traditional TWSVM, there is a singularity in the solution, whereas in our NSTWSVM the singularity is removed by adding a nonzero term to the equation; (2) in order to shorten training time, we use successive overrelaxation (SOR) technique in our NSTWSVM.

This paper is organized as follows: Sect. 2 briefly dwells on the TWSVM and introduces the notation used in this paper. Section 3 proposes our new method for the TWSVM. Section 4 shows experimental results. Section 5 draws a conclusion.

2 Twin Support Vector Machines

For a classification problem, data points belonging to classes +1 and −1 are denoted as matrices $A \in R^{m_1 \times n}$ and $B \in R^{m_2 \times n}$, respectively, where m_1 and m_2 represent the number of patterns of classes.

The aim of TWSVM is to find a pair of nonparallel planes

$$f_1(x) = w_1^T x + b_1 = 0 \tag{1}$$

$$f_2(x) = w_2^T x + b_2 = 0 \tag{2}$$

where $w_1, w_2 \in R^n$, $b_1, b_2 \in R$.

For the above requirements, the TWSVM classifier is obtained by solving the following QPPs

$$\begin{aligned} &\min_{w_1, b_1, \xi_1} \frac{1}{2}(Aw_1 + e_1 b_1)^T (Aw_1 + e_1 b_1) + c_1 e_2^T \xi_1 \\ &s.t - (Bw_1 + e_2 b_1) + \xi_1 \geq e_2 \ \xi_1 \geq 0 \end{aligned} \tag{3}$$

$$\begin{aligned} &\min_{w_2, b_2, \xi_2} \frac{1}{2}(Bw_2 + e_2 b_2)^T (Bw_2 + e_2 b_2) + c_2 e_1^T \xi_2 \\ &s.t \ (Aw_2 + e_1 b_2) + \xi_2 \geq e_1 \ \xi_2 \geq 0 \end{aligned} \tag{4}$$

where $c_1, c_2 > 0$ are parameters, and e_1, e_2 are vectors of ones of appropriate dimensions.

Clearly, (3) and (4) represent two nonparallel planes. The first term in the objective function of (3) or (4) is the sum of squared distances from the hyperplane to points of one class. The second term of the objective function is the sum of a set of error variables. Minimizing the sum of error variables could minimize misclassification due to points belonging to their own classes. The constraints mean that one hyperplane must be at a distance of at least 1 from the points of another class.

In terms of mathematics, TWSVMs are a pair of QPPs. In each QPP, the objective function corresponds to a particular class and the constraints are determined by patterns of the other class. For (3), the points of class +1 are clustered around the plane $w_1^T x + b_1 = 0$. Similarly, the points of class -1 are also clustered around the plane $w_2^T x + b_2 = 0$ for (4). The original QPP has been divided into two small QPPs. Therefore, TWSVM is faster than usual SVM in computation speed.

The Lagrangian corresponding to the problem TWSVM1 is given as follows

$$L(w_1, b_1, \xi_1, \alpha, \beta) = \frac{1}{2}(Aw_1 + e_1 b_1)^T (Aw_1 + e_1 b_1) + c_1 e_2^T \xi_1 + \alpha^T (Bw_1 + e_2 b_1 - \xi_1 + e_2) - \beta^T \xi_1 \tag{5}$$

where $\alpha = (\alpha_1, \alpha_2, \cdots, \alpha_{m_2})^T$ and $\beta = (\beta_1, \beta_2, \cdots, \beta_{m_2})^T$ are the vectors of Lagrange multipliers. The Karush-Kuhn-Tucker (K.K.T) optimality conditions for (5) are given as follows

$$A^T (Aw_1 + e_1 b_1) + B^T \alpha = 0 \tag{6}$$

$$e_1^T (Aw_1 + e_1 b_1) + e_2^T \alpha = 0 \tag{7}$$

$$c_1 e_2 - \alpha - \beta = 0 \tag{8}$$

$$\alpha^T (Bw_1 + e_2 b_1 - \xi_1 + e_2) = 0,\ \beta^T \xi_1 = 0 \tag{9}$$

$$\alpha \geq 0, \beta \geq 0 \tag{10}$$

Since $\beta \geq 0$, we can get the formulation $0 \leq \alpha \leq c_1$ from (8).

Simply, combining (6) with (7), we can obtain

$$[A^T e_1^T][Ae_1][w_1 e_1]^T + [B^T e_2^T]\alpha = 0 \tag{11}$$

Now, defining $H = [Ae_1]$, $G = [Be_2]$ and $u = [w_1, b_1]^T$, (11) can be rewritten as

$$H^T H u + G^T \alpha = 0 \text{ or } u = -(H^T H)^{-1} G^T \alpha \tag{12}$$

Obviously, if and only if $H^T H \neq 0$, (12) exists.

Generally, $H^T H$ is semi-positive definite, so there may be a possible ill-conditioning of $H^T H$. Jayadeva and his co-workers introduced a regularization term to overcome possible ill-conditioning of $H^T H$ in [12]. Therefore, (12) gets modified to

$$u = -(H^T H + \delta I)^{-1} G^T \alpha \tag{13}$$

where I is an identity matrix of appropriate dimensions and δ is a small positive scalar.

By K.K.T conditions, the dual problems of TWSVM are as follows:

$$\begin{aligned} &\underset{\alpha}{Max}\, e_2^T \alpha - \frac{1}{2}\alpha^T G(H^T H)^{-1} G^T \alpha \\ &s.t\ 0 \le \alpha \le c_1 \end{aligned} \tag{14}$$

$$\begin{aligned} &\underset{\gamma}{Max}\, e_1^T \gamma - \frac{1}{2}\gamma^T H(G^T G)^{-1} H^T \gamma \\ &s.t\ 0 \le \gamma \le c_2 \end{aligned} \tag{15}$$

where $Q = [Be_2], P = [Ae_1]$ and the augmented vector $v = [w_2, b_2]^T$ is given by $v = (G^T G)^{-1} H^T \gamma$.

For the nonlinear kernel classifier, the primal programming problems are as follows

$$\begin{aligned} &\min_{u_1, b_1, \xi_1} \frac{1}{2}\left\| \left(K(A, C^T) u_1 + e_1 b_1\right) \right\|^2 + c_1 e_2^T \xi_1 \\ &s.t - (K(B, C^T) u_1 + e_2 b_1) + \xi_1 \ge e_2, \xi_1 \ge 0 \end{aligned} \tag{16}$$

$$\begin{aligned} &\min_{u_2, b_2, \xi_2} \frac{1}{2}\left\| \left(K(B, C^T) u_2 + e_2 b_2\right) \right\|^2 + c_2 e_1^T \xi_2 \\ &s.t\ (K(A, C^T) u_2 + e_1 b_2) + \xi_2 \ge e_1, \xi_2 \ge 0 \end{aligned} \tag{17}$$

where $C^T = [AB]^T$, and $K(\cdot, \cdot)$ is an appropriately chosen kernel.

In the same way, the dual problems of nonlinear TWSVM are inferred as follows:

$$\begin{aligned} &\underset{\alpha}{Max}\, e_2^T \alpha - \frac{1}{2}\alpha^T R(S^T S)^{-1} R^T \alpha \\ &s.t 0 \le \alpha \le c_1 \end{aligned} \tag{18}$$

$$\begin{aligned} &\underset{\gamma}{Max}\, e_1^T \gamma - \frac{1}{2}\gamma^T S(R^T R)^{-1} S^T \gamma \\ &s.t\ 0 \le \gamma \le c_2 \end{aligned} \tag{19}$$

where $S = [K(A, C^T) e_1]$ and $R = [K(B, C^T) e_2]$.

3 TBSVM

In this section, we will recall the TBSVM, and point out its drawbacks. For linear case, two primal problems solved in TBSVM are as follows:

$$\begin{aligned}&\min_{w_1,b_1,\xi_1} \frac{c_3}{2}(\|w_1\|^2+b_1^2)+\frac{1}{2}(Aw_1+e_1b_1)^T(Aw_1+e_1b_1)+c_1e_2^T\xi_1\\&s.t-(Bw_1+e_2b_1)+\xi_1\geq e_2,\ \xi_1\geq 0\end{aligned} \tag{20}$$

$$\begin{aligned}&\min_{w_2,b_2,\xi_2} \frac{c_4}{2}(\|w_2\|^2+b_2^2)+\frac{1}{2}(Aw_2+e_2b_2)^T(Aw_2+e_2b_2)+c_2e_1^T\xi_2\\&s.t-(Bw_2+e_1b_2)+\xi_2\geq e_1,\ \xi_2\geq 0\end{aligned} \tag{21}$$

where $c_i, i=1,2,3,4$ are the penalty parameters and $e_i,\ i=1,2$ are the vectors of ones of appropriate dimensions.

We can obtain their dual problems as

$$\begin{aligned}&\max\ e_2^T\alpha-\frac{1}{2}\alpha^TG(H^TH+c_3I)^{-1}G^T\alpha\\&s.t.0\leq\alpha\leq c_1\end{aligned} \tag{22}$$

$$\begin{aligned}&\max\ e_1^T\gamma-\frac{1}{2}\gamma^TH(G^TG+c_4I)^{-1}H^T\gamma\\&s.t.0\leq\gamma\leq c_2\end{aligned} \tag{23}$$

The nonparallel proximal hyperplanes are obtained from the solution α and γ of (22) and (23) by

$$v_1=-(H^TH+c_3I)^{-1}G^T\alpha \tag{24}$$

$$v_2=(G^TG+c_4I)^{-1}H^T\gamma \tag{25}$$

There are four parameters in (22) and (23), which need more time for parameter selection.

4 NSTWSVM

For (3), there is a similar equation to (12) by combining (6), (7) with (10) as follows:

$$H^THu+G^T\alpha\alpha^T(Gu-\xi_1+e_2)+G^T\alpha=0 \tag{26}$$

or

$$u=-(H^TH+G^T\alpha\alpha^TG)^{-1}(G^T\alpha\alpha^T\xi_1-G^T\alpha\alpha^Te_2+G^T\alpha) \tag{27}$$

And the augmented vector is given by

$$v = (Q^T Q + P^T \gamma\gamma^T P)^{-1}(P^T \gamma\gamma^T \xi_2 - P^T \gamma\gamma^T e_1 + P^T \gamma) \tag{28}$$

Comparing with (13), (27) is more complex in the form, but there is no extra parameters, which avoids not only the trouble of ill-conditioning of matrix, but also the additional selection of parameters.

From (24) and (25), the feature information of the surface can be got. Then the equations of nonparallel surfaces could be described as

$$f_1(x) = w_1^T x + b_1 \tag{29}$$

$$f_2(x) = w_2^T x + b_2 \tag{30}$$

Once the solutions (w_1, b_1) and (w_2, b_2) are obtained from the solutions of (24) and (25), a new point $x \in R^n$ is assigned to class i (i = +1, −1), depending on which one of the two hyperplanes in (29) and (30) it is closer to

$$\min\left|x^T w_i + b_i\right|, i = 1, 2 \tag{31}$$

where i is the class of the point.

In order to extend our results to nonlinear classifiers, we define the Lagrangian function of (16) as follows:

$$L(u_1, b_1, \xi_1, \alpha, \beta) = \frac{1}{2}\left\|(K(A, C^T)u_1 + e_1 b_1\right\|^2 + c_1 e_2^T \xi_1 + \alpha^T(Bw_1 + e_2 b_1 - \xi_1 + e_2) - \beta^T \xi_1 \tag{32}$$

The K.K.T necessary and sufficient optimality conditions for (32) are given by

$$K(A, C^T)^T(K(A, C^T)u_1 + e_1 b_1) + K(B, C^T)^T \alpha = 0 \tag{33}$$

$$e_1^T(K(A, C^T)u_1 + e_1 b_1) + e_2^T \alpha = 0 \tag{34}$$

$$c_1 e_2 - \alpha - \beta = 0 \tag{35}$$

$$\alpha^T(K(B, C^T)u_1 + e_2 b_1 - \xi_1 + e_2) = 0, \ \beta^T \xi_1 = 0 \tag{36}$$

Combining (33) with (34), we can obtain

$$S^T S z_1 + R^T \alpha = 0 \tag{37}$$

where $S = [K(A, C^T) e_1^T]$, $R = [K(B, C^T) e_2^T]$ and $z_1 = [u_1, b_1]^T$.

For (16), there is a similar equation to (26) as follows:

$$S^T S z_1 + R^T \alpha\alpha^T (R z_1 - \xi_1 + e_2) + R^T \alpha = 0 \quad (38)$$

or

$$z_1 = -(S^T S + R^T \alpha\alpha^T R)^{-1} (R^T \alpha\alpha^T \xi_1 - R^T \alpha\alpha^T e_2 + R^T \alpha) \quad (39)$$

Similarly, the augmented vector $z_2 = [u_2, b_2]^T$ is given by

$$z_2 = (R^T R + S\gamma\gamma^T S)^{-1} (S\gamma\gamma^T \xi_2 - S\gamma\gamma^T e_1 + S^T \gamma) \quad (40)$$

Once nonlinear TWSVM are solved to obtain the surfaces from (39) and (40), a new pattern $x \in R^n$ is assigned to class +1 or class −1 in a manner similar to the linear case.

NSTWSVM can avoid the ill-conditioning of matrix, and it doesn't introduce extra parameters compared to TWSVM.

5 Experimental Results

In this section, some experiments are carried out to demonstrate the performance of our method to solve TWSVM. All experiments are implemented by using MATLAB 8.0 on a PC with 2.9 GHz CPU and 2 GB RAM. In order to compare the NSTWSVM with other algorithms, we make experiments on the datasets from the UCI machine learning repository. The NSTWSVM and TBSVM are solved by the SOR technique. TWSVM-a and TWSVM-b represent TWSVM solved by QP and SOR, respectively. The optimal values of $c_i(i = 1, 2)$ in TWSVM-a, TWSVM-b and NSTWSVM are obtained in the same range by using a tuning set comprising of 10% of the dataset. The way to select the optimal values of $c_i(i = 1, 2, 3, 4)$ in TBSVM is same. Once the parameters are selected, the tuning set is returned to learn the final classifier.

The "Accuracy" used to evaluate performance of the methods is defined as follows. Accuracy = (TP+TN)/ (TP+FP+TN+FN), where TP, TN, FP, and FN are the number

Table 1. Test accuracy of linear classifier

Datasets	NSTWSVM	TBSVM	TWSVM-a	TWSVM-b
	Accuracy(%)	Accuracy(%)	Accuracy(%)	Accuracy(%)
	Time(s)	Time(s)	Time(s)	Time(s)
australian	86.96	88.41	87.76	86.23
	0.185	0.919	3.629	0.176
german	73.57	74	74.5	71.61
	0.105	1.644	3.094	0.113
sonar	73.81	73.8	76.5	73.81
	0.101	0.122	1.762	0.036
bupa	79.71	78.91	76.28	78.26
	0.095	0.205	0.314	0.017

(continued)

Table 1. (*continued*)

Datasets	NSTWSVM	TBSVM	TWSVM-a	TWSVM-b
	Accuracy(%)	Accuracy(%)	Accuracy(%)	Accuracy(%)
	Time(s)	Time(s)	Time(s)	Time(s)
wdbc	97.37	96.84	96.79	96.49
	0.075	0.476	4.071	0.077
breast-cancer	87.43	87.12	85.68	83.33
	0.174	0.165	2.087	0.188
heart	87.04	87.037	83.12	86.48
	0.034	0.122	2.043	0.092
diabetes	76.68	76.72	75.32	77.27
	0.124	0.713	2.382	0.179
ionosphere	89.74	92	88.97	89.48
	0.041	0.237	2.748	0.036

of true positive, true negative, false positive, and false negative, respectively. Classification accuracy of each algorithm is measured by the standard tenfold cross-validation methodology.

Table 1 summarizes the experiment results. The performance of the NSTWSVM, TBSVM, TWSVM-a, and TWSVM-b for the linear case is compared in Table 1. We

Table 2. Test Accuracy of Nonlinear Classifier

Datasets	NSTWSVM	TBSVM	TWSVM-a	TWSVM-b
	Accuracy(%)	Accuracy(%)	Accuracy(%)	Accuracy(%)
	Time (s)	Time (s)	Time (s)	Time (s)
australian	86.47	86.06	84.52	85.34
	0.247	1.241	7.6	0.195
sonar	78.75	77.4	77.84	78.58
	0.015	0.167	1.947	0.016
bupa	65.22	64.8	64.91	63.77
	0.047	0.228	2.516	0.031
heart	86.61	85.6	85.45	85.19
	0.016	0.182	2.299	0.015
diabetes	66.12	67.53	64.94	63.06
	0.2942	1.402	8.606	0.312
ionosphere	94.5	93.18	92.36	92.8
	0.047	1.753	3.063	0.046
german	73.5	73.2	70	71.5
	0.14	2.34	13.492	0.141
breast-cancer	77.037	76.36	75.82	76.19
	0.256	1.132	2.115	0.205

can see the accuracy of linear NSTWSVM is significantly better than that of the linear TWSVM-a, and TWSVM-b on most of datasets. The linear NSTWSVM has higher accuracy than the TBSVM on some datasets. It also can be seen that the linear NSTWSVM and TWSVM-b take less computation time than the TWSVM-a, and TBSVM. Table 2 compares the performance of the nonlinear NSTWSVM with that of the nonlinear TBSVM, TWSVM-a, and TWSVM-b using RBF kernel. The results in Table 2 show that the accuracy of nonlinear NSTWSVM is significantly better than the nonlinear TBSVM, TWSVM-a, and TWSVM-b on most of datasets. The computation time of the nonlinear NSTWSVM and TWSVM-b are less than the nonlinear TWSVM-a, and TBSVM.

6 Conclusions

In this paper, we have proposed an approach to solve TWSVM, named NSTWSVM, which avoids the ill-condition by adding a nonzero term to the result of traditional TWSVM. NSTWSVM with SOR technique has advantages as follows: (1) NSTWSVM can avoid the singularity of matrix; (2) it doesn't introduce extra parameters into expression; (3) it has higher classification accuracy and efficiency than TWSVM and TBSVM. The numerical results show that NSTWSVM and higher training speed than other algorithms.

Acknowledgment. This work was supported in part by the National Natural Science Foundation of China under Grants (61472307, 51405387), the Key Research Project of Shaanxi Province (2018GY-018) and the Foundation of Education Department of Shaanxi Province (17JK0713).

References

1. Cortes, C., Vapnik, V.: Support vector networks. Mach. Learn. **20**(3), 273–297 (1995). https://doi.org/10.1007/BF00994018
2. Vapnik, V.: The Nature of Statistical Learning Theory. Springer, New York (1996). https://doi.org/10.1007/978-1-4757-2440-0
3. Vapnik, V.: Statistical Learning Theory. Wiley, New York (1998)
4. Chen, S., Wu, X.: Improved projection twin support vector machine. ACTA Electron. Sinca **45**(2), 408–416 (2017). https://doi.org/10.3969/j.issn.0372-2112.2017.02.020
5. Qi, Z., Tian, Y., Shi, Y.: Robust twin support vector machine for pattern classification. Pattern Recognit. **46**(1), 305–316 (2013). https://doi.org/10.1016/j.patcog.2012.06.019
6. Chen, S., Wu, X.: A new fuzzy twin support vector machine for pattern classification. Int. J. Mach. Learn. Cybern. **3**, 1–12 (2017). https://doi.org/10.1007/s13042-017-0664-x
7. Tanveer, M., Khan, M., Ho, S.: Robust energy-based least squares twin support vector machines. Appl. Intell. **45**(1), 174–186 (2016). https://doi.org/10.1007/s10489-015-0751-1
8. Borgwardt, K.: Kernel methods in bioinformatics. Springer, Berlin Heidelberg (2011). https://doi.org/10.1007/978-3-642-16345-6_15
9. Kumar, M., Gopal, M.: Least squares twin support vector machines for pattern classification. Expert Syst. Appl. Int. J. **36**(4), 7535–7543 (2009). https://doi.org/10.1016/j.eswa.2008.09.066

10. Hao, P., Chiang, J., Lin, Y.: A new maximal-margin spherical-structured multi-class support vector machine. Appl. Intell. **30**(2), 98–111 (2009). https://doi.org/10.1007/s10489-007-0101-z
11. Mangasarian, O., Wild, E.: Multisurface proximal support vector machine classification via generalized eigenvalues. IEEE Trans. Pattern Anal. Mach. Intell. **28**(1), 69–74 (2006). https://doi.org/10.1109/TPAMI.2006.17
12. Jayadeva, R., Khemchandani, R., Chandra, S.: Twin support vector machines for pattern classification. IEEE Trans. Pattern Anal. Mach. Intell. **29**(5), 905–910 (2007). https://doi.org/10.1109/tpami.2007.1068
13. Schölkopf, B., Smola, A.: Learning with Kernels. MIT Press, Cambridge (2002)
14. Zhang, C., Tian, Y., Deng, N.: The new interpretation of support vector machines on statistical. Sci. Chin. Math. **53**(1), 151–164 (2010). https://doi.org/10.1007/s11425-010-0018-6

An Interactive Virtual Reality System for Cardiac Modeling

Haoyang Shi[1], Xiumei Cai[1(✉)], Wei Peng[1], Hao Xu[1], Cong Guo[1], Miao Tian[2], and Shaojie Tang[1]

[1] School of Automation, Xi'an University of Posts and Telecommunications, Xi'an 710121, China
{caixiumei, tangshaojie}@xupt.edu.cn
[2] School of Computer Science and Technology, Xi'an University of Posts and Telecommunications, Xi'an 710121, China

Abstract. To help medical colleges train students better, help physicians achieve a better preoperative preparation, and help patients understand their condition better, an interactive virtual reality (VR) system is proposed in this paper for cardiac modeling. First of all, we processed the cardiac images acquired by the gated single photon emission computed tomography (SPECT) myocardial perfusion imaging (MPI) to generate the MPI color model in Matlab. Secondly, the MPI color model is fused with the computed tomography (CT) surface model in Matlab. Thirdly, the fused model is imported into Unity3D. Finally, we constructed an interactive VR system with the operating room environment and operated on the fused model virtually. The experimental results show that this system can achieve satisfying performance as expected.

Keywords: SPECT · MPI · CT · VR · Matlab · Unity3D · Cardiac modeling

1 Introduction

Nowadays, we find that there are various pain points in medical colleges and hospitals due to the lagging technology. These pain points are distributed in daily teaching activities, training of young doctors, preoperative preparation and disease communication. According to the 2017 China health and family planning statistics yearbook, the total number of medical students was 4,096,819 [1]. According to the forecast, the number of hospital visits in China will reach 35.94 billion, 37.65 billion and 3.935 billion in 2018, 2019 and 2020 [2]. There were 988,000 medical and health institutions in China by the end of April 2016. Medical simulation teaching refers to using medical simulation technology to create a simulated clinical environment, simulate patients and carrying out clinical teaching and practicing under simulation conditions [3]. This technology can help not only medical students, young doctors, but also preoperative preparation and understanding of illness. According to the problems, we constructed a cardiac model with the gated single photon emission computed tomography (SPECT) myocardial perfusion imaging (MPI) data. The data are processed by using a dynamic programming-based automatic quantification method for the gated SPECT MPI [4]. The result obtained in this way is then fused with the CT images acquired from the

P. Krömer et al. (Eds.): ECC 2018, AISC 891, pp. 784–792, 2019.
https://doi.org/10.1007/978-3-030-03766-6_88

same patient. We then import the processed data into Unity3D programming environment for modeling and rendering. Finally, we completed scene construction, model import, model capture, model segmentation and other relevant operations in Unity3D.

2 Processing of Images

2.1 Automatic Quantification of Single Photon Emission Computed Tomography Myocardial Perfusion Imaging Images

In recent years, the SPECT MPI has been increasingly used to detect cardiac diseases. The original data derived from SPECT MPI device is processed and analyzed by using a dynamic programming-based automatic quantification [4]. The flow chart of the method is shown in Fig. 1(a).

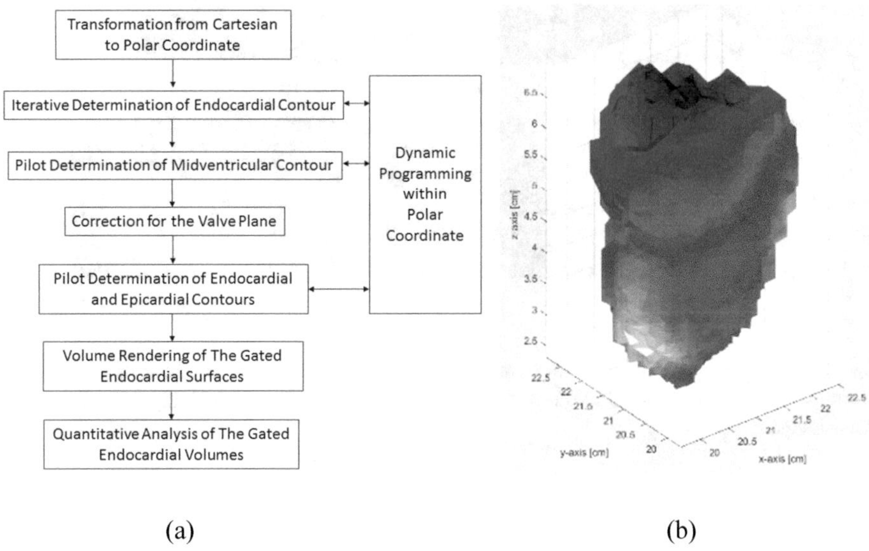

Fig. 1. (a) The flow chart of dynamic programming-based automatic quantification for SPECT MPI, (b) The MPI color model obtained by dynamic programming-based automatic quantification from SPECT MPI

We use long axis slice in the process of modeling. We assume the center of the gated SPECT MPI image volume coincide the center of mass of the left-ventricular. An iterative method is then used to determine the endocardial contour. After the endocardial contour is estimated, the determination of the mid-ventricular contour can be subsequently piloted by the endocardial contour. Finally, both endocardial and epicardial contours can be subsequently piloted by the mid-ventricular contour. In the above process, we all use dynamic programming within polar coordinates. The gray level value of the image (namely the perfusion amount) is incorporated into the endocardial surface as its pseudo-color [4]. The cardiac color model obtained by dynamic programming-based automatic quantification from SPECT MPI (hereafter called MPI color model) is rendered and shown in Fig. 1(b).

2.2 Segmentation of Computed Tomography Images

The CT images acquired from the same patient are segmented by using a simple thresholding (see Fig. 2) [5]. Then Matlab is used to generate a three dimensional (3-D) cardiac surface model from the segmented result (hereafter called CT surface model, see Fig. 3) [5].

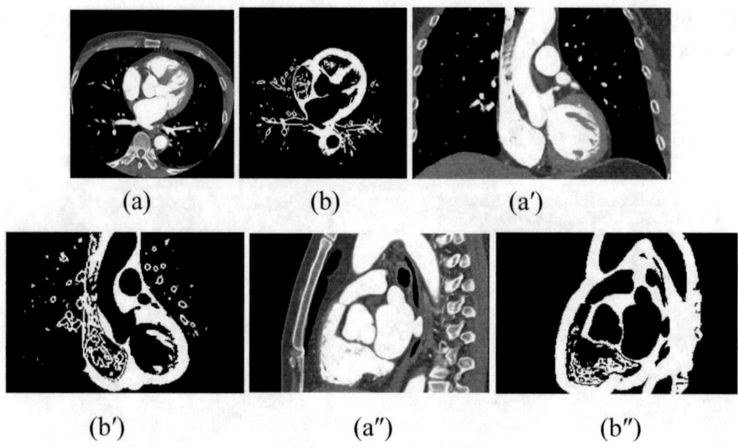

Fig. 2. The 1st, 2nd and 3rd rows correspond to the three orthogonal slices of the patient's thorax, while the left and right columns correspond to the original CTA image and the cardiac tissue image, respectively.

Fig. 3. The CT surface model generated and rendered within Matlab

2.3 Fusion of Myocardial Perfusion Imaging Color Model with Computed Tomography Surface Model

We fused the color information from the MPI color model with the position information from the CT surface model, since the inferior spatial resolution in the MPI color model can be effectively improved by using the CT surface model. The specific way is to add points in the CT model to the MPI model. We ignore coincidence points and add the coordinate information of the remaining points to the MPI model. The nearest five old points around the new point are obtained, and the average grey value of the five old points is assigned to the new point. For example, if the coordinate and grayscale of a point in CT are (x, y, z, a), then the coordinate and grayscale of the new point in MPI are (x, y, z, (a1 + a2 + a3 + a4 + a5)/5). Where, a1, a2, a3, a4 and a5 are the grayscale values of the five nearest old points. The improvement can be clearly appreciated in the VR system proposed below by us.

3 The Construction and Coloration of the Model in Unity3D

Matlab can not output colored 3d models directly. To solve this problem, we used a new rendering method. Using the data processed by Matlab, we directly carried out modeling and rendering operations in Unity3D. The entire process of modeling and rendering in Unity3D will be reported in this section.

3.1 Transmission of Data

We use .txt for data transferring. Due to the large amount of data in this project, we put the .txt into the 'Resources' folder, and use 'resources.load' to read the file. Since the text output by Matlab is in ANSI format, we need to change the format of .txt to UEF8 format for Unity3D reading.

3.2 Point Position and Point Order

Building a model in Unity3D is equivalent to building a mass of triangles. We use 'mesh' component to build the triangle in Unity3D. 'mesh vertices' component corresponds to the position of each triangle vertex, while 'mesh triangle' component corresponds to the join order of vertices. For example, when we create a quadrilateral whose vertices are located at (0, 0, 0), (0, 0, 1), (0, 1, 1) and (0, 1, 0). The specific method is to assign (0, 0, 0, 0, 0, 1, 0, 1, 1, 0, 1, 0) to 'mesh vertices' and assign (0, 3, 2, 0, 2, 1) to 'mesh triangle'. The 12 numbers in the first matrix represent the spatial coordinates of the 4 vertices. The six numbers in the second matrices represent the order of vertices. The quadrilateral is shown in Fig. 4(a).

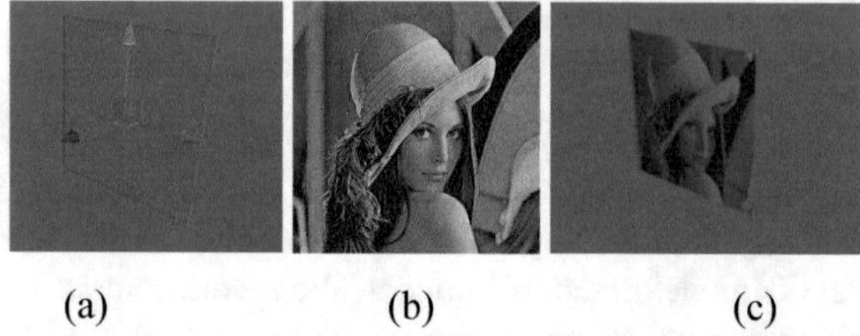

(a) (b) (c)

Fig. 4. (a) The quadrilateral of a fixed-points (0, 0, 0), (0, 1, 0), (0, 1, 1) and (0, 0, 1), (b) is the map (c) shows the mapping effects

After Unity3D reads the txt file, we use the 'length' function to get the length of the matrix directly. After obtaining the matrix, we use the method of cyclic assignment. So, what we do is assigning the $3 \times i$ number, the $3 \times i + 1$ number, and the $3 \times i + 2$ number in the first matrix to the ith digit in 'mesh vertices' and assigning the ith number in the second matrix to the ith digit in 'mesh triangle'. After completing the above steps, the modeling operations can be automated.

3.3 Coloration of Model

Rendering in Unity3D is done by mapping. Specifically, we can set a coordinate for each existing triangle vertex, which corresponds to the specific coordinate position in the map. Taking the quadrangle again as an example, we can assign values to the vertices of all triangles in the quadrangle. For example, we take (0.1, 0.1), (0.9, 0.1), (0.9, 0.9), and (0.1, 0.9) as coordinates of (0, 0, 0), (0, 0, 1), (0, 1, 1), and (0, 1, 0) in the map, respectively. The map is shown in Fig. 4(b) while the effects of the map are also shown in Fig. 4(c).

The biggest problem with rendering described in this paper is the production of texture maps. In the previous subsections, we mentioned the use of grayscale values for pseudo-color rendering, so the focus of this subsection is to create a texture map that can correspond to 256 grayscales. Firstly, we made a color bar with 256 grayscale levels (see Fig. 5(a)). Then, we change the default color mapping of Matlab to the pseudo color mentioned in the previous subsection. Finally, we use color mapping to render grayscale strips.

(a) (b)

Fig. 5. (a) A grayscale bar with 256 grayscale levels, (b) The map contains 256 pseudo-colors

In order to achieve better experimental results, we render 8 gray bars which contain 32 pseudo-colors respectively. The map we made is shown as Fig. 5(b). The order of gray scale growth is from left to right and from bottom to top.

We use a cyclic method to render the model automatically. The concrete method is to determine the grayscale value of each point, and then return the mapping coordinates to each point. For example, when the grayscale range is 156–157, the texture coordinate returned is (0.5625, 0.921875). After using the above method, we built and rendered the model according to the data obtained in Matlab, as shown in Fig. 6.

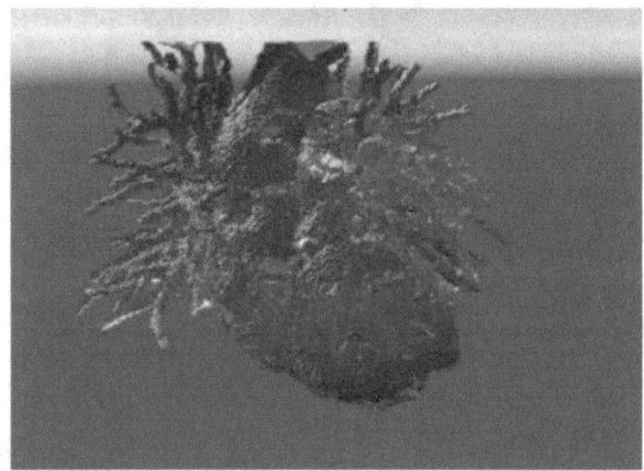

Fig. 6. A cardiac model is built and rendered

3.4 Export of Model

The model needs to be exported because further operations are needed in the subsequent process. The concrete method is to export the model to FBX format using the plug-in in Unity3d. This is a relatively simple process, so it is not described in detail here.

4 Construction of Virtual Reality System

The software platform of this paper is Unity3D, and the hardware platform is HTC Vive. We designed the system, including the construction of the environment and the realization of functions. These two parts are described in detail below.

4.1 Construction of the Environment

The construction of the environment is the most basic part of a VR system. We chose operating room as the theme of simulation environment. The simulation environment of the operation room consists of the following three parts.

The first part is the walls and windows of the operating room. We add collision properties to the walls and windows so that users can be limited to certain areas. The second part is the dynamic character model. There are 3 dynamic doctors in the simulation environment. We add collision, interaction and other functions to these models. The last part is the medical devices and other models. We add collision,

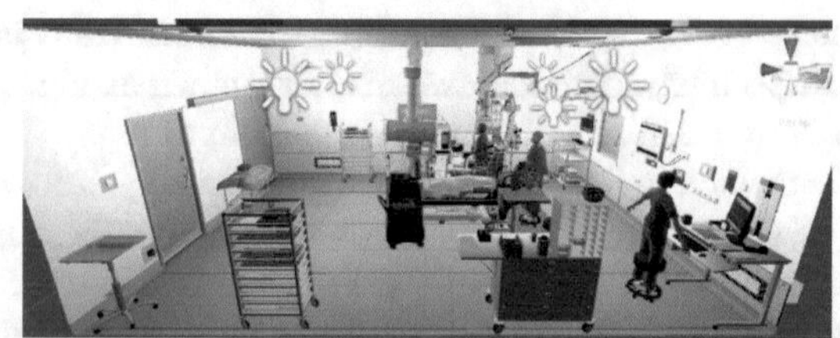

Fig. 7. The overall appearance of the simulation environment

movement and other attributes to the models. After setting up the environment, we imported the cardiac model into the environment and got a complete simulation as shown in Fig. 7.

4.2 Implementation of System Functions

In this section, we added a series of functions to the system. With these functions, a user can implement a series of actions in the scenario. These functions form a complete interactive system with the scenario in the previous section. The following is a description of these functions.

The first function we realized was the walking function of the character. Unlike the complex programming of traditional games, VIVE provides many packaged plug-ins. We use the plug-in to transmit the data of headset device and two handles to Unity3D in real time. In this way, users can move in real time in the simulation environment. In Fig. 8, we show the user's clockwise rotation and the movement to the heart model.

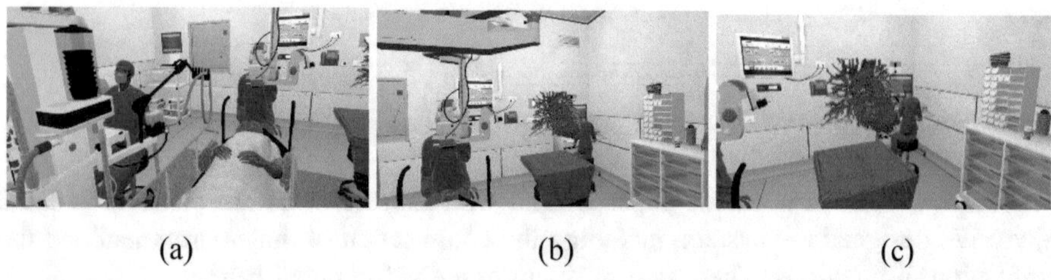

(a) (b) (c)

Fig. 8. (a), (b) show the clockwise rotation, and (c) shows the user's forward operation.

We added the grasping function to the handle in order to make the system more useful. When models collide, if we press a button on the handle, models will be connected by a hinge joint. In this way, objects can be physically connected to achieve the grasping effect. When the button is released, the hinge joint is removed and the object is lowered. In Fig. 9, we show a series of effects achieved by grasping, including rotation function, zoom function, perspective function and so on.

In addition, we designed the model segmentation algorithm. To segment the model is actually to segment the triangular surfaces of the model. Taking Fig. 10(a) as an example, two triangles represent two basic surfaces and horizontal lines represent the

(a) (b) (c)

Fig. 9. (a) shows the scaling function, (b) shows the rotation function, and (c) shows the perspective function.

segmentation track. In triangle abc, each point detects whether the connection line between the point and the cut point is parallel to the edge of the triangle. For example, in triangle abc, ae af are parallel to ab ac, so aef can form a triangle. And be bf are not parallel to ba bc, so points b, e and f cannot directly form a triangle. So the right thing to do is to connect the bf, and generate the triangle bef and the triangle bef. In this way, three triangles are generated in the original triangle abc, namely triangles aef, bef and bfc. We can do the same thing with all triangles and determine the location of all triangles. At this point, all triangles are distributed on both sides of the cutting surface. Two new models are composed of the triangles on both sides. The model that experiences this functionality for segmentation is shown in Fig. 10(b).

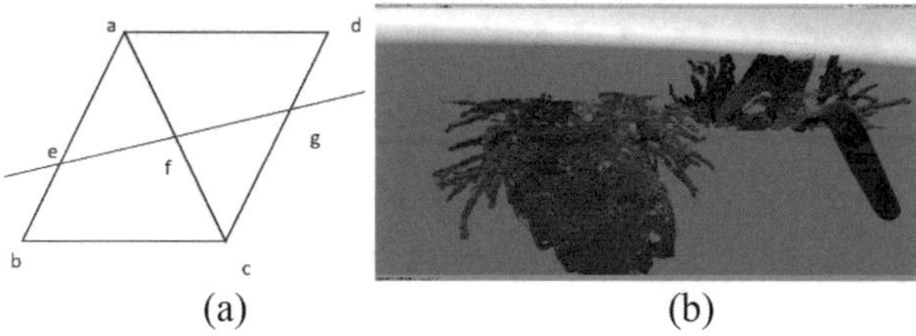

(a) (b)

Fig. 10. (a) shows the segmentation principle, and (b) shows the segmentation results

5 Conclusion

In this work, we firstly used a dynamic programming-based automatic quantification method to process the cardiac images acquired by the gated SPECT MPI to generate the MPI color model. Secondly, the MPI color model was fused with the CT surface model. Thirdly, we proposed a pseudo-color image corresponding to 256 grayscale images, which help us render the fused model accurately within Unity3D. This is not only more feasible than the traditional manual mapping, but also more efficient than the traditional rendering. Finally, we constructed an interactive VR system with the operating room environment and operated on the fused model virtually. By using the proposed system, medical students can learn clinical knowledge better, physicians can analyze the disease and prepare for surgery well, and patients can understand more about their own situations.

Acknowledgments. This work was supported in part by the project for the innovation and entrepreneurship in Xi'an University of Posts and Telecommunications (2018SC-03), the Key Lab of Computer Networks and Information Integration (Southeastern University), Ministry of Education, China (K93-9-2017-03), the Department of Education Shaanxi Province (16JK1712), Shaanxi Provincial Natural Science Foundation of China (2016JM8034, 2017JM6107), and the National Natural Science Foundation of China (61671377, 51709228).

References

1. National Health and Family Planning Commission of PRC: China Health and Family Planning Statistics Yearbook, 1st edn. Peking Union Medical College Press, Beijing (2018)
2. Sun, J., Wen, Q., Chen, F., Wang, Q., Zhu, P.: A study on the prediction of diagnosis and treatment in Chinese hospitals based on the R language ARIMA model. Rural. Econ. Sci.-Technol. **28**(17), 266–269 (2017). https://doi.org/10.3969/j.issn.1007-7103.2017.17.106
3. Zhang, Y.: Discussion on the application of medical simulation teaching in clinical teaching of surgery. Technol. Wind (6), 61 (2018). https://doi.org/10.19392/j.cnki.1671-7341.201806054
4. Tang, S., Huang, J., Hung, G., Tsai, S., Wang, C., Li, D., Zhou, W.: Dynamic programming-based automatic myocardial quantification from the gated SPECT myocardial perfusion imaging. In: The International Meeting on Fully Three-Dimensional Image Reconstruction in Radiology and Nuclear Medicine, Xi'an, China, pp. 462–467 (2017)
5. Tang, S., Zhang, H., Peng, W., Shi, H., Guo, C., Zhao, G., Bian, W., Chen, Y.: A prototype system for three-dimensional cardiac modeling and printing with clinical computed tomography angiography. In: ECC, Spain, pp. 176–186 (2017). https://doi.org/10.1007/978-3-319-68527-4_19

An Improved Method Based on Dynamic Programming for Tracking the Central Axis of Vascular Perforating Branches in Human Brain

Wei Peng[1], Qiuyue Wei[1(✉)], Haoyang Shi[1], Jinlu Ma[1], Hao Xu[1], Tongjie Mu[1], Shaojie Tang[1], and Qi Yang[2]

[1] School of Automation, Xi'an University of Posts and Telecommunications, Xi'an, Shaanxi 710121, China
weiqiuyue123@163.com, tangshaojie@xupt.edu.cn

[2] Department of Radiology, Xuanwu Hospital, Capital Medical University, Beijing 100053, China

Abstract. In the traditional method, generally, it is hard to track the vascular perforating branches in human brain, considering the lower spatial resolution of MRI and involuntary movement of human head. Firstly, the method makes full use of the fuzzy distance transform (FDT) and the local significant factor (LSF) to accurately extract the pivot points in the blood vessel. Then, we improve the original method essentially that the step size is adaptively adjusted according to the curvature of the blood vessel and grayscale information of the vascular perforating branches in MRI data. Thirdly, the central axis of the blood vessel is smoothly and accurately tracked by using the minimum cost path based on dynamic programming. Experiments show that the central axis of vascular perforating branches can be tracked effectively by the improved method.

Keywords: Fuzzy distance transform · Local significant factor
Minimum cost path

1 Introduction

In various analyses of vascular diseases, the effective extraction [1, 2] of the central axis of the blood vessel can reflect the topological structure of the blood vessel well, and at last it is possible to quantitatively analyze blood vessels.

Recently, due to the rapid development of medical image processing technology, the extraction of the central axis of blood vessels has achieved good results. Generally, these methods can be divided into two categories - Topology Preserving Iterative Erosion [3] or Distance Transform based technique [4]. The earliest skeleton extraction was proposed by Blum. Blum's skeleton, or medial axis, is defined using a grassfire transform process [5]. The fire propagates inside the object at an unchanged velocity. The skeleton is the set of extinguish points, where two independent fire-fronts meet. Blum's grassfire transform was later used to extract the central axis or skeleton of an object in the research field of both computer vision and image processing [6]. There are also other methods

P. Krömer et al. (Eds.): ECC 2018, AISC 891, pp. 793–805, 2019.
https://doi.org/10.1007/978-3-030-03766-6_89

based on multi-scale method [7], which makes full use of the flexible frequency bandwidth and ideal enhancement effect of multi-scale real Gabor filtering, and performs central axis enhancement and background noise removal for the vessels of different thicknesses, and then uses the Hessian matrix to calculate the orientation. The information is subjected to non-maximum suppression to obtain the local extreme of the response map, and finally the central axis of the blood vessel is obtained by double threshold segmentation. As well as some scholars' methods based on the minimum-cost path [9], most of the methods have achieved great results. However, for complex medical images and complex human blood vessel topologies, these traditional methods require a large amount of time to enhance and segment the blood vessels and can't fully utilize the grayscale information of the image, and there is no effective method to solve the problem of vascular perforating branches in human brain.

In response to the above problems, we propose a blood vessel central axis tracking method based on optimization theory. This method only requires the user to provide two consecutive initial tracking points, and the method will automatically complete the extraction of the central axis of the blood vessel. The improved method directly uses the information of the original image and selects the pivot point of the blood vessel by combining the fuzzing distance transform (FDT) [8] and the local significant factor (LSF) [4]. The resulting mid-vessel pivot points are then smoothed and accurately centered on the vessel axis through a minimal cost path based on dynamic programming. In the second part, the FDT, LSF and the improved method are described. The last part is on the experiment, where the simulated data and the actual MRI data are used to validate the performance of the improved method.

2 Methodology

2.1 Fuzzy Distance Transformation

The FDT adopted in this work was proposed by scholars such as Punam K. Saha, who calculates the fuzzy distance between two points. The FDT transform is different from the traditional Distance Transformation (DT) [9]. The DT is only suitable for binarized images, while the FDT is suitable for both binarized and grayscale images. Since medical images are displayed in the form of grayscale, they have complex information. Therefore, we prefer to use a distance transform based on grayscale image—FDT, described below.

Supposing that X is a fuzzy set, a fuzzy subset S of X is defined as an ordered set of pairs, i.e., $S = \{(x, \mu_s(x)) | x \in X\}$, where $\mu_s : X \to [0, 1]$ is the membership function of S. A two-dimensional (2D) fuzzy digital object O is a fuzzy subset defined on Z^2, i.e., $O = \{(p, \mu_O(p)) | p \in Z^2\}$, $\mu_O : Z^2 \to [0, 1]$. If $\mu_O(p) > 0$, it means that the pixel p belongs to the support $\Theta_\theta(O)$. Here, $\Theta_\theta(O) = \{p | p \in Z^2, \mu_O(p) \geq \theta\}$ is used to denote the θ support of O, where θ is usually zero.

In a fuzzy object O, the fuzzy distance between pixels p and q is defined as the shortest path length between them. As $p \in \Theta_\theta(O)$ and $q \in Z^2$, the link $\langle p, q \rangle$ representing the length between them is defined as

$$\langle p, q \rangle = \frac{1}{2}(\mu_O(p) + \mu_O(q)) \times \|p - q\| \tag{1}$$

where the $\mu_O(p)$ indicates the membership of pixel p in the fuzzy object O. As $p, q \in O$, $\|p - q\|$ is the Euclidean distance between them.

Let $P(p, q)$ denotes a set of all paths from p to q. For any path $\pi \in P(p, q)$, $\pi = <p = p_1, p_2 \ldots p_m = q>$. The length of π is the sum of all line segments on the path, which is defined as

$$\Pi_O(\pi) = \sum_{i=1}^{m-1} \frac{1}{2}(\mu_O(p_i) + \mu_O(p_{i+1})) \times \|p_i - p_{i+1}\| \tag{2}$$

If $\Pi_O(\pi_{p,q}) \le \Pi_O(\pi)$ and $p, q \in O$, $\pi_{p,q} \in P(p, q)$ is the shortest path, then the fuzzy distance from p to q is expressed as

$$\omega_O(p, q) = \min_{\pi \in P(p,q)} \Pi_O(\pi) \tag{3}$$

Assuming that $p \in Z^2$ and it belongs to $\Theta_\theta(O)$, the fuzzy distance between p and the closest point $q \in \overline{\Theta_\theta(O)}$ is denoted by

$$\Omega_O(p) = \min_{q \in \overline{\Theta(O)}} \omega_O(p, q) \tag{4}$$

2.2 Calculation of Fuzzy Distance Transform Value

The dynamic programming proposed by Punam K. Saha is used to calculate the value of FDT [8]. Punam K. Saha has proved that dynamic programming terminates in a finite number of steps, and as it terminates it produces the desired FDT image. The membership of each pixel is calculated as follows,

$$\mu_O(p) = \begin{cases} G_{m_O,\sigma_O}(f(p)), & \text{if } f(p) \le m_O \\ 1, & \text{otherwise} \end{cases} \tag{5}$$

where f represents a grayscale image, and m_O and σ_O represent the grayscale mean and standard deviation of the image, respectively. G_{m_O,σ_O} denotes a Gaussian function without a normalization.

2.3 Local Significant Factor

Medical image information is relatively complex and noisy, which will affect the accurate extraction of the central axis of the blood vessel. We correct for the center of the blood vessel by adding LSF. $O = \{p \in Z^2 | \mu_O(p) \neq 0\}$ is the support domain, if p satisfies the following inequality,

$$FDT(q) - FDT(p) < (\mu_O(p) + \mu_O(q))|p - q|/2, \ \ p \in O \tag{6}$$

where $|p - q|$ is the Euclidean distance of the two points, and q is a point in the 8 neighborhood of p (Fig. 1).

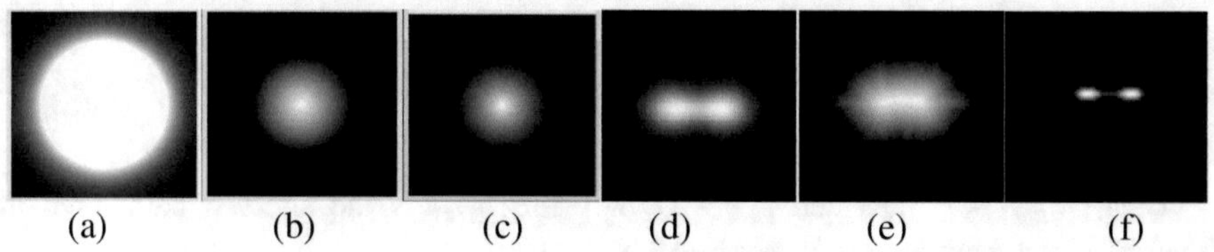

Fig. 1. Comparison of DT and FDT. (a), (d) Simulation model. (b), (e) The result of the DT. (c), (f) The result of the FDT.

If the pixel p satisfies the above formula, its LSF value as defined below can be calculated,

$$LSF(p) = 1 - f_{+}(\max_{q \in N^{*}(p)} \frac{2(FDT(q) - FDT(p))}{(\mu_O(p) + \mu_O(q))|p - q|}) \tag{7}$$

where the function $f_{+}(x)$ returns the value of x if $x > 0$ and zero otherwise. It can be shown that the value of LSF is in the interval [0, 1].

2.4 Improvement Strategy

Figure 2 shows the flow chart of our improved tracking method. The tracking step size can be changed in real time by determining the degree of bending of the blood vessel. At the same time, the branching strategy is used to determine the occurrence of the branch structure, thereby saving the branch point for branch tracking, and the procedure is over until the central axis of all active branches of the blood vessel has been tracked.

Selection of Pivot Points in Blood Vessel. In this paper, the pivot point in the sliced blood vessel is selected by combining FDT with LST. Since in the MRI slice, the gray value of many tissues of the human body is very close to the gray value of the blood vessel center. Therefore, in the process of tracking, the roundness is used to determine whether certain regions in the current slice belong to the blood vessel region. The calculation formula of roundness is as follows:

$$R = 4\pi s/c^2 \tag{8}$$

where s represents the area of the region, i.e., the number of all pixels in the area, and c represents the perimeter of the region, which can be obtained by summing the distances of adjacent points on the edge of the region.

In order to further obtain a more accurate blood vessel area, the anterior-posterior relationship of the slices is utilized. If the vascular area in the current slice varies greatly from that in the previous slice, the tracking in the current slice will be repeated with an increased tracking step. The resulting vascular region is subjected to FDT and LSF to determine the vessel center, and the interpolation range required for the next tracking slice is calculated by

$$extent = \max(FDT) \times reso \times 2 \tag{9}$$

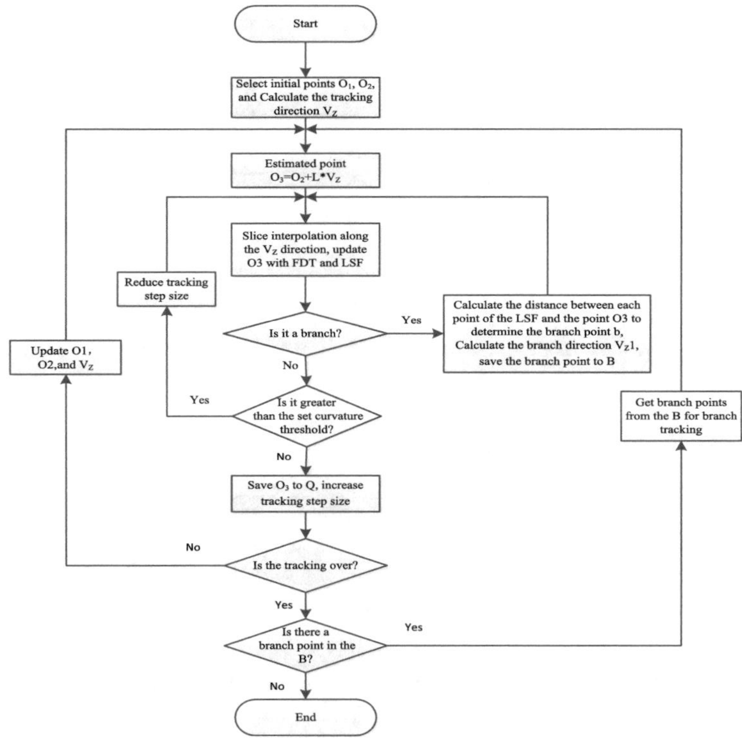

Fig. 2. The flow chart of the algorithmic steps for tracking the vascular central axis.

Vascular Branch Determination. During the tracking process, we calculate the LSF of the slice, keep the LSF of 1, and calculate the distance between them to determine whether the point is a branch point.

The specific algorithmic steps are listed as follows.

1. First select the LSF point corresponding to the maximum point of the slice FDT as the initial point, and set it to O_1, and save it into the cell array A.
2. Calculate the LSF of the slice, and calculate the Euclidean distance between the points where all LSF values are 1 and the initial point O_1, and set distance to D. If $D >$ max (FDT), it is determined that the point belongs to the branch point and is saved in A.
3. Calculate the Euclidean distance between the points with the remaining LSF values of 1 and the points included in A, and set them as D_1 and D_2, respectively. If $D_1 >$ FDT(O_1) and $D_2 >$ FDT(D_2), it is determined that this point belongs to the branch point and is saved in A.
4. Repeat the above procedure until the new branch point does not appear, and finally A contains all the branch points (Figs. 3 and 4).

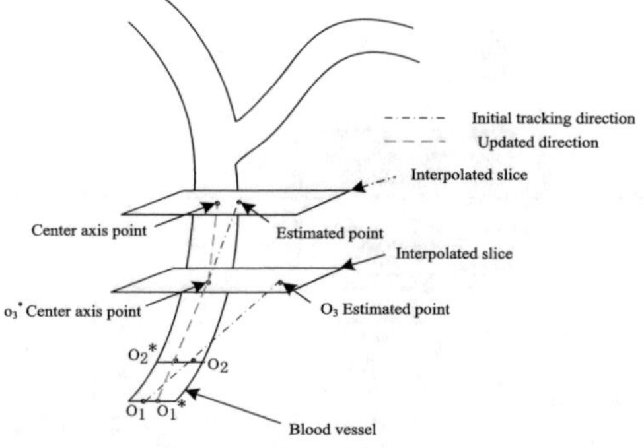

Fig. 3. Schematic of the improved method for tracking the vascular central axis

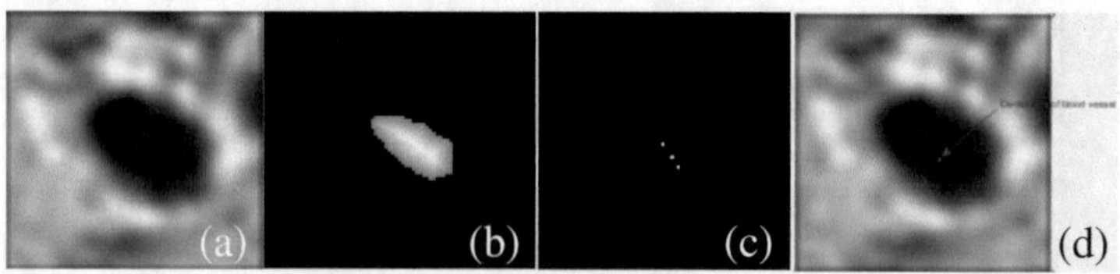

Fig. 4. (a) The original image of the blood vessel slice, (b) the FDT of (a), (c) the LSF of (a), and (d) the central axis point of the blood vessel determined by the improved tracking.

Determination of Tracking Direction. In Fig. 5, the coordinates of the points $O_1(x_1, y_1, z_1)$ and $O_2(x_2, y_2, z_2)$ are shown.

The tracking direction vector is determined as follows,

$$\overrightarrow{Vz} = \overrightarrow{O_1O_2} \Big/ \left|\overrightarrow{O_1O_2}\right| = ((x_2 - x_1)/d, (y_2 - y_1)/d, (z_2 - z_1)/d) \tag{10}$$

where d represents the Euclidean distance between two points,

$$d = \left|\overrightarrow{O_1O_2}\right| = \sqrt{(x_2 - x_1)^2 + (y_2 - y_1)^2 + (z_2 - z_1)^2} \tag{11}$$

The next point O_3 is estimated as

$$O_3 = (x_3, y_3, z_3) = \left(x_2 + \frac{x_2 - x_1}{d}L, y_2 + \frac{y_2 - y_1}{d}L, z_2 + \frac{z_2 - z_1}{d}\right) \tag{12}$$

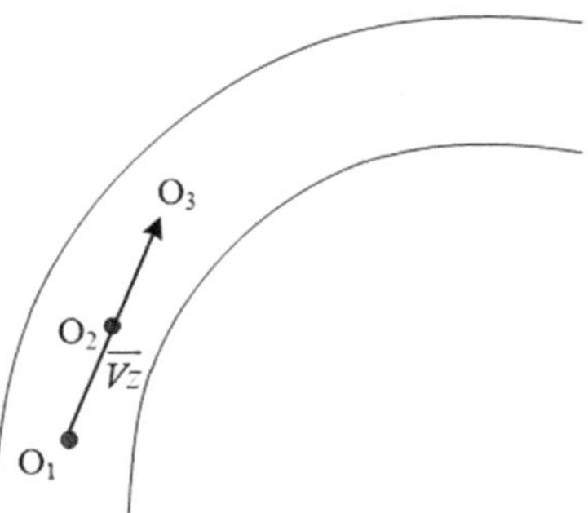

Fig. 5. Schematic of determining the tracking direction

Adaptive Step Size. The general tracking strategy is to fix a small step size, which greatly increases the amount of tracking calculation. If the tracked blood vessel is relatively straight, tracking with a small step size will increase the consumption of computation. If a bended blood vessel is tracked, a large step size will cause the tracking to deviate from the blood vessel, causing the error to increase. In order to balance the two situations encountered in the tracking process, we propose an adaptive step size-based tracking strategy. When the curvature of the blood vessel is relatively small, the method will judge that the blood vessel is relatively smooth. When the curvature of the blood vessel is relatively large, the blood vessel is bent. As the blood vessel is smooth, the tracking step size is increased to improve the tracking efficiency.

While the blood vessel is bent, the tracking step length is reduced to increase the tracking accuracy. As soon as three points ($O_1{*}(x_1, y_1, z_1)$, $O_2{*}(x_2, y_2, z_2)$ and $O_3{*}(x_3, y_3, z_3)$) are known, the curvature is calculated from these three points as follows,

$$\gamma = arc\cos\left(\left(\overrightarrow{r_1}, \overrightarrow{r_2}\right) / \left(\left|\overrightarrow{r_1}\right| . \left|\overrightarrow{r_2}\right|\right)\right) \tag{13}$$

$$k = \gamma / \left(\left|\overrightarrow{r_1}\right| + \left|\overrightarrow{r_2}\right|\right) \tag{14}$$

where $\overrightarrow{r_1} = O_2^* - O_1^*$ and $\overrightarrow{r_2} = O_3^* - O_2^*$. We calculate the curvature from the current three points, thereby adjusting the step size accordingly. The curvature threshold k_{th} is selected, $k_{th} = 0.1$ is taken as a fixed threshold in this work, and L is the step size used for tacking,

$$L = \begin{cases} 2L, & k < k_{th}, \\ L/2, & k \geq k_{th}. \end{cases} \tag{15}$$

Tracking Vascular Perforating Branches. In human brain, tracking vascular perforating branches in MRI data can greatly benefit the diagnosis of diseases caused by the decline of cerebral vascular aging. Unfortunately, vascular perforating branches usually appear not clearly or are broken due to the lower spatial resolution of MRI and involuntary movement of human head. Generally, the traditional methods are hard to track and extract the central axis of these critical branches in human brain.

To solve this problem mentioned above, the step size L will be adaptively updated by using $L_1 = L + count$ with the initial value of *count* being 1. If there is still no vascular area in the interpolated slice, we repeat to adjust L_1 by using $count = count + 1$. In this work, a maximum number is set empirically as 5 for *count*. A concise explanation is given here for this strategy. When the value of *count* is greater than or equal to 5 and the vascular region is not found in the tracked slices, it is determined that the tracking ends for vascular perforating branches. When *count* is less than 5 and the vascular region is found, it is determined that the vascular perforation occurs. The stop criterions for tracking vascular central axis are designed as follows:

1. The coordinate of the estimated next point is beyond the range of the entire image, the tracking is stopped;
2. As the FDT value of the tracked blood vessel is less than a given threshold, the tracking is stopped (Fig. 6).

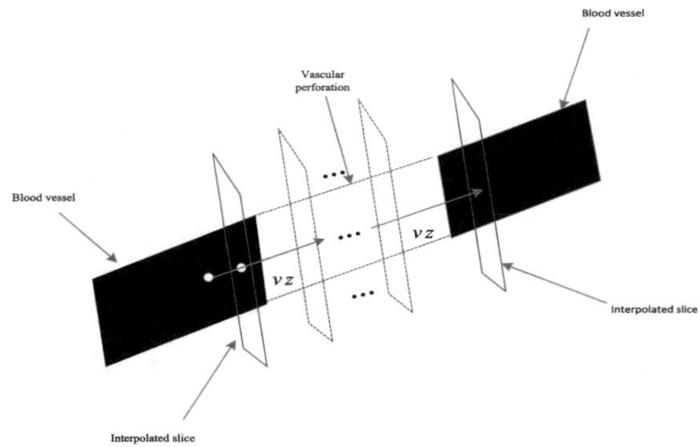

Fig. 6. Schematic of tracking vascular perforating branch, in which the vascular region is a black part and the dotted part is a vascular perforation region. Tracking direction *vz* does not change, along which the tracking step size is adjusted to determine whether or not to pass through the vessel.

2.5 Dynamic Programming for Tracking Vascular Central Axis

The vascular central axis is usually extracted roughly by tracking the blood vessel. Therefore, the central axis of the blood vessel is not smooth enough, and so the error is relatively large. Moreover, in the rough extraction process where the step size is adaptively changed, if a more accurate blood vessel center axis is desired, the dynamic extraction method can be used to optimize the extraction of the central axis. LSF value is used to measure the path cost. For the path π, the energy cost between every two consecutive slices is defined as follows [9],

$$EC = 1/\max(LSF(p) + LSF(q)) \tag{16}$$

where p is the voxel in a slice, and q is the voxel in another slice adjacent to the previous slice. When the LSF value is a maximum of 1, it indicates that the resulting pivot point of the blood vessel is closer to the center of the blood vessel. Therefore, the *EC* takes a minimum so that the Central axis of the tracked blood vessel is closer to the geometric center.

Define Cost(π) as the total energy cost of the path $\pi = <p_1, p_2 \ldots p_m>$, which is the sum of the energy costs of all the slices on the path:

$$\text{Cost}(\pi) = \sum_{i=1}^{m} EC(p_i, p_{i-1}) \tag{17}$$

where tracking is performed from the current initial point until the point p_m that satisfies the tracking stop criterions. If the latest branch is added the current blood vessel branch *S*, the branch of the smallest total cost path is defined as [9],

$$BV_{s,p_m} = \underset{\pi \in \Pi_{s,p_m}}{\arg} \min \mathrm{Cost}(\pi) \tag{18}$$

where Π_{s,p_m} represents all paths from the current branch S to the final tracking point p_m. Minimum cost path branch BV_{s,p_m} was obtained by using dynamic programming.

3 Experimental Results and Analysis

Two experiments were conducted to compare the improved method with the traditional method. The first experiment simulated the vascular structure within MATLAB, and adopted the improved method to obtain the central axis of the blood vessel and to determine the accuracy of the method by calculating the consistency between the evaluated and the actual central axes. The second experiment was performed directly on the MRI data of the blood vessels in human brain to validate the performance of the improved method in the practical clinical application.

3.1 Simulated Vascular Data

We verify the improved method from the simulated vascular data. First, utilizing MATLAB to simulate a tubular structure, we use the sine function as an ideal central axis of the blood vessel, and then use the DT to simulate the transaxial section of blood vessel. As shown in the figures below, a waved blood vessel is simulated. From the

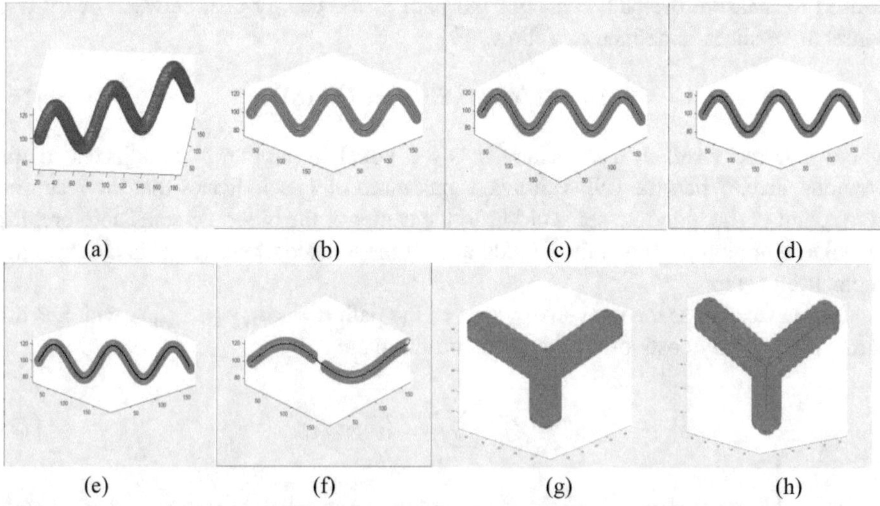

Fig. 7. (a) The blood vessel image simulated and then rendered. (b) The ideal vascular central axis. (c) The vascular central axis obtained by the adaptive step size. (d) The vascular central axis obtained by a fixed step size $L = 1$. (e) The vascular central axis obtained by the dynamic programming with an adaptive step size. (f) The central axis of vascular perforating branch obtained by the dynamic programming with an adaptive step size. (g) the simulated blood vessel with a branch and (h) the central axis of (g) tracked by the improved method.

figures, it can be noticed that the improved method can effectively extract the central axis of the blood vessel.

Due to the branching of the blood vessels, the branching of the blood vessels was also simulated during the experiment. As shown in Fig. 7, the method can well handle the branching of the blood vessels and display the tracking results.

In the experiment, the simulated data were used to evaluate the performance of the improved method. The indexes for assessing the performance of tracking the vascular central axis include the average error between the tracked axis and the ideal axis, the total number of pivot points, and the time of operation. For the average error, we use the Hausdorff distance to measure.

The Hausdorff distance is a measure on the degree of similarity between two sets of points, each set probably having a different number of points. Supposing there are two sets of points $A = \{a_1, a_2, \ldots\}$ and $B = \{b_1, b_2, \ldots\}$, the Hausdorff distance between the two sets of points is defined as,

$$H(A,B) = \max[h(A,B), h(B,A)] \tag{19}$$

$$h(A,B) = \max_{a \in A} \min_{b \in B} \|a - b\|, h(B,A) = \max_{b \in B} \min_{a \in A} \|b - a\| \tag{20}$$

where $\| \ \|$ indicates the Euclidean distance between two points. $h(A, B)$ and $h(B, A)$ are called the one-way Hausdorff distance from A to B and from B to A, respectively. $H(A, B)$ is called the two-way Hausdorff distance. Then, average error is defined as

$$error_{average} = H(A,B)/L_C \tag{21}$$

where L_C represents the length of the simulated blood vessel.

Listed in Table 1 are the results on the indexes for assessing the performance of tracking the vascular central axis with a fixed step size or by our improved method. From the results, we can noticed that, our improved method can effectively reduce the average error as compare to tracking with a fixed step size, while the total numbers of pivot points are almost the same. Meanwhile, our improved method can solve vascular perforating branch problem to certain degree, since it inherits the superior capability of dynamic programming and exploits the grayscale information in MRI data to adjust the step size adaptively.

Table 1. Results on the indexes for assessing the performance of tracking the vascular central axis

Tracking method	The pivot points	Average error (%)	Time (min)
Fixed step size $L = 1$	295	1.36	6.758
The improved method	298	0.93	7.863

3.2 Clinical Vascular Data

We used MRI data of the blood vessel in human brain to evaluate the improved method. For a segment of blood vessel in the MRI data, the central axis obtained by our improved method is compared with that obtained by manually tracking. Our improved method can automatically track the central axis of the blood vessel after just selecting two initial points, whereas manually tracking is time consuming, very tedious for operator, and requires a certain degree of expertise to identify and mark a sequence of pivot points along the blood vessel. It can also be noticed from Fig. 8(a) and (b) that, the central axis of the blood vessel tracked manually is not smooth, whereas that tracked by the improved method is pretty smooth and accurate. Meanwhile, from Fig. 8 (c) and (d), we can observed that the fluctuation extent of grayscale values along the central axis tracked by our improved method is obviously smaller than that acquired by manually tracking.

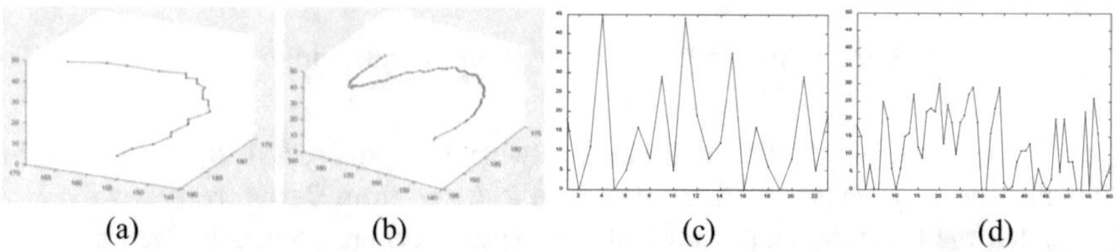

Fig. 8. The central axis of the blood vessel obtained (a) after manually tracking and (b) by the improved method. (c) Grayscale values of the blood vessel obtained after manually tracking and (d) by our improved method.

4 Conclusion

In this paper, tracking the central axis of the blood vessels based on the optimization theory is improved by us. The improved method asks a user to firstly select two pivot points as initial seeds, and then automatically tracks the central axis of the blood vessel. The FDT and LSF values are used to obtain the subsequent pivot points in the blood vessels as in the original method. The central axis can be efficiently tracked by adaptively adjusting the step size according to the curvature of the blood vessel. The grayscale information in MRI data is exploited for tracking the central axis of vascular perforating branches in human brain by adaptively adjusting the step size too. Lastly, the minimum cost path algorithm based on dynamic programming is used for tracking the central axis as in the original method. The experimental results show that the improved method may track the vascular perforating branches in human brain even though these branches usually appear not clearly or are broken due to the lower spatial resolution of MRI and involuntary movement of human head.

Acknowledgments. This work was supported in part by the project for the innovation and entrepreneurship in XUPT (2018SC-03), the Key Lab of Computer Networks and Information Integration (Southeastern University), Ministry of Education, China (K93-9-2017-03), the Department of Education Shaanxi Province (15JK1673), Shaanxi Provincial Natural Science Foundation of China (2016JM8034, 2016JQ5051).

References

1. Yang, G., Kitslaar, P., Frenay, M.: Automatic centerline extraction of coronary arteries in coronary computed tomographic angiography. Int. J. Cardiovasc. Imaging **28**(4), 921–933 (2012). https://doi.org/10.1007/s10554-011-9894-2
2. Akhtar, Y., Mukherjee, D.P.: Reconstruction of three-dimensional linear structures in the breast from craniocaudal and mediolateral oblique mammographic views. IET Image Proc. **11**(11), 1114–1121 (2017). https://doi.org/10.1049/iet-ipr.2016.1063
3. Sadleir, R., Whelan, P.: Fast colon centerline calculation using optimized 3D topological thinning. Comput. Med. Imaging Graph. **29**(4), 251–258 (2005). https://doi.org/10.1016/j.compmedimag.2004.10.002
4. Jin, D., Saha, P.K.: A new fuzzy skeletonization algorithm and its applications to medical imaging. In: 17th International Conference on Image Analysis and Processing (ICIAP). LNCS, Naples, Italy, pp. 662–671 (2013). https://doi.org/10.1007/978-3-642-41181-6_67
5. Blum, H.: A transformation for extracting new descriptors of shape. In: Models for the Perception of Speech and Visual Form, pp. 362–380. MIT Press, Cambridge, (1967)
6. Saha, P.K., Strand, R., Borgefors, G.: Digital topology and geometry in medical imaging: a survey. IEEE Trans. Med. Imaging **34**(9), 1940–1964 (2015). https://doi.org/10.1109/TMI.2015.2417112
7. Lukač, A., Subašić, M.: Blood vessel segmentation using multiscale Hessian and tensor voting. In: 40th International Convention on Information and Communication Technology, pp. 1534–1539. Electronics and Microelectronics (MIPRO), Opatija (2017). https://doi.org/10.23919/mipro.2017.7973665
8. Saha, P.K., Wehrli, F.W., Gomberg, B.R.: Fuzzy distance transform: theory, algorithms, and applications. Comput. Vis. Image Underst. **86**(3), 171–190 (2002). https://doi.org/10.1006/cviu.2002.0974
9. Jin, D., Iyer, K.S., Hoffman, E.A., Saha, P.K.: A new approach of arc skeletonization for tree-like objects using minimum cost path. In: 22nd International Conference on Pattern Recognition, Proceedings of the IAPR International Conference Pattern Recognition, Stockholm, pp. 942–947 (2014). https://doi.org/10.1109/icpr.2014.172

Orientation Field Estimation with Local Information and Total Variation Regularization for Incomplete Fingerprint Image

Xiumei Cai(✉), Hao Xu, Jinlu Ma, Wei Peng, Haoyang Shi, and Shaojie Tang

School of Automation, Xi'an University of Posts and Telecommunications, Xi'an 710121, Shaanxi, China
{caixiumei, tangshaojie}@xupt.edu.cn

Abstract. Orientation field (OF) estimation is an important procedure in fingerprint image preprocessing. As for the hard problem that traditional methods cannot estimate the OF on incomplete fingerprint image accurately and the subsequent recognition will be influenced unavoidably, we propose an algorithm for the OF estimation which combines together the fidelity term of the local information from incomplete fingerprint image and a total variation (TV) regularization term. The local information involves the OFs evaluated by the traditional gradient-based method and the zero-pole model-based method. The experimental results demonstrate that proposed algorithm is effective in reconstructing the OF of incomplete fingerprint image.

Keywords: Incomplete fingerprint · Orientation Field · Total variation Regularization · Estimation

1 Introduction

With the development of science and technology, biometric technology has been widely used [1]. However, in the process of fingerprint acquisition, the images are often low quality or even incomplete because of the influence factors such as wet, scar and molt, as well as the loss and failure of the sensor during the acquisition process. This has caused great troubles for the later fingerprint image processing and recognition, especially in the criminal investigation process, the most of the latent fingerprints collected in the field are incomplete. According to the data, about 10% of the fingerprints in the database are incomplete, so estimation for incomplete fingerprints is quite desired in practice. Shown in Fig. 1 are three typical samples of incomplete fingerprint image. It is difficult to restore incomplete fingerprint image directly. But it is possible to first restore the orientation field (OF) of fingerprint by local information. The restoration of OF can provide more benefit for the restoration of fingerprint image in the later stage.

P. Krömer et al. (Eds.): ECC 2018, AISC 891, pp. 806–813, 2019.
https://doi.org/10.1007/978-3-030-03766-6_90

Fig. 1. Typical samples of incomplete fingerprint image

2 Estimation for Incomplete Region

For an incomplete fingerprint, it is necessary to obtain a coarse OF, regardless of the size of the incomplete region, in order to estimate, synthesize and smooth the OF for the incomplete region. Gradient-based method introduced by Kass and Witkin [2] is used to calculate the coarse OF. After getting the result of coarse OF, we estimate the OF of incomplete region.

2.1 Neighborhood-Based Estimation

Because the ridge flow or the valley flow is evenly distributed in a specific neighborhood, the change is slow and has the characteristics of continuity. The neighborhood-based estimation method [3] can be used to calculate the missing OF. The steps are detailed as follows:

1. The image is divided into 3×3 regions. As shown in Fig. 2, let $\{I, II, IV, V\} \in D_1$, $\{II, III, V, VI\} \in D_2$, $\{IV, V, VII, VIII\} \in D_3$ and $\{V, VI, VIII, IX\} \in D_4$.

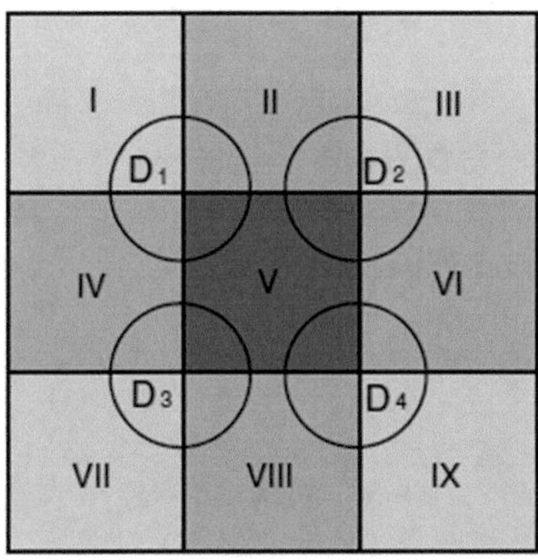

Fig. 2. The block is divided into four parts and V is the target block

2. Calculate the coherence [2] of area D_1, D_2, D_3 and D_4. Let Coh_{max} = Max{Coh_1, Coh_2, Coh_3, Coh_4}. The target block is estimated on the basis of the block which has the largest value of coherence.
3. The direction of the target block is calculated by

$$O_n = \sum_{i=1}^{3} \theta_i w_i \tag{1}$$

Assuming that the coordinates of the block center is (x_m, y_m), w_i is defined as

$$w_i = \left[(x_i - x_m)^2 + (y_i - y_m)^2 \right]^{-\frac{1}{2}} \tag{2}$$

2.2 Minutiae-Based Estimation

Minutiae are one of the most widely used features in fingerprint recognition. In many fingerprint estimation schemes, minutiae provide a lot of fingerprint information. Feng et al. proposed a method to estimate the OF based on the minutiae [4].

The minutiae set is defined as

$$\{x_n, y_n, \alpha_n\}_{n=1}^{N} \tag{3}$$

where x_n and y_n denotes the horizontal and vertical coordinate of the nth minutiae, respectively, while α_n denotes its direction, N denotes the total number of minutiae. For the target block (m, n), we estimate the direction of the block by using the nearest minutiae. The calculation method is given as follows

$$u = \sum_{n=1}^{N} \cos(2\alpha_n)\omega_n \tag{4}$$

$$v = \sum_{n=1}^{N} \sin(2\alpha_n)\omega_n \tag{5}$$

$$O_m(m, n) = \frac{1}{2} \arctan\left(\frac{v}{u}\right) \tag{6}$$

where ω_n is a weight function. Improve the method in [4], we calculate the value of ω_n by the following method:

First of all, the equation of minutiae line obtained according to the information from the minutiae set is

$$y - y_n = \tan\alpha(x - x_n) \tag{7}$$

Supposing that the coordinates of the target block center is (x_O, y_O), the distance d_n from the point to the line on which the detail point located is

$$d_n = \frac{|x_0 \tan\alpha - y_0 + y_n - x_n \tan\alpha|}{\sqrt{1 + \tan^2\alpha}} \tag{8}$$

So, one has

$$\omega_n = \frac{1}{d_n + r} \tag{9}$$

In the upper form, r represents the reliability of the minutiae. If the minutiae are unreliable, r will be assigned a larger value.

2.3 Fused OF

The OF obtained by gradient-based method is coarse. In incomplete region of fingerprint image, the OF result is not accurate. By the method of Sects. 2.1 and 2.2, we replace the results of the low quality region with the following results

$$O_f = \alpha O_n + (1 - \alpha) O_m \tag{10}$$

where α is determined according to the actual situation and fixed to 0.5 in this paper.

3 Estimation Based on Singular Point

Singular point is one of the most important global features in fingerprint image, and it also affects the OF calculation, recognition and matching of fingerprint. Zero-pole model [5] is used in the section to obtain OF. The orientation $O(z)$ at any point z on the complex plane can be regarded as the angle of the complex function $P(z)$:

$$P(z) = \sqrt{e^{2jo_\infty} \frac{(z - z_{c1})(z - z_{c2})\cdots(z - z_{cn})}{(z - z_{d1})(z - z_{d2})\cdots(z - z_{dm})}} \tag{11}$$

$$O(z) = \arg(P(z)) \bmod \pi \tag{12}$$

where z_{ci} and z_{dj} denote the position of the ith core and the jth delta, respectively. O_∞ represents the OF in the ideal state, which is infinitely far from the singular point, where we generally set the value to 0. The combination of the above two formulas leads to

$$O(z) = \left[o_\infty + \frac{1}{2} \times \left(\sum\nolimits_{i=1}^{K} \arg(z - z_{ci}) - \sum\nolimits_{j=1}^{L} \arg(z - z_{dj})\right)\right] \bmod \pi \tag{13}$$

It can be observed that from the above formula that the effect of each core to the orientation is $0.5\arg(z - z_{ci})$, while the delta is $-0.5\arg(z - z_{dj})$. Figure 3 shows the different OFs that calculated by the zero-pole model-based method.

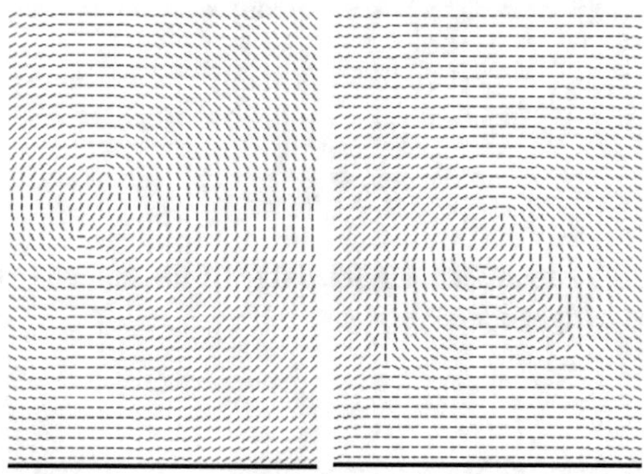

Fig. 3. (left) the OF of two cores, and (right) the OF of two cores and two deltas

4 Orientation Smoothing with TV Regularization

It has been already known that the OFs obtained by the neighborhood- and minutiae-based methods are not reliable. In the section, a new smoothing method is proposed to suppress the noise in the image.

TV model was first proposed by Rudinosher and Fatemi [6]. In the traditional TV regularization, there is only one initial image. Here, we add the zero-pole model to smooth the obtained image and fit the original image. First of all, we improve the TV model as

$$u = \underset{u \in BV(\Omega)}{\arg\inf}\left\{\omega_1\|u_0 - u\|^2_{L^2(\Omega)} + \omega_2\|u_1 - u\|^2_{L^2(\Omega)} + \lambda|u|_{TV}\right\} \tag{14}$$

where ω_1 and ω_2 are two positive scalars used to balance the proportion of the two images in the output image. The sum of ω_1 and ω_2 is 1/2. If the OF of fingerprint obtained in 2.3 is denoted as OF_1, and the OF obtained by the zero-pole model is OF_2, then the following energy functional can be designed following the above mentioned TV model,

$$\begin{aligned}\Phi(OF) = {} & \omega_1 \iint_{\Omega} [OF(x,y) - OF_1(x,y)]^2 dxdy \\ & + \omega_2 \iint_{\Omega} [OF(x,y) - OF_2(x,y)]^2 dxdy + \lambda TV(OF)\end{aligned} \tag{15}$$

where

$$TV(OF) = \iint\limits_{\Omega} |\nabla\ OF(x,y)|dxdy = \iint\limits_{\Omega} \left[(\frac{\partial OF}{\partial x})^2 + \frac{\partial OF}{\partial y})^2 \right] dxdy \quad (16)$$

And the corresponding Euler-Lagrange equation is

$$\omega_1(OF - OF_1) + \omega_2(OF - OF_2) + \lambda div(\frac{\nabla OF}{|\nabla OF|}) = 0 \quad (17)$$

We can use gradient descent method to obtain its solution in an iterative form. the initial of can be a linear combination of OF_1 and OF_2, e.g

$$OF(x,y,0) = 2(\omega_1 OF_1(x,y)\ +\ \omega_2 OF_2(x,y)) \quad (18)$$

5 Experimental Results

In order to validate the proposed algorithm, some fingerprints in FVC2004 database [7] are used for the experiment. First of all, some high-quality fingerprints in the database are selected. The OF of these fingerprints is calculated by the gradient-based method as the ground truth. Then, we manually create different sizes of missing blocks on these high-quality fingerprints. The OFs of these fingerprint images with missing blocks are calculated by the gradient-based method and the method proposed in this paper. We compare the results of the fingerprint OF calculated by the two methods with the ground truth. The error rates of are calculated. Table 1 shows the error rates of the OFs estimated by the gradient-based method and our proposed method. We made different sizes N × N (i.e., 4 × 4, 6 × 6 and 8 × 8) of artificial missing block, the results show that our proposed method can significantly reduce the error rate. Error rates can be calculated as follows:

$$ERR(\%) = \frac{\sum\limits_{B} \left| OF - OF' \right|}{\pi \cdot N^2} \times 100\% \quad (19)$$

A few low-quality fingerprint images are also selected to validate the performance of the proposed method. Figure 4 shows the experimental results. From the results, it is found that the algorithm proposed in this paper can estimate OF well, regardless of the missing block or the noise effect. After smoothed by the proposed algorithm, the OF of the missing block boundary is greatly improved, and its continuity is effectively increased.

Table 1. Comparison of error rate between the gradient-based method and the proposed method

Size of missing block	4×4	6×6	8×8
Gradient-based method	26.85%	23.89%	26.22%
The proposed method	18.82%	11.38%	17.82%

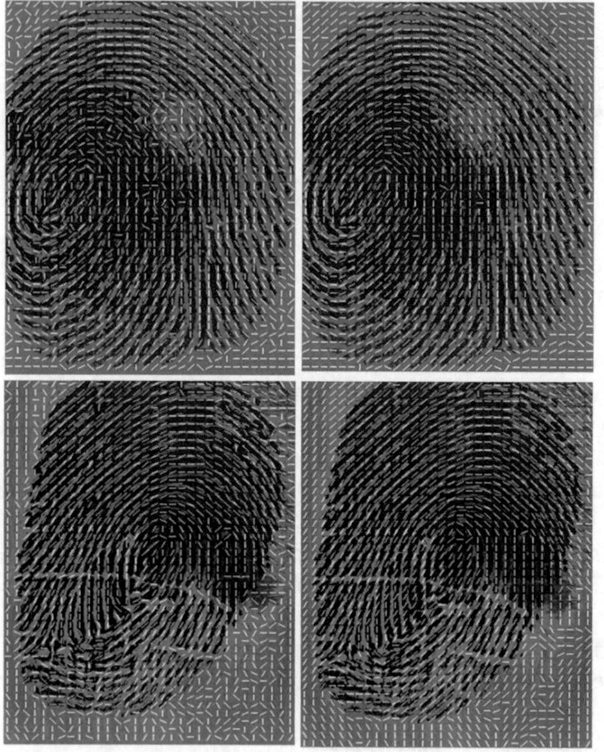

Fig. 4. OFs estimated for the two fingerprint images selected from FVC2004 database. The left column is the result of the gradient-based method and the right is that obtained by the proposed method. Note that $\omega_1 = \omega_2 = 1/4$, $\lambda = 5{,}000$ and the number of iterations is 20.

6 Conclusions

In this paper, an OF estimation algorithm for incomplete fingerprint is proposed, As compared with the gradient-based method, it has the following advantages: (1) For incomplete regions, whether for small or large areas of missing blocks, by introducing the TV model, the fingerprint OF can be well estimated after several iterations. Its continuity and robustness are much better than the traditional method. (2) The local OF can be corrected by the regularization strategy in TV model without relying on external information, thus reducing the error rate of the global OF and improving the matching effect in the later stage. (3) This proposed method can maximize the use of the local information of fingerprint images. Even if part of local information is unavailable, the accuracy of the estimation results can be guaranteed by the TV model.

The experimental results show that the proposed method can effectively estimate the OF of the incomplete fingerprint and has as less error rate as compared to the gradient-based method.

Acknowledgments. This work was supported in part by the project for the innovation and entrepreneurship in Xi'an University of Posts and Telecommunications (2018SC-03), the Key Lab of Computer Networks and Information Integration (Southeastern University), Ministry of Education, China (K93-9-2017-03), the Department of Education Shaanxi Province (16JK1712), Shaanxi Provincial Natural Science Foundation of China (2016JM8034, 2017JM6107), and the National Natural Science Foundation of China (61671377, 51709228).

References

1. Yilong, Y., Xinbao, N., Xiaomei, Z.: Development and application of automatic fingerprint identification technology. J.-Nanjing Univ. Nat. Sci. Ed. **38**(1), 29–35 (2002). https://doi.org/10.3321/j.issn:0469-5097.2002.01.005
2. Kass, M., Witkin, A.: Analyzing oriented patterns. Read. Comput. Vis. 268–276 (1987). https://doi.org/10.1016/b978-0-08-051581-6.50031-3
3. Wang, Y., Jiankun, H., Han, F.: Enhanced gradient-based algorithm for the estimation of fingerprint orientation fields. Appl. Math. Comput. **185**(2), 823–833 (2007). https://doi.org/10.1016/j.amc.2006.06.082
4. Feng, J., Jain, A.K.: Fingerprint reconstruction: from minutiae to phase. IEEE Trans. Pattern Anal. Mach. Intell. **33**(2), 209–223 (2011). https://doi.org/10.1109/TPAMI.2010.77
5. Sherlock, B.G., Monro, D.M.: A model for interpreting fingerprint topology. Pattern Recognit. **26**(7), 1047–1055 (1993). https://doi.org/10.1016/0031-3203(93)90006-I
6. Rudin, L.I., Osher, S., Fatemi, E.: Nonlinear total variation based noise removal algorithms. Phys. D: Nonlinear Phenom. **60**(1-4), 259–268 (1992). https://doi.org/10.1016/0167-2789(92)90242-F
7. Maio, D., et al.: FVC2004: Third fingerprint verification competition. In: Biometric Authentication, pp. 1–7. Springer, Heidelberg (2004) https://doi.org/10.1007/978-3-540-25948-0_1

Minimum Square Distance Thresholding Based on Asymmetrical Co-occurrence Matrix

Hong Zhang[1,2], Qiang Zhi[1,2(✉)], Fan Yang[1,2], and Jiulun Fan[3]

[1] Xi'an University, Xi'an, China
zhmlsa@xupt.edu.cn, 345867389@qq.com, 703905407@qq.com
[2] School of Automation, Xi'an University of Posts and Telecommunications, Xi'an, Shaanxi, China
[3] Xi'an University of Posts and Telecommunications, Xi'an, Shaanxi, China

Abstract. In thresholded image segmentation, correct and adequate extraction of pixel distribution information is the key. In this paper, asymmetrical gray transition co-occurrence matrix is applied to better represent the spatial distribution information of images, and uniformity probability of binarization image is introduced to calculated the deviation information between original and thresholding image. A novel minimum square distance criterion function is proposed to select threshold value, and the vector correlation coefficient is deduced to interpret the reasonable of new criterion. Comparing with relative entropy method, the proposed method is simpler, moreover, it has outstanding object extraction performance.

Keywords: Thresholding method · Co-occurrence matrix
Minimum square distance

1 Introduction

In image preprocessing, segmentation is a critical step, and thresholding is an effective and commonly used segmentation method [1]. Thresholding algorithms can be categorized into histogram shape based, clustering based, entropy based, object attribute based, spatial methods and local methods [2]. Among them, spatial-based methods take full advantage of space statistical information from pixels and neighborhood pixels [3], can obtain more reasonable threshold value.

For a pixel pair with a certain distance, there is gray spatial correlation property. Gray-level co-occurrence matrices were employed to measure the spatial co-occurrence characteristic, and express image feature [4, 5]. Gray-level transition co-occurrence matrix is an application mode of spatial correlation information between pixels, including asymmetric and symmetric [6, 7].

Due to the better statistic information description of pixel and its neighbor, co-occurrence matrix is used widely [8]. We analyzed the construct method of gray-level transition co-occurrence matrix, applied asymmetrical co-occurrence matrix to represent image spatial distribution information. Using prior probability and regional uniformity probability, taking into account the deviation information between original and binarization images, we proposed a new thresholding criterion based on Euclidean

P. Krömer et al. (Eds.): ECC 2018, AISC 891, pp. 814–823, 2019.
https://doi.org/10.1007/978-3-030-03766-6_91

distance function. The new criterion function is more simple to use, and thresholding results show the effectiveness and adaptability of proposed method.

2 Gray Correlation Co-occurrence Matrix

2.1 Co-occurrence Matrix

For an image X of size $M \times N$, the pixel takes values on L gray levels $G = [0, 1, 2, \cdots, L-1]$, i represents the gray level at the pixel (x, y), and j represents the pixel $(x - d\sin\theta, y + d\cos\theta)$ with distance d to pixel (x, y). Generally, the value is $d = 1$, and θ are integer multiples of $\frac{\pi}{2}$. Considering the common frequencies with gray level i of (x, y) and j of $(x - d\sin\theta, y + d\cos\theta)$, a gray level co-occurrence matrix $C_{1,\theta} = (c_{ij}(\theta))_{L\times L}$ of direction θ can be obtained. Here

$$c_{ij}(\theta) = \sum_{x=0}^{M-1}\sum_{y=0}^{N-1} \delta_\theta(x, y) \tag{1}$$

In this formula

$$\delta_\theta(x, y) = \begin{cases} 1, & f(x, y) = i \; and \; f(x - \sin\theta, y + \cos\theta) = j \\ 0, & else \end{cases} \tag{2}$$

$c_{ij}(\theta)$ is the frequency with gray level i and its θ direction pixel level j. If considering the four directions of θ: 0, $\pi/2$, π, $3\pi/2$, the symmetrical matrix C is defined as:

$$C = (c_{ij})_{L\times L} = \frac{1}{4}[C_{1,0} + C_{1,\pi/2} + C_{1,\pi} + C_{1,3\pi/2}] \tag{3}$$

Pal [9] pointed out that the use of grayscale changes in the horizontal direction to the left and in the vertical direction does not provide more information or important improvements. Therefore, in order to reduce the amount of calculation, only adjacent pixels are considered in this paper.

Here, if only two directions $\theta = 0^\circ$ (pixels in the current horizontal right direction) and $\theta = 3\pi/2$ (pixels below the current vertical direction) are taken, an asymmetric co-occurrence matrix $T = (t_{ij})_{L\times L}$ can be formed, as follows:

$$t_{ij} = \sum_{x=0}^{M-1}\sum_{y=0}^{N-1} \delta(x, y) \tag{4}$$

In this formula

$$If \begin{cases} f(x,y) = i, f(x,y+1) = j \\ and/or \\ f(x,y) = i, f(x+1,y) = j \end{cases}, \quad \text{then } \delta(x,y) = 1 \tag{5}$$

Otherwise $\delta(x,y) = 0$.

Normalize the elements of the co-occurrence matrix $T = (t_{ij})_{L \times L}$ to obtain the probability of gray level i to j:

$$p_{ij} = \frac{t_{ij}}{\sum_{i=0}^{L-1} \sum_{j=0}^{L-1} t_{ij}} \tag{6}$$

Let threshold $t \in G$ divide image X into two parts, the object and the background, then t divides matrix T into four quadrants as shown in Fig. 1.

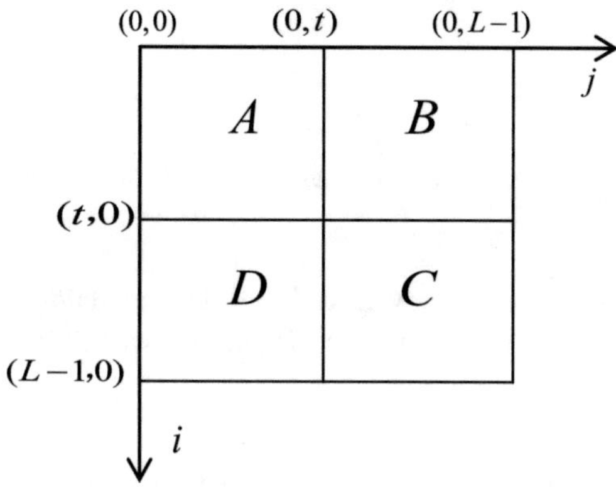

Fig. 1. Quadrant of symbiotic matrix T

If a pixel with a gray value greater than t is assumed to the object, and a pixel with a gray value less than or equal to t belonging to the background, then quadrants A and C correspond to local variations in the object and background respectively, while quadrants B and D represent the changes in boundaries of background and object. The probability of each quadrant is as follows:

$$P_A(t) = \sum_{i=0}^{t} \sum_{j=0}^{t} p_{ij} \tag{7}$$

$$P_B(t) = \sum_{i=0}^{t} \sum_{j=t+1}^{L-1} p_{ij} \tag{8}$$

$$P_C(t) = \sum_{i=t+1}^{L-1} \sum_{j=t+1}^{L-1} p_{ij} \tag{9}$$

$$P_D(t) = \sum_{i=t+1}^{L-1} \sum_{j=0}^{t} p_{ij} \tag{10}$$

2.2 Uniformity Probability

If the selected threshold value is t, we assign that the gray value whose gray level belongs to $G_1 = \{0, 1, \cdots, t\}$ is zero, and the gray level belongs to $G_2 = \{t+1, \cdots, L-1\}$ is $L-1$, a binary image X^* can be obtained. If only the probability uniformity of the divided regions is concerned, for the grayscale within G_1 is treated with equal probability, the grayscale within G_2 is also treated with equal probability, and the uniformity probability p'_{ij} of X^* can be defined as follows:

$$p_{ij}'^{(A)} = q_A(t) = \frac{P_A(t)}{(t+1)\times(t+1)}, \quad 0 \le i \le t, 0 \le j \le t \tag{11}$$

$$p_{ij}'^{(B)} = q_B(t) = \frac{P_B(t)}{(t+1)\times(L-t-1)}, \quad 0 \le i \le t, t+1 \le j \le L-1 \tag{12}$$

$$p_{ij}'^{(c)} = q_C(t) = \frac{P_C(t)}{(L-t-1)\times(L-t-1)}, \quad t+1 \le i \le L-1, t+1 \le j \le L-1 \tag{13}$$

$$p_{ij}'^{(D)} = q_D(t) = \frac{P_D(t)}{(L-t-1)\times(t+1)}, \quad t+1 \le i \le L-1, 0 \le j \le t \tag{14}$$

The variation probability distribution of the co-occurrence matrix containing spatial information can reflect the uniformity within the group (quadrants A and C), and the variation across the boundary (quadrants D and B).

3 Minimum Square Distance Thresholding Method

In thresholding criterion of images, the relative entropy-based method is a simple and effective threshold by describing the deviation between the original image and the binarized image, and which can obtain the optimal matching between original image and the binarized image, then the optimal threshold criterion function is built. In this paper, we describe the deviation of the original image and the binarized image based on the perspective of Euclidean distance, and we can get a more simple thresholding criterion than the relative entropy method. It is defined as:

$$
\begin{aligned}
F(p;p') &= \sum_{i=0}^{L-1}\sum_{j=0}^{L-1}(p_{ij}-p'_{ij})^2 \\
&= \sum_{i=0}^{t}\sum_{j=0}^{t}(p_{ij}-q_A(t))^2 + \sum_{i=0}^{t}\sum_{j=t+1}^{L-1}(p_{ij}-q_B(t))^2 \\
&+ \sum_{i=t+1}^{L-1}\sum_{j=0}^{t}(p_{ij}-q_D(t))^2 + \sum_{i=t+1}^{L-1}\sum_{j=t+1}^{L-1}(p_{ij}-q_C(t))^2
\end{aligned}
\tag{15}
$$

Among them

$$
\begin{aligned}
&\sum_{i=0}^{t}\sum_{j=0}^{t}(p_{ij}-q_A(t))^2 = \sum_{i=0}^{t}\sum_{j=0}^{t}\left(p_{ij}^2 - 2p_{ij}\frac{p_A(t)}{(t+1)(t+1)} + \left(\frac{p_A(t)}{(t+1)(t+1)}\right)^2\right) \\
&= \sum_{i=0}^{t}\sum_{j=0}^{t}p_{ij}^2 - \frac{p_A^2(t)}{(t+1)(t+1)}
\end{aligned}
\tag{16}
$$

$$
\sum_{i=0}^{t}\sum_{j=t+1}^{L-1}(p_{ij}-q_B(t))^2 = \sum_{i=0}^{t}\sum_{j=t+1}^{L-1}p_{ij}^2 - \frac{p_B^2(t)}{(t+1)(L-t-1)} \tag{17}
$$

$$
\sum_{i=t+1}^{L-1}\sum_{j=t+1}^{L-1}(p_{ij}-q_C(t))^2 = \sum_{i=t+1}^{L-1}\sum_{j=t+1}^{L-1}p_{ij}^2 - \frac{p_C^2(t)}{(L-t-1)(L-t-1)} \tag{18}
$$

$$
\sum_{i=t+1}^{L-1}\sum_{j=0}^{t}(p_{ij}-q_D(t))^2 = \sum_{i=t+1}^{L-1}\sum_{j=0}^{t}p_{ij}^2 - \frac{p_D^2(t)}{(L-t-1)(t+1)} \tag{19}
$$

We can obtained

$$
\begin{aligned}
F(p;p') = \sum_{i=0}^{L-1}\sum_{j=0}^{L-1}p_{ij}^2 - \frac{P_A^2(t)}{(t+1)^2} - \frac{P_B^2(t)}{(t+1)\times(L-t-1)} \\
- \frac{P_D^2(t)}{(t+1)\times(L-t-1)} - \frac{P_C^2(t)}{(L-t-1)^2}
\end{aligned}
\tag{20}
$$

In the above formula, the first term $\sum_{i=0}^{L-1}\sum_{j=0}^{L-1}p_{ij}^2$ is a constant, recorded as:

$$
F_{total}(p;p') = \frac{P_A^2(t)}{(t+1)^2} + \frac{P_B^2(t)}{(t+1)\times(L-t-1)} + \frac{P_D^2(t)}{(t+1)\times(L-t-1)} + \frac{P_C^2(t)}{(L-t-1)^2} \tag{21}
$$

The minimum $F(p;p')$ is equal to maximum the value of $F_{total}(p;p')$.

4 Interpretation of Vector Correlation Coefficient

Another interpretation of the above criterion function is the "vector correlation coefficient". The co-occurrence matrix $(p_{ij})_{L\times L}$ of original image may constitute a vector of dimension L^2, and the co-occurrence matrix $(p'_{ij})_{L\times L}$ of threshold image also constitute a L^2 dimension vector. Then the correlation coefficient of this vector is calculated as follows:

$$\begin{aligned} r(t) &= \frac{\sum_{i=0}^{L-1}\sum_{j=0}^{L-1} p_{ij}p'_{ij}}{\sqrt{\sum_{i=0}^{L-1}\sum_{j=0}^{L-1} p_{ij}^2}\cdot\sqrt{\sum_{i=0}^{L-1}\sum_{j=0}^{L-1}(p'_{ij})^2}} \\ &= \frac{\sum_{i=0}^{t}\sum_{j=0}^{t}\frac{p_{ij}P_A(t)}{(t+1)^2}+\sum_{i=t+1}^{L-1}\sum_{j=t+1}^{L-1}\frac{p_{ij}P_C(t)}{(L-t-1)^2}+\sum_{i=0}^{t}\sum_{j=t+1}^{L-1}\frac{p_{ij}P_B(t)}{(t+1)(L-t-1)}+\sum_{i=t+1}^{L-1}\sum_{j=0}^{t}\frac{p_{ij}P_D(t)}{(L-t-1)(t+1)}}{\sqrt{\sum_{i=0}^{L-1}\sum_{j=0}^{L-1} p_{ij}^2}\sqrt{\frac{P_A^2(t)}{(t+1)^2}+\frac{P_C^2(t)}{(L-t-1)^2}+\frac{P_B^2(t)}{(t+1)(L-t-1)}+\frac{P_D^2(t)}{(t+1)(L-t-1)}}} \\ &= \frac{\sqrt{\frac{P_A^2(t)}{(t+1)^2}+\frac{P_C^2(t)}{(L-t-1)^2}+\frac{P_B^2(t)}{(t+1)(L-t-1)}+\frac{P_D^2(t)}{(t+1)(L-t-1)}}}{\sqrt{\sum_{i=0}^{L-1}\sum_{j=0}^{L-1} p_{ij}^2}} \end{aligned} \tag{22}$$

Since $\sqrt{\sum_{i=0}^{L-1}\sum_{j=0}^{L-1} p_{ij}^2}$ is also a constant. Therefore maximizing $r(t)$ is equivalent to maximizing $F_{total}(p;p')$.

It should be noted that in the definition of p'_{ij}, the position of $p_{ij}=0$ is given as a non-zero value, which is unreasonable in some cases. A more reasonable choice of p'_{ij} is that, it is only considered the information at the position of $p_{ij}\neq 0$ [10], that is:

$$q_A(t) = \frac{P_A(t)}{\sum_{i=0}^{t}\sum_{j=0}^{t} s_{ij}} \tag{23}$$

$$q_B(t) = \frac{P_B(t)}{\sum_{i=0}^{t}\sum_{j=t+1}^{L-1} s_{ij}} \tag{24}$$

$$q_C(t) = \frac{P_C(t)}{\sum_{i=t+1}^{L-1}\sum_{j=t+1}^{L-1} s_{ij}} \tag{25}$$

$$q_D(t) = \frac{P_D(t)}{\sum_{i=t+1}^{L-1}\sum_{j=0}^{t} s_{ij}} \tag{26}$$

Here,

$$s_{ij} = \begin{cases} 1, & p_{ij} \neq 0 \\ 0, & p_{ij} = 0 \end{cases} \tag{27}$$

Considering the information of A, B, C, and D regions, $F_{total}(p;p')$ is called the global squared distance threshold criterion [11]. If only the information inside the region [12] is considered, the local distance criterion can be obtained:

$$F_{local}(p;p') = \frac{P_A^2}{(t+1)^2} + \frac{P_C^2}{(L-t-1)^2} \tag{28}$$

If only the information of boundary is considered, the connection distance criterion is obtained:

$$F_{joint}(p;p') = \frac{P_B^2}{(t+1)\times(L-t-1)} + \frac{P_D^2}{(t+1)\times(L-t-1)} \tag{29}$$

When the above criterion function takes the maximum value, the result is the optimal threshold value.

5 Results and Analysis

In order to verify the effectiveness of proposed method, we have thresholding test for some images, Results of four representative images are shown in Figs. 2, 3, 4, 5: Bacteria, Circle, Dot_Blots and Test with sizes 178 × 178, 256 × 256, 576 × 336 and 256 × 256, respectively. Figures 2, 3, 4, 5(a) are original images, Figs. 2, 3, 4, 5 (b) are 1_d Histogram. Figures 2, 3, 4, 5(c)–(e) show the results of proposed Local, Joint, Total Distance method. The results of relative entropy method are shown in Figs. 2, 3, 4, 5(f) for comparison.

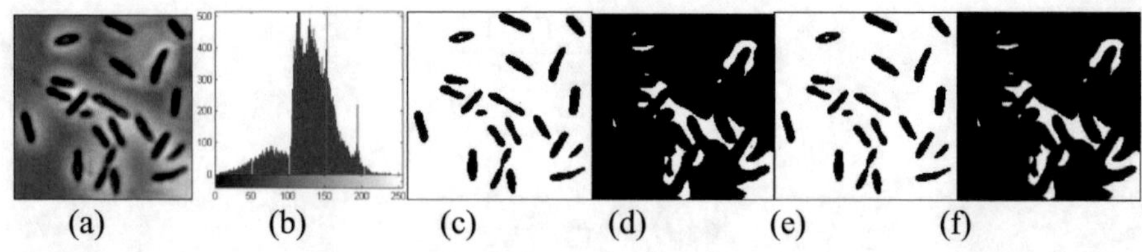

Fig. 2. Bacteria: (a) Original image, (b)1_d Histogram, (c)Local Distance, (d) Joint Distance, (e) Total Distance, (f) Relative Entropy.

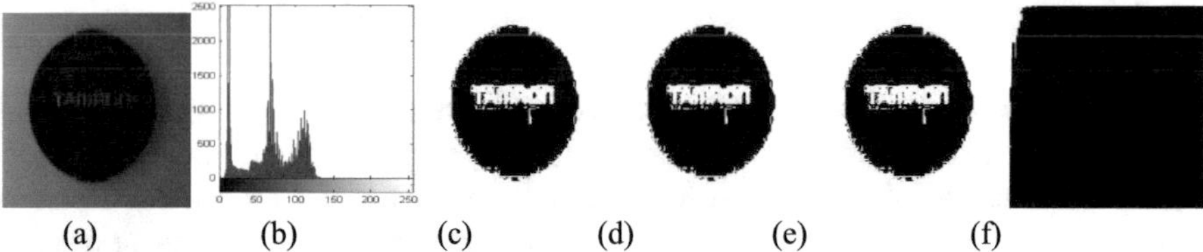

Fig. 3. Circle: (a) Original image, (b) 1_d Histogram, (c) Local Distance, (d) Joint Distance, (e) Total Distance, (f) Relative Entropy.

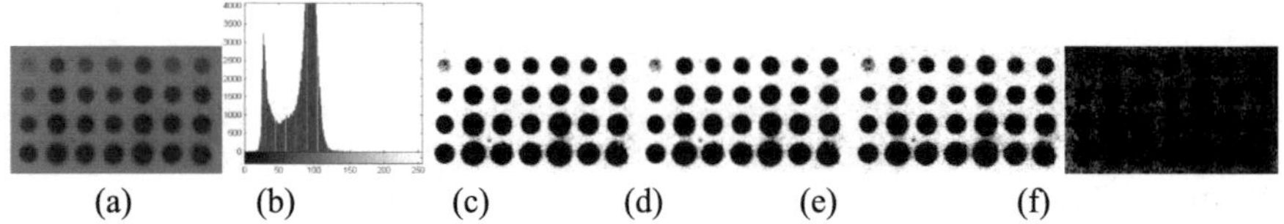

Fig. 4. Dot_Blots: (a) Original image, (b) 1_d Histogram, (c) Local Distance, (d) Joint Distance, (e) Total Distance, (f) Relative Entropy.

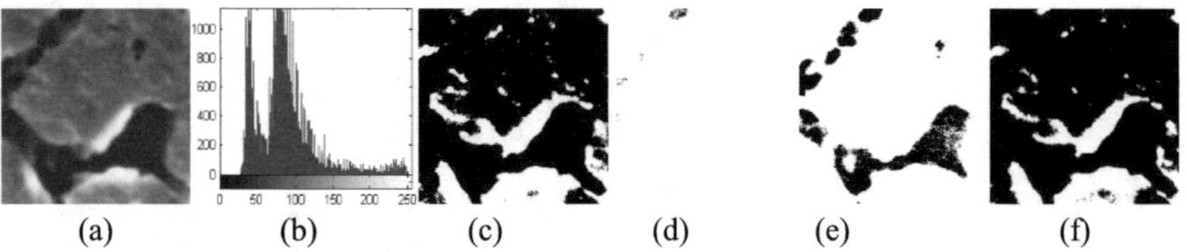

Fig. 5. Test: (a) Original image, (b) 1_d Histogram, (c) Local Distance, (d) Joint Distance, (e) Total Distance, (f) Relative Entropy.

Comparing with the results of four criterions, the best object extraction results are obtained by applying Total Distance method. Especially, for narrower 1_d Histogram as Figs. 3, 4, relative entropy method can hardly extract the object. Table 1 lists the segmentation thresholds of four methods.

Table 1. The results of four methods

Method	Image			
	Bacteria	Circle	Dot_Blots	Test
Local distance	101	13	80	102
Joint distance	166	13	80	31
Total distance	101	13	82	42
Relative entropy	165	121	109	108

6 Conclusion

In this paper, we applied asymmetrical co-occurrence matrix based on pixel and its two neighbor pixels, introduced the uniformity probability of binarization image region to represent image spatial distribution information. For obtaining the deviation information between original and thresholding image, constructed a minimum square distance function, which is used as thresholding criterion. The local and joint criterionare given to measure inside and edge information of segmented region. The vector correlation coefficient interpreted the reasonable of criterion. The results show that, comparing with relative entropy method, our proposed method has outstanding performance on integrity of object and can also obtain best object extraction results.

Acknowledgments. This work is supported by the National Science Foundation of China (No. 61571361, 61671377), and the Science Plan Foundation of the Education Bureau of Shaanxi Province (No. 15JK1682).

References

1. Sang, Q., Lin, Z.L., Acton, S.T.: Learning automata for image segmentation. Pattern Recognit. Lett. **74**, 46–52 (2016). https://doi.org/10.1016/j.patrec.2015.12.004
2. Sezgin, M., Sankur, B.: Survey over image thresholding techniques and quantitative performance evaluation. J. Electron. Imaging **13**(1), 146–168 (2004). https://doi.org/10.1117/1.1631315
3. Chanda, B., Chaudhuri, B.B., Majumder, D.D.: On image enhancement and threshold selection using the gray-level co-occurrence matrix. Pattern Recognit. Lett. **3**(2), 243–251 (1985). https://doi.org/10.1016/0167-8655(85)90004-2
4. Liang, D., Kaneko, S., Hashimoto, M., et al.: Co-occurrence probability-based pixel pairs background model for robust object detection in dynamic scenes. Pattern Recognit. **48**, 1370–1386 (2015). https://doi.org/10.1016/j.patcog.2014.10.020
5. El-Feghi, I., Adem, N., Sid-Ahmed, M.A., et al.: Improved co-occurrence matrix as a feature space for relative entropy-based image thresholding. In: Proceedings of the Computer Graphics, Imaging and Visualization, vol. 49, pp. 314–320 (2007). http://doi.org/10.1109/CGIV.2007.49
6. Fan, J.L., Ren, J.: Symmetric co-occurrence matrix thresholding method based on square distance. Acta Electronica Sinica **39**(10), 2277–2281 (2011). http://en.cnki.com.cn/Article_en/CJFDTOTAL-DZXU201110012.htm. (in Chinese)
7. Fan, J.L., Zhang, H.: A unique relative entropy-based symmetrical co-occurrence matrix thresholding with statistical spatial information. Chin. J. Electron. **24**(3), 622–626 (2015). https://doi.org/10.1049/cje.2015.07.031
8. Subudhi, P., Mukhopadhyay, S.: A novel texture segmentation method based on co-occurrence energy-driven parametric active contour model. Signal, Image Video Process. **12**(4), 669–676 (2018). https://doi.org/10.1007/s11760-017-1206-4
9. Pal, S.K., Pal, N.R.: Segmentation using contrast and homogeneity measure. Patt. Recog. Lett. **5**, 293–304 (1987). https://doi.org/10.1016/0167-8655(87)90061-4

10. Ramac, L.C., Varshney, P.K.: Image thresholding based on ali-silvey distance measures. Pattern Recognit. **30**(7), 1161–1174 (1997). https://doi.org/10.1016/S0031-3203(96)00149-5
11. Chang, C.I., Chen, K., Wang, J., Althouse, M.L.G.: A relative entropy-based approach to image thresholding. Pattern Recognit. **27**(9), 1275–1289 (1994). https://doi.org/10.1016/0031-3203(94)90011-6
12. Lee, S.H., Hong, S.J., Tsai, H.R.: Entropy thresholding and its parallel algorithm on the reconfigurable array of processors with wider bus networks. IEEE Trans. Image Proc. **8**(9), 1242–1299 (1999). https://doi.org/10.1109/83.784435

Cross Entropy Clustering Algorithm Based on Transfer Learning

Qing Wu(✉) and Yu Zhang

School of Automation, Xi'an University of Posts & Telecommunications, Xi'an 710121, China
xiyouwuq@126.com, 543454272@qq.com

Abstract. To solve the problem of clustering performance degradation when traditional clustering algorithms are applied to insufficient or noisy data, a cross entropy clustering algorithm based on transfer learning is proposed. It improves the classical cross entropy clustering algorithm by combining knowledges from historical clustering centers and historical degree of membership and applying them to the objective function proposed for clustering insufficient or noisy target data. The experiment results on several synthetic and four real datasets and analyses show the proposed algorithm has high effectiveness over the available.

Keywords: Cluster · Cross entropy clustering · Transfer learning
Historical clustering center · Historical degree of membership

1 Introduction

Clustering plays an important role in many parts of artificial intelligence, machine learning and pattern recognition [1–3]. Fuzzy C-means clustering algorithm (FCM) [4] is the most widely used clustering algorithm. However, theoretical and experimental studies have shown that the objective function of FCM expresses the degree of ambiguity in exponentially weighted form, which lacks clear physical meaning. Recently, many entropy-based algorithms have been proposed for clustering analysis, where most of them consider containing an entropy term to the objective function. Yang et al. [5] proposed Maximum Entropy Clustering (MEC). Then Fan et al. proposed a fuzzy clustering algorithm based on generalized entropy. But these methods are sensitive to noise, and the interference of the exceptional point often makes the cluster center seriously deviate. Hence Gu et al. [6] introduced the cross entropy into the objective function of the traditional FCM algorithm, and proposed cross-entropy semi-supervised clustering based on pairwise on constraints (CE-SSC), which solved the above problems.

The above clustering algorithms usually divide information based on a large amount of available data. However, the situation where data is insufficient or noisy is often prevalent. If we use the traditional clustering algorithm, the clustering result will be unsatisfactory. How to effectively improve data clustering performance when data is insufficient or noise is one of the research directions of researchers in recent years. The knowledge transfer mechanism [7] is an effective method. It is used to improve the clustering result of the current data.

P. Krömer et al. (Eds.): ECC 2018, AISC 891, pp. 824–832, 2019.
https://doi.org/10.1007/978-3-030-03766-6_92

In this paper, the objective function is modified by adding two transfer learning mechanisms above the cross entropy clustering algorithm and a cross entropy clustering algorithm based on transfer learning (TLCEC) is proposed. By learning historical knowledge the performance of the clustering algorithm can be improved when the sample is insufficient or noisy. Numerical experiments on multiple data sets show that TLCEC can achieve great clustering results.

2 Cross Entropy Clustering

In order to construct the objective function of cross entropy clustering, the definition of cross entropy is given as follows

Definition 1 (Cross Entropy). We define the cross-entropy of x_j with respect to sub-density x_k by

$$H\left(x_j, x_k\right) = -\sum_{i=1}^{c} u_{ij} \ln u_{ik} \tag{1}$$

where u_{ij} denotes membership of the vector x_j belonging to the i-th fuzzy subset. It is required that membership should satisfy

$$\sum_{i=1}^{c} u_{ij} = 1 \tag{2}$$

As we can see that when $i = k$, Eq. (1) is the same as the entropy term in the objective function of the maximum entropy clustering algorithm MEC. It can be seen that cross entropy can be regarded as a generalized form of maximum entropy.

Inference 1. The sample cross entropy can be decomposed into its own cross entropy and relative entropy as follows

$$H\left(x_j, x_k\right) = H\left(x_j, x_j\right) + D\left(x_j, x_k\right) \tag{3}$$

Proof. From Eq. (1), we know

$$\begin{aligned} -\sum_{i=1}^{c} u_{ij} \ln u_{ik} &= -\sum_{i=1}^{c} u_{ij} \ln\left(\frac{u_{ij}}{u_{ij}} \cdot u_{ik}\right) \\ &= -\sum_{i=1}^{c}\left(u_{ij} \cdot \ln\left(u_{ij}\right) + u_{ij} \cdot \ln\left(\frac{u_{ik}}{u_{ij}}\right)\right) \\ &= -\sum_{i=1}^{c} u_{ij} \ln\left(u_{ij}\right) + \sum_{i=1}^{c} u_{ij} \cdot \ln\left(\frac{u_{ij}}{u_{ik}}\right) \\ &= H\left(x_j, x_j\right) + D\left(x_j, x_k\right) \end{aligned}$$

It can be seen that the cross entropy of the sample is a combination of internal and external information. Compared with the maximum entropy function, the entropy function in the clustering algorithm not only represents the sample self-information. We apply the cross entropy to the cluster, which can express data information better and process clusters efficiently. In order to construct the objective function of cross-entropy clustering, the symmetric form of cross-entropy is given below.

Definition 2. The symmetric form of cross-entropy is given as follows

$$H(\mathrm{x_j}, \mathrm{x_k}) = \sum_{i=1}^{c} \sum_{\substack{j=1 \\ k=1}}^{N} u_{ij} \ln \frac{u_{ij}}{u_{ik}} + u_{ik} \ln \frac{u_{ik}}{u_{ij}} \tag{4}$$

As a whole, the objective function of cross-entropy cluster is converted into the following form to be minimized

$$minimize\ J(U, V) = u_{ij} \|x_j - v_i\|^2 - \theta \sum_{i=1}^{c} \sum_{\substack{j=1 \\ k=1}}^{N} \left(u_{ij} \ln \frac{u_{ij}}{u_{ik}} + u_{ik} \ln \frac{u_{ik}}{u_{ij}} \right) \tag{5}$$

The memberships should satisfy as follows

$$\sum_{i=1}^{c} u_{ij} = 1$$

where, θ is a cross entropy adjustment coefficient, which determines the degree of influence of cross entropy. The objective function is composed of two parts. The first part is the objective function of FCM, and the second part is the cross entropy penalty term. It can be seen from the objective function the membership degree of each sample depends on not only the distance factor, but also the cross entropy. Therefore, the algorithm does not fall into local optimum during the iterative process which can form better clustering results.

3 Cross Entropy Clustering Based on Transfer Learning

3.1 Algorithm Objective Function

Although the cross entropy clustering algorithm achieves better clustering effect, when the amount of data is insufficient or noisy, clustering directly using this algorithm will make the cluster center often deviate from the actual cluster center, resulting in poor clustering effect. According to the transfer learning theory, when there is a certain correlation between the data in the source domain and the data in the target domain, and there are certain differences, the beneficial knowledge of the source domain can be used to guide the completion target domain task.

Definition 3 (Transfer Learning). Given a source domain D_S and learning task T_S, a target domain D_T and a learning task T_T, transfer learning aims to improve the learning of the target predictive function $f_T(\cdot)$ in D_T using the knowledge in D_S and T_S, where $D_S \neq T_S$, or $T_S \neq T_T$. In transfer learning, we have the following three important research issues: (1) what to transfer; (2) how to transfer; (3) when to transfer.

In this paper, the membership degree and the cluster center are selected as the transfer knowledge. We propose two transfer rules, the membership transfer rules and the clustering center transfer rules. Each rule is given as follows.

(1) The membership transfer rules of the target domain relative to the source domain is

$$\varphi\left(U, \hat{U}\right) = \alpha \sum_{i=1}^{c} \sum_{j=1}^{N} u_{ij} \left\| x_j - v_i \right\|^2 + (1-\alpha) \sum_{i=1}^{c} \sum_{j=1}^{N} \hat{u}_{ij} \left\| x_j - v_i \right\|^2 \tag{6}$$

where $\hat{u}_{ij}$ is the membership of the sample within the target domain relative to the source domain, and α is the balance factor, which controls the degree of historical membership in the source domain and the degree of influence of the membership in the target domain on the final clustering result. When $\alpha \rightarrow 0$, it indicates that the historical membership degree of transfer is relatively poor, and the clustering result of the target domain is more affected by the membership degree of the source domain. When $\alpha \rightarrow 1$ the historical membership degree of transfer has high reference ability.

(2) The clustering center transfer rules of the target domain relative to the source domain is

$$\varphi\left(V, \hat{V}\right) = \beta \sum_{i=1}^{c} \left\| v_i - \hat{v}_i \right\|^2 \tag{7}$$

where $\hat{v}_i$ represents the i-th center point of the historical class center point of the source domain, β represents the balance factor of the minimum rule of the class center point change and c is the number of clusters. When $\beta \rightarrow 0$ the cluster center point of the target domain and the cluster center point of the source domain need to maintain a small degree of consistency, which indicates that the knowledge of the cluster center point of the source domain is unreliable at this time. When $\beta \rightarrow 1$ the cluster center point of the target domain and the cluster center point of the source domain need to maintain a large degree of consistency, which indicates that the knowledge of the cluster center point of the source domain is high.

3.2 Algorithm Objective Function

Aiming at the advantages of traditional cross entropy clustering algorithm without transfer learning, we propose the KTCEC based on two transfer rules. The algorithm objective function is as follows

$$\min J\left(U,\hat{U},V,\hat{V}\right)=\alpha\sum_{i=1}^{c}\sum_{j=1}^{N}u_{ij}\|x_j-v_i\|^2+(1-\alpha)\sum_{i=1}^{c}\sum_{j=1}^{N}\overset{\wedge}{u_{ij}}\|x_j-v_i\|^2 + \beta\sum_{I=1}^{c}\left\|v_i-\overset{\wedge}{v_i}\right\|^2+\theta\sum_{i=1}^{c}\sum_{\substack{j=1\\k=1}}^{N}\left(u_{ij}\ln\left(\frac{u_{ij}}{u_{ik}}\right)+u_{ik}\ln\left(\frac{u_{ik}}{u_{ij}}\right)\right) \tag{8}$$

The memberships should satisfy as follows

$$\sum_{i=1}^{c}u_{ij}=1$$

The objective function is composed of three parts. It can be seen that when $\alpha=1,\beta=0$, the algorithm degenerates into a classical cross entropy clustering algorithm. In order to obtain the iterative formula of membership degree and clustering center, we construct the Lagrangian function as follows

$$L=\alpha\sum_{i=1}^{c}\sum_{j=1}^{N}u_{ij}\|x_j-v_i\|^2+(1-\alpha)\sum_{i=1}^{c}\sum_{j=1}^{N}\overset{\wedge}{u_{ij}}\|x_j-v_i\|^2+\beta\sum_{I=1}^{c}\left\|v_i-\overset{\wedge}{v_i}\right\|^2 + \theta\sum_{i=1}^{c}\left(u_{ij}\ln\left(\frac{u_{ij}}{u_{ik}}\right)+u_{ik}\ln\left(\frac{u_{ik}}{u_{ij}}\right)\right)+\sum_{j=1}^{N}\lambda_j\left(1-\sum_{i=1}^{c}u_{ij}\right) \tag{9}$$

Setting the derivative of Eq. (9) equal to zero with respect to u_{ij}, we can get the clustering center

$$v_i=\frac{\sum_{i=1}^{c}\left[\alpha u_{ij}+(1-\alpha)\overset{\wedge}{u_{ij}}\right]x_j+\beta\overset{\wedge}{v_i}}{\left[\sum_{j=1}^{N}\alpha u_{ij}+(1-\alpha)\overset{\wedge}{u_{ij}}\right]+\beta} \tag{10}$$

Setting the derivative of Eq. (9) equal to zero with respect to v_i, one can obtain the degree of membership

$$u_{ij}=\frac{\sum_{i=1}^{c}u_{ik}}{W_0\left(\exp\left(-\frac{K}{\theta}\right)\sum_{k=1}^{j}u_{ik}\right)} \tag{11}$$

In Eq. (11), $K=\alpha\|x_j-v_i\|^2-\theta+\theta\sum_{k=1}^{j}\ln u_{ik}-\lambda_j$, and $W_0(\cdot)$ is called the Lambert W function [8], which satisfies $W(Z)\exp(W(Z))=Z$.

3.3 TLCEC Algorithm Description

Our cross entropy clustering algorithm based on transfer learning (TLCEC) algorithm is summarized as follows:

Input: number of categories c, the cross entropy adjustment coefficient θ, source data set X, target data set T, balance factors α and β;

Output: target domain cluster center v_i, membership degree matrix U;

(1) Using the traditional cross entropy clustering algorithm to obtain the historical clustering center of the source domain and the historical membership degree;
(2) Initialize the iteration counter t = 0;
(3) Calculate a new cluster center v_i according to Eq. (10);
(4) Calculate a new membership u_{ij} according to Eq. (11);
(5) When $\|U_{t+1} - U_t\| \leq \zeta$, the algorithm terminates. Otherwise, return to Step (3).

The algorithm flow aims at alternating iterative updating of the membership matrix and the cluster center. In the process of continuous iteration of the algorithm, the iterative formula of membership degree and cluster center uses the knowledge of historical membership degree and historical clustering center.

4 Experiments

To test the performance of our proposed approaches, we compare numerically with k-means, FCM, MECA, CE-SSC and cross entropy fuzzy C-means clustering (CEFCM), respectively, on a synthetic dataset and 4 datasets from UCI Repository. All experiments were implemented by using MATLAB 8.4 on a personal computer with 1.6 GHz and 4 GB RAM. In this paper, we use the normalized mutual information (NMI) and the rand index (RI) [6] to evaluate the experimental results. To ensure the accuracy of the experiment, the experimental result data are averages obtained after running 10 times.

In this paper, we construct a set of source data sets X and 3 sets of target datasets T1, T2, and T3. Datasets are randomly generated by Gaussian probability distribution. In order to ensure sufficient experimental data and to extract data that is instructive for clustering the target dataset, we divide the source domain dataset X into three classes with a total of 600 samples, and each class with 200 samples. The target data set T1 is used to represent a scenario with insufficient data volume, which accounting for only 10% of the source dataset X. The target dataset T2 has a total of 300 data, which represent the case when the amount of data is sufficient. The target dataset T3 add Gaussian noise on T2 to reflect the situation when data is noisy. The distribution of constructed source domain and target domain dataset is shown in Fig. 1.

Table 1 summarizes the experiment results. It can be seen the KTCEC algorithm can also perform clustering well when the amount of data is insufficient on the dataset T1. On the dataset T2, because the data is sufficient, the six algorithms all achieve great clustering results. Especially, CE-SSC and KTCEC have the best clustering effect. On the dataset T3, the clustering performance of other algorithms is significantly worse than the KTCEC when the data is noisy.

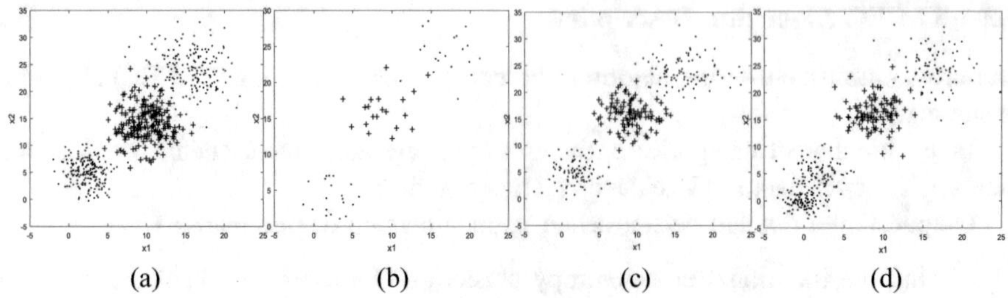

Fig. 1. Synthetic dataset distribution. (a) Source dataset X. (b) Target dataset T1. (c) Target dataset T2. (d) Target dataset T3

Table 1. Experimental result for TLCEC and others on the synthetic datasets

Dataset	Index	K-means	FCM	MEC	CE-SSC	CEFCM	KTCEC
T1	RI	0.792	0.841	0.891	0.934	0.929	**0.943**
	NMI	0.743	0.788	0.806	0.835	0.824	**0.967**
T2	RI	0.854	0.894	0.936	**0.947**	0.941	**0.947**
	NMI	0.793	0.812	0.825	**0.887**	0.847	**0.887**
T3	RI	0.715	0.796	0.846	0.895	0.892	**0.912**
	NMI	0.617	0.685	0.678	0.745	0.723	**0.823**

In order to further verify the performance of the algorithms, we select four real datasets from the UCI database. Each data set extracts 30% from different classes as the target domain data to construct a transfer scenario. The data characteristics of the four datasets are given in Table 2.

Table 2. Experimental results for WLSTWSVM and GLSTSVM on the UCI datasets

Dataset	Datasets types	Number	Dimension	Class
Iris	Source datasets	150	4	3
	Target datasets	45	4	3
Wine	Source datasets	178	13	3
	Target datasets	50	13	3
Seed	Source datasets	210	4	3
	Target datasets	63	4	3
Breast	Source datasets	699	9	2
	Target datasets	210	9	2

Table 3 gives the comparisons of the performance of the KTCEC with K-means, FCM, MEC, CE-SSC, CEFCM. Compared with other algorithms, KTCEC has the best clustering performance and CE-SSC is the second best. It can be seen that the cross entropy clustering has obvious advantages compared with the traditional clustering algorithm. In these four data sets, Iris constructed the least number of transfer scenarios and the number of the target dataset was insufficient. In this case, KTCEC has the best clustering effect. The other algorithms did not introduce any transfer learning mechanism, resulting in the poor final clustering results.

Table 3. Experimental results for TLCEC and others on the UCI datasets

Dataset	Index	K-means	FCM	MEC	CE-SSC	CEFCM	KTCEC
Iris	RI	0.872	0.841	0.879	0.934	0.929	**0.944**
	NMI	0.732	0.748	0.737	0.813	0.810	**0.878**
Wine	RI	0.937	0.934	0.942	0.968	0.964	**0.972**
	NMI	0.844	0.852	0.849	**0.908**	0.901	**0.908**
Seed	RI	0.868	0.872	0.883	0.937	0.937	**0.948**
	NMI	0.676	0.679	0.696	0.841	0.845	**0.865**
Breast	RI	0.927	0.932	0.949	0.973	0.972	**0.976**
	NMI	0.756	0.795	0.817	**0.898**	0.895	0.896

5 Conclusion

In this paper, we introduce the traditional cross entropy clustering algorithms and the knowledge transfer learning mechanism to improve the performance of the algorithm. Then a cross-entropy clustering algorithm based on transfer learning is proposed. KTCEC can automatically adjust the weights of two balance factors in the iterative process, which improves the clustering effect of the algorithm when the number of samples is insufficient or noisy.

Acknowledgments. This work was supported in part by the National Natural Science Foundation of China under Grants (61472307, 51405387), the Key Research Project of Shaanxi Province (2018GY-018) and the Foundation of Education Department of Shaanxi Province (17JK0713).

References

1. Krishnapuram, R., Keller, J.: A possibilistic approach to clustering. IEEE Trans. Fuzzy Syst. **1** (2), 98–110 (1993). https://doi.org/10.3156/jfuzzy.7.5_976
2. Wang, L., Wang, J.D., Li, T.: Cluster's feature weighting fuzzy clustering algorithm integrating rough sets and shadowed sets. Syst. Eng. Electron. **35**(8), 1769–1776 (2013). https://doi.org/10.3969/j.issn.1001-506X.2013.08.31
3. Wang, Y., Peng, T., Han, J.Y., Liu, L.: Density-Based distributed clustering method. J. Softw. **28**(11), 2836–2850 (2017). https://doi.org/10.13328/j.cnki.jos.005343
4. Likas, A., Vlassis, N., Verbeek, J.J.: The globel k-means clustering algorithm. Pattern Recogn. **36**(2), 451–461 (2003). https://doi.org/10.1016/S0031-3203(02)00060-2

5. Zhi, X.B., Fan, J.L., Zhao, F.: Fuzzy linear discriminant analysis guided maximum entropy fuzzy clustering algorithm. Pattern Recogn. **46**(6), 1604–1615 (2013). https://doi.org/10.1016/j.patcog.2012.12.007
6. Li, C.M., Xu, S.B., Hao, Z.F.: Cross-entropy semi-supervised clustering based on pairwise on constraints. CAAI Trans. Intell. Syst. **30**(07), 598–608 (2017). https://doi.org/10.16451/j.cnki.issn1003-6059.201707003
7. Pan, S.J., Yang, Q.: A survey on transfer learning. IEEE Trans. Knowl. Data Eng. **22**(10), 1345–1359 (2010). https://doi.org/10.1109/tkde.2009.191
8. Long, M., Zhou, T.J.: A survey of properties and application of the Lambert W function. J. Hengyang Norm. Univ. **32**(6), 38–40 (2011). https://doi.org/10.13914/j.cnki.cn43-1453/z.2011.06.010

The Design of Intelligent Energy Consumption Acquisition System Based on Narrowband Internet of Things

Wenqing Wang(✉), Li Zhang, and Chunjie Yang

School of Automation, Xi'an University of Posts and Telecommunications, Xi'an 710121, Shaanxi, China
398191005@qq.com

Abstract. This design develop an energy consumption acquisition system based on narrowband Internet of Things using stm32 series low-power microcontroller stm32f407 produced by STMicroelectronics [1]. The system consists of Perception layer, Network layer, and Application layer. The Perception layer includes an intelligent collection terminal. The Network layer upload data to the server through Low Power WAN. The Application layer achieve intelligent monitoring of remote nodes. The system not only meets the requirements of acquisition and transmission in industrial environments, but also can be applied to terminal data collection and transmission requirements in many different networks such as building networks and home networks. The system hardware design adopts the modular circuit design method, which mainly includes: the minimum system board of the microcontroller, the power management module, the clock circuit module, the data storage module, the narrow-band IoT module. The system software design adopts the service-level hierarchical programming idea. For the different modules, write the corresponding driver code; according to different protocols, write the corresponding communication protocol code. Since the design is based on industrial control site requirements, the system uses industrial design standards. The system runs well and is suitable for two-way remote meter reading system in industrial control field.

Keywords: Narrowband Internet of Things · STMicroelectronics Intelligence

1 Introduction

1.1 The Background of the Research

With the development of economy and society, the problem of high energy consumption in buildings in China is becoming increasingly prominent. China's building energy consumption accounts for about 28% of the country's total energy consumption. Most of the newly built 2 billion square meters of buildings in China are high-energy buildings. China's large public buildings have high energy density and serious energy waste problems. Energy-saving space, building energy conservation has become

P. Krömer et al. (Eds.): ECC 2018, AISC 891, pp. 833–841, 2019.
https://doi.org/10.1007/978-3-030-03766-6_93

imperative. The development of narrow-band IoT technology has brought unprecedented opportunities for building energy conservation [2].

The state has successively issued a series of policies, which have pointed out the direction for the implementation and promotion of building energy conservation and consumption reduction work. More and more cities have put together more detailed normative documents around the construction of building energy consumption monitoring systems based on their actual conditions, which has also accelerated the construction and promotion of building energy consumption monitoring systems.

1.2 The Mission of the Research

The energy consumption monitoring system based on the narrowband Internet of Things is designed. The system includes the perception layer, the network layer and the application layer.

The perception layer design the energy consumption collector for data analysis, protocol analysis, data storage from different instruments such as water meter, electricity meter and heat meter.

The network layer completes the NB-IOT network access to the Tianyi cloud platform, and the perception layer data can be uploaded to the platform through the NB-IOT.

The application layer design upper computer to monitor the data from the perception layer and can intelligently control the perception layer device by issuing the command [3].

2 The Introduction of the System

2.1 The Composition of the System

The system adopts layered development, and divides all the designs that need to be completed into perception layer, network layer, and application layer.

The architecture of the system.

2.2 The Design of System

The software design of the system uses a layered design, the entire system runs strictly in accordance with the processes of each layer. The system is mainly composed of the following three parts.

Perception layer. The perception layer completes the design of the energy consumption collector for the collect data from smart meters (electric meters, water meters, gas meters, etc.), various instrument protocol analysis, data storage, and remote transmission.

Network layer. The network layer completes the NB-IOT network access to the Tianyi cloud platform, and implements the perception layer device profile and the codec plug-in development on the platform to implement data storage and command delivery [4]. The network layer is a bridge between the perception layer and the

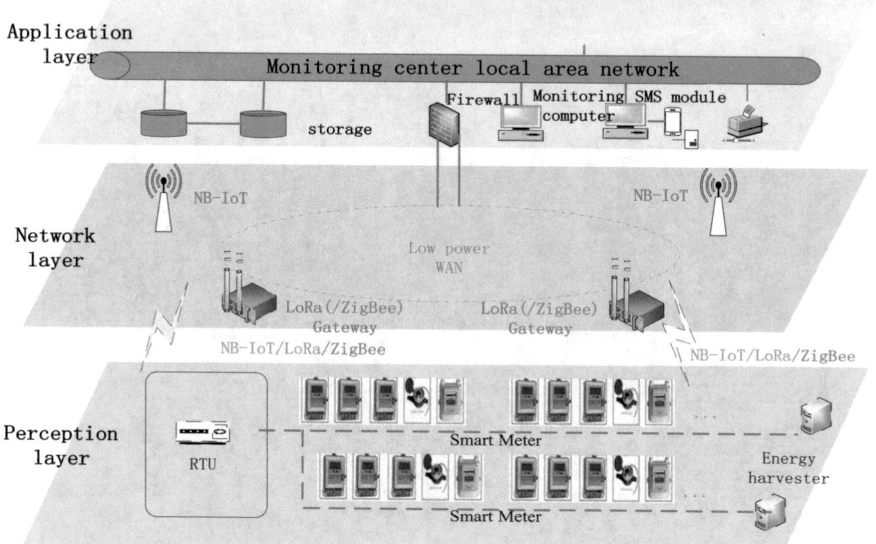

Fig. 1. A figure of system architecture diagram.

application layer. The transmission of data and the issuance of commands require a stable and secure network layer design, so the stability of the network layer is important (see Fig. 1).

Application layer. The application layer is developed by Java, including the upper computer software and mobile APP of the energy consumption collection system. The application layer obtains the perception layer device data through the subscription platform interface, and performs data fusion and data analysis on the data to judge the state of the perception layer device [5]. Users can get all the devices data from the perception layer through the application layer software, and can be remotely controlled through the application layer software [6].

3 The Hardware Design of System

The system hardware design only involves the perception layer. The hardware design of the perception layer involve the microcontroller, power module, data storage module, and narrow-band IoT module.

3.1 The Design of System Circuit

Minimal System Design of Microcontroller. As the core control module of the perception layer, the minimum system of the Microcontroller. The Microcontroller complete all task of the perception layer, and the circuit design is carried out according to the principle of stability and reliability (Fig. 2).

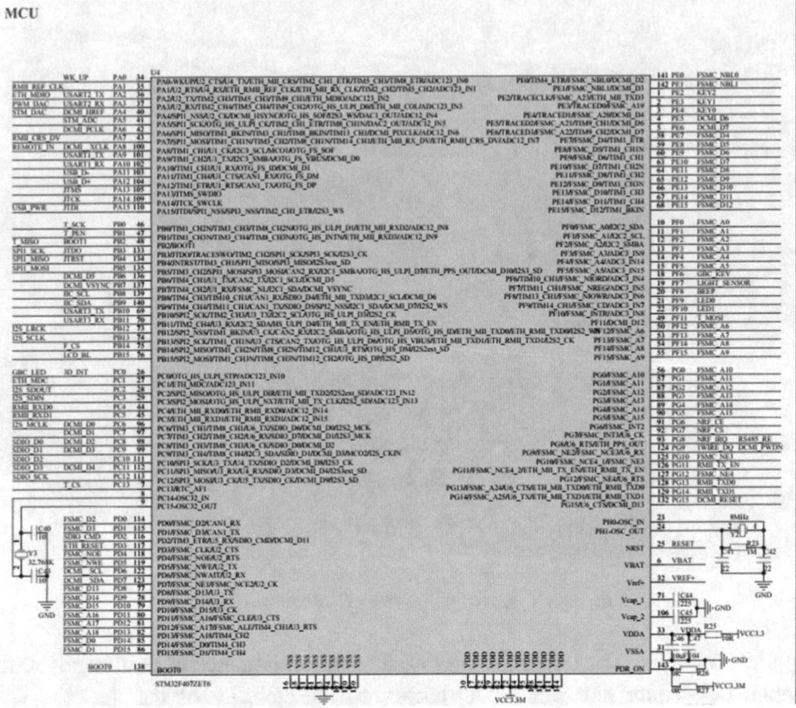

Fig. 2. A minimal system schematic diagram of Microcontroller

The Circuit Design of Power Management. In order to solve the power supply problem of the industrial environment, the system utilizes the voltage regulator chip of the MP2359. The DC input and backup battery input of the 5–20 power supply can be met to achieve a wide voltage range. The backup battery interface has anti-reverse function.

The Circuit Design of Power Management. The filter capacitor is added to the input and output of the MP2359 regulator chip. A diode and a fuse are added to the battery interface to protect the input battery power by blowing the fuse while preventing the power supply from being reversed (Fig. 3).

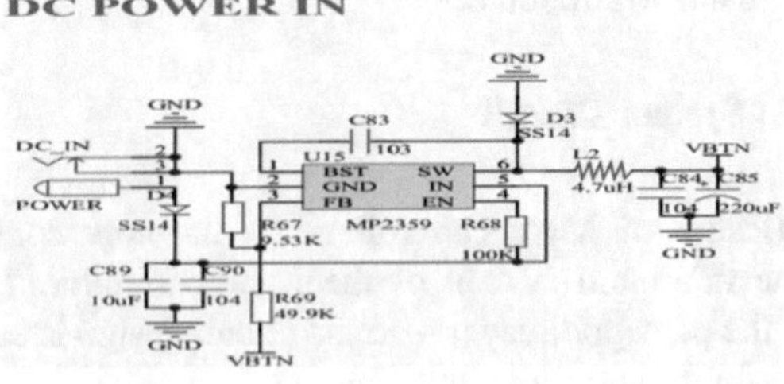

Fig. 3. The schematic of Power switch circuit

3.3 V Voltage Regulator Circuit Design. A series of chips on the periphery of the system are powered by 3.3 V, so 3.3 V regulation is required. So, the AMS1117-3.3 designed by TI is designed to be a 3.3 V regulator circuit (Fig. 4).

3.3V & POWER SWITCH

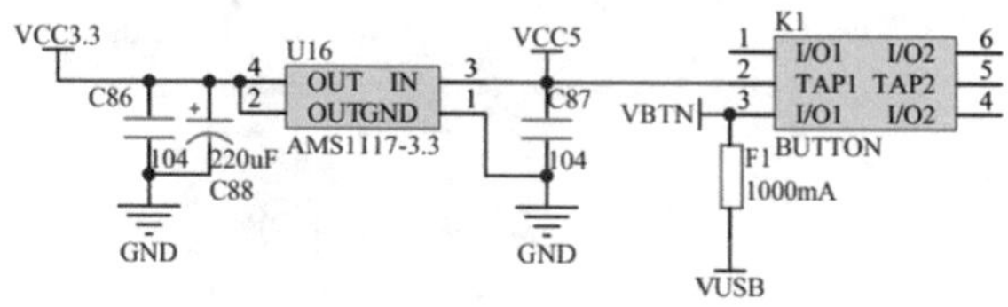

Fig. 4. The schematic of 3.3 V regulator circuit

The Circuit Design of Data Storage. The system design adopts the SD card as a storage device for collecting data by the perception layer. The circuit diagram of the design SD card (Fig. 5).

SD CARD

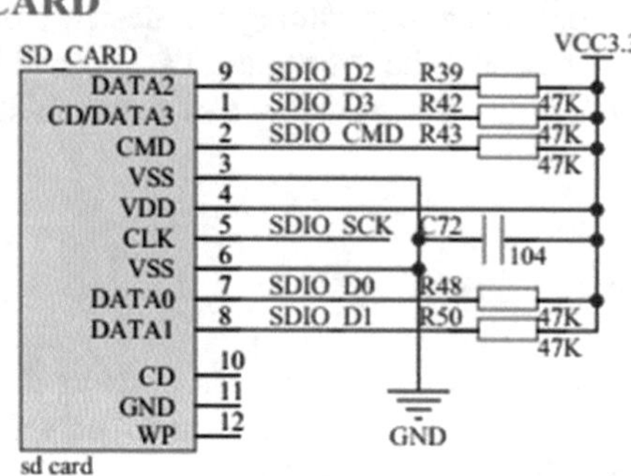

Fig. 5. The schematic of SD card circuit

The Circuit Design of Narrowband IoT. The narrowband IoT module selects BC-95 as the chip as its main control chip, and can transmit data through the MCU (Fig. 6).

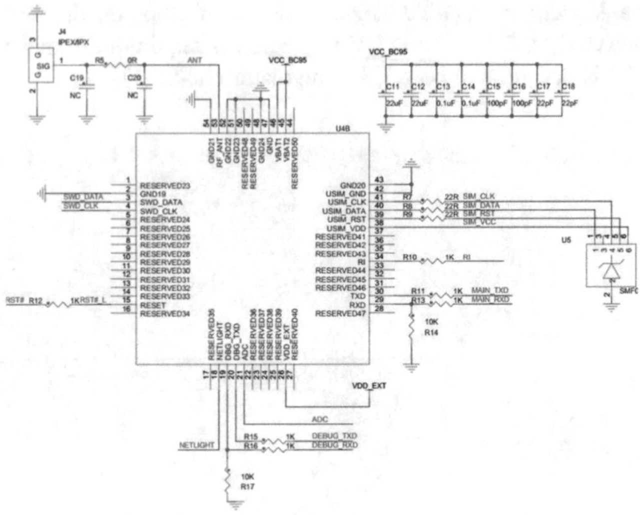

Fig. 6. The schematic of NBIOT

3.2 The Summary of Hardware Design

The hardware design is based on the functions to be implemented by the system. In order to improve the reliability of the system operation, the hardware design, whether in the power supply voltage regulation range or the device's differential selection, has undergone many changes. Moreover, during the design process, a lot of debugging work has been carried out, and the schematic diagram and PCB diagram have been continuously improved, so that the reliability and practicability of the hardware design have been greatly improved.

4 The Software Design of System

The system software design adopts a hierarchical structure of program design. Program design with a well-defined structure is conducive to the writing, reading and modification of the code. Starting from the underlying drive of the sensor layer chip. The lower structure provides a function interface to the upper layer, and the upper structure uses the interface provided by the lower layer to perform control operations.

4.1 The Program Structure of Perceptual Layer

The functions of the perceptual layer program are mainly for data collection, storage and data transmission. These tasks are all required by the MCU (Fig. 7).

The perception layer collects sensor data through the IO port of the MCU, and stores the data to the SD card through the fatfs file system. The UART is sent to the narrowband IoT module to send data to the cloud platform through the narrowband Internet of Things.

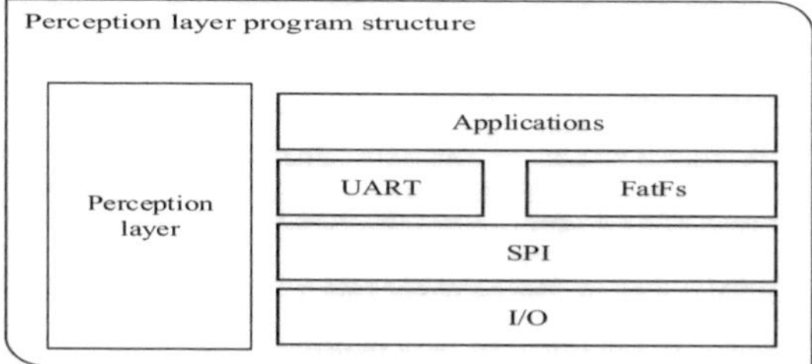

Fig. 7. The program structure of Perceptual layer

4.2 The Program Structure of Network Layer

The main functions of the network layer program include device profile development and device codec plug-in development, and registration of different devices and corresponding development on the cloud platform. The data of the perception layer can be recognized and recorded by the network layer. The network layer and the perception layer implement basic communication through the COAP protocol (Fig. 8).

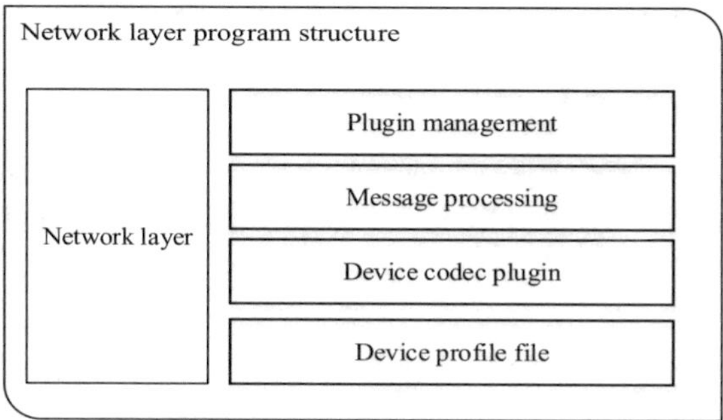

Fig. 8. The program structure of Network layer

As the middle layer, the network layer bears the channel for data transmission at the system design network layer. The cloud platform uses the COAP protocol to communicate with the perception layer NBIOT device, and the application layer communicates with the cloud platform through the http protocol. The system realizes the data transmission from the southbound equipment to the northbound application, and can monitoring and collection the consumption data from perceptual layer and the intelligent control of the southbound equipment.

4.3 The Program Structure of Application Layer

The main function of the application layer program is to subscribe to the interface of the network layer, and the perception layer device can be controlled by signaling (Fig. 9).

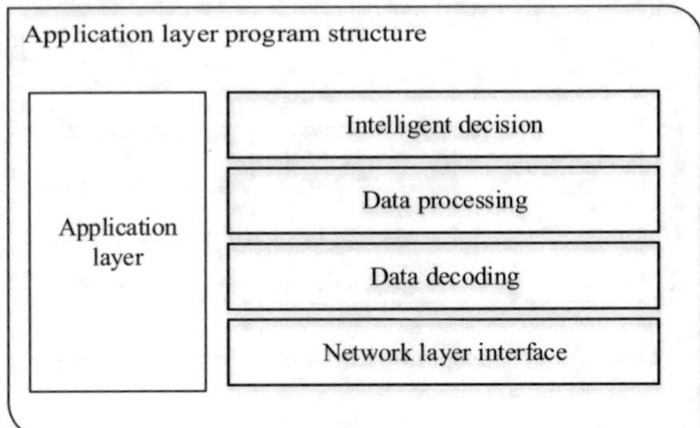

Fig. 9. The program structure of application layer

The application layer first provides an interface with the network layer to ensure communication with the perception layer. And the application layer obtains the perception layer data through the interface, decodes the data, and the decoded data makes corresponding decisions through the data fusion technology. The application layer sends the decision to the perception layer device, and the device performs the corresponding operation [7].

4.4 The Summary of Software Design

The software design is not a simple sequential process, so it cannot be presented in the form of a flowchart, but its core idea is still the processing of data and instructions. The layered interface provides a clear and logical software design.

5 Conclusion

The design starts from the direction of reducing energy consumption in the country, realizes data collection and remote monitoring of smart meters, and adopts the popular narrow-band Internet of Things technology for data transmission. The system collects energy consumption data through the perception layer and transmits the data through the narrowband Internet of Things. The network layer parses and packages the data. The application layer performs data fusion processing and intelligent decision and sends the decision result to the perception layer device through the network layer. Through the design of the energy consumption collection system, the remote collection and monitoring of energy consumption data is realized, and the problem of huge energy consumption in the current society is effectively solved.

Acknowledgments. This work is supported by Design and Development of Intelligent Collection Platform Of Energy Consumption based on Internet of Things, Industrialization project of Shaanxi Provincial Education Department (16JF024) and the Shaanxi Education Committee project (14JK1669) Shaanxi Technology Committee Project (2018SJRG-G-03).

References

1. Zhang, H., Kang, W.: Design of the data acquisition system based on STM32. Procedia Comput. Sci. **17** (2013). https://doi.org/10.1016/j.procs.2013.05.030
2. Wu, J.: Narrowband Internet of Things (NB-IoT): design challenges and considerations. In: IEEE Beijing Section. 2016 13th IEEE International Conference on Solid-State and Integrated Circuit Technology (ICSICT) Proceedings, vol. 1. IEEE Beijing Section (2016)
3. Ming, Z., Wired remote meter reading system based on fieldbus. In: Intelligent Information Technology Application Association. Proceedings of 2011 2nd Asia-Pacific Conference on Wearable Computing Systems (APWCS 2011 V4), vol. 2. Intelligent Information Technology Application Association (2011)
4. Malik, H., Alam, M.M., Moullec, Y.L., Kuusik, A.: NarrowBand-IoT performance analysis for healthcare applications. Procedia Comput. Sci. **130** (2018). https://doi.org/10.1016/j.procs.2018.04.156
5. Xu, J., Li, J., Xu, S.: Data fusion for target tracking in wireless sensor networks using quantized innovations and Kalman filtering. Sci. China (Inf. Sci.) **55**(03), 530–544 (2012)
6. Hoffmann, W.C., Lacey, R.E.: Multisensor data fusion for high quality data analysis and processing in measurement and instrumentation. J. Bionics Eng. **4**(01), 53–62 (2007)
7. Thoudam, S., Buitink, S., Corstanje, A., Enriquez, J.E., Falcke, H., Frieswijk, W., Hörandel, J.R., Horneffer, A., Krause, M., Nelles, A., Schellart, P., Scholten, O., ter Veen, S., van den Akker, M.: LORA: a scintillator array for LOFAR to measure extensive air showers. Nuclear Inst. Meth. Phys. Res. A **767** (2014). https://doi.org/10.1016/j.nima.2014.08.021

A Family of Efficient Appearance Models Based on Histogram of Oriented Gradients (HOG), Color Histogram and Their Fusion for Human Pose Estimation

Yong Zhao[1,2(✉)] and Yong-feng Ju[1]

[1] School of Electronic and Control Engineering, Chang'an University, Xi'an 710064, China
zhaoyong@xupt.edu.cn

[2] School of Automation, Xi'an University of Posts and Telecommunications, Xi'an 710121, China

Abstract. Human pose estimation can be addressed within the pictorial structures framework, where a principal difficulty is to model the appearance of body parts. For solving this difficulty, three new models are proposed in this paper. The appearance model based on Histogram of Oriented Gradients (HOG) and Support Vector Data Description (SVDD) is built by the linear combination of sub-classifiers constructed using the SVDD algorithm, while the mixing weights can be learned by using the maximum likelihood estimation algorithm. Moreover, a human part has a specific location probability in different images, then according to learned location probability from static image to be processed, the corresponding color histogram can be calculated, resulting in the appearance model based on color and location probability. According to the illumination and color contrast between clothes and background of the static image to be processed, the respective mixing weights for two appearance models can be learned and then used to build the combined appearance model. We use our appearance models to human pose estimation based on pictorial structure and evaluate them on two image datasets, experiments results show they can get higher pose estimation accuracy.

Keywords: Human pose estimation · Appearance model
Histogram of Oriented Gradients (HOG) · Color histogram

1 Introduction

Human action and behavior analysis is a research focus in the field of commuter vision because people are typically the dominant objects in images and videos that we encounter every day. Human pose estimation is a process of automatically detecting and estimating the pose of human from a static image or video [1], it has been widely applied in the fields of human-computer interaction, activity recognition and visual surveillance [2]. Although many research achievements have been made, it remains an unsolved hard problem.

P. Krömer et al. (Eds.): ECC 2018, AISC 891, pp. 842–850, 2019.
https://doi.org/10.1007/978-3-030-03766-6_94

The pictorial structure is a probabilistic model for an object given by a collection of parts with connections between certain pairs [3], it has been widely used in human pose estimation [2]. The model consists of unary terms of similarity between image region and part appearance and pairwise terms between adjacent parts and/or joints capturing their preferred spatial arrangement.

Appearance model is a very important part of pictorial structure model, it plays a vital role in human pose estimation, many appearance models have been proposed over the past ten years. An appearance model is built using HOG, shape and color features [4–8].

The existing appearance models mainly use image features such as edge, color and shape and so on, there are some models are built just using one feature, while the others use the fusion of multiple features. Although the existing appearance models have got good effects, there are still having some defects: (i) image features are not used adequately to express the real part accurately; (ii) the existing image feature fusion methods are too simple to implement appropriate fusion.

In addition to appearance model, inference algorithm also plays an important role in human pose estimation, and some effective inference algorithms have been proposed over the past ten years [9–11].

In this paper, three new appearance models using HOG, color and their fusion are proposed and used to estimate the two-dimensional (2D) pose of upper body in static image. This paper is organized as follows: Sect. 2 presents an overview of our approach, three new appearance models are introduced respectively from Sect. 3 to Sect. 5. Experiments and conclusions are then given in Sects. 6 and 7.

2 Overview of Our Approach

A brief flowchart of our appearance models are given in Fig. 1. The appearance model based on HOG and SVDD can be built through the following steps: (1) all of annotated image blocks that belong to the same part are cropped and resized to the same size; (2) a sub-classifier is constructed using the SVDD algorithm for every cell of HOG feature of image blocks, the number of sub-classifiers is equal to the number of cells of one image block; (3) the weight for every sub-classifier is learned; (4) the model is constructed by the linear combination of sub-classifiers with learned weights.

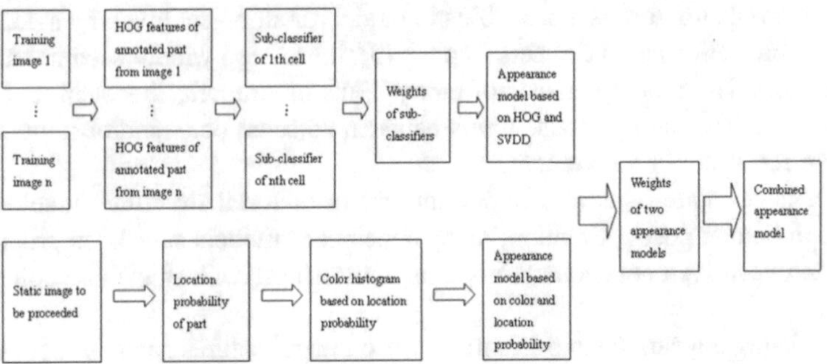

Fig. 1. Flowchart of our approach.

3 Appearance Model Based on HOG and SVDD

HOG descriptor [12–17] can be used to capture effectively appearance of the parts in terms of edge and shading properties while incorporating a controlled level of invariance to lighting and local deformation by quantization and spatial pooling of the image gradients. SVDD [18] is an extended algorithm of SVM, the difference with SVM lies in that it does not tend to find the optimal hyperplane but construct a hypersphere with minimal radius contains all or most of given sample data [18–21]. An appearance model based on HOG and SVDD built by the linear combination of the sub-classifiers with different weights is proposed in this section, which is inspired by the sparse kernel-based ensemble learning (SKEL) algorithm [22].

3.1 Support Vector Data Description

Given a set of training data $\{x_i;\ i = 1,\ 2,\ \cdots,\ N\} \in R_n$, the SVDD algorithm estimates the parameters of the hypersphere by minimizing Eq. (1)

$$\begin{aligned} F(R, c, \xi_i) &= R^2 + C\sum_{i=1}^{N} \xi_i \\ s.t. \quad & \|x_i - c\|^2 \le R^2 + \xi_i \end{aligned} \tag{1}$$

where R and c are the radius and centre of the hypersphere respectively, ξ_i are slack variables that allow some data points out of the hypersphere, C is the penalty parameter used to control the trade-off between the size of the hypersphere and the number of samples that possibly fall outside the hypersphere [23].

3.2 Appearance Model Based on SVDD

The HOG feature of a annotated part from the ith training image can be represented as $X_i = \{x_{i1}, x_{i2}, \ \cdots,\ x_{im}\}$, where m is the number of cells, x_{ij} is a vector corresponding to the jth cell of the part. Given the HOG feature set of the jth cell $\{x_{ij};\ i = 1,\ 2,\ \cdots,\ N\}$, a hypersphere with the parameters (R_j, c_j) can be constructed using the SVDD algorithm, which can be seen as the template of *j*th cell of the part. As mentioned before, the similarity of given vector x_{ij} with the corresponding template can be measured by the distance between the centre of hypersphere. The sub-classifier is constructed by the hypersphere in this section, the output of sub-classifier is the similarity

$$s_{ij} = \begin{cases} 1 & if\ d \leq R_j \\ R_j/d & if\ d > R_j \end{cases} \tag{2}$$

where R_j is the radius of hypersphere, d is the distance between vector x_{ij} and the centre of the hypersphere.

The classifier is constructed by the linear combination of sub-classifiers with different weights, which is the appearance model based on HOG and SVDD. The mixing weights can be learned by maximizing the similarity of corresponding HOG feature of annotated human parts from training images with the appearance model.

$$\begin{aligned} &\max\left(\sum_{i=1}^{N}\sum_{j=1}^{m} w_j s_{ij}\right) \\ &s.t.\quad 0 \leq w_j \leq 1,\ \sum_{j=1}^{m} w_j = 1 \end{aligned} \tag{3}$$

where N is the number of training image, m is the number of cells, w_j and s_j is the mixing weight and the output of the *j*th sub-classifier respectively.

4 Appearance Model Based on Color and Location Probability

Color feature has been extensively used in human pose estimation, color histogram and color symmetry are two main application ways, where color histogram is mainly used to build appearance model [4, 8], color symmetry is mainly used to add constraints to different parts [9, 24]. A new algorithm for learning location probability for the specific static image to be processed is proposed and the appearance model based on color and location probability is built according to the learned location probability in this section.

4.1 Location Probability

For learning location probability of a human part from a static image to be processed, the location region is needed to be determined first. An example of determining the location region of left upper arm is shown in Fig. 2. The detection window of upper body is detected firstly using the upper body detector [12], and then the reduced state space is determined using the approach proposed in [25], the location region is determined finally by all pixels covered by image blocks corresponding to states in the reduced state space. So the location probability for each pixel is learned by using the Eq. (4) in this paper.

$$LP_i(x,y) = \frac{\sum_{l_i \in L_i} p(I|l_i)}{num_i} \quad (4)$$

where (x, y) is the image coordinate of pixel, L_i. is the reduced state space of part i, $p(I|l_i)$ is the similarity for state l_i. num_i is the number of states in L_i.

4.2 Appearance Model Based on Color and Location Probability

The similarity of one state with the appearance model can be calculated using Eq. (5).

$$D_i(x_i, z_i) = \frac{1}{n}\sum_{k=1}^{n}\left(1 - \frac{|x_{ik} - z_{ik}|}{\max(x_{ik}, z_{ik})}\right) \quad (5)$$

where D_i (x_i, z_i) is the normalized Euclidean distance of x_i and z_i, x_i is the color histogram of part i being at a location l_i, n is the dimension of x_i, z_i is the appearance model.

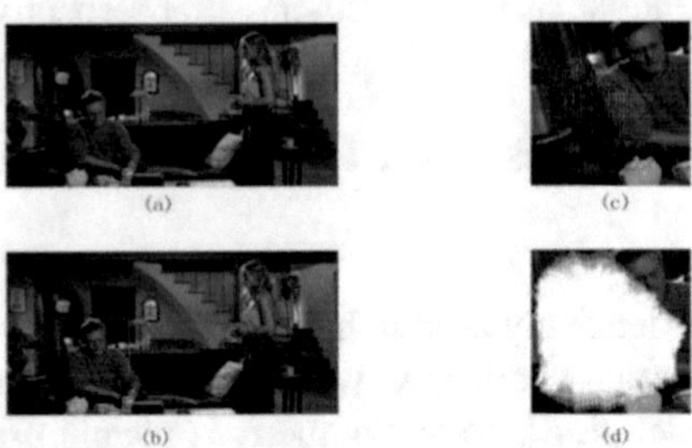

Fig. 2. Example of determining of location region (a) image to be processed; (b) upper body window; (c) reduced state space; (d) location region.

5 Combined Appearance Model

In this section, a new appearance model is built by the linear combination of the two appearance model proposed above, the combined appearance model can be expressed as Eq. (6).

$$s = \sum_{j=1}^{2} w_j s_j \quad (6)$$

where w_j is the mixing weight, s_1 and s_2 is the similarity of the part being present at a location l_i with two appearance models proposed above respectively.

Assume all of the similaritys fit the gauss distribution, the mixing weights can be learned by maximizing the mean similarity

$$\begin{aligned} & \max\left(\frac{1}{N}\sum_{i=1}^{N}\sum_{j=1}^{2} w_j s_{ij}\right) \\ & s.t. \quad 0 \leq w_j \leq 1, \ \sum_{j=1}^{2} w_j = 1 \end{aligned} \tag{7}$$

where N is the state number in reduced state space, w_1 and w_2 is the weight of two appearance models proposed above respectively, s_{i1} and s_{i2} is the similarity of the ith state with two appearance models respectively.

6 Experimental Results and Evaluation

We select the same training and test images from Buffy and PASCAL dataset as [4, 8]. The comparison results are shown in Table 1, where the left and right data of the slack are the mean and standard deviation respectively, "HOG + SVM" and "HOG + SVDD" represent the appearance model using SVM algorithm and proposed in this paper respectively, "color" represents the proposed appearance model based on color and location probability, "HOG + color(1)" represents the appearance model in which the mixing weights of the appearance model based on HOG and SVDD and appearance model based on color and location probability are equivalent, "HOG + color(2)" represents the combined appearance model proposed in this paper.

Table 1. Comparison of similarity

Appearance model	Torso	Head	Upper arms	Lower arms
HOG + SVM	0.79(0.09)	0.84(0.09)	0.72(0.12)	0.68(0.13)
HOG + SVDD	0.85(0.09)	0.88(0.08)	0.77(0.1)	0.73(0.11)
[8]	0.73(0.11)	0.75(0.11)	0.70(0.12)	0.70(0.13)
color	0.77(0.1)	0.80(0.09)	0.75(0.11)	0.72(0.11)
HOG + color(1)	0.79(0.09)	0.81(0.09)	0.74(0.11)	0.70(0.12)
HOG + color(2)	0.88(0.09)	0.90(0.08)	0.79(0.1)	0.75(0.11)

We use the proposed appearance model to human pose estimation based on pictorial structure model and estimate human pose for test images. Table 2 gives the comparison results with the existing human pose estimation algorithm based on pictorial structure model on this two image datasets.

Table 2. Evaluation of pose estimation.

Image dataset	Method	Torso	Head	Upper arms	Lower arms
Buffy	[6]	90.7	95.5	79.3	41.2
	[8]	98.7	97.9	82.8	59.8
	[4]	100	96.2	95.3	63.0
	color	100	100	89.4	60.2
	HOG	100	100	90.7	64.3
	HOG + color	100	100	96.3	65.6
PASCAL	[8]	97.2	88.6	73.8	41.5
	[4]	100	90.0	87.1	49.4
	color	100	100	81.3	48.1
	HOG	100	100	85.0	50.4
	HOG + color	100	100	89.0	52.8

Compared with [6] and [8], all of the proposed appearance models can improve accuracy for any part, "HOG + color" gets the highest estimation accuracy. However, "color" and "HOG" get a slightly lower accuracy for arms and upper arms respectively compared with [4], that is because they use only one image feature to build, while [4] use HOG, color and shape features, while "HOG + color" that use HOG and color features can get higher estimation accuracy for all of parts.

Figure 3 shows a comparison of pose estimation results between our proposed appearance models and [4]. The top line of it are several failure pose estimation results in [4], the second line are pose estimation results used "HOG", the third line are pose estimation results used "color", and the bottom line are pose estimation results used "HOG + color". From these results it appears that "HOG" often works badly in strongly cluttered, the "color" and [4] often works badly in images with worse light conditions or lower color contrast between clothes and background, however, even in those scenes, "HOG + color" can still localize more parts correctly.

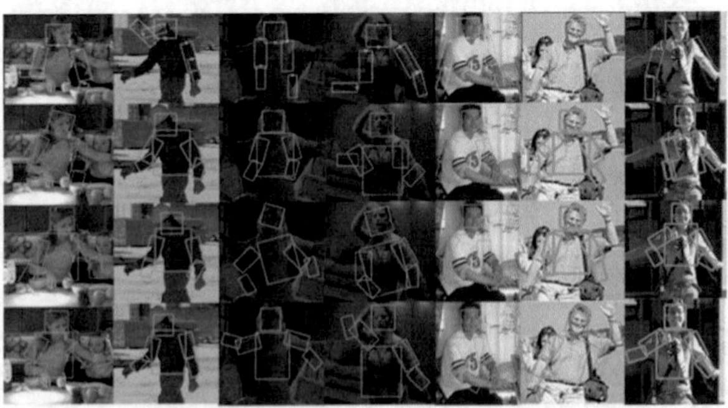

Fig. 3. Comparison of pose estimation results between proposed appearance models and [4].

7 Conclusion and Feature Work

In this paper, three new appearance models are proposed and used to estimate human pose. The first is the appearance model based on HOG and SVDD, which is built by combining multiple sub-classifiers trained using the SVDD algorithm. The second is the appearance model based on color and location probability, which is built using the color histogram calculated according to the learned specific location probability from the static image to be processed. The third is the combined appearance model, which is built by the linear combination of the first and second appearance models. Compared with the existing human pose estimation algorithms based on pictorial structure model, when use our models to human pose estimation based on pictorial structure model get higher estimation accuracy.

References

1. Felzenszwalb, P.F., Huttenlocher, D.P.: Pictorial structures for object recognition. Int. J. Comput. Vis. **61**(01), 55–79 (2005)
2. Thomas, B.M., Hilton, A., Krüger, V.: Visual Analysis of Humans, pp. 131–275. Springer, London (2011)
3. Fischler, M., Elschlagr, R.: The representation and matching of pictorial structures. IEEE Trans. Comput. **22**(01), 67–92 (1973)
4. Sapp, B., Toshev, A., Taskar, B.: Cascaded models for articulated pose estimation. In: Proceedings of the 11th European Conference on Computer Vision, pp. 406–420. Springer, Berlin (2010)
5. Ramanan, D.: Learning to parse images of articulated bodies. In: Proceedings of the 20th Annual Conference on Neural Information Processing Systems, pp. 1129–1136. MIT Press, Cambridge (2006)
6. Andriluka, M., Roth, S., Schiele, B.: Pictorial structures revisited: People detection and articulated pose estimation. In: Proceedings of the 2009 IEEE Conference on Computer Vision and Pattern Recognition, pp. 1014–1021. IEEE, Piscataway (2009)
7. Ukita, U.: Articulated pose estimation with parts connectivity using discriminative local oriented contours. In: Proceedings of the 2012 IEEE Conference on Computer Vision and Pattern Recognition, pp. 3154–3161. IEEE, Piscataway (2012)
8. Eichner, M., Ferrari, V.: Better appearance models for pictorial structures. In: Proceedings of the 20th British Machine Vision Conference, pp. 3.1–3.11. BMVA Press, Dundee (2009)
9. Tian, T.P., Sclaroff, S.: Fast globally optimal 2d human detection with loopy graph models. In: Proceedings of the 2010 IEEE Conference on Computer Vision and Pattern Recognition, pp. 81–88. IEEE, Piscataway (2010)
10. Sapp, B., Jordan, C., Taskar, B.: Adaptive pose priors for pictorial structures. In: Proceedings of the 2010 IEEE Conference on Computer Vision and Pattern Recognition, pp. 422–429. IEEE, Piscataway (2010)
11. Sun, M., Telaprolu, M., Lee, H., et al.: An efficient branch-and-bound algorithm for optimal human pose estimation. In: Proceedings of the 2012 IEEE Conference on Computer Vision and Pattern Recognition, pp. 1616–1623. IEEE, Piscataway (2012)
12. Ferrari, V., Marin-Jimenez, M., Zisserman, A.: Progressive search space reduction for human pose estimation. In: Proceedings of the 2008 IEEE Conference on Computer Vision and Pattern Recognition, pp. 1–8. IEEE, Piscataway (2008)

13. Dalal, N., Triggs, B.: Histograms of oriented gradients for human detection. In: Proceedings of the 2005 IEEE Conference on Computer Vision and Pattern Recognition, pp. 886–893. IEEE, Piscataway (2005)
14. Johnson, S., Everingham, M.: Combining discriminative appearance and segmentation cues for articulated human pose estimation. In: Proceedings of the 12th IEEE International Conference on Computer Vision, pp. 405–412. IEEE, Piscataway (2009)
15. Tran, D., Forsyth, D.: Configuration estimates improve pedestrian finding. In: Proceedings of the Twenty-first Annual Conference on Neural Information Processing Systems. MIT Press, Cambridge (2007)
16. Buehler, P., Everingham, M., et al.: Long term arm and hand tracking for continuous sign language TV broadcasts. In: Proceedings of the 19th British Machine Vision Conference, pp. 1105–1114. BMVA Press, UK (2008)
17. Johnson, S.: Articulated Human Pose Estimation in Natural Images [Ph.D. dissertation]. University of Leeds, UK (2012)
18. Dvaid, M.J., Robert, P.W.: Support vector data description. Mach. Learn. **54**(1), 45–66 (2004)
19. Xue, Z.X., Liu, S.Y., Liu, W.L., et al.: SVDD based learning algorithm with progressive transductive support vector machines. Pattern Recog. Artif. Intell. **21**(6), 721–727 (2008)
20. Zhu, X.K., Yang, D.G.: Multi-class support vector domain description for pattern recognition based on a measure of expansibility. Acta Electronica Sinica **37**(03), 464–469 (2009)
21. Niazmardi, S., Homayouni, S., Safari, S.: An improved FCM algorithm based on the SVDD for unsupervised hyperspectral data classification. IEEE J. Sel. Top. Appl. Earth Obs. Remote Sens. **6**(2), 831–839 (2013)
22. Gurram, P., Kwon, H.: Ensemble learning based on multiple kernel learning for hyperspectral chemical plume detection. In: Proceedings of the SPIE, vol. 7695, pp. 5–9. SPIE Press, Washington (2010)
23. Dvaid, M.J., Robert, P.W.: Support vector domain description. Pattern Recogn. Lett. **20**(11–13), 1191–1199 (1999)
24. Jiang, H.: Human pose estimation using consistent max-covering. IEEE Trans. Pattern Anal. Mach. Intell. **33**(09), 1911–1918 (2011)
25. Han, G.J., Zhu, H., Ge, J.R.: Effective search space reduction for human pose estimation with Viterbi recurrence algorithm. Int. J. Model. Identif. Control **18**(04), 341–348 (2013)

Author Index

P. Krömer et al. (Eds.): ECC 2018, AISC 891, pp. 851–854, 2019.
https://doi.org/10.1007/978-3-030-03766-6

Printed in the United States
By Bookmasters